TURBULENT TRANSPORT IN MAGNETIZED PLASMAS

Dedicated to
Elisabeth (Libby) Becker Horton (1943–2007)
who made this research possible.

Foreword

For over thirty years the Institute for Fusion Studies at The University of Texas at Austin has been teaching plasma physics with an emphasis on laboratory plasma experiments and the problems associated with magnetic confinement of high temperature plasmas in the laboratory. This book results from the lectures and research projects of numerous scientists, postdoctoral and graduate students working with the author during this period. The Institute for Fusion Studies was designated in 1980 as the home base for organizing official US-Japan exchanges and workshops under an agreement for exchanges in energy research signed by President Carter and Prime Minister Fukuda in 1978. Thus, the material and publications in this book are enriched by many contributions from our colleagues in Japan. Since 2000 the author has worked at the Cadarache Laboratory where the next large fusion magnetic confinement device is being built for operation in 2017. The author is indebted to scientists at Cadarache Laboratory and throughout the European Union for many important contributions to the book.

The subject of plasma physics is now a rich and mature discipline of basic physics. Plasma physics combines the areas of mechanics, electromagnetic field theory, and statistical mechanics. The subject is a rich and rewarding course of study for physics students who have finished basic courses in these three disciplines of classical physics. Plasma science shows in a natural and beautiful setting how the laws of mechanics, electricity and magnetism, and statistical theory work together to predict the behavior of high temperature gases which occur in the presence of externally applied magnetic fields and magnetic fields produced by electric currents arising from the differing trajectories of the ions and electrons in the plasma.

For students without a knowledge of magnetohydrodynamics (MHD) physics, we recommend an alternative way to learn the basic principles used here for plasmas by studying one of the numerous books on geophysical fluid dynamics. The Coriolis force for fluids on a rotating planet has the same mathematical form as the Lorentz force on charged fluids so there are numerous analogous problems in the geophysical fluid dynamics and magnetized plasmas. The monograph of Horton and Ichikawa (1996) develops this hydrodynamic analogy in some detail in a manner that introduces many of the concepts and laws that govern drift waves and temperature gradient driven instabilities in magnetized plasmas. For a start in geophysical fluid dynamics the new book *Waves in the Ocean and Atmosphere* with subtitle Introduction to Wave Dynamics by J. Pedlosky, 2003, is a valuable concise text from lectures given by Pedlosky. Classical books on geophysical flows would then be accessible to the reader which include *Atmosphere-Ocean Dynamics*, A. E. Gill, 1982, and *Geophysical Fluid Dynamics*, J. Pedlosky, 2nd Edition, 1987.

The author would like to express his gratitude to his colleagues and students who have contributed to the knowledge and work presented in this book. Owing to the forty-year history of the research presented here, however, the author must confine the list to his closest colleagues who worked at the Institute for Fusion Studies. The author wishes to thank and acknowledge the contributions in this work from A. Aydemir, H. Berk, D. W. Ross, K. Gentle, P. J. Morrison, T. Tajima, D-I. Choi, J. W. Van Dam, S. M. Mahajan, F. L. Waelbroeck, B. N. Breizman, H. V. Wong, D. Brower, R. D. Bengtson, W. Rowan, M. Kotschenreuther and L. Leonard. He also benefited from including work developed in collaboration with D. Biskamp, S. Hamaguchi, H. Sugama, I. Holod, A. Ishizawa, Y. Kishimoto, and A. Beklemishev. Many useful suggestions were contributed from The Director of the Fusion Research Center, W. Drummond, and the Institute for Fusion Studies Directors, M. N. Rosenbluth, D. Baldwin, R. D. Hazeltine, and J. W. Van Dam.

The extensive magnetic fusion research in the United States was enabled by talented leaders in the Department of Energy who managed The University of Texas group from 1967. These DoE leaders include S. Dean, B. Hirsch, A. Trivelpiece, D. Nelson, R. Hunter, W. Sadowski, M. L. Crisp, C. Bolton, and E. Synakowski.

As with all modern science and engineering fields, fusion plasma physics has many specialized terms and acronyms. A good glossary that defines most of the terms used in this book can be found at `http://www.iter.org/glossary`. A widely-used handbook for the formulas and constants used in plasma physics can be found at `http://www.ppd.nrl.navy/nrlformulary`. Chapter 1 has an abbreviated list of the most frequently used physical constants in this book. The plasma literature has a mixture of cgs and mks units.

The typing and LaTeX formatting of the book was performed by the long-time contributor to The University of Texas Fusion Research group, Suzy Mitchell.

Contents

Chapter 1

Basic Concepts and Historical Background

Plasmas are created by heating gases through the absorption of ionizing radiation in the visible through the ultraviolet part of the electromagnetic spectrum and through collisional ionization of molecules with energetic electrons. The absorption and heating processes are typically localized in space so that large spatial gradients occur in the plasma density, temperature and other characterizing parameters. Due to the high electrical conductivity of ionized gases, large electrical currents typically flow and produce substantial self-generated magnetic fields in the plasma. In other plasmas strong magnetic fields exist, produced by external currents such as those flowing in the cores of the planets or those in the conducting coils designed to produce plasma confinement in laboratory machines. Thus, the focus of this book is on the transport processes in magnetized plasmas.

Due to the 40 years of intense international pursuit of the laboratory confinement of high temperature ($T > 10^7\,\mathrm{K}$) hydrogenic plasmas for the purpose of obtaining controlled nuclear fusion, there is an enormous resource to be found in the scientific literature on the experiments, the theory, and now, the numerical simulations of the plasma transport processes. While there are some books, most notably those by Mikhailovskii (1992), Wakatani (1998), Weiland (1999), and Wesson (2004), developing the subject, and numerous review articles on the transport problem, there is need for a book suitable for the first stages of transport research. Thus, in this work we attempt to provide a general presentation starting from elementary considerations, while including the most recent results that one may view as firmly established in the understanding of the nonlinear transport processes in ionized gases. The collisional transport produced by the binary Coulomb interactions between the charges in a plasma is a well-developed subject treated in the works of Balescu (1988) and Helander and Sigmar (2002). When the plasma temperature is high and density is low, the distance between these nearby encounters is long and the orbits of the particles between the collisions are complicated. The mean distance between Coulomb collision is called the mean-free-path and complex orbits have names like "banana" and "potato" particle orbits [Helander and Sigmar (2002)]. In general, the turbulent transport greatly exceeds the collisional transport in tokamaks. The exceptions are discussed in Chapter 14 of

Helander and Sigmar (2002).

A clear and reliable way to understand transport in fluids and plasmas is to follow the evolution of test particles, or a passive field, typically called a *dye* or *trace impurity*, in the system. Test particle transport theory, simulations, and experiments are emphasized here, perhaps more than is traditional in the existing plasma literature. The limitations of the test particle method are serious, however, which accounts for their being less well developed in earlier works on plasma transport. The limitations arise because the motion of the charged particles modifies the electric charge and current distributions in the plasma. These changes in the electric charge density $\rho_q(\mathbf{x}, t)$ and current density $\mathbf{j}(\mathbf{x}, t)$ in turn modify the electromagnetic fields $\mathbf{E}(\mathbf{x}, t)$ and $\mathbf{B}(\mathbf{x}, t)$ in the plasma. Thus, the orbits of the particles are themselves changed again due to the collective effect of the accumulated charges and currents from the fields, due to their motions in these same fields. This problem is known as the self-consistent field problem. The self-consistent field problem forms the essence of the theoretical and simulation modeling of plasma dynamics. It is this circular sequence of interactions of the motion of the charged particles in the self-consistent electromagnetic fields that contains both the externally-applied fields and those generated by the particle motions that determine the actual transport of the plasma. Thus, the large body of works on plasma transport theory and modeling have their complexity rooted in the problem of determining, as clearly as possible, the solutions of the orbital equations with the self-consistent plasma electric and magnetic fields. Obviously, many different theoretical approaches can be and have been made to extract key transport information from this problem. Often, the approach is set by *a priori* assumptions about the nature of the fields; coherent waves or vortices, weak turbulent fields or states of strong turbulence, for example.

While the most practical information desired may be contained in turbulent transport coefficients such as the particle diffusivity D and the electron and ion thermal diffusivities χ_e and χ_i, respectively, there are important processes where no such simple diffusivities exist. In such regimes there can be bursty in time large transport events, the dominance of coherent structures such as streamers, zonal flows, the existence of long flights of the particles through small gaps in transport barriers, and other exotic transport processes.

1.1 Space and Astrophysics

Early plasma physics research began in about 1900 when scientists puzzled over the high altitude lights in the northern sky and the systematic deflections of the magnetic compass needle in careful measurements taken across the northern latitudes of the Scandinavian countries. Birkeland, whose name is attached to the large plasma currents known to flow into and out of the northern ionosphere from the magnetosphere, pioneered this research by building models of the magnetosphere in the laboratory, in an effort to reproduce the auroral observations. The properties of the Birkeland plasma currents are now measured with spacecraft that are equipped with high-resolution particle and field instruments [Kivelson and Russell (1995)]. Birkeland set up a laboratory magnetic confinement experiment and reproduced the effect of the auroral rings, but uncovered many more problems of complicated physics than could be answered with the knowledge and equipment of that era. Alfvén and Falthammer, along with Stoermer, pursued these complex issues and realized that these were generic problems for the solar system and what they called the cosmos in their subsequent books that give many of the basic plasma physics concepts used today. Two of these concepts developed by Alfvén that are used extensively in the drift wave transport of plasmas are the low frequency electromagnetic Alfvén plasma waves and the guiding center drift of the guiding centers for describing the trajectories of charged particles in the confining magnetic fields in plasmas. For these and other important discoveries, Hannes Alfvén received the Nobel Prize in 1970 in plasma physics.

The plasma physics of the solar corona [Aschwanden (2006)] has similarities to that of the toroidally confined in tokamaks. The flux ropes observed with high resolution solar observatories such as SOHO, TRACE, and SDO show both smaller ($L < 10\,\mathrm{Mm}$) structures in the chromosphere to lower corona and large ($L < 100\,\mathrm{Mm}$) structures that extend into the upper corona and last for many hours. These large flux ropes are called prominences and appear as EUV radiation lines, particularly from ionized (Fe) atoms. The images from solar-observing spacecraft, including TRACE, SOHO and STEREO show magnetic flux loops with plasma currents that are embedded in the photosphere. In these large coherent structures one can see strands of more or less dense plasma trapped in smaller ($< 10\,\mathrm{Mm}$) helical flux tubes that show the winding number of the helical magnetic field. In terms of the tokamak magnetic field twist number q, these solar magnetic fields appear as structures with $q < 0.5$. The corresponding plasma currents are then of order 10^{11} Amps. The solar corona has an electron density n_e and temperature T_e such that the pressure $p_e = n_e T_e$ is less than the magnetic field pressure $P_M = B^2/2\mu_0$. These currents are driven by the injection of magnetic flux from vertical convection of the plasma up through the photosphere from the subsurface solar convection zone. The injected magnetic flux is of order $10^8\,\mathrm{Wb/s}$ producing the accelerating toroidal electric field. The foot-points of the flux tubes are usually brighter in EUV and soft X-ray emission than the surrounding plasma. Thus, we see that a tokamak-like toroidal structure may occur in nature as a self-organized natural state of the plasma. This observation may be consistent with the fact that across the spread of laboratory magnetic confinement geometries the simple ohmic heated tokamak has the highest energy confinement time τ_E and the largest Lawson product $n_e\tau_E$ of any of the man-made confinement devices.

In the fusion experiments the auxiliary heating, P^{aux}, is applied with values far above the natural ohmic heating power, P_{oh}, and the confinement product of $n_e\tau_E$

decreases with approximately the inverse square root of the ratio of $P^{\rm aux}/P_{oh}$. The physics behind this trend is that the drift wave turbulence and turbulent transport increase as the temperature profiles become steeper with increased auxiliary power. The core plasma temperature must reach temperatures in the 10-20 KeV range needed to produce the nuclear reactivity sufficient for a net power gain in the driven laboratory confinement system. The fusion reactivities for the D-T and

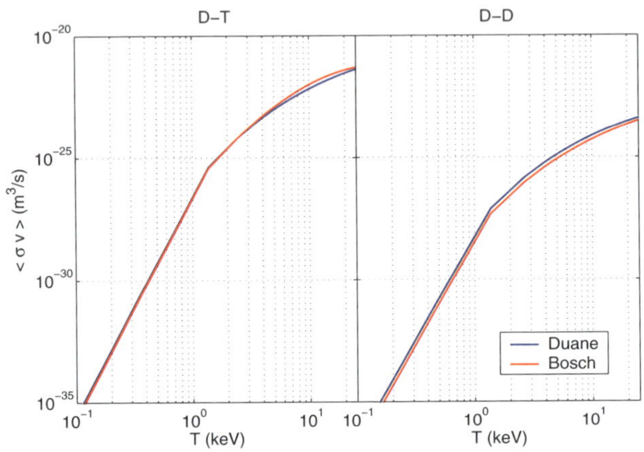

Fig. 1.1 The nuclear fusion reactivity $\langle \sigma v \rangle$ of deuterium D with tritium T in frame (a) and deuterium alone in frame (b) as the temperature of the Maxwellian distribution functions increases from 100 eV to 20 KeV [Duane (1972); Bosch and Hale (1992)].

D-D collisions as a function of temperature are shown in Fig. 1.1.

The average in $\langle \sigma v \rangle$ shown in Fig. 1.1 is for a Maxwell-Boltzmann distribution function at the temperature given on the horizontal axis in KeV. A database from nuclear physics is used for the fusion-reaction cross-sections $\sigma(v)$, which are in the range of a few barns, where one barn is defined as $10^{-24}\,\mathrm{cm}^2$. In contrast, the typical cross-sections for atomic processes including ionization of the neutral atoms are of order $10^{-16}\,\mathrm{cm}^2$.

The principal difficulty in achieving fusion in the laboratory arises from the small value of the nuclear cross-section for fusion reactions compared with much larger cross-sections for atomic and Coulomb scattering of the ions. Thus, it becomes essential to use tritium as the target and fuel material for the experiments and thermonuclear reactors, owing to the much larger fusion cross-section, as seen by comparing the two reactivities shown in Fig. 1.1. Tritium is not naturally occurring, owing to its 12.6 year half-life for decay into deuterium. Thus, the tritium is made from lithium in other nuclear reactors, principally as a byproduct from fission reactors. For fusion power reactors the tritium consumption for a 1 GW steady state reactor would be a few kilograms of tritium per year. The tritium fuel would be produced by the fusion system itself by surrounding the reacting plasma with a blanket of lithium, which is naturally occurring, and "breeding" the tritium through the interaction of the escaping fusion neutrons with the lithium. The reaction of the fusion neutron with the naturally occurring Li($A = 7, Z = 3$) nucleus yields a helium nucleus He($A = 4, Z = 2$) plus tritium T($A = 3, Z = 1$). The cross-section for this reaction is large, so the fusion power system will produce its own fuel from lithium. Alternatively, one may say that lithium is the primary fuel for the D-T fusion reactor, following Maisonnier (1989). The issues of the breeding of tritium, the neutronics of fusion reactors and the radiological aspects of the problem, are described in Kammash (1975).

1.2 World War II, Teller 1952

The path followed in this book comes from a second major lines of plasma research developed directly as a natural following up of the success in producing the hydrogen bomb from the nuclear fusion reactions in deuterium and tritium after World War II. The project led by Edward Teller was pushed aggressively by a cadre of his co-workers and there are many interesting books that describe the developments of nuclear fusion research. The recent book, *The Brotherhood of the Bomb* (2002) by Herken, describes the political and scientific developments of this era. After this military science project was concluded, the developers turned their effort to producing a controlled thermonuclear fusion reaction in the laboratory. The work took place under the project name Sherwood supported by the Atomic Energy Commission in the United States. Similar projects were developed in the Soviet Union (by the Kurchatov group) and in Great Britain. In 1958 the projects were declassified and a general science meeting was held in Geneva where all three countries showed their results. The results from the three countries were similar and showed that the plasma was typically unstable to strong electromagnetic-particle instabilities, the description of which is a large part of this book.

The fastest and largest length scale of these plasma instabilities is also common to space and astrophysical plasmas. Due to their large scale, the instabilities can often be described by a truncated description of the mean particle motion called the magnetohydrodynamic or fluid description of the plasma. There are different levels of the approximations used in the reduction of the particle orbits to the fluid models. The most extreme level of reduction, which gives the most tractable model and can be likened to the Bohr model, which is not a quantum mechanics theory, in that the single fluid ideal magnetohydrodynamic model is not derived from a systematic deduction from the many-body (MHD) problem. Rather, the MHD model is an extension of the ideal gas behavior and the inviscid (no viscosity) limit of the fluid equations with infinite electrical conductivity of the fluid. The ideal MHD model, like the Bohr atom, is never correct, but always useful. There are well-known books by Bateman (1978), Friedberg (1987, 2007), and Hazeltine and Meiss (1992) on the MHD models and their generalizations to take into account the real, or non-ideal, aspects of the plasma. We assume the reader is familiar with these descriptions.

1.3 Controlled Nuclear Fusion

The search for controlled thermonuclear fusion is dominated by the approach that uses strong magnetic fields to confine high-temperature laboratory plasmas. From the 1970s to the 1990s the prevailing thought was that the low-amplitude drift wave turbulence would not seriously impede reaching the critical Lawson condition in MHD stable machines when heated with sufficient auxiliary (external) plasma heating power, P_{aux}. Exceeding the Lawson condition, $n_e\tau_E \sim 2\times 10^{20}\,\text{m}^{-3}\text{s}$ at 15-25 KeV, is required for D-T ignition. In the Lawson condition on $n_e\tau_E$, the time $\tau_E = W_p/(P_{\text{in}} - dW_p/dt)$ is the energy confinement time, where W_p is the stored plasma energy, P_{in} is the total power input and dW/dt is the time variation of the stored plasma energy (dW/dt).

The plasma fusion community now realizes that merely increasing the auxiliary heating is not sufficient to reach ignition. As external heating is added to the plasma, the turbulence level increases dramatically and more coherent structures appear so that the confinement time is seriously degraded in comparison to that at lower levels of auxiliary heating.

1.4 Magnetic Confinement Conditions for Nuclear Fusion

A less stringent condition is derived from the fusion power P_f amplification factor defined by $Q_f = P_f/P_{\text{loss}}$. The fusion Q_f is maximized at a lower temperature between $10-20\,\text{KeV}$ where the nuclear reaction rate is $\langle\sigma v\rangle_{\text{DT}} \propto T_i^2$ so that $Q_f \sim n_e\tau_E T \gtrsim 10^{21}\,\text{m}^{-3}s\,\text{KeV}$. Here, to reduce the amplification Q_f to the fusion triple product $n_e\tau_E T$, we use $P_f = (1/4)E_f n_e^2\langle\sigma v\rangle_{\text{DT}}$ where $E_f = 17.6\,\text{MeV}$ is the fusion energy release by a D-T reaction and $P_{\text{loss}} = nT/\tau_E$. The reactivity function $\langle\sigma v\rangle_{DT}$ is shown in Fig. 1.1, where at lower temperatures the increase is proportional to the square of the ion temperature T_i^2.

For a mixture of deuterium of density n_D and tritium of density n_T, the reactivity is the product of $n_D n_T$ and $\langle\sigma v\rangle$. For charge balance between the electrons and ions we have $n_e = n_D + n_T$ and, thus, the maximum reactivity occurs when $n_D = n_T = n_e/2$. The inevitable presence of impurity ions from the plasma-wall interactions reduces the reactivity from this ideal rate as shown in Chapter 11.5.

Reference values for the drift wave turbulence parameters in the International Thermonuclear Experimental Reactor (ITER) device are given in Table 9.1 currently under construction. Coherent structures are now understood to produce intermittent plasma turbulence. There is now a broader level of recognition in the fusion community that an understanding of the chaos and coherent structures created by plasma drift wave instabilities is required to reach practical fusion energy producing regimes [Horton and Ichikawa (1996)]. In this monograph we examine in detail how the basic drift wave mechanism works.

The condition for the onset of fusion power is set by the values of the cross-sections for atomic scattering and nuclear fusion reactions. Atomic scattering cross-sections are of order $10^{-16}\,\mathrm{cm}^2$ and the nuclear cross-sections are of order $10^{-24}\,\mathrm{cm}^2$. The nearly elastic scattering of the hot electrons from the ions results in the intrinsic loss of electron energy by radiation of high frequency electromagnetic waves in the process called Bremsstrahlung radiation [Jackson (1999)]. Both processes increase with the square of the plasma density so that the ignition temperature is set by a temperature around 7 KeV for the fusion power to dominate. Chapter 16 contains a derivation of the condition for ignition.

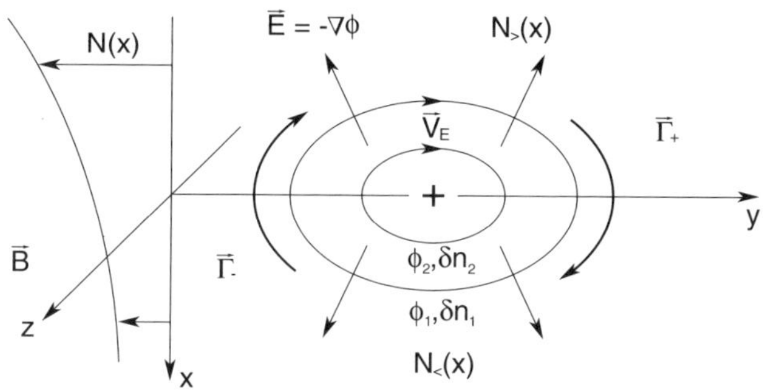

Fig. 1.2 Drift-Wave mechanism showing $\mathbf{E} \times \mathbf{B}$ convection in a nonuniform, magnetized plasma.

1.5 Nature of Plasma Turbulence

Plasmas with electrically interacting ions and electrons naturally have a variety of collective modes of oscillations occurring at low frequencies and long wavelengths compared with the more familiar oscillations of electron plasma waves. In the spatially nonuniform plasma in a magnetic field the electric fields associated with the low frequency collective modes produce drifts of the guiding centers of the charges. A characterization is given of the spatial scale of the collective modes by the gyroradius ρ_s of the ions $s = i$ and electrons $s = e$ and by the scale length L_p of the inhomogeneity of the plasma. Typically, but not always, the gyroradius scale is small compared to the inhomogeneity scale length L_n for the density n and L_T for the temperature T fields. Mathematically this is expressed by introducing epsilon as a small parameter where $\epsilon = \rho_i/L_n \ll 1$. The magnetized plasma will $\mathbf{E} \times \mathbf{B}$ convect as shown schematically in Fig. 1.2 around the maximum and minimum of the electric potential variations in the plane perpendicular to the magnetic field.

The parallel (along z) motion of the particles allows the density variations parallel to the magnetic field to align with the potential variations. The plasma ion and electron densities adjust to the electric potential $\varphi(\mathbf{x}, t)$ to be close to the local thermodynamic equilibrium Boltzmann relation, where the density, for example, becomes close to a function of the potential with $n_s(\mathbf{x}, t) = n_0 \exp\left[-q_s\varphi(x, t)/k_B T_s\right]$.

This situation is defined as the "adiabatic regime" in the plasma physics literature and the electrons in particular will move to form this situation as shown in Fig. 1.2. This adiabatic density response, while not exact, is often taken as a useful model especially when forces like the temperature gradients are driving the plasma turbulence. When the density response is adiabatic, we see from Fig. 1.2 that there is no net density convection integrated across a given radial surface. The density flux Γ_+ brought in on the right-hand side of the structure is balanced by the plasma density flux Γ_-, returned on the left side of the structure, in both the ion temperature gradient [ITG] gradient turbulence and the electron temperature gradient [ETG] driven turbulence described in Chapters 13 and 14.

In more general situations, however, the electron density is not able to stay in phase with the electric potential fluctuation as in the case with the strong temperature gradient-driven instabilities. This general case is shown in Fig. 1.3 where there is a phase shift that develops between the density and potential minima and maxima. Owing to the phase shift $\Psi_{\delta n,\phi}$ shown in the figure there is not a net transport of the density down the density gradient from the convection of higher density region that is shifted from positive potential region. The phase shift of the density from the potential is a general property analyzed in Chapters 3 and 9. The phase shifts determine the extent to which electric potentials produce a net cross-field transport, as opposed to vortex or wave motion without a net plasma transport. When the temperature gradient is driving the fluctuations, the phase shift may be positive, as shown, or negative and give rise to outward or inward convection of the plasma density.

Calculating the local particle flux, $\delta n_e v_r$, gives Γ and averaging over y or time gives $\Gamma = \sum_{\mathbf{k}} [k_y \phi_k \delta n_k / B] \sin \Psi_k$. The electron response to the potential determines the phase shift Ψ_k and amplitude ratio $\delta n_e/\phi_k$. The special case where the phase shift is zero gives $\delta n_e/n_e = e\phi_k/T_e$ and is called the adiabatic response, owing to its being the linearization of the equilibrium Boltzmann relation.

This adiabatic response is used in many of the classical drift wave models, e.g., when the temperature gradient drives the drift waves unstable rather than the density gradient ($n \to T$) as in Fig. 1.2. In turbulent plasmas this convective flux occurs at each $\mathbf{k}$ with the total transported flux being a sum over all scales of k.

The typical size of the turbulent cells is $\lambda_\perp = 2\pi/k_\perp \sim 6\rho_s$ and the frequency is approximately $\omega_k = k_y v_{de}$, where $v_{de} = -(T_e/eBn_e)(dn_e/dr) = T_e/eBL_n = c_s\rho_s/L_n$. The scale length, $\rho_s = c_s/\omega_{ci}$, is called the ion inertial scale length and determines the scale of the waves and vortex structures. In the quasi-geostrophic flow in planetary atmospheres the equivalent quantity is the Rossby radius [Horton and Ichikawa (1996)]. For reference, a present-day tokamak, e.g., Tore Supra [Hoang, *et al.* (2003)] has $R/a = 2.4\,\mathrm{m}/0.72\,\mathrm{m}$, $B = 2\,\mathrm{T}$, $n_e = 3\times 10^{19}\,\mathrm{m}^{-3}$ and $T_e = 1\,\mathrm{KeV}$. For these parameters $\rho_s/L_n = 0.1\,\mathrm{cm}/80\,\mathrm{cm}$, the sound speed $c_s = 3\times 10^7\,\mathrm{cm/s}$ and thus $v_{de} = 3\times 10^4\,\mathrm{cm/s}$ and the drift wave frequency for $k_y = 20\,\mathrm{cm}^{-1}$ is about $6\times 10^5\,\mathrm{rad/s}$ or 100 kHz. Reference values for ITER are included in Table 1.1. These

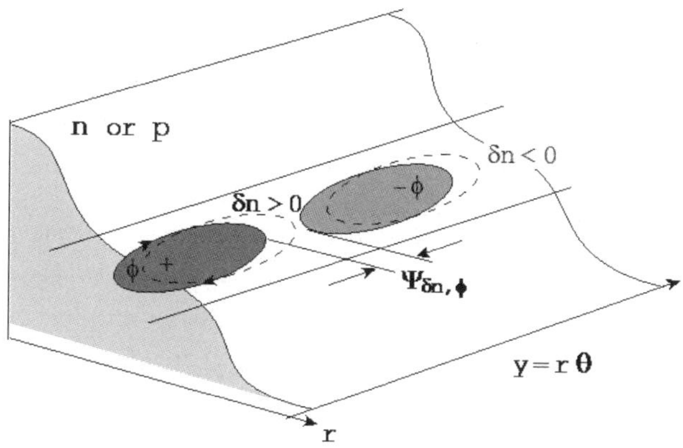

Fig. 1.3 The general case where the drift wave density fluctuations are shifted in phase from the potential fluctuations by Ψ. This results in a net flux of density down the density gradient when the particle flux is integrated over a fixed radial surface r = const outside of the annulus of the fluctuation zone. This is the general situation that is measured in the early drift-wave experiments analyzed in Chapters 3 and 4.

drift wave fluctuations have been studied in detail in a wide variety of laboratory machines.

Tore Supra, constructed during 1981-1988, reported successful long pulse (7 seconds) Ohmic discharges at the 1988 International Conference on Plasma Physics and Controlled Fusion in Nice, France [Equipe Tore Supra (1989)]. The machine's chief architect, R. Aymar (1989), reported for the Equipe Tore Supra, the results for plasma parameters up to $T_e = 1.5\,\text{KeV}$ and $n_e = 4 \times 10^{13}/\text{cm}^3$ for Ohmic discharges with loop voltage $V_{\text{loop}} = 1\,\text{V}$ Volt and plasma currents $I_p = 600\,\text{kAmp}$ maintained for four seconds. The machine parameters are given in Table 1.2. A twenty-year perspective on the research from Tore Supra is given in Aymar (2001).

Table 1.1 ITER parameters with typical drift wave frequencies and wavenumbers.

Installed auxiliary heating	73 MW
Fusion power	500 MW
Plasma major radius (R)	6.2 m
Plasma minor radius (a)	2.0 m
Plasma volume	830 m^3
Plasma current(I_p)	15 MA
Central solenoid flux	120 Wb
Toroidal field at 6.2 m radius	5.3 T
Density gradient length (L_n)	1 m
Electron temperature (T_e)	10 KeV
Electron Diamagnetic Drift Velocity (v_{de})	1.9 km/s
Ion inertial radius (ρ_s)	0.3 cm
Gyro-Bohr diffusivity ($D_{gB} = \rho_s v_{de}$)	5.7 m^2/s
Drift wave frequency ($\omega = k_y v_{de}$)	190 KHz at $k_y \rho_s = 0.3$

After several years of operation, it was reported at the 1988 IAEA in Nice that the Tokamak Fusion Test Reactor (TFTR) [Bell, *et al.* (1989); Hawryluk (1998)], with 30 MW of balanced deuterium neutral beam heating, achieved an ion temperature of 32 KeV and the triple product of $n_e(0)\tau_E T_i(0) = 4\times10^{20}\,\mathrm{m^3\,s\,KeV}$. Since the electron temperature remained at about 8-10 KeV, this type of hot ion mode of operation is called the "hot ion supershot" regime, and is one of the promising approaches to fusion power. The Bell, *et al.* (1989) TFTR article describes how deuterium-beams into deuterium plasma converts the measured neutron flux of 3×10^{16} /s into what it would be with deuterium-tritium fuel fusion power. The conversion predicts the tritium experiments should yield a fusion of $Q_{\rm fus} = 0.5\pm0.13$. In 1998 the tritium experiments in TFTR were concluded. The highest fusion power amplification achieved was $Q_{\rm fus} = 0.5\pm0.13$.

Smaller, but important, tokamaks reported encouraging results at the Nice 1988 IAEA meeting including Doublet III, PDX, ASDEX, Alcator C-Mod, TEXTOR, ISX-B, PBX-M and then new, large Japanese tokamak JT60. The encouraging results from these experiments led to formation of groups to study a design for the NEXT-STEP machine that eventually became ITER. An important step towards demonstrating steady-state plasma confinement was obtained by the Tore Supra tokamak with parameters given in Table 1.2. In this experiment the superconductivity coils allow confinement for 10^3 s with a gigajoule of plasma energy passing through the system at KeV plasma temperatures. Currently, ITER is several years under construction at the Cadarache Laboratory in the Luberon Mountains outside of Aix-en Provence.

Table 1.2 Tore Supra super conducting Tokamak in Cadarache, France 1988-present.

major radius R	2.37 m
minor radius a	0.8 m
toroidal magnetic field	4.5 T
plasma current	1.7 MA
discharge duration	60 s
solenoid flux swing	21 Wb
real-time feedback control of plasma	
niobium-titanium windings	1.8 K

Drift waves are sufficiently slow so that the electrons adjust to the potential fluctuations according to

$$E_\parallel + \frac{1}{en}\nabla_\parallel p_e = 0. \tag{1.1}$$

Equation (1.1) states that the forces acting on the electron fluid parallel to the magnetic field line are in a state of balance. Out of balance, the electrons will adjust rapidly by motion parallel, or along the magnetic field lines. Thus, the waves have the opposite polarization of the MHD waves, which have $E_\parallel = 0$. The parallel electric field, measured in detail in the experiments, is a key part of the drift waves and the related trapped particle mode instabilities. The fluctuating electron current, $\delta j_\parallel$, driven by the parallel electric field takes the energy $\langle \delta j_\parallel E_\parallel \rangle < 0$ out of the nonuniform electron distributions and produces the wave growth. The effect is as though the plasma had a negative resistivity/conductivity $\sigma_\parallel(\omega, \mathbf{k})$ for the drift wave.

When there is zero or negligible phase shift between the density and the potential fluctuation, the $\mathbf{E} \times \mathbf{B}$ convection produces vortex structures shown in Fig. 1.2. These plasma structures can have large amplitudes forming nonlinear coherent vortices in the plasma. The study of these nonlinear structures and their propagation speeds is a rich area that is explored in Chapter 10. Under the presence of unfavorable field line magnetic curvature these vortices are accelerated out of the plasma down the effective gravity created by the charge polarization from the magnetic curvature drifts of the ion and electron guiding centers. The effect is to produce a bursty, intermittent transport that adds to the background of turbulent wave transport.

Detailed measurements and comparisons with theory suggest that, in fact, one of the most serious limiting factors for the success of ITER and similar machines is the control of drift wave turbulent transport. The seriousness of the turbulence becomes more evident as machines with increasing injected power $P_{\rm aux}$ reach core power densities of $1\,\mathrm{MW/m^3}$. The plasma temperature rises only as a fractional power law, like $P_{\rm aux}^{1/2}$, owing to the associated increased turbulence levels, and the energy confinement time, τ_E, thus decreases even though the temperature increases. Thus, the triple fusion power product may remain essentially constant.

As a consequence of this stronger heating power, $P_{\rm aux}$, the turbulence is amplified, which shortens the plasma energy confinement time, $\tau_E = W/P_{\rm aux}$, in agreement with the ITER database [I.P.E. Group (1999); Kaye, *et al.* (1997); L-H Mode Database (1994)]. The drift wave turbulence has explained the empirical database laws for both helical systems [Yamada, *et al.* (2005)] and Tokamaks [I.P.E. Group (1999); Kaye, *et al.* (1997)], showing that $\tau_E = \tau_E^0 (P_0/P_{\rm aux})^{2/3}$. Thus, doubling the $P_{\rm aux}$ shortens the confinement time, $\tau_E \rightarrow \tau_E^0/1.59 = 0.64\tau_E^0$. For example, the reference design for ITER (see Table 1.1) has $P_{\rm ICRF} = P_{\rm ECE} = 20\,\mathrm{MW}$ and $P_{\rm NBI} = 16\,\mathrm{MW}$ injected into the plasma volume of $V = 2\pi^2\,\mathrm{Rab} \simeq 800\,\mathrm{m^3}$. The average power density is then $0.07\,\mathrm{MW/m^3}$. To reach the power density of $1\,\mathrm{MW/m^3}$ at the core, as in current large tokamaks, the ICRH and ECE power needs to be focused in the central core volume of $40\,\mathrm{m^3}$, corresponding to the core plasma radius $r_c = 0.6\,\mathrm{m}$.

In toroidal magnetic geometry there needs to be mechanisms for stabilizing the plasma from interchange motions that are analogous to the interchange instability that occurs when a dense liquid or gas is supported above a light, less dense gas or liquid. This instability is called the Rayleigh-Taylor instability in hydrodynamics. In plasmas there is analogous instability called the magnetic Rayleigh-Taylor insta-

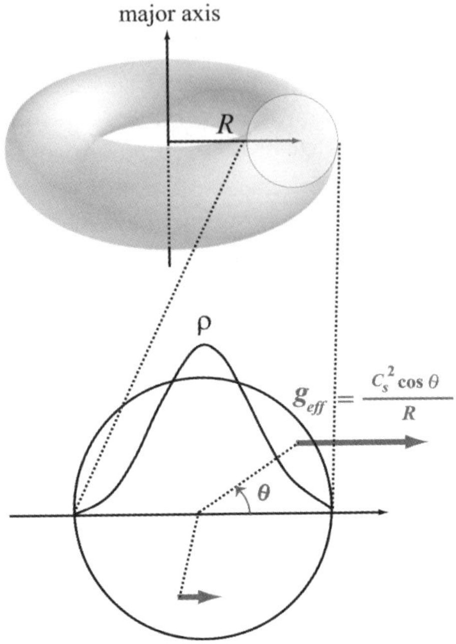

Fig. 1.4 Schematic showing the effective gravity that arises from separation of electric charge in a plasma in a toroidal confinement magnetic field. The effective acceleration represented by $g_{\rm eff}$ in the lower figure arises from the motion of the particles in the curved magnetic field lines of the torus. The sound speed is $c_s = \sqrt{[(T_e + T_i)/m_i]}$ and R is the distance from the axis of symmetry to the observation point. The profile ρ is the mass density.

bility. The effective plasma acceleration $g_{\rm eff}$ shown in Fig. 1.4 arises from electric forces created by the separation of the ion and electron charges.

The detailed description of the plasma interchange modes is complicated and derived in Chapter 5. The direction of the effective gravity is horizontal and directed outward from the axis of symmetry. Thus, for the peaked plasma mass density profile ρ shown in the lower diagram, there is a strong acceleration determined by the sound speed c_s squared divided by the major radius R of the torus. This configuration is unstable to the onset of an interchange of plasma from the higher density core to the lower density outer region. The methods of maintaining the stability are described in Chapter 6 in the context of the MHD model. There it is shown that the tension in the magnetic field attached to the plasma provides the restoring force. This balance between the magnetic tension force and the acceleration from $g_{\rm eff}$ give the fundamental limit on the maximum plasma pressure that a torus with a specified magnetic field B_ϕ and major radius R can contain. This pressure limit is the called the beta limit and is given by an expression called the Tryon formula [Troyon, *et al.* (1984)]. The limit is thoroughly validated by the extensive tokamak database accumulated from the many well-documented tokamak experiments performed over the past forty years. The details are found in Chapters 5 and 6.

Figure 1.5 gives a few of the key equations that will be derived in Chapter 5 along with the analogous equations that describe stratified neutral fluids. The physical constants used in the calculations for the plasma parameters of the plasmas such as the frequencies, growth rates and diffusivities are given in Table 1.3.

1.6 Breakthrough with Tokamak Confinement

In 1968, there was a breakthrough in the confinement of hot plasmas by the creation of the plasma with ohmic resistive heating in a toroidal chamber where the plasma itself forms the effective secondary coil of the transformer. By driving magnetic flux $\Phi_1(t)$ in the primary solenoid in one direction to a maximum number of volt-seconds, a toroidal voltage $2\pi R E_\phi$ is induced from the toroidal electric field E_ϕ that ionizes and heats the plasma. The successful creation of a thermal plasma with T_e approaching a $T_e \sim 10^7\,\mathrm{K} \sim 1\,\mathrm{KeV}$ temperature with energy confinement times τ_E of order milliseconds was announced by Artsimovich, *et al.* (1969a,b) and the Tokamak Team at the Kurchatov Laboratory in Moscow.

The result of such high electron temperature measured by the diamagnetic loop that gives the total plasma energy content $W_p = 3/2 \int_V n(T_e + T_i) d^3x$ rather than T_e directly was met with much skepticism. The argument was settled in favor of the measured energy W_p being a true thermal temperature by the invited British team of Peacock, *et al.* (1969). The British team used the new technique of laser scattering off the electrons by the Thomson cross-section (discussed in Chapter 2.8) to verify directly that the electron velocity spectrum was close to a Maxwell-Boltzmann distribution. The impact of the discovery of toroidal confinement times much longer than the prevalent Bohm confinement times led many laboratories to change to the tokamak device as their leading fusion confinement experiment. Steady progress was made on an international basis in advancing the plasma confinement toward the condition given by Lawson for ignition for fusion.

For the subject of this book of turbulent transport, it is interesting to note that the two key formulas written on the blackboard in Artsimovich's lecture on the experiment while visiting the Princeton Plasma Physics Laboratory in 1968, were the effective electrical conductivity σ^* from the ohmic current driven by the transformer and the effective electron thermal diffusivity K^* from the energy confinement time. The type of formula Artsimovich wrote was called pseudo-classical transport scaling with

$$K^* = \nu_{\text{eff}}\, \rho_{e\theta}^2 \tag{1.2}$$

where ν_{eff} is the effective electron collision frequency defined by the measured electrical conductivity $\sigma^* = ne^2/m_e\nu_*$ and $\rho_{e\theta} = (m_e T_e)^{1/2}/eB_p(a)$ where $B_p(a)$ is the maximum poloidal (B_p) magnetic field produced by the ohmic plasma current I_p.

The thermal diffusivity formula Eq. (1.2) is an example of the general principle that the confinement in the tokamak depends on the self-consistent magnetic field B_p created by the plasma current. In formula Eq. (1.2) K^* is inherently limited by the strength of external applied magnetic fields. The external field B_T sets the stability limit for the tokamak system. For plasma currents I_p such that the poloidal field $B_p > aB_T/R$ a power kink instability occurs and the plasma current is lost from the core. This rapid event called a major or minor current disruption occurs when the central twist of the magnetic field becomes too strong. The critical twisting is measured by the "safety factor" $q = aB_T/RB_p$ and occurs when q drops well below unity or hovers just below unity.

Pressure Gradient Interchange

$p' \equiv -\frac{p}{L_p}$ and $\boldsymbol{E}_\perp = -\boldsymbol{v}\times\boldsymbol{B}$

$$\gamma_{\mathrm{MHD}}^2 = -\frac{p'}{\rho}\frac{2}{R}$$

blobs accelerate out

$$g_{\mathrm{eff}} = 2c_s^2/R \sim 2(3\times 10^5\,\mathrm{m/s})^2/(3\mathrm{m}) \sim 6\times 10^{10}\,\mathrm{m/s^2}$$

over outside of torus

g_{eff} sets limits on three scales:
MHD p', ρ_i-ITG and ρ_e-ETG.

Rayleigh-Taylor

gravity g and
density gradient

$$\gamma_{\mathrm{RT}}^2 = -\frac{g}{\rho}\frac{dp}{dr} = -\frac{\boldsymbol{g}\cdot\nabla\rho}{\rho}$$

Fig. 1.5 Table with key formulas showing the relationship of the pressure gradient driven interchange instability in a magnetized plasma with the classical neutral fluid gravitation instability in an inverted mass density profile. Key equations for the magnetized plasma instability are on the left and the neutral fluid analog on the right. The role of gravity is played by the centrifugal force the charged particles experience moving along the curved magnetic field lines, and the role of the mass density gradient is played by the plasma pressure gradient. The outer half of a torus has an unstable stratification of the pressure gradient, and the instability operates in the drift waves on the micro-scales of the ion and electron gyroradius as explored in Chapters 13 and 14.

1.7 Confinement Records Set in Early Tokamaks

The evolution of the tokamaks to their present state as in the currently operational Joint European Torus (JET) may be thought of as occurring in three generations, although at the time there were few dramatic changes in the approach and the evolution was more continuous than generational. The first generation would be those built in the 1970s to test the breakthrough described in Section 1.6 an-

Table 1.3 Physical constants in mks units.

Physical Quantity	Symbol	Value	mks Units
Boltzmann constant	k_B	1.3807×10^{-23}	$\mathrm{JK^{-1}}$
Elementary charge	e	1.6022×10^{-19}	C
Electron mass	m_e	9.1094×10^{-31}	kg
Proton mass	m_p	1.6726×10^{-27}	kg
Gravitational constant	G	6.6726×10^{-11}	$\mathrm{m^3s^{-2}kg^{-1}}$
Planck constant	h	6.6261×10^{-34}	Js
	$\hbar = h/2\pi$	1.0546×10^{-34}	Js
Speed of light in vacuum	c	2.9979×10^{8}	$\mathrm{ms^{-1}}$
Permittivity of free space	ϵ_0	8.8542×10^{-12}	$\mathrm{Fm^{-1}}$
Permeability of free space	μ_0	$4\pi \times 10^{-7}$	$\mathrm{Hm^{-1}}$
Proton/electron mass ratio	m_p/m_e	1.8362×10^{3}	
Electron charge/mass ratio	e/m_e	1.7588×10^{11}	$\mathrm{Ckg^{-1}}$
Bohr radius	$a_0 = \epsilon_0 h^2/\pi m e^2$	5.2918×10^{-11}	m
Atomic cross-section	πa_0^2	8.7974×10^{-21}	$\mathrm{m^2}$
Classical electron radius	$r_e = e^2/4\pi\epsilon_0 mc^2$	2.8179×10^{-15}	m
Thomson cross-section	$(8\pi/3)r_e^2$	6.6525×10^{-29}	$\mathrm{m^2}$
Compton wavelength	$h/m_e c$	2.4263×10^{-12}	m
of electron	$\hbar/m_e c$	3.8616×10^{-13}	m
Fine-structure constant	$\alpha = e^2/2\epsilon_0 hc$	7.2974×10^{-3}	
	α^{-1}	137.04	

nounced by Artsimovich and the Kurchatov Nuclear Laboratory group, including many famous Soviet scientists. The second generation would be designs with shaping, such as the doublet design of Ohkawa that evolved into the current DIII-D machine, and in Europe the building of the ASDEX and the Tokamak Configuration Variable. Next came the construction of the super conducting tokamak for steady-state operation in the machine, Tore Supra, that still may be included as the steady-state superconducting machine. Tore Supra holds the world record for the longest discharge of 15 min with more than one gigajoule of energy passing through the well-confined plasma. The third generation would then be the large machines built to attempt to reach the Lawson conditions for break-even, or to at least have a few energy confinement times during which the fusion power produced by the 17.6 MeV released energy per fusion event would equal the injected heating power. These machines, in historical order of approaching this $Q^{\mathrm{DT}}_{\mathrm{fusion}} = 1$, are the Princeton Plasma Physics Laboratories' TFTR (Tokamak Fusion Test Reactor), the JET (Joint European Tokamak), and the Japanese Atomic Energy Research Institute (JAERI) machine called JT60-U where the 60 designates that the plasma chamber volume is $60\,\mathrm{m}^3$. A complete account of the TFTR experiments leading up to their record shot number 76778 yielding approximately 9 MW of fusion power from $P_{\mathrm{aux}} = 34\,\mathrm{MW}$ of injected power is described in the educational review article of Harwyluk (1998). The record fusion power projection is described at `http://www.jet.efda.org/jet/history-anniversaries/`.

1.7.1 *First generation tokamaks: Ormak, PLT, Alcator, ATC and TFR*

ORMAK was the first neutral beam heated tokamak to achieve $T_i > T_e$ with $T_i \sim 1.5\,\text{KeV}$. The PLT, ORMAK and Alcator machines found that the energy confinement time, τ_E, increased with density, $\overline{n}_e$. Using variable machine sizes (a, R), the energy confinement scaling law was inferred to be a multiplicative constant times the formula $\tau_E = \overline{n}_e\, a^2 R$ with the constant such that for the high density $\overline{n}_e = 3 \times 10^{14}\,\text{cm}^{-3}$ the Alcator plasma reached the confinement time reached 20 ms. The projections based on this Alcator scaling law were that second-generation large Tokamaks would reach a confinement time $\tau_E \lesssim 1\,\text{s}$. The results, however, would turn out to be that the maximum achieved energy confinement time in TFTR was $\tau_E \leq 0.1\,\text{s}$. The Soviet Union at Kurchatov Laboratory in Moscow achieved similar results [Razumova (1984)] shown in Table 1.4in a machine designated as T-11.

The Adiabatic Toroidal Compressor tokamak was an early air core transformer tokamak with 15 KeV neutral beam injection with up to 100 kW of power. The plasma rotation was measured by the Doppler shifts of the kink modes measured with soft x-rays and magnetic loops. The NBI heating increased the ion temperature from 200 eV to 350 KeV and produced a plasma rotation speed of about 5 km/s. The core electron temperature reached 1.2 KeV. Drift waves were measured in ATC as described in Chapter 2.9.5 by scattering of electromagnetic waves.

Compression experiments decreased the major and minor radius simultaneously and showed the adiabatic increase of the T_i and less than adiabatic increase in T_e. Nominal parameters were $B_t \sim 1.6\,\text{T}$ and $I_p \sim 60\,\text{kamp}$ with $n_e \sim 3 \times 10^{13}/\text{cm}^3 = 3 \times 10^{19}/\text{m}^3$.

At that time (in the 1970s), the MIT group was getting high density discharges with $n_e \sim 3 \times 10^{14}$ to $10^{15}\,\text{cm}^{-3}$ in the Alcator machine with $R = 54\,\text{cm}$ and the limiter radius $a_L = 10\,\text{cm}$ with $B_t = 8\,\text{T}$ and $I_p = 160\,\text{kA}$ and loop voltage of 2 to 3 volts. The core $T_e \sim 1100\,\text{eV}$ at core $n_e = 3 \times 10^{14}/\text{cc}$ from Ohmic power of 480 kW.

As described in Chapter 2.9, Mazzucato carried out microwave scattering experiments on ATC [Mazzucato (1976)] that showed the frequency and wavenumber spectrum consistent with those expected from drift waves. Horton (1976) gave a theoretical derivation of the drift wave spectrum explaining the peaking of the fluctuation spectrum at $k_\perp \rho_s \sim 0.3$ at the amplitude consistent with the scattering data with $\delta n_e / \langle n_e \rangle \sim 0.01$ to 0.03.

The Princeton Large Torus (PLT) experiments reached $I_p = 600\,\text{kA}$ with peak electron temperatures of 2 KeV for intervals up to $\delta t = 1\,\text{s}$.

ORMAK (Oak Ridge Tokamak) achieved ion temperatures of 1.5 KeV with neutral beam heating. This machine was the first tokamak to achieve $T_i > T_e$.

The energy confinement time $\tau_E \sim 20\,\text{ms}$ in Alcator at $n_e = 3 \times 10^{14}$ so that a record for the Lawson product $n_e \tau_E \sim 6 \times 10^{12}\,\text{cm}^{-3}\,\text{s}$ was reached. This confinement was within a factor of 20-30 of the Lawson condition. All these machines

Table 1.4 Six First-Generation Tokamak Parameters.

Device	R/a	B_T/I_p	W	τ_E
PLT	1.34 m/.4 m	3.2 T/400 kA	10 kJ	10-30 ms
Alcator	0.54 m/0.1 m	8.2 T/160 kA	10 kJ	20 ms
ATC	0.86 m/.12 m	.16 T/60 kA	5 kJ	3 ms
TFR	1 m/.2 m	6 T/360 kA	50 kJ	30 ms
ORMAK	0.8 m/.23 m	2.6 T/230 kA	10 kJ	10-15 ms
T-11	1.5 m/.34 m	4.5 T/430 kA	20 kJ	40 ms(OH)-30 ms (ECH)

showed the Alcator scaling law for the energy confinement time with

$$n_e\tau_E \sim n_e^2 a^2 R. \tag{1.3}$$

So the path to ignition by increasing the physical dimensions seemed clear.

1.7.2 *TFTR and the D-T fusion plasmas*

From these results the design of the TFTR machine was made with the expectation of an energy confinement time of $\tau_E \sim 1$ s and Lawson product of $n_e\tau_E \sim 6 \times 10^{13}$ to $10^{14}\,\text{cm}^{-3}\,\text{s}$. It took over ten years to complete the experiments on TFTR and the highest Lawson product achieved was $n_e\tau_E \sim 2 \times 10^{13}\,\text{cm}^{-3}\,\text{s}$. The reason is now clear that the confinement time degrades with increasing plasma temperature and does not continue to increase linearly with density, as given in Eq. (1.3). The result is due to the drift wave turbulence increasing the thermal diffusivity. The Tokamak Fusion Test Reactor (TFTR) operated at the Princeton Plasma Physics Laboratory (PPPL) from 1982 to 1997. TFTR set records for the core plasma temperature with $T_i(0) = 45$ KeV equivalent to 500 million degrees centigrade [Hawryluk (1998)]. While not achieving break-even as originally planned, the TFTR tokamak achieved all of its hardware design goals and thus made substantial contributions in many areas of fusion technology development (`http://www.pppl.gov//tokamakfusiontestreactor.cfm`).

The parameters of key tokamaks discussed in some detail in this book are given in Table 1.4, Table 1.5, and Table 1.6. An overview of the history of the progress towards the design of the ITER machine from the forerunner design committee for the INTOR device is found in Stacey (2010). In an appendix this book gives a list of the parameters of some hundred tokamaks starting with the T-1 device in the USSR in 1957 and going through to ITER with expected operation in 2020. The list includes EAST and KSTAR new superconducting tokamaks in China and South Korea.

1.7.3 *Third-generation tokamaks with international growth*

The new results on the effect of optimizing the plasma cross-sectional shape did not come in time to influence the design of TFTR. These shaping results eventually came from Doublet II and Doublet III from General Atomics at San Diego. The PDX (Princeton Divertor Experiment) experiment at Princeton showed that higher confinement and larger plasma currents can be achieved for given major radius and toroidal magnetic field by using an elliptically elongated cross-section. The TCV machines in Ecole Polytechnique Lausanne extensively documented the effect of plasma shaping. In the 1980s the new machines JT60-U, JET and ITER were designed with elliptical cross-sections to take advantage of the higher plasma current limits. The final design parameters of ITER were settled on in a "Final Design Report," [ITER-EDA (1998)]. The world record was achieved for the Q-value, the ion temperature, and the fusion triple product.

The TCV machines in Ecole Polytechnique Lausanne extensively documented the effect of plasma shaping [Goodman, *et al.* (2003)].

The JT60-U tokamak did not operate with tritium (T). When the neutron yield from the D-D collisions is measured and converted to what the yield would have been if the working gas were 50% D and 50% T. The record value of the fusion multiplication would have been achieved with the equivalent $Q_{DT} = 1.25$ according to the JAERI publications [Fujita, *et al.* (1999)]. These record discharges had neutron emission rates of 3.6×10^{16} n/s at the end of a 4 s current ramp-up of I_p, starting at 0.5 MA and rising to 2.5 MA. Feedback control using the measured neutron production to control the neutral beam power injection rate was used to get the plasma stably through the unstable region where the minimum value of the safety factor q reached the value 3. The profiles of density ion and electron temperature all showed the "box" shape with a steep gradient occurring at the magnetic surface, where the safety factor $q(r)$ reached its minimum value, which was just above 2 at the end of the discharge. The record discharge was shot number E31872 and at the final time $n_e(0) = 8.5 \times 10^{19}/\text{m}^3$, $T_i(0) = 16.8$ KeV and $T_e(0) = 7.2$ KeV. The energy confinement time was reported as $\tau_E = 1.07$ s, giving a multiple over the emperical tokamak scaling law for energy confinement, called the ITER89P formula for τ_E of $H_{\text{ITER89P}} = 3.2$. The actual fusion power produced from the D-D nuclear reactions divided by the 10 MW of neutral beam power is $Q_{DD} = 5.6 \times 10^{-3}$ which when multiplied by 220 for the ratio of the energy released in the two different nuclear reactions gives the published value of the equivalent performance of $Q_{DT} = 1.25$. There are issues as to how reliable this conversion between two different working gases, along with much higher energy products from the D-T reactions, can be carried out. More energetic nuclear decay products produce more influx of impurities from the walls of the vessel to the plasma volume.

The JET tokamak with $R/a = 2.88\,\text{m}/0.9\,\text{m}$ with $B/I_p = 4\,\text{T}/4.5\,\text{MA}$ produced record discharges by 1995-1999 with what has become a standard regime of the ITER design called the H-mode for high confinement. During the 1997 campaign with DT experiments the best results were obtained in the H-mode with an edge transport barrier or an edge pedestal [Cordey, *et al.* (1999); Gibson and the JET Team (1998)].

The H-mode plasma has bursty releases of stored plasma energy across the edge pedestal that are called Edge Localized Modes or ELMs for short. The ELMy H-mode in the D-T plasma had the longest energy confinement time now reaching $\tau_E \sim 700\,\mathrm{ms}$ at densities of $\langle n_e \rangle \gtrsim 10^{19}\,\mathrm{m}^{-3}$. By the late 1990s extensive confinement databases were assembled and elaborate scaling laws with long names were published. The scaling laws will not be presented in this book, but the interested reader is referred to the Nuclear Fusion articles on ITER design. For example, the Cordey, *et al.* (1999) article compares their DT shots with the ITERH-EPS 97(y) and the ITERH-93P scaling laws given in their Eq. (1.1) and Eq. (1.2). The first detailed engineering and physics design studies for the original international test reactor are given in Aymar, *et al.* (2001), where the device was called ITER-FEAT with larger major radius ($R = 8\,\mathrm{m}$) than the final machine under construction as specified in Table 1.1 with a major radius of $R = 6.2\,\mathrm{m}$.

In the 1970s and 80s, the Soviet Union physicists proceeded with both the tokamak machines building T-4, T-10 and proposing a large T-20 device. The Kurchatov Laboratory in Moscow had difficulty maintaining a steady research program and so the information available from these machines is too limited to repeat here. The Soviet leaders realized that they would not have sufficient financial strength to carry out the design and construction of the fusion burn device and proposed already in 1978 at the US-USSR Joint Fusion Power Coordinating Committee Meeting to have an international effort at building an international tokamak similar in size to the current ITER device. After the collapse of the Soviet Union in 1991, the Russian Federation President, President Gorbachev, continued at a much-reduced level, to participate in the international effort to design a "proof of principle" or POP tokamak device. The Academician, Boris Kadomtsev, a brilliant theoretical physicist, was designated the Russian representative to the international tokamak design group. The understanding of the sawteeth and minor disruptions in the core of the tokamak starts with the Kadomtsev models of magnetic reconnection in the core of the tokamaks as explained in Chapter 15. For those seeking more information on the Soviet Union and the Russian Federation's contribution to the magnetic fusion program we suggest reading the reports in the bi-annual publications from the International Fusion Energy Conferences organized by the International Atomic Energy Agency in Vienna, Austria. Many important concepts and models included in this book have their origins from the research in the Soviet Union. In addition to the tokamak confinement device, the high-energy neutral beam-driven mirror machines developed at the Budker Institute in the Soviet science city of Novosibirsk and the development of the neoclassical theory of particle transport in toroidal confinement systems by R. Sagdeev and his young associate Alex Galeev are mentioned. In the 1980s, Professors Sagdeev and Galeev became leaders of the Space Research Institute located just outside Moscow. This institute made many important contributions to space plasma physics.

The large shaped cross-section tokamak called JT60-U for the Japanese Tokamak with approximate volume of 60 m^3 was operated at the JAEA near Mito with parameters given in Table 1.5 from 1990 until 2007. In 1997, outstanding results from the JT60 Team were reported in Koide, *et al.* (1997). All JT60-U experiments were performed in deuterium plasmas so an extrapolation of order 200 is required to convert the neutron yield from the D-D plasma to the D-T plasma assuming the only difference is with replacing the nuclear cross-sections.

Table 1.5 Parameters of the third generation large tokamak JT60-U built at the Japan Atomic Energy Research Institute, Naka, Japan.

A. Device concept and components	
Concept	tokamak[1]
Vacuum vessel: Plasma facing material:	carbon[1]
Divertor/limiter facing material	carbon-fiber composites[1]
Working gas	D2[1]
Boundary configuration:	divertor[1]
Toroidal magnetic field (on axis), $B(T)$:	4.2[1]
B. Plasma geometry	
Major radius, $R(m)$:	~ 3.4[1]
Minor radius, $a(m)$:	~ 1.0[1]
Plasma volume (m^3):	< 100[1]
Remarks:	Plasma elongation = 1.2-1.8[1]
C. Plasma parameters (maximum values)	
Central ion temperature, T_i (KeV):	45[2]
Central electron temperature, T_e (KeV):	26[3]
Line averaged plasma density, $\langle n \rangle$ (10^{20} m^{-3}):	-
Central electron density, n (10^{20} m^{-3}):	2.8[4]
Plasma current, I (MA):	5[13]
Discharge pulse duration (s):	15[1]
Plasma beta, $\langle \beta \rangle$(%):	2.7[5]
Stored energy, (MJ):	10.9[6]
Energy confinement time, $\tau)E(s)$:	1.1[7]
D. Plasma heating and current drive (device values)	
Ohmic heating, (MW):	-
NBI heating, (MW):	40[8]
RF heating: -IC: (MW):	7[9]
RF heating: -EC: (MW):	3[10]
RF heating: -LH: (MW):	8[11]
E. Reactor relevant plasma performance parameters	
Triple fusion product, (10^{20} KeV sm^{-3}):	15.3[1]

[1]N. Hosogane, *et al.* "Development and operational experiences on the JT60-U tokamak and power supplies," Fusion Sci. and Tech., 42, 368 (2002).

[2]S. Ishida, *et al.* "High performance experiments in JT60-U high current divertor discharges," Proc. 16th Inter. Conf. on Fusion Energy, Montreal, Vol. 1, 315 (1996).

[3]S. Ishida and the JT60 Team, "Recent results and future plan on JT60-U," to be published in Proc. in 19th IEEE/NPSS Symposium on Fusion Engineering, Atlantic.

[4]Y. Kamada, *et al.* "Pellet injection with improved confinement in JT60," Proc. 13th Inter. Conf. on Plasma Phys. and Control. Nucl. Fusion Res., Washington, Vol. 1, 291 (1991).

[5]Y. Kamada, *et al.* "High triangularity discharges with improved stability and confinement in JT60-U," Proc. 16th Inter. Conf. on Fusion Energy, Montreal, Vol. 1, 247 (1996).

[6]S. Ishida, *et al.* "Achievement of high fusion performance in JT60-U reversed shear discharges," Phys. Rev. Lett., 79, 3917 (1997).

[7]T. Fujita, *et al.* "High performance reversed shear plasmas with a large radius transport barrier in JT60-U," Nucl. Fusion, 38, 207 (1998).

[8]M. Kuriyama, *et al.* "Operation and development on the positive ion-based neutral beam injection system for JT60 and JT60-U," Fusion Sci. and Tech., 42, 410 (2002).

[9]H. Kimura, *et al.* "ICRF Heating and TAE Modes in Reactor Relevant JT60-U Discharges," Proc. 16th Inter. Conf. on Fusion Energy, Montreal, Vol. 3, 295 (1996).

[10]Y. Ikeda, *et al.* "The 110 GHz electron cyclotron range of frequency system on JT60-U: design and operation," Fusion Sci. and Tech., 42, 435 (2002).

[11]Y. Ikeda, *et al.* "Development and operation of JT60 LHRF launchers plasma devices and operations," Vol. 1, 155 (1991).

[12]T. Fujita, *et al.* "High performance experiments in JT60-U reversed shear discharges," Nucl. Fusion, 39, 1627 (1999).

[13]JT60 Team (Presented by M. Kikuchi), "Recent JT60-U result towards steady state operation of tokamaks," Proc. 15th Inter. Conf. on Plasma Phys. and Control. Nucl. Fusion Res., Seville, Vol. 1, 31 (1994).

Under this assumption on the equivalence of the D-D experiments to an equivalent and D-T plasmas, the Koide and the JT60-U team [Koide and JT60 Team (1997)] published that they achieved the equivalent reactivity of a break-even experiment with $Q_{\text{fusion}} = 1$. In this measure of performance, the Q_{fusion} is the energy release per second expressed as a multiplication factor defined by the ratio of the nuclear power P_{fus} from nuclear fusion reactions to the injected heating power P_{aux}.

The high performance of the JT60-U plasmas was achieved by having reversed magnetic shear (RMS) plasmas with internal transport barriers in both the electron and ion thermal flux channels.

Wolf (2003) reviews the various types of internal transport barriers in different plasma confinement modes in detail. The regimes with higher ion than electron temperature are analyzed separately from those with higher electron than ion temperatures. The types of confinement regimes such as the RS regime with reversed magnetic shear are presented. The types of MHD oscillations are discussed by Wolf (2003). The edge-localized modes called ELMs that are categorized in the literature as Type-I, Type-II and Type-III are described and the confinement modes such as ELMy H-mode free and ELM-free H-mode are characterized. The avoidance of large type I ELMs has become a high priority for the operation of ITER so numerous ideas have been advanced to explain the onset of these large ELMs and methods of controlling the ELMs with external coils has grown in to a major research area.

A sudden convection of hot core plasma into the cool edge plasma results in a radiation spike in the visible (red) line radiation from the neutral hydrogen or deuterium gas as the readily measured diagnostic for the edge plasma localized modes. The red line radiation with wavelength 656 nm arises from the $n = 3$ to $n = 2$ electron transition in a neutral hydrogen gas.

This allowed the NBI driven plasma with $P_{\text{abs}} = 17\,\text{MW}$ and $I_p = 2.8\,\text{MA}$ in the chamber with major radius $R = 3.1\,\text{m}$, minor radius at the mid-plane of $a = 0.73\,\text{m}$, and toroidal magnetic field $B_t = 4.3\,\text{T}$ plasma to achieve the high energy confinement time of order one second with a stored plasma energy $W_p = 10.9\,\text{MJ}$.

In more detail, the energy confinement time for this plasma was estimated to be $\tau_E = 0.97\,\text{s}$ and the D-D plasma neutron emission is $S_n = 4.5 \times 10^{16}\,\text{n/s}$ corresponding to $Q_{\text{DD}} = 4.7 \times 10^{-3}$. The types of reversed q-profiles and temperature profiles are shown Chapter 10.

In Chapter 2 the dispersion relations are developed for the drift waves, ion acoustic waves, and the Alfvén waves that form the low-frequency collective oscillations that lead to turbulent transport of the plasma across the confining magnetic fields.

1.8 JET Record Fusion Power Experiments

The Joint European Tokamak called JET was designed in the 1970s and has produced significant fusion power from deuterium-tritium plasmas. The JET experiment, approved in 1975, was a major step up from the second-generation tokamaks that had plasma currents $I_p \simeq 300\,\text{kA}$ and volumes of order $1\,\text{m}^3$. JET was designed and did operate with plasma currents $I_p = 4\,\text{MA}$. The plasma chamber was designed to have a small aspect ratio $R/a \leq 3$ and height $b/a = 1.7$ allowing a large plasma current I_p for the given magnetic field $B_T \leq 4\,\text{T}$ and machine size $R = 3\,\text{m}$, $a = 1.15\,\text{m}$. The plasma volume is of order $100\,\text{m}^3$.

Eventually, with an upgraded plasma heating system with 22 MW of neutral beam injection (NBI), 22 MW of ion cyclotron wave heating (ICRH), and 7.3 MW of lower hybrid current device (LHCD) the plasma current reached 7 MA which is approximately one half the designed current value (15 MA) of the ITER device currently under construction.

The JET machine was designed with a poloidal divertor and the divertor configuration has gone through numerous changes since 1980. All modes of plasma

Table 1.6 JET and ITER main parameters comparison [Shimada, *et al.* (2007)] of ITER Physics Basis).

	Maximum values achieved on JET separately	ITER Design values
R and a	3 m 1.25 m	6.2 m 2 m
Elongation	1.8	1.7
Plasma volume	100 m^3	840 m^3
Magnetic field on axis	4 T	5.3 T
Plasma current in D-shaped plasma	7 MA	16 MA
Plateau current	60 s (1 MA)	15 MA
Modes of operation	L, H and ELMy	ELMy
Plasma contact	Carbon/beryllium limiters-pumped divertor	Pumped divertor
Neutral injection to the plasma	22 MW	33-50 MW
Coupled ICRH	22 MW	20-40 MW
ECRH	0	20-40 MW
Current drive	3 MA (LH) (5 MA) external	
Central density	2×10^{20} m^{-3}	10^{20} m^{-3}
Electron temperature	20 KeV	21 KeV
Ion temperature	40 KeV	18 KeV
Q value in DT plasma	0.6 (0.9 net)	10
Fusion power	16 MW	500 MW
Fusion energy	22 MJ in 4 s	120 GJ in 200

confinement states, the L-mode, the H-mode, and the ELMY-H-mode have been achieved in JET.

Prior to 1990, the metallic walls of the vacuum vessel were coated with carbon to withstand the plasma-surface interaction. At the highest heating power, the plasma became highly diluted with fully ionized carbon ions. Thus, in a rather inspired leap of judgment, the Director, Henri Rebut, changed the coating material to the lower-Z material of beryllium. The result was plasma with lower dilution from the wall material and record values of the metrics used to measure the machine performance [Dietz, *et al.* (1990)]. Thus, the ITER experiment is equipped with first wall coatings of beryllium rather than carbon.

In Table 1.6 we compare the values achieved in the JET experiment with those in the designs of the ITER experiment currently under construction at the Cadarache Nuclear Laboratory in the Luberon Mountains 100 km north of Marseilles, France. The ITER project is the International Thermonuclear Experimental Reactor with the European Union and Japan playing the key roles and with five other partners at the current time. The 13 to 15 billion dollar project is scheduled to start plasma confinement experiments in 2019. The foundation and numerous buildings are now finished in 2012.

A new generation of tokamaks just commencing operation now are superconducting for steady state operation with results just now arriving. These new machines are the KSTAR tokamak in South Korea and EAST in China. Both machines are producing important new results with radio frequency methods of driving the plasma current as described in Chapter 14. A superconducting upgrade of the JT60U machine is under construction at the time of this writing. This new tokamak is called JT60SuperAdvanced and will test the integrity of the first walls to the intense neutron fluxes from the fusion reactions. The JT60SA is a new joint Japan-Euro project called the Broader Approach which is in reference to a step beyond the ITER machine science. The new machine would test reactor design issues including the lifetime of the walls of the vacuum vessel. Tore Supra is now being modified through the WEST project from a circular limiter tokamak to a monoblock W divertor test bed facility for plasma-wall interaction research [Bucalossi (2011)].

Worked Problem 1.1 Gyro-Averaged Fields

In the lab or observer's reference frame, the charge q with mass m moves in the constant electromagnetic fields $\boldsymbol{E}$ and $\boldsymbol{B}$ with velocity $\boldsymbol{v}$.

(1) Use the Lorentz force to show that in the frame moving with the $\boldsymbol{E}\times\boldsymbol{B}$ velocity there is no perpendicular electric field so the particle moves in a helical orbit with gyroradius $\rho = v_\perp/\Omega$ where $\Omega = qB/m$ is the gyrofrequency. In the local frame with $\boldsymbol{B} = B\widehat{\boldsymbol{e}}_z$ the motion is $x(t) = x_{\rm gc}+\rho\cos(\zeta-\Omega t)$, $y(t) = y_{\rm gc}+\rho\sin(\zeta-\Omega t)$. Check that $\dot{v}_x = \Omega v_y$, $\dot{v}_y = -\Omega v_x$ and sketch the orbit for a 1 KeV proton in a 1 T magnetic field.

(2) For a wave $E = E_k\cos(ky-\omega_k t)$ find the electric force on the charge averaged over one gyro-period $T = 2\pi/\Omega$. Show that the effective electric field $\langle E_k\rangle$ acts on the guiding center $(x_{\rm gc}, y_{\rm gc})$ reduced by the factor $J_0(kv_\perp/\Omega)$ where J_0 is the Bessel function.

Solution:

(1) The motion perpendicular to the magnetic field has the Lorentz force

$$\boldsymbol{F} = q(\boldsymbol{E}+\boldsymbol{v}\times\boldsymbol{B}) \tag{1.4}$$

in the lab frame. Transforming to the $\boldsymbol{E}\times\boldsymbol{B}$ frame with

$$\boldsymbol{v} = \frac{\boldsymbol{E}\times\boldsymbol{B}}{B^2}+\boldsymbol{v}' \tag{1.5}$$

removes the electric force and the equation of motion is

$$\frac{d\boldsymbol{v}'}{dt} = \Omega\boldsymbol{v}'\times\widehat{\boldsymbol{e}}_z \tag{1.6}$$

with solutions given in the statement of the problem. Note if the Lorentz transformation is made the magnetic field is reduced by $B' = B/\gamma$ where $\gamma = (1-v_E^2/c^2)^{-1/2}$ and the cyclotron frequency is lowered to $\Omega = eB/m\gamma$.

(2) Figure 1.6 shows the cyclotron orbit in the $\cos(ky-\omega t)$ wave with $\rho = v_\perp/\Omega > 1/k_\perp$ to see the canceling direction of th electric force over a

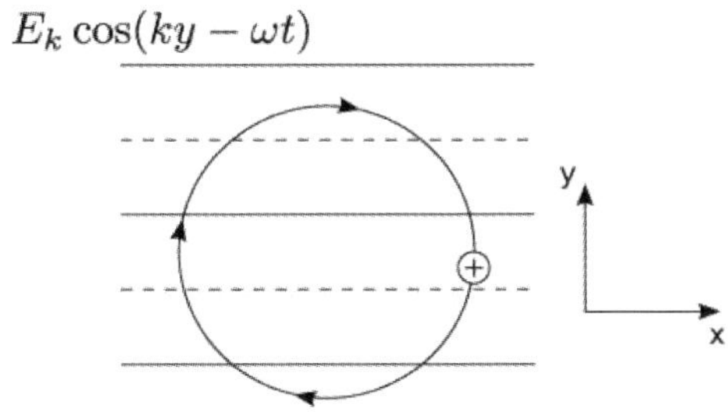

Fig. 1.6 The gyro orbit of a charged particle in a plasma wave.

gyro-period. The average of $\cos[k_y y(t) - \omega t]$ with $y(t) = y_{\rm gc} + \rho \sin(\zeta - \Omega t)$ requires the expression

$$e^{ix \sin\theta} = \sum_{n=-\infty}^{\infty} J_n(x) e^{in\theta} \tag{1.7}$$

from the generating function for Bessel function $J_n(x)$. Now assuming $\omega \ll \Omega$, the average over the cyclotron period of the electric force gives

$$\begin{aligned} \langle F \rangle &= q \int_0^{2\pi/\Omega} dt E_k \cos\left[k\left(y_{\rm gc} + \rho \sin(\zeta - \Omega t)\right) - \omega t\right] \\ &= q E_k J_0(k\rho) \cos\left(k y_{\rm gc} - \omega t\right) \end{aligned} \tag{1.8}$$

after using the real part of Eq. (1.7) and taking the integral over one cyclotron period. Thus, plasma waves give charge and currents with the fields reduced by the Bessel function as form factors.

References

Artsimovich, L. A., Bobrovskii, G. A., Gorbunov, E. P., Ivanov, D. P., Kirillov, V. D., Kuznetsov, E. I., Mirnov, S. V., Petrov, M. P., Razumova, K. A., Strelkov, V. S., and Shcheglov, D. A. (1969). *Plasma Phys. Control, Nucl. Fusion* **1** (International Atomic Energy Agency), p. 157, *Nucl. Fusion Suppl.* **17**.

Artsimovich, L. A., Anashin, A. M., Gorbunov, E. P., Ivanov, D. P., Petrov, M. P., and Strelkov, V. S. (1969), *Zh. Eksp. Teor. Fiz. Pis'ma* **10**, p. 130; *Sov. Phys. JETP Letts.* **10**, p. 82.

Aschwanden, M. J. (2006). *Physics of the Solar Corona: An Introduction with Problems and Solutions* (Springer-Verlag) (March 21), ISBN-10:3540307656.

Aymar, M. (1989). Two-electron Rydberg law for the ns2 1S0 Wannier states of alkaline-earth atoms, *J. Phys. B: At. Mol. Opt. Phys.* **22**, p. 2359, doi:10.1088/0953-4075/22/15/011.

Aymar, R., Chuyanov, V. A., Huguet, M., Shimomura, Y., ITER Joint Central Team and ITER Home Teams. (2001). Overview of ITER-FEAT - The future international burning plasma experiment, *Nucl. Fusion* **41**, p. 1301, doi:10.1088/0029-5515/41/10/301.

Balescu, R. (1988). *Transport Processes in Plasmas* (North-Holland-Amsterdam), Vols. 1-2.

Bateman, G. (1978). *MHD Instabilities* (MIT Press, Cambridge, MA), p. 270.

Bell, M. G. and the TFTR Team. (1989). *Plasma Phys. and Control. Nucl. Fusion Res., 1988* (International Atomic Energy Agency, Vienna), Vol. 1, pp. 27-40.

Bosch, H.-S. and Hale, G. M. (1992). Improved formulas for fusion cross-sections and thermal reactivities, *Nucl. Fusion* **32**, 611, doi:10.1088/0029-5515/32/4/107.

Herken, G. (2002). Brotherhood of the Bomb: The Tangled Lives and Loyalties of Robert Oppenheimer, Ernest Lawrence, and Edward Teller (Henry Holt and Co.).

Bucalossi, J., Argouarch, A., Basiuk, V., Baulaigue, O., Bayetti, P., Bécoulet, M., Bertrand, B., Brémond, S., Cara, P., Chantant, M., Corre, Y., Courtois, X., Doceul, L., Ekedahl, A., Faisse, F., Firdaouss, M., Garcia, J., Gargiulo, L., Gil, C., Grisolia, C., Gunn, J., Hacquin, S. Hertout, P., Huysmans, G., Imbeaux, F., Jiolat, G., Joanny, M., Jourdheuil, L., Jouve, M., Kukushkin, A., Lipa, M., Lisgo, S., Loarer, T., Maget, P., Magne, R., Marandet, Y., Martinez, A., Mazon, D., Meyer, O., Missirlian, M., Monier-Garbet, P., Moreau, P., Nardon, E., Panayotis, S., Pégourié, B., Pitts, R. A., Portafaix, C., Richou, M., Sabot, R., Saille, A., Saint-Laurent, F., Samaille, F., Simonin, and A. Tsitrone, E. (2011). Feasibility study of an actively cooled tungsten divertor in Tore Supra for ITER technology testing, *Fusion Engineering and Design* **86**, pp. 684688.

Cordey, J. G., Balet, B., Bartlett, D. V., Budnya, R. V., Christiansen, J. P., Conway, G. D., Eriksson, L.-G., Fishpool, G. M., Gowers, C. W., de Haasb, J. C. M., Harbour, P. J., Horton, L. D., Howman, A. C., Jacquinot, J., Kerner, W., Lowry, C. G., Monk, R. D., Nielsen, P., Righi, E., Rimini, F. G., Saibene, G., Sartori, R., Schunke, B., Sips, A. C. C., Smith, R. J., Stamp, M. F., Startc, D. F. H., Thomsen, K., Tubbing, B. J. D., and von Hellermann, M. G. (1999). Plasma confinement in JET H-mode plasmas with -, D, DT and T isotopes, *Nuclear Fusion* **39**, p. 301, `http://iopscience.iop.org/0029-5515/39/3/301`.

Dietz, K. J. and The JET Team. (1990). Effect of beryllium on plasma performance in JET, *Plasma Phys. Control. Fusion* **32**, pp. 837-852, `http://dx.doi.org/10.1088/0741-3335/32/11/002`.

Duane, B. H. (1972). Fusion Cross-Section Theory, Rept. BNWL-1685 (Brookhaven National Laboratory).

Equipe Tore Supra. (1989). *Plasma Phys. Control. Nucl. Fusion Res. 1988* **1**, pp. 9-26, (International Atomic Energy Agency, Vienna).

Friedberg, R. (1987). *Ideal Magnetohydrodynamics* (Cambridge University Press).

Friedberg, R. (2007). *Plasma Physics and Fusion Energy* (Cambridge University Press) ISBN:13-978-0-521-85107-7.

Fujita, T., Kamada, Y., Ishida, S., Neyatani, Y., Oikawa, T., Ide, S., Takeji, S., Koide, Y., Isayama, A., Fukuda, T., Hatae, T., Ishii, Y., Ozeki, T., Shirai, H., and JT-60 Team. (1999). High performance experiments in JT60-U reversed shear discharges, *Nucl. Fusion* **39**, p. 1627, `http://iopscience.iop.org/0029-5515/39/11Y/302`.

Fujita, T., *et al.* (1998). High performance reversed shear plasmas with a large radius transport barrier in JT60-U, *Nucl. Fusion* **38**, p. 207.

Gibson, A. (1998). *Phys. Plasmas* **5**, p. 1839, doi:10.1063/1.872854.

Goodman, T.P., Ahmed, S. M., Alberti, S., Andr'ebe, Y., Angioni, C., Appert, K., Arnoux, G., Behn, R., Blanchard, P., Bosshard, P., Camenen, Y., Chavan, R., Coda, S., Condrea, I., Degeling, A., Duval, B. P., Etienne, P., Fasel, D., Fasoli, A., Favez, J.-Y., Furno, I., Henderson, M., Hofmann, F., Hogge, J.-P., Horacek, J., Isoz, P., Joye, B., Karpushov, A., Klimanov, I., Lavanchy, P., Lister, J. B., Llobet, X., Magnin,

J.-C., Manini, A., Marlétaz, B., Marmillod, P., Martin, Y., Martynov, An., Mayor, J.-M., Mlynar, J., Moret, J.-M., Nelson-Melby, E., Nikkola, P., Paris, P. J., Perez, A., Peysson, Y., Pitts, R. A., Pochelon, A., Porte, L., Raju, D., Reimerdes, H., Sauter, O., Scarabosio, A., Scavino, E., Seo, S. H., Siravo, U., Sushkov, A., Tonetti, G., Tran, M. Q., Weisen, H., Wischmeier, M., Zabolotsky, A., and Zhuang, G. (2003). An overview of results from the TCV tokamak, *Nucl. Fusion* **43**, pp. 1619-1631.

Hawryluk, R. J. (1998). Results from deuterium-tritium tokamak confinement experiments, *Rev. Mod. Phys.* **70**, 2, pp. 537-587, doi:10.1103/RevModPhys.70.537.

Hazeltine, R. D. and Meiss, J. D. (1992). *Plasma Confinement* (Addison-Wesley, Redwood City).

Helander, P. and Sigmar, D. J. (2002). *Collisional Transport in Magnetized Plasmas* (Cambridge Monographs on Plasma Physics).

Hoang, G. T., Horton, W., Bourdelle, C., Hu, B., Garbet, X., and Ottaviani, M. (2003). *Phys. Plasmas* **10**, pp. 405-413, doi:10.1063/1.1534113.

Horton, W. (1976). Spectral Distribution of Drift-Wave Fluctuations in Tokamaks, *Phys. Rev. Lett.* **37**, pp. 1269-1272, `http://link.aps.org/doi/10.1103/PhysRevLett.37.1269`.

Horton, W. and Ichikawa, Y. (1996). *Chaos and Structures in Nonlinear Plasmas* (World Scientific) ISBN:81-7764-234-0.

I. P. E. Group. (1999). *Nucl. Fusion* **39**, p. 2175.

ITER-EDA. (1998). Final Design Report (published in *Nuclear Fusion* **39**, p. 2775, 1999) (International Atomic Energy Agency, Vienna), `http://www-jt60.naka.jaea.go.jp/english/index-e.html`.

Jackson, J. D. (1999). *Classical Electrodynamics* (Wiley, New York) 3rd Ed., pp. 708-724.

Kammash, T. (1975). *Fusion Reactor Physics* (Ann Arbor Science).

Kaye, S. M., Greenwald, M., Stroth, U., Kardaun, O., Kus, A., Schissel, D., DeBoo, J., Bracco, G., Thomsen, K., Cordey, J., Miura, Y., Matsuda, T., Tamai, H., Takizuda, T., Hirayama, T., Kikuchi, H., Naito, O., Chudnovskij, A., Ongena, J., and Hoang, G. T. (1997). *Nucl. Fusion* **37**, pp. 1303-1328, doi:10.1088/0029-5515/37/9/I10.

Kivelson, M. G. and Russell, C. T. (1995). *Introduction to Space Physics* (Cambridge University Press) ISBN:0-521-45714-9.

Koide, Y. and JT60 Team. (1997). *Progress in confinement and stability with plasma shape and profile control for steady-state operation in the Japan Atomic Energy Research Institute Tokamak-60 Upgrade, Phys. Plasmas* **4**, p. 1623, `http://link.aip.org/link/?PHPAEN/4/1623/1`.

L-H Mode Database Working Group. (1994). *Nucl. Fusion* **34**, pp. 131-167.

Maisonnier, C. (1989). *Artsimovich Memorial Lecture, Plasma Phys. Control. Nucl. Fusion Res. 1988* **1**, pp. 3-8 (International Atomic Energy Agency, Vienna).

Mazzucato, E. (1976). *Phys. Rev.* **36**, p. 792.

Mikhailovskii, A. B. (1992). *Electromagnetic Instabilities in an Inhomogeneous Plasma* (Institute of Physics Publishing) ISBN:0-7503-0182-1.

Peacock, N. J., Robinson, D. C., Forrest, M. J., Wilcock, P. D., and Sannikov, V. V. (1969). *Nature* **224**, pp. 488-490.

Razumova, K. A. (1984). Results from T-7, T-10, T-11 and TM-4 Tokamaks, *Plasma Phys. Control. Fusion* **26**, 37, pp. 1984, `http://iopscience.iop.org/0741-3335/26/1A/304`.

Shimada, M., Campbell, Mukhovatov, V., Fujiwara, M., Kirneva, N., Lackner, K., Nagami, M., Pustovitov, V. D., Uckan, N., Wesley, J., Asakura, N., Costley, A. E., Donné, A. J. H., Doyle, E. J., Fasoli, A., Gormezano, C., Gribov, Y., Gruber, O., Hender, T. C., Houlberg, W., Ide, S., Kamada, Y., Leonard, A., Lipschultz,

B., Loarte, A., Miyamoto, K., Mukhovatov, V., Osborne, T. H., Polevoi, A., and Sips, A. C. C. (2007). Chapter 1: Overview and summary, *Nucl. Fusion* **47**, p. S1, doi:10.1088/0029-5515/47/6/S01.

Stacey, W. M. (2010). *The Quest for a Fusion Energy Reactor* (Oxford University Press), ISBN:978-0-19-973384-2.

Troyon, F., Gruber, R., Saurenmann, H., Semenzato, S., and Succi, S. (1984). MHD limits to plasma confinement, *Plasma Phys. Control. Fusion* **26**, pp. 209-215.

Wakatani, M. (1998). *Stellarator and Heliotron Devices* (Oxford University Press) ISBN:0-19-507831-4.

Weiland, J. (1999). (CRC Press).

Wesson, J. A. (2004). (Oxford University Press) ISBN:0198509227.

Wolf, R. C. (2003). Internal Transport Barriers in Tokamaks Plasmas, *Plasma Phys. Control. Fusion* **45**, pp. R1-R91.

Yamada, H., Harris, J. H., Dinklage, A., Ascasibar, E., Sano, F., Okamura, S., Talmadge, J., Stroth, U., Kus, A., Murakami, S., Yokoyama, M., Beidler, C. D., Tribaldos, V., Watanabe, K. Y., and Suzuki, Y. (2005). *Nucl. Fusion* **45**, 12, pp. 1684-1693, doi:1088/0029-5515/45/12/024.

Chapter 2

Alfvén and Drift Waves in Plasmas

In this chapter we analyze the dispersion relations that describe the linear wave propagation of drift waves in inhomogeneous plasmas. We show how drift waves relate to plasma waves of the infinite, uniform plasma. The objective of this chapter is to develop an overview of the more elementary and universal aspects of drift waves, before studying their realization in particular plasma geometries. For clarity, in this chapter we recall from available references [Krall and Trivelpiece (1973); Mikhailovskii (1967); Mikhailovskii (1974); Mikhailovskii (1992)] the required plasma electrical conductivities and charge susceptibilities, rather than interrupting the presentation here with their lengthy derivation from the Vlasov-Maxwell system of equations. Refer to Chapter 11 for the derivations.

2.1 Low-Frequency Wave Dispersion Relations

The study of drift waves, as with other plasma waves, begins with the study of small amplitude sinusoidal solutions of the self-consistent field equations for a system of ions and electrons interacting through their mutual Coulomb forces. The self-consistent plasma electric fields $\mathbf{E}(\mathbf{k},\omega)$ are determined by the solutions of Maxwell's equations with the sources of charge density $\rho(\boldsymbol{k},\omega)$ and current density $\mathbf{j}(\mathbf{k},\omega)$ calculated from the plasma dynamics due to these same electromagnetic fields. Of course, the time varying electric fields and plasma currents induce time varying magnetic fields $\mathbf{B}(\mathbf{k},\omega)$ which are often of a relatively small amplitude and produce relatively weak transport compared with that from the time varying electric fields.

The plasma dynamics in the presence of the oscillating electric field is described by the linearized kinetic equations for the perturbed distribution functions $\delta f_j(\mathbf{v},\mathbf{k},\omega)$. In certain cases, particularly when collisions dominate the particle motion along the magnetic field, the fluctuating charge and current densities are adequately described by the fluid equations with parallel transport coefficients. This hydrodynamic regime is studied in Chapter 3. In both cases the theoretical results required for the study of small amplitude motion are conveniently expressed as a frequency and wavenumber dependent plasma conductivity tensor $\sigma_{\alpha\beta}(\mathbf{k},\omega)$ and the scalar charge density susceptibility function $\mathcal{X}(\mathbf{k},\omega)$. The theoretical results for these quantities and their derivations are well known from the widely available reference works on plasma waves in nonuniform media [Braginskii (1965); Stix (1973); Mikhailovskii (1967); Krall and Trivelpiece (1973)].

We begin here by simply introducing these conductivity $\sigma_j(\mathbf{k},\omega)$ formulas with the appropriate explanatory material rather than repeating their lengthy derivations from the kinetic equation. We shall give the first principles kinetic derivations in later chapters (cf. Chapter 11) where particular confinement geometries are introduced after the relevant estimates from elementary considerations have been established.

Let us begin very simply by considering plasma that is nonuniform only because of a density variation in a direction perpendicular to that of the confining magnetic field. We introduce a coordinate system with $\widehat{\mathbf{e}}_z$ in the direction of the magnetic field $\boldsymbol{B}$, $\widehat{\mathbf{e}}_x$ in the direction of the density gradient, and $\widehat{\mathbf{e}}_y$ in the direction orthogonal to $\widehat{\mathbf{e}}_x$ and $\widehat{\mathbf{e}}_x$. The orthogonal coordinate system is taken so as to have $\widehat{\mathbf{e}}_x$, $\widehat{\mathbf{e}}_y$, $\widehat{\mathbf{e}}_z$ right-handed. The plasma pressure gradient ∇p produced by the density gradients is balanced by the $\mathbf{j}\times\mathbf{B}$ force where the current density is $\mathbf{j}=j_y\widehat{\mathbf{e}}_y$ is the sum of the partial currents from the drift velocities v_{dj} of each charge species

$$j_y = \sum_j e_j n_j v_{dj}. \tag{2.1}$$

Here, the jth charge species has charge e_j, mass m_j, density n_j and temperature T_j. The diamagnetic drift velocity v_{dj} is given by

$$v_{dj} = \frac{T_j}{e_j B}\frac{1}{n_j}\frac{dn_j}{dx} \tag{2.2}$$

when the temperature is uniform. More generally, from force balance $\mathbf{j}\times\mathbf{B}=\nabla\Sigma p_j$ we find that v_{dj} is proportional to the sum of partial pressure gradients. The positive and negative charges drift in opposite directions and combine in Eq. (2.1) to give the net current across the magnetic field. This current j_y is called the diamagnetic current and v_{dj} is called the diamagnetic drift velocity. The direction of the plasma current is such as to produce a magnetic field component ΔB that reduces the given $B\widehat{\mathbf{e}}_z$ which accounts for the meaning of the name diamagnetic. The change in the magnetic field B from Ampére's law with the diamagnetic plasma current then gives $d(p+B^2/8\pi)/dx=0$ in absence of magnetic curvature. Here the total pressure $p=\sum_j n_j T_j$ as follows from Eq. (2.1), Eq. (2.2), and Ampére's law.

The measurement of ΔB is one of the basic diagnostic signals used to determine the total stored plasma energy $W_p(t)$. The flux from $\Delta B(t)$ is called the diamagnetic signal and is measured as the voltage induced in a coil surrounding the plasma column.

Clearly, the current density in Eq. (2.1) satisfies the force balance equation

$$\mathbf{F} = \left(j_y B - \frac{dp}{dx}\right)\mathbf{e}_x = 0 \tag{2.3}$$

after using Eq. (2.2). The electric force $\rho_q E_x$ does not enter force balance since the equilibrium is charge neutral with $\rho_q = \sum_j e_j n_j(x) = 0$ which requires that $\sum_i Z_i n_i(x) = n_e(x)$. As a consequence the drift velocities v_{dj} are related by

$$\sum_j \frac{n_j e_j^2}{T_j} v_{dj} = 0. \tag{2.4}$$

Equation (2.4) follows by substituting Eq. (2.2) into Eq. (2.1) using $d/dx(\sum_j e_j n_j) = 0$. For a pure electron-ion plasma, Eq. (2.4) reduces to

$$\sum_{j=i,e} \frac{e_j}{T_j} v_{dj} = 0. \tag{2.5}$$

Thus, the higher temperature species have the larger diamagnetic drift velocities. Often the drift waves propagate in the direction of the higher temperature species.

Let us consider a single Fourier component $\mathbf{k}\omega$ of an oscillating electric field perturbation of the form

$$\mathbf{E}(x,t) = \widehat{\mathbf{E}}_{\mathbf{k}\omega} e^{i\mathbf{k}\cdot\mathbf{x} - i\omega t} + \widehat{\mathbf{E}}^*_{\mathbf{k}\omega} e^{-i\mathbf{k}\cdot\mathbf{x} + i\omega t} \tag{2.6}$$

with the vector wavenumber

$$\mathbf{k} = k_y \widehat{\mathbf{e}}_y + k_z \widehat{\mathbf{e}}_z. \tag{2.7}$$

Here we allow the angular frequency ω to be complex $\omega = \omega_r + i\gamma$. The linearized Vlasov equation for the perturbation $\delta f_j(\mathbf{k}, \omega, \mathbf{v})$ of the phase space density function f_j is induced by the electric field Eq. (2.6) in the particle distributions. The linear fluctuations can be calculated exactly and used to calculate the complex dispersive plasma conductivities $\sigma_j(\mathbf{k}, \omega, \mathbf{v})$. Maxwell's equations then yield the following dielectric tensor defined by

$$\varepsilon_{\alpha\beta}(\mathbf{k}, \omega) = I_{\alpha\beta} + 4\pi i \frac{\sigma_{\alpha\beta}(\mathbf{k}, \omega)}{\omega} \tag{2.8}$$

where for low frequency drift-Alfvén waves ($\omega \ll \omega_{ci}$) the values of $\varepsilon_{\alpha\beta}$ are given by

$$\varepsilon_{xx} \cong 1 + \frac{\omega_{pi}^2}{\omega_{ci}^2}\left(1 - \frac{\omega_{*i}}{\omega}\right)\frac{1 - \Gamma_0(b)}{b} \tag{2.9}$$

$$\varepsilon_\perp \equiv \varepsilon_{yy} \cong 1 + \frac{\omega_{pi}^2}{\omega_{ci}^2}\left(1 - \frac{\omega_{*i}}{\omega}\right)\frac{1 - \Gamma_0(b)}{b} \tag{2.10}$$

$$\varepsilon_\parallel \equiv \varepsilon_{zz} \cong 1 - \frac{\omega_{pe}^2}{k_\parallel^2 v_e^2}\left(1 - \frac{\omega_{*e}}{\omega}\right) Z'\left(\frac{\omega}{k_\parallel v_e}\right) - \frac{\omega_{pi}^2}{\omega^2}\left(1 - \frac{\omega_{*i}}{\omega}\right)\Gamma_0(b) \tag{2.11}$$

where the diamagnetic drift frequencies and gyroradius averaging are

$$\omega_{*j} \equiv k_y v_{dj} \qquad \text{and} \qquad \Gamma_0(b) \equiv e^{-b} I_0(b) \tag{2.12}$$

with I_0 being the modified Bessel function of zeroth order. In these equations we use b to denote the square of the wavenumber perpendicular to the magnetic field $k_\perp$ measured with respect to the thermal ion gyroradius; i.e.

$$b = \frac{1}{2}\frac{k_\perp^2 v_i^2}{\omega_{ci}^2} = \frac{k_\perp^2 T_i}{m_i \omega_{ci}^2}. \tag{2.13}$$

The physical origin of the Bessel functions in Eq. (2.9) to Eq. (2.12) is made clear by considering the average over one gyro or cyclotron orbit period of the electric field in Eq. (2.6). The average is performed using the vector from the guiding center to the particle $\mathbf{X}(t) = \mathbf{X}_{gc} + \mathbf{b} \times \mathbf{v}/\omega_c$. Integrating around one gyro circle at fixed guiding center position $\mathbf{X}_{gc}$ gives J_0 and then computing the integral over the Maxwellian velocity distribution yields the modified Bessel function I_0. The calculation is carried out explicitly in Chapter 5.1. In Eq. (2.9)-Eq. (2.11) the complex plasma dispersion function $Z(\zeta)$ and its derivative $Z'(\zeta)$ appear from the parallel velocity integrals over the wave-particle propagator $g_{k\omega} = (\omega - k_\parallel v_\parallel + i0^+)^{-1}$. The properties of the dispersion function are given in Fried and Conte (1961) and the derivation is given in Chapter 2.1.

Note that Eq. (2.8) is written in cgs units. In mks units the 4π is replaced with $\mu_0 = 4\pi \times 10^{-7}\,\mathrm{mT/A}$ where m is meters, T is Tesla, A is amperes.

The dielectric components from the electrical conductivities in Eq. (2.9)-Eq. (2.11) are valid in the frequency-wavelength regime defined by

$$\begin{aligned} &k_\parallel v_i \ll \omega \ll \omega_{ci} \ll \omega_{ce} \\ &k_\perp v_e \ll \omega_{ce}, \end{aligned} \tag{2.14}$$

which is sufficient for present purposes.

To see that conditions Eq. (2.14) are satisfied for plasmas with weak density gradients over the scale of the ion gyroradius we make the following estimates. It is standard to define the density gradient scale length L_n by

$$\frac{1}{n}\frac{dn}{dx} = \frac{-1}{L_n} \tag{2.15}$$

which gives from Eq. (2.2) and Eq. (2.12)

$$\omega_{*e} = k_y \frac{T_e}{eBL_n} = k_y \rho_s \frac{c_s}{L_n} \tag{2.16}$$

where we define the ion sound-inertial scale length $\rho_s = c_s/\omega_{ci} = c(m_i T_e)^{1/2}/eB$ with the ion sound, or acoustic speed c_s defined by electron temperature and the ion mass by $T_e = m_i c_s^2$. For $T_e \sim T_i$ the spatial scale length ρ_s is similar to the ion thermal gyroradius ρ_i. For $k_y \rho_i \leq 1$ the drift wave frequency is bounded above by

$$\omega_k \leq \max_k(\omega_k) = \frac{c_s}{L_n} \tag{2.17}$$

which gives

$$\frac{\omega_k}{\omega_{ci}} \leq \frac{\rho_s}{L_n}. \tag{2.18}$$

Thus for $\rho_s/L_n \ll 1$, as is typical of magnetized plasmas, the drift wave frequencies are well below the ion cyclotron frequency ω_{ci}.

After eliminating the perturbed magnetic field $\mathbf{B}_{\mathbf{k}\omega}$ in Ampére's law by substituting $\mathbf{B}_{\mathbf{k}\omega} = c\mathbf{k} \times \mathbf{E}_{\mathbf{k}\omega}/\omega$ from Faraday's law, we obtain the system of self-consistent field equations

$$\left(k_\parallel^2 - \frac{\omega^2}{c^2}\,\varepsilon_{yy}\right) E_y - k_y k_\parallel E_z = 0 \tag{2.19}$$

$$-k_y k_\parallel E_y + \left(k_\perp^2 - \frac{\omega^2}{c^2}\,\varepsilon_{zz}\right) E_z = 0 \tag{2.20}$$

$$\left(k_y^2 + k_\parallel^2 - \frac{\omega^2}{c^2}\,\varepsilon_{xx}\right) E_x = 0 \tag{2.21}$$

which are written here in the approximation that $\sigma_{xz} = \sigma_{xy} = \sigma_{yz} = 0$.

From the field equations Eq. (2.19)-Eq. (2.21) we find that there are compressional modes that have $B_{z\mathbf{k}\omega} = k_y E_x/\omega$ with the dispersion relation

$$k^2 = \frac{\omega^2}{c^2}\,\varepsilon_{xx} \qquad \text{and} \qquad E_y = E_z = 0 \tag{2.22}$$

separate from the approximately incompressional modes with $B_{z\mathbf{k}\omega} \equiv 0$. These modes with $E_y, E_z \neq 0$ from Eq. (2.19) and Eq. (2.20) are the coupled Alfvén-drift waves given by

$$k_\parallel^2 \varepsilon_\parallel + k_y^2 \varepsilon_\perp = \frac{\omega^2}{c^2}\,\varepsilon_\parallel \varepsilon_\perp \qquad \text{and} \qquad E_x = 0 \tag{2.23}$$

as derived from Eq. (2.19)-Eq. (2.21). The compression here refers to the change in $B_z \equiv B_0 + \delta B_z$ which produces a large perturbed magnetic pressure force $\mathbf{F}_B = -\nabla(B^2/2\mu_0)$.

Observe that the incompressible modes follow from the parallel component of Ampére's law and Poisson's equation. For the incompressible mode, Faraday's law gives only a perpendicular component of the perturbed $B_{x\mathbf{k}\omega}$ field perpendicular to the ambient magnetic field $B\widehat{\mathbf{e}}_z$. The parallel component of Ampére's law, consequently, reduces to

$$\begin{aligned}(\nabla \times \mathbf{B})_\parallel = -ik_y B_x &= \frac{ick_y}{\omega}\,(k_\parallel E_y - k_y E_\parallel) \\ &= \frac{4\pi\sigma_\parallel}{c}\,E_\parallel - \frac{i\omega}{c}\,E_\parallel = -i\,\frac{\omega}{c}\,\varepsilon_\parallel E_\parallel .\end{aligned} \tag{2.24}$$

From Poisson's equation we have $\rho_q = (i/4\pi)(k_\parallel E_\parallel + k_y E_y)$, and thus from charge conservation it follows that

$$\begin{aligned}\frac{\partial \rho_q}{\partial t} + \nabla\cdot\mathbf{j} = &+\frac{\omega}{4\pi}\,(k_y E_y + k_\parallel E_\parallel) \\ &+ i(k_y\sigma_\perp E_y + k_\parallel \sigma_\parallel E_\parallel) = 0.\end{aligned} \tag{2.25}$$

In the limit where $\mathbf{E}_{k\omega} = -i\mathbf{k}\phi_{k\omega}$ the charge conservation Eq. (2.5) reduces to

$$\left(k_{\parallel}^2\epsilon_{\parallel} + k_{\perp}^2\epsilon_{\perp}\right)\phi_k = 0 \tag{2.26}$$

gives the low frequency limit of Eq. (2.23).

Worked Problem 2.1:
Low Frequency Energy Conservation Theorem

Problem:

Find the electric current j_p produced by a low-frequency oscillation of an electric field $E(t)$ in a plasma that accelerates the charged particles across the ambient magnetic field $B_z = \text{const}$. From the formula for the plasma current, which is called the polarization current, show that the rate of change dW_E/dt of the kinetic energy W_E in the $\mathbf{E}\times\mathbf{B}$ drift velocities of the particles equals the electric power deposited in the plasma through the polarization current.

Solution:

Low frequency is defined in the (drift) wave literature as frequencies $\omega \ll \omega_c = qB/m$ where the motions are then given by the $\mathbf{E}\times\mathbf{B}$ drift velocity $\mathbf{v}_E$. The acceleration of the charge from the increase of $\mathbf{v}_E$ in time then gives rise to a smaller drift velocity $\mathbf{v}_{p,j}$ of the particle through Newton's law

$$m_j\frac{d\mathbf{v}_E}{dt} = q_j\mathbf{v}_p \times \mathbf{B}. \tag{2.27}$$

We solve for the polarization velocity $\mathbf{v}_{p,j}$ by crossing Eq. (2.27) with $\mathbf{B}$ and working out the double vector cross product to show that

$$\mathbf{v}_p = \frac{m_j}{q_j B^2}\frac{d\mathbf{E}}{dt}. \tag{2.28}$$

The drift velocity $\mathbf{v}_p$ is the polarization drift velocity which, owing to the factor m_j/q_j, is in the opposite directions for ions and electrons so as to create an electric field polarization of the plasma. This polarization of the plasma gives the dielectric constant $\epsilon_\perp$ used in deriving the Alfvén wave from the electromagnetic waves in dielectric materials from electromagnetic wave theory.

The polarization drift is in opposite directions for positive and negative charges and thus produces the plasma polarization current $\mathbf{j}_p$ given by

$$\mathbf{j}_p = \sum_j \frac{n_j m_j}{B^2}\frac{d\mathbf{E}}{dt}. \tag{2.29}$$

The energy conservation theorem follows by computing the rate of change of particle kinetic energy

$$W_E = \frac{1}{2}\sum_j m_j n_j\, v_E^2 \tag{2.30}$$

in the $\mathbf{E} \times \mathbf{B}$ drift motion of the particles

$$\frac{dW_E}{dt} = \frac{n_j m_j \mathbf{v}_E}{B^2} \cdot \frac{d\mathbf{E}}{dt} \tag{2.31}$$

$$= \mathbf{j}_p \cdot \mathbf{E}(t) \tag{2.32}$$

where to obtain the identity we used the dot product of Eq. (2.29) with $\mathbf{E}$ for the electric power transfer from the electric field to the particle motion.

The form of the energy conservation theorem, Eq. (2.32), is closely related to the Poynting theorem when we introduce the plasma dielectric function $\epsilon_\perp$. As in the electrodynamics of materials the polarization current $\mathbf{j}_p$ and the polarization charge density ρ_p are related by the conservation of charge as in Eq. (2.25). For magnetized plasmas the polarization at low frequencies is high which gives a high index of refraction to electromagnetic waves slowing the velocity of the transverse waves from the speed of light to the Alfvén speed.

We see that we can interpret $\mathbf{j}_p$ in terms of the perpendicular dielectric function $\epsilon_\perp$ given by

$$\epsilon_\perp = \epsilon_0 + \sum_j \frac{n_j m_j}{B^2} \tag{2.33}$$

where ϵ_0 is the dielectric of the vacuum in mks units. The student will want to show that the units of the plasma density contribution are the same as the vacuum contribution which is usually expressed in Farad/meter. At a relatively low plasma density, the plasma contribution exceeds the vacuum contribution and slows the electromagnetic waves through the standard dispersion relation

$$\epsilon_\perp \omega^2 = c^2 k^2 \tag{2.34}$$

as we will see in greater detail in Chapter 4. The form of the Poynting theorem in a dielectric MKS now called SI material $\mathbf{E} \cdot \mathbf{D}$ now applies with $\mathbf{D} = \epsilon \mathbf{E}$.

There are two systems of units used in the plasma physics literature: the cgs units of centimeters, grams and seconds and the mks units of meters, kilograms, seconds. The electrical units corresponding to the cgs system are E(statvolt/cm), B(gauss), j(abamperes) and to the mks units E(volts/m), B(Tesla) and j(amps/m^2). The speed of light enters in the cgs units for Maxwell's equations and the Lorentz force whereas in the mks units the fundamental constants $\epsilon_0 = 8.854 \times 10^{-12}\,\mathrm{F/m}$ and $\mu_0 = 4\pi \times 10^{-7}\,\mathrm{T \cdot m/A}$ enter the corresponding mks equations. The speed of light determines μ_0 in the mks system through $c^2 = 1/\epsilon_0 \mu_0$. It is important for plasma physicists to know both systems of units since the astrophysical and early fusion literature is written in the cgs units while laboratory and space plasma physics is written in terms of the mks system of units.

The NRL formulary found at `http://www.ppd.nrl.navy.mil/nrlformulary/` is a valuable tool for going between the units and finding reference values in both systems of units for many typical plasma systems. Errors are often made by the novice in evaluating electric and magnetic phenomenon in either cgs or mks units. As an example, we give the reference values for a typical laboratory plasma of the drift speed and electric polarization current for a plasma with a radial electric field E_r of $1\,\mathrm{kV/m}=(1/300\times 100\mathrm{cm/m})$ statvolt/cm increasing at the rate of 1 kHz in a cylindrical plasma with $\mathbf{B} = B\mathbf{e}_z$ of $B = 1T = 10^4\,\mathrm{G}$. The value of the $\mathbf{E}\times\mathbf{B}$ drift velocity is $1000\,\mathrm{m/s}= 10^5\,\mathrm{cm/s}$ and for a density of $n = 10^{20}$ protons/m^3 $= 10^{14}$ protons/cm^3, the plasma polarization current is $j_p = [10^{20}/\mathrm{m}^3\,1.67\times 10^{-27}\,\mathrm{kg}/(1\mathrm{T}^2)]\times 10^6\,\mathrm{V/ms}= 0.167\,\mathrm{A/m}^2$. By the energy conservation theorem, the power deposited in the particle kinetic drift velocities is $0.167\,\mathrm{kW/m}^3$, which is a typical Alfvén wave heating power in laboratory plasmas.

Worked Problem 2.2

Problem:

Find the collisionless damping rate γ for (1) electron plasma oscillations $\omega = \omega_{pe}k_\parallel/k$ and (2) ion acoustic waves $\omega = k_\parallel c_s$ in a strongly magnetized plasma with temperature T_e.

Solution:

Strongly magnetized means that the cyclotron frequency $\omega_{ce} = eB/m_e \gg \omega_{pe}$, the plasma frequency, so the electron motion is one-dimensional. We describe the motion in the wave with electron Vlasov equation for the electron phase space density $f(x, v_\parallel, t)$, which satisfies

$$\frac{\partial f}{\partial t} + v_\parallel \frac{\partial f}{\partial z} + v_E \frac{\partial f}{\partial x} - \frac{eE_\parallel}{m_e}\frac{\partial f}{\partial v_\parallel} = 0. \tag{2.35}$$

(1) electron plasma oscillations

The electron density $n_e = \int f dv_\parallel$ and we solve for the part of $f = f_0 + \delta f$ that oscillates with the wave electric field $\mathbf{E} = \mathbf{E}_{k\omega}\exp(i\mathbf{k}\cdot\mathbf{x} - i\omega t)$. The fluctuating δf follows from solving

$$\left(-i\omega + ik_\parallel v_\parallel\right)\delta f = \frac{eE_\parallel}{m_e}\frac{\partial f_0}{\partial v_\parallel} - \frac{E_y}{B}\frac{\partial f_0}{\partial x} \tag{2.36}$$

and the associated oscillating density in Poisson's equation

$$\nabla\cdot\mathbf{E} = -k^2\phi_{k\omega} = -\frac{e}{\varepsilon_0}\int_{-\infty}^{+\infty} dv_\parallel\,\delta f_{k\omega}(v_\parallel). \tag{2.37}$$

Substituting δf from Eq. (2.36) into Eq. (2.37) gives the dispersion relation

Kinetic Dispersion Relation

$$k^2\varepsilon_k(\omega)\phi_{k\omega} = \left[k^2 - \frac{ne^2}{\varepsilon_0 m_e}\int_{-\infty}^{+\infty} dv_\| \frac{\left(k_\| \frac{\partial f_0}{\partial v_\|} + \frac{k_y}{eB}\frac{\partial f_0}{\partial x}\right)}{\omega - k_\| v_\| + i0^+}\right]\phi_{k\omega} = 0 \tag{2.38}$$

where the wave-particle resonance at $\omega - k_\| v_\| = 0$ is resolved by taking the adiabatic growth of the wave from $E_k \exp(-i(\omega + i0^+)t)$ or equivalently from using the Landau initial value problem solution with a Laplace transform. The residue of the complex, analytic density response function in Eq. (2.38) gives the collisionless damping of the wave when $df_0/dx = 0$. The resonance $\omega_k - k_\| v_\| = 0$ provides the drift wave growth rate when df_0/dx is large.

For waves with $\omega_k \gg \gamma_k$ we solve Eq. (2.38) by a Taylor series expansion about the real root $\omega = \omega_k$. For electron plasma waves with $\omega_k \gg k_\| v_e$ we take the high phase velocity limit of Eq. (2.38) $\varepsilon_k(\omega) = 1 - (\omega_{pe}/k^2)(k_\|^2/\omega^2) = 0$ to obtain the plasma waves $\omega_k = \pm\omega_{pe} k_\|/k$. Now, with this frequency we can evaluate the residue from complex part of the response function to obtain from Eq. (2.38)

$$\frac{2i\gamma\omega_{pe}^2}{\omega^3} - \frac{i\pi\omega_{pe}^2}{k^2}\int dv_\| \delta\left(\omega - k_\| v_\|\right) k_\| \frac{\partial f_0}{\partial v_\|} = 0.$$

This Landau resonance problem occurs throughout plasma wave theory so it is important to define the associated plasma dispersion functions $Z(\zeta)$ and $dZ/d\zeta$ that occur over and over. The Maxwellian velocity distribution function gives

$$\int_{-\infty}^{+\infty} \frac{dv_\| f_M(v_\|)}{\omega - k_\| v_\|} = \int_{-\infty}^{+\infty} \frac{dx}{\pi^{1/2}} \frac{\exp\left(-x^2/2\right)}{x - \omega/k_\| v_e + i0^+} \equiv Z(\zeta)$$

which defines the analytic complex dispersion function $Z(\zeta)$ with $\zeta = \omega/|k_\||v_e$ being the complex variable continued by analytic continuation from the upper half of the complex frequency or ζ plane.

Thus we arrive at the electron plasma wave damping rate γ_k^{el} as

$$\gamma_k^{el} = -\left(\frac{\pi}{8}\right)^{1/2} \frac{\omega_{pe}^2}{k^3 v_e^3}\omega_k^2 \exp\left(\frac{-\omega_k^2}{k_\|^2 v_e^2}\right).$$

(2) ion-acoustic waves

For the second question, the low-frequency ion acoustic waves have $\zeta = \omega/|k_\||v_e \ll 1$ and the expansion of the plasma response function requires some analysis skipped here. One finds that the result is $Z(\zeta) \cong -2\zeta - i\pi^{1/2}$. Using this small phase velocity response function in Eq. (2.38) gives

$$\omega^2 = \frac{\omega_{pi}^2}{1 + \frac{\omega_{pe}^2}{k_\|^2 v_e^2}\left(1 + i\pi^{1/2}\frac{\omega}{|k_\||v_e}\right)}.$$

Expanding up the small resonant electron contribution gives the damping rate of the ion-acoustic (ia) wave as

$$\gamma_k^{ia} = -\left(\frac{\pi}{8}\right)^{1/2} |\omega_k| \left(\frac{T_e}{T_i}\right)^{3/2} \exp\left(-\frac{\omega_k^2}{2k_\parallel^2 v_e^2}\right).$$

We note that the plasma waves $\omega_k = \pm\omega_{pe}(k_\parallel/k) = \pm\omega_{pe}\cos(\theta)$ have the same kinematic structure as internal gravity waves with $\omega_{pe} \to N-$ the Brunt-Väisäla frequency and the angle θ is between the **k**-vector and the horizontal. The Brunt-Väisäla or buoyancy frequency for a stable stratification of the mass density $\rho(z)$ is given by $N(z) = \sqrt{(-gd\rho/pdz)}$.

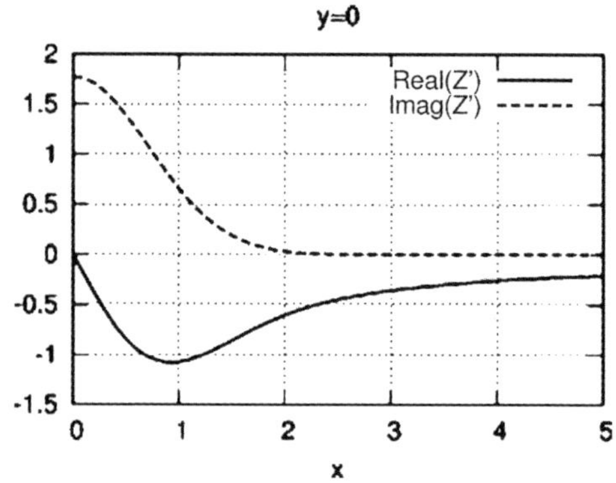

Fig. 2.1 The real and imaginary parts of the plasma dispersion function $Z(x)$ in the range $x = [0, 5]$ for the dimensionless phase velocity $x = \omega/k_\parallel v_e$ for real ω in the Maxwellian plasma velocity distribution $v_e = \sqrt{[2T_e/m_e]}$ with temperature T_e. The real part of $Z(x)$ is anti-symmetric in $x \to -x$ and the imaginary part of $Z(x)$ is symmetric in $x \to -x$.

The plasma dispersion function $Z(\zeta)$ is shown in Fig. 2.1 for real values of $\zeta = x = \omega/kv_e$. For real $\zeta = x$, the response from the $\mathrm{Re}\, Z'$ gives the continuous change from adiabatic shielding in the low frequency dynamics with $|x| \ll 1$ to the hydrodynamic reactive oscillatory response for $|x| \gg 1$. Plasma waves or plasma oscillations occur when $|x| \gg 1$ and the (exponentially) small imaginary part of Z' then describes the response of the resonant particles with $\omega = k_\parallel v_e$.

In the adiabatic regime defined by $|x| \ll 1$ the imaginary resonant response is order of $|x|$ and thus there is strong dissipation or growth from these wave-particle interactions. This is the regime of the electrons in the drift wave and thus when $\omega < \omega_*$ there is a strong growth of the drift wave from the resonant electrons.

When the perturbed parallel electron current is calculated and the Poynting theorem is used to calculate the averaged transfer of power from the particles to the waves as $-\langle \mathbf{j} \cdot \mathbf{E} \rangle$ we find that this is positive or negative and of order of $\omega/k_\parallel v_e$ for the electron drift wave. The condition for the negative value of $\langle \mathbf{j} \cdot \mathbf{E} \rangle$ is that $\omega < \omega_*$, which is then the condition for the instability to occur. The proof of this statement is most carefully given by construction of aNyquist diagram that maps

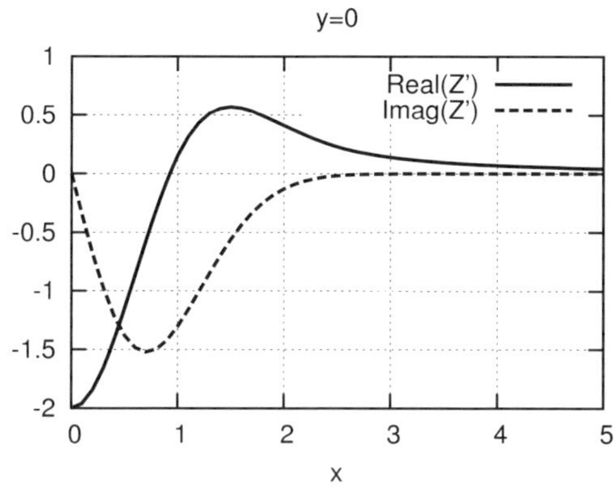

Fig. 2.2 The real and imaginary part of the plasma dispersion function $Z'(x) = dZ/dx$ in the range $x = [0,5]$ for the dimensionless phase velocity $x = \omega/k_{v_e}$ for real ω in the Maxwellian plasma velocity distribution $v_e = \sqrt{[2T_e/m_e]}$ with temperature T_e. The real part of $Z'(x)$ is symmetric in $x \to -x$ and the imaginary part of $Z(x)$ is anti symmetric in $x \to -x$. The response at $|x| \ll 1$ is $Z''(x) = -2$ which gives the Debye shielding of charges in the plasma. This regime in the fluid description is called the adiabatic response of the species. The regime $|x| \gg 1$ is $Z''(x) = 1/x^2$ and is called the hydrodynamic or cold plasma response giving electron plasma oscillations.

the upper-half of the complex ω-plane with the dispersion function $\epsilon(\omega, k)$ on the complex ϵ plane. When the mapped image encircles the origin of the ϵ plane, the Cauchy theorem guarantees there is a root of the $\epsilon(\omega, k) = 0$ inside the contour in the ω-plane. This provides one method of checking the stability of the system and deriving the critical conditions for the onset of an instability.

In Fig. 2.2 the derivative $Z' = dZ/d\zeta$ of the Z-function is shown for the dimensionless phase velocity $\zeta = \omega/kv_e$ in the range $[-5, +5]$.

2.2 Reduction of the Kinetic Dispersion Relation

For the low frequency and long parallel wavelength modes of interest the vacuum term (displacement current) in $\varepsilon_\parallel$ is negligible compared to the large parallel electron currents produced by the fluctuating electric field. Consequently, we make the following reduction:

$$\begin{aligned}\varepsilon_\parallel &= 1 - \frac{\omega_{pe}^2}{k_\parallel^2 v_e^2}\left(1 - \frac{\omega_{*e}}{\omega}\right) Z'\left(\frac{\omega}{k_\parallel v_e}\right) - \frac{\omega_{pi}^2}{k_\parallel^2 v_i^2}\left(1 - \frac{\omega_{*i}}{\omega}\right)\Gamma_0(b) Z'\left(\frac{\omega}{k_\parallel v_i}\right) \\ &\cong - \frac{\omega_{pi}^2}{2k_\parallel^2 c_s^2}\left(1 - \frac{\omega_{*e}}{\omega}\right) Z'\left(\frac{\omega}{k_\parallel v_e}\right) - \frac{\omega_{pi}^2}{k_\parallel^2 v_i^2}\left(1 - \frac{\omega_{*i}}{\omega}\right)\Gamma_0 Z'\left(\frac{\omega}{k_\parallel v_i}\right).\end{aligned} \tag{2.39}$$

Likewise, the perpendicular ion current, and at high densities $\omega_{pe} > \omega_{ce}$ the perpendicular electron current, dominates the vacuum displacement current in most cases of practical interest. That is the particle currents dominate the vacuum displacement current. This dominance of the plasma current allows the approximation

$$\begin{aligned}\varepsilon_\perp &= 1 + \frac{\omega_{pi}^2}{\omega_{ci}^2}\left(1 - \frac{\omega_{*i}}{\omega}\right)\frac{1-\Gamma_0(b)}{b} + \frac{\omega_{pe}^2}{\omega_{ce}^2}\left(1-\frac{\omega_{*e}}{\omega}\right) \\ &\cong \frac{\omega_{pi}^2}{\omega_{ci}^2}\left(1 - \frac{\omega_{*i}}{\omega}\right)\left(\frac{1-\Gamma_0(b)}{b}\right) + \frac{\omega_{pe}^2}{\omega_{ce}^2}\left(1-\frac{\omega_{*e}}{\omega}\right)\end{aligned} \tag{2.40}$$

to be made. With these reductions in Eq. (2.39) and Eq. (2.40) substituted into the field equations, Eq. (2.24) and Eq. (2.25), we have derived a simpler dispersion relation.

$$(\nabla \times \mathbf{B})_\parallel = -ik_y B_x = \frac{ick_y}{\omega}(k_\parallel E_y - k_y E_\parallel) = \frac{4\pi\sigma_\parallel}{c} E_\parallel \tag{2.41}$$

$$\nabla \cdot \mathbf{j} = ik_y \sigma_\perp E_y + ik_\parallel \sigma_\parallel E_\parallel = 0. \tag{2.42}$$

Substituting Eq. (2.39) and Eq. (2.40) into Eq. (2.41) and Eq. (2.42), and canceling the overall plasma density factor ω_{pi}^2/c_s^2, we obtain the drift-Alfvén wave dispersion relation

Low β Drift- Alfvén Waves

$$\begin{aligned}&\frac{T_e}{T_i}[1-\Gamma_0(b)]\left(1-\frac{\omega_{*i}}{\omega}\right) + \left[1 - \frac{1-\Gamma_0(b)}{b}\frac{\omega(\omega-\omega_{*i})}{k_\parallel^2 v_A^2}\right] \\ &\quad\times\left[-\frac{1}{2}\left(1-\frac{\omega_{*e}}{\omega}\right)Z'\left(\frac{\omega}{k_\parallel v_e}\right) - \frac{k_\parallel^2 c_s^2}{\omega^2}\left(1-\frac{\omega_{*i}}{\omega}\right)\Gamma_0(b)\right] = 0.\end{aligned} \tag{2.43}$$

For $b \ll 1$ the first term in Eq. (2.43) is small since $\Gamma_0(b) \simeq 1-b$. Consequently, at sufficiently small b there are two modes that are given by the two-second factors within the square brackets in Eq. (2.43). The ion drift-Alfvén mode is defined by

$$\omega(\omega - \omega_{*i}) = k_\parallel^2 v_A^2 \tag{2.44}$$

arising from the first factor in Eq. (2.43). The electron drift-ion acoustic mode with $\omega < k_\parallel v_e$ defined by

$$\omega(\omega-\omega_{*e}) - k_\parallel^2 c_s^2\left(1-\frac{\omega_{*i}}{\omega}\right) = 0 \tag{2.45}$$

arises from the second factor in Eq. (2.43). Equation (2.45) is a cubic equation with the ion acoustic waves and a thermal mode. The roots of the ion acoustic wave become unstable when ω_{*i}/ω_{*e} is large and the instability is called the ion temperature gradient mode. Chapter 13 develops the theory and application of this instability. Clearly, the first term proportional to $1-\Gamma_0(b)$ in Eq. (2.43) couples these two modes given in Eq. (2.44) and Eq. (2.45).

The precise description of the complex $\omega(\mathbf{k})$ roots of Eq. (2.43) is a difficult problem requiring numerical analysis. Rather than pursue such an analysis at this point in the presentation let us simply write the four modes that are described by Eq. (2.44) and Eq. (2.45) in the small b limit. The motivated reader can independently pursue both the problems (1) of determining the wavelength domains in $k_y - k_\parallel$ where the following approximation solutions are valid and (2) the form of the roots $\omega_k^\alpha(\alpha = 1,2,3,4)$ in regions outside the range of the $k_y - k_\parallel$ domains where the modes have the following simple forms. With these cautions, then, we may write the following formulas for the four modes of propagation that follow from Eq. (2.43) in the $k_\parallel v_e > \omega$ domain where electron Landau damping is the dominant dissipation mechanism, except for the ion-acoustic mode for which ion Landau damping dominates unless T_e/T_i is large. Here we assume that $c_s \ll v_A \ll v_e$.

Shear Alfvén Waves

$$\omega_k^a = \pm k_\parallel v_A - i\left(\frac{\pi}{8}\right)^{1/2} \frac{k_\parallel^2 v_A^2 k_\perp^2 \rho_s^2 (1 + T_i/T_e)}{|k_\parallel| v_e}. \tag{2.46}$$

Ion Acoustic Waves

$$\omega_k^s = \pm k_\parallel c_s - i\left(\frac{\pi}{8}\right)^{1/2} \frac{k_\parallel^2 c_s^2}{|k_\parallel| v_e}\left[1 + \left(\frac{m_i}{m_e}\right)^{1/2}\left(\frac{T_e}{T_i}\right)^{1/2} \exp\left(-\frac{T_e}{2T_i}\right)\right]. \tag{2.47}$$

Drift Waves

$$\omega_{\mathbf{k}}^{\mathrm{dw}} = \omega_{*e} + i\left(\frac{\pi}{2}\right)^{1/2} \frac{\omega_{*e}^2 k_\perp^2 \rho_s^2 (1 + T_i/T_e)}{|k_\parallel| v_e}. \tag{2.48}$$

Ion Drift Modes

$$\omega_{\mathbf{k}}^{id} = \omega_{*i} - i\left(\frac{\pi}{2}\right)^{1/2} \frac{\omega_{*i}^2 (1 + T_e/T_i)}{|k_\parallel| v_e}. \tag{2.49}$$

The formulas Eq. (2.46)-Eq. (2.49) apply where $k_\parallel v_e > \omega$. On the other hand in the electron hydromagnetic limit where $k_\parallel v_e < \omega$ the plasma modes of propagation obtained from Eq. (2.43) are as follows:

Inertial Shear Alfvén Waves

$$\omega_{\mathbf{k}}^a = \pm \frac{k_\parallel v_A}{(1 + k_\perp^2 c^2/\omega_{pe}^2)^{1/2}}. \tag{2.50}$$

Ion Drift Modes

$$\omega_{\mathbf{k}}^{id} = \omega_{*i} \tag{2.51}$$

where both the electron and ion damping are now exponentially small and hence not included in these equations. The damping is small since the phase velocity $\omega/k_{\|}$ is faster than the electron thermal velocity. The ion acoustic waves do not occur in this regime and are replaced by the lower hybrid waves and electron acoustic waves.

2.3 Drift Waves

The drift wave formulas in Eq. (2.47), Eq. (2.48), Eq. (2.49), Eq. (2.50) and Eq. (2.51) are the first of a large number of drift wave formulas needed to understand the wave-induced transport in magnetized, inhomogeneous plasma. Both hydrodynamic and kinetic descriptions will be used in the following chapters for the drift waves.

2.4 Kinetic Alfvén Waves

Alfvén waves are ubiquitous in space and laboratory plasmas. There are two kinds of wave dispersion that are important for Alfvén waves. The form in the cold plasma arises from the electron inertial skin depth $\delta = c/\omega_{pe}$. In hot plasmas the dispersion comes from the ion sound-inertial scale length ρ_s and the ion thermal gyroradius ρ_i for hot ion plasmas.

2.5 Coupling of the Drift Wave, Ion-Acoustic and Shear Alfvén Waves

A precise description of the coupling of the modes defined in Eq. (2.46)-Eq. (2.48) requires that the full dispersion relation be retained. Here we give the fifth order polynomial in ω that describes the coupling of these modes in the $k_{\|}v_e > \omega$ regime where the electron Landau damping is negligible. Rationalizing the fractional expression in Eq. (2.43), we obtain the polynomial dispersion

5th Order Dispersion Reduction

$$D_5(\mathbf{k}, \omega) = \sum_{n=0}^{5} a_n \omega^n = 0 \tag{2.52}$$

where polynomial coefficients are

$$a_0 = -\frac{k_\parallel^2 c_s^2}{\tau}\,\omega_* \Gamma_0(b)$$

$$a_1 = k_\parallel^2 c_s^2 \Gamma_0 - \frac{c_s^2}{v_A^2}\,\frac{1-\Gamma_0(b)}{b}\,\Gamma_0(b)\,\frac{\omega_*^2}{\tau^2}$$

$$a_2 = -\omega_* \Gamma_0 \left(1 - \frac{2c_s^2}{v_A^2 \tau}\right)$$

$$a_3 = -\frac{\omega_*^2}{\tau k_\parallel^2 v_A^2}\left(\frac{1-\Gamma_0}{b}\right) - 1 - \tau(1-\Gamma_0) - \frac{c^2}{v_A^2}\,\Gamma_0$$

$$a_4 = \frac{\omega_*(\tau - 1)}{\tau k_\parallel^2 v_A^2} \cdot \frac{1-\Gamma_0(b)}{b} \tag{2.53}$$

$$a_5 = \frac{1-\Gamma_0(b)}{k_\parallel^2 v_A^2 b}. \tag{2.54}$$

In Fig. 2.3 we show the disposition of the roots of Eq. (2.52) as a function of the $k_\parallel$ for $k_y \rho_s \ll 1$. We notice that the roots are complex along the dashed curves where the waves resonate with the thermal ions. The fluid expansions of the dispersion relation are not capable of describing the waves accurately in the regions where $\omega \simeq k_\parallel v_j$ where $v_j = (T_j/m_j)^{1/2}$ is the thermal speed. For the limit of a uniform plasma $\omega_* = 0$ and the polynomial reduces due to $a_0 = 0$ to a fourth degree with the standard forward and backward ion acoustic and shear Alfvén waves.

From the analysis presented in this chapter, and particularly from Fig. 2.3, it is evident that for typical laboratory plasmas where $c_s \ll v_A \lesssim v_e$ the decoupling of the ion drift-Alfvén modes given in Eq. (2.44) from the electron drift-acoustic modes given in Eq. (2.45) is a good approximation. The relative error made in

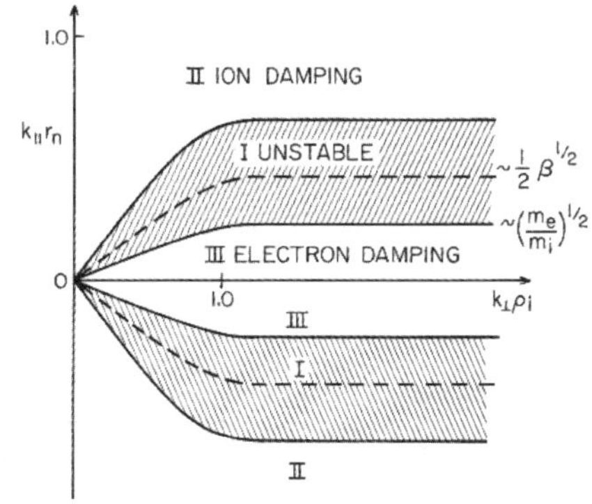

Fig. 2.3 The stable and unstable (shaded) regions of **k**-space for the electron drift wave driven by density gradient specified by $L_n = r_n$.

neglecting the Alfvén wave effects on the drift wave-acoustic modes is of order

$$\widehat{\beta} = \frac{\omega_{pi}^2 \omega^2}{(c^2 c_s^2 k_\parallel^2 k_\perp^2)} \simeq \frac{\omega_x^2}{(k_\parallel^2 v_A^2 b)} = \frac{\beta}{2k_\parallel^2 L_n^2} \tag{2.55}$$

where we use $\omega \simeq \omega_*$ and $c_s^2/v_A^2 = \beta/2$. The ratio $\widehat{\beta}$ follows comparing the terms $a_5\omega^5$ and $a_3\omega^3$ in Eq. (2.52). For electrostatic motion $\widehat{\beta}$ is small compared to unity. Neglecting these Alfvén coupling effects altogether reduces the dispersion relation to the electrostatic modes have the polarization $\mathbf{E} = -\nabla\phi = \Sigma - i\mathbf{k}\phi_{k\omega}$ with $\delta\mathbf{B} = 0$.

2.5.1 *Electrostatic drift waves*

For $\widehat{\beta} < 1$ the plasma modes are electrostatic (ES) with

$$D_{\text{ES}} = k_y^2 \varepsilon_\perp + k_\parallel^2 \varepsilon_\parallel = 0 \tag{2.56}$$

with $\varepsilon_\parallel$ and $\varepsilon_\perp$ given by Eq. (2.39) and Eq. (2.40), respectively. In this case the electric field is purely electrostatic with

Electrostatic Polarization

$$E_y = \left(\frac{k_y}{k_\parallel}\right) E_\parallel .$$

As is apparent from Eq. (2.48) the drift wave branch is intrinsically unstable. The approximate solution given in Eq. (2.48) is valid for $k_\parallel c_s < \omega_{*e} < k_\parallel v_A$. From these bounds on $|\omega_{*e}|$, it follows that there is instability only for $k_\parallel L_n < k_y \rho_s$ for $k_y \rho_s \leq 1$, and since $\omega_k \leq c_s/L_n$ for larger $k_y \rho$ the requirements that $k_\parallel c_s < \omega_k$ becomes $k_\parallel L_n < 1$ at $k_y \rho > 1$.

In simple geometries where the length $L_\parallel$ of the magnetic field line determines $k_\parallel$, the stability diagram in Fig. 2.3 gives unstable drift waves with $L_\parallel < 2\pi L_n$ for high $k_y \rho_s \lesssim 1$. Ion drift mode wavelength waves $k_y \rho_s < 1$ require even longer magnetic field lines to be drift wave unstable. The condition is approximately $L_\parallel < \frac{2\pi L_n}{k_y \rho_s}$.

In Fig. 2.3 we show the stability boundary implied by the electrostatic dispersion relation Eq. (2.56) computed numerically illustrates these conditions on $k_\parallel r_n$ versus $k_y \rho$ for instability. In Fig. 2.3 the symbol $r_n = L_n$ is used for the density gradient scale length as in the early drift wave literature.

The unstable region shown in Fig. 2.3 becomes wider for values of the parallel electron drift velocity $u = -j_\parallel / e n_e$ since confined plasmas often contain parallel currents with u/c_s of the order unity or so. The case shown is for $T_e = T_i$ and the plasma is sufficiently collisionless to neglect finite $\lambda_{\text{mfp}} = v_e/\nu_e$ effects altogether. Clearly, the physical limit of instability implied by the $k_\parallel - k_y$ stability boundary is that once $k_\parallel L_n$ exceeds the critical value given by the curve $k_{\parallel\text{crit}} = k_\parallel(k_y \rho)$, the parallel phase velocity of the wave is sufficiently low that ion Landau damping dominates the electron destabilization as given by the growth rate in Eq. (2.48).

We notice that although electron drift $u/c_s \sim 1$ is destabilizing, the effect is rather weak.

We see from Fig. 2.3 that a principal method of controlling the stability of the drift wave is through control of the parallel wavelength $\lambda_\| = 2\pi/|k_\||$. If, in the example given, the parallel wavelengths longer than approximately $(2\pi/0.15)L_n = 42L_n$ can be excluded from the system (by a finite system length or by magnetoshear) then unstable modes will not arise.

In toroidal confinement systems this limiting of the parallel wavelength to smaller than critical values is not a possible solution. In fact in toroidal systems with a nonuniform rotational transform in the magnetic field there always occur small radial regions where the parallel wavelengths $\lambda_\|$ can approach infinity. Thus, we may expect the toroidal system to have many unstable drift waves.

For magnetic fields with a substantial gradient in the rotational transform, a property which is defined as magnetic shear, the regions of long parallel wavelengths is spatially localized. In such systems the locally unstable region is consequently connected to neighboring regions of stable $k_\|(\mathbf{x})$ due to the spatial nonuniformity of the helical magnetic field. The net stability of the modes is determined by the strength of the connection between the stable and unstable regions due to the wave propagation. We now investigate this stability problem of the drift wave in a sheared magnetic field.

2.6 Drift Wave Eigenmodes in a Sheared Magnetic Field

As a simple example of the localizing and stabilizing effect of the magnetic shear on drift waves we consider a local radial region at $r = r_0$ with $x = r - r_0$ forming a slab of plasma in which the magnetic field varies locally according to

$$\mathbf{B}(\mathbf{x}) = B_0 \left(\widehat{\mathbf{e}}_z + \frac{x}{L_s} \widehat{\mathbf{e}}_y \right) \tag{2.57}$$

where L_s is the magnetic shear length. Clearly the magnetic field, which is along the z axis at $x = 0$, rotates by $\pm 45°$ in going a distance $x = \pm L_s$ from the $x = 0$ plane. This magnetic field is created by external currents for $B_0 \widehat{\mathbf{e}}_z$ and a constant plasma current $\nabla \times \mathbf{B} = B_0/L_s \, \widehat{\mathbf{e}}_z = (4\pi j_z/c)\widehat{\mathbf{e}}_z$ for $B_y(x)$ in Eq. (2.57). Thus the model applies where dj_z/dx is negligible for the waves of interest. For modes associated with dj_z/dx see Chapter 15.

Due to the infinite uniform medium property of the system in the y- and z-directions the fields for the plasma modes must vary according to

$$E_\alpha(\mathbf{x}, t) = E_{\alpha,k}(x) \exp\left(ik_y y + ik_z z - i\omega t\right) + \text{c.c.} \tag{2.58}$$

with all allowed k_y and k_z values considered. The derivation of the equations for $E_\alpha(x)$ from first principles is lengthyB. The results are perhaps apparent from Eq. (2.39) with the observation that b in that equation is now the square of the local perpendicular wavenumber which, due to the x dependence of the magnetic field vector in Eq. (2.39), is $k_\perp^2 = k_y^2 + k_x^2$ and, furthermore, the k_x wavenumber is properly defined from the fields by $\partial E_\alpha/\partial x = ik_x E_\alpha$. Thus, we expect that the first principle calculation should simply yield Eq. (2.39) with

$$b \to \frac{v_i^2}{2\omega_{ci}^2}\left(k_y^2 - \frac{\partial^2}{\partial x^2}\right). \tag{2.59}$$

Making the Ansätz that the relevant modes $E_{\alpha,k}$ have wavelengths in the x-direction that are long compared with the ion gyroradius, the operator $\rho_i^2\partial^2/\partial x^2$ is small compared with $k_y^2\rho_i^2$ and is Taylor series-expanded out of the Bessel functions in Eq. (2.39) to obtain the second-order differential equation governing the field amplitude variation for the nonlocal mode. Performing the small $\rho_i^2\partial^2/\partial x^2$ expansion in Eq. (2.39) yields the wave equation

Drift Wave Eigenmodes

$$\begin{aligned}\rho_s^2\frac{\partial^2 E_y(x)}{\partial x^2} &+ \left\{-k_y^2\rho_s^2 + \left[1 - \frac{\omega(\omega-\omega_{*i})}{k_\parallel^2 v_A^2}\right]\right.\\ &\times\left.\left[\frac{1}{2}\left(1-\frac{\omega_{*e}}{\omega}\right)Z'\left(\frac{\omega}{k_\parallel v_e}\right) - \frac{k_\parallel^2 c_s^2}{\omega^2}\right]\right\}E_y(x) = 0.\end{aligned} \tag{2.60}$$

Here $\rho_s^2 = \tau\rho_i^2$ with $\tau = T_e/T_i$ is the ion inertia scale length. In Eq. (2.60) the parallel wavenumber is a function of x due to the rotation of the magnetic field vector with changing position in x. In fact, we have from Eq. (2.57) that $k_\parallel(x) = k_z + k_y(x/L_s)$ so that for given k_y and k_z the different x positions define a set of straight lines on the marginal stability plane in Fig. 2.3 which passes from the unstable region to the stable region as x is varied. From the marginal stability diagram we should expect the unstable mode attempts to localize its amplitude to the small $k_\parallel(x)$ region. To look for such unstable localized modes we shift the origin of the x coordinate to the point where $k_\parallel = 0$ or $x' = x - x_0$ where $x_0 = -k_z L_s/k_y$. Or equivalently, for the given origin defined by Eq. (2.57), the k_y, k_z modes which are most localized to the unstable region are those modes which have $k_z = 0$.

The physical meaning of this shift $x \to x - x_0$, or the choice of k_y/k_z for a given x_0, is to bring the pitch of the magnetic field twist to equal the pitch of the phase fronts $k_y y + k_z z =$ constant of the wave.

For the modes with $k_\parallel(x) = k_y x/L_s$ we have the wave equation

Sheared Slab Model

$$\begin{aligned}\rho_s^2\frac{\partial^2}{\partial x^2}E_y(x) &+ \left\{-k_y^2\rho_s^2 + \left[1 - \frac{\omega(\omega-\omega_{*i})L_s^2}{k_y^2 x^2}\right]\right.\\ &\times\left.\left[\frac{1}{2}\left(1-\frac{\omega_{*e}}{\omega}\right)Z'\left(\frac{\omega L_s}{k_y v_e|x|}\right) + \frac{c_s^2 k_y^2 x^2}{\omega^2 L_s^2}\right]\right\}E_y(x) = 0\end{aligned} \tag{2.61}$$

where for sufficiently large x the hydrodynamic approximation for the ions $\omega \gg k_\parallel v_i$ fails and the mode enters a region of strong ion Landau damping which exponentially damps the amplitude of the mode. For large values of x the mode equation Eq. (2.61) reduces to

$$\left(\rho_s^2 \frac{\partial^2}{\partial x^2} + \frac{c_s^2 k_y^2}{\omega^2 L_s^2}\, x^2\right) E_y(x) \cong 0 \tag{2.62}$$

with the two linearly independent asymptotic solution given by

$$E_y^{\pm}(\mathbf{x}, t) = \frac{C_{\pm}(\omega)}{|x|^{1/2}} \exp\left(ik_y y \pm \frac{ik_y c_s}{2\omega L_s}\, x^2 - i\omega t\right) + \text{c.c.} \tag{2.63}$$

For a wave packet with given $C_{\pm}(\omega)$ the group velocity of the wave in the x-direction is clearly opposite to the phase velocity of the waves. The constant wave fronts follows from Eq. (2.63).

Phase Front Propagation

$$S_{k,\omega}^{\pm}(x, t) = \pm \frac{k_y c_s^2}{2\omega L_s}\, x^2 - \omega t = \text{const.} \tag{2.64}$$

where $S^{\pm}$ is called the eikonal of the wave function.

The stationary phase point of the wave packet follows from the derivative of the phase in Eq. (2.64) with respect to ω giving

Group Velocity Propagation

$$\frac{dS_{k,\omega}^{\pm}}{d\omega} = \mp \frac{k_y c_s^2}{2\omega^2 L_s}\, x^2 - t = 0. \tag{2.65}$$

Equation (2.65) follows by considering $\int d\omega C_{\pm}(\omega) \exp\left(iS_{k,\omega}^{\pm}(x, \omega, t)\right)$ evaluated by the method of stationary phase [Jackson (1998)] Section 10.12.

2.7 Symmetries of the Drift Wave Eigenmodes

In the derivation of Eq. (2.61) we assume that the flow from $v_E = -E_r/B$ is constant as to the Doppler shift wave frequency and that there is no gradient of the drift wave frequencies $\omega_{*i,e}$. In general, this is not the case and these background gradients break the alternating even and odd symmetries of the eigenmodes given by Eq. (2.61). When the shearing effects are weak we treat them with perturbation theory as in quantum mechanics finding second order shifts of the eigenvalues.

The symmetry breaking due to the shearing has a second more subtle effect that is important at finite plasma pressure. The modes with odd reflection symmetry for the electrostatic potential have an even symmetry of the associated magnetic perturbation of the $\delta B_r(x)$ field arising from the oscillating electron parallel currents in the drift waves. These magnetic perturbations from the odd symmetry potential modes create small magnetic islands that can then grow from nonlinear driving terms and energy associated with the shear in the ambient magnetic field. The calculations for this effect are postponed until Chapter 15 where the instabilities that arise from magnetic reconnection of twisted magnetic fields are analyzed. Simulations showing the transformation at finite plasma pressure $\beta_e \geq 0.005$ percent of the magnetic pressure from the electrostatic drift waves to the magnetic tearing mode parity are shown in Muraglia, *et al.* (2009, 2011).

2.8 Outgoing Wave Boundary Conditions

In the time interval from t to $t + dt$ the wave phase fronts move from x by $dx = \pm(\omega^2 L_s / x k_y c_s^2)dt$ whereas the wave packet maximum moves from x by $dx = \mp(\omega^2 L_s / x k_y c_s^2)dt$. Consequently, the asymptotic solution that represents a propagation of the wave energy in a wave packet from the unstable region about $k_\parallel(x) \simeq 0$ to the regions of wave absorption at large $|k_\parallel|$ is the negative branch (lower signs) in Eq. (2.63). This negative branch solution has outgoing wave energy and incoming phase fronts [Pearlstein and Berk (1969)].

The explicit form of the wave absorption at large $|k_\parallel|$ is from the ion-wave resonance given by Im $Z'(\omega/k_\parallel v_i)$ in Eq. (2.25). In writing Eq. (2.60) this imaginary part of the ion Z'-function has been dropped. Eigenvalue calculations with the function $Z'(\zeta_i)$ where $\zeta_i = \omega/|k_\parallel|v_i$ are given in Gladd and Horton (1973). The component of the wavenumber vector **k** parallel to the magnetic field is critical in determining the response of the plasma to the wave.

With these results we are able to determine the physically acceptable boundary conditions (B.C.) on the solutions of Eq. (2.61) for large positive and negative values of x. With these outgoing wave boundary conditions which we may write as

Outgoing Energy B.C.

$$\lim_{|x|\to\infty} \left[\rho_s \frac{dE_y}{dx} + i \frac{k_y c_s x}{\omega L_s} E_y \right] = 0 \tag{2.66}$$

a unique solution of Eq. (2.62) is specified. The unique nontrivial solution only occurs for the eigenvalues of ω. In the approximation where the electron response is $Z' = -2$ and $v_A/c_s \to \infty$ the differential equation reduces to such an extent that we can write down the well-known eigenvalues and eigenfunctions. Retaining the full complex potential in Eq. (2.62) requires lengthy analytical perturbation theory calculation or numerical shooting method analysis.

For modes sufficiently broad that $\langle |k_\parallel(x)| v_e/\omega \rangle > 1$ we may approximate $Z'(\omega/k_\parallel(x)v_e) \simeq -2$ and reduce Eq. (2.60) to

$$\rho_s^2 \frac{\partial^2 E_y}{\partial x^2} + \left[-k_y^2 \rho_s^2 - \left(1 - \frac{\omega_{*e}}{\omega}\right) + \frac{k_\parallel^2 c_s^2}{\omega^2} \right] E_y(x) = 0. \tag{2.67}$$

The spectrum $n = 0, 1, 2, 3, ...$ of solutions of the wave equation Eq. (2.67) satisfying the boundary condition given in Eq. (2.66) are

Eigenmodes

$$\varphi_{k_y,n}(x) = H_n \left[\left(\frac{ik_y \rho_s c_s}{\omega_k L_s} \right)^{1/2} x \right] \exp\left(-\frac{ic_s k_y x^2}{2\omega_k \rho_s L_s} \right) \tag{2.68}$$

where the eigenfrequency $\omega_{k,n}$ is shifted from $\omega_k = \omega_{*e}/(1 + k_y^2 \rho_s^2)$ to the spectrum of waves

Eigenfrequencies

$$\omega_{k,n} = \frac{\omega_{*e}[1 - i(L_n/L_s)(2n+1)\mathrm{sg}(k_y)]}{1 + k_y^2 \rho_s^2}. \tag{2.69}$$

Here $n = 0, 1, 2, ...$ is radial mode number and $H_n(\xi)$ are the Hermite polynomials. The physical effect of the shear is manifested by the damping at the rate $|\omega_*|(L_n/L_s)$ produced by the outward propagation of energy from the locally unstable region. The damping in Eq. (2.69) from the propagation of the wave energy from unstable region to the stable region is the basis of the important principle of magnetic shear stabilization of the drift waves. The boundary condition Eq. (2.66) assumes that all outgoing wave energy is lost. In reality the wave energy is absorbed by the ion Landau damping in the regions where $|k_\parallel| v_i > |\omega_k|$. This radial region is in the tails of the eigenmode wave functions far from the mode rational surface where $k_\parallel = 0$.

The numerical solutions of the drift wave eigenvalue problem with a magnetic shear is shown in Fig. 2.4. The growth with rate decreases with $S = 0.06, 0.8$ and stable for $S = 0.10$. In this figure from Gladd and Horton (1973) the symbol $a = \rho_s = c_s/\omega_{ci}$ and $r_n = L_n$ and $T_i = T_i$.

Let us consider the nature of the eigenfunction produced by the shear stabilization and determine the implications of the mode structure. In Chapter 10.1.1 we will see that electron dissipation $i\delta_k$ shifts the denominator of Eq. (2.69) to $1 + k_y^2 \rho_s^2 - i\delta_k$. The electron dissipation then gives $\gamma = \mathrm{Im}(\omega_{k,n}) > 0$ corresponding to exponential growth in time for the wave for the k_y, n eigenmodes with

Unstable Growth

$$\delta_k > \frac{r_n}{L_s}(2n+1)\mathrm{sg}(k_y) \tag{2.70}$$

where $n = 0, 1, 2, \ldots$ are the radial eigenvalue mode numbers in Eq. (2.68).

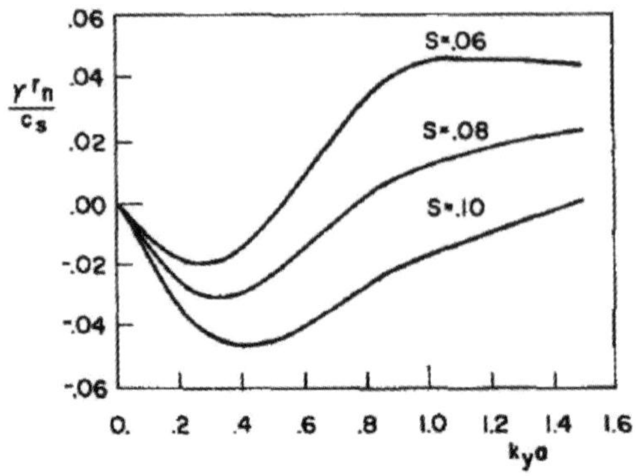

Fig. 2.4 Eigenmode growth rate γ as a function of wavenumber k_y for three values of the magnetic shear $S = L_n/L_s$ parameters. In this figure $a = \rho_s$ and $r_n = L_n$.

2.8.1 *Localized ion drift modes*

From the mode equation given in Eq. (2.60) we see that there also occurs the possibility of narrower modes localized to regions such that $\langle |k_\parallel(x)| v_e \rangle \ll \omega$. Such hydrodynamic modes would encompass the FLR ion drift mode with

$$\left(k_\perp^2 \rho_i^2\right) \omega(\omega - \omega_{*i}) + k_y^2 \sum_j v_{d_j} v_{D_j} = 0 \tag{2.71}$$

or the plasma slab where $v_{D_j} = cT_j/eBR_c$ is the particle guiding center drift velocity induced by curvature of the magnetic field line. The strength of the magnetic curvature is measured by the radius of curvature R_c. These modes given by Eq. (2.71) are extremely important in system with weak or no magnetic shear such as mirror machines and field reversed confinement (FRC) machines. The modes are called "flute modes."

Often the analysis is clearer in dimensionless variables. Thus we introduce the dimensionless variables $\xi = x/\rho_s, \Omega = \omega/\omega_{*e}, K = k_y\rho_s, S = L_n/L_s$, and $\beta = 8\pi p/B^2$.

The ion mode equation in such regions is

$$\frac{\partial^2 \varphi}{\partial \xi^2} + \left[-K^2 + \frac{S^2\xi^2}{\Omega^2} - \frac{1}{2}\beta + \left(1 - \frac{1}{\Omega}\right)\left(\frac{m_i}{m_e}\right)\left(\frac{S^2\xi^2}{\Omega^2} - \frac{1}{2}\beta\right)\right]\varphi = 0 \tag{2.72}$$

where $E_y(x) = -ik_y\varphi(x)$, $\xi = x/\rho$, $\Omega = \omega/\omega_{*e}$ and $S = L_n/L_s$. The eigenvalue Ω is a function of four parameters: $K = k_y\rho_s$, $S = r_n/L_s$, m_i/m_e, $\beta = 8\pi p/B_0^2$.

The ground state eigenmode of Eq. (2.72) is

$$\varphi = \exp\left(-\frac{\sigma\xi^2}{2}\right) \tag{2.73}$$

with

$$\sigma = \left(\frac{m_i}{m_e}\right)^{1/2} \frac{|S|}{\Omega} \left(\frac{1}{\Omega} - 1\right)^{1/2}. \tag{2.74}$$

The eigenvalue Ω for the ground state is given by

$$\begin{aligned} -\frac{S}{\Omega}\left(\frac{1}{\Omega}-1\right)^{1/2}\left(\frac{m_i}{m_e}\right)^{1/2} - K^2 - \frac{1}{2}\beta \\ +\frac{1}{2}\left(\frac{m_i\beta}{m_e}\right)\left(\frac{1}{\Omega}-1\right) = 0. \end{aligned} \tag{2.75}$$

For $0 < \Omega_k < 1$ there is one mode with real Ω: outside this frequency range there are growing and decaying solutions. All roots of Eq. (2.75) must satisfy the localization criterion $|k_\parallel' v_e \Delta x/\omega| < 1$ to satisfy the validity condition of reducing Eq. (2.61). Typically, only the root close to $\Omega = 1/[1 + K^2 2m_e/m_i\beta]$, which in dimensional form, is

$$\omega_k = \frac{\omega_{*e}}{(1 + k_\perp^2 c^2/\omega_{pe}^2)} \tag{2.76}$$

satisfies the localization requirement. The dispersion of the waves from the electron inertial scale length or skin depth $\delta_s = c/\omega_{pe}$ is important for small-scale drift-wave turbulence. We will see the shielding effect of this electromagnetic field in the case of tearing modes and electron temperature gradient driven modes in Chapters 14 and 15.

2.9 Ion Acoustic Wave Turbulence

Many space and laboratory plasmas have spectra of ion acoustic waves. When a small density gradient is present across an ambient magnetic field the forward and backward ion acoustic waves become split in to a drift wave with a phase velocity faster than the ion acoustic speed and another slower wave. For the density gradient small or vanishing, the solutions of electrostatic dispersion given ion oscillations with adiabatic shielding $\delta n_e = n_0(e\varphi/T_e)$ behavior of the electrons. This ion oscillation is defined as the ion acoustic waves and the ion fluid is compressed and rarefracted during the oscillations. The electric field $\mathbf{E} = -\nabla\varphi$, however, dominates the ion pressure gradient giving $c_s^2 = T_e/m_i$. It is useful to study the measurements of the ion acoustic wave spectra as an introduction to measuring the spectrum of the drift waves.

Renormalized kinetic theory for the long range collective oscillations of turbulent plasmas is developed through infinite order summations of the interactions in the Vlasov-Poisson self-consistent field equations. Three levels of renormalized turbulence theory are known and the simply renormalized equations are solved for the problem of ion acoustic instability. The physical predictions of renormalized turbulence theory are used to interpret several experiments on ion acoustic turbulence.

2.9.1 *Electromagnetic scattering measurements of ion acoustic waves*

The spectral distribution of ion acoustic fluctuations and drift wave fluctuations are measured by electromagnetic scattering from the electron density fluctuations δn_e from the waves. In laboratory plasmas the theoretically predicted wavelengths are sufficiently short that the method of scattering light and infrared electromagnetic radiation lends itself to the detection of the electron density fluctuations in the plasma. Strong scattering maximum occur at the Bragg angle where the $\Delta\mathbf{k}$ of the scattered electromagnetic wave equals the wavenumber vector of the fluctuation in the electron density $n_e(\Delta\mathbf{k})$. From these scattering experiments an understanding of the wave vector distribution of the density fluctuations in ion acoustic and drift wave turbulence is obtained. The measured spectral distributions agree with the theoretical distributions.

Non-perturbative measurements of the plasma fluctuation are carried out with Bragg scattering, $\mathbf{k} = \mathbf{k}_{\rm out} - \mathbf{k}_{\rm in}$, where an in coming microwave or lower wave, $\mathbf{E}_{\rm in}\exp(i\mathbf{k}_{\rm in}\cdot\mathbf{x} - i\omega_{\rm in}t)$, sets the electron into quiver motion with $\delta\mathbf{v}_e = -e\mathbf{E}_{\rm in}/im_e\omega_{\rm in}$ that radiates $\mathbf{E}_{\rm out}\exp(i\mathbf{k}_{\rm out}\cdot\mathbf{x} - i\omega_{\rm out}t)$, with current $\delta\mathbf{j} = -e\delta n_{\mathbf{k}} \times \delta\mathbf{v}_e(\mathbf{k}_{\rm in})$. The scattered wave, $\mathbf{E}_{\rm out}\exp(i\mathbf{k}_{\rm out}\cdot\mathbf{x} - i\omega_0 t)$ has $\mathbf{k}_{\rm out} = \mathbf{k} + \mathbf{k}_{\rm in}$ and $\omega_{\rm out} = \omega_{\mathbf{k}} + \omega_{\rm in}$. Since $\omega_{\mathbf{k}} \ll k_{\rm in}c$, the frequency shift is often too small to be resolved. The scattering cross-section,

$$\frac{d\sigma}{d\Omega} = \left(\frac{e^2}{m_e c^2}\right)^2 \left\langle |\delta n_e(\mathbf{k})|^2 \right\rangle \frac{\Omega_{\rm scat}}{n_e}$$

where $\Omega_{\rm scat}$ is the scattering volume defined by the insertion of the two electromagnetic waves giving a non-perturbative measurement of $\langle|\delta n_e(\mathbf{k})^2\rangle$. Both X- and O-mode polarizations are useful. Here the dimensions of the cross-section are given by the square of the classical radius of the electron $r_0 = e^2/m_e c^2 = 2.8\times10^{-13}\,$cm.

In the scattering experiments, the enhancement of the scattered power due to correlations in the electron density is measured by the dimensionless form factor $S(\mathbf{k})$ where the wavenumber $\mathbf{k}$ of the fluctuation is determined by the scattering geometry. For scattering geometries where there are a large number of wavelengths in the scattering volume V_s, knowledge of the form factor $S(\mathbf{k})$ is equivalent to determining the spectral distribution of $N(\mathbf{k})$ of the electron density correlation function.

For a specified type of small amplitude fluctuations, theory is used to relate the measured density correlation function $N(\mathbf{k})$ to the potential fluctuation spectrum $I(\mathbf{k})$ in nonlinear theory. Here we review the formulas required for the analysis of the scattering data.

The differential cross-section for scattering the incident electromagnetic wave $\widehat{\varepsilon}_0 E(x)\exp(i\mathbf{k}_0\cdot\mathbf{x} - i\omega_0 t)$ from the scattering volume V_s into the direction $\widehat{\mathbf{n}}$ of observation is

$$\frac{d\sigma}{d\boldsymbol{\Omega}} = r_0^2 F(\mathbf{q})\,|\widehat{\varepsilon}_0 \times \widehat{\mathbf{n}}|^2 \tag{2.77}$$

where $\mathbf{q} = (\omega/c)\widehat{\mathbf{N}} - \mathbf{k}_0$, $r_0 = e^2/m_e c^2$ and $F(\mathbf{q})$ is the scattering form factor defined by

$$F(\mathbf{q}) = \left\langle \left| \int_{V_s} d\mathbf{x}\, n_e(\mathbf{x}) e^{-i\mathbf{q}\cdot\mathbf{x}} \right|^2 \right\rangle. \tag{2.78}$$

For randomly distributed electrons the form factor $F(\mathbf{q})$ clearly reduces to $F = N_s = n_e V_s$, the number of electrons in the scattering volume. The dimensionless scattering form factor $S(\mathbf{q})$, also called the static form factor, is defined relative to this incoherent scattering value of F, that is

$$S(\mathbf{q}) = \frac{F(\mathbf{q})}{n_e} V_s. \tag{2.79}$$

For fluctuations with $q^3 V_s \gg 1$ it is evident from Eq. (2.78) that $S(\mathbf{q})$ is simply related to the spectral distribution of the electron density fluctuations $N(\mathbf{k})$ defined by

$$\langle n_e(\mathbf{x}') n_e(\mathbf{x}) \rangle = \int_{V_s} d\mathbf{k}\ N(\mathbf{k}) e^{i\mathbf{k}\cdot(\mathbf{x}'-\mathbf{x})}. \tag{2.80}$$

Upon using the relations for the Fourier transforms in the limit $k^2 k_\| V_s \to \infty$ we obtain

$$S(\mathbf{k}) = (2\pi)^3 \frac{N(\mathbf{k})}{n_e} \tag{2.81}$$

from Eq. (2.78), Eq. (2.79), and Eq. (2.80). When the electron density fluctuations are due to ion acoustic or drift wave fluctuations we have the simple relation

$$\frac{N(\mathbf{k})}{n_e^2} = \frac{e^2 I(\mathbf{k})}{T_e^2} \tag{2.82}$$

between the spectral distribution of the electron density fluctuations and the potential fluctuations.

The total cross-section for high frequency electromagnetic wave scattering from a single electron is the Thompson cross-section $\sigma_T = (8\pi/3) r_0^2 \simeq 2/3 \times 10^{-24}\,\mathrm{cm}^2$ and is valid for photon energies less than the rest mass of the electron.

2.9.2 *Laser scattering experiment in Helium plasma*

Experimental measurements by Slusher, *et al.* (1976) of the spectral distribution of the ion acoustic fluctuations were performed in a low temperature helium discharge plasma. In this experiment a constant electrical current with drift velocity $u/v_e \approx 0.1 - 0.15$ in a partially ionized helium plasma produces a broad spectrum of ion acoustic fluctuations whose distribution in wave vectors k is measured by scattering a 10.6 micron wavelength CO_2 laser beam from the plasma. The spectrum reported by Slusher, *et al.* (1976) is shown in Fig. 2.5. The observed spectrum is peaked at $k\lambda_{De} \approx 0.6$ and has an angular half width of approximately 15°-20°.

The observed spectrum agrees with that obtained in the mode simulations based on renormalized turbulence theory as shown in Fig. 2.3. The level of the turbulence is found experimentally to be a sensitive function of the presence of light hydrogen ions which occurs as an impurity in the helium plasma.

Fluid simulations with three nonlinear pressure equations for p_e, p_i and p_z and parallel flow velocities $v_{\parallel i}$ and $v_{\parallel e}$ are given in Dubuit, *et al.* (2007). The Dubuit *et al.* work uses parameters from a typical L-mode Tore Supra discharge.

According to the theory, these light ions are rapidly accelerated to form a fast ion tail. Although the fractional hydrogen density is known only approximately, Slusher, *et al.* (1976) conclude that there is approximately a balance between the linear electron growth rate and the Landau damping of the waves on the fast hydrogen ions. The excess of the electron growth rate over the linear hydrogen ion tail damping rate determines the level of the nonlinear wavenumber spectrum. This excess is not known with sufficient accuracy to permit a comparison of the theory with the experimental turbulence level. The observed turbulence level W is reported as $W/nT_e \lesssim 10^{-4}$.

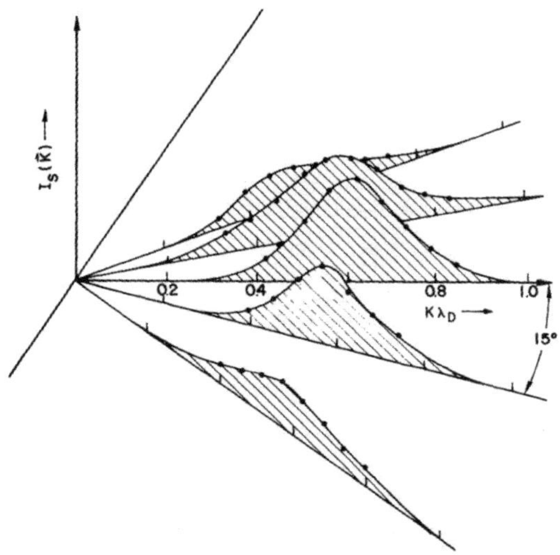

Fig. 2.5 Spectral distributions of current driven ion acoustic fluctuations observed by Slusher, *et al.* (1976) from a far-infrared laser scattering experiment in a partial ionized unmagnetized low-temperature helium plasma. The scattered intensity is shown on an arbitrary linear scale versus the $k\lambda_{De}$ and the angle between the fluctuation vector $\mathbf{k}$ and the direction of the electron drift velocity that carries the current driving the turbulence.

The principal result from this experiment is a detailed mapping of the relative intensity of the spectral distribution $I(\mathbf{k})$ of the fluctuations throughout the $\mathbf{k}\lambda_{De}$ region where theory predicts the modes to be significant. Comparing the theoretical distribution from renormalized turbulence theory gives the observed spectrum in Fig. 2.5.

In a similar helium positive column experiment the spectral distribution of ion acoustic fluctuations is reported by Ilic (1977) from probe measurements. The steady-state turbulent energy density is reported as $W/nT \approx 10^{-4}$ at low neutral gas pressure and $W/nT \approx 10^{-5}$ at higher neutral gas pressures. It is only at higher turbulence levels, $W/nT > 10^{-3}$, that theory predicts significant nonlinear deformations of the spectral distribution $S(\mathbf{k},\omega)$ from $\exp(2\gamma_k \tau)$ where τ is the collisional decay time from ion-neutral collisions. The figures of Ilic for $S(\mathbf{k},\omega)$ are consistent with this linear theory amplification factor.

2.9.3 *Probe measurements of the two-point correlation functions*

In a stationary uniform fully ionized krypton plasma in which ion acoustic turbulence is driven by a constant return current [Stenzel (1978)] has measured the characteristics of the turbulence by two-point correlation function data obtained with probes. The correlation data verify that the frequency and wavelength relations of the maximum of $I(\mathbf{k},\omega)$ for the observed signals satisfy the real part of the linear dispersion relation for ion acoustic modes in the presence of the turbulence. The waves are shown to have a finite lifetime due to the presence of the turbulence as predicted by theory through the line width ν_k. In space the waves first grow exponentially and then decay due to the finite lifetime $1/\nu_k$ determined by the turbulence.

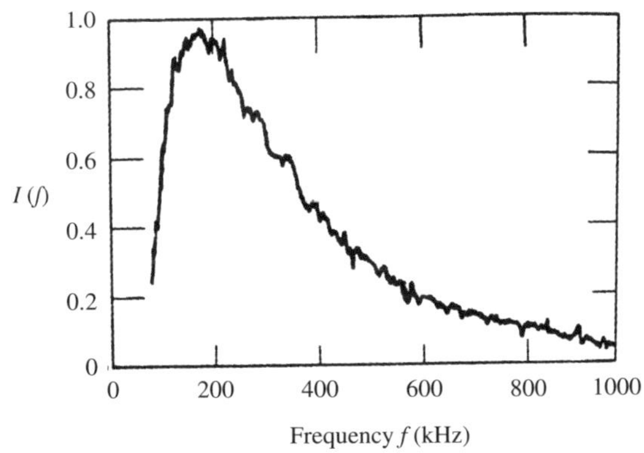

Fig. 2.6 Current driven ion acoustic turbulence measured in a large volume collisionless Krypton plasma experiment by Stenzel (1978). The turbulent spectrum is produced by a uniform plasma return current with $u \approx 40c_s \approx 0.08v_e$ in a steady state where $T_e/T_i \approx 20(m_{Kr} = 84m_H)$.

The experiments report that the observed frequency spectrum is consistent with that obtained from the wavenumber spectrum reported by Horton, *et al.* (1977) transformed from wavenumber to frequency through the linear dispersion relation. A typical frequency spectrum measured by Stenzel for which this comparison is made is shown in Fig. 2.6. The Fourier transform of the two-space correlation function shows a broad **k** space spectrum that is consistent with the modified Kadomtsev, spectrum peaked at $k\lambda_{De} \approx 0.2$ and having an angular width that approaches $90°$. The level of the turbulent energy density $W/nT_e \simeq 10^{-2}$ compared to the electron thermal energy density is reported to be a few percent.

The experiment shows that the process of induced wave scattering is a relatively strong process. The process involves observed wave amplitudes when the beat wave phase velocity is appropriately resonant with either the thermal ions or electrons. The wave scattering is verified by test wave propagation in the experiments.

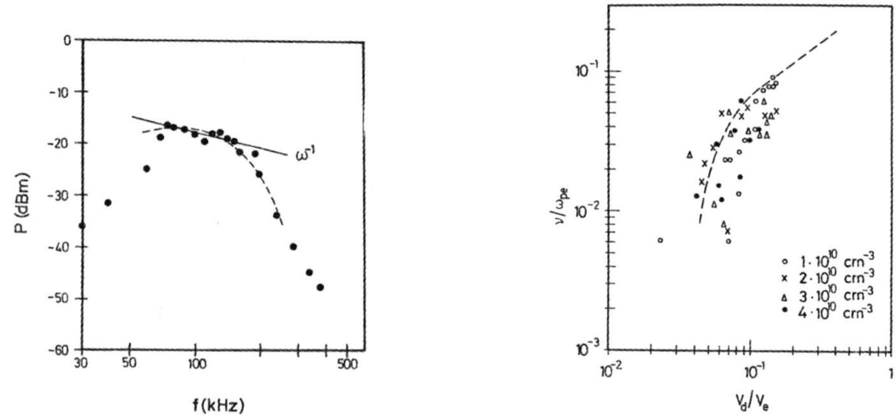

(a) (b)

Fig. 2.7 (a) The frequency dependence of the power spectrum for $V_g = 218\,\text{V}$, where $\omega_{pi}/2\pi \simeq 2.2\,\text{MHz}$ and $v_d/v_e \simeq 0.1$. The dashed curve is the theoretical wave spectrum of Horton, *et al.* (1977) [Kawai and Guyot (1977); Kawai, *et al.* (1978)]; (b) The effective collision frequency as a function of v_d/v_e at the pressure of 2×10^{-4} Torr, where $T_e = 1\,\text{eV}$ and $L = 10\,\text{cm}$. The dashed line is the theoretical values of the turbulent electron scattering frequency ν [Kawai and Guyot (1977)].

In comparing experimental data taken with probes to particle data taken by electromagnetic wave scattering one should be aware that appreciable distortion of wavenumber spectrum is potentially a problem with the probe data.

2.9.4 *Probe measurements of the spectrum and anomalous resistivity*

A large volume, steady-state collisionless argon plasma is investigated by Kawai and Guyot (1977) and Kawai, *et al.* (1978) with Langmuir probes, double probes and an electron energy analyzer. The mean free path in the plasma is determined by neutral collisions and is much longer than the dimensions of the system. Ion heating is negligible and electron heating is weak. The electron to ion temperature ration, $T_e/T_i = 15$, determined experimentally from the propagation speed of test waves, the ion acoustic instability is observed to occur at the expected value of the electron drift velocity u. Direct measurements of the time averaged electric field $\langle \mathbf{E} \rangle$, the electron drift velocity $\mathbf{u}$ and the frequency spectrum of the electron density and the electrostatic potential fluctuations are reported for parameters in the range $0.03 < u/v_e < 0.2$. The associated turbulent resistivity collision frequency ν is observed to lie in the range of $7 \times 10^{-3} < \nu/\omega_{pe} < 9 \times 10^{-2}$. The frequency spectrum and the resistivity are analyzed by Kawai, *et al.* (1978) using the formulas given by Choi and Horton (1974) from renormalized turbulence theory. In Fig. 2.7 their analysis of the observed frequency spectrum is shown for the case $u/v_e = 0.1$. The dashed curve is the result that Kawai, *et al.* (1978) obtain from applying the formulas of renormalized turbulence theory to their experiment. In obtaining the frequency spectrum the wavenumber spectrum $I(k)$ is converted to a frequency spectrum making use of the linear dispersion relation. For comparison with the renormalized frequency spectrum, the $1/\omega$ dependence obtained from the Kadomtsev spectrum is also shown. The shape of the measured frequency spectrum agrees with the renormalized spectrum except possibly in the wings of the spectrum where there is only a small fraction of the net energy density.

The turbulent energy density as a fraction of the electron thermal energy density follows from the frequency integral of the fluctuation spectrum. On the other hand, the effective resistivity collision frequency is derived from the time-averaged electric field $\langle E \rangle$ and the electron drift velocity $u = \langle j \rangle / en$. These quantities are observed to be related through $\nu/\omega_{pe} = (0.30 - 0.33)(W/nT_e)$ over the range of parameters accessible to the experimenters. This result follows quite generally from the turbulence theory since both quantities are determined by similar integrals of the $\mathbf{k}$ spectrum. The numerical coefficient in this relationship depends on the shape of the spectrum, and the observed value of 0.30-0.33 is to be compared with the value of 0.37 obtained in the mode simulations.

In the parametric study of the observed resistivity collision frequency, which by the preceding formula is equivalent to the parametric dependence of W/nT_e, is investigated for the drift velocity varying from $u/v_e = 0.03$ to 0.2 with the electron density varying from $(1 - 4) \times 10^{10}$ cm^{-3}. The result of the parametric analysis of the data along with the collision frequency computed with the renormalized spectrum is shown in Fig. 2.7 as reported by Kawai, *et al.* (1978). The collision frequency increases abruptly with increasing u/v_e in the region only slightly above the critical drift velocity and then increases proportional to u/v_e in the range about $u/v_e \simeq 0.1$. For still larger drift velocities there is little further increase in ν/ω_{pe}. The dashed curve, which appears to interpret the data adequately is calculated by Kawai, *et al.* (1978) from the formulas in Choi and Horton (1974).

2.9.5 *Drift wave spectral distributions*

Since drift waves are closely related in their properties to the ion acoustic waves, further understanding of the present problem concerning the predictions of renormalized turbulence theory is obtained from studying the recent experiments on the stationary spectrum of drift wave fluctuations. Drift waves are the fast wave branch of the ion acoustic waves in a nonuniform magnetized plasma (Eq. (2.45)-Eq. (2.48) and Chapter 5.4).

For drift modes, a relatively simple formula [Horton (1976)] for the spectral distribution $I(k_\perp)$ of the turbulence at a given radial position is obtained from simply renormalized theory of plasma turbulence applied to localized drift wave normal modes. In the derivation of the spectrum the linear electron growth rate $\gamma^\ell(k_\perp)$ of the drift waves is balanced by the induced wave scattering arising through the $\mathbf{E}\times\mathbf{B}$ nonlinearities in the ion dynamics. The nonlinear ion response is obtained from renormalized turbulence theory. In this ion response the renormalized particle propagator becomes $[\omega-\omega'-(k_\parallel-k'_\parallel)v_\parallel+i\nu_{k-k'}]^{-1}$ where ν_k is determined through the diagrams equivalent to those in Eq. (29) in Choi and Horton for the nonlinear $\mathbf{E}\times\mathbf{B}$ convective flux. For this problem one finds that even at the low fractional turbulence levels observed in the experiments and predicted by the theoretical formula for $I(k_\perp)$ we have $\langle\omega_\mathbf{k}\rangle \gtrsim \langle\nu_{_k}\rangle \gg \langle k_\parallel\rangle v_i$ so that the wave scattering resonance is dominated by the turbulent $\mathbf{E}\times\mathbf{B}$ motions $\nu_k = \langle(\mathbf{k}\cdot\mathbf{v}_E)\rangle^{1/2}$. In this turbulence a balance occurs between the rate of nonlinear transfer of energy from the mode $k_\perp, \omega_{k\perp}$ and the rate of energy transfer into the mode through the electron growth rate $\gamma^\ell(k_\perp)$. The resulting spectral distribution is given by the formula in Horton (1976).

In the ATC tokamak experiment at Princeton, where the drift wave spectrum was measured, the fastest growing linear modes are predicted by theory to have wavelengths across the magnetic field of approximately one centimeter. Resolution of electron density fluctuations on this spatial scale is obtained by scattering a 90 GHz ($\lambda_0 = 1/3\,\mathrm{cm}$) microwave beam from the plasma in the experiments of Mazzucato (1976, 1978). In the ATC experiment shown in Fig. 2.8, scattered intensities from six values of $k_\perp$ are measured along with the frequency spectrum of the scattered signal at each $k_\perp$. The frequency spectrum is observed to be broadly peaked at the $\omega_{k\perp}$ corresponding to drift wave theory, and the width of the spectrum is consistent with the ν_k formula of renormalized theory. Changing the direction of the magnetic field produces the predicted shift of the frequency by $\omega_{k_\perp} \to -\omega_{k_\perp}$ and $\nu_{k_\perp} \to +\nu_{k_\perp}$ and provides a null measurement of $2\omega_{k_\perp}$. The frequency spectrum and the determination of $2\omega_{k_\perp}$ are shown in Goldston, *et al.* (1977).

In these rather collisionless plasmas the drift wave normal modes are predicted by theory to be partially localized to the outside region of the tokamak where the trapped electron density is maximum. In Fig. 4 of Goldston, *et al.* (1977) the increase in the scattered signal observed as the scattering volume is moved from the inside to the outside position of the tokamak is shown. The relative increase in the scattered power expected from theory is observed in the experiment.

The theoretical spectral distribution was calculated in some detail for the radial plasma profiles used in the experiment. With these parameters and profiles the

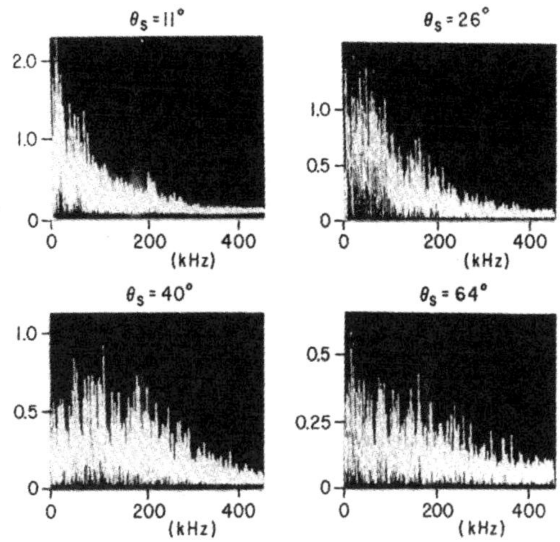

Fig. 2.8 Spectra of microwaves scattered by density fluctuations with wavelengths of $\lambda(11°) = 2.3\,\text{cm}$, $\lambda(26°) = 1.2\,\text{cm}$, $\lambda(40°) = 0.6\,\text{cm}$, $\lambda(64°) = 0.4\,\text{cm}$. θ_s is the scattering angle; the ordinate is in arbitrary units and is proportional to the electric field of the scattered microwaves [Mazzucato (1976) and Goldston, *et al.* (1977)].

dispersion relation and the nonlinear terms used in the energy transfer rate were evaluated to obtain $I(\mathbf{k}_\perp)$ in the scattering volume. The dimensionless scale for the wavenumber is determined by the ion inertia through $\rho_s = c(m_i T_e)^{1/2}/eB$, which is approximately one millimeter in this experiment.

Subsequent to the analysis of the ATC spectrum and prior to further reports on microwave scattering at the Sixth Conference on Plasma Physics and Nuclear Fusion in Berchtesgaden, additional theoretical analysis was performed to predict the spectral distribution of the fluctuations to be expected in the tokamak at Fontenay-aux-Roses called TFR. The theoretical prediction published by Horton (1977) explains the spectrum obtained by microwave scattering and published by the TFR Equipe in the proceedings of that conference. The theoretical prediction is shown in Fig. 1 of Horton, *et al.* (1977) and the observed scattering experiment the level of the fluctuations was predicted to be $\widetilde{n}_e/\overline{n}_e \approx 10^{-2}$ and the associated turbulent energy density is $W/nT_e \approx 10^{-4}$. This level is somewhat higher than the level of $\widetilde{n}_e/\overline{n}_e = (3\text{-}5)\times 10^{-3}$ reported by the TFR Equipe. The fixed scattering volume, however, is located in azimuth at 105° around from the outside of the torus where the normal mode theoretically has its maximum amplitude. Taking into account the theoretical variation of the amplitude with azimuthal angle, the theoretical prediction and the experimentally inferred amplitude for the electron density fluctuation consistent to within the accuracy of the amplitude measurements.

In conclusion, turbulence theory applied to the problem of ion acoustic instability and the related problem of drift wave instability in an inhomogeneous magnetized plasma explains the principal features of several turbulence experiments.

The nonlinear theory allows accurate calculations sufficiently complete to permit the calculation of the spectral distribution of the correlation functions and the associated nonlinear dispersion relations. With knowledge of the spectral distributions the turbulent transport processes such as the low frequency electrical resistivity and the thermal conductivity are readily evaluated. For drift waves the important turbulent transport processes are the cross-field particle and thermal fluxes produced by the turbulent $\mathbf{E} \times \mathbf{B}$ drifts of the particles.

2.9.6 *Microwave scattering experiments in PLT*

The drift wave spectrum measured in PLT showed a monotonically decreasing wavenumber spectrum rather than the peaked wavenumber spectrum of the ATC experiment. The PLT tokamak experiments were at $B = 0.3\,\mathrm{T}$ with $n_e = 5\times10^{13}\,\mathrm{cm}^{-3}$ and a variable T_e profile. Density fluctuations with k wavenumber from $2\,\mathrm{cm}^{-1}$ to $23\,\mathrm{cm}^{-1}$ were measured. For comparison the earlier ATC spectrum is shown in the filled circles. The two spectra are consistent with drift wave turbulence with $(\widetilde{n}/n)_{\mathrm{PLT}} \approx (0.5 \text{ to } 1.0)\times10^{-2}$ while the smaller ATC device has a higher level with $(\widetilde{n}/n)_{\mathrm{ATC}} \gtrsim 10^{-2}$.

Table 2.1 ATC-PLT Scattering Experiment Parameters.

	PLT	ATC
R/a	132/35	80-90/15-17 cm
B/I_p	0.32 T/400 kA	$0.16 - 0.19$ T
n_e	$4 \times 10^{13}\,\mathrm{cm}^{-3}$	$3 \times 10^{13}\,\mathrm{cm}^{-3}$
T_e	$300 - 800$ eV	1000 eV
T_i	$500 - 900$ eV	150-200 eV
k	$10\,\mathrm{cm}^{-1}$	$15\,\mathrm{cm}^{-1}$
$\omega_* \pm \nu_k$	100 ± 100 kHz	300 ± 300 kHz

Table 2.1 gives the details of the comparison of the PLT and ATC scattering experiments from Mazzucato (1976, 1978).

The PLT scattering data is also of interest with regard to the issue of the possible presence of the trapped electron mode. The dissipative trapped electron mode growth rate varies roughly as $\gamma_k = c_1(\omega_{*e}^2/\nu_{\mathrm{eff}})\,\eta_e$ where $\eta_e = d\ell n\; T_e/d\ell n\, n_e$ in sharp contrast to the sheared slab collisionless drift wave which has $\gamma_k = c_2(\omega_{*e}^2/|k_\parallel|v_e)[\langle k_\perp^2\rho_s^2\rangle - \eta_e/2]$ where positive η_e is stabilizing. The constants c_1 and c_2 are near unity.

In the PLT discharge presented in Mazzucato (1978), the heavy impurities on axis resulted in a radiative cooling of the core plasma producing a hollow T_e profile with a fixed parabolic density profile. Simultaneous with the inversion of the temperature profile to a system with $\eta_e < 0$ the fluctuation level increases substantially. The increase of the fluctuation level is consistent with the presence of the sheared-slab collisionless drift waves and not with the trapped particle instability. Since $\epsilon = r/R \leq 18\,\text{cm}/132\,\text{cm}$ the fraction of trapped electrons is small $f_T < (18/132)^{1/2} = 0.3$. The inverted profile was reported to be macroscopically stable throughout the discharge.

2.10 Drift Waves and Transport in the TEXT Tokamak

In the TEXT tokamak at The University of Texas, far-infrared laser scattering experiments with the six wavenumbers $k_\perp = 1, 2, 4.5, 7, 9$, and $12\,\text{cm}^{-1}$ were carried out and correlated with the electron thermal diffusivity in an ohmic tokamak with $R = 1\,\text{m}$, $a = .27\,\text{m}$, $B = 2.8\,\text{T}$ for $I_p = 150\,\text{kA}$ to $350\,\text{kA}$. The density increased from $n = 2 \times 10^{13}\,\text{cm}^{-3}$ to $9 \times 10^{13}\,\text{cm}^{-3}$ with the reference discharge $I_p = 300\,\text{kA}$ and $\overline{n}_e = 3.5 \times 10^{13}\,\text{cm}^{-3}$. The discharges have sawtooth oscillations except at the lowest currents. The reference discharge had $T_{e0} = 10^3\,\text{eV}$, $T_{i0} = 550\,\text{eV}$, $Z_{\text{eff}} = 2$ and $\tau_E = 12\,\text{ms}$.

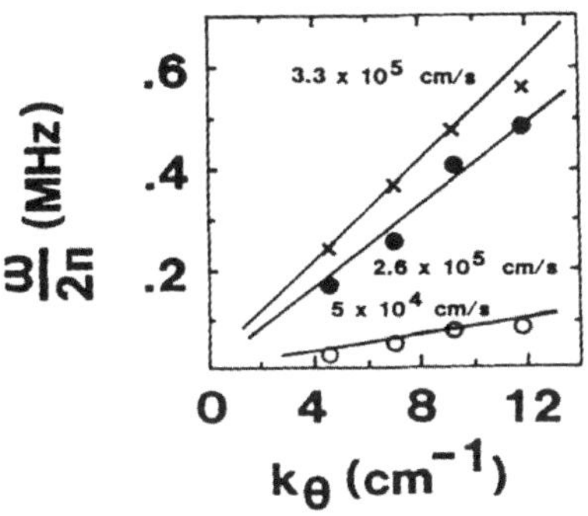

Fig. 2.9 Microturbulence statistical dispersion for electron components, $n_e = 2 \times 10^{13}\,\text{cm}^{-3}$ (crossed), $2 \times 10^{13}\,\text{cm}^{-3}$ (filled circles) and ion component $n_i = 8 \times 10^{13}\,\text{cm}^{-3}$ (open circles) [Brower, *et al.* (1987)].

The typical frequency spectrum for each wavenumber is shown in Fig. 2.9 from Brower, *et al.* (1987).

The inferred density fluctuation level $\widetilde{n}(r)$ and the centroid of the frequency spectrum $\omega(k)$ are shown in Fig. 2.10. The fluctuation level $\widetilde{n}$ is found to linearly increase with mean density with $\widetilde{n} \simeq 0.02\overline{n}$.

The role of the Doppler shift from the radial electric field was accurately known in the TEXT tokamak due to the HIBP, or Heavy Ion Beam Probe, from University of Renassler [Hallock, *et al.* (1994)]. The HIBP probe diagnostic uses the change in the heavy ion particle orbit at the point of the charge exchange collision to determine electrostatic potential of the plasma at this time and point of the charge exchange with the ambient working gas ion. This HIBP diagnostic gave profiles of the $E_r(r,t)$ electric field from which the Doppler shift $k_\theta v_{E_r}$ was determined as

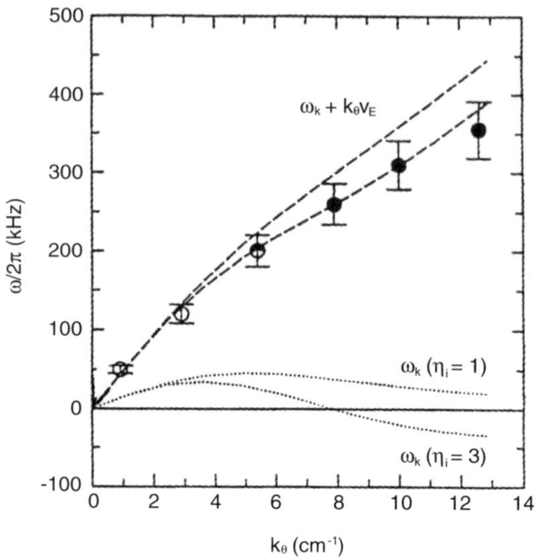

Fig. 2.10 Frequency spectrum from far-infrared laser scattering from the electron density fluctuations associated with drift waves in the TEXT tokamak [Bravenec, *et al.* (1992)]. The dotted line is the frequency in the plasma rest frame and the dashed line is the frequency in the laboratory frame.

shown in Fig. 2.10 from Bravenec, *et al.* (1992). For an inward pointing electric field the plasma as a whole rotates in the electron diamagnetic direction increasing the frequency measured in the laboratory frame for drift waves propagating in the electron diamagnetic direction. For the ion temperature gradient driven drift waves rotating in the ion diamagnetic direction in the plasma rest frame, the Doppler shift was large enough to change the direction to the electron diamagnetic direction in the laboratory frame of the TEXT tokamak. Figure 2.10 shows the Doppler shifted drift wave frequency for the discharge taking into account the ion temperature gradient parameter $\eta_i = 1$ and 3 for the far-infrared laser scattering experiment of Brower, *et al.* (1987). Including the E_r-Doppler shift is clearly important to identify the drift waves in the presence of the ion temperature gradients in TEXT and other tokamaks.

Bravenec, *et al.* (1992) note that these variations of the fluctuation levels with parameters are consistent with turbulent $\chi_e(n)$ formulas. The theoretical turbulence works of Koch and Tang (1978) and Hasegawa and Wakatani (1983) were used to explain the correlations between the fluctuations and confinement.

2.11 Drift Waves in Stellarators

The inverse variation of the plasma energy and particle confinement times τ_E and τ_p

are sometimes difficult to establish in the magnetic confinement experiments. One early example where the inverse correlation τ_p and $\tau_E \propto 1/\langle \delta n^2/n^2 \rangle$ was reported used far-infrared laser scattering from the electron density fluctuations in a stellarator equipped with both ohmic heating and neutral beam heating. A discharge where the initial ohmic heating was gradually reduced over 100 ms and the neutral beam heating was ramped up while the scattering data was continuously measured showed a clear inverse correlation of confinement energy W_p with fluctuation level [Meyer and Mahn (1981)].

The experiments were performed in the Wendelstein VIIA Stellarator at the Max-Planck Institute in Garching. The VIIA stellarator has $B = 3T$, $R = 2\,\text{m}$, $a = 0.1\,\text{m}$ and negligible plasma current except when ohmic heating is applied. There were three neutral H_2 beam injections operating at 27 kV at $P_N = 1\,\text{MW}$.

For both D_2 and He working gases the confinement goes up with increasing density from the low density limit and the fluctuation $\langle \delta n^2/n^2 \rangle$ level goes down. Above the maximum confinement time the variation $\tau_E \propto 1/\langle \delta n^2/n^2 \rangle$ continues for the confinement time.

The time history of the plasma stored energy W_p and the fluctuation energy density $W(t)$ show this inverse relation when the ohmic power is ramped down and the compensating neutral beam power is ramped up over period of 100 ms. Finally, comparison of the fluctuation levels with stored plasma energy for two versus three neutral beams also showed this inverse correlation.

References

Braginskii, S. I. (1965). *Reviews of Plasma Physics* **1**, p. 205,[Authorized translation from the Russian by Herbert Lashinsky, University of Maryland, USA.] Edited by M. A. Leontovich (Consultants Bureau, New York).

Bravenec, R. V., Gentle, K. W., Richards, B., Ross, D. W., Sing, D. C., Wootton, A. J., Brower, D. L., Luhmann, N. C., Peebles, W. A., Yu, C. X., Crowley, T. P., Heard, J. W., Hickok, R. L., Schoch, P. M., and Yang, X. Z. (1992). Core turbulence and transport studies on the Texas Experimental Tokamak, *Phys. Fluids B* **4**, p. 2127, `http://dx.doi.org/10.1063/1.860019`.

Brower, D. L., Peebles, W. A., Kim, S. K., Luhmann, N. C., Tang, W. M., and Phillips, P. E. (1987). Observation of a high-density ion mode in tokamak microturbulence, *Phys. Rev. Lett.* **59**, pp. 48-51, `http://link.aps.org/doi/10.1103/PhysRevLett.59.48`.

Choi, Duk-In and Horton, W. (1974). *Phys. Fluids* **17**, 2048.

Dubuit, N., Garbet, X., Parisot, T., Guirlet, R., and Bourdelle, C. (2007). Fluid simulations of turbulent impurity transport, *Phys. Plasmas* **14**, p. 042301.

Fried, B. D. and Conte, S. S. (1961). *The Plasma Dispersion Function* (New York: Academic Press).

Gladd, N. T. and Horton, W. (1973). *Phys. Fluids* **16**, 879, doi:10.1063/1.1694440, ISSN:0031-9171.

Goldston, R. J., Mazzucato, E., Slusher, R. E., and Surko, C. M. (1977). *Plasma Phys. Control. Nucl. Fusion* (International Atomic Energy Agency, Vienna) **I**, p. 371.

Hallock, G. A., Hickok, R. L., and Hornady, R. S. (1994). The TMX heavy ion beam probe, Plasma Science, *IEEE Transactions on Plasma Science* **22**, 4, pp. 341-349, ISSN:0093-3813, doi:10.1109/27.310639.

Hasegawa, A. and Wakatani, M. (1983). Plasma Edge Turbulence, *Phys. Rev. Lett.* **50**, pp. 682-686 `http://link.aps.org/doi/10.1103/PhysRevLett.50.682`, doi:10.1103/PhysRevLett.50.682.

Horton, W. (1976). `http://link.aps.org/doi/10.1103/PhysRevLett.37.1269`.

Horton, W., Okuda, H., Cheng, C. Z., Kuo, Y. Y., Lee, W. W., Matsuda, Y., and True, M. (1977). *Plasma Phys. Control. Nucl. Fusion* (International Atomic Energy Agency, Vienna) **II**, pp. 467-479.

Ilic, D. B. (1977). `http://link.aip.org/link/PFLDAS/v20/i10/p1717/s1`, ISSN:0031-9171.

Jackson, J. D. (1998). *Classical Electrodynamics* 3rd Ed., (John Wiley and Sons, Inc.), ISBN:047130932X.

Kawai, Y. and Guyot, M. (1977), `http://link.aps.org/doi/10.1103/PhysRevLett.39.1141`.

Kawai, Y., Hollenstein, Ch., and Guyot, M. (1978). *Phys. Fluids* **21**.

Koch, R. A. and Tang, W. M. (1978). *Phys. Fluids* **21**, p. 1236, doi:10.1063/1.862362, ISSN:0031-9171.

Krall, N. A. and Trivelpiece, A. W. (1973). *Priciples of Plasma Physics* (McGraw-Hill Inc.), ISBN:0-070035346-8.s

Mazzucato, E. (1976). `http://link.aps.org/doi/10.1103/PhysRevLett.36.792`.

Mazzucato, E. (1978). `http://link.aip.org/link/PFLDAS/v21/i6/p1063/s1,ISSN:0031-9171`.

Meyer, J., and Mahn, C. (1981). `http://link.aps.org/doi/10.1103/PhysRevLett.46.1206`.

Mikhailovskii, A. B. (1967). *Review of Plasma Physics* **2**, p. 159, ed. M.A. Leontovich (Consultants Bureau, New York).

Mikhailovskii, A. B. (1974). *Theory of Plasma Instabilities* **1** (Consultants Bureau, New York).

Mikhailovski, A. B. (1992). *Electromagnetic Instabilities in an Inhomogeneous Plasma* (Institute of Physics Publishing) ISBN:0-7503-0182-1.

Muraglia, M., Agullo, O., Yagi, M., Benkadda, S. Beyer, P., Garbet, X., Itoh S.-I., Itoh, K., and Sen, A. (2009). Effect of the curvature and the β parameter on the non-linear dynamics of a drift tearing magnetic island, *Nucl. Fusion* **49**, p. 055016, doi:10.1088/0029-5515/49/5/055016.

Muraglia, M., Agullo, O., Benkadda, S., Yagi, M., Garbet, X., and Sen, A. (2011). Generation and Amplification of Magnetic Islands by Drift Interchange Turbulence, *Phys. Rev. Lett.* **107**, p. 095003, `http://link.aps.org/doi/10.1103/PhysRevLett.107.095003`.

Pearlstein, L. D. and Berk, H. L. (1969). Universal Eigenmode in a Strongly Sheared Magnetic Field, *Phys. Rev. Lett.* **23** pp. 220-222. `http://link.aps.org/doi/10.1103/PhysRevLett.23.220`.

Slusher, R. E., Surko, C. M., Moler, D. R., and Porkolab, M. (1976). `http://link.aps.org/doi/10.1103/PhysRevLett.37.1747`.

Stenzel, R. L. (1978). Experiments on current-driven three-dimensional ion sound turbulence. Part I: Return-current limited electron beam injection, *Phys. Fluids* **21**, 93, doi:10.1063/1.862084

Stix, T. H. (1973). Decay of poloidal rotation in a tokamak plasma, *Phys. Fluids* **16**, No. 8, p. 1260, `http://link.aip.org/link/PFLDAS/v16/i8/p1260/s1`.

Chapter 3

Mechanisms for Drift Waves

3.1 Drift Wave Turbulence

Drift waves occur universally in magnetized plasmas producing the dominant mechanism for transport of particles, energy and momentum across magnetic field lines. A wealth of information obtained from laboratory experiments for plasma confinement is reviewed for drift waves driven unstable by density gradients, temperature gradients and trapped particle effects. The modern understanding of origin of the scaling laws for Bohm and gyro-Bohm transport fluxes measured in tokamaks arises from drift wave turbulence. The role of sheared flows and magnetic shear in reducing the transport fluxes is discussed and illustrated with large scale computer simulations. Plasmas turbulence models are derived with reduced magnetized fluid descriptions. The types of theoretical descriptions include weak turbulence theory and anisotropic Kolmogorov-like spectral indices, and the mixing length. A number of standard turbulent diffusivity formulas are given for the various space-time scales of the drift wave turbulent mixing.

Worked Problem 3.1

Problem:

Find a Hamiltonian description of the drift waves with $d = 2$ degrees of freedom for the parallel motion $p_1 = mv_\parallel$, $q^1 = z$ and the cross-field guiding-center motion $dp_2 = Bdx$, $dq^2 = dy$. Use the cross-field polarization dielectric $\epsilon_\perp$ derived in Worked Problem 2.1 when writing the Poisson equation.

Solution:

The guiding-center motion derived in plasma textbooks follows from Hamilton's equations of motion $\dot{p}_\alpha = -\partial H/\partial q^\alpha$, $\dot{q}^\alpha = \partial H/\partial p_\alpha$ with the Hamiltonian

$$H(p_\alpha, q^\alpha) = \frac{1}{2}mv_\parallel^2 + q\phi(x,y,z,t) + \mu B(x,y) \qquad (3.1)$$

where μ is the first adiabatic invariant that is constant to the first order in the small parameters $\epsilon = \omega/\Omega_i$ and $k_\perp^2 \rho_i^2$. In the potential energy q is the particle charge, which we take as $q = e$ to avoid the use of the symbol q for both Hamiltonian coordinate and the electric charge.

The particle phase space density $f(x, y, z, v_\parallel, t)$ is conserved by the volume preserving Hamiltonian flow, so that

$$\frac{\partial f_j}{\partial t}(p, q, t) + [H, f_j] = 0 \tag{3.2}$$

for each species j compute the guiding-center motion from

$$\frac{dz}{dt} = v_\parallel \quad \frac{dv_\parallel}{dt} = -\frac{e}{m}\frac{\partial \phi}{\partial z} \tag{3.3}$$

$$\frac{dx}{dt} = \frac{E_y}{B} = -\frac{1}{B}\frac{\partial \phi}{\partial y} \tag{3.4}$$

$$\frac{dy}{dt} = -\frac{E_y}{B} = \frac{1}{B}\frac{\partial \phi}{\partial x} \tag{3.5}$$

and re-expressing the motion in terms of p_α, q^α is defined in the statement of the problem gives

$$[H, f] = v_\parallel \frac{\partial f}{\partial z} - \frac{e}{m}\frac{\partial \phi}{\partial z}\frac{\partial f}{\partial v_\parallel} + \frac{1}{B}\frac{\partial \phi}{\partial x}\frac{\partial f}{\partial y} - \frac{1}{B}\frac{\partial \phi}{\partial y}\frac{\partial f}{\partial x} \tag{3.6}$$

for the directional or Lie derivative of the Hamiltonian flow in the phase space [Jose and Saletan (2002)].

The self-consistent electric potential satisfied Poisson's equation taking into account the polarization charge density $\rho_p = -\nabla \cdot \boldsymbol{j}_p$ from Worked Problem 2.1. The field equation in configuration space is

$$\nabla_\perp \cdot (\epsilon_\perp \nabla_\perp \phi) + \partial_z^2 \phi = -4\pi \sum_j q_j \int f_i d^3v \tag{3.7}$$

where $\epsilon_\perp = 1 + \sum_j \omega_{pi}^2/\Omega_j^2$ is the dielectric function. In the space of the fluctuations where

$$\phi(\boldsymbol{x}, t),\ \delta f_j\left(\boldsymbol{x}, v_\parallel, t\right) \sim \phi_{\boldsymbol{k}}, \delta f_j\left(\boldsymbol{k}, v_\parallel, \omega\right) \times \exp(i\boldsymbol{k}\cdot\boldsymbol{x} - i\omega t) + \text{c.c.}$$

the partial differential equations Eq. (3.6) and Eq. (3.7) become algebraic with complex dispersion function

$$k^2 \epsilon_{\text{ES}} = k_\perp^2 \epsilon_\perp + k_\parallel^2 + \sum_j \frac{4\pi n_j e_j^2}{m_j} \int \frac{d^3v \left(\omega \dfrac{\partial f_j}{\partial \epsilon} + \dfrac{k_y}{e_j B}\dfrac{\partial f_j}{\partial x}\right)}{\omega - k_\parallel v_\parallel - k_y v_D + i0^+} \tag{3.8}$$

where $\epsilon = mv^2/2$ is the particle energy, $\mu = mv_\perp^2/2B$ is the particle magnetic moment, and $v_{Dj} = (\mu/e_j)\partial_x \ell n\, B(x)$ is the guiding-center drift velocity.

Show how the dispersion relation in Eq. (3.8) reduces to that of Worked Problem 2.2 when the plasma becomes uniform in space.

3.2 Drift Wave Mechanism

One of the most intellectually interesting problems in plasma physics is the problem of turbulence and the associated transport of the plasma properties including density, temperature and momentum. Plasma turbulence can be rather different from that in neutral fluids due to the numerous waves in the plasma. Plasma turbulence is closer to that of geostrophic flows on rotating planets where the Rossby wave is closely analogous to the plasma drift wave. At finite amplitude the wave dispersion often balances the wave nonlinearity to form solitary structures and coherent vortex structures. This aspect of plasma turbulence is explored in numerous workshop proceedings [Horton and Reichl (1984)], the Marseilles Workshops published as Benkadda, *et al.* (1995) and Horton and Ichikawa (1996). Drift waves are the most widely studied form of plasma turbulence in magnetized, confined plasmas so this first chapter of the ITER School Proceedings is devoted to the study of these drift waves.

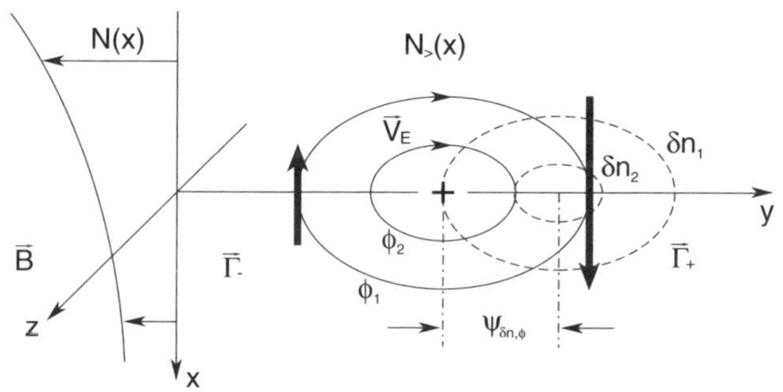

Fig. 3.1 A segment of a drift-wave fluctuation showing the variation of the electrostatic potential perpendicular to the magnetic field at a given instant of time. The contours of $\varphi = \text{const}$ in the plane perpendicular to $B\widehat{z}$ are the stream lines of the $\boldsymbol{E} \times \boldsymbol{B}$ particle motion, and a segment of the correlated but phase-shifted density variation δn. The potential and density variation are out of phase by $\psi_{\delta n,\varphi}$ with the density fluctuation leading the potential fluctuation for propagation along the positive $\widehat{y}$ axis.

In Fig. 3.1 we show the profiles of electron density, n_e, and pressure, p_e, along with a drift wave localized over a radial region $\Delta r = \ell_c$, between the two cells in the correlation length. The electrostatic potential that gives rise to the electric field, $\vec{E} = -\nabla\phi$, is shown by the positive and negative ovals. Ahead of the potential there are contours of enhanced density ($\delta n_e > 0$) and lower density ($\delta n_e < 0$) phase shifted by $\psi_{\vec{k}} = \psi(\delta n, \phi)$. The magnetic field is taken to be in the horizontal $\boldsymbol{z}$ direction. Thus, the positive potential cell gives a clockwise rotation and the negative cell a counter-clockwise rotation of $\boldsymbol{E} \times \boldsymbol{B}$ convected plasma. Clearly, due the phase shift there is a greater transport of plasma down rather than up the density gradient. Thus, there is a net outward flux of density or pressure given by the y-integrated or time averaged flux $\langle n_e v_r \rangle = \langle n_e E_y/B \rangle$. To be explicit, the model in Fig. 3.1 and calculation is characterized by the electric potential wave $\phi_k \sin(k_y y - \omega_k t)$ and by the density wave $\delta n_e \sin(k_y y - \omega_k t - \psi_k)$.

3.3 Energy Bounds for Turbulence Amplitudes

3.3.1 *Density gradients*

The electron drift wave and trapped electron instability (trapped particle mode) [Kadomtsev and Pogutse (1970)] are driven by the square of the plasma density gradient in plasmas with no background temperature gradient. The wave with a temperature gradient is analyzed in Section 3.3.2. In the early literature the electron drift wave with

Growth Rate of Density Gradient Driven Drift Wave

$$\gamma_k = \left(\frac{\pi}{2}\right)^{1/2} \frac{\omega_{*e}^2 k_\perp^2 \rho_s^2}{|k_\parallel| v_e} \propto \left(\frac{dn}{dx}\right)^2$$

was called the universal instability owing to the drive from $(\nabla n)^2$, which is always present in confined plasmas.

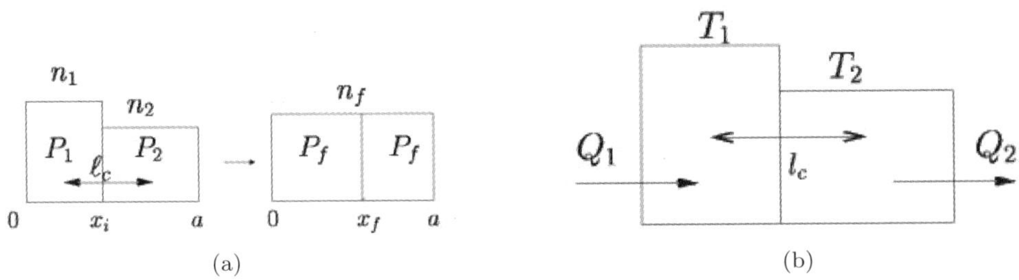

Fig. 3.2 Expansion diagram under density gradient $n_1 > n_2$ for calculating the released plasma energy in the left figure. The right figure is the energy diagram for the temperature gradient driven turbulence. The regions are one radial correlation length ℓ_c apart.

Figure 3.2 shows the model for the energy release for the density gradient. The left higher density n_1 (and pressure $p_1 = n_1 T$) box from $x = [0, x_i]$ is allowed to expand to $[0, x_f]$ and compress the gas in the right box until the force balance is reached with

$$P_f = P_1 \left(\frac{x_i}{x_f}\right)^\Gamma = P_2 \left(\frac{a - x_i}{a - x_f}\right)^\Gamma. \tag{3.9}$$

Solving for the new position x_f and calculating the work, $W = 1/a \int_{x_i}^{x_f} (P_1 - P_2) dx$ per unit volume done by the gas during the expansion gives the free energy. For

$\Delta x/a \ll 1$, the turbulent energy per unit volume is

$$W_{\rm turb} \leq W_{\rm max} = \frac{(\Delta P)^2 x_i(a - x_i)}{\Gamma P a^2} \leq \frac{(\Delta P)^2}{4\Gamma P}. \tag{3.10}$$

The limit $\Gamma \to \infty$ corresponds to an incompressible gas for which the partition between the pressures P_1 and P_2 will not move.

If the drift wave energy $W_{\rm turb}$ is almost entirely in the $\boldsymbol{E} \times \boldsymbol{B}$ kinetic energy of the ions, then we have the bound for the average turbulent convection velocity,

$$\left\langle v_E^2 \right\rangle^{1/2} \leq \left(\frac{\ell_c}{L_n}\right) c_s. \tag{3.11}$$

When the correlation length ℓ_c is limited by the ion inertial scale length $\ell_c = \rho_s$, the bound for the mean square turbulent plasma velocity is

$$\left\langle v_E^2 \right\rangle \leq \left(\frac{\rho_s}{L_n}\right)^2 c_s^2 \equiv v_{de}^2. \tag{3.12}$$

The limit of $\left\langle v_E^2 \right\rangle \leq v_{de}^2$ follows from another argument called the mixing length limit. This argument states that when the gradient of the fluctuating density, $\nabla \delta n$, reaches the value $dn/dx = -n/L_n$ of the ambient density gradient, the gradient source, on average, is eliminated and growth stops. Then $\delta n/n = 1/k_x L_n$, with $\ell_c = 1/k_x = \rho_s$. The same amplitude limit applies and the scale of the turbulent diffusion $D_{gB} = \left\langle v_E^2 \right\rangle^{1/2} \ell_c = (\rho_s/L_n)(T_e/eB)$ is defined as the gyro-Bohm diffusivity. To derive the famous Bohm diffusivity, $D_B = T_e/eB$, one instead assumes ℓ_c to be the geometric mean of L_n and ρ_s [Bohm, *et al.* (1949)].

3.3.2 *Temperature gradients*

For auxiliary heated plasmas that develop a temperature gradient ∇T (either ion or electron) from input power $P_{\rm aux}$, the relevant thermodynamic argument is the Carnot engine. As shown in Fig. 3.3, there is an input of energy Q_1 in the left box. The vortices and streamers cycle the plasma over the correlation length ℓ_c, connecting T_1 and T_2. Now, for each cycle, the temperature gradient drives the drift wave to release the energy,

$$W_{\rm turb} \leq \Delta W_{\rm carnot} = \Delta S \Delta T \tag{3.13}$$

where ΔS is the change in the entropy of the gas. From kinetic theory, the value of the gas constant, Γ, depends on the type of drift wave. $\Gamma = 5/3$ where the wave-particle resonance is $\omega_k = \omega_{\nabla B} + \omega_{\rm curv}$ and $\Gamma = 3$ where $\omega_k = k_\parallel v_\parallel$ and $k_\perp \rho_i \ll 1$. For toroidal ITG and ETG, a simple calculation for modes with $\Gamma = 5/3$ yields

$$\Delta S = n k_B \left(\frac{3}{2}\frac{\Delta T}{T} - \frac{\Delta n}{n}\right) \tag{3.14}$$

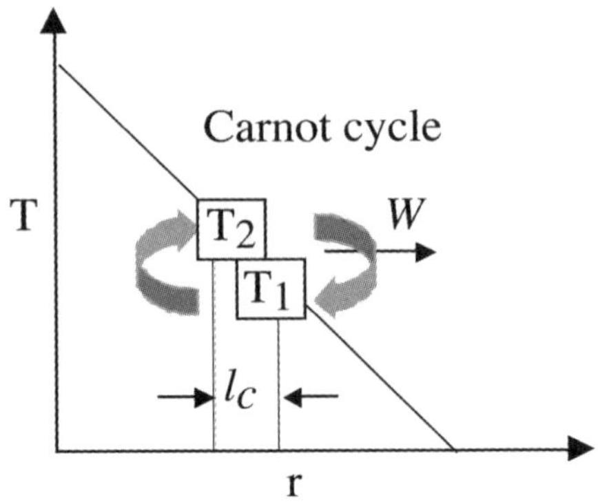

Fig. 3.3 Carnot cycle diagram for temperature-gradient driven drift waves. The maximum energy W released to the plasma turbulence is $W = Q_1 - Q_2 = n\Delta T(\Delta T/T - (\Gamma - 1)\Delta n/n)$.

which in Eq. (3.13) together with $\Delta T = -\ell_c dT/dr$ give

$$\Delta W_{\text{carnot}} = \frac{3}{2} nT\ell_c^2 \frac{\partial \ln T}{\partial r}\left[\frac{\partial \ln T}{\partial r} - \frac{2}{3}\frac{\partial \ln n}{\partial r}\right] \tag{3.15}$$

where $\epsilon = E/T_i$ and μ denotes system parameters such as η_i and T_e/T_i. The average $< \cdot >$ is over the local Maxwell phase-space density $f(\epsilon)$.

Figure 3.4 shows the roots of the Vlasov-Poisson dispersion relation, $D_{\vec{k}}(\omega, \mu) = 0$, with increasing $\eta \equiv \nabla \ln T / \nabla \ln n$. Thus, for the kinetic dispersion relation,

$$D_{\vec{k}}(\omega, \mu) = \frac{1}{T_e} + \frac{1}{T_i}\left[1 - \left\langle \frac{\omega - \omega_{*i}(\eta, \epsilon)}{\omega - \omega_{Di}\epsilon + i0^+}\right\rangle\right] \tag{3.16}$$

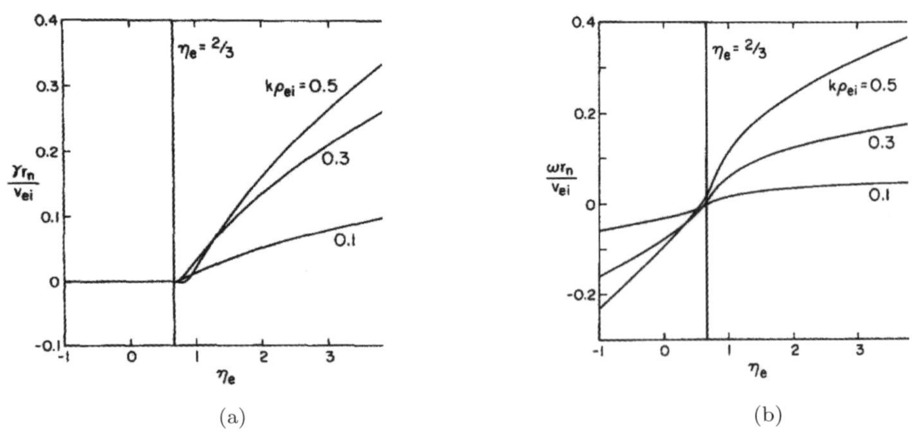

Fig. 3.4 The Vlasov (GKE) equation growth rate and frequency as a function of the temperature gradient divided by the density gradient [Horton, *et al.* (1988)].

The critical temperature gradient from the Nyquist analysis is given by

$$\eta_{\mathrm{crit}} = \frac{\partial \ln T}{\partial \ln n} = \frac{2}{3} \tag{3.17}$$

as shown in Fig. 3.4. The agreement between the thermo-dynamical formula and the stability condition of the kinetic dispersion relations can be understood when one sees from Fig. 3.4 that the wave frequency, $\omega_{\vec{k}} \simeq 0$ at the critical point. Here the relevant gradient of the phase-space density function $f(r)$ is proportional to $\partial/\partial r[n/T^{d/2}]$ where

$$\frac{\partial f}{\partial r} = \frac{\partial}{\partial r}\left(\frac{n(r)}{T^{d/2}(r)}\right) \quad \text{for} \quad E \ll T \tag{3.18}$$

with $\Gamma = (d+2)/d$ for a plasma with d degrees of freedom. For one-dimensional waves like Langmuir waves, $d = 1$ and $\Gamma = 3$, and for toroidal drift waves $d = 3$ and $\Gamma = 5/3$.

The key, universal result given by Eq. (3.15) for W_{carnot} is that there is only a window of η values, $0 < \eta \leq \Gamma - 1$, which are generally stable. Drift waves are unstable for both inverted gradients $\nabla n \cdot \nabla T < 0$, and strong temperature gradients $\nabla T \gg \nabla n$.

3.3.3 *Drift wave eigenmodes in toroidal geometry*

The first effect of toroidal geometry that changes the stability of the drift waves is the unfavorable magnetic field line curvature vector, $\boldsymbol{\kappa} = -(\widehat{\mathbf{b}} \cdot \nabla)\widehat{\mathbf{b}} \simeq (\widehat{\boldsymbol{r}} \cos\theta + \widehat{\boldsymbol{\theta}} \sin\theta)/R$ which covers the outer half ($\cos\theta >$) of the toroidal surfaces. This magnetic curvature creates small scale interchange turbulence with waves that are only partially stabilized by the magnetic shear, $s = rq'/q$, where $q(r) = rB_\phi/RB_\theta$ gives the helical pitch of the field lines. The E_r-shear can also stabilize the modes as discussed in Section 3.3.4. In addition, and closely related, the weakening of the magnetic field with increasing radius (see Fig. 3.5) leads to an outward acceleration given by the force, $F = -\mu \nabla B$ acting on the plasma.

In the toroidal drift wave equations the local gradients of the density and pressure are normalized with respect to the gradient length, R, of the magnetic field. Thus, the key parameters are

$$\epsilon_n = \frac{L_n}{R} \quad \text{and} \quad \epsilon_p = \frac{L_p}{R} \tag{3.19}$$

throughout the toroidal equations.

The faster radial flows shown in Fig. 3.5 give the plasma compression,

$$\boldsymbol{\nabla} \cdot \vec{v}_E = \frac{2v_x \cos\theta}{R} \tag{3.20}$$

compared with the plasma convection,

$$\vec{v}_E \cdot \nabla n = -\frac{n v_x}{L_n} = -\frac{n}{L_n B}\frac{\partial \Phi}{\partial y} \tag{3.21}$$

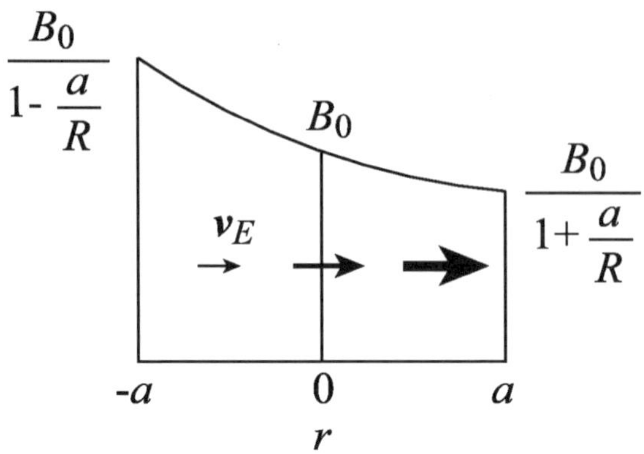

Fig. 3.5 Compression/Rarefraction of the plasma by $\boldsymbol{E} \times \boldsymbol{B}$ convection in the toroidal magnetic field.

gives rise to the two terms proportional to $(1 - 2\epsilon_n \cos\theta)\omega_{*e}$ in the drift wave eigenmode equation. For flat density profiles toroidal compression dominates.

The density and potential fluctuations set up along the magnetic field line give rise to ion acoustic waves with $\omega^2 = k_\parallel^2 c_s^2$, where

$$k_\parallel^2 \to -\frac{1}{q^2R^2}\left(\frac{\partial}{\partial\theta} + q(r)\frac{\partial}{\partial\phi}\right)^2 \tag{3.22}$$

is the differential operator $\nabla_\parallel^2$ in the nonuniform plasma.

Similarly, the shear-Alfvén wave $\omega^2 = k_\parallel^2 v_A^2$ propagates with the $k_\parallel^2$ operator. The periodicity of the torus means that a physical field is the same at $(\theta + 2\pi m, \phi + 2\pi n)$ for all integer values of m, n. These rational surfaces appear when the pitch of the magnetic field is the same as that of the mode $\exp(im\theta - in\phi)$, which occurs when the right-hand side of Eq. (3.22) vanishes. The compressional term in Eq. (3.20) and convection term in Eq. (3.21) combine with the ion acoustic waves to make the drift wave eigenmode equation for waves of the form,

$$\begin{aligned}\Psi(x,t) &= \sum_m \Phi_{m,n}(r)e^{im\theta - in\phi - i\omega t} + \text{c.c.} \\ &= \Phi(r,\theta)e^{-in\phi - i\omega t} + \text{c.c.}\end{aligned} \tag{3.23}$$

The toroidal mode solution, $\Phi(r,\theta)$ of this Mathieu-type equation is for each given r determined by

$$\begin{aligned}&\frac{c_s^2}{\omega^2 q^2 R^2}\left(\frac{\partial}{\partial\theta} - inq(r)\right)^2 \Phi \\ &+ \left[1 - i\delta_{\vec{k}} - \rho_s^2\nabla_\perp^2 - \frac{\omega_{*e}}{\omega}(1 - 2\epsilon_n\cos\theta)\right]\Phi = 0.\end{aligned} \tag{3.24}$$

Here we used the simple phase-shifted, non-adiabatic electron response, $\delta n_e = n_e(e\phi/T_e)(1 - i\delta_{\vec{k}})$. Introducing the phase variation $\Phi \to \Phi(\theta)e^{inq(r)\theta}$, required to keep $k_\parallel^2\Phi$ small, we find that $\nabla_\perp^2\Phi = (\partial_r^2 + r^{-2}\partial_\theta^2)\Phi e^{inq\theta} = -(k_r^2 + k_\theta^2)\Phi = -k_\theta^2(1+s^2\theta^2)\Phi$, where $k_r = sk_\theta\theta$ and $k_\theta = nq/r$.

The drift-eigenmode equation for the one-dimensional ansatz $\Phi(\theta)$ is called the "ballooning mode" equation, which from Eq. (3.22)-Eq. (3.24) is

$$\begin{aligned} &\frac{c_s^2}{\omega^2 q^2 R^2}\frac{\partial^2\Phi}{\partial\theta^2} + \left[1 - i\delta_{\vec{k}} + k_\theta^2\rho_s^2\left(1+s^2\theta^2\right)\right. \\ &\left. - \frac{\omega_{*e}}{\omega}(1 - 2\epsilon_n\cos\theta)\right]\Phi = 0. \end{aligned} \tag{3.25}$$

The solution of Eq. (3.25) with a shooting code requires as a boundary condition such that when $\theta^2 \to \infty$ the solution $\Phi \to \exp i\sigma/2\,\theta^2$, where $\sigma = \omega qR|sk_\theta|\rho_s/c_s$ with $\mathrm{Im}\,\sigma \propto \mathrm{Im}\,\omega > 0$. When the mode is strongly localized to the small θ^2 region, we may use $\cos\theta \simeq 1-\theta^2/2$ to reduce Eq. (3.25) to the analytic harmonic oscillator problem with eigenvalues,

$$\omega_{k_\theta,n} = \frac{\omega_{*e}(1-2\epsilon_n) - ic_s|sk_\theta|\rho_s/(qR)(2n+1)}{1 + k_\theta^2\rho_s^2 - i\delta_{\vec{k}}} \tag{3.26}$$

and eigenfunctions, $\Phi_n(\theta) = \exp(i\frac{\sigma}{2}\theta^2)H_n\left((-i\sigma)^{1/2}\theta\right)$ with $n = 0, 1, 2, \ldots$ The $n = 0$ mode is a symmetric ground state eigenmode with flow vortex center on the rational surface. This symmetry is called the interchange. The $n = 1$ mode is $\theta\exp(i\sigma\theta^2/2)$ with odd parity and corresponds to the varicose mode in neutral fluids with the flows into and out of the rational surface. For finite $\beta_e = 8\pi\, p_e/B^2$ the odd mode develops a reconnection magnetic field component. Small magnetic islands are formed. Thus, the magnetic shear, $|s|$, introduces a damping of the waves due to the coupling to the ion acoustic waves that the propagate wave energy to region of Landau damping. Marginal stability conditions follow from Eq. (3.26) once the details of the dissipative $\delta_{\vec{k}}(\mu)$ function are given.

The structure of the eigenmode equation 3.25 can be transformed to a structure like the sheared slab eigenmode equation in Eq. (2.59) in Chapter 2.6 by taking the Fourier transform with respect to the variable θ defined on the real line $[-\infty \le \theta \le +\infty]$. The first term becomes the ion acoustic wave term $k_\parallel^2$ of Eq. (2.59) and the term $(k_\theta^2\, s\theta)^2$ becomes the radial derivative term in Eq. (2.59). The toroidal term is absent so that an important destabilization mechanism is absent in the slab drift waves compared with the toroidal drift waves. The two eigenmode wave functions are clearly related when the toroidal mode is localized to the outside of the torus with $|\theta| < \pi$.

As noted in Chapter 2.10 the presence of sheared flows particular from nonuniform radial electric field E_r in the plasma and from sheared diamagnetic frequencies breaks the even and odd symmetries of the eigenmodes and shifts the wave function off the rational magnetic surfaces. At finite plasma electron pressure this results in driving up a magnetic fluctuation δB_r and coupling the tearing and ballooning mode symmetries described in Chapters 11-15.

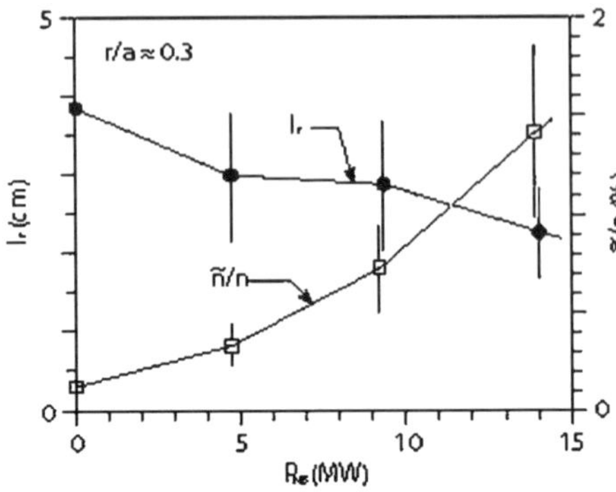

Fig. 3.6 Microwave reflectometer measurements of the radial correlation length l_r and the fractional fluctuation level in the core of TFTR [Mazzucato (1976)].

For the TFTR machine Fig. 3.6 gives the estimate of the density fluctuation level $\delta n/n$ as a function of increasing neutral beam (NB) heating power $P_{\rm NB}$. At the left, the data for $P_{\rm NB} = 0$ is for the ohmic discharge with low $\delta n/n \simeq 0.001$. As the beam heating power increases, the fractional fluctuations rise to $\delta n/n \simeq 0.003$ at 5 MW to 0.006 at 9 MW and 0.015 at 14 MW. This increase of $\delta n/n$ is consistent with power balance $\langle nT\rangle/\tau_E = (P_{\rm NB} + P_{\rm Oh})/{\rm Vol}$ with $1/\tau_E \propto \chi_{gB}/a^2 \propto T^{3/2}$. Here Vol is the mean plasma volume determined by the mean value theorem,

$$\int \frac{3}{2} nT{\rm d}^3x = \frac{3}{2}\langle nT\rangle \cdot {\rm Vol}. \tag{3.27}$$

In tokamaks with higher temperatures and thus stronger radial temperature gradients, the drift wave changes character and is destabilized by the ion temperature gradient, ∇T_i. In this regime, called the ion temperature gradient (ITG) instability, the electron dissipation $i\delta_{\vec{k}}$ can be dropped and the electron response taken as adiabatic and is therefore defined by $\delta n_e/n_e = e\phi_{\vec{k}}/T_e$.

Simpler basic experiments validating the ITG model were carried out in the Columbia Linear Machine (CLM) as given in Sen, *et al.* (1991). These plasmas, produced by RF heating, have $n_e \sim 2\times 10^{14}\,{\rm m}^{-3}$ and $T_i \sim 10\,{\rm eV}$. There is a temperature anisotropy $T_{i\|} \neq T_{i\perp}$ that complicates the linear dispersion but is useful for both NBI or RF heated toroidal experiments and the central cell of the tandem mirror. In the GAMMA-10 tandem mirror resonant ion cyclotron heating in the central cell results in $T_{i\perp} \sim 6.5\,{\rm KeV}$ and $T_{i\|} \sim 2.5\,{\rm KeV}$ [Cho (2005,2006,2008)]. The details of the two temperature components of the ITG theory with the simple cylindrical central-cell plasma data can be found in Hojo (1995) and Sen, *et al.* (1991).

In Section 3.3.2, we discuss the free energy of the ITG modes that limits their intensity. Simulations with gyrokinetic Vlasov codes confirm these limits. The literature concerning the numerical modeling of ITG modes is extensive due to its role as the standard model for explaining the anomalous ion thermal diffusivity χ_i.

The ion thermal diffusivity is strongly influenced by the sheared mass flows in the background plasma [Hamaguchi and Horton(1990)]. Thus we now consider the dynamics of a plasma with a strongly nonuniform $\boldsymbol{E}\times\boldsymbol{B}$ sheared flow in the equilibrium. The dimensionless shear-flow parameter Υ_s derived here will be important in Chapter 18 for analyzing the formation of internal transport barriers in tokamaks.

Worked Problem: KH Instability

Find the eigenvalue equation for electrostatic potential waves in a plasma with $\epsilon_\perp \gg 1$ that has a sheared $\boldsymbol{E}\times\boldsymbol{B}$ flow velocity $\boldsymbol{V}_0 = v_0(x)\widehat{\boldsymbol{e}}_y$ across the constant $\boldsymbol{B} = B\widehat{\boldsymbol{e}}_z$.

Solution:

From Problem 2.1 we know that the net plasma current is the polarization current

$$\boldsymbol{j}_p(\boldsymbol{x},t) = \frac{\rho}{B^2}\frac{d\boldsymbol{E}_\perp}{dt} \tag{3.28}$$

which must close, in the absence of drift wave parallel currents, in loops across $\boldsymbol{B}\widehat{\boldsymbol{e}}_z$ according to

$$\nabla\cdot\boldsymbol{j}_p = \nabla\cdot\left[\frac{\rho}{B^2}\frac{d(-\nabla_\perp\varphi)}{dt}\right] = 0 \tag{3.29}$$

for electrostatic fluctuations $\boldsymbol{E} = -\nabla\varphi(\boldsymbol{x},t)$.

Using the convective derivative with $\mathbf{v}_E\cdot\nabla\varphi$, one uses

$$\frac{d}{dt} = \frac{\partial}{\partial t} + \frac{1}{B}\left(\frac{\partial\varphi}{\partial x}\frac{\partial}{\partial y} - \frac{\partial\varphi}{\partial y}\frac{\partial}{\partial x}\right) \tag{3.30}$$

from the $\boldsymbol{E}\times\boldsymbol{B}$ flow, to reduce Eq. (3.29) to the nonlinear scalar equation

$$\frac{\rho}{B^2}\frac{d}{dt}\nabla_\perp^2\varphi = 0. \tag{3.31}$$

Equation (3.31) is equivalent to the equation of 2D incompressible flow in hydrodynamics whose properties are extensively explored in fluid turbulence. In the uniform plasma the vorticity from the $v_x v_y$-flow is in the origin of equation vorticity

$$\omega = \partial_x v_y - \partial_y v_x = \frac{\nabla_\perp^2\varphi(\boldsymbol{x},t)}{B} \tag{3.32}$$

so that Eq. (3.31) states that the vorticity, and all functions of the vorticity $F(\omega)$, are constant in the frame moving with the $\boldsymbol{E}\times\boldsymbol{B}$ flow.

Waves $(\boldsymbol{k},\omega)$ in the one-dimensional sheared flow $v(x)$ are of the form

$$\varphi(x,y,t) = \varphi_0(x) + \delta\varphi_k(x)e^{ik_y y - i\omega t} + \text{c.c.} \tag{3.33}$$

Linearizing Eq. (3.31) yields the eigenvalue equation

$$(\omega - k_y v(x)) \left[\partial_x^2 \delta\varphi_{k_y} - k_y^2 \delta\varphi_{k_y}\right] + k_y v''(x)\varphi_{k_y}(x) = 0. \qquad (3.34)$$

Equation (3.34) is isomorphic with the famous equation for the Kelvin-Helmholtz instability in hydrodynamic.

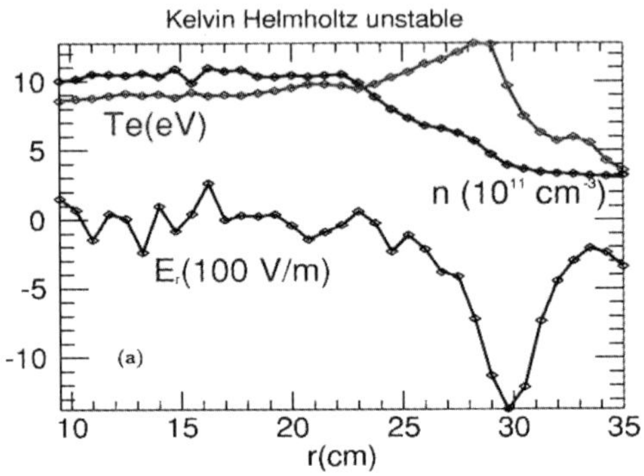

Fig. 3.7 Mean plasma profiles obtained as an average in 25 experimental shots during the stationary turbulent state in the wall-biased [Perez, *et al.* (2006)] configuration of LAPD.

The simplest analytic solution showing the stable and unstable wave spectrum is given by the model of $v(x)$ relevant to the "vee-shaped" radial electric $E_r(r)$ occurring in H-mode edge plasmas. For the vee-flow profile we use the model

$$v(x) = v_0 \left(1 - \frac{|x|}{a}\right) \text{ for } |x| \leq a \text{ and } v(x) = 0 \text{ for } |x| > a \qquad (3.35)$$

and follow Chandrasekhar (1961) to write out the quadratic equation dispersion relation with unstable modes for $|k_y|a < 1$ and stable modes for higher wavenumbers $k_y a > 1$. Note that the Rayleigh theorem requires an inflection point ($v'' = 0$) in the flow profile for linearly unstable eigenmodes. However, the K-H operator is not self-adjoint so that there can be substantial convective instability from the initial value problem with $v(x) = v'(x)$ profiles without inflection points.

The nonlinear dynamics of the Kelvin-Helmholtz instability results in the formation of a chain of tilted vortices, as shown in Fig. 3.9 from a simulation of Horton, *et al.* (1987). There is a phase shift between the two cross-field velocity components so that the average along the chain of vortices produces a net transport of the sheared velocity through the Reynolds stress tensor $\langle v_x v_y \rangle$. The radial divergence of this stress tensor then gives the rate of transport of the positive moment on the right-hand side to the negative momentum on the left-hand side. In other examples in Chapters 13-17 we show how the pressure gradient instabilities can generate this Reynolds stress flux and thus create the sheared flows. The creation of sheared flow from density and temperature gradient is of key importance in the hot confined plasma.

3.3.4 *The effect of magnetic and E_r shear on drift waves*

The amplitude of the ITG turbulence is controlled by the magnetic shear, $s = rq'/q$, and the shear in the azimuthal (or poloidal) flow velocity, $v_E = -E_r/B$. For example, Hamaguchi and Horton (1990) show, with an FLR fluid model, the variation of the amplitudes and the resulting turbulent ion heat flux, $q_i = \langle n_i T_i v_r \rangle$, with magnetic and E_r shears. Kishimoto, *et al.* (1996) showed similar results with a PIC code. In the toroidal code, the flow shear, v_E', from E_r stifles the χ_i transport, whereas suppression of χ_i in the local slab model is given by

$$\chi_i(v_E') = \frac{\chi_i}{1 + (L_s v_E'/c_s)^2}. \tag{3.36}$$

The transport supression factor in the denominator of Eq. (3.36) is defined as $\Upsilon_s =$

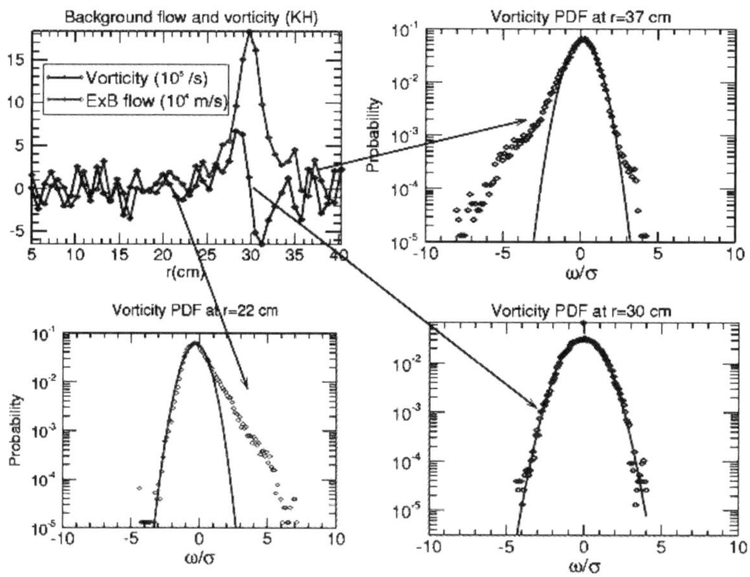

Fig. 3.8 Vorticity probability distribution functions for representative radius in strong theE_r-shear regime [Perez, *et al.* (2006)].

$L_s v_E'/c_s$ arises from the eigenmode calculation of the driftwaves in the sheared flows [Hamaguchi and Horton (1992)]. Here $L_s = qR/s(r)$ is the magnetic shear length defined by the distance spanned by the magnetic field when the $\vec{B}$-vector rotates by approximately $\pi/4$. The physical picture of the velocity-shear effect is given in Biglari, *et al.* (1998), where the large eddies or vortices of size $\Delta r = \rho_s(L_s/L_T)$ in the absence of shear flows are reduced in the shearing time, $\tau_{\text{sh}} = 1/v_E'$, to small eddies. The details of the shear flow stabilization are complicated [Diamond, *et al.* (2005)]. A kinetic theory calculation is given by Dong, *et al.* (1992) and the role of shear flow in magnetized plasma is reviewed by Itoh and Itoh (1996), Terry (2000), and Diamond, *et al.* (2005).

Experiments for validation of drift wave theory were performed on LAPD. Variable shear-flow regimes were created by varying the radial electric field, E_r, from weak to moderate and then to strong. In the moderate and strong E_r regime, the electron-density transport barrier was measured as shown in Fig. 3.7. The main characteristic of the drift wave was identified clearly in the weak E_r-shear regime and as the mixed Kelvin-Helmholtz drift wave regime in the strong E_r-shear regime.

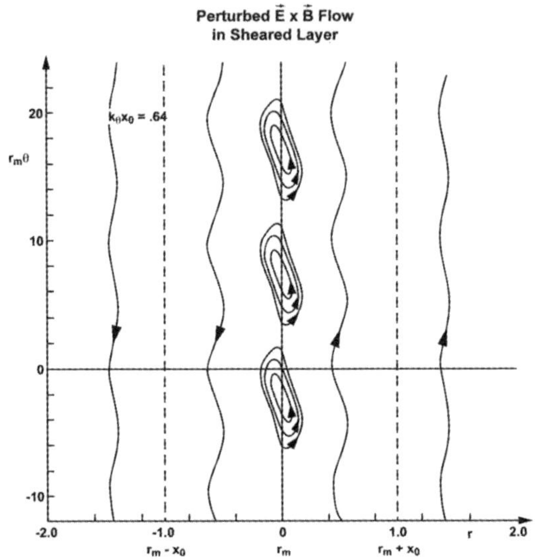

Fig. 3.9 Simulation for LAPD ($L = 17\,\mathrm{m}$, $a = 0.5\,\mathrm{m}$) with the location of the vortices in the shear-flow layer as confirmed with the vorticity probe driven on a movable shaft taking 10 shots per $\Delta r = 0.8\,\mathrm{cm}$ through the region of drift wave and Kelvin-Helmholtz turbulence.

In the strong E_r-shear regime there is a deep well of negative E_r, as occurs in the edge of H-mode tokamak plasmas. The vorticity probe measures $\omega_z = \widehat{\mathbf{b}} \cdot \nabla \times \vec{v}_E$ as shown in Fig. 3.8.

As the probe sweeps across the shear layer, the vorticity changes from counterclockwise to clockwise rotating vortices, just as expected from simple considerations and computer simulations. The region between these two oppositely spinning vortices forms the partial transport barrier as suggested by Biglari, *et al.* (1998) and Terry (2000).

The details of the effects of both magnetic and E_r shear are subtle since, like for the KH instability, the entire shape of flow profile, including its curvature, needs to be analyzed [Balmforth and Morrison (1999)]. In some circumstances shear can be destabilizing [Chen and Morrison (1991)]. Also, the actual region of transport suppression, the location of the transport barrier, sometimes occurs where the shear is a minimum, as suggested by Horton, *et al.* (1998) and Marcus, *et al.* (2008).

The strong shear flow in a plasma creates temporal and spatial intermittency

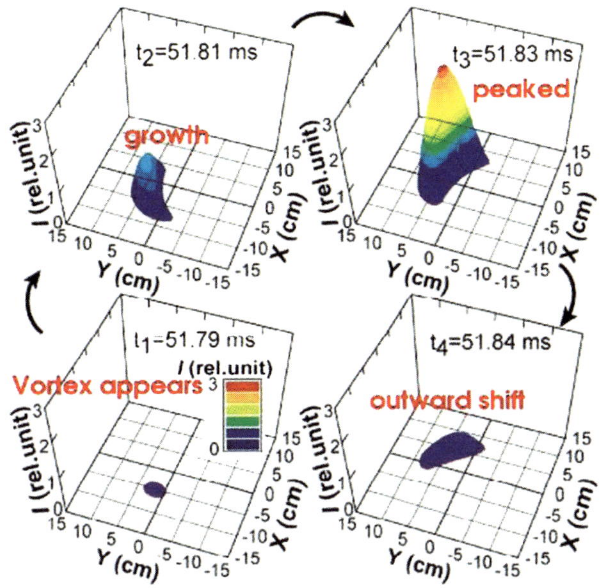

Fig. 3.10 X-ray tomography of a detailed vortex structure of an intermittent coherent structure showing its typical lifetime ranging around a hundred microseconds along with the rotational motion of the $\boldsymbol{E} \times \boldsymbol{B}$ drift [Cho, *et al.* (2005)].

as shown by the probability distribution function with large skewness and excess probability levels over the Gaussian level (parabola) in Fig. 3.8. In the central cell of the GAMMA-10 tandem mirror, where the electron temperature $T_e \sim 800\,\mathrm{eV}$, there is sufficient soft-X-ray emission to image the growth and decay of the vortices using two micro-channel plates and tomographic image reconstruction [Cho, *et al.* (2005,2006)], .

3.4 Weak Turbulence Theory for Drift Waves

For turbulent states with a broad range of wavenumbers, $\Delta\mathbf{k}$, and frequencies, $\Delta\omega_{\mathbf{k}}$, in the fluctuation spectrum the weak turbulence theory is developed by assuming that the deviation of the fluctuations from Gaussian statistics is weak. Under this condition the chain of coupled correlation functions is broken off at fourth order with what is called the "quasi-normal closure approximation." This means that the ensemble average of the product of four $\phi_{\mathbf{k}}$ fields is factorized as

$$\begin{aligned}\langle \phi_{\mathbf{k}_1}\phi_{\mathbf{k}_2}\phi_{\mathbf{k}_3}\phi_{\mathbf{k}_4=-\mathbf{k}_1-\mathbf{k}_2-\mathbf{k}_3}\rangle &= \langle \phi_{\mathbf{k}_1}\phi_{\mathbf{k}_2}\rangle\langle \phi_{\mathbf{k}_3}\phi_{\mathbf{k}_4}\rangle \\ &+ \langle \phi_{\mathbf{k}_1}\phi_{\mathbf{k}_3}\rangle\langle \phi_{\mathbf{k}_2}\phi_{\mathbf{k}_4}\rangle + \langle \phi_{\mathbf{k}_1}\phi_{\mathbf{k}_4}\rangle\langle \phi_{\mathbf{k}_2}\phi_{\mathbf{k}_3}\rangle\end{aligned} \tag{3.37}$$

which closes the equations for the dynamics of $I_{\mathbf{k}}(t) = \langle \phi^*_{\mathbf{k}}\phi_{\mathbf{k}}\rangle$.

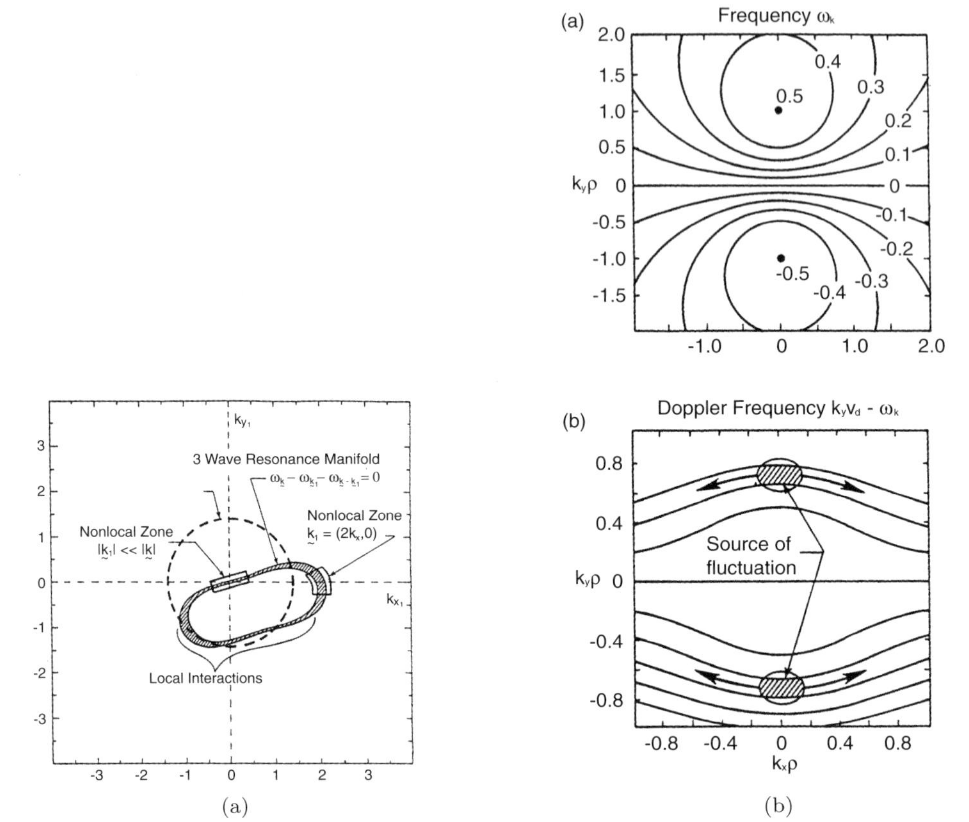

Fig. 3.11 (a) The resonant three-wave manifold for the drift-wave interaction calculations in weak turbulence theory. (b) The flux of wave energy from the source region to the damped modes given by the wave-kinetic equation.

For drift waves the turbulent energy spectrum,

$$W_{\mathbf{k}} = \left(1 + k_\perp^2 \rho_s^2\right) I_{\mathbf{k}}(t) \tag{3.38}$$

and the resulting weak-turbulence wave kinetic equation is

$$\begin{aligned}\frac{\mathrm{d}W_{\mathbf{k}}}{\mathrm{d}t} &= 2\gamma_{\mathbf{k}} W_{\mathbf{k}} + \sum_{\mathbf{k}_1, \mathbf{k}_2 = \mathbf{k} - \mathbf{k}_1} \frac{(\mathbf{k} \times \mathbf{k}_1 \cdot \widehat{\mathbf{z}})^2}{B^2} (k_2^2 - k_1^2) R_{\mathbf{k},\mathbf{k}_1,\mathbf{k}_2} \cdot \\ &\quad \left[\left(k_2^2 - k_1^2\right) W_{\mathbf{k}_1} W_{\mathbf{k}_2} - \left(k^2 - k_1^2\right) W_{\mathbf{k}} W_{\mathbf{k}_1} + \left(k^2 - k_2^2\right) W_{\mathbf{k}} W_{\mathbf{k}_2}\right]\end{aligned} \tag{3.39}$$

with the three-wave resonance function given by

$$\begin{aligned}R_{\mathbf{k},\mathbf{k}_1,\mathbf{k}_2} &= -\operatorname{Im}\left[\frac{1}{\omega_{\mathbf{k}} - \omega_{\mathbf{k}_1} - \omega_{\mathbf{k}_2} + i(\nu_{\mathbf{k}} + \nu_{\mathbf{k}_1} + \nu_{\mathbf{k}_2})}\right] \\ &\simeq \pi\delta(\omega_{\mathbf{k}} - \omega_{\mathbf{k}_1} - \omega_{\mathbf{k}_2})\end{aligned} \tag{3.40}$$

for $\nu_{\mathbf{k}} \ll \max(\omega_{\mathbf{k}})$. In the classical model with adiabatic electrons, the wave-kinetic equation obeys the conservation laws of energy, $d/dt \sum_{\mathbf{k}} W_{\mathbf{k}} = 0$, and enstropy, $d/dt \sum_{\mathbf{k}} (1 + k_\perp^2) W_{\mathbf{k}} = 0$. Detailed accounts of the derivation of Eq. (3.38) and Eq. (3.40) are given in Horton and Ichikawa (1996).

These two invariants lead to the cascade of enstropy, $\Omega = \sum_{\mathbf{k}} k_\perp^2 W_{\mathbf{k}}(t)$, to high $\mathbf{k}_\perp$ and the inverse cascade of energy W to low $\mathbf{k}_\perp$. The equilibrium solution of the wave-kinetic equation with no growth/damping, i.e. $\gamma_{\mathbf{k}} \equiv 0$, is the equipartition of the energy $W_{\mathbf{k}} = W_{\text{total}}/N_{\text{modes}}$, where N_{modes} is the total number of modes.

In reality, there are limited regions of $\mathbf{k}$-space where $\gamma_{\mathbf{k}} > 0$ and hence where modes can grow from the gradient mechanisms. In all other $\mathbf{k}$-space regions modes are damped. While some approximate analytic solutions to Eq. (3.38)-Eq. (3.40) are known in terms of anisotropic Kolmogorov-type spectra $1/|k_x|^{m_x}|k_y|^{m_y}$, the general solution to the problem can only be found numerically.

Figure 3.11 shows by the shaded loop in $\mathbf{k}$-space the region of resonant three-wave interaction defined by the subspace $\omega(\mathbf{k}) - \omega(\mathbf{k}_1) - \omega(\mathbf{k}_2) = 0$. There are two special parts of this resonant manifold: one where $\mathbf{k}_1 = (2k_x, 0)$, which is a zonal flow type of state with $u_y(x) = k_x \phi_{\mathbf{k}_1} \sin(2k_x x)$ and $u_x = 0$, and the second region is the quasi-adiabatic part of the resonant manifold where two neighboring waves $(\mathbf{k}_1, \mathbf{k}_2)$ resonate, i.e. $(\mathbf{k}, \omega_{\mathbf{k}}) - (\mathbf{k}_2, \omega_{\mathbf{k}_2}) = (\mathbf{k}_1, \omega_{\mathbf{k}_1})$ with small values of $\mathbf{k}_1$ and $\omega_{\mathbf{k}_1}$. This second class gives a scale separation where the symbols $\boldsymbol{q} = \mathbf{k}_1 \ll |\mathbf{k}|$ and $\Omega = \omega_{\mathbf{k}_1} \ll |\omega_{\mathbf{k}}|$ are often introduced for the large-scale part of the spectrum. These cross-scale $\mathbf{k} \leftrightarrow \boldsymbol{q}$ interactions give a long-range energy transfer in $\mathbf{k}$-space. These two regions of special interest are marked by rectangles in Fig. 3.11.

Figure 3.12 shows an example of the anisotropic Kolmogorov spectrum applied, where there is only an unstable ring for $0.35 < (k_x^2 + k_y^2)^{1/2} \leq 0.4$ with small $\gamma_0 > 0$ and damping of $-\gamma_0$ at both high $\mathbf{k}$ and low $\mathbf{k}$. The steady-state spectra of $W(k_x) = \sum_{k_y} W(k_x, k_y) \simeq 1/|k_x|^{m_x}$ and $W(k_y) = \sum_{k_x} W(k_x, k_y) = 1/|k_y|^{m_y}$ are shown in Fig. 3.12. A majority of the turbulent energy has accumulated in $\mathbf{k}$'s below the growth region ($k_\perp \sim 0.35$), while the region above the growth region has a power-law decay with $m_x = 3.7 \pm 0.4$ and $m_y = 4.1 \pm 0.4$.

These results are typical but not universal for drift wave turbulence. The adiabatic electron model leads to that the nonlinear $\boldsymbol{E} \times \boldsymbol{B}$ convection of the density by $\mathbf{v}_E \cdot \nabla n = 0$ vanishes. In general however, there is an appreciable non-adiabatic part of the density response to ϕ. The nonadiabatic part, for example, $\delta n = (1 + \delta_0 \partial/\partial y)\phi \rightarrow (1 + i\delta_0 k_y)\phi_{\mathbf{k}}$ leads to a nonlinear $\boldsymbol{E} \times \boldsymbol{B}$ drift $\mathbf{v}_E \cdot \nabla n_e \neq 0$ and turbulent transport for δn_e. Waltz (1983) has shown that the additional nonlinear term from $\mathbf{v}_E \cdot \nabla n$ leads to a lower amplitude as shown in Fig. 3.13. The values of $\phi_{\text{rms}} = \left\langle \phi^2 \right\rangle^{1/2}$ and the diffusivity D are shown as a function of the non-adiabatic parameter δ_0. Many dissipative processes can lead to the phase shift represented by the operator $ik_y \delta(k_\perp, \mu)$ in the electron density response [Horton (1976)] where $\{\mu\}$ is the list of system and plasma parameters.

Experiments in the Brazilian tokamak at Sao Paulo identified the three-wave

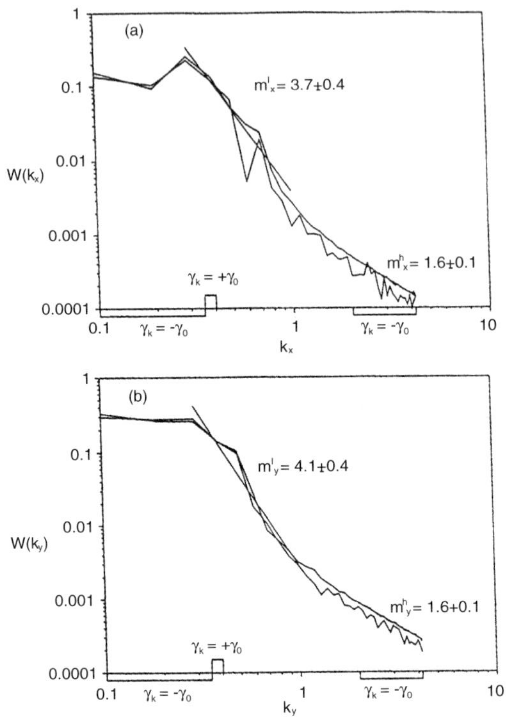

Fig. 3.12 The anisotropic spectral distribution of wave energy for the driven-damped Hasegawa-Mima equation. Spectral indices are computed for comparison with weak turbulence theory.

coupling of drift waves using sophisticated signal processing of the probe data in the edge region of the plasma [Batista, *et al.* (2006)].

3.5 Ion Temperature Gradient Mode

There are numerous nonlinear formulations of the ion temperature gradient mode instability in the literature. The models range from one-field models (e.g. Hasegawa-Mima) with one nonlinearity $\mathbf{v}_E \cdot \nabla$, to two field-models with the fluctuating electrostatic potential ϕ and fluctuatiing ion temperature. The one-field model, however, uses the adiabatic electron response $\delta n_e/n_e = e\phi/T_e$ for the density fluctuation whereas the two-field model uses either the fluctuating ion pressure, δp_i, or the fluctuating ion temperature, δT_i, dynamical equation. These gyro-fluid models assume a closure that can be either an equation of state with adiabatic gas constant, Γ, for the ion fluid or closure formulas for the parallel ion-thermal-heat flux, $\delta q_{i\,\|}$ [Hammett and Perkins (1990); Mattor and Parker (1997); Mattor (1999); Sugama, *et al.* (2001)].

The value of the adiabatic gas constant Γ should be chosen to be relevant to the specific regime of the plasma and the geometry for the wave-particle interactions.

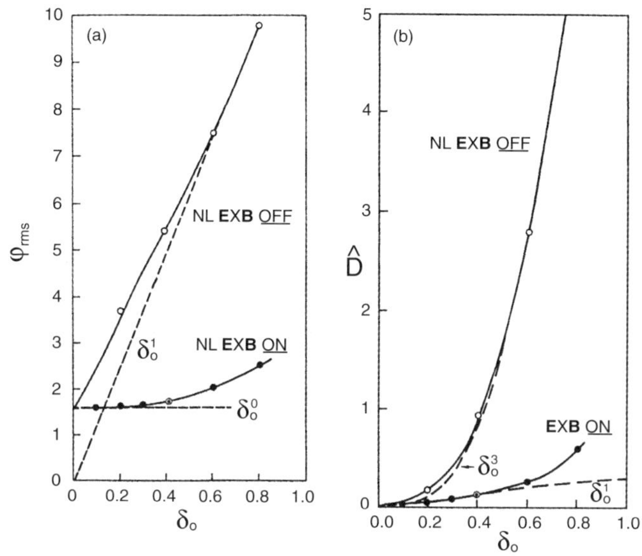

Fig. 3.13 Scaling of the drift-wave turbulence with δ_0. In frame (a) the scaling of the rms potential fluctuation level and (b) the normalized diffusivity $D/D_{\rm dw}$ for the Hasegawa-Mima nonlinearity (open circles) and the self-consistent nonlinear drift-wave Eq. (99) (solid circles) containing the $\boldsymbol{E} \times \boldsymbol{E}$ nonlinearity in the convection of the density [Waltz (1983)].

For example, in the collisionless slab the appropriate value for the ITG modes is that $\Gamma = 3$ due to the one-degree-of-freedom ($f_{\rm dof} = 1$) motion of the ions along the magnetic field lines with the resonance $\omega = k_\| v_\|$. In the torus, however, the resonance is of the ions with the waves has $f_{\rm dof} = 3$ so the appropriate choice of Γ is $\Gamma = 5/3$. In 3D systems the ITG modes become critical in the simplest kinetic dispersion relations when η_i exceeds $\Gamma - 1 = 2/3$. This is the point in the Carnot cycle without dissipation where no network is done in cycling the plasma from the high to low-temperature regions. Similar considerations apply to the ETG and other instabilities. The use of a local or effective Γ is also required for astrophysical gases by Shapiro and Tuekolsky (1983) in their Chapter 3 where the symbol Γ_1 for this locally relevant adiabatic gas constant. For electrons there is fast thermal transport along the magnetic field lines so the relevant value of Γ_e is often barely exceeding unity. Unity applies when the gas is strictly isothermal. A clear example for the need of a electron gas constant near unity is given by the temperature of the solar wind at the Earth (1 AU) where $T_e \sim 10\,\text{eV}$ for the greatly-expanded flux tube volumes that are connected to the solar corona. The value of 5/3 for the electron gas would have the electron temperature below the recombination value at one astronomical unit rather than 10 eV.

References describing different models are Horton, *et al.* (1980), Hamaguchi and Horton (1990), and Dimits, *et al.* (1991, 2000) for the slab ITG mode and Horton, *et al.* (1981), Waltz, *et al.* (1994), and Beer and Hammett (1999) for the toroidal ITG mode. The toroidal ITG mode is typically stronger than the slab ITG mode and has a mathematical structure similar to a small scale ($\lambda_\perp \gtrsim \rho_i$) Rayleigh-Taylor instability. The Rayleigh-Taylor MHD instability,

$$\omega^2 = -\gamma_{\mathrm{MHD}}^2 = \frac{k_y^2}{k_\perp^2}\left(\frac{1}{\rho_s}\frac{\mathrm{d}p}{\mathrm{d}x}\frac{1}{R_c}\right) \gg k_\parallel^2 v_A^2 \tag{3.41}$$

is changed in drift wave theory to a slower growing mode with $\gamma(k_y = 0) = 0$ and a well-defined maximum growth rate, $\gamma_{\max}$, at a finite $k_y\rho_s \simeq [(1-2\epsilon_n)/(1+\eta_i)]^{1/2}$. Here $\epsilon_n = L_n/R$ and $\eta_i = \mathrm{dln}T_i/\mathrm{dln}n_i$. The electrostatic shielding by the adiabatic electron fluid introduces a dielectric constant, $\epsilon = 1 + 1/k_\parallel^2\lambda_{De}^2 \gg 1$. Since $k_\parallel\lambda_{De} = k_\parallel v_e/\omega_{pe} \ll 1$, since $k_\parallel\lambda_{De} = k_\parallel v_e/\omega_{pe} \ll 1$. Thus, the dielectric $\epsilon \gg 1$ reduces the electric fields in the drift wave-interchange instability.

The energy source of the instability is explained in Section 3.3.2. The typical fastest-growing mode has $\gamma_{\max} = 0.1c_s/(L_{Ti}R)^{1/2}$ for a toroidal system with a magnetic radius of curvature, $R_c = R$, on the outside of the torus where the mode balloons in amplitude. The ballooning-mode equation and a simple nonlinear model for the toroidal ITG are given in Horton, *et al.* (1981).

Balescu (2005) gives a version of the Hamaguchi-Horton slab ITG model as an example of an alternative model. Balescu (2005) describes the ITG mode with the four partial differential equations,

$$\frac{\partial \delta n_i}{\partial t} = -v_{de}\frac{\partial \delta\phi}{\partial y} - \nabla_\parallel \delta v_\parallel - \frac{c_s^2}{\Omega_i}[\delta\phi, \delta n_i] \tag{3.42}$$

$$\frac{\partial}{\partial t}\left(\delta n_i - \rho_s^2\nabla_\perp^2\delta\phi\right) = -v_{de}\frac{\partial}{\partial y}\left(1 - \tau(1+\eta_i)\rho_s^2\nabla_\perp^2\right)\delta\phi - \nabla_\parallel\delta v_\parallel$$
$$-\frac{c_s^2}{\Omega^i}\left[\delta\phi, (\delta n_i - \rho_s^2\nabla_\perp^2\delta\phi)\right] \tag{3.43}$$

$$\frac{\partial \delta v_\parallel}{\partial t} = -\frac{1}{m_i n_0}\nabla_\parallel\delta p_i - c_s^2\nabla_\parallel\delta\phi - \frac{c_s^2}{\Omega_i}\left[\delta\phi, \delta v_\parallel\right] \tag{3.44}$$

$$\frac{\partial \delta p_i}{\partial t} = -v_{de}\tau(1+\eta_i)\frac{\partial\delta\phi}{\partial y} - \Gamma p_{i0}\nabla_\parallel\delta v_\parallel - \frac{c_s^2}{\Omega_i}[\delta\phi, \delta p_i] \tag{3.45}$$

where $\tau = T_i/T_e$, $v_{de} = (\rho_s/L_n)c_s$ and Γ is the ion adiabatic gas constant. Each equation has a convective nonlinearity.

The earliest example of the slab ITG turbulence is given in Horton, *et al.* (1980) with three partial differential equations. Two versions are given: 1) with full compressional nonlinearity as in MHD for the ion pressure and 2) with reduced nonlinearity of purely $\boldsymbol{E} \times \boldsymbol{B}$ convection of nonlinear pressure as in Eq. (3.44) and Eq. (3.45) in the Balescu model. The energy integral for the system takes two different forms for these two versions of the ITG model. The Horton, *et al.* (1980) work solved the nonlinear convection terms with upwind finite difference formulas and found fast relaxation of the pressure profile that was fixed with Dirichlet boundary conditions at the core and the edge of the simulation box. The CRAY-1 computer was used to make the earlier 3D simulations of drift wave turbulence.

Currently there are a number of global and flux-tube codes that simulate the full gyro-kinetic equations for the ITG turbulence [Lin, *et al.* (1998); Dimits, *et al.* (2000); Jenko and Dorland (2001)]. Here we show the distribution of the electrostatic potential fluctuations in all three dimensions in Fig. 3.14. In this figure where saturation has occurred, the tendency for the fluctuations to establish a zonal-flow pattern is evident from the ringed structure of the potential.

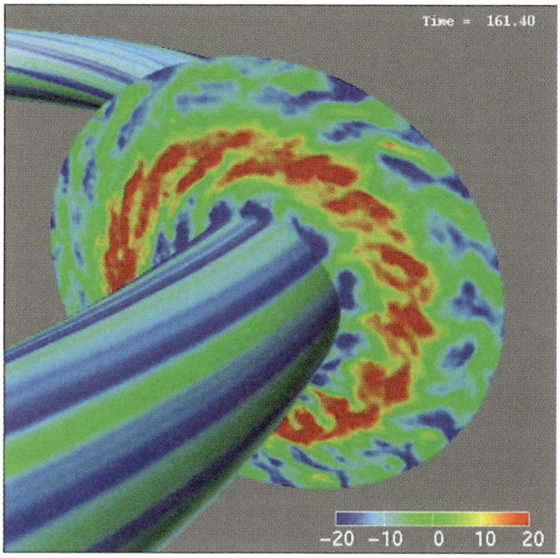

Fig. 3.14 ITG simulation from Watanabe GKV showing the contours of constant electrostatic potential and density ($\delta n/n = e\delta\varphi/T_e$) in the saturated state.

The adiabatic closure in Eq. (3.45) allows the turbulence to be computed with well-developed spectral codes that produce high accuracy in the fluid conservation laws. Dissipation in ITG turbulence occurs through ion-wave resonances and ion-ion collisions ν_i.

In short-mean-free-path $\lambda_{\rm mfp}$ plasmas, the heat flux is calculated by an expansion in the small parameter $k_\parallel \lambda_{\rm mfp}$. There is a well-accepted and widely-used set of transport formulas given by Braginskii (1965) for the collisional dominated regime with small $k_\parallel \lambda_{\rm mfp}$. Thus, in this book we discuss some problems in terms of the Braginskii fluid equations. In the collisional regime the parallel ion transport is given by

$$q_\parallel = -3.9 \frac{n_i T_i}{m_i \nu_i} \nabla_\parallel T_i. \tag{3.46}$$

In the problem of nonlinear simulations of ITG turbulence [Hamaguchi and Horton(1990)] used this formula with a small value of ν_i such that $\max(k_\parallel v_i) = \nu_i$.

This cut-off method is known to be an adequate approximation for the nonlocal thermal flux.

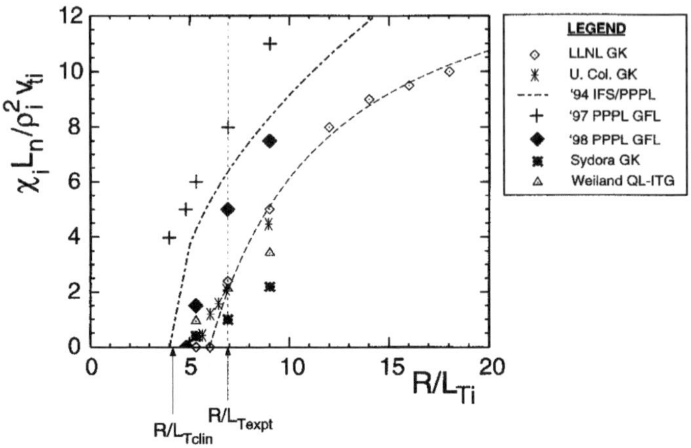

Fig. 3.15 Comparisons of the turbulent ion thermal diffusivity χ_i in units of $\rho_i^2 v_{ti}/Ln$ versus the ion temperature gradient parameter R/L_{T_i} from nonlinear simulation codes with different architectures with the developers listed in the box. The reference plasma parameters such as $a, R, B, q, \nabla T_i, \nabla n, T_e$ were adopted from mid-radius value of a typical tokamak discharge [Dimits, *et al.* (2000)] in a SciDAC project. The electron response was modeled as adiabatic and other simplifications introduced so that the seven codes could simulate the same plasma. The critical gradient inferred from the experimental data is indicated.

The critical gradient for the onset of the ITG instability depends on the exact nature of the wave-ion resonance at the marginal stability wave frequency $\omega_{\rm crit}(\boldsymbol{k})$. Nyquist diagrams are used to derive these conditions from a complex dispersion relation $D_k(\omega)$ in both fluid and kinetic theory. In Chapters 13 and 14 the collisionless kinetic dispersion relations are analyzed with Nyquist diagrams. Here we show in Fig. 3.15 the results for the critical ion temperature expressed as R/L_{T_i} for a particular set of plasma parameters representative of the mid-radius of tokamak. The parameters and modeling are described in Dimits, *et al.* (1991) from the team project to compare the results of seven simulation groups listed in the box. The codes used by the groups varied from two-component fluid codes to different types of kinetic theory codes. These simulation codes are described in detail in Chapter 19.

Improved closure models, which give rise to the residual zonal flows predicted by Rosenbluth and Hinton (1998), are described by Beer and Hammett (1999) and recently by Sugama, *et al.* (2007).

3.6 Drift Waves Paradigms: Hasegawa-Mima and Hasegawa-Wakatani Models

The basic physical mechanisms of drift wave instabilities and the associated plasma transport are described here with reduced nonlinear models or paradigms. Drift waves develop into a mixture of wave turbulence and coherent structures with the details of the mixture depending on the driving forces ∇n_e, ∇T_e and ∇T_i and the amount of magnetic shear and background velocity shear in the system. The mixture of zonal flows, waves and vortices has certain similarities to the Rossby waves in planetary atmospheres and the equatorial-to-polar temperature gradient driven turbulent heat transport. The principal difference is in the replacement of the gravitational force on planets with a electric force from charge separation in the curved magnetic fields. The Coriolis force has the same mathematical form as the Lorentz force in the momentum balance equations.

To see the reduction of the ITG turbulence to the simple one-field Hasegawa-Mima model, the density fluctuation δn_e is taken as adiabatic so that with quasineutrality $\delta n_i = \delta n_e = n_0(e\delta\varphi/T_e)$ in Eq. (3.43) and the ion acoustic waves are dropped with $\nabla_\parallel \delta v_\parallel \to 0$. The vorticity Eq. (3.43) then reduces to a single, simple nonlinear partial differential equation for $d\omega/dt = 0$ with $\omega = \delta\varphi - \rho_s^2 \nabla_\perp^2 \delta\varphi - v_d x$ and the convective derivative given by $\boldsymbol{v}_E \cdot \nabla\omega = [\delta\varphi, \omega] = 0$. In this limit the drift wave equation is isomorphic to the Charney equation of geostrophic flows. This subject is developed in detail in Horton and Hasegawa (1994) and Horton and Ichikawa (1996).

The free energy sources created by the radial gradients of the plasma density and the temperature are useful to characterize the strength of the drift wave turbulence. Real time diagnostic signals from microwave interferometers and Thompson scattering data be used to compute the thermodynamic formulas for the upper bounds of the turbulence given in Section 3.3 for the density and temperature fluctuations. The derivations of the free energy formulas treat the plasma as an ideal gas with a compressibility gas constant, $\Gamma = (d+2)/d$, where d is the number of degrees of freedom associated with the type of instability under consideration. The Carnot-cycle description of the auxiliary heated plasma is particularly relevant. The method is used in Horton, *et al.* (2005) for the turbulence driven by the electron temperature profiles in Tore Supra and is being used for real-time data analysis in NSTX.

The structure of the drift wave eigenmodes in the double-periodic toroidal geometry is described after a brief review of the early experiments on drift waves in steady state cylindrical Q-machines. Experiments in the Columbia Linear Machine [Sen, *et al.* (1991)] were devoted to the ion temperature gradient (ITG) mode and the six-channel CO_2 laser scattering experiments in the TEXT tokamak [Brower, *et al.* (1987)] clearly show the role of the hybrid of the ITG mode and trapped electron mode TEM. An example of the isolines of the electrostatic potential contours for the Hasegawa-Mima equation from direct numerical simulations is shown in Fig. 3.16. The spectral method of fluid simulations is described in detail in Chapters 6 and 7 of Horton and Ichikawa (1996).

The ITG mode has become the standard model for describing ion transport in basically all toroidal experiments with strong ion temperature gradients. The

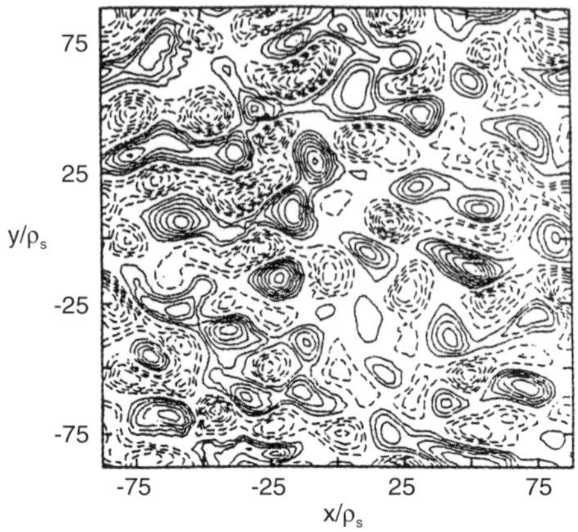

Fig. 3.16 Isopotential contours computed from direct numerical simulations of the driven-damped Hasegawa-Mima equation.

basic toroidal mechanism and nonlinear models are given in Horton, *et al.* (1981). Nowadays large-scale Tera-flop computers commonly compute the ITG turbulence from direct numerical solutions of the gyrokinetic-Vlasov-Poisson system. For example, the state of the turbulence is shown in Section 3.5 and Chapter 19 for the state-of-the art simulations. The standard article on the comparison of direct numerical simulation for the ITG turbulence is Dimits, *et al.* (1991). Review of the state of the numerical simulations of the ETG turbulence is Nevins, *et al.* (2006). The electron turbulent transport is more demanding to compute but shows large enhancements over the quasilinear fluxes due to the formation of extended radial structures with correlation lengths, $\ell_c \gg \rho_e$. The structures are insensitive to the radial electric-field shear but do depend sensitively on the magnetic shear profile.

The empirical energy confinement scaling laws for both helical system [Yamada, *et al.* (2004)] and tokamaks [Kaye and ITER Group (1997)] follow the drift wave formulas and simulations to within the accuracy of the experimental data. Some of the more detailed comparisons of theoretical models [Kinsey, *et al.* (1996a); Kinsey and Bateman (1996); Redd, *et al.* (1998); Bateman, *et al.* (1998)] with tokamak databases described in Chapters 18 and 19.

The validation of a gyrokinetic code for temperature gradient driven turbulence in a cylindrical plasma with constant B is shown in Fig. 3.17. The simulation agrees with the steady-state 10 eV hydrogen ETG experiments carried out in the Columbia Linear Machine by Sen, *et al.* (1991) with the GTC codes developed by Lin, *et al.* (1998). The validation of the GTC code with the CLM experiment was a SciDAC project [Fu, *et al.* (2012)]. The steady turbulence with the dominant $m = 11, 12, 13$ modes in Fig. 3.17 were measured in CLM validating the ETG model and simulation code.

A practical gyrofluid code named GS-2 is available for modeling and interpret-

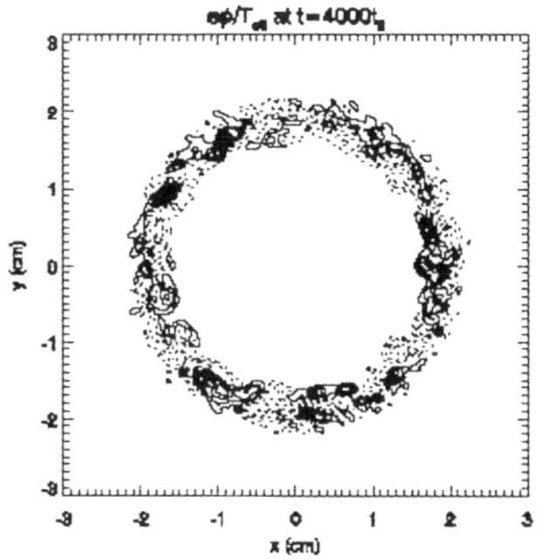

Fig. 3.17 Example of an electron temperature gradient driven drift wave spectrum in the non-linear state with a spectrum of vortex structures with outward extending plumes that produce an anisotropy in the radial and poloidal drift wave fluctuation spectrum [Fu, *et al.* (2012)].

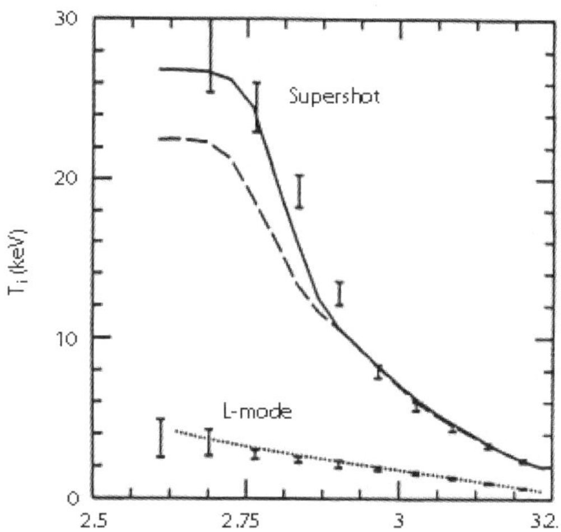

Fig. 3.18 Comparison of the IFS-PPPL model for an L mode and a supershot discharge in TFTR. The complex χ_l formula is able to account for the large increase in the core ion temperature from the experimental values of T_l at $\rho = 0.8$ taken as the boundary condition [Kotschenreuther (1995a,b)].

ing tokamak data with the ITG and TEM instabilities. An early success of these gyrofluid simulations is shown in Fig. 3.18, where neutral beam driven discharges in the TFTR tokamak were shown to be interpreted by using nonlinear gyrofluid simulations to construct a complicated, but comprehensive interpolating formula $\chi_i(\{\mu\})$ for the ion thermal conductivity. The χ_i formulas with a large parameter set $\{\mu\}$ taken from the TRANSP transport code were able to reproduce the bifurcation of the plasma to the supershot regime shown in Fig. 3.18. The formula derived from a large number of nonlinear simulations by Kotschenreuther, *et al.* (1995a,b) is useful.

A closely-related problem with dropping the divergence of the finite parallel heat flux is explained in some detail by Kulsrud in the Handbook of plasma physics. Kulsrud shows that double adiabatic versions of the MHD problem give rise to a large discrepancy in the threshold condition for the onset of the mirror mode instability when compared to the correct Vlasov equation formulas. In this case the kinetic mirror mode is unstable for conditions where the double adiabatic MHD equations predict stability. This error again arises from the parallel phase velocities not being fast compared with the ion thermal velocities as required for a valid local fluid description with an adiabatic closure.

The topic of submarginal turbulence and self-organized criticality is surveyed by Krommes (1999, 2002).

The GS-2 [Kotschenreuther (1995b)] is currently fully electromagnetic with three ion species designed for running on massively parallel computers. The electron temperature profile in NSTX is resilient [Gates, *et al.* (2002)] and consistent with the fast ETG turbulence.

References

Balescu, R. (2005). Aspects of anomalous transport in plasmas, *Series in Plasma Physics* (Institute of Physics Publishing).

Balmforth, N. J. and Morrison, P. J. (1999). A necessary and sufficient instability condition for inviscid shear flow, *Studies Appl. Math.* **102**, pp. 309-344.

Batista, A. M., Calderas, I. L., Lopes, S. R., Viana, R. L., Horton, W., and Morrison, P. J. (2006). Nonlinear three-mode interaction and drift-wave turbulence in a tokamak edge plasma, *Phys. Plasmas* **13**, Article No. 042510.

Bateman, G., Kritz, A. H., J. E. Kinsey, and Redd, A. J. (1998). *Phys. Plasmas* **5**, pp. 2355-2362.

Beer, M. A. and Hammett, G. W. (1996). Toroidal gyrofluid equations for simulations of tokamak turbulence, *Phys. Plasmas* **3**, 11, pp. 4046-4064, doi:10.1063/1.871538.

Beer, M. A. and Hammett, G. W. (1999). *Proceedings of the Joint Varenna-Lausanne Intl. Workshop on Theory of Fusion Plasmas, Varenna, 1998*, eds. J. W. Connor, E. S. Indoni, and J. Vaclavik (Bologna, Societa Italiana de Fisca).

Benkadda, S., Elskens,Y., and Doveil, F., eds. (1995). *Transport, Chaos and Plasma Physics: Institut Mediterranean De Technologie, Marseilles, France July 10-21* Advanced Series in Nonlinear Dynamics, (World Scientific Pub. Co. Inc).

Biglari, H., Diamond, P. H., and Terry, P. W. (1998). *Phys. Rev. Lett.* **60**, pp. 200-203.

Bohm, D., Burhop, E. H. S., and Massey, H. S. W. (1949). The use of probes for plasma exploration in strong magnetic fields, *The characteristics of electrical discharges in magnetic field* eds. A. Guthrie, and R. K. Wakerling, (New York, McGraw-Hill).

Braginskii, S. I. (1965). *Reviews of Plasma Physics* ed. M. A. Leontovich, (Consultants Bureau, New York) I, p. 205.

Brower, D. L., Peebles, W. A., Kim, S. K., Luhmann, N. C., Tang, W. M., and Phillips, P. E. (1987). *Phys. Rev. Lett.* **59**, pp. 48-51.

Chandrasekhar, S. (1961). *Hydrodynamic and Hydromagnetic Stability* (Oxford University

Press), pp. 481-498, ISBN:0-486-64071-X.

Chen, X. L. and Morrison, P. J. (1991). A sufficient condition for the ideal instability of shear flow with parallel magnetic field, *Phys. Fluids B* **3**, pp. 863-865.

Cho, T., Pastukhov, V. P., Horton, W., Numakura, T., Hirata, M., Kohagura, J., Chudin, N. V., and Pratt, J. (2008). Active control of internal transport barrier formation due to off-axis electron-cyclotron heating in GAMMA-10 experiments, *Phys. Plasmas* **15**, p. 056120, `http://dx.doi.org/10.1063/1.2906262`.

Cho, T., Kohagura, J., Numakura, T., Hirata, M., Higaki, H., Hojo, H., Ichimura, M., Watanabe, O., Yoshikawa, M., Kojima, A., Miyake, Y., Miyata, Y., Shimizu, K., Tomii, Y., Yoshida, M., Sakamoto, K., Imai, T., Pastukhov, V. P., Miyoshi, S., and G. Group. (2006). *Phys. Rev. Lett.* **97**, p. 055001.

Cho, T., Kohagura, J., Hirata, M., Numakura, T., Higaki, H., Hojo, H., Ichimura, M., Ishii, K., Islam, K. M., Itakura, A., Katanuma, I., Nakashima, Y., Saito, T., Tatematsu, Y, Yoshikawa, M., Takemura, Y., Kojima, A., Kobayashi, T., Yamaguchi, Y., Miyata, Y., Yokoyama, N., Tomii, Y., Miyake, Y., Kiminami, S., Shimizu, K., Kubota, Y., Saimaru, H., Higashizono, Y., Mase, A., Yasaka, Y., Ogura, K., Sakamoto, K., Yoshida, M., Pastukhov, V. P., Imai, T., Miyoshi, S., and G. Group. (2005). *Nucl. Fusion* **45**, pp. 1650-1657.

Diamond, P. H., Itoh, S-I., K. Itoh, and Hahm, T. S. (2005). Zonal flows in plasma: A review, *Plasma Phys. Control. Fusion* **47**, pp. R35-R161.

Dimits, A. M., Drake, J. F., Guzdar, P. N., and Hassam, A. B. (1991). *Phys. Plasmas* **3**, pp. 620-626.

Dimits, A. M., Bateman, G., Beer, M. A., Cohen, B. I., Dorland, W., Hammett, G. W., Kim, C., Kinsey, J. E., Kotschenreuther, M., Kritz, A. H., Lao, L. L. Andrekas, J. M., Nevins, W. M., Parker, S. E., Redd, A. J., Shumaker, D. E., Sydora, R., and Weiland, J. (2000). *Phys. Plasmas* **7**, pp. 969-983.

Dong, J.-Q., Horton, W., and Kim, J-Y. (1992). *Phys. Plasmas* **4**, pp. 1867-1876.

Fu, X., Horton, W., Sen, A. K., Sokolov, V., Lin, Z., and Xiao, Y. (2012). *Phys. Plasmas* **19**, p. 032303, doi:1063/1.3686143.

Gates, D., *et al.* (2002). *Phys. Plasmas* **10**, p. 1659.

Hamaguchi, S. and Horton, W. (1992). *Phys. Fluids B* **4**, p. 319.

Hamaguchi, S. and Horton, W. (1990). *Phys. Fluids B* **2**, pp. 1833-1851.

Hammett, G. W. and Perkins, F. W. (1990). *Phys. Rev. Lett.* **64**, pp. 3019-3022.

Hojo, H. (1995). *Jpn. J. Appl. Phys.* **34**, p. 6501.

Horton, W., Park, H-B., Kwon, J-M., Strozzi, D., Morrison, P. J. and Choi, D-I. (1998). Drift wave test particle transport in reversed shear profile, *Phys. Plasmas* **5**, pp. 3910-3917.

Horton, W., Choi, D.-I., and Tang, W. M. (1981). *Phys. Fluids* **24**, pp. 1077-1085.

Horton, W. and Hasegawa, A. (1994). Quasi-two-dimensional dynamics of plasmas and fluids, *Chaos* **4**, 2, p. 227.

Horton, W. and Reichl, L. E., eds. (1984). *Statistical Physics and Chaos in Fusion Plasmas* (John Wiley and Sons, New York).

Horton, W. and Ichikawa, Y. (1996). *Chaos and Structures in Nonlinear Plasmas* (World Scientific), ISBN:81-7764-234-0.

Horton, W., Hong, B. G., and Tang, W. M. (1988). *Phys. Fluids* **31**, p. 2971.

Horton, W., Tajima, T., and Kamimura, T. (1987). Kelvin-Helmholtz instability and vortices in magnetized plasma, *Phys. Fluids* **30**, pp. 3485-3495, `http://link.aip.org/link/doi/10.1063/1.866429`.

Horton, W., Estes, R., and Biskamp, D. (1980). *Plasma Phys.* **22**, pp. 663-678.

Horton, W., Wong, H. V., Morrison, P. J., Wurm, A., Kim, J. A., Perez, J. C., Pratt, J.,

Hoang, G. T., LeBlanc, B. P., and Ball, R. (2005). *Nucl. Fusion* **45**.
Horton, W. (1976). *Phys. Rev. Lett.* **37**, p. 1269.
Itoh, K. and Itoh, S.-I. (1996). *Plasma Phys. Control. Fusion* **38**, pp. 1-49.
Jenko, F. and Dorland, W. (2001). *Plasma Phys. Control. Fusion* **43**, pp. A141-A150.
Jose, J. V. and Saletan, E. J. (2002). Classical Mechanics (Cambridge UP, Cambridge), p. 46.
Kadomtsev, B. and Pogutse, O. (1970). Turbulence in Toroidal Systems, *Rev. Plasma Phys.* **5**, ed. A. Leontovich, pp. 249-400.
Kaye, S., Greenwald, M., Stroth, U., Kardaun, O., Kus, A., Schissel, D., DeBoo, J., Bracco, G., Thomsen, K., Cordey, J., Miura, Y., Matsuda, T., Tamai, H., Takizuda, T., Hirayama, T., Kikuchi, H., Naito, O., Chudnovskij, A., Ongena, J., and Hoang, G. T. (1997). *Nucl. Fusion* **32**, pp. 1303-1328.
Kinsey, J. E., G. Bateman, Kritz, A. H., and A. Redd. (1996). *Phys. Plasmas* **3**, pp. 561-570.
Kinsey, J. E., and G. Bateman. (1996). *Phys. Plasmas* **3**, pp. 3344-3357.
Kishimoto, Y., T. Tajima, Horton, W., M. J. LeBrun, and J.-Y. Kim. (1996). *Phys. Plasmas* **3**, pp. 1289-1307.
Kotschenreuther, M., Dorland, W., Beer, M. A., Hammett, G. W. (1995). Quantitative predictions of tokamak energy confinement from first-principles simulations with kinetic effects, *Phys. Plasmas* **2**, p. 6, ISSN:1070-664X.
Kotschenreuther, M., Rewoldt, G., and Tang, W. M. (1995). *Comp. Phys. Comm.* **88**, pp. 128-140, `http://dx.doi.org/10.1016/0010-4655(95)00035-E`
Krommes, J. A. (2002). Fundamental statistical description of plasma turbulence in magnetic fields, *Phys. Reps.* **360**, pp. 1-353, doi:10.1016/S0370-1573(01)00066-7.
Krommes, J. A. (1999). *Plasma Phys. Control. Fusion* **41**, pp. A641-A652.
Lin, Z., Hahm, T. S., Lee, W. W., W. M. Tang, and R. B. White. (1998). *Science* **281**, pp. 1835-1837.
Marcus, F. A., Caldas, I. L., Guimarẽs-Filho, Z. O., Morrison, P. J., Horton, W., Kuznetsov, Yu. K., and Nascimento, I. C. (2008). Reduction of chaotic particle transport driven by drift waves in sheared flows, *Phys. Plasmas* **15**, p. 112304-1-9.
Mattor, N. (1999). *Phys. Plasmas* **6**, p. 1065.
Mattor, N., and Parker, S. E. (1997). *Phys. Rev. Lett.* **79**, pp. 3419-3422.
Mazzucato, E. (1976). *Phys. Rev. Lett.* **36**, pp. 792-794.
Nevins, W. M., Candy, J., Cowley, S., T. Dannert, Dimits, A., W. Dorland, C. Estrada-Milla, Hammett, G. W., Jenko, F., M. J. Pueschel, and D. E. Shumaker. (2006). *Phys. Plasmas* **13**, p. 122306.
Pastukhov, V. P. (2005.), Equations for nonlinear MHD convection in shearless magnetic systems, *Plasma Phys. Reps.* **31**, 7, pp. 577-590, doi:10.1134/1.1992585.
Perez, J., Horton, W., Gentle, K., Rowan, W., Lee, K., and Dahlburg, R. (2006). *Phys. Plasmas* **13**, pp. 2101-2110.
Redd, A. J., Kritz, A. H., Bateman, G., and Horton, W. (1998). *Phys. Plasmas* **5**, 1369-1379.
Rosenbluth, M. N. and Hinton, F. L. (1998). *Phys. Rev. Lett.* **80**, pp. 724-727.
Sen, A. K., Chen, J., and Mauel, M. (1991). *Phys. Rev. Lett.* **66**, pp. 429-432.
Shapiro, S. L. and Teukolsky, S. A. (1983). (Cambridge Press) ISBN:0-471-87316-0.
Sugama, H., Watanabe, T. H., and Horton, W. (2001). *Phys. Fluids* **8**, p. 26172628.
Sugama, H., Watanabe, T. H., and Horton, W. (2007). *Phys. Plasmas* **14**, p. 022502.
Terry, P. W. (2000). *Rev. Mod. Phys.* **72**, pp. 109-165.
Waltz, R. E. (1983). Numerical study of drift wave turbulence with simple models for wave-wave nonlinear coupling, *Phys. Fluids* 26, p. 169, doi:10.1063/1.864004.

Waltz, R. E., Kerbel, G. D., and Milovich, J. (1994). Toroidal gyro-Landau fluid model turbulence simulations in a nonlinear ballooning mode representation with radial modes, *Phys. Plasmas* **1**, pp. 2229-2244.

Yamada, H., Harris, J. H., Dinkalge, A., Ascasibar, E., Sano, F., Okamura, S., Talmadge, J., Stroth, U., Kus, A., Murakami, S., Yokoyama, M., Beidler, C. D., Tribaldos, V., Watanabe, K. Y., and Suzuki, Y. (2004). *Nucl. Fusion* **45**, pp. 1684-1693.

Chapter 4

Two-Component Magnetohydrodynamics

4.1 Collisional Transport Equations

The two component fluid equations are used to describe the low-frequency-large-scale dynamics of plasmas. The two-component adjective means that the ion and electrons are described separately as fluids with their own characteristic properties. This formulation allowing for the ions and electrons to behave differently is of key importance making the dynamics richer and more accurate than possible with the one-component MHD model. Many strong assumptions must be made to reduce the two-component fluid theory to the MHD model. Both ion acoustic waves and drift waves are lost when using the MHD model. Generally, lower frequency waves including magnetic reconnection dynamics require the two-component fluid theory.

The two-component fluid equations are valid for disturbances with frequencies ω much less than the ion cyclotron frequency ω_{ci} and wavelengths perpendicular $\lambda_\perp$ and parallel $\lambda_\parallel$ to the magnetic field such that $\lambda_\perp \gg \rho_i$ and $\lambda_\parallel \gg \lambda_{\rm mfp}$ where ρ_i is the average ion gyroradius and $\lambda_{\rm mfp}$ is the mean-free-path determined by the collision frequency $\nu_{\rm coll}$. The plasma equilibrium is determined by the cross-field transport processes of diffusion and convection which lead to a collisional decay at a rate proportional to $\nu_{\rm coll}/B^2$. The electrostatic modes are basically drift waves and sound waves with frequencies $\omega \simeq k_y v_{de}$ and $\omega \simeq k_\parallel c_s$ which are fast compared with the equilibrium time scales and slow compared with the electromagnetic fast modes that establish the equilibrium plasma. The corrections due to the coupling of the electrostatic modes to the faster Alfvén dynamics is of order $\omega^2/k_\parallel^2 v_A^2 \simeq \beta$ where the toroidal geometry determines the scale of the parallel wavelengths. The neglect of the electromagnetic coupling compared to the toroidal effects restricts the plasma beta to $\beta < (r/R)^2$. Here the plasma $\beta = 8\pi p/B^2[\text{cgs}] = 2\mu_0 p/B^2[\text{mks}]$ is the ratio of the plasma pressure to the magnetic pressure.

In this chapter the domains of plasma instability are delineated from the dispersion relation derived from the linearization of the fluid equations given in Eq. (4.1)-Eq. (4.7) in toroidal systems. In the analysis the equilibrium radial gradients are weak over the scale of the gyroradius and the dispersion relations are correct to within small corrections arising from the radial normal mode problem. The different drifts of the ions and electrons from their different temperatures and variation in ion and electron coupling to the electromagnetic fields is responsible for the rich behavior of the plasma.

The plasma fluid is described by the electron and ion number densities n_e, n_i, the velocities $\boldsymbol{v}_e, \boldsymbol{v}_i$, and temperatures T_e, T_i. The transport equations for these moments are

$$\frac{\partial}{\partial t} n_i + \nabla \cdot (n_i \boldsymbol{v}_i) = 0 \tag{4.1}$$

$$\frac{\partial}{\partial t} n_e + \nabla \cdot (n_e \boldsymbol{v}_e) = 0 \tag{4.2}$$

$$m_i n_i \left(\frac{\partial}{\partial t} + \boldsymbol{v}_i \cdot \nabla \right) \boldsymbol{v}_i + \nabla \cdot \boldsymbol{\pi}_i = e n_i \left(\boldsymbol{E} + \frac{\boldsymbol{v}_i}{c} \times \boldsymbol{B} \right) - \nabla p_i - \boldsymbol{R} \tag{4.3}$$

$$m_e n_e \left(\frac{\partial}{\partial t} + \boldsymbol{v}_e \cdot \nabla \right) \boldsymbol{v}_e + \nabla \cdot \boldsymbol{\pi}_e = -e n_e \left(\boldsymbol{E} + \frac{\boldsymbol{v}_e}{c} \times B \right) - \nabla p_e + \boldsymbol{R} \tag{4.4}$$

$$\frac{3}{2} n_i \left(\frac{\partial}{\partial t} + \boldsymbol{v}_i \cdot \nabla \right) T_i + n_i T_i \nabla \cdot \boldsymbol{v}_i = -\nabla \cdot \boldsymbol{q}_i - \boldsymbol{\pi}_i : \frac{\partial \boldsymbol{v}_i}{\partial \boldsymbol{x}} + Q_i \tag{4.5}$$

$$\frac{3}{2} n_e \left(\frac{\partial}{\partial t} + \boldsymbol{v}_e \cdot \nabla \right) T_e + n_e T_e \nabla \cdot \boldsymbol{v}_e = -\nabla \cdot \boldsymbol{q}_e + Q_e \tag{4.6}$$

and the quasineutrality condition

$$n_i = n_e \tag{4.7}$$

for singly charged ions. The case of two ion species is treated in Chapter 9.

The effects of particle collisions are contained in the frictional force $\boldsymbol{R}$, the momentum flux $\boldsymbol{\pi}$, the heat flux $\boldsymbol{q}$ and the heat generation Q.

The momentum equation Eq. (4.4) for the electrons will often be written in the limit that the electron mass $m_e \to 0$. This is a good approximation for low-frequency waves when $k_\perp^2 c^2/\omega_{pe}^2 \ll 1$. There are exceptions where the electron inertia is important in both Alfvén waves and tearing modes, as described in detail in Chapter 15.

As is well known from collisional transport theory [Balescu (1988); Helander and Sigmar (2002)], the form of the collisional transport $\boldsymbol{R}, \boldsymbol{q}, \boldsymbol{\pi}$ fluxes depends strongly on the character of the particle orbits between collisions. In the collisional regime defined by $\omega_b < \nu_{\rm col} < \omega_{ci}$ where the collision frequency is intermediate to the average particle bounce frequency ω_b and the cyclotron frequency ω_{ci}, the collisional fluxes are taken from the approximation that $\boldsymbol{B} = \text{const}$. The bounce frequency ω_b arises from the reflection of the parallel motion of the particles by the mirror force $F_\parallel = -\mu \nabla_\parallel B$. In the constant B limit, where the orbits are helical gyro-orbits with constant parallel speeds, the collisional fluxes [Braginskii (1965)] parallel to $\boldsymbol{B}$ and the finite gyroradius fluxes (FLR) across $\boldsymbol{B}$ are

$$\boldsymbol{R} = en\left(\eta_\| \boldsymbol{j}_\| + \eta_\perp \boldsymbol{j}_\perp\right) - \beta n \nabla_\| T_e \tag{4.8}$$

$$\boldsymbol{\pi} = \boldsymbol{\pi}_{\text{FLR}} - \frac{nT_i}{\nu_i} \nabla_\| v_{\|i} \tag{4.9}$$

$$\boldsymbol{q}_e = -\beta \frac{T_e}{e} \boldsymbol{j}_\| - \alpha \frac{nT_e}{m_e \nu_e} \nabla_\| T_e - \frac{5}{2} \frac{cnT_e}{eB} \widehat{\boldsymbol{b}} \times \nabla T_e \tag{4.10}$$

$$\boldsymbol{q}_i = -\gamma \frac{nT_i}{m_i \nu_i} \nabla_\| T_i + \frac{5}{2} \frac{cnT_i}{eB} \widehat{\boldsymbol{b}} \times \nabla T_i \tag{4.11}$$

$$Q_i = \Delta Q = 5.9 \frac{m_e}{m_i} n \nu_e (T_e - T_i) \tag{4.12}$$

$$Q_e = \frac{\boldsymbol{R} \cdot \boldsymbol{j}}{n_e} - \Delta Q = \eta_\| j_\|^2 + \eta_\perp j_\perp^2 - \beta \frac{j_\| \nabla_\| T_e}{e} - \Delta Q \tag{4.13}$$

where α, β, γ are numerical coefficients of order unity. The parallel resistivity is given by $\eta_\| = m_e \nu_e / ne^2$ and $\eta_\perp = 2\eta_\|$. For reference, the collision frequency ν_e defined here is 0.51 times the collision frequency defined in Braginskii (1965) and $\alpha = 1.6$, $\beta = 0.71$ and $\gamma = 3.0$. Note that this transport coefficient $\gamma = 3.0$ defined by Braginskii is distinct from the growth rate symbol conventionally taken as γ and distinct from the adiabatic gas constant γ. Capital Γ is used for the adiabatic gas constant and seldom need the constant symbol γ in the Braginskii ion heat flux equation occurring in Eq. (4.11).

Before linearizing the equations, some reductions are made in the transport equations appropriate for the low frequency modes. For the modes considered we have $\omega/\omega_{ci} \sim \varepsilon^2$ and $k_\perp a_i \sim \varepsilon \ll 1$, and the perpendicular particle fluxes are solved for by iteration in ε. The first-order ion and electron particle fluxes are

$$n\boldsymbol{v}_{\perp i,e}^{(1)} = \frac{nc\boldsymbol{E} \times \boldsymbol{B}}{B^2} \pm \frac{c\boldsymbol{B} \times \nabla p_{i,e}}{eB^2}. \tag{4.14}$$

The divergence of the next order ion particle flux is given by

$$\nabla \cdot \left(n\boldsymbol{v}_{\perp i}^{(3)}\right) = \nabla \cdot \left[\frac{nm_i c^2}{eB^2}\left(\frac{\partial}{\partial t} + \boldsymbol{v}_i^{(1)} \cdot \nabla\right) \boldsymbol{E}_\perp\right] \tag{4.15}$$

where in calculating this term the convective derivative and the finite gyroradius part of the momentum tensor $\boldsymbol{\pi}_{\text{FLR}}$ in Eq. (4.3) are combined as discussed in detail in Smolyakov (1998) and for drift waves in Hinton and Horton (1971). In a constant magnetic field similar reductions occur in the finite gyroradius contributions to the heat balance equations. To the lowest order in the aspect ratio, where $|\boldsymbol{B}| =$const. the equations governing the perturbations can be written as follows:

$$\frac{\partial n_i}{\partial t} + \frac{c\boldsymbol{E} \times \boldsymbol{B}}{B^2} \cdot \nabla n_i + \nabla \cdot \left(\frac{n_i m_i c^2}{eB^2}\left(\frac{\partial}{\partial t} + \boldsymbol{v}_i^{(1)} \cdot \nabla\right) \boldsymbol{E}_\perp\right)$$

$$+ \frac{\partial}{\partial x_\|}(n_i v_{\|i}) = 0 \tag{4.16}$$

$$\frac{\partial n_e}{\partial t} + \frac{c\boldsymbol{E} \times \boldsymbol{B}}{B^2} \cdot \nabla n_e + \frac{\partial}{\partial x_\parallel}(n_e v_{\parallel e}) = 0 \tag{4.17}$$

$$m_i n_i \left(\frac{\partial v_{\parallel i}}{\partial t}\right) + \left(\boldsymbol{v}_E + v_{\parallel i}\widehat{\boldsymbol{b}}\right) \cdot \nabla v_{\parallel i} = -\nabla_\parallel (p_i + p_e)$$

$$+ \frac{\partial}{\partial x_\parallel}\left(\frac{n_i T_i}{\nu_i}\frac{\partial v_{\parallel i}}{\partial x_\parallel}\right) \tag{4.18}$$

$$j_\parallel = \frac{1}{\eta_\parallel}\left(E_\parallel + \frac{T_e}{e n_e}\frac{\partial n_e}{\partial x_\parallel} + \frac{(1+\beta)}{e}\frac{\partial T_e}{\partial x_\parallel}\right) \tag{4.19}$$

$$\frac{3}{2} n_i \left(\frac{\partial T_i}{\partial t} + (\boldsymbol{v}_E + v_{\parallel i}\widehat{\boldsymbol{b}}) \cdot \nabla T_i\right) + n_i T_i \frac{\partial v_{\parallel i}}{\partial x_\parallel}$$

$$= \frac{\partial}{\partial x_\parallel}\left(\frac{\gamma n_i T_i}{m_i \nu_i}\frac{\partial T_i}{\partial x_\parallel}\right) + \Delta Q \tag{4.20}$$

$$\frac{3}{2} n_e \left(\frac{\partial T_e}{\partial t} + (\boldsymbol{v}_E + v_{\parallel e}\widehat{\boldsymbol{b}}) \cdot \nabla T_e\right) + n_e T_e \frac{\partial v_{\parallel e}}{\partial x_\parallel} = \frac{\partial}{\partial x_\parallel}\left(\alpha \frac{n_e T_e}{m_e \nu_e}\frac{\partial T_e}{\partial x_\parallel}\right)$$

$$+ \frac{\beta}{e} T_e \frac{\partial j_\parallel}{\partial x_\parallel} + \eta_\parallel j_\parallel^2 - \Delta Q. \tag{4.21}$$

The self-consistent field problem is closed by the condition of quasineutrality:

$$n_i = n_e \tag{4.22}$$

where n_i and n_e are computed from Eq. (4.16)-Eq. (4.21) with $\boldsymbol{v}_E = c\boldsymbol{E} \times \boldsymbol{B}/B^2$ and $\widehat{\boldsymbol{b}} = \boldsymbol{B}/|\boldsymbol{B}|$.

The linear solution of the self-consistent field equation gives the dispersion relations and the polarizations of the waves $\omega_{k\alpha} = \omega_{\boldsymbol{k}}(\alpha)$ and instabilities of wavenumber $\boldsymbol{k}$ and type α. For electromagnetic disturbances the field closure includes Ampere's law with the plasma electrical currents $j_\perp$ and $j_\parallel$ giving rise to the wave magnetic field components.

The derivation of the two-component fluid equations is complicated. These equations are widely known as the Braginskii equations from their publication in Braginskii (1965). A careful description and derivation is given in Chapter 4 of Helander and Sigmar (2002), along with a proof of the Onsager symmetries possessed by the transport matrix. A useful physical picture of the content of the transport equations along with references to other works on this problem, particularly those of Balescu, is given in Helander and Sigmar (2002).

The fundamental view of what these fluid equations mean in terms of the reduction of the many-body Coulomb interaction in statistical mechanics is developed in Elskens and Escande (2003). The relationship to the Hamiltonian formulation of the particle dynamics and the reduction to the Vlasov-Poisson self-consistent field equations is analyzed in detail by Elskens and Escande (2003) (Chapters 3,4 and 5). Their analysis is concerned with simpler uniform plasmas but shows the nature of the Landau damping to the singular eigenfunctions of the Vlasov-Poisson field equations and the role of Coulomb interactions on the dynamics in a fundamental way. Such a detailed formulation, however, is not practical for use in the study of drift waves in tokamak plasmas. Thus, in this chapter we reduce the particle simulations following the guiding-center orbits to a closed chain of fluid equations. These reduced partial differential equations for the fluid moments are found to be generally accurate for a variety of drift wave instabilities and turbulent transport. The complete analysis of the forces from the divergence of the FLR momentum stress tensor and the dynamics of the corresponding heat flux $\boldsymbol{q}$ is an intricate problem which is best left to the literature. A simple example is found in the drift wave work of Hinton and Horton (1971) for the Q-machine experiments. The general analysis including the FLR deduction from the dynamical equations for both the momentum stress tensor Π and the heat flux tensor q is given in Smolyakov (1998).

4.2 Current, Density and Temperature Gradient Driven Drift Modes

The density gradient, the temperature gradients, and the plasma current $j_\parallel$ in the equilibrium are driving forces for instability in the drift modes of the plasma. In addition to the driving forces, the form of instability is determined by the dissipative processes in the plasma. The role of the dissipative processes can be viewed as providing constraints in the dynamics of the plasma. For example, the degree of thermal transport determines the variation in the ion and electron temperatures in the wave. A fast thermal transport rate constrains the wave motion to occur at nearly constant temperature and eliminates some of the terms causing instability. In the limit of zero resistivity another constraint of zero net parallel force on the electron fluid is provided which eliminates the instability driven by the density gradient.

At high densities and low temperatures both ions and electrons behave adiabatically with $\omega \gg (k_\parallel^2 T_e/m_e\nu_e)$, $(k_\parallel^2 T_i/m_i\nu_i)$. At the higher temperatures of interest in fusion and space physics, however, the electron transport of heat becomes rapid compared with the wave frequency, and electrons behave approximately isothermally. For sound waves, for example, the wave frequency is $\omega \simeq k_\parallel v_i$, and the ions respond adiabatically for $k_\parallel \lambda_{\text{mfp}} < 1$, and the electrons respond approximately isothermally for $k_\parallel \lambda_{\text{mfp}} > (m_e/m_i)^{1/2}$. In the following analysis this domain of slightly collisional waves defined by $(m_e/m_i)^{1/2} < k_\parallel \lambda_{\text{mfp}} < 1$ is examined in detail.

In the limiting case of perfect electron transport and negligible thermal ion transport along the magnetic field, the dispersion relation obtained is

Reactive Limit of Dispersion Relation

$$D_R(\omega)=\omega^2-\frac{k_\parallel^2}{m_i}\left(T_e+\frac{5}{3}T_i\right)-\omega\left[k_\perp v_{de}-(k\rho_s)^2\left(\omega-k_\perp v_{di}\right)\right]$$
$$+\frac{2}{3}\frac{k_\parallel^2T_i}{m_i}\frac{k_\perp v_{de}}{\omega}\left(1-\frac{3}{2}\eta_i\right)-\frac{2}{3}\frac{k_\parallel^2T_i}{m_i}(k_\perp\rho_s)^2\left(1-\frac{k_\perp v_{di}}{\omega}\right)=0. \tag{4.23}$$

The first two terms determine the form of the compressible sound wave in the magnetized plasma. with isothermal electrons and adiabatic ions. The three modes in Eq. (4.23) arise from the time dependence of $\widetilde{n},\widetilde{v}_{\parallel i}$, and $\widetilde{T}_i$. The fast electron thermal transport makes $\widetilde{T}_e$ negligible for these lower frequency modes.

4.2.1 *Ion acoustic waves and the thermal mode*

In the absence of a temperature gradient ($\eta_i = d\ell n\, T_i/d\ell n\, n_i = 0$) the three roots of Eq. (4.23) are real. For cold ions the modes are the coupled drift-ion acoustic waves with

$$\omega=\frac{1}{2}\left[k_y v_{de}\pm\left((k_y v_{de})^2+4k_\parallel^2c_s^2\right)^{1/2}\right]. \tag{4.24}$$

For finite ion temperature an additional stable mode arises from $\partial\widetilde{T}_i/\partial t$ and has a frequency $\omega\simeq k_y v_{di}$ for small $k_\parallel$ and a frequency $\omega\simeq\frac{2}{5}k_\perp v_{de}/(1+\frac{3}{5}T_e/T_i)$ for large $k_\parallel$.

4.2.2 *Ion temperature gradient instability*

In the presence of an ion temperature gradient such that $\eta_i>2/3$, an unstable mode appears in Eq. (4.23). For small $k_\parallel$ the mode is purely growing with

$$\gamma\sim|k_\parallel|\,v_i\left(\eta_i-\frac{2}{3}\right)^{1/2}. \tag{4.25}$$

As $k_\parallel$ increases the mode has a growth rate

$$\gamma\sim\left(|k_\parallel|\,v_i\right)^{2/3}(k_\perp v_{de}\eta_i)^{1/3}\quad\text{with}\quad\gamma_{\max}\lesssim k_\perp v_{di}\eta_i. \tag{4.26}$$

The compressional effect of the ion-sound waves stabilizes the ion temperature gradient mode for

$$k_\parallel c_s>k_\perp v_{di}\eta_i. \tag{4.27}$$

The use of magnetic shear to achieve the necessary $k_\parallel$ to stabilize the mode as given by condition Eq. (4.27) is found [Coppi, *et al.*, 1968a; Coppi, *et al.* (1968b)] to require an order unity value for the magnetic shear parameter $S=L_n/L_S$. Such a large magnetic shear is outside the range of the perturbation expansion in power of $(k_x\rho_i)^2$. This shear stabilization problem leads to the development of integral eigenvalue formulations of the problem by Dong, *et al.* (1992) and Dong and Horton (1993). Thus, the ion temperature gradient mode is one of the fundamental means of driving the low-frequency plasma turbulence universally seen in laboratory confinement devices. Part of the slowing down in the progress between 1970 to the present time towards reaching a fusion reactor can be attributed to the fact that as more and more ion heating is applied to driving up the ion nuclear fusion rate, the ion temperature gradient turbulence becomes stronger and stronger. One solution to this problem seems to be to keep the parameter $\eta_i < 2/3$ by peaking the density profile. Experiments in C-Mod Alcator [Greenwald, *et al.* (2007)] with peaked density profiles from pellet injection seem to confirm this method for achieving improved thermal confinement. Recall that $\eta_i = \frac{d\ln T_i}{d\ln n_i}$ and the nature of these unstable modes is analyzed in detail in Chapter 13.

4.3 Closure Models for Coupled Chain of Fluid Moments

The use of partial differential equations rather than phase space densities from the Vlasov-Maxwell system of kinetic equations as used in Chapter 2, depends on a truncation of the chain of derived fluid moments from the kinetic equations. This is a well-known problem explored for example by Balescu (1988) and many other authors. The problem is universal to both laser plasma for inertial confinement, magnetic confinement and space-astrophysical problems. Here we discuss the methods used for the drift wave turbulence. Other methods apply to shocks and inertial confinement problems.

4.3.1 *Closure models for the chain of the fluid moments*

4.3.1.1 *Examples of heat flux problem in fluid closures*

In short-mean-free-path $\lambda_{\rm mfp}$ plasmas the heat flux $\boldsymbol{q}_i(x,t)$ is calculated by an expansion in the small parameter $k_\parallel \lambda_{\rm mfp}$. Even in this regime this calculation of the heat flux $q_\parallel$ is a complicated problem owing to the differential velocity space operators describing the Coulomb collisions of the charges. In the presence of waves there are boundary layers in phase space associated with the differential operators describing the Coulomb collisions. The degree of collisionality depends on the energy of the particle. The response problem is treated in Chapter 13 for the trapped electron drift waves. There is, however, a well-accepted and widely-used set of transport formulas given by Braginskii (1965) for the collisional dominated regime with small $k_\parallel \lambda_{\rm mfp}$. Thus, in this chapter we discuss some problems in terms of the Braginskii fluid equations. In the collisional regime the parallel ion transport is given by

$$q_{\|} = -3.9\frac{n_i T_i}{m_i \nu_i}\nabla_{\|} T_i. \tag{4.28}$$

For electron thermal transport there is a similar formula for the thermal flux with coefficient α in Eq. (4.10), but there is the additional term in showing thermal-electric effect given by the coefficient $\beta = 0.71$ in Eq. (4.10). The conjugate effect appears in Eq. (4.13).

The problem of nonlinear simulations of ITG turbulence [Hamaguchi and Horton(1990)] uses a generalized form of the heat flux in Eq. (4.28) with a limit on the smallness of the collision frequency ν_i, such that that its minimum value is given by the validity condition formula with a small value of ν_i such that $\max(k_{\|} v_i) = \nu_i$. This cut-off method is known to be an approximation for a more systematic determination of the fluctuating parallel thermal flux.

Hammett and Perkins (1990) presented the following simple long mean free path generalization

$$\text{Div } q_{\boldsymbol{k}\|} = -2\left(\frac{2}{\pi}\right)^{1/2} n_i \left|k_{\|}\right| v_i \delta T_i(\boldsymbol{k}). \tag{4.29}$$

The transform of Eq. (4.29) back to x-space from k-space is a Hilbert transform of the temperature fluctuation. The formula in Eq. (4.29) was derived by considering those truncations of the linear Vlasov-Poisson eigenvalue problem with the corresponding fluid equations that were accurate representations of the linear Vlasov growth rates. Effectively, the full plasma dispersion function $Z(\phi)$ is approximated by first-order rational polynomial called a Pade approximate. The H-P formula Eq. (4.29) can also be understood from making a series of reasonable approximations from a more general non-Markovian transport theory developed by Chang and Callen (1992) and Held, *et al.* (2001). This formula Eq. (4.29) has the advantage of providing a simple computational algorithm that accounts for the kinetic effects. Due to these desirable practical properties this closure model is widely used and the method generalized to toroidal fluid equations. In the generalization the method is called the Landau fluid system of equations. Waltz, *et al.* (1990), Dorland and Hammett (1993), Beer, *et al.* (1997), Beer and Hammett (1999) and others, have reported numerous ITG and trapped electron mode fluid simulations using the Landau fluid model.

The validity of this approach with regard to the nonlinear problem was the focus of a debate in which Mattor and Parker (1997, 1999) raised serious issues regarding the use of the Landau fluid model. Their argument started with the assertion that the Eq. (4.29) formula does not represent the weak turbulence formulas for the nonlinear mode coupling processes-especially the process called shielded Compton scattering of the fluctuations by the particles. The arguments then expanded to note that the equations have a high level of dissipation throughout $k_{\|}$-space and thus the entropy balance equations are not well represented. The closure issues is taken

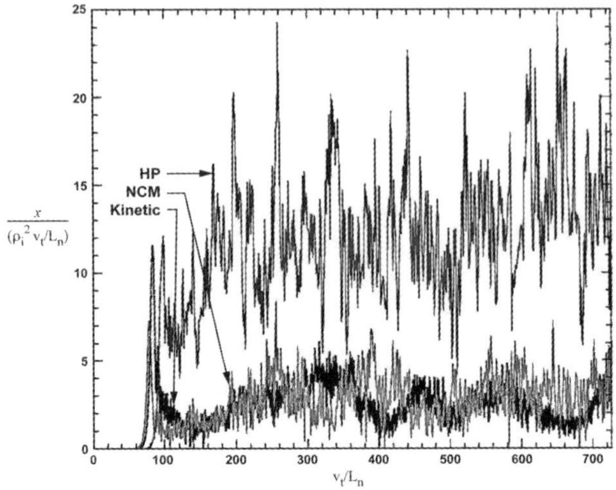

Fig. 4.1 Normalized turbulent ion heat diffusivity $\chi/(\rho_i^2 v_t/L_n)$ as a function of normalized time $v_t t/L_n$ obtained by the two-dimensional slab ITG turbulence simulation [Sugama, *et al.* (2007)]. Lines correspond to results from the kinetic simulation, the fluid simulation using the NCM, and that using the Hammett-Perkins (HP) model, respectively.

up by Watanabe, *et al.* (2000) and Watanabe and Sugama (2002) who performed numerical comparisons with exact solutions of the full phase space distribution functions to test the accuracy of Eq. (4.29) for the closure problem. In the course of that research they developed an alternative closure model that more faithfully reproduces the conservation of entropy intrinsic to the collisionless Vlasov equation. This closure model is called the Nondissipative Closure Model or NCM to denote the capturing of key features of the collisionless Vlasov-Poisson system of field equations.

Figure 4.1 shows a comparison of the development of the slab ITG turbulent ion thermal diffusivity χ_i for the two closure models compared with values derived from the solution of the continuum Vlasov equation. The higher level of dissipation in the HP closure mode results in an artificially high and steady level of the thermal flux and diffusivity. The NCM closure model produces a more intermittent thermal flux with inward and outward bursts of larger convection events. The NCM evolution for χ_i is closer to the time trace of χ_i computed from the exact solution of the Vlasov equation.

In Fig. 4.1 we show some of the simulations of Watanabe and Sugama (2002) and that address these issues. Watanabe uses a high resolution direct numerical simulation of the drift kinetic equation to compute the ITG nonlinear turbulence. Watanabe compares the results of the simulation with the Landau fluid result based on the Hammett-Perkins formula and the actual simulation results. Sugama and Watanabe also introduces an improved closure formula that captures the time reversibility of the underlying kinetic equation and at the same time gives a good representation of the linear growth rate of the ITG modes. The key difference is to retain the damped modes of the plasma dispersion function defined and analyzed in Worked Problem 2.2.

When the velocity moments of the original gyrokinetic equation are taken to derive the gyrofluid equations, we need to construct closure models which describe high-order moment variables arising from kinetic effects such as parallel streaming, finite gyro-radii, and toroidal magnetic drift. Various aspects of the closure models for low collisionality regimes are discussed in a series of works by Hammett and Perkins (1990), Dorland and Hammett (1993), and Beer, *et al.* (1997). In these models, high-order moment variables are written in forms given by linear combination of low-order moment variables so as to reproduce well the linear kinetic dispersion relation. In the Hammett-Perkins model, the parallel heat flux $q_{\|k}$ with the wavenumber vector k is given in terms of the temperature fluctuation T_k as

$$q_{\|k} = -2\sqrt{\frac{2}{\pi}}\, n_0 v_t \, i\frac{k_\|}{|k_\||} T_k \tag{4.30}$$

where $k_\|$ is the wavenumber in the direction parallel to the magnetic field. The inverse Fourier transform of $ik_\|/|k_\|| = i\text{sg}(k_\|)$ is the Hilbert integral transform operator. More detailed discussion of the Hilbert operator for Landau clamping is in Chapter 5 of Horton and Ichikawa (1997).

The Hammett-Perkins model is applied to calculation of frequencies and growth rates of linear ITG modes while it does not describe time-reversal symmetric behaviors of the three-mode nonlinear kinetic ITG system [Mattor (1997, 1999)] which consists of two unstable slab ITG modes and a stable mode representing a quasi-linear flattening of the background temperature profile. Mattor and Parker (1997, 1999) presented a nonlinear closure model and successfully reproduced the solution of the three-mode ITG problem.

Realizing that the time-reversal symmetric solution of the three-mode ITG problem can be represented by the superposition of an unstable normal-mode distribution function and its conjugate damped mode, Sugama, *et al.* (2001, 2003, 2007) derived the nondissipative closure model (NCM) which relates the parallel heat flux $q_{\|k}$ in the linearly unstable wavenumber region to the temperature fluctuation T_k and the parallel flow u_k in terms of real-valued coefficients,

$$q_{\|k} = C_{T_k} n_0 v_t T_k + C_{u_k} n_0 T_i u_k \quad \text{(for linearly unstable modes)} \tag{4.31}$$

where the real-valued coefficients C_{Tk} and C_{uk} are determined by requiring that the kinetic dispersion relation should be derived from the closure relation. The NCM gives the correct relation between $q_{\|k}$, T_k, and u_k both for the unstable normal-mode solution and its complex-conjugate solution as well as for any linear combination of these solutions.

Since the time reversal symmetry of the original collisionless kinetic equation is retained in Eq. (4.31), the NCM can exactly reproduce the solution of the three-mode ITG problem. The three-wave problem is used by Watanabe, *et al.* (2000) to verify the NCM closure method.

Subsequently, the NCM was used in two-dimensional fluid simulations of the strong turbulence driven the slab ITG modes and its validity for quantitative prediction of the anomalous ion thermal transport in the slab ITG turbulence was confirmed by comparison to collisionless kinetic simulation results as shown in Fig. 4.1 [Sugama, *et al.* (2003)].

In the slab ITG turbulence shown in Fig. 4.1, the zonal-flow component was not included. With zonal flow only quiet steady states with turbulence transport suppressed by large zonal-flow generation were obtained by both kinetic and fluid simulations when including the zonal flow component [Watanabe and Sugama (2002)]. In toroidal configurations, zonal-flow generation is weaker than in slab cases although zonal flows are still an important factor to regulate turbulent transport. Rosenbluth and Hinton (1998) identified the zonal flows as a cause for the difference between gyrokinetic and gyrofluid simulation results in the toroidal ITG turbulence.

Additional closure models, which give residual zonal flows derived by Rosenbluth and Hinton (1998), were analyzed by Beer and Hammett (1999) and Sugama, *et al.* (2007).

4.4 Pressure Gradient Driven Instabilities

As shown in Fig. 1.5 in Chapter 1 there is a pressure gradient driven instability which has the curved magnetic field producing an effective gravity from the associated charge dependent drifts of the ion and electron guiding centers in magnetized plasmas. This instability is known as the resistive-g instability since the mathematical structure and the physics is closely analogous to the gravitation instabilities known as the Rayleigh-Taylor instability. In the toroidal geometry the effective gravity is given by

$$\boldsymbol{g} = -\left(\left\langle v_\parallel^2 \right\rangle + \frac{1}{2}\left\langle v_\perp^2 \right\rangle\right)\boldsymbol{\kappa} \tag{4.32}$$

where $\boldsymbol{\kappa}$ is the magnetic curvature vector given by

$$\boldsymbol{\kappa} = (\widehat{\boldsymbol{b}}\cdot\nabla)\widehat{\boldsymbol{b}} = -\frac{\widehat{\boldsymbol{n}}}{R_c} \tag{4.33}$$

where $\widehat{\boldsymbol{b}} = \boldsymbol{B}(\boldsymbol{x})/|B|$ is the unit vector along the magnetic field. This effective g-vector has a large horizontal component from the toroidal magnetic field and a small radially outward component from B_θ^2/r. Both components produce the effect of driving the convection cells that produce the interchange of hot core plasma with outer cooler plasma. The acceleration $\boldsymbol{g}$ is produced by the particle drifts in the nonuniform magnetic fields. As shown in Fig. 1.5 in Chapter 1, the resulting time scale for growth of the convective cells is the geometric mean of the pressure gradient scale length L_p and the radius of curvature R_c of the magnetic field divided by the sound speed in the plasma. This time for growth of a small fluctuation is on the scale of microseconds in MHD unstable plasmas. Thus, for the establishment of a confined plasma state in which drift waves exists the magnetic geometry must have MHD stability to the interchange or magnetic Rayleigh-Taylor instability. In Chapter 5 we show examples where there are regions outside a critical layer which are MHD unstable so the qualitative differences in the confined interior plasma and MHD unstable exterior plasma are clearly manifested.

Here we consider what happens after the system is stable to the interchange instability but the plasma is described by the transport equations in this chapter, namely the resistive-viscous two-component plasma transport equations. The plasma has a remnant of the Rayleigh-Taylor instability but with a growth rate controlled by a hybrid exponent of the Rayleigh-Taylor growth rate and the resistive diffusion rate for the magnetic field to diffuse a localized plasma current density $j_{\|}$. The parallel plasma current controls the rate of change of the plasma vorticity and thus the creation and decay of the vortex structures in the stabilized Rayleigh-Taylor system. Drift wave mode structures spanning the region from the hot core plasma to the cooler edge plasma remain in the plasma.

Essentially all methods of stabilizing MHD instabilities rely on the restoring force from bending the magnetic field exceeding the destabilizing force from the perturbed pressure gradient. This force condition translates directly into an energy condition. The MHD stability is often presented in terms of an energy principle for the dynamics since this gives information without solving the eigenvalue-eigenvector problem for the linear wave equations. For drift waves the energy principle is not so useful and for transport calculations one needs the eigenvalues and eigenvectors.

The basic method used to produce stability to interchange instabilities on the fast time scale is to use the principle that for fast motions the plasma carries the magnetic field with it from the frozen in flux condition (Kulsrud, 2005) of

$$\boldsymbol{E} + \boldsymbol{v} \times \boldsymbol{B} = 0 \tag{4.34}$$

which together with Faraday's law leads to

$$\frac{\partial \boldsymbol{B}}{\partial t} = \nabla \times (\boldsymbol{v} \times \boldsymbol{B}). \tag{4.35}$$

By introducing the displacement vector $\boldsymbol{\xi}(x,t)$ such that $\boldsymbol{v} = \partial \boldsymbol{\xi}/\partial t$, we can solve for $\delta \boldsymbol{B}$ from Eq. (4.35) to get

$$\delta B_r = (\boldsymbol{B} \cdot \nabla)\xi_r = iB\, k_{\|} \xi_r. \tag{4.36}$$

The radial acceleration of the plasma is given by

$$\rho \frac{\partial^2 \xi_r}{\partial t^2} = F_r = \frac{(\boldsymbol{B} \cdot \nabla)^2 \xi_r}{\mu_0} - \frac{2\delta p}{R_c} \tag{4.37}$$

giving Alfvén waves in the absence of the pressure fluctuation δp and the field line curvature $2/R_c$. The pressure fluctuation is given by convection $\xi_r\, dp_0/dr$ and compression $p_0\nabla\cdot\boldsymbol{\xi}$. For incompressible motion with $\nabla\cdot\boldsymbol{\xi}=0$, the radial force reduces to

$$F_r = \left(-\frac{k_\parallel^2 B^2}{\mu_0} - \frac{2}{R_c}\frac{dp_0}{dr}\right)\xi_r \tag{4.38}$$

where the first term is a restoring force producing Alfvén oscillations at the frequency

$$\omega_A^2 = \frac{k_\parallel^2 B^2}{\mu_0 \rho} \tag{4.39}$$

and the second force drives the interchange instability with

$$\omega^2 = -\gamma^2 = \frac{2}{R_c}\frac{1}{\rho}\frac{dp_0}{dr} \tag{4.40}$$

where $\xi_r = \xi_r(0)\exp(\gamma t)$.

For radially distributed motions we construct the energies from the plasma motion

$$W_{\rm alf} = \frac{1}{2\mu_0}\langle\delta B_\perp^2\rangle = \frac{B^2}{2\mu_0}\left\langle\left(k_\parallel\xi_r\right)^2\right\rangle \tag{4.41}$$

$$W_{\rm int} = -\left(\frac{2}{R_c\rho}\frac{dp_0}{dr}\right)\langle\xi_r^2\rangle \tag{4.42}$$

and kinetic energy

$$W_{KE} = \frac{1}{2}\left\langle\rho_0\left(\frac{\partial\xi_r}{\partial t}\right)^2\right\rangle. \tag{4.43}$$

The interchange motion releases the energy $W_{\rm int}$ so the system is stable when the increase in the magnetic energy $W_{\rm alf}$ is greater than the energy released by the convection across the pressure gradient $W_{\rm int}$. This, is the basic method by which all magnetic confinement systems toroidal and mirror geometries provide MHD stability. The $W_{\rm int} < W_{\rm alf}$ is satisfied when the pressure gradient, and thus the maximum pressure itself p, is less than the magnetic energy density $W_{\rm mag}$ times the geometrical factors controlling the minimum parallel wave number $k_\parallel$ and the pressure gradient scale length L_p and the radius of curvature R_c of the magnetic field lines. The minimum $k_\parallel$ is can be controlled in number of ways. In the tokamak the value is set by the major radius and the safety factor $q(r)$ and the magnetic shear $s(r)$ defined by

$$\begin{aligned} q(r)R &= \text{length to go once around the torus following the magnetic field line} \\ &= \frac{rB_\phi}{B_p} \end{aligned} \tag{4.44}$$

and the magnetic shear function

$$s(r) = \frac{r}{q}\frac{dq}{dr}. \tag{4.45}$$

where B_ϕ and B_p are the toroidal and poloidal magnetic fields at minor radius r from the magnetic axis. The connection length is defined as $L_c = qR/2$ which separates the region of favorable curvature on the inside of the torus and the unstable region on the outside of the torus. This formula is widely used to estimate the parallel wavenumber $k_\parallel = 1/qR$ in both drift wave and MHD calculations. The formula is adequate on average but fails to recognize the role of the special rational magnetic surfaces at $r_{m/n}$ where $k_\parallel = 0$. Of course, this applies only at a point in $r = r_{m/n}$ and thus one needs a measure δ_r for the weight of the effect over the radial zone surrounding the rational surface. The weight is the width of the eigenmode that exists from the dynamical equations for the unstable motion. For the drift waves we found the mode width in Chapters 2.6 and 2.7. The width was proportional to the inertial scale length ρ_s and inversely proportional to the magnetic shear parameter s. Thus, for low magnetic shear the mode width is wide and there is a large radial width to the interchange motion. While for strong magnetic shear the mode is narrow and the interchange motion is restricted to be close to the rational surface. When these considerations and factors are used in the estimates for the released energies in the equation $W_{\rm int} < W_{\rm alf}$, we arrive at the limit on the plasma pressure-to magnetic pressure given by

$$\beta_{\rm tor} = \frac{2\mu_0 p}{B^2} = \frac{L_p R_c}{L_c^2} = \frac{L_p}{q^2 R} \tag{4.46}$$

for the toroidal confinement system.

For the mirror confinement geometry the situations changes as the minimum $k_\parallel$ is determined by the length of the field line set either by the distances to the conducting end walls or the distance to the stabilizing mirror cells. The radius of curvature R_c of the field lines also changes becoming infinite on the axis of the mirror cell and decreases with the radius r as one moves off the axis to give

$$\frac{1}{R_c} = \frac{r}{L_m^2} \tag{4.47}$$

as the curvature of the magnetic field lines. Repeating the balance condition in the $W_{\rm int} < W_{\rm alf}$ now yields the pressure limit

$$\beta_{\rm mirror} = \frac{L_p R_c}{L_m^2} = \frac{L_p}{r}. \tag{4.48}$$

This behavior is borne out in the Earth's magnetosphere where the plasma beta exceeds unity in much of the nightside plasma sheet. When the magnetic fields lines become stretched, the curvature is the strongest in the region where the plasma $\beta \sim 1$ becomes unstable [Crabtree, *et al.* (2003); Horton, *et al.* (2005)] with the interchange motions as observed by spacecraft plasma instruments.

For helical toroidal systems like the LHD, the pitch of the helical magnetic field is controlled by the external coils and fixed. There is a rational surface corresponding to the $q_{\rm ext} = 2$ winding pitch of the magnetic field. In this case the plasma pressure limit given by Eq. (4.46) reduces to the simple result of

$$\beta_{\rm LHD} = \frac{a}{2R}. \tag{4.49}$$

For LHD this limit is a volume averaged $\beta = 0.05$ which seems to be a firm limit in the plasma pressure achieved in LHD [Yamada, *et al.* (2004)].

Now, we consider that the plasma pressure is well below the limits set by Eq. (4.46) and see how the effect of finite resistivity produces the resistive-g instabilities. These resistive-g modes are important in the outer layers of the tokamak where the plasma is weakly collisional. The multi-mode code used in Chapters 16 and 18 for transport includes transport formulas for the resistive-g and these formulas and their generalizations are now used to predict the presence of edge localized modes, called ELMs for short, in toroidal confinement machines.

In Chapters 13 and 14 the stability analysis shows that magnetized plasmas with a significant temperature gradients across the magnetic field are a common situation owing to the thermal insulating properties of the magnetic field. Such temperature gradients occur in the magnetosphere, the solar corona and with particularly high gradients in the laboratory experiments aimed at magnetic confinement for nuclear fusion. Thus, we often have situations where the pressure gradient is below the critical value for the onset of the MHD modes, however, the temperature gradients are above the critical values required for the onset of intense thermal convection from the drift wave scale turbulence. The situations differ according to whether the gradient is the ion temperature gradient or the electron temperature gradient is driving the turbulent thermal convection. The ion temperature gradient case is known as the ITG mode and treated in Chapter 13 and the case of the electron temperature gradient driving the turbulence is called the ETG modes and treated in Chapter 14.

The simulations for transport must consider the stable MHD equilibrium with the resistive-g modes in the edge plasma, the ITG and ETG turbulence in the core plasma as intrinsic to determining the performance of a fusion machine like JET, and ITER. For reference Table 4.1 gives for JT60U some of the key parameters.

In the large tokamaks experiments reviewed in Chapter 1 we see that the core ion temperatures have reached 30 KeV in the TFTR device and 38 KeV in the JT60-U machine with $R/a = 3.1\,{\rm m}/0.7\,{\rm m} \simeq 4$. Thus, we know that the magnetic field can provide strong thermal insulation withstanding gradients greater than 30 KeV/m under ideal conditions of the closed irrational magnetic flux tubes. Table 4.1 gives the parameters for a well-known JT60-U shot that has a core ion temperature of 38 KeV and core electron temperature of 12 KeV.

The ion (deuterium) quantities T_i and v_ϕ are inferred from spectroscopic measurement of carbon ions. There are few direct measurements of the fully ionized

Table 4.1 JT60-U high-β_p experiment with internal transport barrier: parameters in Phase I of shot 17110.

R/a	$3.1\,\mathrm{m}/0.7\,\mathrm{m}$
B_ϕ	$4.4\,\mathrm{T}$
I_p	$2\,\mathrm{MA}$
P_{NBI}	$27\,\mathrm{M}$
$n_D(0)$	$4.1 \times 10^{19}\,\mathrm{m}^{-3}$
$T_i(0)/T_e(0)$	$38\,\mathrm{KeV}/12\,\mathrm{KeV}$
$n_D \tau_E T_i$	$1.1 \times 10^{21}\,\mathrm{m}^{-3} \cdot s \cdot \mathrm{KeV}$
$v_\phi(0)$	$-100\,\mathrm{km/s}$

hydrogen isotopes in the tokamak plasma other than the neutron yields and the gamma rays from the nuclear reactions.

The whole concept of magnetic confinement for thermonuclear fusion of plasmas depends on this thermal insulating principle. From the deuterium-tritium reactivity function shown in Fig. 1.1 one finds that the core ion plasma temperatures of order $10^8\,\mathrm{K}$ are required for significant fusion nuclear reactivity. This high temperature plasma must be partially insulated from the vessel walls which must remain below $5000\,\mathrm{K}$ to retain their integrity. Thus, steep temperature gradients of order $10\,\mathrm{KeV/m}$ must be stably maintained in the thermonuclear fusion device. The associated ion thermal diffusivity χ_i is then of order or less than $1\,\mathrm{m}^2/\mathrm{s}$ in devices with minor radius of a few meters.

Such magnetic thermal insulation is consistent with the results from the latest generation of tokamak and spherical tokamak experiments under favorable conditions. The thermal flux across the magnetic surfaces is large $q_i = -n_i \chi_i dT_i/dr$ reaching $0.1\,\mathrm{MW/m}^2$ and requiring large heating rates of many MW to maintain the high temperature plasma against turbulent losses. For the alpha particle decay products from the fusion reactions to provide the power to maintain the ion temperature the core electron temperature T_e must also be sufficiently high in the $10\,\mathrm{KeV}$ range. Chapter 14 is devoted to instabilities driven by the electron temperature gradient and the associated turbulent transport of electron thermal energy across the magnetic flux tubes. This electron thermal transport problem is more difficult to control in the magnetic confinement experiments than the ion thermal turbulent transport as emphasized by Kadomtsev in his monograph [Kadomtsev (1992)].

In both laboratory plasmas and in astrophysical plasmas where a strong local heating takes place there is clearly a potential source of plasma instability. There can be MHD (single component conducting fluid in Chapter 6) instabilities controlled by the constraints of frozen-in magnetic flux and the adiabatic equation of state. These fast instabilities limit the plasma pressure gradients with respect to the confining magnetic pressure. These fast, single fluid instabilities have be eliminated in the magnetic confinement devices proposed for nuclear fusion. Here, we consider that the plasma is stable to ideal MHD motions and study the remaining smaller scale drift wave instabilities. The drift waves have dispersive frequencies and finite parallel electric fields which means the associated waves and vortices have resonant wave-particle interactions.

The drift wave dispersion relation in a collisionless plasma does not assume a local thermal Maxwell-Boltzmann distribution function. In the high temperatures of fusion and geospace plasmas the particle distributions in energy typically have high energy tails and different effective temperatures across and parallel to the magnetic field. In low beta $\beta = p/(B^2/2\mu_0)$ plasmas the ion temperature gradient instability, known as the ITG mode, has nearly an electrostatic polarization for the coupled drift waves and the ion acoustic waves that gives rise to the collective instability. In higher pressure plasmas with $\beta > m_e/m_i$ the ITG mode couples with the shear Alfvén wave becoming a dispersive oscillation with an electromagnetic polarization. As the plasma pressure increases the inductive electric field from the time changing δB_x begins canceling part of the electrostatic component of the parallel electric field. This cancelation reduces the energy transfer rate $\langle j_\parallel E_\parallel \rangle$ and reduces the growth rate of the instability. There are numerous studies of these weakly electromagnetic ITG modes including Rewoldt, *et al.* (1999), Hong, *et al.* (1989a,b) that detail how the ITG modes changes with increasing plasma beta. At certain intermediate values of plasma beta both the ITG and the kinetic ballooning mode of the MHD polarization are present with different mode frequencies and growth rates.

4.4.1 *Scale invariance properties arising from an Ohm's law with electron inertia*

The dynamics for electrostatic pressure gradient driven turbulent fluctuations for a large aspect ratio tokamak (in a coordinate system $r, \theta, \phi = \zeta$ where r and θ are cylindrical coordinates about the magnetic axis and ζ is the toroidal angle) can be written: vorticity equation

$$\frac{\rho}{B_0}\frac{d}{dt}\boldsymbol{\nabla}_\perp^2 \delta\phi - \boldsymbol{B}\cdot\boldsymbol{\nabla}\delta j_\parallel = \frac{\widehat{e}_\zeta \cdot \boldsymbol{\nabla} R^2 \times \boldsymbol{\nabla}\delta p}{R^2} \tag{4.50}$$

with incompressible convection

$$\frac{d}{dt}\delta p = 0 \tag{4.51}$$

of the pressure fluctuation. Here $\widehat{e}_\zeta = \widehat{e}_\phi$ is the unit vector in the toroidal direction. In the right-hand side of Eq. (4.50) we use $\nabla \times \boldsymbol{E} = 0$ to reduce the divergence of the diamagnetic current.

For low-collision frequencies, the electron inertia becomes important so the collisionless Ohm's law

$$\frac{\boldsymbol{B}\cdot\boldsymbol{\nabla}\delta\phi}{B_0} = \frac{m_e}{ne^2}\frac{d}{dt}\delta j_\parallel \tag{4.52}$$

where

$$\frac{d}{dt} = \frac{\partial}{\partial t} + \frac{\boldsymbol{e}_\zeta \times \boldsymbol{\nabla}\delta\phi \cdot \boldsymbol{\nabla}}{B_0} \tag{4.53}$$

$$\boldsymbol{B} \cdot \boldsymbol{\nabla} = \frac{B_0}{R}\left(\frac{\partial}{\partial \zeta} + \frac{1}{q}\frac{\partial}{\partial \theta}\right) \qquad R = R_0 + r\cos\theta. \tag{4.54}$$

Here $\delta\phi, \delta j_{\|}$ and δp are the fluctuations in electrostatic potential, parallel current and pressure, respectively.

Considering fluctuations with short-perpendicular wavelengths comparable to the collisionless skin depth $\delta = c/\omega_{pe}$, we introduce

$$x = \frac{r - r_0}{\delta} \qquad y = \frac{r_0\theta}{\delta} \tag{4.55}$$

as local coordinates perpendicular to the magnetic field at a minor radius r_0. Defining dimensionless forms for the turbulent fluctuations $\delta\phi, \delta j_{\|}$ and δp by

$$\frac{\delta\phi}{B} = \frac{\delta^2}{Rq} v_A \delta\phi \qquad \delta j = \frac{B_0}{Rq\mu_0} j \qquad \delta p = \delta\left(-\frac{d\delta p_0}{dr}\right)\delta p \tag{4.56}$$

where the dimensionless time variable is $\tau = v_A t/Rq$, these nonlinear resistive-g equations can be simplified to

$$\frac{d}{d\tau}\nabla_\perp^2 \delta\phi = \left(\frac{\partial}{\partial\theta} + sx\frac{\partial}{\partial y}\right)\delta j + \alpha\left(\cos\theta\frac{\partial}{\partial y} + \sin\theta\frac{\partial}{\partial x}\right)\delta p \tag{4.57}$$

$$\frac{d\delta p}{d\tau} + \frac{\partial\delta\phi}{\partial y} = 0 \tag{4.58}$$

$$\left(\frac{\partial}{\partial\theta} + sx\frac{\partial}{\partial y}\right)\delta\phi = \frac{d}{d\tau}\delta j \tag{4.59}$$

where

$$\frac{d}{d\tau} = \frac{\partial}{\partial\tau} + \frac{\partial\delta\phi}{\partial y}\frac{\partial}{\partial x} \qquad \nabla_\perp^2 = \frac{d^2}{dx^2} + \frac{d^2}{dy^2}. \tag{4.60}$$

and the dimensionless pressure gradient is

$$\alpha = -2\mu_0 R_0 \left(\frac{dp_0}{dr}\right)\frac{q^2}{B_0^2}. \tag{4.61}$$

Scale invariance techniques [Connor (1993)] can be applied to these equations to derive the scaling of the thermal diffusivity. The invariance scaling gives that the diffusivity by χ is

$$\chi \sim \frac{\langle (r - r_0)^2 \rangle}{t} \sim \frac{v_A}{Rq}\left(\frac{c}{\omega_{pe}}\right)^2 \frac{\langle x^2 \rangle}{\tau}. \tag{4.62}$$

The solutions of Eq. (4.57), Eq. (4.58), and Eq. (4.59) must have the property

$$\frac{\langle x^2 \rangle}{\tau} = F(\alpha, s) \tag{4.63}$$

where F is, as yet, an arbitrary function.

The equations Eq. (4.57), Eq. (4.58), and Eq. (4.59) are invariant under a single scale transformation

$$\begin{array}{llll} \delta\phi \to \lambda^2 \delta\phi & \delta j \to \lambda \delta j & \delta p \to \delta p & x \to \lambda x \\ y \to \lambda y & \tau \to \lambda^{-1} \tau & \alpha \to \lambda^2 \alpha & \end{array} \tag{4.64}$$

which allows us to show that

$$F = \alpha^{3/2} f(s). \tag{4.65}$$

Following Connor (1993) we can use periodic equilibrium variation in θ and the assumption 'flute'-like so that a two-scale analysis along the field line to find the form of the function $f(s)$ can be determined. The analysis shows the $f(s) = \mathrm{const}/s^{1/2}$ and the function $F(\alpha, s)$ in Eq. (4.63) must be

$$F = \frac{c\alpha^{3/2}}{s^{1/2}} \tag{4.66}$$

in Eq. (4.63) with a constant c.

In terms of the g-modes the effect of unfavorable curvature can be modeled by the replacement

$$\alpha \left(\cos\theta \frac{\partial}{\partial y} + \sin\theta \frac{\partial}{\partial x} \right) \to g \frac{\partial}{\partial y} \tag{4.67}$$

where g is the analogue of α in Eq. (4.57). In the resistive-g (rg) limit ($m_e = 0$) scale invariance allows one to derive

$$F_{rg} = C_{rg} \frac{g^{3/2}}{s^2}. \tag{4.68}$$

With electron inertia ($m_e \neq 0$), the turbulent thermal diffusivity is

$$\chi \sim \frac{v_A}{qR} \frac{c^2}{\omega_{pe}^2} \frac{\alpha^{3/2}}{s^{1/2}}. \tag{4.69}$$

Formula Eq. (4.69) agrees with that obtained by Itoh and Fukuyama (1992).

4.4.2 *Scaling of correlation length and time*

From scale invariance we can also deduce scalings for the perpendicular correlation length ℓ_c and time τ_c of the turbulence from the scaling results Eq. (4.64). Thus

$$\ell_c \sim \left(\frac{\alpha}{s}\right)^{1/2} \frac{c}{\omega_{pe}} \qquad \tau_c \sim \left(\frac{\alpha}{s}\right)^{-1/2} \frac{Rq}{v_A} \tag{4.70}$$

in agreement with Itoh and Fukuyama (1992). This supposes that the fluctuations are not coupled by toroidal effects to generate radially extended structures and longer correlation lengths which increases the transport coefficients.

4.4.3 *Magnetic flutter thermal transport*

For a high density of rational surfaces, the perturbed magnetic fields $\boldsymbol{B}_0 + \delta\boldsymbol{B}$ produce stochastic magnetic field lines. The electron thermal transport is due to the stochastic magnetic fields produced by the resistive-g turbulence in the strong turbulence limit.

$$\chi_e^{\rm st} \sim v_e \left(\frac{\delta B_r}{B}\right) \ell_c \tag{4.71}$$

where v_e is the electron thermal velocity and ℓ_c is the correlation length along the magnetic field from $\ell_c \Delta k_\parallel = 1$. Here δB_r is the rms value of the magnetic fluctuation. From Ampére's law for δB_r one shows that δj leads to

$$\chi_e^{\rm st} \sim \frac{v_e}{Rq}\left(\frac{c}{\omega_{pe}}\right)^2 \frac{\alpha^2}{s} \tag{4.72}$$

for the electron thermal diffusivity for pressure gradient (α) driven magnetic field line stochasticity.

4.4.4 *Electron inertia Ohm's law*

The use of the electron equation of motion, Eq. (4.52), is appropriate for 'cold' electrons, i.e. when $v_e \tau_c / Rq \ll 1$. In the opposite limit [Kadomtsev and Pogutse (1984)] introduced an Ohm's law involving Landau damping. Ignoring diamagnetic effects, valid for turbulence with $\omega_* \tau_c \ll 1$ where ω_* is a diamagnetic frequency, the collisionless Ohm's law is

$$\boldsymbol{B} \cdot \boldsymbol{\nabla}\phi = \frac{m_e v_e}{ne^2} \boldsymbol{B} \cdot \boldsymbol{\nabla} j \tag{4.73}$$

or, using the field line following the variables, Eq. (4.59),

$$\left(\frac{\partial}{\partial\theta} + sx\frac{\partial}{\partial y}\right)\left(\phi - \frac{v_e}{V_A} j\right) = 0. \tag{4.74}$$

Scaling arguments can be applied to Eq. (4.57), Eq. (4.58), and Eq. (4.74), which replaces Eq. (4.59) as an Ohm's law. The result for χ is

$$\chi \sim \left(\frac{c}{\omega_{pe}}\right)^2 \frac{v_e}{Rq} \alpha g(s) \tag{4.75}$$

where $g(s)$ = constant in the two-scale approximation. Connor argues that the result in Eq. (4.75) remains valid even if the Ohm's law Eq. (4.73) is replaced by a drift-kinetic description of electron dynamics, provided the same conditions on v_e and ω_* hold.

The correlation length and time are given by

$$\ell_c \sim \left(\frac{v_e}{v_A}\right)^{1/2} \alpha^{1/4} \left(\frac{c}{\omega_{pe}}\right) \sim \alpha^{1/4} \left(\frac{c}{\omega_{pe}} \rho_s\right)^{1/2} \qquad \tau_c \sim \frac{Rq}{v_A} \alpha^{-1/2} \tag{4.76}$$

where $\rho_s = (m_i T_e)^{1/2}/eB$ is the 'ion sound' Larmor radius. In Chapter 15.5 the length ρ_s is invoked to explain the sawtooth reconnection process in a collisionless plasma as given by Porcelli (1992).

For typical tokamak conditions,

$$\frac{v_e}{Rq} \tau_c \sim \frac{v_e}{v_A} \alpha^{-1/2} \gg 1 \tag{4.77}$$

so that the result Eq. (4.75) is more appropriate than Eq. (4.69). The associated stochastic magnetic field electron thermal diffusivity is now given by

$$\chi_e^{\mathrm{st}} \sim \left(\frac{c}{\omega_{pe}}\right) \rho_s \frac{v_e}{Rq} \frac{\alpha^{3/2}}{s^{1/2}}. \tag{4.78}$$

4.5 Momentum Stress Tensor Stability Analysis

The contribution to the plasma currents from the force arising from the divergence of the off-diagonal momentum stress tensor or the gyro-orbits of the particles gives a key stabilizing term to the magnetic ballooning and interchange modes and to the rotational instabilities. The effect of this term in the dispersion relation is given by

$$\omega^2 \to \omega\,(\omega - \omega_{*i}) \qquad \text{where} \qquad \omega_{*i} = \frac{k_y}{eBn_i} \frac{dp_i}{dr}. \tag{4.79}$$

The effect of this ion diamagnetic pressure gradient term on the interchange-ballooning mode is to stabilize the higher wavenumbers. This effect is called the finite-Larmor radius stabilization of MHD modes in the early literature. The effect is strong and important for geometries with large ion gyroradius as in the Field Reversed Configurations.

Worked Problem: FLR stabilization of Interchange Modes

Show that the interchange instability dispersion relation is modified by the divergence of the current from the force $\boldsymbol{F}_{\rm gyro} = -\nabla\cdot\boldsymbol{\Pi}_{\rm FLR}$ to become $k_\perp^2 \rho^2\, \omega(\omega - \omega_{*i}) + \omega_D\, \omega_{*p} = 0$. Solution in Hinton and Horton (1971), Kim, *et al.* (1991) and Smolyakov (1998).

4.6 Kinetic Ballooning Mode Instability

The coupled drift-shear Alfvén mode including the complete Bessel function gyroradius effect and the $\nabla_\perp \boldsymbol{B}$-curvature guiding center drift resonance of kinetic theory is solved for the toroidal ballooning mode eigenvalues and eigenfunctions.

The basic physical mechanisms of drift wave instabilities and the associated plasma transport are now known and understood in detail. Drift waves develop into a mixture of wave turbulence and coherent structures with details of the mixture depending on the driving forces ∇n_e, ∇T_e and ∇T_i and the amount of magnetic shear and background velocity shear in the system. The mixture of zonal flows, waves and vortices has certain similarities to the waves in planetary atmospheres and the equatorial-to-polar temperature gradient driven turbulent heat. The principal difference is in the replacement of the gravitational force on planets with a electric forces from charge separation in the curved magnetic fields. The role of the Coriolis force on planets for waves with periods longer than the rotation period of the planet is analogous mathematically to the Lorentz force from the ambient magnetic field.

The free energy sources created by the radial gradients of the plasma density and the temperature are used to characterize the strength of the drift wave turbulence. Diagnostic signals from microwave interferometers and Thompson scattering data described in Chapter 2 can be used to compute the thermodynamic formulas for the upper bounds of the turbulence as shown in Chapters 13 and 14 for the density and temperature fluctuation levels. The derivations of the free energy formulas treat the plasma as an ideal gas with a compressible gas constant, $\Gamma = (d+2)/d$, where d is the number of degrees of freedom associated with the type of instability under investigation.

The Carnot-cycle description of the auxiliary heated plasma is particularly relevant for the instabilities driven by the temperature gradients. The electron temperature profiles in Tore Supra drive the temperature gradient instabilities, as given in Horton, *et al.* (2005). The structure of the drift wave eigenmodes in the double-periodic toroidal geometry is described in detail in Chapters 13 and 14.

Experiments for the ion temperature gradient (ITG) mode were carried out in the Columbia Linear Machine [Sen, *et al.* (1991)] and in TEXT [Brower, *et al.* (1987)]. Both experiments show the role of the ITG drift mode in turbulent transport.

The ITG mode has become the standard model for describing ion thermal transport in basically all toroidal magnetic confinement experiments. The basic toroidal instability mechanism and nonlinear models are given in Horton *et al.* (1981). High-ion temperatures can be stable by simultaneously maintaining sufficient density profile peaking.

In Chapter 19 we discuss how large-scale Tera-flop and Peta-flop computers are used to compute the ITG turbulence from direct numerical solutions of the gyrokinetic-Vlasov-Poisson system. These codes allow the computation of the drift wave turbulence in complicated systems including the nonaxisymmetric systems like the Large Helical Device at the National Institute for Fusion Science (NIFS). The NIFS codes are called GKV for gyrokinetic Vlasov.

The standard article on the comparison of direct numerical simulation for the

ITG turbulence is Dimits *et al.* (1991). Review of the state of the numerical simulations of the ETG turbulence is Nevins *et al.* (2006).

The electron turbulent transport is more demanding to compute but shows large enhancements over the quasilinear fluxes due to the formation of extended radial convective structures with correlation lengths, $\ell_c \gg \rho_e$. The convective structures are insensitive to the radial electric-field shear but do depend sensitively on the magnetic shear profile.

The empirical energy confinement time scaling laws for both helical system [Yamada, *et al.* (2004)] and tokamaks [Kaye and ITER Group (1997)] follow the drift wave formulas and simulations to within the accuracy of the experimental data. Some of the more detailed comparisons of theoretical models [Kinsey, *et al.* (1996a); Kinsey and Bateman (1996b); Redd, *et al.* (1998); Bateman, *et al.* (1998)] with tokamak databases are referenced but a detailed discussion of the agreement between the data and simulations is beyond the scope of this chapter.

References

Balescu, R. (1988). *Transport Processes in Plasmas* (North-Holland, Amsterdam and New York), vol. 2, ISBN:0444870938.

Bateman, G., Kritz, A. H., Redd, A. J., *et al.* (1998). *Proceedings of the Seventeenth International Atomic Energy Agency Fusion Energy Conference* (Yokohama, Japan, 19-24 October, 1998, International Atomic Energy Agency, Vienna, 1999), paper IAEA-F1-CN-69/THP2-19.

Beer, M. A., Hammett, G. W., Rewoldt, G., Synakowski, E. J., and Zarnstorff, M. C. (1997). *Phys. Plasmas* **4**, pp. 1792-1799.

Beer, M. A., and Hammett, G. W. (1999). *Proceedings of the Joint Varenna-Lausanne Intl. Workshop on Theory of Fusion Plasmas, Varenna, 1998*, eds. J. W. Connor, E. S. Indoni, and J. Vaclavik, Bologna, Societa Italiana de Fisca.

Braginskii, S. I. (1965). *Rev. Plasma Phys.*, ed. M. A. Leontovich, (Consultants Bureau, New York) I, p. 205.

Brower, D. L., Peebles, W. A., Kim, S. K., Luhmann, N. C., Tang, W. M., and Phillips, P. E. (1987). *Phys. Rev. Lett.* **59**, pp. 48-51.

Chang, Z. and Callen, J. D. (1992). Unified fluid/kinetic description of plasma microinstabilities. Part I: Basic equations in a sheared slab geometry, *Phys. Fluids B* **4**, p. 1167, doi:10.1063/1.860125.

Connor, J. W. (1993). *Plasma Phys. and Control. Fusion* **35**, pp. 757-763.

Coppi, B., Laval, G., Pellat, R., and Rosenbluth, M. N. (1968). *Plasma Phys.* **10**, p. 1.

Coppi, B., Rosenbluth, M. N., and Yoshikawa, S. (1968). *Phys. Rev. Lett.* **20**, p. 190.

Crabtree, C., Horton, W., Wong, H. V., and Van Dam, J. W. (2003). Bounce-averaged stability of compressional modes in geotail flux tubes, *J. Geophys. Res.* **108**, p. 10.1029. Published.

Dimits, A. M., Drake, J. F., Guzdar, P. N., and Hassam, A. B. (1991). *Phys. Plasmas* **3**, 6, pp. 20-626.

Dong, J.-Q., Horton, W., and Kim, J-Y. (1992). Toroidal kinetic η_i-mode study in high-temperature plasmas, *Phys. Fluids B* **4**, p. 1867, `http://dx.doi.org/10.1063/1.860040`.

Dong, J.-Q. and Horton, W. (1993). Kinetic quasi-toroidal ion temperature gradient instability in the presence of sheared flows, *Phys. Fluids B* **5**, p. 1581, `http://dx.doi.org/10.1063/1.860898`.

Dorland, W. and Hammett, G. W. (1993). *Phys. Fluids B* **5**, pp. 812-835.

Elskens, Y. and Escande, D. (2003). *Microscopic Dynamics of Plasmas and Chaos* (Institute of Physics Publishing, Bristol and Philadephia) ISBN:0-750306122.

Greenwald, M., *et al.* (2007). *Nucl. Fusion* **47**, p. L26.

Hamaguchi, S. and Horton, W. (1990). *Phys. Fluids B* **2**, pp. 1833-1851.

Hammett, G. W. and Perkins, F. W. (1990). *Phys. Rev. Lett.* **64**, pp. 3019-3022.

Helander, P. and Sigmar, D. (2002). *Collisional transport in magnetized plasmas* (Cambridge University Press) ISBN:0521807980.

Held, E. D., Callen, J. D., Hegna, C. C., and Sovinec, C. R. (2001). Conductive electron heat flow along magnetic field lines *Phys. Plasmas* **8**, p. 1171, doi:10.1063/1.1349876.

Hinton, F. L. and Horton, Jr., C. W. (1971). Amplitude limitation of a collisional drift wave instability, *Phys. Fluids* **14**, pp. 116-123.

Hong, B. G., Horton, W., and Choi, D-I. (1989a). *Phys. Fluids B* **1**, p. 1589.

Hong, B. G., Horton, W., and Choi, D-I. (1989b). *Plasma Phys. Control. Fusion* **31**, p. 1291 doi:10.1088/0741-3335/31/8/006.

Horton, W., Choi, D-I., and Tang, W. M. (1981). *Phys. Fluids* **24**, pp. 1077-1085.

Horton, W., Wong, H. V., Morrison, P. J., Wurm, A., Kim, J. H., Perez, J. C., Pratt, J., Hoang, G. T., LeBlanc, B. P., and Ball, R. (2005). *Nucl. Fusion* **45**.

Ito, K., Itoh, S.-I., and Fukuyama, A. (1992). *Phys. Rev. Lett.* **69**, 1050.

Rev. Plasma Phys., ed. B. B. Kadomtsev. (1992). (Consultants Bureau).

Kadomtsev, B. B., and Pogutse, O. P. (1984). *Plasma Phys. Control. Fusion* 24th Conf., 2, Vienna, IAEA, p. 69.

Kaye, S. M., Greenwald, M., Stroth, U., Kardaun, O., Kus, A., Schissel, D., DeBoo, J., Bracco, G., Thomsen, K., Cordey, J., Miura, Y., Matsuda, T., Tamai, H., Takizuda, T., Hirayama, T., Kikuchi, H., Naito, O., Chudnovskij, A., Ongena, J., and Hoang, G. T. (1997). *Nucl. Fusion* **32**, pp. 1303-328.

Kim, C-B., Horton, W., and Hong, B-G. (1991). Fluctuation and thermal energy balance for drift-wave turbulence, *Plasma Phys. Control. Fusion* **33**, p. 469.

Kinsey, J. E., Bateman, G., Kritz, A. H., and Redd, A. (1996). *Phys. Plasmas* **3**, pp. 561-570.

Kinsey, J. E. and Bateman, G. (1996). *Phys. Plasmas* **3**, pp. 3344-3357.

Kulsrud, R. (2005). *Plasma Physics for Astrophysics*, ISBN:9780691120737.

Mattor, N. (1999). *Phys. Plasmas* **6**, p. 1065.

Mattor, N., and S. E. Parker. (1997). *Phys. Rev. Lett.* **79**, pp. 419-3422.

Nevins, W. M., Candy, J., Cowley, S., Dannert, T., Dimits, A., Dorland, W., Estrada-Milla, C., Hammett, G. W., Jenko, F., Pueschel, M. J., and Shumaker, D. E. (2006). *Phys. Plasmas* **13**, p. 122306.

Porcelli, F. (1992). *Phys. Rev. Lett. A* **168**, p. 59.

Redd, A. J., Kritz, A. H., Bateman, G., and Horton, W. (1998). *Phys. Plasmas* **5**, pp. 1369-1379.

Rewoldt, G., Ku, L.-P., Tang, W. M., and Cooper, W. A. (1999). Drift mode calculations in nonaxisymmetric geometry *Phys. Plasmas* **6**, p. 4705, doi:10.1063/1.873757.

Rosenbluth, M. N., and Hinton, F. L. (1998). *Phys. Rev. Lett.* **80**, pp. 724-727.

Sen, A. K., Chen, J., and Mauel, M. (1991). Production and identification of the ion-temperature-gradient instability, *Phys. Rev. Lett.* **66**, pp. 429-432, `http://link.aps.org/doi/10.1103/PhysRevLett.66.429)`.

Smolyakov, A. (1998). *Can. J. Phys.* **76**, p. 321.

Sugama, H., Watanabe, T.-H., and Horton, W. (2001). *Phys. Fluids* **8**, pp. 2617-2628.

Sugama, H., Watanabe, T.-H., and Horton, W. (2003). *Phys. Plasmas* **10**, pp. 726-736.

Sugama, H., Watanabe, T.-H., and Horton, W. (2007). *Phys. Plasmas* **14**, p. 022502.

Waltz, R. E., DeBoo, J. C., and Rosenbluth, M. N. (1990). *Phys. Rev. Lett.* **65**, pp. 2390-2393.

Watanabe, T.-H. and Sugama, H. (2002). *Phys. Plasmas* **9**, pp. 3659-3662.

Watanabe, T.-H., Sugama, H., and Sato, T. (2000). Nondissipative kinetic simulation and analytical solution of three-mode equations of the ion temperature gradient instability *Phys. Plasmas* **7**, p. 984, `http://dx.doi.org/10.1063/1.873897`.

Yamada, H., Harris, J. H., Dinkalge, A., Ascasibar, E., Sano, F., Okamura, S., Talmadge, J., Stroth, U., Kus, A., Murakami, S., Yokoyama, M., Beidler, C. D., Tribaldos, V., Watanabe, K. Y., and Suzuki, Y. (2004). *Nucl. Fusion* **45**, pp. 1684-1693.

Chapter 5

Laboratory Experiments for Drift Waves

The experimental study of the small amplitude waves in steady state, low temperature cylindrical plasmas led to the discovery of the ion acoustic waves and the drift waves described in Chapter 3. These low temperature, cylindrical plasmas had a low level of fluctuations in them compared with the toroidal plasmas so they were called Q-machines where Q designated "quiet" plasmas meaning low noise or ambient fluctuation levels. In Q-machine plasma waves could be launched with antennas or produced by instability mechanisms and studied in detail with Langmuir probes owing to the steady, low temperature nature of the plasma. The plasmas were singly ionized gases produced from easily ionized alkali metals of cesium and potassium [D'Angelo and Motley (1963)]. Figure 5.1 shows a schematic of the plasma fields in

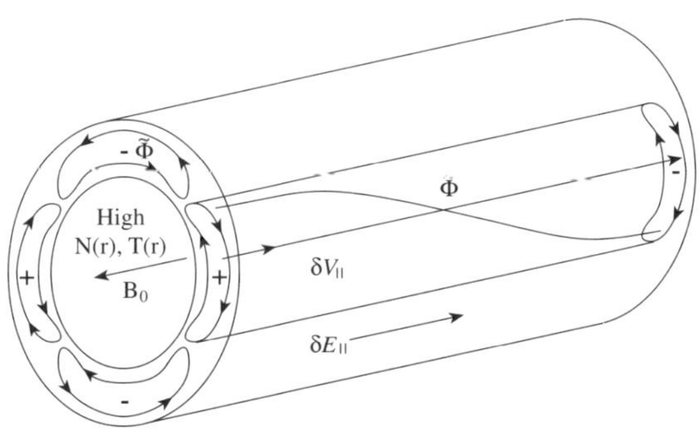

Fig. 5.1 Three-dimensional configuration of the drift-wave fields in a cylinder.

a typical Q-machine used to study drift waves and ion acoustic waves. There were many of these devices in the US and Europe during the 1960s and 1970s. Figure 5.2 shows in detail the design of one such Q-machine. There is an oven at the right end with a tungsten hot plate at a temperature of 5000 K that ionizes the cesium atoms that strike the plate in what is known as contact ionization. The entire magnetized cylinder then fills with singly ionized cesium ions with a temperature of about $T = 0.5\,\text{eV} = 5000\,\text{K}$. The plasma diffuses by collisions to the region beyond

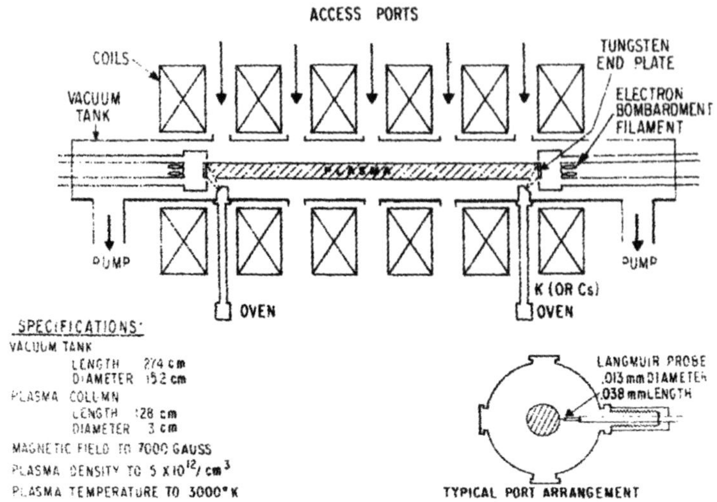

Fig. 5.2 Schematic and key parameters of the Q-machine used in the research of the linear and nonlinear properties of drift waves. Note the spray indicated from the ovens for the potassium (K) or cesium (Cs) atoms on to the tungsten hot plates at both ends of the 1.8 m plasma column [MATT-586 (1968) and Hendel, *et al.* (1968)].

the hot plate radius of $r_p = 12\,\mathrm{cm}$ in this particular device developing a peaked density profile and a uniform temperature profile. This situation spontaneously produces the drift wave instability described in Chapter 3.2. The strength of the instability can be controlled by changing the density and the magnetic field in the system. Figure 5.1 shows a typical set of fluctuating density and potential profiles associated with the drift waves. The measured phase shift with fluctuating density leading the potential is given by the dispersion relation in Chapter 3 and produces the net convection by the $n\boldsymbol{E}\times\boldsymbol{B}$ flux from the core radii of higher density to the lower density edge region. The plasma flux was measured in detail and agreed with drift wave theory. The nonlinear calculations for the wave amplitude and the particle flux are complicated so we sketch the key points of the theories rather than giving the entire calculations.

The plasma equations used in the theory of these ion-acoustic and drift waves must be taken from a two-component fluid theory in which the ions and electron components of the fluid behave differently. This is in sharp contrast to the MHD model where the ions and electrons are combined to form a single fluid and both ion and electron components move together with the $\boldsymbol{E}\times\boldsymbol{B}$ drifts and there is no parallel electric field or either a small resistivity supported parallel electric field. In contrast, the drift waves have an intrinsic parallel electric field.

The parallel electric field $E_\parallel$ is supported by the electron pressure gradient almost independent of the dissipation introduced by the parallel plasma current driven by the plasma electrical conductivity and the parallel electric field. The ion acoustic wave and the drift wave have a parallel electric field $E_{\|}$ supported by the parallel force from the electron pressure gradient $-\nabla p_e/n_e$. For isothermal electron behavior we have $-\nabla p_e/n_e = -T_e \nabla \ln n_e$. This parallel electric field is measured in the Q-machines and the axial wave function is found to be a standing wave of approximately the form of a $\sin(\pi z/L)$ where L is the length of the Q-machine. The ends of the wave functions form a sheath region in which the plasma adjusts its electron and ion densities to satisfy the sheath plasma conditions [Hendel (1969)]. These sheath boundary conditions need not concern us here and the wave function to a good approximation is the lowest standing wave with zero potential boundaries conditions (Dirichlet boundary conditions) in the long cylinder. The radial eigenmode wave function is determined by the radial density and temperature profiles and the difference of the ion and electron cross field drifts. The symmetry of the cylinder dictates the azimuthal wave functions are of the form $f_m(z)\cos(m\theta - \omega t +$ phase shift). The situation is shown in Fig. 5.1.

The parallel electrical resistivity, $\eta_{\|}$, led to the first particular forms of the drift wave instabilities. The other transport coefficients of viscosity and thermal diffusivity also enter into the accurate form of the drift wave dispersion relations and were needed to obtain detailed agreement with the laboratory measurements. Thus the full confirmation of the details of the drift waves occurred two decades of Q-machine experiments and two decades of research. The two-component FLR-fluid equations in Chapter 4 are used to give the full dynamics of drift waves in the collisional plasma.

Further investigations into various plasma regimes revealed the omnipresence of the drift wave modes over a variety of machines and operating conditions. The corresponding theories confirmed that the modes were unstable under a variety of conditions. Thus, the drift waves are often called the "universal mode instability" indicating that it is potentially present in almost all nonuniform, magnetically confined plasmas. The variants of the drift waves that are currently limiting the achievement of the required confinement in fusion devices are often called the ion and electron temperature gradient driven drift waves where as the name suggests the large radial temperature gradients are the dominant driving forces for instability. These forms of the drift waves are analyzed in Chapters 13 and 14.

To keep the analysis clear, we first consider instabilities in which the plasma temperature profile is flat or $dT/dr = 0$. This corresponds precisely to the Q-machine systems since the hot end plates used in the device keep the temperature uniform over the volume defined by surface of the hot plate and the length of cylinder. The most comprehensive experimental studies of the drift wave instability were also carried out for the collision-dominated form of the drift waves. In this chapter we investigate the nature of the collision-dominated drift waves. The study in this chapter embraces both the linear and nonlinear aspects of these collective modes as described by the collisional transport equations. The linear study of the collisional drift waves with a nonlinear study follows the historical development.

5.1 Basic Laboratory Experiments for Drift Waves with Uniform Temperature Profiles

Under a wide range of circumstances a confined, MHD stable plasma can exhibit motions through which particles are convected across the magnetic field by $\boldsymbol{E} \times \boldsymbol{B}$ drifts. The drift waves or the universal drift wave modes are an important group of these instabilities of a confined plasma which are being studied extensively both theoretically and experimentally. An important period of the development of the understanding of drift waves comes from the stability of the two-component fluid equations in MHD stable plasmas in cylindrical geometry with a constant axial magnetic field. Owing to their low first ionization potential, the Alkali elements are most easily used to form the plasma with hot plates being sufficient to produce a steady-state plasma at a sufficient density to be considered fully ionized. Simple plasma waves emerged in these plasmas which were easily studied with probes in a steady state. In the 1960s these low temperature plasma experiments were named Q-machines because of the "quiet" plasma they produced.

The low temperature ($T \sim 3000\,\mathrm{K} \sim 0.3\,\mathrm{eV}$) alkali plasma experiments the conditions for instability as a function of the plasma parameters were thoroughly explored and shown to be in agreement with the theory [Chen (1965a,b,c); Politzer (1971), Kadomtsev and Pogutse (1966)]. In this way the properties of the basic drift wave became well known so they were recognized, after some difficulties due to their transformation to turbulent waves in magnetic confinement geometries owing to the large number of unstable drift waves in the larger machines. The schematic of the Q-machine used by the Hendel group in the 1960s at the Princeton Plasma Physics Laboratory is shown in Fig. 5.2.

The higher mode numbers with smaller wavelengths are damped by the ion collisional viscosity. Thus, as the steady state magnetic field is increased the viscous damping decreases and higher azimuthal mode numbers appear in the density and potential fluctuations. Figure 5.3 shows the amplitude of the fluctuations versus the plasma radius $r(mm)$ for the mean plasma density $n_{os}(r)$ before instability and $n_{ow}(r)$ with the drift wave. The relevant drift wave dispersion relation and nonlinear equations for the experiment are given in Hinton and Horton (1971). From the dispersion relation applied to the parameters of a reference experiment the marginal stability diagram of the form shows contours of constant growth rate in the space of B/k_y versus the parallel wavelength $\lambda_{\|}$.

In Fig. 5.3 the three panels show (a) the decrease in the core electron density and flattening of the density profile in the presence of the drift wave, (b) the radially localized electrostatic potential and density fluctuation from the drift wave and (c) the phase shift ψ by which the density wave leads the potential wave with the data points from their error bars. In the region of large amplitude the density wave leads the potential wave by a phase shift of approximately $\pi/4$ radians. The phase shift is due principally to the electron-ion collision frequency but the complete

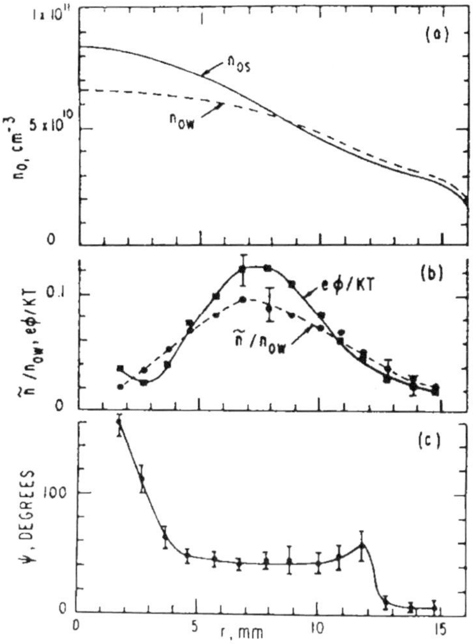

Fig. 5.3 The three panels show the data for (a) the change in the electron density potential and density fluctuation from the drift wave and (c) the phase shift ψ by which the density wave leads the potential wave with the data points from their error bars [Hendel (1969)].

dispersion relation from the two-component drift wave equations has a number of transport contributions including those associated with the electron temperature fluctuations. The variation of the amplitude and frequency with increasing magnetic field B is shown in Fig. 5.4.

The relative density fluctuation amplitude measured for the magnetic field increased from 1.5 to 6.5 kG. At $B \sim 4\,\text{kG}$ there is a transition from the $m = 2$ to $m = 3$ mode as the dominant amplitude signal. At $B \sim 5.5\,\text{kG}$ there is a transition from the $m = 3$ to $m = 4$ fluctuation as dominant. The dashed lines are the growth rates from the dispersion relation. Later experiments by Politzer (1971) explored the transitions to still higher mode numbers and the onset of broadband drift wave turbulence.

In magnetic confinement experiments with hydrogen plasmas such as tokamaks, stellarators, multipoles, and the spherator, density fluctuations with characteristics of drift wave frequencies are universally observed. The multipole experiments commonly show fluctuations of the drift wave type.

In toroidal octupole experiments [Coppi, *et al.* (1968)] the plasma density fluctuations were identified as the higher frequency drift-cyclotron mode and a low frequency flute ($k_\| = 0$) microinstability rotating in the ion diamagnetic direction. In linear quadrupole experiments [Coppi, *et al.* (1968)] the drift-ballooning mode

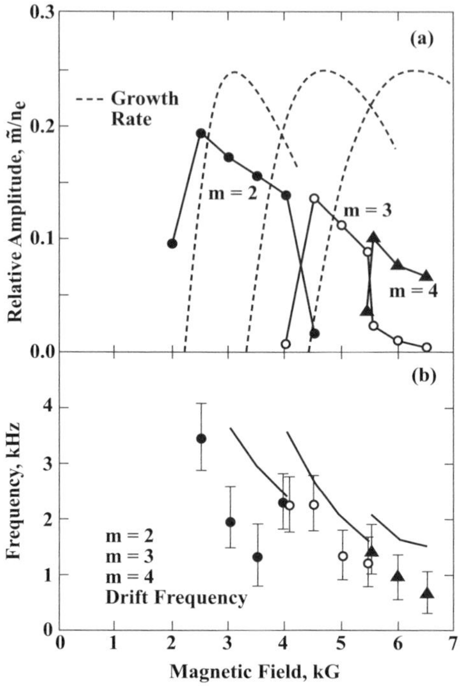

Fig. 5.4 The amplitude and the linear growth rate versus the magnetic field strength. The transitions to higher mode numbers with the increasing magnetic field is due to the decrease of the ion collisional viscosity with square of the ion gyroradius [Hendel, *et al.* (1968)].

are observed and studied in some detail. In addition, fluctuations in the spherator [Johnson and Greene (1967)] are identified as collisional drift waves and were studied by varying the magnetic shear, the ion mass and the plasma density. In the spherator experiments the amplitude and presence of the density turbulence was controlled by the variable magnetic shear in qualitative agreement with the critical magnetic shear requirement.

In addition to the density and temperature gradient driven instabilities, a parallel plasma current provides a source of plasma instability. Thus, an important problem is to determine limits of stability in each geometry including the cross-field gradients and the parallel plasma current. The study in this section is limited to collision-dominated modes where the electron mean-free-path $\lambda_{\rm mfp}$ is less than the parallel wavelength $\lambda_{\|}$. In general, drift waves occur with long parallel wavelengths, and the collision-dominated waves can occur in regimes where $\lambda_{\rm mfp} > R$ where R is the major radius of the toroidal device. Restricting consideration to long wavelength, collision-dominated waves permits a fluid description of the waves.

For highly turbulent plasmas the effect of the smallest scales in the turbulence on the large scales dynamics can be estimated by using the transport coefficients in large-scale equations of motion. In this case, one assumes that the average effect of the small scale-length plasma turbulence is adequately represented at large scale lengths by effective transport coefficients. The most common instance of this multi-scale modeling occurs in applying resistive MHD equations in high temperature plasmas. Here the Coulomb collisional resistivity is negligible and the resistivity used in the MHD equations represents the effect of the small scale plasma instabilities driven by the plasma current. The current driven drift wave instabilities are one source of this anomalous and turbulent resistivity.

In other regimes, for instance, short scale length ($\lambda_{\|} \lesssim 10\lambda_{De}$) ion sound instability provides a large effective momentum transfer collision frequency at small fractional turbulent energy densities $W = \langle E^2 \rangle / 8\pi\, \mathrm{nT}$. Only a small amount of turbulent energy at these scale lengths is required to make the effective mean-free-path much less than the major radius R. In fact, for consistency numerous early toroidal experiments with substantial parallel currents required the introduction of an effective collision frequency considerably enhanced over the classical Spitzer resistivity to explain the measured toroidal loop voltages.

5.2 Discovery of Drift Waves in Early Q-Machine Experiments

An early line of investigation that lead to the discovery of resistive drift waves was performed by D'Angelo and Motley (1963) and Chen (1964), when spontaneous oscillations appeared in the Q-machines. The oscillations were at lower frequencies than anything known at that time. The Kadomtsev-Nedospasov helical instability [Kadomtsev and Pogutse (1966)] of in a partially ionized gas called a positive column had an appealing mechanism, but it depends on the electron-neutral gas collisions. Trying to find a similar mechanism in a fully ionized plasma, Chen and D'Angelo included the electrical resistivity. The Coulomb collisional resistivity is different from the electron-neutral collisions that apply in weakly ionized plasmas. In fully ionized plasma the Coulomb resistivity is independent of the plasma density and decreases strongly with increasing electron temperature. In the weakly ionized plasma the resistivity is determined by the electrons colliding with the neutral atoms.

Chen realized that the ions must drift slower than the $\boldsymbol{E} \times \boldsymbol{B}$ drift of the electrons and that the mechanism for this slower drift is the larger gyroradius of the ions averaging the E-field over the ion orbit. This effect is called the finite Larmor radius effect, or FLR effect, for brevity and is a key effect in both the density gradient driven drift waves and in the stability of the MHD-like interchange modes. The reduction factor from the finite Larmor or gyroradius is calculated in Worked Problem 1.1 at the end of Chapter 1. This FLR effect is discussed further in Chapters 11-14 and 19 on gyrokinetic simulations.

Chen (1967) thought of including the ion viscosity stress tensor terms as a way of representing the FLR effects. These FLR effects make the ion drifts slower than the electrons producing a charge separation. The averaging effect gives a downward shift in the drift wave frequency proportional to $k_\perp^2 \rho_i^2$ where ρ_i is the ion gyroradius.

Chen (1963) basically reported on the resistive drift instability though the author was apparently reluctant to call the waves an instability at the time. This first work on drift waves is not well known, because it came before conventional notation was established and did not claim an instability. At the same time Rudakov and Sagdeev (1961) published their theoretical discovery of the drift wave instability. The Rudakov-Sagdeev theoretical work is widely known as the discovery article on drift wave instabilities.

Chen (1965a) applied this instability to anomalous diffusion and suggested that the "diffusion" need not follow Fick's Law. The physical mechanism of drift instabilities was explained in Chen (1965a,b) and again in Chen (1965c). Chen applied the theory to Q-machine experiments, showing how parallel wavelengths longer than the machine length could be spontaneously excited. The plasma sheaths on the metal end plates were accounted for properly.

In Chen (1966) the drift wave experiment and analysis was repeated in cylindrical geometry with the addition of a Rayleigh-Taylor instability due to the centrifugal force $F_r = mv_\theta^2/r$ from the ambient radial electric field E_r that produces the azimuthal rotational velocity $v_\theta = -E_r/B$ in the cylindrical geometry.

Subsequently, Chen and Mosher (1967) performed experiments with a solid metal current carrying axial rod passing through holes in the center of the hot tungsten end plates of a Q-machine. The current carrying rod produced a sheared helical magnetic field in the plasma. Vortex structures similar to those for the drift wave coupled drift waves and ion acoustic waves are developed in Su, *et al.* (1991, 1992).

An asymmetry in the plasma waves between right ($E_r < 0$) and left ($E_r > 0$) rotations was found to arise due to the Coriolis force in the rotating plasma. In Chen (1966) the plasma viscosity due to the ion-ion Coulomb collisions was considered. At the Princeton Plasma Physics Laboratory the second Q-machine shown in Fig. 5.2 was set up by Hendel, *et al.* (1968) which confirmed the Chen results. The nonlinear calculations of Hinton and Horton (1971) included the Coulomb ion-ion collisions as the physical mechanism that balanced the resistive growth rates determining the sequence of mode numbers found in the experiments and shown in Fig. 5.3 and Fig. 5.4. The finite ion gyroradius viscosity was also included in the model along with the electron temperature fluctuations. The Hinton-Horton model can be reduced to the Hasegawa-Wakatani model by taking the ion temperature to be negligible and dropping the electron temperature fluctuations δT_e.

Hendel's group measured the thresholds in the magnetic field $B_{\rm crit}$ for the onset of various drift wave modes in the Q-machine. The measurements were explained in the Hinton-Horton model by the mechanisms generating higher harmonics that were damped by the ion viscosity in the nonlinear steady state. The inclusion of ion cross-field viscosity is now the standard procedure for controlling the high k-modes amplitudes in computer simulations of drift wave turbulence. The wave growth rate could be made slow in these experiments so that not only the wave frequency, $\omega(m)$, but the growth rate, $\gamma(m)$, was shown to agree with the theoretical model. The radial eigenvalue problem corresponds to a plasma physics version of the famous Orr-Sommerfeld problem in hydrodynamics and is thus a difficult fourth-order boundary value problem.

The drift wave structures are now being captured with high speed CCD cameras that image the UV and soft X ray radiation from the hot plasma. An example of the growth and decay of a coherent drift wave structure is shown in Fig. 3.10 taken in the central cell plasma of the GAMMA 10 plasma at the University of Tsukuba. Chapter 10 is devoted to the study of the coherent structures that arise from the drift wave equations.

References

Chen, F. F. (1963). Low Frequency Instabilities of a Fully Ionized Gas, *Proc. Sixth Conf. Ionized Phenomena in Gases* (Paris, France) **2**, pp. 435-451.

Chen, F. F. (1964). Normal modes for electrostatic ion waves in an inhomogeneous plasma, *Phys. Fluids* **7**, p. 949.

Chen, F. F. (1965a). Resistive over-stabilities and anomalous diffusion, *Phys. Fluids* **8**, p. 912.

Chen, F. F. (1965b). Excitation of drift instabilities in thermionic plasmas, *J. Nucl. Energy* Pt. C7, p. 399, doi:10.1088/0368-3281/7/1/306.

Chen, F. F. (1965c). Effects of sheaths on drift instabilities in thermionic plasmas, *Phys. Fluids* **8**, p. 752.

Chen, F. F. (1966). Microinstability and shear stabilization of a low-beta rotating, resistive plasma, *Phys. Fluids* **9**, p. 965.

Chen, F. F. and Mosher, D. (1967). Shear stabilization of a potassium plasma, *Phys. Rev. Lett.* **18**, p. 639, doi:10.1103/PhysRevLett.18.639.

Coppi, B., Rosenbluth, M. N., and Yoshikawa, S. (1968). *Phys. Rev. Lett.* **20**, p. 190, doi:10.1103/PhysRevLett.20.190.

D'Angelo, N. and Motley, R. W. (1963). *Phys. Fluids* **6**, p. 422, doi:10.1063/1.1706749.

Hendel, H. W. (1969). *Phys. Rev. Lett.* **23**, pp. 568.

Hendel, H. W., Chu, T. K., and Politzer, P. A. (1968). Collisional Drift Waves-Identification, Stabilization, and Enhanced Plasma Transport, *Phys. Fluids* **11**, p. 2426, `http://dx.doi.org/10.1063/1.1691833`.

Hinton, F. L. and Horton, W. (1971). *Phys. Fluids* **14**, pp. 116-123.

Johnson, J. L. and Greene, J. G. (1967). *Plasma Phys.* **9**, p. 611.

Kadomtsev, B. B. and Pogutse, O. (1966). *Zh. Eksp. Teor. Fiz.* **51**, p. 1734, [(1967). *Sov. Phys. JETP* **24**, p. 1172].

Politzer, P.A. (1971). Drift instability in collisionless alkali metal plasmas, *Phys. Fluids* **14**, p. 2410.

Rudakov, L. I. and Sagdeev, R. Z. (1961). On the instability of a nonuniform rarefied plasma in a strong magnetic field, *Sov. Phys. Doklady* **6**, p. 415.

Su, X. N., Horton, W., and Morrison, P. J. (1991). Drift wave vortices in inhomogeneous plasmas, *Phys. Fluids B* **3**, pp. 921-930.

Su, X. N., Horton, W., and Morrison, P. J. (1992). Drift wave vortices in nonuniform plasmas with sheared magnetic fields, *Phys. Fluids B* **4**, p. 1238.

Chapter 6

Magnetohydrodynamics and Magnetic Confinement Geometries

6.1 The MHD-Magnetohydrodynamic Model

A universal model that plays a key role in plasma physics is the MHD model. From the perspective of Chapter 2 that gives the kinetic theory and Chapter 4 that develops the two-component (ion and electron) fluid equations, the MHD model is a further step down in the accuracy and detailed description of the plasma dynamics. In fact, it is not possible to derive precisely the conditions under which the MHD model is valid. It is not unusual in science, and physics in particular, that the simpler less detailed description, is more useful. In physics two famous examples of this rule come to mind: Ohm's law and the Bohr atom. From the perspective of the more complex models introduced in Chapters 2 and 4, one can say that the ideal MHD model is to plasma physics as the Bohr atom model is to quantum atomic physics. One concludes that the MHD model is rarely precisely correct, but is always useful just as with the Bohr model of the atom. There are phenomena, of course, that are well-described by the idealized Bohr atomic and MHD plasma models. For plasma MHD the key properties that make the model so useful, and often adequate, are the conservation laws the model embodies. The MHD model is a set of seven partial differential equations advancing in time a single perfectly conducting fluid of mass density ρ, flow velocity $\boldsymbol{u}$, and isotropic pressure p that obeys the adiabatic gas law with a user-specified adiabatic gas constant Γ. The MHD model is needed for various purposes. Essentially all laboratory plasma confinement experiments are designed to have large parameter regions that are MHD stable. The drift wave and Kelvin-Helmholtz instabilities are of interest in MHD stable plasmas.

First and foremost, the study of drift waves and anomalous transport assumes that there already exists a stable MHD equilibrium. The point is that, even if the details of the MHD stability description are not precisely known, it is unlikely that drift waves and microinstabilities are of interest, or even well-posed mathematically, if the plasma is already unstable in the MHD description. There are some exceptions to this rule at marginal stability, and we discuss such a case below. But, the exceptions are rare. The very formulation of the equilibrium, or even the evolving ambient plasma state, requires the MHD equations be satisfied locally on the time scale of transport. The reconstruction of the density and temperature profiles used in the analysis of drift wave turbulent transport will use the conditions of local MHD equilibrium in the inversion of the interferometer and Thomson scattering data, for example.

6.1.1 *MHD equations*

To obtain the MHD equations from the kinetic equations or even the two-component FLR-fluid equations requires a long list of approximations. For those interested in such an effort, see the discussion in Krall and Trivelpiece (1968) and Braginskii (1965). Here we write down the ideal equations and then discuss some of the key physical processes missing from the description with respect to the subject of drift waves and tearing modes.

The conservation laws for a single ideal, perfectly conducting fluid with mass density ρ and center of mass flow velocity $\boldsymbol{u}$ are

$$\frac{\partial \rho}{\partial t} + \nabla \cdot (\rho \boldsymbol{u}) = 0 \tag{6.1}$$

$$\rho \left(\frac{\partial \boldsymbol{u}}{\partial t} + \boldsymbol{u} \cdot \boldsymbol{\nabla} \boldsymbol{u} \right) = \boldsymbol{j} \times \boldsymbol{B} - \nabla p + \rho \boldsymbol{g} \tag{6.2}$$

$$\frac{\partial \boldsymbol{B}}{\partial t} = \nabla \times (\boldsymbol{u} \times \boldsymbol{B}) \tag{6.3}$$

$$\frac{\partial p}{\partial t} + \boldsymbol{u} \cdot \boldsymbol{\nabla} p + \Gamma p \nabla \cdot \boldsymbol{u} = 0 \tag{6.4}$$

where the current density follows from $\mu_0 \boldsymbol{j} = \nabla \times \boldsymbol{B}$ and the electric field is determined by idealization of a plasma gas with infinite electrical conductivity

$$\boldsymbol{E} + \boldsymbol{u} \times \boldsymbol{B} = 0. \tag{6.5}$$

Equation (6.1) and Eq. (6.4) are combined to show that the specific entropy of the plasma gas $S = p/\rho^{\Gamma}$ called the specific entropy is constant in the frame $\boldsymbol{u}$ of the moving plasma. The equation of state (EOS) of the ionized gas is $S =$ const in the rest frame of the gas. Combining this statement with the mass conservation law one defines the entropy density $s(\boldsymbol{x}, t) = n k_B \ln(p/\rho^{\Gamma})$ is a conserved density satisfying

$$\frac{ds}{dt} = \frac{\partial s}{\partial t} + \nabla \cdot (\boldsymbol{u} s) = 0. \tag{6.6}$$

Here n is the number density, k_B the Boltzmann constant and Γ is the adiabatic gas constant.

Equation (6.5) is the condition that the electric field $\boldsymbol{E}'$ vanishes in the local rest frame of the moving plasma. Equation (6.6) is the statement that in a fluid element following the fluid flow, there is no change in the entropy. These two equations are the strong statements that fail in most of the plasma dynamics studied in this book. They become reasonable approximation only for very fast, and thus strongly out of force balance dynamics.

Equation (6.5) then produces the motional EMF in the laboratory frame that is the fixed initial frame in which the velocity $\boldsymbol{u}$ is measured with respect to. For example, in the frame attached to Earth, the solar wind plasma has the velocity $\boldsymbol{u}_{\mathrm{sw}}$ that is directed radially outward from the Sun with speeds that range from 200 km/s to 700 km/s. This plasma contains the interplanetary magnetic field $\boldsymbol{B}_{\mathrm{IMF}}$ and then in magnetosheath surrounding the Earth's magnetosphere has an additional $\Delta\boldsymbol{B}$ component. In the Earth's frame of reference this solar wind flowing across the ambient magnetic field $\boldsymbol{B}(\boldsymbol{x},t)$ drives the electric field

$$\boldsymbol{E}_{\mathrm{sw}} = -\boldsymbol{u}_{\mathrm{sw}} \times \boldsymbol{B} \tag{6.7}$$

that maps down the magnetic field lines, since $\boldsymbol{B}\cdot\boldsymbol{E} = 0$, to impose the measured electric field in the ionospheric polar caps. For this discussion, the polar cap is defined as the region where a field line has only one end attached to Earth and the other end goes to the interplanetary magnetic field $\boldsymbol{B}_{\mathrm{IMF}}$. These solar wind generated potentials range from 40 kV in quiet times to hundreds of kilovolts during magnetic storms. The electric fields are measured by spacecraft surrounding the Earth, by spacecraft in the ionosphere and inferred by ground based radar measurements of ionospheric plasma convection [Kelley, 2009; Kivelson and Russell (1983)].

In the laboratory plasma the most important example of Eq. (6.5) is of the radial electric field $E_r(r,t)$ that is produced by the high-speed toroidal rotational velocities (200 to 600 km/s) driven by the neutral beam injection (NBI) into tokamak plasmas. For toroidal plasma flow velocities u_ϕ, these rotation driven electric fields are then radial with

$$E_r^{(0)} = u_\phi B_\theta \tag{6.8}$$

and are known to be very important in influencing the plasma confinement.

The laboratory measurements of E_r are complex and indirect. Thus, the errors in the radial profile of E_r are rather large. The zero superscript indicates the dominant component of E_r for large toroidal flows $\boldsymbol{u} = \boldsymbol{u}_\phi = R\Omega\widehat{\boldsymbol{e}}_\phi$. However, what is known is that, even at the highest velocities, the MHD approximation $E_r = E_r^{(0)}$ given by Eq. (6.8) is incomplete in rotating tokamaks. There are carefully-documented measurements that show that E_r is controlled not only by u_ϕ but by the flows driven by the ion pressure $\nabla p_i/en_i$ and the impurity pressure gradients as well as the parallel force $F_{\|ab}$ between the working gas ions and the impurity ions. Grierson, *et al.* (2010) develop the measurements of E_r and correlate the profiles of E_r with the predictions of neoclassical theory Hirshman and Sigmar (1997).

6.1.2 *Conservation form of MHD*

The success of the MHD model in describing a multitude of plasma dynamics is due to the conservation of mass, momentum, energy and charge. Numerical implementations are most reliable when those properties are built into the time advancement scheme as emphasized by Barnes and Brackbill (1977) and Boris and Book (1973). For example, shocks are a universally-observed, complex phenomenon that are well described outside the dissipation layer by the conservation laws of MHD. In conservation form the equations are

$$\frac{\partial \rho}{\partial t} + \nabla \cdot (\rho \boldsymbol{u}) = 0 \tag{6.9}$$

$$\frac{\partial}{\partial t}(\rho \boldsymbol{u}) + \nabla \cdot \left[\rho \boldsymbol{u}\boldsymbol{u} + \left(p + \frac{B^2}{2\mu_0}\right)\boldsymbol{I} - \frac{\boldsymbol{B}\boldsymbol{B}}{\mu_0}\right] = 0 \tag{6.10}$$

and

$$\frac{\partial}{\partial t}\left(\frac{1}{2}\rho u^2 + \frac{B^2}{2\mu_0} + \frac{p}{\Gamma - 1}\right) + \nabla \cdot \left[\rho \boldsymbol{u}\left(\frac{u^2}{2} + \frac{\Gamma}{\Gamma - 1}\frac{p}{\rho}\right)\right. \tag{6.11}$$

$$\left. + \boldsymbol{u}\frac{B^2}{\mu_0} - \boldsymbol{B}\,\frac{\boldsymbol{u} \cdot \boldsymbol{B}}{\mu_0}\right] = 0 \tag{6.12}$$

where $(I)_{ij} = \delta_{ij}$. Care must be taken to maintain $\nabla \cdot \boldsymbol{B} = 0$ exactly in using Eq. (6.10) since a residual (numerical) error creates a large magnetic monopole force $\boldsymbol{B}(\boldsymbol{\nabla} \cdot \boldsymbol{B})$ in Eq. (6.10). Charge is conserved as long as $\nabla \cdot \boldsymbol{j} = \nabla \cdot (\nabla \times \boldsymbol{B}) \equiv 0$ is maintained. Morrison and Greene (1980) show that the conservation form is naturally Hamiltonian and Kulsrud (1983) gives generalizations of MHD for astrophysics where the gravity field is included, for analogous momentum conservation law and the virial theorem.

In advancing Eq. (6.9) through Eq. (6.12) one uses the ideal conductivity law in Eq. (6.3) which adds two degrees of freedom. Thus, the system has $5 + 2 = 7$ degrees of freedom. Linearization about a uniform state yields a cubic equation in ω^2 and a zero frequency $\omega = 0$ mode.

The great difficulty, as well as beauty, of the system Eq. (6.9)-Eq. (6.12) is that finite amplitude disturbances steepen as the disturbances propagate leading to finite time singularities. These singularities are the formation of the MHD shocks [Burgess (1995)]. Methods for handling the shock problem are the Godunov method and the flux corrected transport FTC problem. In terms of the physics, however, the MHD description clearly fails for the transition region of shocks. Here, we know that kinetic particle descriptions are required to describe the reflected ions, the turbulent fluctuations and anomalous resistivity and viscosity. The lower hybrid drift waves are measured with CO_2 laser scattering experiments in numerous laboratory shock experiments. Spacecraft passing through the bow shock measure many types of plasma wave waves and complex particle dynamics. For shock physics the reader is referred to Kennel (1969), Sagdeev and Galeev (1969), Burgess (1995), and Kivelson and Russell (1983, p. 129).

Worked Problem: Grad-Shafranov Equilibrium

In cylindrical geometry (r, ϕ, z) with symmetry (axisymmetric equilibria) about the vertical z-axis

(1) Find the magnetic flux function $\psi(r, z)$ that describes the specified magnetic fields $B_r(r, z)$ and $B_z(r, z)$ arising both from external coils and the plasma currents.

(2) Show that the contours of $\psi(r, z) = \text{const}$ describe their "poloidal" magnetic field vectors.

(3) When there is no flow ($u = 0$) and no external force ($g = 0$), express the force balance in terms of magnetic flux ψ. The resulting equation is Grad-Shafranov equation. Note that the pressure p is a function of the flux function $p = p(\psi)$ and the toroidal magnetic field B_ϕ can be expressed in terms of ψ through $I = rB_\phi = I(\psi)$.

Solution:

(1) From the vanishing divergence of the magnetic field vector $\boldsymbol{B}$ and the axisymmetry, we have

$$\frac{1}{r}\frac{\partial}{\partial r}(rB_r) + \frac{\partial B_z}{\partial z} = 0. \tag{6.13}$$

This constraint is satisfied by a scalar function $\psi(r, z)$ and is a sufficient condition for the existence of the flux function, such that

$$B_r = \frac{-1}{r}\frac{\partial \psi}{\partial z} \qquad B_z = \frac{1}{r}\frac{\partial \psi}{\partial r}. \tag{6.14}$$

(2) Consider the equations for trajectories tangent to the magnetic field vector

$$\frac{dr}{du} = B_r = \frac{-1}{r}\frac{\partial \psi}{\partial z} \tag{6.15}$$

$$\frac{dz}{du} = B_z = \frac{1}{r}\frac{\partial \psi}{\partial r}. \tag{6.16}$$

The trajectories are along the level contours defined by

$$\psi(r, z) = \text{const}. \tag{6.17}$$

In fact, the trajectories are Hamiltonian with $H(p_1 = r^2/2,\ q^1 = z) = \psi(r^2, z)$.

The situation is isomorphic to that of the stream function $\psi(r, z)$ for 2D incompressible $\nabla \cdot \boldsymbol{u} = 0$ fluid flows. Furthermore, the vorticity $\nabla \times \boldsymbol{u}$ in the symmetry direction $\widehat{e}_\phi$ is isomorphic with the plasma current $\mu_0\, \boldsymbol{j} = \nabla \times \boldsymbol{B}$ in the toroidal direction.

The current now has a poloidal component $\boldsymbol{J}_p$ in addition to the toroidal component J_ϕ. The total current is

$$\mu_0\,\boldsymbol{J} = \nabla \times \boldsymbol{B} = -\left[\frac{\partial}{\partial r}\left(\frac{1}{r}\frac{\partial \psi}{\partial r}\right) + \frac{1}{r}\frac{\partial^2 \psi}{\partial z^2}\right]\widehat{\boldsymbol{e}}_\phi + \nabla\left(rB_\phi\right) \times \nabla\phi. \quad (6.18)$$

The second-order operator acting on the flux function ψ is defined in the literature as $\Delta^*\psi$ and called the Grad-Shafranov operator. The definition of Δ^* is

$$\Delta^* = r\,\frac{\partial}{\partial r}\frac{1}{r}\frac{\partial}{\partial r} + \frac{\partial^2}{\partial z^2} \quad (6.19)$$

with which the toroidal component of Eq. (6.18) becomes

$$\Delta^*\psi = -\mu_0\, rJ_\phi. \quad (6.20)$$

In addition to the toroidal J_ϕ there is the poloidal current from the last term in Eq. (6.18)

$$\mu_0\,\boldsymbol{J}_p = \nabla\left(rB_\phi\right) \times \nabla\phi \equiv \nabla I \times \nabla\phi \quad (6.21)$$

where the literature defines the scalar function $I \equiv rB_\phi(r,z)$.

(3) Now the force resulting from these currents $\boldsymbol{J}_p$ and J_ϕ is

$$\boldsymbol{f} = \boldsymbol{J} \times \boldsymbol{B}. \quad (6.22)$$

Carrying out the cross products one arrives at the force

$$\boldsymbol{f} = \frac{J_\phi}{r}\,\nabla\psi - \frac{I}{\mu_c r^2}\,\nabla I. \quad (6.23)$$

The force $\boldsymbol{f}$ is perpendicular to the level lines of the flux function when it can be balanced by a pressure gradient. When balanced by the plasma pressure, $\boldsymbol{B}\cdot\nabla p = 0$ gives $p = p(\psi)$. And $J \cdot \nabla p = 0$ gives $I = I(p) = I(\psi)$. Then Eq. (6.23) becomes

$$\frac{dp}{d\psi}\nabla\psi = \frac{J_\phi}{r}\nabla\psi - \frac{J}{\mu_\theta r^2}\frac{dI}{d\psi}\nabla\psi.$$

Thus, for MHD equilibria without flow ($\rho\boldsymbol{u} = 0$) we have two function that determine the static equilibrium $p(\psi)$ and $I(\psi)$ for the pressure and toroidal magnetic field function. Calculating $\boldsymbol{J} \times \boldsymbol{B} = \nabla p$ then gives that

$$\Delta^*\psi = -\mu_0 r^2\,\frac{dp}{d\psi} - I\,\frac{dI}{d\psi} \quad (6.24)$$

called the Grad-Shafranov equation. There is extensive computer software to solve the Grad-Shafranov equation for specified profile functions $p(\psi)$ and $I(\psi)$. The turbulent transport is the mechanism by which these are established from the externally applied heating and fueling sources.

More comprehensive analysis and derivations of the Grad-Shafranov equation are given in Goedbloed, *et al.* (2010) on advanced magnetohydrodynamics with general methods for solving the G-S equation. We return to the G-S equation for transport calculations in Chapter 17 and note that the collisional transport theory in tokamaks is also formulated in terms of the G-S equation in Helander and Sigmar (2002), Chapter 7. There is extensive software for solving the G-S equation both for theory and in real-time for shot analysis.

6.1.3 *MHD stable plasmas*

Plasmas are created by various atomic and molecular ionization methods in the laboratory and in nature. Often the creation process is rather violent, forming plasmas with fast flows, large electric currents and/or with high energy particles. After the passage of some time, typically measured by the number of Alfvén wave oscillations, the plasma will settle into a local, slowly evolving quasi-equilibrium state. In this second stage we define the characteristic time scale $\tau_{\rm eq}$ and space scale $L_{\rm eq}$ for the plasma time and space scale of evolution. When the quasi-equilibrium state contains a significant magnetic field B such that $\beta \equiv$ energy density ratio of thermal particle energy density $P = k_B nT$ to the magnetic energy density $B^2/2\mu_0$ is of order unity, the plasma scale length $L_\perp \ll L_\parallel$. This inequality $L_\perp \ll L_\parallel$ is also satisfied in laboratory magnetic confinement experiments.

For each problem some modeling judgment is required to determine when the profiles have entered into this slowly varying quasi-equilibrium phase of evolution. Even if the system is slightly above the critical state according to stability analysis of ideal MHD for the corresponding frozen state it is still important to determine the microinstabilities of the evolving system. This stability analysis can be carried out with sufficient rigor on the evolving background for wave frequencies $\omega(\boldsymbol{k})$ and wavenumbers $\boldsymbol{k}$ such that $\omega\tau_{\rm eq}$ and $kL_{\rm eq}$ are large compared with unity.

For example, the shock fronts structures are often analyzed locally for the small scale lower hybrid drift wave turbulence. The transport bifurcations in the tokamak system also are sufficiently slow to allow drift wave stability analysis to apply locally in radius and time in the transport barrier region for determining the turbulent transport in the barrier region. A key point is that the MHD equilibrium analysis alone will yield a large (infinite) class of equilibrium solutions even when the boundary conditions are fixed. It is the turbulent transport coupled to the sources and sinks of particles, energy and momentum that determine the particular MHD equilibrium that the plasma settles into. The large equilibrium forces of $\boldsymbol{j} \times \boldsymbol{B}$ and ∇p will be equilibrated first through compressional motions and maintain the plasma close to force balance equilibrium throughout the transport evolution. This is a self-consistency loop called the problem of equilibrium, stability and transport. Transport codes such as CRONOS, BALDUR, and TRANSP take the coupled evolution dynamics into account. What has become clear with running these codes is that the coupling of the dynamical loops of back-reaction of the transport on the profiles and thus the turbulence is a fundamental part of dynamical feedback and feed-forward loop. These loops have critical points in which the fractional flows of energy-momentum from the input to the output take different pathways through the system. Many names are associated with the different plasma states such as L-mode, H-mode, ELMy H-mode, ELM-free H-modes, for the different plasma states. For many purposes, transitions between these plasma states can be viewed as second-order phase transitions.

The importance of tracking the self-consistent interactions in the transport problem has only been fully realized and taken into account in the past decade. A key laboratory experiment helped start this revolution in transport studies as well as the maturing of the transport simulation models that tracked the stability and transport effects of the many different instabilities that occur in tokamaks. The laboratory experiment that made clear the occurrence of bifurcations in the transport was the discovery of the improved high confinement mode, called the H-mode, in the ASDEX device. The pre-existing state is then called the L-mode for low confinement. The L-mode state is the base confinement level of standard tokamak operation and is the most thoroughly documented confinement state of high temperature plasmas.

The tokamak plasma data describing this confinement state is available at `http://erie.ornl.gov/iter/vendors/` and `http://orion.ph.utexas.edu/~starpower/`. The original $R = 8$ ITER machine design was based on L-mode confinement and only when it became clear that the factor of roughly two-fold confinement improvement from the H-mode would produce large costs savings in the designed size of the machine was the L-mode scaling law abandoned and the H-mode confinement state adopted as the baseline performance model for the ITER plasma. The projected confinement times for the L- and H-modes in the original and reduced-sized ITER experiment are shown in Table 1.1. To achieve the L-to-H bifurcation, the injected heating power must exceed a critical value defined as $P_{L\to H}$. In this way we see that the change of confinement phase is analogous to a phase transition.

The L-mode data is studied and parameterized to empirical scaling laws for baseline tokamak operation for more than a decade. Tokamaks can be routinely designed and brought on line with the performance given by the L-mode scaling laws. To finish this brief survey we note that there are many means known for enhancing the confinement over that given by the L-mode and this enhancement is quantified as the H-factor. For example, with the use of reversed magnetic shear the JT60-U experiment, and many others subsequently, obtained H-factors of 2 to 2.5 meaning that the global energy confinement time was 2 to 2.5 times longer than the time predicted by the L-mode formula for the same 9 system parameters. These predictions are studied in detail in Chapter 18 since they are some of the most thoroughly understood of the laboratory plasma confinement studies.

Bifurcations of plasma states are to be expected of course as discussed already by

Kadomtsev. The phenomena of bifurcations is studied in detail in neutral fluid experiments for many years. The most important correspondence between the plasma L- to H-mode bifurcation and those in neutral fluids is for the Rayleigh-Bénard convection system. As described qualitatively in Fig. 1.5 and shown mathematically in Chapter 4, there is a close correspondence between the resistive-g plasma model and the Rayleigh-Bénard system. In toroidal plasmas there is an effective gravity from $g = 2c_s^2/R$ where R is the major radius of the torus. The ion temperature gradient ITG system is the collisionless limit of the resistive $-g$ system, and thus is also related the Rayleigh-Bernard convection system. In terms of drift waves, the ITG system applies when there is significant coupling of the vorticity equation to the ion thermal balance equation through the magnetic curvature ($\kappa = 1/R_c = 2/R$) and $\nabla_\perp B$ drift particle currents produced by the toroidal magnetic field geometry. In a straight system there is also a weak ITG mode but now the coupling comes through an inverted parallel pressure gradient $\nabla_\| \delta p_i$ that drives the parallel ion mass flow R.

The Rayleigh-Bernard experiments of Krishnamurti and Howard (1981) and Howard and Krishnamurti (1986) show that as the temperature gradient is turned up or the fluid viscosity is lowered there is a bifurcation of the fluid flows to new states with large a scale shear flow. In higher flow system there is a global circular flow around the entire box of the experiment. This global sheared mass flow then interferes with and thus reduces the convection in the pre-existing temperature gradient driven vortices. The vortices become tilted in the presence of the sheared flows. The fluid simulations of the plasma ITG show exactly this behavior even in the presence of magnetic shear [Hamaguchi and Horton (1991)]. Furthermore, Howard and Krishnamurti (1986) develop a low order model for the onset of the flow with the tilted vortices interacting with the sheared mass flow. The low-order model is such that in below the bifurcation point, or what is called in the subcritical state, the system reduces to the Lorenz system of the three odes to describe the nonlinear cellular convection.

The same bifurcation phenomenon happens in plasmas but in more varied and complex ways. There are simulations of the process in two and three dimensions with fluid models in Carreras, *et al.* (1987), Newman, *et al.* (1998), and Wakatani (1998). There are gyrokinetic simulations of the process. Analytic models of the L to H bifurcations are given in Diamond, *et al.* (2005), Newman, *et al.* (1998), Sugama, *et al.* (2001), Sugama, *et al.* (2003), and Sugama, *et al.* (2007).

The universality of the rotational shear flow coupling to the wave spectrum has now been extended to stellarators and large helical devices. Recently, studies in non-neutral plasma show that the wave that plays the role of the called the diocotron mode can also influenced by and coupled to a background sheared flow.

6.1.4 *Interchange stability condition*

In this section we will consider the energy released from the interchange of two magnetic flux-tubes in mirror magnetic field geometry. If the potential energy of the final state, upon interchange, is lower, then it is assumed a suitable plasma motion, $\boldsymbol{v} = d\boldsymbol{\xi}/dt$ exists to enable the interchange. In MHD it is easy to justify that such a motion exists because the equations of motion follow from minimizing the potential energy, as in classical mechanics.

The potential energy is the sum of the magnetic field energy $B^2/2\mu_0$ and the internal energy of the plasma $p/(\Gamma - 1)$. The internal plasma energy in a flux-tube is given by

$$E_p = \int \frac{p}{\Gamma - 1} dV = \int \frac{p}{\Gamma - 1} ds dA = \int \frac{p}{\Gamma - 1} \Phi \frac{ds}{B} = \frac{p\Phi}{\Gamma - 1} \int \frac{ds}{B} \tag{6.25}$$

where s is the length along the flux-tube, dA is the cross-sectional area of the flux-tube, and $\Phi = BdA$ is the magnetic flux which is a constant for each flux-tube. The most familiar regime is a gas with three degrees of freedom for which $\Gamma = 5/3$ and $1/(\Gamma - 1) = 3/2$.

In any adiabatic motion we have $PV_\perp^\Gamma$ = const so that on interchange of flux-tubes labeled 1 and 2 we have

$$P_1' = P_2 \left(\frac{V_2}{V_1} \right)^\Gamma \quad \text{and} \quad P_2' = P_1 \left(\frac{V_1}{V_2} \right)^\Gamma \tag{6.26}$$

where the prime signifies the final state. Thus, the change in internal plasma energy due to the interchange of two flux-tubes is given by

$$\Delta E_p = \frac{1}{\Gamma - 1} \left[\left\{ P_1 \frac{V_1^\Gamma}{V_2^\Gamma} V_2 + P_2 \frac{V_2^\Gamma}{V_1^\Gamma} V_1 \right\}_{\text{final}} - \{P_1 V_1 + P_2 V_2\}_{\text{init}} \right]. \tag{6.27}$$

The magnetic energy in a flux-tube is

$$E_M = \int \frac{B^2}{2\mu_0} dV = \frac{\Phi^2}{2\mu_0} \int \frac{ds}{dA} \tag{6.28}$$

and upon interchanging two flux-tubes labeled 1 and 2 the change in the magnetic energy is given by

$$\Delta E_M = \frac{1}{2\mu_0} \left[\left\{ \Phi_1^2 \int_1 \frac{ds}{dA} + \Phi_2^2 \int_2 \frac{ds}{dA} \right\}_{\text{final}} - \left\{ \Phi_1^2 \int_2 \frac{ds}{dA} + \Phi_2^2 \int_1 \frac{ds}{dA} \right\}_{\text{initial}} \right]. \tag{6.29}$$

Thus, if we exchange two flux-tubes with the same amount of magnetic flux, i.e., $\Phi_1 = \Phi_2$, then there is no change in magnetic energy. Thus, the $\Delta E_p > 0$ stability condition reduces to the condition that is called negative V'' - in the fusion literature and is given by

$$\frac{dp}{d\psi} \frac{d}{d\psi} \int \frac{ds}{B} + \frac{\Gamma p}{V} \left(\frac{d}{d\psi} \int \frac{ds}{B} \right)^2 \geqq 0$$

where the volume of the flux tube per unit of magnetic flux $d\psi$ is given by $V = \int(ds/B)$ where we use that $dV = \Delta A ds$ where ΔA is the cross-sectional area of the flux tube such that $\Delta\psi = B\Delta A = \text{const}$ along the tube.

6.1.5 *Plasma energy functional of MHD*

The plasma energy functional of MHD is given by

$$\delta W_p^{\mathrm{MHD}} = \frac{1}{2}\int \frac{d\Psi dyds}{B}\left[\Gamma p(\nabla\cdot\boldsymbol{\xi})^2 + \boldsymbol{Q}_\perp^2 + \boldsymbol{Q}_L^2 - 2\boldsymbol{\xi}\cdot\boldsymbol{\kappa}\boldsymbol{\xi}\cdot\nabla p\right] \tag{6.30}$$

as derived in Friedberg (2007). The functional in Eq. (6.30) is the potential energy for the displacement ξ of the system from an MHD equilibrium. The equilibrium is stable if all possible displacements ξ give a positive value of $\delta W_p^{\mathrm{MHD}}$. This method is highly developed in the plasma physics community in both magnetic confinement studies [Troyon, 1987] and in the astrophysics community [Shapiro and Teukolsky, 1983].

6.1.6 *Limitations of the MHD model*

For the MHD model to give (i) accurate growth rates for an optimized displacement vector ξ and (ii) marginal stability boundaries in agreement with the first principle stability calculations places severe constraints on the system parameters.

Namely, the motions predicted from the MHD model must be sufficiently unstable, and thus fast in the dynamics, for the MHD model to be a valid description of the processes. The statement of marginal stability in the context of MHD means the plasma is stable with respect to fast-large scale motions. This constraint on applicability gives rise to the notion of FAST MHD.

When the MHD model predicts motions that are fast compared with the drift mode frequencies and the transit or bounce time of the particles in their motion along the field lines, then the MHD description is confirmed by the kinetic theory. Often this condition is not satisfied in a relaxed system. The condition is especially not satisfied in the case of confinement geometries that are carefully constructed to be MHD stable. When such a system is operated in the MHD stable window of system parameters, the remaining instabilities are of the kinetic resonant particle variants of the MHD oscillations. These oscillations and instabilities require kinetic theory to describe adequately. Some of these instabilities are entirely missed by the MHD description with a single fluid and a vanishing parallel electric field.

For motion of frequency ω and wavenumber k there are natural collective oscillations at the diamagnetic drift frequencies $\omega_* = k_y v_d$ as described in Section 2.1 associated with each species ω_{*i} and ω_{*e}. The MHD growth must be fast compared to both the ω_*'s and the transit/bounce frequency for the kinetic description to reduce to MHD. In addition, there are the requirements that (1) the parallel electric field be negligible and that (2) the distinction between the ion and electron pressure gradients is not important. These conditions are difficult to meet unless the motion is rather violent.

In the intermediate state where MHD gives a good guide, but not an accurate value for the growth rate and stability condition, the kinetic theory may give a maximum growth rate $\gamma_{\max}$, for example, that is one half the value $\gamma_{\max}^{\text{mhd}}$ of the MHD calculation. More significantly, however, is that the kinetic growth rate $\gamma(k)$ will be non-vanishing and with a significant magnitude for wavenumber k values where the MHD growth rate is stable $\gamma_k^{\text{mhd}} = 0$. There are both low and high wavenumber values for which the first principle calculations give unstable oscillations whereas the MHD oscillations are only unstable in the window. This phenomenon is well known in the branch of the MHD literature that deals with the resistive/viscous fluid equations.

A typical result, for example, is that below the critical beta of $\beta_{\text{crit}} = L_p R_c / L_c^2 \simeq 5 I_p / a B_T$ for the onset of the ballooning MHD mode, there is a resistive interchange mode with growth rate given by $\gamma^{rg} \leq c_s/(L_p R_c)^{1/2}$ [Sugama and Horton (1994); Carreras, *et al.* (1987)].

The fast internal kink instability that limits the maximum toroidal current I_p that a tokamak can contain also has this same limitation. Below the MHD critical current there are important kinetic-drift wave instabilities associated with the toroidal current density profiles. These slower modes are seen in the precursor oscillations before the current disruptions and are discussed in Chapter 15.

Laboratory imaging by Park, *et al.* (2004) shows movies of the precursor motions and then the final disruption dynamics in the TEXTOR plasmas. The movies are made from tomographic reconstruction of the plasma electron temperature from soft-X ray emission with high resolution spectra. Namely, for currents that are below the MHD onset of the internal kink $m = 1/n = 1$ mode there is the resistive tearing mode unstable growth rate. The calculations for this mode are famous and elaborate since they provide one of the most intrinsic limits to high-performance operation of tokamaks.

A detailed calculation of the stability theory of the 1/1-mode known as the internal kink mode is given by Crew and Ramos (1983). They calculate $\delta W(\xi)$ for each toroidal mode number n and the internal kink mode growth rates, $\gamma = n v_{A1}/R_0$, for shifted circular magnetic surfaces by an expansion in low $\epsilon = r/R_0$ for $\beta_p = 2\mu_0 \langle p \rangle / B_\theta^2(a) > 1$ provided $\epsilon \beta_p < 1$. Here v_{A1} is the Alfvén velocity calculated with poloidal magnetic field from the plasma current at the rational surface $q(r_{1/1}) = 1$ where the field line twist matches the magnetic perturbation $\delta B_r = \delta B_{1,1} \cos(\theta - \phi)$ so that a magnetic island is formed. This special magnetic surface is where the field line is returning to exactly the same point, thus being the shortest such length in the torus, after one revolution in both the toroidal (long way) and poloidal (short way) around the torus. Experimentally, from the beginning of the tokamak experiments, strong oscillations and disruptions of the toroidal plasma current originate from the vicinity of this surface. Images of the plasma from X-ray tomography from H. Park are available at `http://orion.ph.utexas.edu/~starpower`. Analysis and simulations [Aydemir (1990)] show there is an unstable window for $\beta_{p1} < \beta_p < \beta_{p2}$ with the internal kink. The nonlinear effect of this mode is the displacement of the core plasma inside the $q = 1$ rational surface into a helical $\times$-point. At the $\times$-point magnetic reconnection occurs as given by the Kadomtsev theory. The core plasma temperature drops abruptly in what is called the sawtooth of instability in MHD stable domains.

6.2 Double Adiabatic Pressure Tensor for Anisotropically Heated Plasmas

The use of the scalar pressure in the MHD model is a severe restriction for plasmas with strong external heating when the mean-free-path is long compared to the parallel scales $L_\parallel$ of the background plasmas. Two important effects and their associated instabilities are missed with the isotropic pressure model. Heating methods such as neutral beam injection and ion cyclotron resonance heating drive up a temperature anisotropy with $T_\parallel > T_\perp$ for NBI heating and with $T_\perp > T_\parallel$ for the cyclotron resonance heating. To include these anisotropies in the wave dynamics the scalar pressure is replace in the momentum equation with a pressure tensor

$$\boldsymbol{p} = p_\perp(\boldsymbol{I} - \boldsymbol{bb}) + p_\parallel \boldsymbol{bb} \tag{6.31}$$

where the force from the pressure tensor in the momentum equation becomes

$$\boldsymbol{F}_p = -\nabla\cdot\boldsymbol{p} = -\nabla p_\perp - \left(p_\parallel - p_\perp\right)(\boldsymbol{b}\cdot\nabla)\boldsymbol{b} - [\boldsymbol{B}\cdot\nabla\left(p_\parallel - p_\perp)/B\right))]\widehat{\boldsymbol{b}}. \tag{6.32}$$

The parallel component of the force $\boldsymbol{F}$ is used in the collisional transport theory in Chapter 17. So we give here the form $F_\parallel$ that is used in that chapter and in general for the bootstrap current calculations. The parallel part of the force can be written in several different forms, one of which is

$$\boldsymbol{B}\cdot\boldsymbol{F} = -\left(p_\perp - p_\parallel\right)(\boldsymbol{b}\cdot\boldsymbol{\nabla}B) - B\boldsymbol{b}\cdot\nabla p_\parallel. \tag{6.33}$$

In the low-collisionality regime where the trapped particles dominate the transport the fluxes are computed from the average over the magnetic surface of the parallel force from the anisotropic pressure tensor. The divergence of the anisotropic pressure tensor produces a parallel force given by

$$F_\parallel = -\left(p_\perp - p_\parallel\right)\boldsymbol{b}\cdot\boldsymbol{\nabla}\ln B - \boldsymbol{b}\cdot\nabla p_\parallel = -\frac{p_\perp}{B}\boldsymbol{b}\cdot\nabla B - \boldsymbol{B}\cdot\nabla\left(\frac{p_\parallel}{B}\right). \tag{6.34}$$

When this force is averaged over the entire flux surface with periodicity in the toroidal coordinates the net parallel force in the between ψ and $\psi + \delta\psi$ is given by

$$\langle\boldsymbol{B}\cdot\boldsymbol{\nabla}\cdot\boldsymbol{\pi}\rangle = \left\langle\left(p_\perp - p_\parallel\right)\nabla_\parallel B\right\rangle. \tag{6.35}$$

In conclusion, the banana-plateau flux is driven by parallel variation of magnetic field strength, $\nabla_\parallel B \neq 0$, in combination with a difference between the parallel and perpendicular pressures. This difference can be written in terms of the Legendre polynomial $P_2(\xi) = (3\xi^2 - 1)/2$.

$$p_{\|} - p_{\perp} = \int mv^2 P_2(\xi) f d^3v. \tag{6.36}$$

Here ξ is the cosine of the pitch angle with $\xi = v_{\|}/v$.

Thus we see that the various heating methods as NBI, ICRH and ECH contribute to the anisotropy driving a parallel force that produces transport. The simplest cases of this are familiar ones in the collisional regime where the force due the parallel resistivity and thermoelectric effects determine the Pfirsch-Schlüter diffusion, derived in Chapter 9.7 and the generalizations from thermoelectric effects. This name "diffusion" is a misnomer in that these fluxes are actually from cross field convective flows that are set up by the forces acting parallel to the magnetic fields. The calculations for the flows and the parallel plasma flows are then the first part of the calculation for these parallel forces that drive these net fluxes. When the parallel flow is related to the cross field pressure gradients and perpendicular electric field this calculation is called the calculation for the parallel viscosity coefficients of the plasma. The works of Hirshman and Sigmar (1997) and Helander and Sigmar (2002) derive lengthy formulas for the transport effects in the absence of the fluctuations. The works of Sugama and Horton (1994), Sugama, *et al.* (1996) and Balescu (2005) derive the transport formulas in the presence of the fluctuations.

The value of this form of the parallel pressure gradient force is that when the surface average use transport theory is taken, the second term proportional to $B\boldsymbol{b} \cdot \boldsymbol{\nabla} F$ vanishes for any F, leaving the anisotropy in the pressure as the net driving force. These results are used extensively in both collisional and turbulent transport calculations.

Now the force is understood by introducing the volume element of the flux magnetic tube $dAds = d\psi ds/B$ where $d\psi$ is the element of the flux tube volume defined by the product of its area and the local magnetic field vector $d\psi = \boldsymbol{B} \cdot \boldsymbol{dA}$. Here $\boldsymbol{dA}$ is the vector element of area and thus $d\psi$ is the magnetic flux through this surface element. For plasma in the given flux tube the cross-sectional area decreases as the magnetic field strength increases. The plasma becomes compressed in the perpendicular direction and the perpendicular pressure increases strongly. To describe the microscopic kinetic physics correctly one takes into account that the particle's magnetic moments $\mu = mv_{\perp}^2/2B$ is conserved if the dynamics is slow on the space-time scale of the cyclotron orbit. This means that the fluid motion must be such as to have

$$\frac{dp_{\perp}}{dt} + 2p_{\perp}\nabla \cdot \boldsymbol{u} - p_{\perp}\boldsymbol{b} \cdot (\boldsymbol{b} \cdot \boldsymbol{\nabla})\boldsymbol{u} = 0 \tag{6.37}$$

which is consistent with the conservation of the average magnetic moment expressed in fluid variables as $p_{\perp}/\rho B = \text{constant}$.

For the evolution of the parallel pressure we must use a weaker adiabatic invariant associated with the parallel motion, namely the second particle adiabatic invariant from the action $J = \int v_{\|} ds$. Averaged over the particles motion this statement is that $p_{\|}B^2/\rho^3 = $ constant. Thus the partial differential equation derived from the ordering yields the evolution of the parallel pressure as

$$\frac{dp_{\|}}{dt} + p_{\|}\nabla\cdot\boldsymbol{u} + 2p_{\|}\left(\boldsymbol{b}\cdot(\boldsymbol{b}\cdot\nabla)\boldsymbol{u}\right) = 0. \tag{6.38}$$

To understand that Eq. (6.37) and Eq. (6.38) are consistent with $p_\perp$ and $p_{\|}$ equations of state, one uses that the $\boldsymbol{b}$-component of $\partial_t\boldsymbol{B} = \nabla\times(\boldsymbol{u}\times\boldsymbol{B})$ gives $\boldsymbol{b}(\boldsymbol{b}\cdot\boldsymbol{\nabla}\boldsymbol{u}) = B\boldsymbol{\nabla}\cdot\boldsymbol{u} + dB/dt$.

These equations are known as the Chew-Goldberger-Low equations and are described in the usual text books and in Volkov (1966). The linearization of these equations gives the quadratic equation in ω^2 with

$$\omega^4 - B\omega^2 - C = 0 \tag{6.39}$$

with $B(k)$ and $C(k)$ given by

$$B(k) = \left(\frac{2p_\perp}{\rho} + v_A^2\right)k^2 + \frac{(2p_{\|} - p_\perp)}{\rho}k_{\|}^2 \tag{6.40}$$

and

$$C(k) = \left(\frac{p_\perp^2}{\rho^2} - 3\left(\frac{p_{\|}}{\rho_o}\right)\left(\frac{2p_\perp}{\rho} + v_A^2\right)\right)k^2k_{\|}^2 + \left(\frac{3p_{\|}^2 + p_\perp(3p_{\|} - p_\perp)}{\rho^2}\right)k_{\|}^4. \tag{6.41}$$

For small anisotropies these waves give the kinetically correct dispersion relations without ad hoc choices for the adiabatic gas constant. For larger anisotropies and plasma pressures comparable to the magnetic pressure the discriminant of Eq. (6.39) goes negative for certain directions of propagation. The instabilities are (i) the firehose mode for parallel propagation and (ii) the mirror mode for perpendicular propagation. Figure 6.1 and Figure 6.2 show the unstable k-space domains for the firehose and mirror modes. Both instabilities are important and common in the magnetospheres and solar wind plasmas. The mirror mode limits the maximum beta $\beta = 2\mu_0 P_\perp/B^2$ for a stable mirror trapped plasma for a specified mirror ratio $R_M = B_{\max}/B_{\min}$.

To understand the nonlinear dynamics of either the firehose or the mirror mode one needs to introduce the kinetic theory dispersion terms. This requirement can be seen from the fact that once dispersion relation Eq. (6.39) gives a growing mode, all modes with $k\to\infty$ are unstable in the model given in Eq. (6.39). Kinetic theory shows that the firehose mode becomes a right-hand polarized wave that connects to the whistler dispersion relation, while the mirror mode becomes a compressional wave with FLR and ω_* corrections.

The pressure anisotropies change the MHD equilbrium from that given by the Grad-Shafranov equation to a complex pair of partial differential equations. The MHD stability analysis now becomes complicated with numerous special cases with the mirror mode stability limiting the high beta minimum-B mirror equilibrium

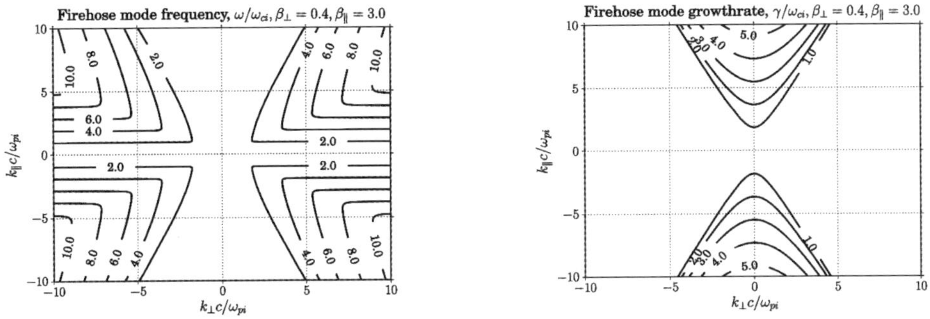

(a) (b)

Fig. 6.1 The left figure above shows the frequency contours when $\beta_\perp = 0.4$, $\beta_\parallel = 3.0$. ($\beta_\perp = 8\pi P_\perp/B^2$, $\beta_\parallel = 9\pi P_\parallel/B^2$). The blank regions, which form two triangle-like shapes, are the firehose instability regions. The criterion for firehose mode instability is $\beta_\parallel - \beta_\perp > 2$. The right figure above shows the growth rate contours for the same parameters. Firehose mode mainly propagates along the magnetic field.

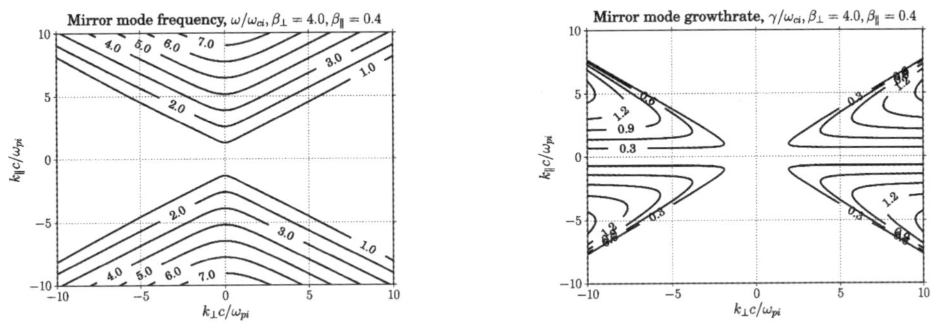

(a) (b)

Fig. 6.2 The left figure above shows the frequency contours when $\beta_\perp = 4, \beta_\parallel = 0.4$. In contrast to firehose instability, we observe another kind of instability which mainly propagates perpendicular to the magnetic field. The double adiabatic criterion for mirror mode instability is $\beta_\perp^2 > 6\beta_\parallel(1+\beta_\perp)$. The right figure shows the growth rate contours. In kinetic theory the factor 6 becomes unity.

and the firehose mode limiting the parallel pressure in toroidal systems with parallel NBI heating. The variational formulation of the MHD stability analysis for the anisotropic pressure equilibrium is developed in the Nelson, *et al.* (1978). In that analysis one finds the firehose and mirror mode criterion limit further the maximum pressure that can be stability confined in a magnetic confinement system. A comprehensive extension of the MHD to include anisotropy and subsconic flows is developed in Ito, *et al.* (2007) and (2011) where the anisotropic MHD is extended by calculating the electron parallel heat flux, whose moment is then closed in such away to reproduce a corresponding dispersion relation derived by kinetic theory. The ions are taken as cold for the closure of their momentum flux. The authors call the resulting single fluid model anisotropic Hall MHD. In an equilibrium with poloidal flow V_p, an analysis of the two coupled equilibrium equations shows that the system is elliptic provided the poloidal flow is subsonic by a margin that depends on the degree of anisotropy in the pressures $p_\perp$ and $p_\parallel$.

Kinetic theory gives a lower threshold for the mirror mode and is required for finding the maximum growth rate $\gamma_{\max}$ and corresponding wavenumber $k_{\max}$. The kinetic theory stability parameter is $K = T_\perp/T_\parallel - 1 - 1/\beta_\perp > 0$ and $\gamma_{\max}/\omega_{ci} \propto \beta_\perp^{1/2} K^{3/2}$ [Pokhotelov, *et al.* (2004)].

6.3 Ballooning Interchange Modes and the Trapped Particle Instability

Once the spatial variation of the magnetic field vector $\boldsymbol{B} = B(x)\boldsymbol{b}(\boldsymbol{x})$ where $\boldsymbol{b}$ is the unit tangent vector is taken into account the force accelerating the plasma becomes a parallel mirror force and a perpendicular field line bending or curvature force. The mirror force traps particles with large magnetic moments in the local minimum of the magnetic field strength which separates their dynamics from the high parallel velocity particles that pass over the maxima of the magnetic field. This separation of the trapped and passing particles exists until a Coulomb collision rotates the velocity vector sufficiently to allow the particle to change its class from trapped or passing. This difference means that the trapped particles may behave as though they were in flute modes with $k_\parallel = 0$ and thus become interchange unstable for magnetic field line curvature concave towards the higher pressure plasma.

The lowest-order eigenmode of the plasma with a changing sign of the curvature of the magnetic field from concave to convex and remain flute mode with $k_\parallel = 0$ and still be unstable when the condition $(dp/d\psi)(dV/d\psi) < 0$ is satisfied. This condition expresses the appropriate averaging of the magnetic field line curvature κ through the gradient $\frac{dV}{d\psi}$ of the flux tube volume. This is the stability condition for fast growing interchange modes that are relatively constant along the magnetic field line the displacement ξ_r is measured in magnetic flux. That is the contravariant displacement $\xi^\psi = B\xi_r(s)$.

Here $V(\psi)$ is the flux tube volume given by

Flux Tube Volume

$$V(\psi) = \int \left[\frac{ds}{B(s,\psi)}\right].$$

This instability condition can be understood from either thermodynamics or kinetic theory. In thermodynamics the condition follows from requiring the energy released in a compressional motion down the pressure gradient exceed the work required to compress the plasma. In kinetic theory the condition follows from the flux tube average of the microscopic particle wave resonance condition of

Line Average Interchange

$$\int \frac{ds}{B(s)} [\omega_* \omega_D] > 0$$

which is the condition for the guiding center drift velocity to separate charge from over the wave in such a wave to produce an $\boldsymbol{E} \times \boldsymbol{B}$drift of the plasma down the pressure gradient. This interchange condition was understood early by Longmire and Rosenbluth (1956).

Including the stabilizing term from plasma compression with an ideal gas constant of Γ we get the condition for instability is

Gradient of Entropy

$$\frac{dpV^\Gamma}{d\psi} < 0.$$

This is a famous stability condition and applies to a wide variety of confinement geometries including magnetospheres and solar flux tubes. The condition is used to understand how the dipole magnetic fields with unfavorable magnetic curvature and guiding center drifts are stable until the pressure gradient becomes very steep. This is because for a dipole field the volume per unit of flux

$$V(r) = \int \frac{ds}{B(s,\psi)} = \text{const}\, \frac{r^4}{B_{dp}} \tag{6.42}$$

so the pressure gradient must exceed $-d \ln p/dr > 4\Gamma/r$ for the MHD motion to release plasma thermal energy.

6.4 Experimental Discovery of the Trapped Particle Instabilities

Dipole magnetic fields with everywhere unfavorable magnetic curvature for interchange stability were not considered for laboratory plasma confinement. The simple extension to a kind of "double dipole" with two concentric loops with parallel currents, formally called the quadrupole, gives rise to alternating regions of favorable and unfavorable magnetic curvature inside the critical flux surface ψ_c. Inside the surface ψ_c the field lines have average favorable magnetic curvature where the condition in Eq. (6.42) is satisfied, and now the system is one suitable for magnetic confinement. That there are regions of favorable curvature near the field null is clear from geometry of the field lines as well as the fact that B^2 increases away from the magnetic $\times$-point. In fact, moving away from the magnetic null or $\times$-point the system has local minimum $B(\boldsymbol{x})$ surfaces not just average minimum-B properties. Thus the separatrix region is one of strong single particle and plasma confinement.

A number of linear and toroidal quadrupole and octupole experiments were built and operated in the late 1960s and early 1970s. These machines showed for the first time that confinement better than the Bohm value by more than an order of magnitude could be achieved. They also clearly demonstrated that drift wave fluctuations are universal in most regions of an MHD stable plasma.

An important example of the quadrupole experiments was the 5-meter long, 0.45 meter diameter cylindrical device. The plasma was produced by inductively coupling 200 kHz of RF power to the helium working gas. The typical after-glow plasma density was decaying on the time scale of a 1 msec with T_e dropping from 10 to 1 ev and $T_i \sim T_e/5$.

The fluctuations in the density and the potential were measured by the ion saturation current and the floating potential measurements with Lagrangian probes. The probe is movable and has a variable bias to determine the plasma fluctuating δn_e and $\delta\varphi$. Common to all the quadrupole and octupole experiments there is an abrupt drop from the high fluctuation level $\delta n/n \sim$ 50-100% in the flute mode unstable region where

MHD Unstable

$$\frac{dV}{d\psi}\cdot\frac{dp}{d\psi} < 0 \tag{6.43}$$

to a small amplitude 10-15% turbulence in the MHD stable region when the sign of Eq. (6.43) reverses. The fluctuations with frequencies $f \sim 60$ to 200 kHz are drift waves. These drift wave oscillations occur in the $V(\psi) = \oint ds/B$-stable regions where $(dV/d\psi)(dp/d\psi) > 0$. The drop in the fluctuation level occurs right at the critical magnetic surface ψ_c where the derivative of the flux tube volume $dV/d\psi$ changes sign.

The $ds/v_{\parallel}$-averaged sum of the ∇B and curvature drift also changes sign at ψ_c. The average of the guiding center drifts over the flux tube is given by the gradient of the flux tube volume $V(\psi) = \int ds/B$. Thus, the critical surface where the average drift changes from stabilizing to destabilizing is where

$$\frac{dV}{d\psi} = \frac{d}{d\psi}\int \frac{ds}{B}\bigg|_{\psi=\psi_c} = 0. \tag{6.44}$$

Inside of ψ_c the average magnetic curvature is favorable, and the fluctuations are of the drift wave type with $k_{\parallel} \neq 0$. Thus, $E_{\parallel} = -\boldsymbol{b}\cdot\nabla\phi = \mathrm{Re}[-ik_{\parallel}\widetilde{\phi}_k \exp(i\boldsymbol{k}\cdot\boldsymbol{x} - i\omega t)]$ is finite for the fluctuations inside ψ_c.

6.5 Discovering the Trapped Particle Instability

Meade and Yoshikawa (1967) were the first to report the details of this transition from large amplitude flute mode with $\delta n/n \approx 0.5$ outside ψ_c with frequencies in the 1-20 kHz, to the ballooning drift waves with $\delta n/n \lesssim 0.15$ with frequencies $f \geq 60\,\mathrm{kHz}$ inside the ψ_c surface. The word "flute" is used to describe the topology of the $k_\| = 0$ fluctuations. The topology of the flute motion is analogous to the fluted columns in Greek architecture.

The structure is found to be an odd parity ballooning mode with its nodes at the points of maximum favorable curvature and anti-nodes at the point of maximum bad curvature. This parallel variation means that there is one parallel wavelength $2\pi/k_\|$ along the closed field line. For the He plasma at the condition of $n_e = 2 \times 10^{12}\mathrm{cm}^{-3}$ and $T_e = 5\,\mathrm{ev}$ the mean-free-path $\lambda_{\mathrm{mfp}} = 5\,\mathrm{m}$ and the field line length is $0.7\,\mathrm{m}$. Thus, $k_\| \lambda_{\mathrm{mfp}} \simeq 5$ and the drift wave is essentially the collisionless trapped electron mode. The local Boltzmann relation $\delta n_e/n \simeq e\delta\phi/T_e$ was satisfied to within 50%. The transverse (axial) wavelength was $6 - 10\,\mathrm{mm}$ so that $k_\perp \rho_i = 0.5$ to 1.0 with $\rho_i = (m_i T_i)^{1/2}/eB_{\mathrm{min}}$. There are approximately $10\rho_i$ between the separatrix ψ_s surface and the critical surface ψ_c for the baseline operation with 45 kA in the internal conductors that are 20 cm apart. The internal conductors had a stainless steel shield that was grounded and supplied with up to 60 kA for 200 ms. The subsequently published work of Coppi (1968a,b) refer to the Meade and Yoshikawa (1967) experiment and a closely related experiment of Ohkawa and Yoshikawa (1968) in the GA octupole. In the GA octupole the fluctuations in the MHD stable region are described as having $k_\perp \rho_i > 1$ and interpreted as high-frequency drift waves that are a FLR modified pressure gradient ballooning interchange mode [Horton, *et al.*, 2002].

In summary the multipole experiments clearly demonstrated the existence of flute-type interchange modes with high amplitude fluctuation levels in the MHD unstable region beyond the critical surface ψ_{crit} defined by

$$\frac{dV}{d\psi}\frac{dp}{d\psi}\bigg|_{\mathrm{crit}} = 0 \tag{6.45}$$

and ballooning drift wave turbulent fluctuations in the MHD stable regions. In terms of the bounce-averaged guiding-center drift velocities $\omega_{Dj}(\lambda)$ the condition Eq. (6.45) is equivalent to

$$\langle \overline{\omega_{D_j}}(\lambda)\omega_{*p_j} \rangle_{\mathrm{MB}} \geq 0 \tag{6.46}$$

where the average is over the Maxwell-Boltzmann (MB) distribution functions.

In the inner region, inside ψ_{crit}, of flute stability where $V'p' > 0$, the experiments showed a wide spectrum of high frequency oscillations that were partially localized to the regions of locally unfavorable curvature. These localized modes were called ballooning modes in the title of the Meade and Yoshikawa (1967) article. Both groups report that the confinement time was up to ten times the value expected from the tradition Bohm formula. To beat the Bohm value by an order magnitude was a major breakthrough at the time. The stellarator experiments had for years produced Bohm confinement times. Bohm confinement would preclude building a reasonably-sized nuclear fusion reactor.

6.5.1 *Parallel particle dynamics*

The condition that the growth time, $1/\gamma_{\max}$, be shorter that the time for the particles to travel over the inhomogeneous of the perturbed fields, $1/k_{\|}v_T$, and the mirror time of the single particles is required for the fluid model to be valid. In the case of no mirror force, if this condition is not satisfied then the time scale is comparable to the particle transit time over the parallel wavelength. In this regime there is phase mixing and Landau growth or damping from the interaction of the particles with the k,ω mode. These are effects that require the kinetic equation to describe. In the case that the parallel wavelength is very long, but the equilibrium magnetic field increases from some minimum value $B_{\min}$, then the bouncing motion of the mirroring particles averages the perturbed fields over the bounce motion. The self-consistent field equations are then integral equations over this bounce motion. When the integral field equation yields unstable motions with growth times shorter than the bounce time of the particles, the integral equations reduce to the local differential equations of the fluid models.

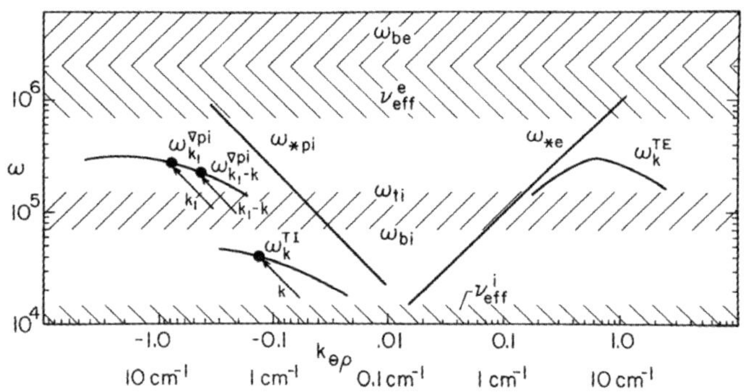

Fig. 6.3 A schematic of the frequency ranges given by shaded zones for the electron and ion bounce frequencies ω_b and the collisional detrapping frequencies $\nu_{\rm eff}$ for the electrons and the ions in a typical high-temperature tokamak. Between these shaded regions are the unstable drift waves propagating in the electron diamagnetic direction on the right and in the ion diamagnetic direction on the left. On the left is a typical three-mode interaction between wavenumbers $\mathbf{k}, \mathbf{k}_1$ and $\mathbf{k}_2 = \mathbf{k} - \mathbf{k}_1$. The three frequencies may be resonant or nonresonant in such three-mode interactions. The transit frequencies have lower frequency given by the bounce frequency ω_b and upper frequency given by the transit frequency ω_t for the small pitch-angle particles.

The first theory that extended the MHD stability formulation of the perturbed plasma potential energy $\delta W(\boldsymbol{\xi})$ to take these effects in to account is the Kruskal-Oberman stability theory [Kruskal and Oberman, 1958]. To correctly calculate the perturbed plasma energy in this regime, the theory introduces the bounce averaging operator for particles using that their energy and magnetic moment are constant. The magnetic moment, $\mu = mv_\perp^2/2B(\boldsymbol{x})$, is a constant of the motion in both the equilibrium and perturbed electromagnetic fields and that the energy E is constant $E = \frac{mv^2}{2}$ in the equilibrium (but not the perturbed state). Then the response of the plasma is determined by the bounce averaging of the perturbed field, for example, the perturbed electrostatic potential $\phi_k(s)$ and compressional change in the magnetic field $\delta B_\|$ through

Bounced Averaged

$$\overline{\phi}_k = \frac{1}{\tau}\int_{s_1}^{s_2} \frac{ds'\ \phi_k(s')}{v_\|(E,\mu,s')} \tag{6.47}$$

$$\overline{\delta B_\|(k)} = \frac{1}{\tau}\int_{s_1}^{s_2} \frac{ds'\ \delta B_\|(k,s')}{v_\|(E,\mu,s')} \tag{6.48}$$

where $\tau = \int_{s_1}^{s_2} ds'/v_\|(E,\mu,s')$. The parallel velocity is given by $v_\| = [2/m(E - \mu B(s) - q\Phi(s))]^{1/2}$. In Eq. (6.47) and Eq. (6.48) where $q_s\Phi$ is often small compared with μB. The turning points s_1, s_2 are then given by the roots of $E = \mu B(s_{1,2})$. Often the pitch angle-like variable $\lambda = \mu B_{\min}/E$ is used in toroidal geometry for the calculations of $\overline{\phi}$ and $\delta\overline{B}_\|$ and the other bounce averaged fields occurring in the low-frequency dynamics.

The mirror bounce frequencies $\omega_{be} \gg \omega_{bi}$ introduce new structure to the frequency spectrum as shown in Fig. 6.3. The upper shaded region is that of the electron bounce frequency and the collisional detrapping ν_{eff} destabilizes the waves.

Clearly, this orbit-averaged response can be very different from the local response depending on the symmetry and number of spatial oscillations of the fields $\phi_k(s)$ and $\delta B_{\|k}(s)$ along the magnetic field line. The bounce averaged stability theory equations are developed in Chapter 5. The problem is particularly important for determining the stability of the magnetic dipole trapped plasmas in nature and the multipole magnetic plasma confinement systems.

In Eq. (6.47) and Eq. (6.48) the s_1 and s_2 are the mirroring points of the particle where $v_\| = 0$. The weight $ds/v_\|$ is the time the orbit spends in ds at s and this integral is then divided by $\tau = \int_1^2 ds/v_\|$ the total time to move from the turning point s_1 to that at s_2. The particle is decelerated and stopped at s_1 and s_2 by the combination of the parallel decelerating force from $F_\| = -\mu\frac{dB}{ds} - qd\Phi/ds$. In the absence of an equilibrium potential $\Phi(s)$ the mirror points are only a function of the particle pitch angle α through $\sin^2\alpha = \lambda = \mu B_{\min}/E$.

In this case it is apparent that the bounced-averaged field depends only on the dimensionless pitch angle variable $\lambda = \mu B_{\min}/E$. Thus, the bounce averaged fields are

$$\overline{\Phi}(\lambda) = \frac{1}{\tau}\int_1^2 \frac{ds\ \phi(s)}{\sqrt{q(1-\lambda B(s))}} \tag{6.49}$$

$$\overline{\delta B}(\lambda) = \frac{1}{\tau}\int_1^2 \frac{ds\delta B_\|(s)}{\sqrt{2(1-\lambda B(s))}}. \tag{6.50}$$

This averaging operator cannot be inverted to give $\phi(s)$ from $\overline{\phi}(\lambda)$ uniquely. To be able to make an inversion one needs to compute the whole sequence of pitch angle averaged fields weighted by $\sin(n\tau(s)/2\tau(\lambda))$ and $\cos(n\tau(s)/2\tau(\lambda))$ where the half-period $\tau(s) = \int_1^s ds'/v_{\|}(E, \mu, s')$ maps s to particle time τ. The presence of a weak rotation of the plasma Doppler shifts the frequency in this equations from the plasma rest frame to the laboratory frame. For a rotation with significant shear involved, the situation becomes much more complicated. The problem of the stability of the rotating plasma is the subject of Chapter 17.

References

Aydemir, A.Y. (1990). *Phys. Fluids B* **2**, p. 2135.

Balcscu, R. (2005). Aspects of anomalous transport in plasmas, *Series in Plasma Physics* (Institute of Physics Publishing).

Barnes, D. C. and Brackbill, J. U. (1977). *Nucl. Sci. Eng. (US)* **64**, p. 1.

Boris, J. P. and Book, D. L. (1973). *J. Comp. Phys.* **11**, pp. 38-69, doi:10.1016/0021-9991(73)90147-2.

Braginskii, S. I. (1965). *Rev. Plasma Phys.* ed. M. A. Leontovich (Consultants Bureau, New York), **I**, p. 205.

Burgess, D. (1995). Chap. 5, Collisionless Shocks, *Introduction to Space Physics* eds. M. G. Kivelson and C. T. Russell, (Cambridge, MA) p. 129.

Carreras, B. A., Garcia, L., and Diamond, P. H. (1987). Theory of Resistive pressure-gradient driven turbulence, *Phys. Fluids* **30**, p. 1388.

Coppi, B., Laval, G., Pellat, R., and Rosenbluth, M. N. (1968a). *Plasma Phys.* **10**, p. 1.

Coppi, B., Rosenbluth, M. N., and Yoshikawa, S. (1968b) *Phys. Rev. Lett.* **20**, p. 190.

Crew, G. B. and Ramos, J. J. (1983). *Phys. Fluids* **26**, 9, pp. 2621-2634.

Diamond, P. H., Itoh, S-I., Itoh, K., and Hahm, T. S. (2005). *Plasma Phys. Control. Fusion* **47**, pp. R35-R161.

Friedberg, R. (2007). *Plasma Physics and Fusion Energy* (Cambridge University Press) ISBN:13-978-0-521-85107-7.

Goedbloed, J. P., Keppens, R., and Poedts, S. (2010). *Advanced Magnetohydrodynamics* pp. 269-304 (Cambridge Univ. Press) ISBN:978-0-521-87957-5.

Grierson, D. A., Burrell, K. H., Solomon, W. M., and Pablant, N. A. (2010). Deuterium velocity and temperature measurements on the DIII-D tokamak, *Rev. Sci. Instrum* **81**, p. 10D735, doi:10.1063/1.3491222.

Hamaguchi, S. and Horton, W. (1991). *Phys. Fluids B* **4**, p. 319.

Helander, P. and Sigmar, D. J. (2002). *Collisional Transport in Magnetized Plasmas* (Cambridge Monographs on Plasma Physics).

Hirshman, S. P. and Sigmar, D. J. (1997). *Phys. Fluids* **20**, p. 518.

Horton, W., Wong, H. V., Van Dam, J. W., and Crabtree, C. (2002). *J. Geophys. Res.* **106**, pp. 18803-18822.

Howard, L. N. and Krishnamurti, R. (1986). *J. Fluid Mech.* **170**, pp. 385-410.

Ito, A., Ramos, J. J., and Nakajima, N. (2007). Ellipticity of axisymmetric equilibria with flow and pressure anisotropy in single-fluid and Hall magnetohydrodynamics, *Phys. Plasmas* **14**, p. 062502.

Ito, A. and Nakajima, N. (2011, Equilibria of toroidal plasmas with toroidal and poloidal Flow in high-beta reduced magnetohydrodynamic models, *Nucl. Fusion* **51**,

p. 123006.

Kelley, G. (2009). (Elsevier) ISBN:13-978-0-12-088425-4.

Kennel, C. F. (1969). *Rev. Geophys.* **7**, pp. 379-419, doi:10.1029/RG007i001p00379 .

Kivelson, M. A. and Russell, C. T. (1983). *J. Geophys. Res.* **88**, pp. 49-57, doi:10.1029/JA088iA01p00049.

Krall, N. A. and Trivelpiece, A. W. (1968). *Advances in Plasma Physics, I*, eds. A. Simon and W. B. Thompson (Interscience Publishers, New York), p. 153.

Krishnamurti, R. and Howard, L. N. (1981). *Proc. Nat. Acad. Sci., USA* **78**, p. 4.

Kruskal, M. D. and Oberman, C. R. (1958). *Phys. Fluids* **1**, p. 275, doi:10.1063/1.1705885.

Kulsrud, R. M. (1983). *Handbook of Plasma Physics*, eds. M. N. Rosenbluth and R. Z. Sagdeev (New York: North Holland).

Longmire, C. L. and Rosenbluth, M. N. (1956). *Phys. Rev.* **103**, pp. 507-510, doi:10.1103/PhysRev.103.507.

Meade, D. M. and Yoshikawa, S. (1967). *Phys. Fluids* **10**, pp. 2649-2652.

Morrison, P. J. and Greene, J. M. (1980). Noncanonical Hamiltonian density formulation of hydrodynamics and ideal magnetohydrodynamics, *Phys. Rev. Lett.* **45**, pp. 790-794; **48**, p. 569 (1982).

Nelson, G. O., Spies, and Hedrick, C. L. (1978). Necessary stability criterions for plasma equilibrium with tensor pressure, *Phys. Fluids* **21**, p. 1742.

Newman, D., Carreras, B., Lopez-Bruna, D., Diamond, P. H., and Lebedev, V. (1998). *Phys. Plasmas* **5**, p. 938.

Ohkawa, T. and Yoshikawa, M. (1968). *Phys. Fluids* **11**, pp. 2039.

Park, H., Mazzucato, E., Munsat, T., Domier, C. W., Johnson, M., Luhmann, Jr., N. C., Wang, J., Xia, Z., Classen, I. G. J., Donné, A. J. H., and van de Pol, M. J. (2004). Simultaneous microwave imaging system for densityand temperature fluctuation measurements on TEXTOR, *Rev. Scien. Instrum.*, **75**, 10, pp. 3787-3792, ISSN:0034-6748.

Pokhotelov, O. A., Sagdeev, R. Z., Balikhin, M. A., and Treumann, R. A. (2004). Mirror instability at finite ion-Larmor radius wavelengths. *J. Geophys. Res.* **109**, pp. A09213-A09222, doi:10.1029/2004JA010568.

Sagdeev, R. Z. and Galeev, A. A. (1969). *Nonlinear Plasma Theory* (Benjamin, New York).

Shapiro, S. L. and Teukolsky, S. A. (1983). (Cambridge University Press) ISBN:0-471-87316-0.

Sugama, H. and Horton, W. (1994). Transport suppression by shear flow generation in multiplicity resistive-g turbulence, *Phys. Plasmas* **1**, p. 2220.

Sugama, H., Okamoto, M., Horton, W., and Wakatani, M. (1996). Transport processes and entropy production in toroidal plasmas with gyrokinetic electromagnetic turbulence, *Phys. Plasmas* **3**, 6, pp. 2379-2394, doi:10.1063/1.871922.

Sugama, H., Watanabe, T. H., and Horton, W. (2001). *Phys. Fluids* **8**, p. 2617-2628.

Sugama, H., Watanabe, T.-H., and Horton, W. (2003). *Phys. Plasmas* **10**, pp. 726-736.

Sugama, H., T.-H. Watanabe, T.-H., and Horton, W. (2007). *Phys. Plasmas* **14**, p. 022502.

Troyon, F. (1987). Theoretical studies of MHD stability, *Phil. Trans. R. Soc. Lond. A* **322**, pp. 163-171, doi:10.1098/rsta.1987.0045.

Volkov, T. F. (1966). Hydrodynamic Description of a Collisionless Plasma, *Rev. Plasma Phys.* **4**, pp. 1-21.

Wakatani, M. (1998). Stellarators and heliotron devices, (Oxford University Press) ISBN0-19-507831-4.

Chapter 7

Laboratory Plasma Experiments for Waves and Transport

7.1 Laboratory Plasma Drift Waves in Cylinders

Let us now examine how these drift waves arise in simple cylindrical geometries and what is measured with probes in university-scale drift wave experiments.

The first detailed study of the drift wave growth, saturation and anomalous transport was in a series of Q-machine experiments [Hendel, *et al.* (1968); Chen and Hasegawa (1974)] describe the identification of the mode frequencies and wavelengths in detail with variation of the system parameters in the Q-machine. The linear growth rate, $\gamma_{\vec{k}} = \mathrm{Im}\,\omega$, is produced by the phase shift, $\psi_{\delta n,\phi}$, from the electron collisions, ν_e, resulting in a density response,

$$\frac{\delta n_e}{n_e} = \frac{\omega_{*e} + ik_{\parallel}^2 v_e^2/\nu_e}{\omega + ik_{\parallel}^2 v_e^2/\nu_e} \frac{e\phi_{\vec{k}}}{T_e} \tag{7.1}$$

from the potential wave $\phi_{\vec{k}}$. The growth rate, $\gamma_{\vec{k}}$, is controlled in the experiments by varying the ion viscosity μ, where $\mu/\rho_m = (3/10)\nu_i\rho_i^2$, ν_i is the ion collision frequency, ρ_i the ion gyroradius and ρ_m the mass density. Since $\mu/\rho_m \propto 1/B^2$, the experimentalists varied the $\boldsymbol{B}$-field through the critical value, $\vec{B}_{\mathrm{crit}}$, to measure directly the growth rate as well as the wave frequency ω_k in the laboratory. The nonlinear saturation occurs when the nonlinearity in the ion dynamics balances the growth rate from the electron dynamics in Eq. (7.1). The growth is from ω_{*e}^2, which is the driving term of the instability. The $\boldsymbol{v}\cdot\nabla f$ convective nonlinear saturation is generic to drift waves since the rapid parallel electron motion leaves the electron-wave phase as in the linear response theory.

Formula Eq. (7.1) shows the key feature for density transport. When $\omega = \omega_{*e}$ the complex fraction is unity and the density is the adiabatic Boltzmann response. For $0 < \omega < \omega_{*e}$ the density wave leads the potential wave and there is net $\boldsymbol{E}\times\boldsymbol{B}$ transport down the density gradient. The high-frequency regime $\omega > \omega_{*e}$ is characterized by evanescent wave structures that become the vortex structures described in Chapter 10, when the amplitude of the potential fluctuation exceeds a critical level.

7.2 Helimak Confinement

7.2.1 *Helimak geometry and plasma dynamics*

The Helimak system produces laboratory simulations of space and astrophysical plasma with combined mass flow and sheared magnetic flux tubes as in the solar corona. The Helimak geometry is shown in Fig. 7.1 The electron temperature is similar to the corona at $T_e \sim 10\,\mathrm{eV}$ and the electron density is $n_e \sim 10^{11}\mathrm{cm}^{-3}$ a factor of 100 higher than the lower solar coronal. The magnetic field is strong so the plasma has a low plasma pressure-to-magnetic pressure ratio. By combined flow and magnetic shears, the Helimak creates a situation in which a sheared flow field and a sheared magnetic field are both present. This type of system is perhaps best illustrated by providing examples of the two canonical cases: the magnetized jet [Dahlburg, *et al.* (1998)] and the plane current vortex sheet [Dahlburg, *et al.* (1997)]. Both of these systems have sheared magnetic field structures. In the case of the magnetized jet, there is a fluid jet superimposed on the magnetic field structure. The magnetized jet is used to model a variety of physical systems, including solar surges [Carbone, *et al.* (1987); Dahlburg and Karpen (1994)] and the fast flows in the Earth's magnetotail [Sato and Walker (1982)]. The magnetized jet can be turned into a magnetized wake by application of a Galilean transform. This configuration is used to model the formation and acceleration of the slow solar wind [Einaudi and Rubini (1986); Einaudi and Rubini (1989); Einaudi, *et al.* (1999)], the heliospheric current sheet [Wang, *et al.* (1988)], and nonthermal galactic jets [Dahlburg (2001,2002)]. This plasma system with sheared magnetic fields and fast flows is frequently used to model solar surges [Carbone, *et al.* (1987)] and tokamak plasmas with equilibrium flows [Paris and Sy (1988)].

Laboratory plasmas experiments with combined flows and magnetic shear are the Helimak [Perez, *et al.* (2006)] and the LAPD [Horton, *et al.* (2005)]. In general, these plasmas share the characteristic that they are bounded, in contrast to astrophysical plasmas in which boundaries are often so far away from the region of interest that they may be ignored. Hence it is of interest to determine how the presence of walls alter the stability of magnetofluids with combined flow and magnetic shears. Are walls a stabilizing influence, or do they serve as regions for destabilization and turbulence production? For example, many of the unstable modes found in earlier studies are ideal, dissipationless instabilities. These modes are generally of large spatial extent in the cross-stream direction. Walls could disrupt the structure of these modes and alter their stability properties. Conversely, it is found in both fluid and fusion confinement systems that walls can provide a source region for instabilities, since large stresses can develop in response to the velocity or magnetic field striving to meet the imposed boundary condition.

For example, in magnetized plane Poiseuille flow, large Reynolds and Maxwell stresses develop at the channel walls for the case of a DC magnetic field [Nachtsheim

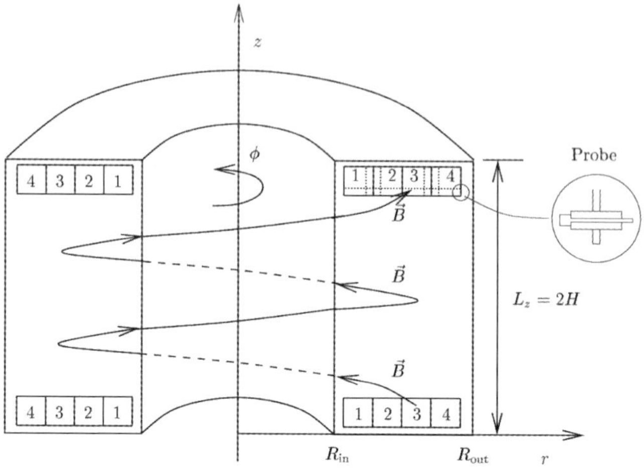

Fig. 7.1 A schematic drawing of the Helimak illustrating the laboratory coordinate system. The mathematical analysis uses a rotated coordinate system that brings $\widehat{e}_z$ parallel to the magnetic field at $r = (R_{\rm in} + R_{\rm out})/2$ and $\widehat{e}_y$ as the binormal direction perpendicular to $\widehat{e}_z$ and $\widehat{e}_r$. The probes labeled 1, 2, 3, 4 are biased to control the radial electric field and thus the plasma flow velocity $W_0(x)$.

and Reshotko (1965)]. Kent (1996) shows that sheared magnetic fields also strongly influenced the stability of MHD plane Poiseuille flow.

Early studies of magnetized channel flows indicate that there can be a significant interaction between sheared flows and magnetic fields in magnetofluids bounded in channels [Stern (1963); Kent (1966)]. Interest in MHD plane Poiseuille flow [Stuart (1954); Velikhov (1959)] has waned, so we study a representative system of contemporary interest: the Helimak experiment at The University of Texas at Austin. The Helimak is a toroidal magnetic confinement device designed to realize in the laboratory the sheared magnetic field slab geometry used by theorists in the analysis of localized modes in tokamaks [Perez, *et al.* (2006)]. When small resistive boundary layers are important it is not necessary for theory to keep the global radial profiles to study the plasma instabilities and turbulence. By virtue of the helical magnetic field structures, modes with long parallel wavelengths $\lambda_\parallel > 50 - 100\,\mathrm{m}$ are allowed in the experiment. The Alfvén wave dynamics is thus a natural mode of the system and the fluctuations of the magnetic field are an important aspect of the system's dynamics as opposed to experiments with short parallel wavelengths.

In this subsection we focus on the linearized theory and calculation of MHD instabilities associated with a sheared, bounded magnetofluid with flow and magnetic fields characteristic of the Helimak. The flow velocity is externally induced by the radial electric fields applied by segmented endplate electrodes numbered 1,2,3,4 in Fig. 7.1. Included are the effects of resistivity and viscosity, as well as no-slip walls [Dahlburg, *et al.* (1998)]. Linear theory provides a guide to the parameter

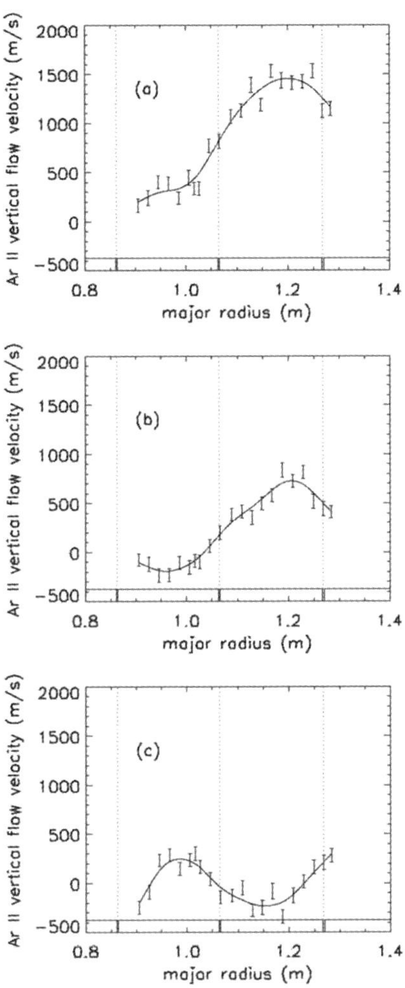

Fig. 7.2 Vertical flow profiles of a Helimak plasma at bias voltage values (a) $V_{\text{bias}} = 0$ (b) $V_{\text{bias}} = -10\,\text{V}$, and (c) $V_{\text{bias}} = -15\,\text{V}$. The error bars correspond to the statistical uncertainty in spectroscopic Doppler shift measurements of spectral atomic emission lines from a working gas of singly ionized argon.

space of the Helimak. In particular the stress analysis gives an understanding of the energetics of Helimak flow and fluctuating magnetic fields.

Figure 7.2 shows three examples of the vertical cross-field flow $v_z = w(x)$ configurations created in the Helimak plasma with the data points and error bars from spectroscopic Doppler shift measurements of the spectral lines from the working gas of singly ionized Argon. The flow profiles are controlled by the bias voltage V_{bias} given in the figure which changes the cross-field electric field E_r. The plasma responds in a complex manner to give the measured velocity profiles shown in frames a, b and c. The profiles have shear and the associated vorticity dV_z/dr of order

5×10^3 to 10^4/s changes sign at the inflection points in the profile. By virtue of the long lengths L of the magnetic field lines given by $L = B_\phi H/B_z > 50\,\mathrm{m}$ the Alfvén wave frequency is low $k_\| v_A/2\pi \sim 10\,\mathrm{kHz}$.

Figure 7.3 shows the fluctuation amplitudes of the electric potential and the electron density measured with Langmuir probes in the 10 ev Argon plasma. The probes show that $e\phi/T_e \gg \delta n_e/n_e$ over much of the flow profile which means that we may drop the drift wave terms $|\omega_{*e}/\omega|$ to first order. The Ohm's law then simplifies to $E_\| = \eta j_\|$ in the modeling of the experiment. The key fluctuations are the electric potential and the perpendicular component of the magnetic field in these flow dominated plasmas. Owing to the large mass of the Argon ions ($m_{Ar}/m_p = 40$) these low temperature plasmas have $m_i\beta_e/m_e \sim 1 - 2$ where β_e is the ratio of the electron thermal energy density to the magnetic energy density.

7.2.2 *Magnetized jet: the unbounded case*

A first point to note is that a fluid jet can be transformed into a wake field by a Galilean transformation. This transformation can be achieved by moving to a frame of reference moving at the maximum jet speed in the original frame of reference. Unbounded magnetized jets and wakes are being studied in a variety of contexts. In general, researchers have investigated systems which are either magnetically dominated or flow dominated. It turns out that this is significant because the Alfvén number ($A = v_A/v_0$ where v_A is the characteristic Alfvén velocity and v_0 is the characteristic flow velocity) is important in determining the evolution of the system. Note that the inverse of the Alfvén number A, is the Alfvén Mach number, or $M_A = v_0/v_A$. For example, for the low A, or flow dominated plasmas, researchers have looked at the heliospheric current sheet, the Earth's magnetotail, and at nonthermal galactic filaments. For the high A, or magnetically dominated case, researchers have looked at solar surges and tokamaks. An interesting problem that requires investigation of both the high A and low A regimes is that of slow

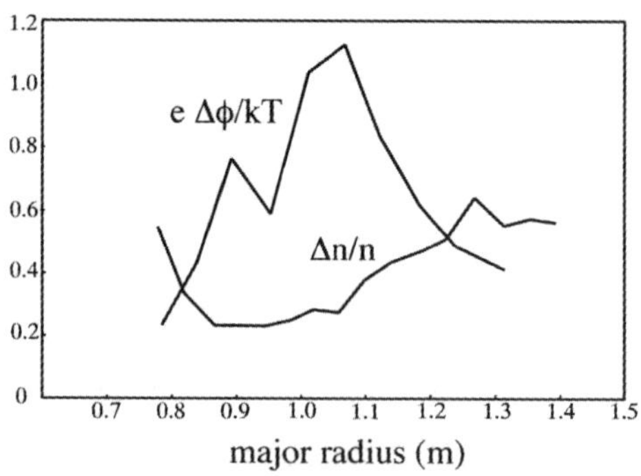

Fig. 7.3 Fluctuation amplitudes of the electric potential and the electron density measured with Langmuir probes in a 10 eV argon Helimak plasma.

solar wind formation and acceleration. In this case the plasma originates in a magnetically dominated region below the solar Alfvén point, and then moves to a flow dominated region above the Alfvén point.

Varicose and Sinuous Modes

In the linear analysis, three unstable modes are found [Dahlburg, *et al.* (1998)]. For flow dominated systems, with $0 \leq A \leq 1$, two unstable modes were found, an ideal varicose mode and a sinuous mode. A varicose mode is one in which the cross-stream flow is antisymmetric, and a sinuous mode is one in which the cross stream flow is symmetric. As A is increased these modes are stabilized by the stiffness of the magnetic field.

A generalization of the Howard semicircle theorem to the case of the magnetized jet shows that a strong enough magnetic field suppresses the ideal (vanishing resistivity) instabilities. As A increases above one, the system becomes magnetically dominated. In this case a resistive varicose instability occurs. In the fusion literature the varicose mode is called the tearing mode parity.

Analysis of the fluctuation energy-stress balance [Dahlburg (1997, 1998)] shows that there are three stress components: (i) the Reynolds stress, (ii) the Maxwell stress, and (iii) the cross-field stress. The values of the stresses and direction of energy flow are sensitive to the Alfvén number. The Reynolds stress moderates the transfer of energy between the basic flow field and the perturbed flow field. The Maxwell stress moderates the transfer of energy between the basic flow field and the perturbed magnetic field. The cross-stress is the dynamo term that moderates the transfer of energy between the basic magnetic field and the perturbed flow and magnetic fields.

The resistive varicose mode is characterized by a large positive cross-stress contribution to the perturbed energy in the high-A regime. For low-A the cross-stress contribution is negative. An increase in a positive Maxwell stress contribution is not enough to compensate for increased Ohmic and viscous dissipation for low-A. The ideal varicose mode has a large positive Maxwell stress contribution for all values of A. Again, there is increased dissipation for high-A, resulting in damping.

For the sinuous mode, the Reynolds stress is very large for low-A. There is also a more modest contribution to the perturbed energy due to the Maxwell stress. Both processes are strongly damped as A becomes large leading to increased stability. Research on the nonlinear incompressible problem focused on the magnetized wake, due to its use as a model for several solar physics, space physics, and astrophysical situations. In two spatial dimensions, it is found that moving plasmoids form and accelerate [Einaudi, *et al.* (1999)]. Of great interest is finding how rapidly the wake (or jet) spreads in the cross stream direction. When the span-wise direction is included, secondary instabilities occur, followed by a transition to turbulence [Dahlburg and Karpen (1994)].

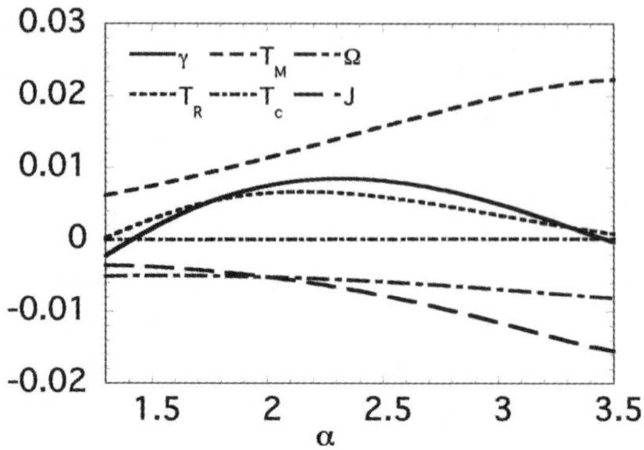

Fig. 7.4 Variation of perturbation energy balance terms with respect to streamwise wavenumber or binormal wavenumber, $k_\perp L = \alpha$. On the y-axis, γ is the growth rate in units of v_A/L and T_R, T_M, T_C, Ω, and J are the normalized values of the fluctuation energy contributions due to the Reynolds stress, Maxwell stress, cross stress, viscous dissipation, and Ohmic dissipation, respectively. For these calculations the parallel wavenumber $k_\| L = \beta = 0.0$. The magnetic and viscous dissipation numbers are $S = S_v = 5000$.

Compressible Dynamics

Compressibility introduces a significant increase in complexity for this problem. The linearized equations are far more extensive and difficult to solve. The three unstable modes still occur [Dahlburg, *et al.* (2001)]. Increasing the Mach number leads to decreased growth rates. In contrast with the current-vortex sheet, no second modes are found at high Mach numbers [Dahlburg and Einaudi (2000)]. An interesting feature is that, for certain situations, the three-dimensional ideal varicose modes grew at a faster rate than their two-dimensional counterparts [Dahlburg, *et al.* (2001)]. In the nonlinear compressible problem, massive accelerated plasmoids are observed [Einaudi, *et al.* (2001)]. These are similar to the incompressible plasmoids mentioned earlier, although the details of the acceleration are altered. Rappazzo, *et al.* (2005) analyze the diamagnetic and expansion effects on the formation of plasmoids. Even the incompressible problem has proven to be a rather complex physical problem.

7.2.3 *Magnetized jets in Helimak*

The Helimak Experiment

The Helimak geometry shown in Fig. 7.1, is similar to the magnetohydrodynamic (MHD) Poiseuille flow geometry [Stuart (1954); Velikhov (1959)] with concentric conducting walls at $r = a = 0.6\,\mathrm{m}$ and $r = b = 1.6\,\mathrm{m}$ and a height of $H = 2\,\mathrm{m}$. The dominant toroidal field, $B_\phi \simeq 0.1\,\mathrm{T} = 1\,\mathrm{kG}$ is produced by a set of 16 toroidal field coils around the vacuum chamber. Three poloidal field coils are used to produce a

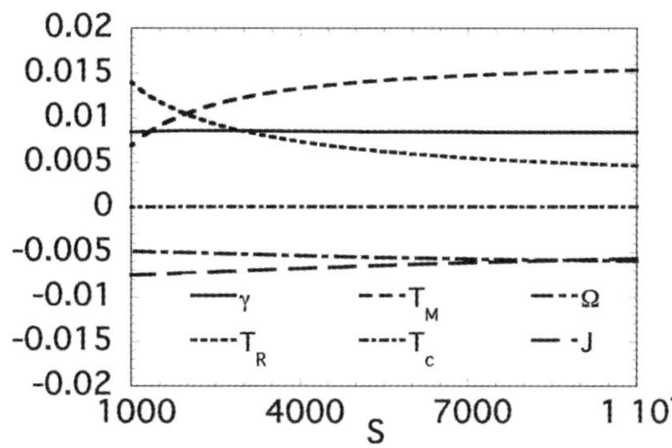

Fig. 7.5 Variation of perturbation energy balance terms with respect to Lundquist number (S). Here γ is the growth rate and T_R, T_M, T_C, Ω, and J represent normalized versions of the perturbed energy contributions due to the Reynolds stress, Maxwell stress, cross stress, viscous dissipation, and Ohmic dissipation, respectively. For these calculations $\alpha = 2.3, \beta = 0.0$ and $S_v = 5000$. The mode at mid-radius has $k_\perp = 4.6\,\mathrm{m}^{-1}$ and $k_\| = 0$.

weaker vertical field, B_z, which may be varied up to 10% of the toroidal field by changing the ratio of the current flowing through the toroidal coils to the current flowing through the poloidal field coils. The magnetic field lines are thus helices spiraling from bottom to top and whose pitch varies with the radius as the toroidal field decreases as $1/r$, where r is measured from the vertical symmetry axis of the device. The field line length $L_\|$ may be varied from less the 20 m to more than 1 km by reducing the B_z field.

The jet is a localized flow $v_z = w(x)$, induced by the biased endplates at $z = \pm H/2$. Plasma flows across the magnetic field from the steady-state radial electric field E_r. Probe measurements of the radial electric field show localized E_r structures, which means that there is a localized high speed vertical plasma stream $v_z = v_0(r)$ in the vertical direction $\widehat{e}_z$.

The plasma has toroidal $B_\phi = 0.1\,\mathrm{T}$ and vertical or poloidal field from $B_z = 0$ to $0.05\,T = 500\,\mathrm{G}$. The ionized Argon plasma is produced with ECH heating with temperatures of up to $T_e = 10\,\mathrm{eV}$ at the peak of the temperature profile. The plasma density is approximately $2 \times 10^{17}/\mathrm{m}^3 = 2 \times 10^{11}/\mathrm{cm}$. From these parameters the Alfvén velocity is $v_A(r) = 1 - 3 \times 10^8\,\mathrm{cm/s}$, the resistivity is $\eta = 10^{-5}\Omega\,\mathrm{m}$ and the kinetic viscosity $\nu = 10^4\mathrm{cm}^2/s$. Details on the plasma parameters are given in Table 7.1.

For this plasma we define

Magnetic Reynolds Number

$$S = S_M = \frac{V_A a}{(\eta/\mu_0)} \tag{7.2}$$

Table 7.1 Helimak viscoresistive MHD parameters.

Toroidal Magnetic Field	$B_\phi = 0.1\,\mathrm{T}$
Poloidal Magnetic Field	$B_z = 0.01\,\mathrm{T}$
Half Height	$H = 1.0\,\mathrm{m}$
Half Width	$L = 0.5\,\mathrm{m}$
Aspect ratio	$H/L = 2$
Magnetic Field Line Length	$L_\parallel = 20\,\mathrm{m}$
Argon Density	$n_e = 10^{11}\,\mathrm{cm}^{-3}$
Electron Temperature	$T_e = 10\,\mathrm{eV}$
Chamber Volume	$V = 13.8\,\mathrm{m}^3$
Space-Time Scale	$(L,T) = (L, L/V_A)$ $= (0.5\,\mathrm{m}, 0.6\mu\,\mathrm{s})$
Alfvén velocity	$v_A = 1 - 3 \times 10^8\,\mathrm{cm/s}$
Sound speed	$c_s = 4 \times 10^5\,\mathrm{cm/s}$
Dimensionless pressure	$\frac{m_i}{m_e}\beta_e = 3$
Ion sound gyroradius	$\rho_s = 2\,\mathrm{cm}$
Magnetic Diffusivity	$\eta/\mu_0 = 30\,\mathrm{m}^2/\mathrm{s}$
Magnetic Reynolds Number	$S_M = V_A L/(\eta/\mu) = 10^4$
Ion kinematic viscosity	$\nu = 0.3\rho_i^2\nu_i = 0.3\,\mathrm{m}^2/\mathrm{s}$
Viscous Reynolds Number	$S_v = 10^4$
Bohm Diffusivity	$\chi_e = 1/16\ T_e/B = 6\,\mathrm{m}^2/\mathrm{s}$

and

Viscous Reynolds Number

$$S_v = \frac{V_A a}{\nu}. \tag{7.3}$$

Thus, we take the magnetic Reynolds number as $S = 5000$ and the viscous Reynolds number as $S_v = 5000$ for the baseline case and consider variations from these values in the analysis. The velocity of the singly-ionized Argon atoms Ar^{+1} is measured spectroscopically

Singly ionized Argon has emission lines in the blue part of the visible spectrum. The Doppler shift of these lines is measured to give accurate, well-resolved measurements of the ion velocity $v_z = w(r)$. The ion velocity is of order $1\,\mathrm{km/s}$ and thus larger than the ion thermal velocity, comparable to the ion acoustic speed, and a significant fraction of the Alfvén velocity. Due to the large mass of the argon ions, $m_i\beta_e/m_e \sim 1$, significant electromagnetic fluctuations are measured with magnetic probes. The reference time is L/v_A of order one microsecond.

7.2.4 *The slab model for the Helimak*

Now we explore the behavior of a magnetofluid contained in a channel which corresponds to a slab model of the Helimak geometry. A schematic drawing of the Helimak and the slab coordinate systems is given in Fig. 7.1. In the laboratory frame of reference, we let x correspond to the radial direction $\widehat{\boldsymbol{e}}_r$; y' correspond to the toroidal direction $\widehat{\boldsymbol{e}}_\phi$; and z' correspond to the axial direction $\widehat{\boldsymbol{e}}_z$. Since most of the calculations are performed in a rotated frame of reference aligned with magnetic field, we use primes to denote the laboratory frame of reference.

Periodic boundary conditions are imposed in the y' and z' directions and no-slip boundary conditions in the radial direction.

In the simulations we parameterize the radius r with dimensionless x such that

$$r = [a + b + (b - a)x]/2 \tag{7.4}$$

so that $x = -1$ is $r = a$ and $x = +1$ is $r = b$. Thus, $dx = 2dr/(b - a) = dr/L$ with L given in Table 7.1.

The natural time units and velocity units $\boldsymbol{v} = \boldsymbol{V}\, v_A$ and $t = L/v_A$ where L is the half width of the plasma annulus. The field lines spiral upward (downward) so that modes with a helical pitch close to that of the magnetic field line have very long ($> 100\,\mathrm{m}$) parallel wavelength. Thus $\omega_A = k_\parallel v_A \lesssim 100\,\mathrm{kHz}$ is easily achieved. Magnetic probes show that there is a significant δB_x produced in these plasmas with $\omega_{*e} \sim |k_\parallel| v_A \leq 100\,\mathrm{kHz}$.

The boundary conditions are

$$u'(x = \pm 1) = v'(x = \pm 1) = w'(x = \pm 1) = 0 \tag{7.5}$$

where u' is the perturbed flow in the x-direction, v' is the perturbed flow in the y'-direction, and w' is the perturbed flow in the z'-direction.

In the Helimak, most of the magnetic field is due to the currents flowing through the sixteen external toroidal field coils, which varies as

$$B_\phi = \frac{B_0 R_0}{r} = \frac{2aB_0}{b + a + (b - a)x} \tag{7.6}$$

where the x coordinate corresponds to the scaled radial variable with $x = 0$ located at the center of the channel and $x = \pm 1$ to the walls. The magnetic field scaled to the inner wall value gives

$$\boldsymbol{B}'_{0y'}(x) = \frac{1.2}{2.2 + x}\widehat{\boldsymbol{e}}_{y'}, \qquad \boldsymbol{B}'_{0z'} = 0.1\widehat{\boldsymbol{e}}_{z'}. \tag{7.7}$$

Probe measurements of the radial electric field show a localized radial E_r well which means that there is a localized high speed vertical plasma stream in the vertical direction $W' = E_r/B_\phi = W'_0(x)$. The $W(x)$ with a hyperbolic cosine function $\mathrm{sech}(r - r_0/w) \to \mathrm{sech}(3x)$ is modeled. For the baseline model we take

$$\boldsymbol{V}'_0 = 0.0\widehat{\boldsymbol{e}}_{y'}, \qquad \boldsymbol{W}'_0(x) = 0.1\frac{\mathrm{sech}(3x) - \mathrm{sech}(3)}{\mathrm{sech}(0) - \mathrm{sech}(3)}\widehat{\boldsymbol{e}}_{z'} \tag{7.8}$$

where $\boldsymbol{V}'_0$ is the basic flow in the y'-direction and $\boldsymbol{W}'_0$ is the basic flow in the z'-direction. Here in the lab coordinates y' is toroidal direction and z' is vertical in Fig. 7.1.

Figure 7.4 shows the growth rate γ (solid line) versus wavenumber for $S = s_v = 5000$ as a function of $\alpha = k_\perp L$. The maximum growth $\gamma_{\max} = 0.07\, v_A/L$ occurs at $\alpha = 2.3$ corresponding to $k_\perp \rho_s = 2.4\rho_s/L \simeq 0.05$. The measure $\gamma_{\max}/k_\perp^2 \simeq 10\,\mathrm{m}^2/\mathrm{s}$.

Field-line Following Coordinates

In many toroidal plasma theories and codes a coordinate transformation is introduced to follow magnetic field line at each radius $r = r(x)$. Here we introduce this coordinate system by a rotation of the y', z' coordinates through the angle $\theta(x)$ given by $\tan\theta = -B_\phi(r)/B_z$ that is a tipping of the z' axis by about $80°$ to bring the new z coordinate parallel to the local $\boldsymbol{B}(x)$. Locally the plasma has translational invariance in the azimuthal direction $\widehat{e}_\phi$ and vertical direction $\widehat{e}_z$. Thus, we are able to define wavenumbers α and β corresponding to the bi-normal and parallel direction in the fusion literature and are called the stream-wise (α) and the span-wise (β) in the hydrodynamic literature. In the reduced resistive MHD equations, in this section on the Helimak, we have

$$F = \alpha V_0(x) + \beta W_0(x) = \vec{k} \cdot \vec{V}_0(x) \tag{7.9}$$

$$G = \alpha B_y(x) + \beta B_z(s) = \vec{k} \cdot \vec{B}(x). \tag{7.10}$$

Boundaries and Broken Symmetry

There are rigid boundaries in the channel problem. Boundaries can have both a stabilizing and a destabilizing influence. Another significant difference between the unbounded magnetized jet and the present fields is the lack of symmetry in the magnetic field. In particular, while the stream-wise flow field is symmetric in x, the magnetic field is about three times larger at $x = -1$ than at $x = 1$. This broken symmetry influences the details of our numerical method, in that we must retain all of the Chebyshev polynomials to account for the asymmetry. The numerical method is described in Section 7.2.5. In terms of the total kinetic and magnetic energies, the system is magnetically dominated. As we shall see, the linear dynamics appear to be best described as flow dominated. However, the sheared magnetic field has a large effect on the plasma dynamics.

7.2.5 *Linear dynamics in Helimak*

Linearized Resistive-Viscous Magnetohydrodynamic Equations

The vector fields for the incompressible plasma dynamics are

$$\boldsymbol{v}(\boldsymbol{x}, t) \equiv \text{ flow velocity} \tag{7.11}$$

$$\boldsymbol{\zeta}(\boldsymbol{x}, t) = \nabla \times \boldsymbol{v} \equiv \text{ vorticity} \tag{7.12}$$

$$\boldsymbol{B}(\boldsymbol{x},t) \equiv \text{ magnetic field} \tag{7.13}$$

$$\boldsymbol{j}(\boldsymbol{x},t) = \nabla \times \boldsymbol{B} \equiv \text{ electric current density} \tag{7.14}$$

$$\Pi(\boldsymbol{x},t) \equiv \text{ mechanical kinetic pressure} \tag{7.15}$$

$$+ \text{ thermal pressure per unit mass} \tag{7.16}$$

$$S \equiv \text{ Lundquist number} \tag{7.17}$$

$$S_v \equiv \text{viscous Lundquist number.} \tag{7.18}$$

The resistivity and viscosity are assumed to be constant and uniform. The velocities are measured in units of the Alfvén speed (v_A), and time is measured in units of the Alfvén transit time, L/v_A, where the characteristic distance, L, is defined by the channel half-width.

The nonlinear magnetohydrodynamic (MHD) equations:

$$\frac{\partial \boldsymbol{v}}{\partial t} = \boldsymbol{v} \times \boldsymbol{\zeta} - \nabla\Pi + \boldsymbol{j} \times B + \frac{1}{S_v}\nabla^2 \boldsymbol{v} \tag{7.19}$$

and

$$\frac{\partial \boldsymbol{B}}{\partial t} = \nabla \times (\boldsymbol{v} \times \boldsymbol{B}) + \frac{1}{S}\nabla^2 \boldsymbol{B} \tag{7.20}$$

together with the constraints: $\nabla \cdot \boldsymbol{v} = 0$ and $\nabla \cdot \boldsymbol{B} = 0$.

After linearizing Eq. (7.19) and Eq. (7.20), we decompose the first-order terms in the following manner:

$$a_1(x,y,z,t) = a(x)\; e^{i\alpha y + i\beta z - i\omega t} + \text{ c.c.} \tag{7.21}$$

where α is the stream-wise wavenumber, β is the span-wise wavenumber, and ω is the complex frequency and growth rate. Here ∇y is perpendicular to the radial gradient and the mid-radius magnetic field $\vec{B}_0$. This direction is called bi-normal in the fusion literature.

After eliminating the pressure, the differential equations for the linear modes are found. To write the equations compactly we define the operators $D \equiv d/dx$ and $\widehat{\xi} = D^2 - (\alpha^2 + \beta^2)$, then the cross-stream (or radial) velocity (u) and magnetic induction (b_x) equations reduce to:

$$\widehat{\xi}^2 u - iS_v F\widehat{\xi}u + iS_v(D^2F)u = -i\omega S_v\widehat{\xi}u + iS_v\left[(D^2G)b_x - G\widehat{\xi}b_x\right] \tag{7.22}$$

and

$$\left(\widehat{\xi} - iSF + i\omega S\right) b_x = -iSGu \tag{7.23}$$

with the boundary conditions: $u(x=\pm1) = Du(x=\pm1) = b_x(x=\pm1) = 0$. Here u and b_x are the x (radial) components of the perturbed velocity and magnetic field, respectively. The profiles enter the mode equations through

$$F = \alpha V_0 + \beta W_0 = \boldsymbol{k} \cdot \boldsymbol{V}_0 \tag{7.24}$$

and the local parallel wavenumber $G(x,\alpha,\beta) = k_{\|}B$ with

$$G = \alpha B_{0y} + \beta B_{0z} = \boldsymbol{k} \cdot \boldsymbol{B}. \tag{7.25}$$

The fundamental equations Eq. (7.22) and Eq. (7.23) form a sixth-order system of coupled first-order ordinary differential equations with six boundary conditions.

Alfvén wave resonances occur from Eq. (7.22) and Eq. (7.23) with $\omega^2 = G^2(x) = k_{\|}^2 v_A^2$ in the local limit with flow $F = 0$ and no dissipation $1/S = 1/S_v \to 0$. To get the 3D eigenfunction, we also solve perturbed vorticity ($\zeta = \beta v - \alpha w$) and electric current ($\chi = \beta b_y - \alpha b_z$) equations:

$$\left(\widehat{\xi} - i\alpha S_v F\right)\zeta + i\alpha S_v G \chi = -i\omega S_v \zeta - \beta S_v \left[(DQ)v - (DP)b_y\right] \tag{7.26}$$

and

$$\left(\widehat{\xi} - i\alpha S F\right)\chi + i\alpha S G \zeta = -i\omega S \chi - \beta S \left[(DQ)b_y - (DP)v\right]. \tag{7.27}$$

with the boundary conditions $\beta v(x=\pm 1) - \alpha w(x=\pm 1) = \beta b_y(x=\pm 1) - \alpha b_z(x=\pm 1) = 0$. Here v and w denote the stream-wise and span-wise perturbed velocities, b_y and b_z denote the stream-wise and span-wise components of the perturbed magnetic field, and

$$P = \beta B_{0y} - \alpha B_{0z} \tag{7.28}$$

and

$$Q = \beta V_0 - \alpha W_0. \tag{7.29}$$

In the ideal inviscid limit the eigenmode equations reduced to the pair

$$\left[\omega\left(\partial_x^2 - k_\perp^2\right) + F'' + F\partial_x^2\right]u + \left[G\left(\partial_x^2 - k_\perp^2\right) - G''\right]b_x = 0,$$

$$G(x)u + (\omega - F(x))b_x = 0$$

giving the standard MHD Alfvén wave eigenmode equation ([e.g. Chen and Morrison (1990, 1991)] $\{\partial_x[(\omega - F)^2 - G^2]\partial_x - k_\perp^2[(\omega - F)^2 - G^2]\}\xi = 0$ which is now expressed in the MHD Lagrangian displacement field $\xi(\omega - F) = u$.

Decomposition in Chebyshev-τ Method

The linear equations with boundary conditions are solved with the Chebyshev-τ method [Orszag (1971); Dahlburg (1983, 1986)]. The equilibrium and the perturbed fields are expanded in truncated Chebyshev series. The expansions are defined by

$$F(x) = \sum_{n=0}^{N} \widetilde{F}_n T_n(x) \quad \text{and} \quad G(x) = \sum_{n=0}^{N} \widetilde{G}_n T_n(x) \tag{7.30}$$

for the equilibrium fields and by the perturbed fields:

$$u(x) = \sum_{n=0}^{N} \widetilde{v}_n T_n(x) \quad \text{and} \quad b_x(x)) = \sum_{n=0}^{N} \widetilde{b}_n T_n(x). \tag{7.31}$$

The perturbed velocity field $\widetilde{u}_n$ is the expansion coefficient, and $T_n(x)$ is the nth Chebyshev polynomial of the first kind.

Now we use a spectral expansion in Chebyshev polynomials to solve the stability problem. The method is described by Goedbloed, *et al.* (2010) and Dahlburg and Einaudi (2000).

The equations satisfied by the unknown expansion coefficients are obtained by substituting the $N \to \infty$ expansions of Eq. (7.30) and Eq. (7.31) into Eq. (7.22) and Eq. (7.23), each of which produces a countable infinite number of equations in the expansion coefficients for $n = 0, 1, 2, \ldots$ when the orthogonality and recursion relations are used [Orszag (1971)]. All coefficients are set beyond $n = N$ to zero and use the $n = 0$ to $N - 4$ equations from Eq. (7.22) and the $n = 0$ to $N - 2$ equations from Eq. (7.23). The boundary conditions using $G_n(\pm 1) = (\pm 1)^n$ gives the conditions

No-slip Boundaries

$$\sum_{n=0}^{N} \widetilde{v}_n = \sum_{n=0}^{N} (-1)^n \widetilde{v}_n = \sum_{n=0}^{N} n^2 \widetilde{v}_n = \sum_{n=0}^{N} (-1)^n n^2 \widetilde{v}_n = 0 \tag{7.32}$$

and for δB_x the boundary condition is

$$\sum_{n=0}^{N} \widetilde{b}_n = \sum_{n=0}^{N} (-1)^n \widetilde{b}_n = 0. \tag{7.33}$$

This method of truncation is called the τ-approximation [Lanczos (1956)].

The spectral discretization process yields a generalized eigenvalue problem that can be written as $AX = \omega BX$ where the vector $X = (\widetilde{v}_0, \widetilde{v}_1 \ldots \widetilde{v}_N, \widetilde{b}_0, \widetilde{b}_1 \ldots \widetilde{b}_N)$ and A and B are nonsymmetric $(2N + 2)$ by $(2N + 2)$ square matrices.

As is customary for this type of stability problem, either global or local methods are used to determine the eigenvalues.

Global Method

For the global method we first transform the problem into the standard eigenvalue problem $CX = \omega X$ and then use the QR algorithm [Gary and Helgason (1970)]. This produces a full spectrum of complex eigenvalues $\omega_n(\alpha, \beta)$.

Local Method

When a good approximation or guess is available for the eigenvalue, local solutions are found using inverse Rayleigh power iteration. This iteration converges to the eigenvalue (and its associated eigenfunction) closest to the initial guess for the eigenvalue.

Generally we use the global method to identify the eigenvalue with the largest imaginary part (this is the most unstable mode or the least stable mode), and then use the local method to refine the eigenvalue and to compute the form of the eigenfunction. The local method is also used for examining how the solution changes as a parameter, for example the Lundquist number, is varied.

Solutions of the linearized equations Eq. (7.22) and Eq. (7.23) are now examined. For the baseline Helimak case we use the following parameters: the dimensionless stream-wise wavenumber $\alpha = k_\perp L = 2.3$, the span-wise wavenumber $\beta = 0.0$, the resistive Lundquist number $S = 5000$ and the viscous Lundquist number $S_v = 5000$. The corresponding drift wave normalized wavenumber is $k_\perp \rho_s = 0.1$ and growth rate is $\gamma L/c_s \simeq 0.8$.

This choice of α and β produces the largest growth rate for the specified values of S and S_v. See the curve γ in Fig. 7.4. There are a large number of damped, traveling modes for these particular parameters. There is also one unstable traveling mode. The form of the perturbed fields show that this is a global instability with comparable magnetic and flow velocity amplitudes.

Sinuous, Varicose and Resistive Varicose Modes

Three types of unstable modes are identified for the magnetized jet: an ideal sinuous mode, an ideal varicose mode, and a resistive varicose mode. The sinuous mode magnetic field is kink-like in appearance, while for the varicose modes the magnetic field is sausage-like (ideal) or tearing-like (resistive). For the Helimak fields the unstable mode is a somewhat distorted version of the sinuous mode reported for the magnetized jet [Dahlburg, *et al.* (1998)]. In contrast with the magnetized jet sinuous mode, the Helimak mode is suppressed near the inner wall at $x = -1$, where the magnetic field is enhanced. There are no unstable ideal or resistive varicose unstable modes, as are seen in the magnetized jet, found for the Helimak fields. Note that the resistive varicose mode is the a tearing-like mode which is expected to be stable in the Helimak.

The dispersion relation and the variation of the growth rates with respect to α, parameterized by the span-wise wavenumber (β) are now considered. Calculations for $S_v = S = 5000$ are now set. A nested family of curves is produced as the β is increased. All of the curves have the same form. This dispersion relation shows evidence of both low- and high-α cutoffs, as is typical for dissipative fluids confined to channels [Drazin and Reid (1981)]. The lower cutoff is due to the finite channel width and the upper cutoff is due to the finite dissipation. The variation of β has several effects on the α dispersion relation. First, an increase in β generally leads to a decrease in the growth rate $\gamma = \text{Im}(\Omega)$. Second, increasing β moves the maximum growth rate value to a larger value of α.

The variation of the growth rate $\gamma(\beta) = \text{Im}(\omega)$, with respect to the span-wise wavenumber β, is parameterized by α, For all cases computed there is a cutoff value (β_c) above which the instability is stabilized. There is a decrease in the magnitude of the growth rate as β is increased. This implies that a Squires theorem [Squire (1933); Dahlburg, *et al.* (1985); Seehafer and Schumacher (1998); Hughes and Tobias, (2001)] might be proved for the Helimak-like configuration, which we now proceed to consider.

A Magnetohydrodynamic Squire Theorem

A comparison theorem is proven that states that for the same wavenumbers α and β and the same dissipation parameters S and S_v, the 2D system is more unstable than the 3D system. In hydrodynamics this theorem is known as the Squires theorem [Drazin and Reid (1981)]. Thus, we show how the Squires theorem is generalized to a dissipative magnetohydrodynamical system. Hence, for the same S and S_v, the growth rate of the two-dimensional unstable mode will be greater than that of the three-dimensional unstable eigenmode.

3D System

After linearizing Eq. (7.19) and Eq. (7.20) about the basic magnetic and flow fields, and then applying the eikonal decomposition, we are left with the following six equations:

$$\begin{aligned}&\left\{D^2 - \left(\alpha^2 + \beta^2\right) - iS_v\left(\alpha V_0 + \beta W_0\right) + i\alpha S_v c\right\} u \\ &\quad = S_v Dp - iS_v\left(\alpha B_{0y} + \beta B_{0z}\right) b_x\end{aligned} \tag{7.34}$$

$$\begin{aligned}&\left\{D^2 - \left(\alpha^2 + \beta^2\right) - iS_v\left(\alpha V_0 + \beta W_0\right) + i\alpha S_v c\right\} v \\ &\quad = i\alpha S_v p + S_v\left(DV_0\right) v - iS_v\left(\alpha B_{0y} + \beta B_{0z}\right) b_y - S_v\left(DB_{0z}\right) b_x\end{aligned} \tag{7.35}$$

$$\begin{aligned}&\left\{D^2 - \left(\alpha^2 + \beta^2\right) - iS_v\left(\alpha V_0 + \beta W_0\right) + i\alpha S_v c\right\} w \\ &\quad = i\beta S_v p + S_v\left(DW_0\right) u - iS_v\left(\alpha B_{0y} + \beta B_{0z}\right) b_z\end{aligned} \tag{7.36}$$

$$\begin{aligned}&\left\{D^2 - \left(\alpha^2 + \beta^2\right) - iS\left(\alpha V_0 + \beta W_0\right) + i\alpha S c\right\} b_x \\ &\quad = -iS\left(\alpha B_{0y} + \beta B_{0z}\right) u\end{aligned} \tag{7.37}$$

$$\begin{aligned}&\left\{D^2 - \left(\alpha^2 + \beta^2\right) - iS\left(\alpha V_0 + \beta W_0\right) + i\alpha S c\right\} b_x \\ &\quad = S\left(DB_{0y}\right) u - S\left(DV_0\right) b_x - iS\left(\alpha B_{0y} + \beta B_{0z}\right) v\end{aligned} \tag{7.38}$$

and

$$\begin{aligned}&\left\{D^2 - \left(\alpha^2 + \beta^2\right) - iS\left(\alpha V_0 + \beta W_0\right) + i\alpha S c\right\} b_z \\ &\quad = S\left(DB_{0z}\right) - S\left(DW_0\right) b_x - iS\left(\alpha B_{0y} + \beta B_{0z}\right) w\end{aligned} \tag{7.39}$$

where $c = \omega/\alpha$ is the cross-field phase velocity.

The two-dimensional problem is found by setting $\beta = 0$ and $w = b_z = 0$, which gives the following four differential equations

2D System

$$\left\{D^2 - \alpha^2 + i\alpha S_v c\right\} u = S_v Dp - i\alpha S_v B_{0y} b_x + i\alpha S_v V_0 u \qquad (7.40)$$

$$\begin{aligned}\left\{D^2 - \alpha^2 + i\alpha S_v c\right\} u \\ &= i\alpha S_v p + i\alpha S_v V_0 v + S_v\left(DV_0\right) u - S_v\left(DB_{0y}\right) b_x \\ &\quad -i\alpha B_{0y} b_y\end{aligned} \qquad (7.41)$$

$$\left\{D^2 - \alpha^2 + i\alpha Sc\right\} b_x = i\alpha S V_0 b_x - i\alpha S B_{0y} u \qquad (7.42)$$

and

$$\begin{aligned}\left\{D^2 - \alpha^2 + i\alpha S_v c\right\} b_y &= i\alpha S B_{0y} u - i\alpha S V_0 b_y \\ &\quad + S\left(DB_{0y}\right) v - S\left(DV_0\right) b_x.\end{aligned} \qquad (7.43)$$

One asks can a transformation be found which casts equations Eq. (7.34)-Eq. (7.39) in the same form as Eq. (7.40)-Eq. (7.43)? The answer is yes. Multiply Eq. (7.35) and Eq. (7.38) by α, and multiply Eq. (7.36) and Eq. (7.39) by β. Then add the new Eq. (7.36) to the new Eq. (7.35), and the new Eq. (7.39) to the new Eq. (7.38). This gives the following result:

$$\begin{aligned}&\left\{D^2 - \left(\alpha^2 + \beta^2\right) - iS_v\left(\alpha V_0 + \beta W_0\right) + i\alpha S_v c\right\}(\alpha v + \beta w) \\ &iS_v\left(\alpha^2 + \beta^2\right) p + S_v D\left(\alpha V_0 + \beta W_0\right) u + S_v D b_x + i\left(\alpha B_{0y}\right. \\ &\quad \left.+\beta B_{0z}\right)(\alpha v + \beta w)\end{aligned} \qquad (7.44)$$

$$\begin{aligned}&\left\{D^2 - \left(\alpha^2 + \beta^2\right) - iS\left(\alpha V_0 + \beta W_0\right) + i\alpha S_v c\right\}(\alpha b_y + \beta b_z) \\ &\quad = S_v D\left(\alpha B_{0y} + \beta B_{0z}\right) u - iS\left(\alpha B_{0y} + \beta B_{0z}\right)(\alpha v + \beta w)\end{aligned} \qquad (7.45)$$

Then apply an MHD version of the Squire transformation by defining $\widetilde{\alpha}\widetilde{v} = \alpha v + \beta w, \widetilde{\alpha}\widetilde{b}_y = \alpha b_y + \beta b_z, \widetilde{u} = u, \widetilde{b}_x = b_x, \widetilde{\alpha} = \left(\alpha^2 + \beta^2\right)^{\frac{1}{2}}, \widetilde{p}/\widetilde{\alpha} = p/\alpha, \widetilde{c} = c, \widetilde{\alpha}\widetilde{S}_v = \alpha S_v$, and $\widetilde{\alpha}\widetilde{S} = \alpha S$. First apply this transformation to Eq. (7.34), Eq. (7.37), Eq. (7.44), and Eq. (7.45), then after dividing out a factor of $\widetilde{\alpha}$ we are left with:

$$\left\{D^2 - \widetilde{\alpha}^2 + i\widetilde{\alpha}\widetilde{S}_v\widetilde{c}\right\}\widetilde{u} = i\widetilde{\alpha}\widetilde{S}_v\widetilde{p} - i\widetilde{\alpha}\widetilde{S}_v B_0\widetilde{b}_x - \widetilde{S}_v\left(DB_0\right)\widetilde{b}_z \qquad (7.46)$$

$$\left\{D^2 - \widetilde{\alpha}^2 + i\widetilde{\alpha}\widetilde{S}_v\widetilde{c}\right\}\widetilde{w} = \widetilde{S}_v D\widetilde{p} - i\widetilde{\alpha}\widetilde{S}_v B_0\widetilde{b}_z \qquad (7.47)$$

$$\left\{D^2 - \widetilde{\alpha}^2 + i\widetilde{\alpha}\widetilde{S}_v\widetilde{c}\right\} b_x = \widetilde{S}_v\left(DB_0\right)\widetilde{w} - i\widetilde{\alpha}\widetilde{S} B_0\widetilde{u} \qquad (7.48)$$

and

$$\left\{D^2 - \widetilde{\alpha}^2 + i\widetilde{\alpha}\widetilde{S}_v\widetilde{c}\right\} b_z = -i\widetilde{\alpha}\widetilde{S}B_0\widetilde{w}. \tag{7.49}$$

These equations have the same form as Eq. (7.40)-Eq. (7.43) and hence define an equivalent two-dimensional problem. Since $\widetilde{\alpha} > \alpha$ for the three-dimensional problem, the growth rate for the three-dimensional problem is equal to the growth rate for a two-dimensional problem at lower S and S_v. Hence, for the same S and S_v, the growth rate of the two-dimensional unstable mode will always exceed that of the three-dimensional unstable eigenmode Q.E.D.

7.3 Toroidal Octupoles and Field Reversed Configurations

The first toroidal machines to show high plasma confinement were the toroidal octupoles [Ohkawa and Yoshikawa (1968)]. Octupole experiments [Meade (1996)] show orders of magnitude, improved confinement over that found in the ohmic heated model-C stellarator [Young (1967); Pease and Yoshikawa (1962)].

7.3.1 *Toroidal geometry with helical magnetic fields*

In the case of open field lines as in the mirror machine or a dipolar magnetosphere geometry, the calculation of the particle bounce averages of the wave fields is straightforward once the equilibrium $B(s)$ and Φ(s) fields are parameterized by the distance s along the field line. For the toroidal system with both toroidal and poloidal field components the calculation becomes more subtle. Namely, in the toroidal systems there are two types of magnetic field lines: those that close on themselves after one or several passages around the major axis and those that are ergodic and never close precisely on themselves. These field lines have an irrational value for their winding number or "safety factor" q. Electrons following these irrational winding number field lines tend to have a small, out-of-phase density response to the potential response.

The magnetic field component the long way around the torus called the toroidal B_t magnetic field and there is the magnetic field component the short way around the torus called the poloidal B_p magnetic field. The result of the vector sum of the two field (poloidal and toroidal) components produces is a helical magnetic field structure. The rate of change of the pitch of the structure is defined as the magnetic shear in the field lines.

On the special **rational** surfaces where the field line returns on its self after a given number n of transits around the major symmetry axis, the calculation proceeds as in the mirror geometry in each section of the field line between the $B_{\min}$ and $B_{\max}$. These closing field lines are called the rational field lines or field lines with rational rotational transforms or safety factor $q(r_{m,n}) = n/m$.

Charged particles with small pitch angles move over the whole, periodic field line. In the presence of an drift wave with mode numbers m and n, that is varying on the magnetic surface as $\cos(m\theta - n\phi)$, the result is a radial motion is proportional to $1/(m - nq)$. For other radial positions a field line covers the magnetic surface ergodically and the integration of the fluctuation along the infinite field line length returns zero for the average as the motion ergodically covers the magnetic surface. The only exception is for modes that are strictly constant on the magnetic surface, that is for the $m = n = 0$ modes. Thus, charged particles on these ergodic field lines behave very differently depending on whether they are mirror trapped in regions of the minimum magnetic field $B_{\rm min}$ or if they are at small pitch angles to the magnetic field and move over $B_{\rm max}$ to average the wave field over the entire magnetic surface.

In the literature the pitch is typically designated by α with $\cos\alpha = \vec{v} \cdot \vec{B}/vB$ and not to be confused with the bi-normal wavenumber $\alpha = k_\perp L$ used in Sec. 7.2. The condition on the pitch angle is $\sin\alpha < \sqrt{(1 - B_{\rm min}/B_{\rm max})}$. Thus the larger pitch angle particles are trapped in the region between $B_{\rm max}$ and $B_{\rm min}$ and respond more strongly to the perturbing fields than do the passing particles with small $\sin\alpha$ which average the fields over the entire magnetic surface. Here we define the angle pitch angle α at the local minimum of the magnetic field $B_{\rm min}$.

Mirror Trapped Particles

$$\sin^2\alpha \geq \frac{B_{\rm min}}{B_{\rm max}}. \tag{7.50}$$

When condition Eq. (7.50) is satisfied on the pitch angle, then the particle is reflected by the mirror force along the field line. For smaller pitch angles the particle passes over the magnetic field maxima and continually averages the wave perturbation over the lifetime of the fluctuation. If the fluctuation is frozen over this time, the average value is zero. If the fluctuation begins to change phase over the transit time of the particle then a nonzero average is returned. Clearly, the passing electrons will rapidly move along the field and average the perturbation to zero while, in contrast, the trapped electron respond with a value of the potential fluctuation characteristic of that at the $B_{\rm min}$.

This difference in the response is the origin of the trapped electron instability, where the passing electrons behave one way and the trapped electron behave very differently. This difference in response will also mean that in the trapped electron driven turbulence only the trapped electrons will undergo the turbulent transport in the system.

This trapped particle response situation is investigated through the study of electron orbits in the presence of models for the drift wave fields. In the limit of vanishing amplitude for the fields the collisional scattering of the guiding centers of the trapped and passing electrons make a random walk out of the system. This collisional random walk is called the neoclassical transport [Helander and Sigmar (2002)].

The mirror force trapped electron component of the plasma dominates the loss since (1) the trapped particles are more sensitive to the collisions and (2) the trapped particles have a larger radial excursion called the banana orbit. These characteristics apply both to the trapped ions and trapped electrons. In many experiments only the electrons are in the trapped particle regime. The trapped particle orbits are called banana orbits owing to their shape when displayed over one full bounce period in a plane perpendicular to the toroidal magnetic field. The radial width of the banana orbit vanishes as $1/B_p$ for the large magnetic field strength limit. Thus, it is the weaker poloidal magnetic field rather than the strong toroidal field component that confines the particle orbits in the tokamak. High plasma currents I_p are required for good plasma confinement in tokamaks. For example, $I_p \leq 7\,\mathrm{MA}$ in JET and JT60U with ITER designed to carry 15 MA of plasma current. In contrast, the Large Helical Device, LHD, has plasma currents of order 50 kA.

In the presence of the drift waves with amplitudes greater than a small critical level the trapped electrons begin diffusing out of the system while the passing electrons have only small radial oscillations about their smaller their flux surface position. Calculations for the neoclassical transport in the presence of fluctuations have been unified with the drift wave induced fluxes for both the electrostatic and electromagnetic turbulence.

The attentive reader will now have realized that there is even a more fundamental problem (issue) for this toroidal case with alternating surfaces with **rational** and **irrational** rotational transforms for the winding of the magnetic field line around the toroidal surfaces. Discussion on this issue follows.

7.3.2 *Density of rational surfaces and the KAM tori*

Consider a fast electron with zero pitch angle that maps out the magnetic field lines in the torus. If the electron is launched on a surface that has a rational value m/n for the rotational transform then the orbit is periodic: the electron returns to its initial position after some time passes for n revolutions around the torus. After n turns around the major axis the electron has made m turns around the minor axis. If, instead, the electron is launched on a neighboring surface with an **irrational** rotational transform it comes close, but never, returns precisely to its initial position. This orbit is called Moser orbit or a quasi-period orbit. The effect of the perturbation to these two orbits is very different when calculated by perturbation theory. In the first case their arises a vanishing denominator describing a secular change in the particle's orbit with time. (From perturbation theory the original orbit appears unstable to the presence the perturbation [Jose and Saletan (2002)].

The KAM (Kolmogorov-Arnold-Moser) theory solved this problem for the effect of these resonant orbits in the problem of celestial mechanics and then in Hamiltonian systems in general. The result is that there is the opening up of an "island" in the $r-\theta$ plane of radial width δr proportional to the square root of the amplitude of the perturbation strength. The island is constructed by the Poincare map by recording the puncture or crossing point on a reference surface for each passage of the electron around the torus. The perturbation varying with the fluctuation $\cos(m\theta - n\phi)$ creates magnetic islands with m-fold fixed-point inside islands formed on the rational surface $r(m/n)$ defined by the roots of $q(r) = m/n$. The orbit at the center of the islands is then a stable fixed point of the nonlinear map created by the electron equations of motion.

The fixed points $r(m/n)$ have the periodicity of the rational surface. The orbital stability question is now a famous result [Jose and Saletan (2002); Frisch (1996)] and numerous monographs at different levels of mathematical sophistication are available. The effect of the island is most easily computed if one substitutes an area preserving nonlinear map $M(x, y) \to x^*, y^*$ for original nonlinear flow generated by the orbital differential equations. The orbital motion is that from a Hamiltonian system of equations and thus is equivalent to a symplectic area preserving map on the surface of section. Generally, this map must be computed by numerical integration of the differential equations for the orbits. The topology and general properties, however, can often be seen by replacing the flow with a suitable nonlinear map. Many examples are given in detail in Lichtenberg and Liberman (1992). The original mathematical solutions and theorems are clearly presented in the lectures of Moser (1973) in a Princeton University Monograph, Stable and Random Motions in Dynamical Systems.

The density and size of the islands around the rational points in the dynamical system is controlled by the twist in the rotational transform profile. For the simple example given above of a zero pitch angle electron the orbital twist is the same as the twist of the magnetic field lines which is given by the magnetic shear parameter $s(r)$ given by

Magnetic Shear

$$s = \frac{rq'(r)}{q(r)}.$$

This magnetic shear profile is of fundamental importance in the tokamak. In the following chapters many uses of both the rotational rate $1/q(r)$ and the magnetic shear $s = rq'(r)/q(r)$ profiles will be uncovered.

For ion orbits the phase space twist rotational transform can be dominated by the shear in the $\boldsymbol{E} \times \boldsymbol{B}$ flow velocity. The radial electric field E_r and since the parallel velocity determine the surface of section plots in general. For low values of $v_\parallel/q(r)$ however the shear in the E_r field often is the dominant influence on the ion orbits. The plasma fluid will rotate from the E_r field and thus the drift waves themselves be propagating in a sheared plasma flow velocity $v_\theta = -E_r/B(r)$. In general this E_r shearing can add to or partially cancel the magnetic shear influence on the ions and ion fluid dynamics. The role of the fluid shearing is developed in Chapters 13 and 17. Sufficiently strong E_r shear can produce a drift wave version of the Kelvin-Helmholtz instability in magnetically confined plasmas. In other cases a localized shear flow region can decouple the edge from the core plasma giving rise to a transport barrier and the high confinement mode called the H-mode regime. The H-mode confinement is analyzed in Chapters 13 and 14.

Sheared $\boldsymbol{E} \times \boldsymbol{B}$ flows in plasmas can strongly change the nature of the instabilities by creating smaller cross-field variations as the different elements of the plasma gas flow at different speeds $\delta v = \delta x \times dv_E/dx$ where the width $\delta x = \ell_c$ is the radial correlation of the mode in the absence of sheared flows. A table of the effects for the level of the shear flowed required to effect a variety of plasma instabilities is given in Tajima, *et al.* (1991). The strong instabilities like the Rayleigh-Taylor are less disrupted by the sheared flows than the weaker instabilities like the ITG modes. The effects are well known in laser plasma physics as well as magnetic confinement systems physics.

Determining the number of and the density of the rational surfaces is a famous problem in number theory. The solution for high-order n/m rational numbers is expressed in terms of the Euler quotient function as described in Beklemishev and Horton (1992). Here we give the approximation for the density of the rational surfaces for the limit of large n and m values. This density is given by

Density of Rational Surfaces

$$\rho_m(\text{rational surface density}) = m\,\frac{6|q'|}{\pi^2 q^2} \tag{7.51}$$

for all modes up to mode number m. Thus we see that the density increases rapidly and the total number of rational surfaces becomes very large. The spacing between the rational surfaces of order n is $dr = 1/nq'$ for non-vanishing shear. At low values of the magnetic shear $rq'/q \to 0$ the density of rational surfaces vanishes and there maybe only one or no rational surfaces in a given region. Given a measure for the radial width of the eigenmodes, or magnetic islands widths, on each rational surface, it is useful to construct the pictures shown in Fig. 7.6 from the summation over all rational surfaces. From Fig. 7.6 the outer regions $\rho > 0.6$ have a high density of strongly overlapping drift wave functions. In this region the ballooning mode theory applies with the summation over the outer regions and the rational density is such that the modes overlap significantly, as can be determined by counting the number of dark stripes appearing above a given radial position.

In Fig. 7.6 with reversed magnetic shear profile, we see immediately that in the reversal layer the gaps between the mode rational surfaces are large and the overlap is weak. Chapters 14 and 18 shows experiments where there are sharp changes in the plasma transport across regions of reversed magnetic shear. These are called RS discharges for reversed magnetic shear.

Even before making the mode calculations or the orbital calculations of particles in these system one senses that by having a reversed magnetic shear RS system there is a region of reduced transport. In the confinement literature this effect is

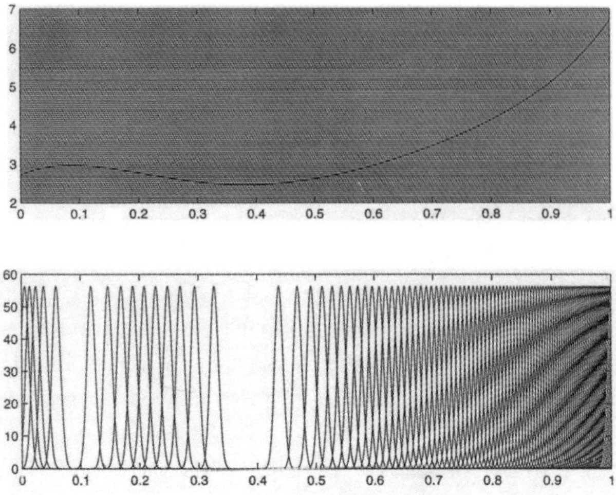

Fig. 7.6 The density of states for a typical reversed shear profile. The $q(r)$ profile is shown by the underlying solid curve in the top panel with resonant lines for toroidal mode numbers up to 20. The lower panel shows the density of rational surfaces with each given the same gaussian wave function. The local maximum at $q = 3$ gives a gap at $\rho = 0.1$ and the local minimum at $q = 5/2$ gives a large gap centered at $\rho = 0.37$ [Beklemishev and Horton (1992)].

referred to as a transport barrier produced by reversed shear. In reality there is still transport across this region but the transport flux is greatly reduced as borne out by detailed calculations and simulations.

The motivation for the MHD stability problem is clear. The MHD stability needs to be considered in some detail as recognized at the beginning of the thermonuclear fusion program. Briefly, there is an MHD potential energy δW from which it is clear that to not have the stabilizing perturbed magnetic energy overwhelm the destabilizing interchange energy that is released by convective motion, we must have an almost vanishing $k_{\|}$ on each toroidal surface. This vanishing $k_{\|}$ then conflicts with the condition for the wave function being doubly period on the torus except for the uninteresting case of the perturbation being constant on the surface. To see this more clearly suppose the function is doubly periodic with given m and n. Then one sees that the parallel wavenumber cannot stay small but increases rapidly as one deviates from the mode rational surface where $q = m/n$. In the neighborhood of the rational surface $q = m/n$, the $k_{\|}$ increase as $k_{\|} = (s/qR)(r - r_{m,n})$ where $q(r_{m,n}) = m/n$ and $s = rdq/dr$ is the magnetic shear parameter.

The magnetic shear is the rate of change of the twist in the helical magnetic field structure as it moves across the magnetic surfaces. Later in this chapter and Chapters 14 and 15 we see that special modes occur on the reversed shear (RS) surfaces where the shearing of the field changes direction. Special considerations apply the reversed shear surface and there can be a partial barrier to the anomalous transport set up at the shear reversal layer.

The ballooning mode solution to this problem is given in Connor, *et al.* (1979) and is explained and amplified in the works of Dewar and Glasser (1983) and Hazeltine and Waelbroeck (1988).

Later on the method for the drift waves, which is somewhat more intricate than for the linear MHD stability problem, was developed. The drift wave literature since 1980 abounds with calculations of the surface eigenvalue problem for the wave functions expressed in the extended poloidal angle of the ballooning mode representation. In some of the flux tube gyrokinetic simulations the local twisted field line following coordinates use the ballooning mode formulation to return the information back to the (global) laboratory coordinates. Some of the early nonlinear calculations were also done in ballooning mode coordinates although it is now recognized that the severe truncations made in some of those calculations are not justified in view of the high-resolution numerical solutions of the nonlinear problem.

7.3.3 *Dynamical limitations of the MHD model and the fluid moments closure problem*

Serious problems arise in attempts to use the MHD model for drift waves, tearing modes and the mirror mode instability. The common origins of these problems is the failure of two key assumptions in the MHD model. First, the motions in these instabilities do not satisfy the condition of being FAST so that the adiabatic closure assumption that

Adiabatic Closure

$$\nabla \cdot \boldsymbol{q} = 0 \tag{7.52}$$

is not satisfied. Secondly, the drift waves and the tearing modes both contain, in essential ways, a finite parallel electric field. A more general way to express the failure of the MHD model for these problems, is to state that many plasma waves require a **generalized Ohm's law**.

The MHD model in the electric field in the plasma is much too simple. As the field lines move in the plasma wave oscillation, the electrons rapidly move along the field line so as to set up a parallel electric potential near to a local Boltzmann distribution of the density and pressure along the magnetic field that follows the wave oscillations. In this case the parallel electric field follows the electron density variation and is given by $E_\| = (k_\| T_e/e)(\delta n_e/n_e)$. The parallel current is largely nondissipative and given by $j_\| = \sigma_\| E_\|$ with $\sigma_\| = i n_e e^2(\omega - \omega_{*e})/m_e k_\|^2 v_e^2$. The parallel current-voltage phase relation changes sign at $\omega = \omega_*$. At $\omega = \omega_{*e}$ the approximate wave polarization is $\delta j_\| = 0$ rather than $\delta E_\| = 0$.

More generally for tearing modes, resistive-g modes, and shorter wavelength electron temperature gradient driven drift waves the Ohm's law required is of the form

Ohm's Law

$$E_{\|} + \frac{\nabla_{\|} p_e}{e n_e} = \eta_{\|} j_{\|} \tag{7.53}$$

where $\eta_{\|}$ is the electrical resistivity along the magnetic field line. Note that electromagnetic waves at the gradient $\nabla_{\|} = (\widehat{b} + \delta \vec{B}/B) \cdot \nabla$ has a wave-induced component from $\delta \boldsymbol{B}$.

The parallel electron pressure gradient brings in the ion inertial scale length term

Ion-Sound Gyroradius

$$\rho_s = \frac{(m_i T_e)^{1/2}}{eB}.$$

The $\boldsymbol{j} \times \boldsymbol{B}$ Hall term brings in the scale length of the collisionless ion skin depth

Ion Inertial Scale Length

$$\delta_i = \frac{c}{\omega_{pi}}.$$

This δ_i-term eliminates the singularity of the electromagnetic waves at the ion cyclotron frequency ($\omega = \omega_{ci}$) and is important for the collisionless reconnection process [Galeev (1984); Galeev and Zeleny (1979); Drake, *et al.* (1992)].

For sufficiently small scale waves the electron inertial $\delta_e = c/\omega_{pe}$ becomes critical and is given by the keeping the electron mass in the parallel acceleration equation. This term plays a key role in the electron physics of magnetic reconnection and is described in the Alfvén wave experiments of Gekelman (1999) in the laboratory simulations of the Alfvén waves as they occur in the ionosphere.

Partly because of the very large convergence of the magnetic field lines in the dipole geometry, the cross-field scale lengths due to the large scale Alfvén waves generated in the equatorial region of the magnetosphere converge to cross-field wavelengths of order 10 km which is comparable to the electron skin depth δ_e. These Alfvén waves where $k_\perp \delta_e$ is order unity are slowed down and dispersed due to the electron mass effect. Another source of Alfvén wave dispersion is the ion mass effect occurring through the parallel electron pressure gradient fluctuations. These so-called inertial Alfvén waves [Lysak and Lotko (1996)] have the dispersion relation

Inertial Alfvén Waves

$$\omega^2 = \frac{k_{\|}^2 v_A^2}{1 + k_\perp^2 c^2/\omega_{pe}^2}. \tag{7.54}$$

In magnetic reconnection simulations in Chapter 15 the electrons stay attached to the magnetic field lines until the spatial gradient scale of the electron skin depth is reached. Then the electrons separate from the field lines forming high current densities that produce lower hybrid drift waves and in strong bursty events the Buneman two-stream instability [Buneman (1958)].The two-stream instability heats the electrons very rapidly and the plasma then rapidly evolves to two-stream stable with $T_e \gg T_i$ stable. The hot electron plasma with a plasma current or electron drift velocity leaves the ion acoustic and current driven drift wave turbulence to provide the anomalous resistivity.

The details of the magnetic reconnection processes are currently an active area of research. Detailed computer simulations and laboratory experiments are clarifying the complexity of these events. Tearing mode theory in Chapter 15 describes well many aspects of the dynamics but the full complexity of the multiscale nonlinear dynamics requires computer simulations.

There are, in fact, many different regimes to this process of magnetic field line reconnection. Tools to study the microturbulence used to analyze these reconnection processes and their computer simulations are provided. A closely-related nonlinear dynamics is that of nonlinear waves that form shock events. Resources for the physics of magnetic reconnection is *Nonlinear Magnetohydrodynamics* by Biskamp (1993) and for plasma shocks [Burgess (1995)] and slow shocks are used to construct magnetic reconnection model.

7.3.4 *Reduction of the toroidal plasma dynamics*

For the slow (subsonic and sub-Alfvénic) evolution of the confined plasma, the perpendicular part of the momentum equation should not be solved forward in time as performed in deriving Eq. (7.54) since the inertia acceleration term, $Mn\,d\boldsymbol{V}/dt$ is typically the small difference between the large, nearly balancing forces of the pressure gradient ∇p and $\boldsymbol{j} \times \boldsymbol{B}$ forces. Instead we solve the cross-field momentum equation for the current density $\boldsymbol{j}_\perp$ necessary to balance the perpendicular forces as

$$\boldsymbol{j}_\perp = \frac{\boldsymbol{B}}{B^2} \times \left[Mn\left(\frac{\partial \boldsymbol{v}}{\partial t} + \boldsymbol{v}\cdot\nabla\boldsymbol{v}\right) + \nabla p \right]. \tag{7.55}$$

Likewise from Ohm's law we solve for the cross-field velocity field $\boldsymbol{v}_\perp$ necessary to satisfy the perpendicular part of Ohm's law

$$\boldsymbol{v}_\perp = \frac{\boldsymbol{B}}{B^2} \times (-\boldsymbol{E} + \eta_\perp \boldsymbol{j}_\perp) \tag{7.56}$$

which is the $\boldsymbol{E} \times \boldsymbol{B}$ motion plus a resistive pressure gradient driven flux

$$\boldsymbol{v}_{\perp\eta} = -\frac{\eta_\perp}{B^2}\nabla_\perp p \tag{7.57}$$

where one uses $\boldsymbol{J} \times \boldsymbol{B} = \nabla_\perp p$.

At high temperature and strong magnetic field the flux Eq. (7.57) is small compared to the radial flux from $\boldsymbol{E} \times \boldsymbol{B}$ convection that arises from the electric force due to $\eta_\parallel$. This $\eta_\parallel$ driven particle flux is called the Pfirsch-Schlüter "diffusion" even though the flux is from $\boldsymbol{E} \times \boldsymbol{B}$ convection. Thus, we have the following three equations to advance the system in time:

$$\frac{\partial n}{\partial t} = -\nabla \cdot (n\boldsymbol{v}) \tag{7.58}$$

$$\frac{\partial v_{\|}}{\partial t} = -S^2 \frac{B_\theta}{Br}\frac{\partial}{\partial \theta} \ell n(n) - \widehat{\boldsymbol{e}}_{\|} \cdot (\boldsymbol{v} \cdot \nabla)\boldsymbol{v} \tag{7.59}$$

$$\nabla \cdot \left[\frac{Mn}{B}\widehat{\boldsymbol{e}}_{\|} \times \left(\frac{\partial \boldsymbol{v}_\perp}{\partial t} + (\boldsymbol{v} \cdot \nabla)\boldsymbol{v}\right)\right] + \nabla \cdot \left(\frac{\boldsymbol{B} \times \nabla p}{B^2}\right) \tag{7.60}$$

$$+\nabla \cdot (\widehat{\boldsymbol{e}}_{\|} j_{\|}) = 0 \tag{7.61}$$

where for the toroidal geometry with $\boldsymbol{B} = B_\theta \widehat{\boldsymbol{e}}_\theta + B_\phi \widehat{\boldsymbol{e}}_\phi$, the flow fields are

$$\boldsymbol{v}_\perp = \frac{\widehat{\boldsymbol{e}}_{\|}}{B} \times \left(\nabla\varphi - \frac{R_0 E_0(t)}{R}\widehat{\boldsymbol{e}}_\phi + \eta_\perp \boldsymbol{j}_\perp + \eta_\perp j_r \widehat{\boldsymbol{e}}_r\right) \tag{7.62}$$

and

$$j_{\|} = \frac{1}{\eta_{\|}}\left(-\frac{B_\theta}{Br}\frac{\partial \phi}{\partial \theta} + \frac{B_\phi R_0}{BR}E_0(t)\right). \tag{7.63}$$

The sound speed squared in $S^2 = k_B(T_i + T_e)/m_i$ in Eq. (7.59). The system is coupled in that $\boldsymbol{j}$ in Eq. (7.60) depends on $\boldsymbol{v}$ which is being solved for in Eq. (7.59) and Eq. (7.60). For small η this term is small and is sufficient to use the previous value of $\boldsymbol{j}$ to compute $\boldsymbol{v}_\perp$ and iterate if necessary to find the plasma currents and flows.

A fair amount of vector manipulation is required to write Eq. (7.57), Eq. (7.58), Eq. (7.59), Eq. (7.60) and Eq. (7.61) out in component form. Here we only give the reduction of one term. The vectors are defined as

$$\widehat{\boldsymbol{e}}_{\|} = \frac{\boldsymbol{B}}{B} = \frac{B_\theta}{B}\widehat{\boldsymbol{e}}_\theta + \frac{B_\phi}{B}\widehat{\boldsymbol{e}}_\phi \tag{7.64}$$

$$\widehat{\boldsymbol{e}}_\perp = \widehat{\boldsymbol{e}}_{\|} \times \widehat{\boldsymbol{e}}_r = \frac{B_\phi}{B}\widehat{\boldsymbol{e}}_\theta - \frac{B_\theta}{B}\widehat{\boldsymbol{e}}_\phi \tag{7.65}$$

and note that B_ϕ/B, B_θ/B are only functions of r. Then we have the following partial differential equation results:

$$\boldsymbol{v}_\perp = \frac{\widehat{\boldsymbol{e}}_\perp}{B}\frac{\partial \varphi}{\partial r} - \frac{\widehat{\boldsymbol{e}}_r}{B^2}\frac{B_\phi}{r}\frac{\partial \varphi}{\partial \theta} - \frac{R_0}{BR}E_0(t)\frac{B_\theta}{B}\widehat{\boldsymbol{e}}_r + \frac{\widehat{\boldsymbol{e}}_\perp}{B}\eta j_r - \widehat{\boldsymbol{e}}_r\frac{\eta j_\perp}{B} \tag{7.66}$$

and

$$\begin{aligned}\widehat{\boldsymbol{e}}_{\|} \times \frac{\partial \boldsymbol{v}_\perp}{\partial t} = &-\frac{\widehat{\boldsymbol{e}}_r}{B}\frac{\partial^2 \varphi}{\partial r \partial t} - \frac{\widehat{\boldsymbol{e}}_\perp}{B}\frac{B_\phi}{Br}\frac{\partial^2 \varphi}{\partial \theta \partial t} - \frac{R_0 \dot{E}_0(t)}{BR} \cdot \frac{B_\theta}{B}\widehat{\boldsymbol{e}}_\perp \\ &- \frac{\eta_\perp}{B}\frac{\partial j_\perp}{\partial t}\widehat{\boldsymbol{e}}_\perp - \frac{\eta_\perp}{B}\frac{\partial j_r}{\partial t}\widehat{\boldsymbol{e}}_r.\end{aligned} \tag{7.67}$$

The equation for $\nabla \cdot \boldsymbol{j} = 0$ reduces to

$$
\begin{aligned}
& -\frac{1}{r}\frac{\partial}{\partial r}\left(\frac{Mn_r r}{B^2}\frac{\partial^2 \varphi}{\partial r \partial t}\right) - \frac{1}{r}\frac{\partial}{\partial \theta}\left(\frac{R_0 \dot{E}_0}{BR} \cdot \frac{B_\theta B_\phi}{B^2}\right) \\
& -\frac{1}{r}\frac{\partial}{\partial r}\left(\frac{r\eta_\perp \dot{j}_r}{B^2}\right) - \frac{1}{r}\frac{\partial}{\partial \theta}\left(\frac{\eta_\perp \dot{j}_\perp B_\phi}{B^3}\right) + \nabla \cdot \frac{Mn}{B^2}\left(\widehat{\boldsymbol{e}}_\parallel \times (\boldsymbol{v}\cdot\nabla)\boldsymbol{v}\right) \\
& +\nabla \cdot \left(\frac{\boldsymbol{B} \times \nabla p}{B^2}\right) - \frac{1}{\eta_\parallel}\left(\frac{B_\theta}{Br}\right)^2 \frac{\partial}{\partial \theta}\left(\frac{1}{B}\frac{\partial \varphi}{\partial \theta}\right) = 0
\end{aligned}
\tag{7.68}
$$

where

$$
j_r = -\frac{Mn}{B}\frac{\partial v_\perp}{\partial t} + \frac{Mn}{B}\left(\widehat{\boldsymbol{e}}_r \cdot \widehat{\boldsymbol{e}}_\parallel \times (\boldsymbol{v}\cdot\nabla)\boldsymbol{v}\right) - \frac{B_\phi}{B^2 r}\frac{\partial p}{\partial \theta} \tag{7.69}
$$

$$
j_\perp = \frac{Mn}{B}\frac{\partial v_r}{\partial t} + \frac{Mn}{B}\left(\widehat{\boldsymbol{e}}_\parallel \times (\boldsymbol{v}\cdot\nabla)\boldsymbol{v}\cdot\widehat{\boldsymbol{e}}_\perp\right) + \frac{1}{B}\frac{\partial p}{\partial r}. \tag{7.70}
$$

An important remaining task is to compute the components of $(\boldsymbol{v}\cdot\nabla)\boldsymbol{v}$ for the nonlinear inertial acceleration.

All functions must be single valued in θ, φ and thus periodic in θ, φ and several constraints are derived from Eq. (7.68)-Eq. (7.70) by integrating over θ. Hazeltine, *et al.* (1971) analyze the equations by solving in the limit of small r/R_0 for the first order θ-variation and then computing the rate of change (∂t) in $\overline{n}(r,t), \overline{\phi}(r,t), \overline{v}_\perp$ where the bar denotes some type of average over θ $[\int(d\theta/2\pi) \times (R(\theta)/R_0).]$ Another approximation made in the Hazeltine, *et al.* (1971) analysis is that the radial gradients are sufficiently weak. The electric field is

$$
\boldsymbol{E} = -\nabla\varphi(r,\theta,t) + \frac{R_0}{R}E_0(t)\widehat{\boldsymbol{e}}_\phi. \tag{7.71}
$$

The plasma is ohmically-heated by the applied external toroidal electric field $E_0(t)\widehat{\boldsymbol{e}}_\phi$.

There are eight field variables determined by the eight equations Eq. (7.66), Eq. (7.67), Eq. (7.68), and Eq. (7.69). Equation (7.69) determines n, Eq. (7.70) determines φ, Eq. (7.71) determines $\boldsymbol{j}_\parallel$, and $\boldsymbol{u}_\perp$. The electric field is given as curl-free, and the magnetic field is taken to be the approximate tokamak field

$$
\boldsymbol{B} = \frac{R_0}{R}\left(0, B_\theta(r),\ B_\phi^{(0)}\right) \tag{7.72}
$$

where $R = R_0 + r\cos(\theta)$, $B_\phi^{(0)}$, R_0 are constants, and a toroidal current is specified by the form of B_θ by Amperé's law

$$
\frac{1}{r}\frac{\partial}{\partial r}\left(rB_\theta(r)\right) = \mu_0 j_\phi(r). \tag{7.73}
$$

The system of equations is third order in time from the time dependence of n and $\boldsymbol{v}$.

The time scales associated with the dynamics in the equations are as follows:

1. Fast ion-sound oscillations with

$$\omega \simeq \pm \frac{m}{r} \frac{B_\theta}{B} c_s \tag{7.74}$$

where $c_s = [(T_i + T_e)m_i]^{1/2}$ is the speed of sound in the isothermal equations and m is the approximate mode number of the θ-variation. There are density variations with $n = n_0 + n_1 \cos\theta$ and potentials $\varphi = \varphi_1 \sin\theta$ from the toroidal currents in the $\nabla B \neq 0$ toroidal magnetic field. These $m = 1$ ion acoustic waves are called geoacoustic modes and abbreviated as GAMS.

2. Unstable axisymmetric ($m = n = 0$) rotation with the slow growth rate

$$\gamma = \frac{\eta \beta c^2}{\mu_0 r^2} \cdot \frac{R_0^2}{r^2} \tag{7.75}$$

where $\beta = 8\pi p/B^2(\text{cgs}) = 2\mu_0 p/B^2(\text{mks})$. Large $m = 0$ potential oscillations are measured in Hamada, *et al.* (2005) where the fluctuations are associated with flows and the GAMS. The GAMS can reduce the turbulent flux and are strong in rotating plasmas [Casson, *et al.* (2010); Kendl and Scott (2005)].

3. Diffusion time scale τ_D in the rotating state with

$$\frac{1}{\tau_D} = \frac{1}{n}\frac{\partial}{\partial r}\langle n v_r \rangle \approx \frac{8\pi^2}{L^2}\frac{\eta \beta c^2}{r^2} \approx \left(\frac{B_\phi r}{B_\theta R}\right)^2 \frac{\eta_\| \beta c^2}{r^2} = \frac{q^2 \eta_\| \beta}{r^2}. \tag{7.76}$$

The theoretical analysis can be done in the limit of small resistivity η where the diffusion time $\tau_D = 4\pi L^2/c^2\eta$ is long compared to sound wave period qR/c_s.

Pfirsch-Schlüter transport drives a poloidal spin-up of the plasma [Stringer (1969); Hassam, *et al.* (1991)]. Hassam, *et al.* (1991) show that anomalous transport in the toroidal geometry with $\cos\theta$-variations on the flux surfaces also has a rotational spin-up. The collaboration of Hassam and Antonsen (1994) show the details of the spontaneous rotation in tokamaks. The competition between the sheared rotation created by the Reynolds stress tensor from the drift wave turbulence and the Stringer spin-up mechanism is analyzed by McCarthy, *et al.* (1993). The analysis becomes intricate with the scrape-off layer SOL becoming important as known from the experiments. A thorough review of the issues is given by Connor and Wilson (2000) in an article on the L-H transition problem.

References

Beklemishev, A. D. and Horton, W. (1992). *Phys. Fluids B* **4**, p. 200.

Biskamp, D. (1993), *Nonlinear Magnetohydrodynamics* (Cambridge University Press).

Buneman, O. (1958). *Phys. Rev. Lett.* **1**, 8; *Phys. Rev.* **115**, p. 50, doi:10.1103/PhysRevLett.1.8.

Burgess, D. (1995). Chap. 5, Collisionless Shocks, *Introduction to Space Physics* eds. M. G. Kivelson and C. T. Russell, (Cambridge, MA) p. 129.
Carbone, V., Einaudi, G., and Veltri, P. (1987). *Sol. Phys.* **111**, p. 31.
Casson, F. J., Peeters, A. G., Angioni, C., Camenen, Y., Hornsby, W. A., Snodin, A. P., and Szepesi, G. (2010). Gyrokinetic simulations including the centrifugal force in a rotating tokamak plasma, *Phys. Plasmas* **17**, p. 102305.
Chen, L., and Hasegawa, A. (1974). Plasma heating by spatial resonance of Alfvén waves. *Phys. Fluids* **17**, p. 1399, `http://dx.doi.org/10.1063/1.1694904`.
Chen, X. L. and Morrison, P. J. (1991). A sufficient condition for the ideal instability of shear flow with parallel magnetic field. *Phys. Fluids B* **3**, pp. 863-865.
Chen, X. L. and Morrison, P. J. (1990). Resistive tearing instability with equilibrium shear flow, *Phys. Fluids B* **2**, pp. 495-507.
Connor, J. W. and Wilson, H. R. (2000). *Plasma Phys. Control. Fusion* **42**, p. R1, doi:10.1088/0741-3335/42/1/201.
Connor, J. W., Hastie, R. J., and Taylor, J. B. (1979). *Proc. R. Soc. Lond. A* **365**, p. 1, `http://www.jstor.org/stable/79805`.
Dahlburg, R. B., Zang, T. A., Montgomery, D., and Hussaini, M. Y. (1983). *Proc. Nat. Acad. Sci USA* **80**, p. 5798.
Dahlburg, R. B. and Karpen, J. T. (1985). *J. Geophys. Res.* **100**, p. 23489.
Dahlburg, R. B., Zang, T. A., and Montgomery, D. (1986). *J. Fluid Mech.* **169**, p. 71.
Dahlburg, R. B. and Karpen, J. T. (1994). *Astrophys. J.* **434**, p. 766.
Dahlburg, R. B., Boncinelli, P., and Einaudi, G. (1997). *Phys. Plasmas* **7**, p. 1356. **4**, 1213.
Dahlburg, R. B., Boncinelli, P., and Einaudi, G. (1998). *Phys. Plasmas* **5**, p. 79.
Dahlburg, R. B. and Einaudi, G. (2000). *Phys. Plasmas* **7**, p. 1356.
Dahlburg, R. B., Keppens, R., and Einaudi, G. (2001). *Phys. Plasmas* **8**, p. 1697.
Dewar, R. L. and Glasser, A. H. (1983). *Phys. Fluids* **26**, p. 3038.
Drake, J. F., Finn, J. M., Guzdar. P., Shapiro, V., Shevchenko, V., Waelbroeck, F. L., Hassam, A. B., Liu, C. S., and Sagdeev, R. Z. (1992). *Phys. Fluids B* **4**, p. 488, doi:10.1063/1.860246.
Drazin, P. G. and Reid, W. H. (1981). *Hydrodynamic Stability* (Cambridge University Press).
Einaudi, G. and Rubini, F. (1986). *Phys. Fluids* **29**, p. 2563.
Einaudi, G. and Rubini, F. (1989). *Phys. Fluids B* **1**, p. 2224.
Einaudi, G., Boncinelli, P., Dahlburg, R. B., and Karpen, J. T. (1999). *J. Geophys. Res.* **104**, p. 521.
Einaudi, G., Chibbaro, S., Dahlburg, R. B., and Velli, M. (2001). *Astrophys. J.* **547**, p. 1167.
Frisch, U. (1996). *Turbulence* (Cambridge University Press, January 1996), ISBN:0-52145-713-0.
Galeev, A. A. (1984). Spontaneous reconnection of magnetic field lines in a collisionless plasma, *Handbook of Plasma Physics* **2**, eds. M.N. Rosenbluth and R. Z. Sagdeev (North-Holland, New York) pp. 305-335.
Galeev, A. A. and Zeleny, L. M. (1979). *Pis'ma Az. Eksp. Teor. Fis.* **29**, 669; *Space Res. Institute Report Pr-501* (Academy of Sciences, Moscow).
Gary, J. and Helgason, R. (1970). *J. Comp. Phys.* **5**, p. 169.
Gekelman, W. (1999). Review of laboratory experiments on Alfvén waves and their relationship to space observations, *J. Geophys. Res.*, **104**, A7, pp. 14,417-14,435, doi:10.1029/98JA00161.
Goedbloed, J. P., Keppens, R., and Poedts, S. (2010). *Advanced Magnetohydrodynamics with Applications to Laboratory and Astrophysical Plasmas*, pp. 269-304 (Cambridge

University Press) ISBN:978-0-521-87957-5.

Hamada, Y., Nishizawa, A., Ido, K, Watari, T., Kojima, M., Kawasumi, Y., Narihara, K., Toi, K., and JIPPT-IIU Group. (2005). Zonal flows in the geodesic acoustic mode frequency range in the JIPP T-IIU tokamak plasmas, *Nucl. Fusion* **45**, pp. 81-85, doi:10.1088/0029-5515/45/2/002.

Hassam, A. B. and Antonsen, T. M. (1994). *Phys. Plasmas* **1**, p. 337.

Hassam, A. B., Antonsen, T. M., Drake, J. F., and Liu, C. S. (1991). *Phys. Rev. Lett.* **66**, p. 309.

Hazeltine, R. D. and Waelbroeck, F. L. (1988). *The Framework of Plasma Physics* (Perseus Books, Reading, Massachusetts).

Hazeltine, R. D., Lee, E. P., and Rosenbluth M. N. (1971). *Phys. Fluids* **14**, 361, doi:10.1063/1.1693436.

Helander, P. and Sigmar, D. J. (2002). *Collisional Transport in Magnetized Plasmas* (Cambridge Monographs on Plasma Physics).

Hendel, H. W., Chu, T. K., and Politzer, P. A. (1968). Collisional drift waves–Identification, stabilization, and enhanced plasma transport, *Phys. Fluids* **11**, 2426, `http://dx.doi.org/10.1063/1.1691833`.

Horton, W., Perez, J. C., Carter, T., and Bengtson, R. (2005). *Phys. Plasmas* **12**, pp. 022303.

Hughes, D. W. and Tobias,S. M. (2001). *Proc. R. Soc. London Ser. A* **457**, p. 1365.

Jose, J. V. and Saletan, E. J. (2002). Classical Mechanics (Cambridge UP, Cambridge), p. 46.

Kendl, A. and Scott, B. D. (2005). Shear flow reduction by the geodesic transfer mechanism in tokamak edge turbulence, *Phys. Plasmas* **12**, p. 064506, `http://dx.doi.org/10.1063/1.1938147`.

Kent, A. (1966). *Phys. Fluids* **9**, p. 1286.

Lanczos, C. (1956). *Applied Analysis* (Prentice-Hall, NY).

Lichtenberg, A. J. and Lieberman, M. A. (1992). Regular and Chaotic Dynamics, *Appl. Math. Sci.*

Lysak, R. L., and Lotko, W. (1996). *J. Geophys. Res.* **101**, pp. 5085-5094.

McCarthy, D. R., Drake, J. F., Guzdar, P. N., and Hassam, A. B. (1993). *Phys. Fluids B* **5**, p. 1188.

Meade, D. M. (1996). *Phys. Rev. Lett.* **17**, p. 685.

Moser, J. (1973). On a nonlinear problem in differential geometry. *Dynamical Systems* (New York: Academic Press).

Nachtsheim, P. R. and Reshotko, E. (1965). NASA Technical Note d-3144 (National Aeronautics and Space Administration, Washington D.C.).

Ohkawa, T. and Yoshikawa, M. (1968). *Phys. Fluids* **11**, p. 2039.

Orszag, S. A. (1971). *J. Fluid Mech.* **50**, p. 689.

Paris, R. B., and W. N-C. Sy. (1988). *Phys. Fluids* **26**, p. 2966.

Pease, R. S. and Yoshikawa, S. (1962). *Phys. Fluids* **9**, p. 2059.

Perez, J. C., Horton, W., Gentle, K., Rowan, W. L., Lee, K., and Dahlburg, R. B. (2006). *Phys. Plasmas* **7**, p. 1356.

Rappazzo, A. F., Velli, M., Einaudi, G., and Dahlburg, R. B. (2005). *Astrophys. J.* **633**, p. 474.

Sato, T. and Walker, R. J. (1982). *Geophys. Res.* **87**, p. 7453.

Seehafer, N. and Schumacher, J. (1998). *Phys. Plasmas* **7**, p. 1356.

Squire, H. B. (1933). On the stability for three-dimensional disturbances of viscous fluid flow between parallel walls, *Proc. R. Soc. A* **142**, p. 621.

Stern, M. E. (1963). *Phys. Fluids* **6**, p. 636.

Stringer, T. E. (1969). *Phys. Rev. Lett.* **22**, p. 770.
Stuart, J. T. (1954). *Proc. R. Soc. London Ser. A* **221**, p. 189.
Tajima, T., Horton, W., Morrison, P. J., Schutkeker, J., Kamimura, T., Mima, K., and Abe, Y. (1991). Instabilities and vortex dynamics in shear flow of magnetized plasmas, *Phys. Fluids B* **3**, 4, pp. 938-954.
Velikhov, E. P. (1959). *Sov. Phys. JETP* **36**, p. 848.
Wang, S., Lee, L. C., Wei, C. Q., and Akosofu, S.-I. (1988). *Sol. Phys.* **117**, p. 157.
Young, K. M. (1967). *Phys. Fluids* **10**, p. 213, doi:10.1063/1.1761978.

Chapter 8

Turbulence Theory for Drift and Alfvén Waves

8.1 General Considerations: Analogs with Geophysical Fluids

Chapters 2-4 describe the physical processes that give rise to the low-frequency drift waves and temperature-gradient-driven instabilities in magnetized plasma. In those descriptions we used the simplest possible descriptions of the plasma motions to derive the waves and vortices. Now to understand these drift instability processes in detail, as they occur in the large literature on the subject, we must develop two lines of investigation: one is the FLR-fluid description which is a two-component set of fluid equations that are to be solved in limit that $m_a n_a \partial_t \boldsymbol{u}_a \ll e_a n_a \boldsymbol{u}_a \times \boldsymbol{B}$ so that the inertial acceleration is a small, but critical component of the wave vortex dynamics. To carry out this procedure is complicated as is also true for the corresponding geophysical fluid dynamics problem. In fact, it will be useful for the student to see how the geostrophic limit of fluids on rotating planets corresponds to the magnetized plasma limit where the dominant flow is

$$n_a \boldsymbol{u}_a = \frac{n_a \boldsymbol{E} \times \boldsymbol{B}}{B^2} + \frac{\boldsymbol{B} \times \nabla p_a}{e_a B^2}. \tag{8.1}$$

The books of Pedlosky (1987) and Gill (1982) on geophysical fluid dynamics define the geostrophic flow as follows. The basic idea is that the pressure gradient force balance is balanced by the Coriolis force

$$-\nabla p + 2\rho \boldsymbol{u} \times \boldsymbol{\Omega} = 0 \tag{8.2}$$

where p is the pressure of column, ρ mass density, and $\boldsymbol{\Omega}$ the angular velocity of the planet. Solving Eq. (8.2) for $\boldsymbol{u}$ gives the geostropic flow velocity which is the analog of Eq. (8.1).

Then, the next level of geophysical fluid dynamics where the inertial acceleration $\rho_m\, d\boldsymbol{u}/dt$ is retained for a finite value of ω/Ω is called the quasi-geostrophic fluid dynamics. This is the level that gives the all-important reduced dynamical equations of the Rossby waves and vortices. In plasma physics the corresponding flows are those arising from

$$n_a \boldsymbol{u}_a = \frac{\boldsymbol{B}}{e_a B^2} \times \left(m_a n_a \frac{d\boldsymbol{u}_a}{dt} + \nabla \cdot \boldsymbol{\pi}_a \right). \tag{8.3}$$

In Eq. (8.3) the accelerating forces on the right side are evaluated using the first-order flows in Eq. (8.1). That this iteration is valid is seen by comparing the two $n_a \boldsymbol{u}_a$-terms and finding their ratio is $\omega/\Omega_a = |(m_a/e_a B)d/dt| \ll 1$. In the general case the force from the divergence of the collisionless FLR momentum stress tensor has contributions that cancel the terms in

$$\frac{d\boldsymbol{u}_a}{dt} = \frac{\partial \boldsymbol{u}_a}{\partial t} + \boldsymbol{u}_a \cdot \nabla \boldsymbol{u}_a \tag{8.4}$$

that arise from the part of $\boldsymbol{u}_a \cdot \nabla \boldsymbol{u}_a$ from Eq. (8.1) due to the pressure gradient drift.

These cancellations are complicated and are developed in Chapter 4.22 for the case of electrostatic waves in a constant $\boldsymbol{B}$ field. The details of the cancellation are best left to the original literature. The cancellation, in a way, is a clue that the kinetic description may yield a compact, complementary method of deriving the dynamical equations. This, in fact, is found to be the case. Thus, the second major approach that is widely used is to reduce the full Vlasov or kinetic equation to its low-frequency limit is the use of gyrokinetic equations. This is the subject of the gyrokinetic analysis of drift waves developed in Chapters 11 and 12.

Now, the reader is wondering why not learn and use one description or the other? What happens is that each approach clearly reveals certain features, such as the conservation laws in the fluid description, and the drift-orbital motions and the critical conditions for marginal stability in the kinetic description, but tends to obscure other features. For this reason it seems that researchers wanting a full understanding of the low-frequency transport processes must be willing to use both descriptions. The comparisons and contrasts between the two descriptions is very revealing, and we will attempt to draw attention to these features as we develop the roughly parallel formulations of the problem with both the kinetic and FLR fluid mathematical descriptions.

One question may arise in the minds of those familiar with the stability analysis in fluid systems from what has already been stated: namely that the kinetic analysis is used for finding the conditions of marginal stability. Clearly, the dispersion relations derived from the fluid systems give critical conditions for instability and for neutral fluid. Neutral fluid instability threshold formulas are found in Chandrasekhar (1961), for example, and are well verified in neutral fluids experiments. The problem is that for the plasma fluid description the truncations that occur in the thermal energy or pressure equation that neglects the divergence of the flow of heat flux $q_\parallel$ along the magnetic field lines given by $\nabla_\parallel q_\parallel$. This is called the "closure problem." While for sufficiently fast modes this neglect of $\nabla_\parallel q_\parallel$ is a tolerable approximation and will even lead to reasonable approximations for the critical condition for an intrinsically fast MHD instability, the approximation will completely fail for slower instabilities. One famous example in the MHD context is explained in detail by Kulsrud (1983) is the mirror mode important in space plasmas. The threshold in the temperature anisotropy $T_\perp/T_\parallel$ for instability changes by a factor of six between the MHD [Chew, *et al.* (1956)] dissipationless and the Vlasov-Maxwell description. The problem is severe in general for drift waves since they are low frequency ($\omega \ll \omega_{ci}$), slow growing ($\gamma \sim v_i/L$) modes with relatively small wave lengths. Thus, while we want to know the fluid description's prediction for the stability threshold, we will generally not want to rely on this fluid-based critical condition until the corresponding kinetic theory answer is known.

In both the fluid and kinetic theory cases the problem of determining the critical condition and the number of roots and even their values is greatly aided by knowledge of the Cauchy theorem for analytic functions. The dispersion relations are always of the form of analytic function

$$D(\omega, \boldsymbol{k}, \mu) = 0 \tag{8.5}$$

where μ denotes the parameters of the problem. In Eq. (8.5) the fluctuations of f are assumed to vary as

$$f = \delta f_{\boldsymbol{k}\omega} e^{i\boldsymbol{k}\cdot\boldsymbol{x}-i\omega t} + \delta f^*_{\boldsymbol{k}\omega} e^{-i\boldsymbol{k}\cdot\boldsymbol{x}+i\omega^* t} \tag{8.6}$$

so that the response functions, such as the polarization function,

$$P(\omega, \boldsymbol{k}, \mu) = \left\langle \left[\frac{\omega - \omega_* \left(1 + \eta(v^2 - 3/2)\right)}{\omega - \omega_D\left(v_\perp^2, v_\parallel^2\right) - k_\parallel v_\parallel} \right] J_0^2 \right\rangle \tag{8.7}$$

are made analytic by finding their analytic condition from the $\mathrm{Im}(\omega) > 0$ region of the complex ω-plane. The analytic continuation of Eq. (8.7) is made by deforming the real $v_\perp$ and $v_\parallel$-integrals into the complex plane.

Once the analyticity of $D(\omega, \boldsymbol{k}, \mu)$ is in hand, then the Nyquist theorem is an important tool for finding the critical conditions and then for locating the roots of Eq. (8.5). The Nyquist theorem and the application to plasma waves is a topic developed in every textbook on plasma physics so we do not develop the procedure here. Since we will frequently use or quote the result, the reader may wish to review the procedure when reading Section 13.2.

A typical linear spectrum of frequencies $\omega(k_x, k_y)$ for the drift waves is shown in Fig. 8.1. There is a maximum of the frequency in the $k_x - k_y$-plane. The circle in the figure is contour of constant frequency at about one-half the maximum frequency. The units of k_x and k_y are the ion inertial scale length ρ_s. The imaginary part of the complex variable ω is defined as γ and the corresponding contours of growth and damping γ are shown in part (b) of Fig. 8.1.

The ring at $\gamma^\ell(k_x, k_y, \mu) = 0$ divides the growing modes from the damped modes. In the nonlinear mode coupling equations that follow, the growing modes feed energy into the turbulence and the damped modes absorb the energy. Here μ is a list (vector) of the system parameters. Thus, the amplitude and phases of each fluctuation component will evolve to a final turbulent spectrum. The condition for there to exists a well-defined turbulent spectrum are given in detail in Horton and Ichikawa (1996). Qualitatively, the condition is that there is sufficient damping from negative

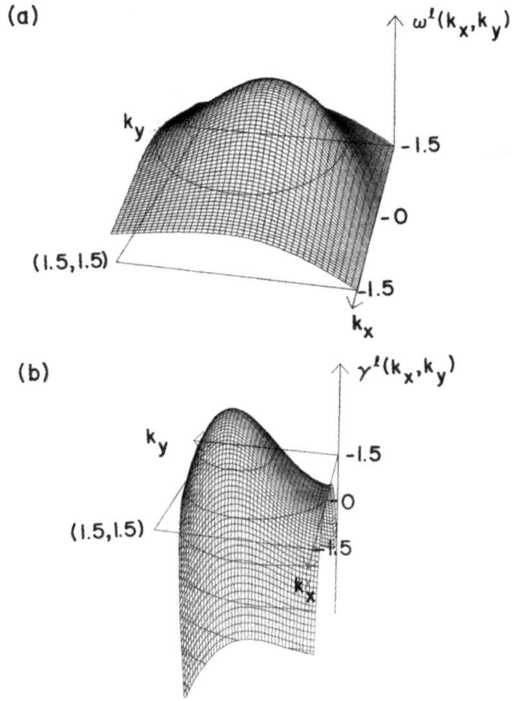

Fig. 8.1 Spectrum of typical drift wave frequencies $\omega(\mathbf{k})$ and growth-damping rate $\omega(\mathbf{k})$ for density gradient or trapped electron mode drift wave turbulence. The wavenumbers are normalized as $k\rho_s$ and the frequency normalized by $\omega L_n/c_s$. There is a well-defined maximum for the angular frequency and the growth rate at different points in the $\mathbf{k}$ space. The lower panel (b) shows that there are a large number of strongly-damped eigenmodes that will absorb wave energy scattered by mode coupling or wave-particle interactions to high $\mathbf{k}$ regions.

$\gamma(\mathbf{k})$ modes at high wavenumbers from the small scale fluctuations. In some simulation problems this requires introducing a hyperviscosity in the vorticity equation. The energy absorbed at these smallest scales is then associated with other higher frequency modes such as lower hybrid waves and ETG modes in Chapter 14.

In fully developed fluid turbulence, universal wavenumber spectra of the energy density $E(k)$ is of the power-law form $E(k) = \varepsilon k^{-m}$, developed from scale invariance and observed in fluid simulations, provided the stirring force is sufficient to prevent the formation of large-scale vortices. For 3D neutral fluid turbulence the Kolmogorov energy spectrum widely applies in what is called the "inertial range of k" where both $\omega_k = \gamma_k = 0$ [Frisch (1995)].

The presence of a significant number of vortex structures changes the energy spectrum as shown in Chapter 5 of Horton and Ichikawa (1996). In plasmas, however, there is no real inertial region with scale invariance due to the various dispersion scales e.g. $\rho_s = c(m_i T_e)^{1/2}/eB)$ and the wave emission/absorption processes. The growth rates γ_k driving turbulence are typically peaked at $k_y \rho_s \approx 0.5 - 0.7$ and

$k_x \rho_s \approx 0$. Thus, we investigate two drift wave models to determine the dependence of the spectral indices m_x, m_y defined by $E(k_x) = E_1 k_x^{-m_x}$ and $E(k_y) = E_2 k_y^{-m_y}$ on the system parameters and to examine the conditions for weak turbulence to be valid. For the weak turbulence measure we use the kurtosis $K = \langle \varphi^4 \rangle / \langle \varphi^2 \rangle^2$ which must be near three for homogeneous, space filling turbulence. A larger kurtosis $K \sim$ 6-10 in the long-time steady state is often found, which means that there are strong vortex structures characterized by radius r_0 and line density $n_v = N_v / L_x L_y$. The kurtosis is related to the packing fraction $f_p = n_v \pi r_0^2$ by $K \simeq 1/f_p$ where N_v is the number of well-defined vortices in the area $L_x L_y$ across the field $B \widehat{\boldsymbol{e}}_z$.

Some aspects of drift wave turbulence are related to the turbulence in two-dimensional fluid dynamics collected in the volume edited by Moreau (1983) on the experiments, theory and simulations of 2D neutral fluid turbulence.

8.2 Nonlinear Drift Wave Models

For the nonlinear dynamics we use the model given by Terry and Horton (1982) and the Hasegawa and Mima (1977) model as interpreted by Novakovskii, *et al.* (1988) for the unstable plasma.

The fundamental equations for both the rotating fluid and the magnetized plasma are

$$\frac{\partial n}{\partial t} + \boldsymbol{v} \cdot \nabla n + n \nabla \cdot \boldsymbol{v} = 0 \tag{8.8}$$

and

$$\frac{d\boldsymbol{v}}{dt} = -\nabla U + \boldsymbol{v} \times \boldsymbol{\Omega} \tag{8.9}$$

where the potential is electric $U = e\Phi/m$ or gravitational $gh(x, y, t)$. Introducing the frequency ω and scale L ordering that $\varepsilon = \omega/\Omega \sim v/L\Omega \ll 1$, the $\boldsymbol{E} \times \boldsymbol{B}$ or geostrophic flow velocity is $\boldsymbol{v} = \widehat{\boldsymbol{z}} \times \nabla U/\Omega$ and the inertial acceleration is given $\boldsymbol{v}_p = -\Omega^{-2}(d/dt)\nabla U$.

The rotational part of Eq. (8.9) with $\boldsymbol{\omega}_z = \widehat{\boldsymbol{e}}_z \cdot \nabla \times \boldsymbol{v}$ yields Ertel's theorem by taking the $\boldsymbol{\Omega} = \Omega \widehat{\boldsymbol{e}}$ component and neglecting the parallel compression $\nabla_\parallel v_\parallel$ that gives coupling to ion-acoustic waves. From Eq. (8.9) we have

$$\frac{d}{dt}\left(\frac{\Omega + \omega_z}{n}\right) = 0 \tag{8.10}$$

where $d_t = \partial_t + (c/B)[\Phi,\ ]$. One uses that $\omega_z = (c/B)\nabla^2\Phi$, $\Omega = eB/mc$ and the dissipative electron response $\mathcal{L}^{\mathrm{ah}}$ to derive

$$\widetilde{n}_e = \frac{eN}{T_e}\left(\Phi + \delta_0 \left(c_1 + \nabla^2\right) \frac{\partial \Phi}{\partial y}\right) \equiv N(1 + \mathcal{L}^{\mathrm{ah}})\Phi. \tag{8.11}$$

Using Eq. (8.11) in Eq. (8.8), we obtain the following dissipative drift wave equation:

$$\mathcal{L}\frac{\partial\varphi(x,y,t)}{\partial t}+v_d\frac{\partial\varphi}{\partial y}+\left[\frac{\partial\varphi}{\partial x}\frac{\partial}{\partial y}(\mathcal{L}\varphi)-\frac{\partial\varphi}{\partial y}\frac{\partial}{\partial x}(\mathcal{L}\varphi)\right]+\nu\nabla^4\varphi=0 \tag{8.12}$$

where $\mathcal{L}=1-\nabla^2+\mathcal{L}^{\rm ah}$ and ν is the ion-ion collisional viscosity. Balancing the wave growth $\gamma_{\boldsymbol{k}}=\mathrm{Im}[k_y\,v_d/\mathcal{L}(\boldsymbol{k})]\simeq k_y^2\,v_d\,\delta_0(k_\perp^2-c_1)/(1+k_\perp^2)^2-\nu k^4/(1+k^2)$ with the nonlinearity shows, that the amplitudes first increase with δ_0 but then saturates near the mixing length level as δ_0 becomes larger. The values of δ_0 and c_1 required for modeling the collisional, plateau and trapped electron regimes of the drift waves are given in Horton (1986).

For $\delta_0=1/4$, $c_1=-1/4$, and $\nu=0.15$ we show the turbulent steady state in Fig. 8.2. These vortices have rotation rates $\Omega_E(\boldsymbol{k})$ greater than the linear wave frequencies ω_k, and thus they behave as a gas of vortices rather than as linear, dispersive waves. The dissipative Reynolds number $R_{en}=\widetilde{v}_EL/\nu\simeq(2)(176)/(.15)=2.3\times10^3$, the kurtosis is $\langle\varphi^4\rangle/\langle\varphi^2\rangle^2\simeq6$ and the skewness near zero. The dissipationless $\boldsymbol{E}\times\boldsymbol{B}$ Reynolds number, or the Kubo number, Eq. (8.12) is $R_E\simeq2$.

Positive c_1 gives the ordinary drift wave and negative c_1 represents the dissipative trapped electron mode which is a prime candidate for the turbulence measured in tokamaks. For these, δ_0, c_1, ν parameter values $\gamma_{\rm max}=0.0163$ at $\boldsymbol{k}=(0,0.71)$ and $\gamma_{\rm min}=-0.472$ at $\boldsymbol{k}=(\pm1.5,\pm1.5)$. Here the wavenumbers are in units of $1/\rho_s$, the frequencies and growth rates in units of c_s/L_n.

The wavenumber spectrum from the turbulence in Fig. 8.2 is found to be similar to that of the drift wave vortex gas considered by Meiss and Horton (1983). The normalization in Fig. 8.3 is that the total energy is $E=\sum_{k_x}E(k_x)=\sum E(k_y)=25.6$. The k_x spectrum is monotonic decreasing while the k_y spectrum has a weak maximum well below the growth rate maximum. The enstrophy is $U=2.1$ and $\overline{k}=(U/E)^{1/2}=0.28$.

The strength of a vortex can be characteried by the mean rotation frequency compared to wave frequency for the given scale $k_\perp$ of the fluctuation. One may define $R_E=\overline{k}_\perp\langle v_E^2\rangle^{1/2}/|\omega_k|$ and call this the Kubo number for the drift wave turbulence. From simulations and experiments one finds $E_E\sim$1-2 in drift wave turbulence.

For comparison we show the weak turbulence spectrum computed with the highly idealized model given by Novakovskii, *et al.* (1988), in which the Hasegawa and Mima (1977) Eq. (8.12) with $\mathcal{L}=1-\nabla^2$ and $\nu=0$, is added $\gamma(\boldsymbol{k})\phi_k$ with the values $\gamma(\boldsymbol{k})=0$ for $0.4<|k|<1.2$ (inertial range) and $\gamma(\boldsymbol{k})=+0.005$ for $0.35\le|k|\le0.4$ (generation range) and $\gamma(\boldsymbol{k})=-0.005$ elsewhere. Under these conditions Fig. 8.3 shows that the spectral indices $W(k_x,k_y)=W_0/k_x^{m_x}\,k_y^{m_y}$ with $m_x=3.7$ and $m_y=4.1$ close to theoretical results of $m_x=4\sim m_y$ obtained from turbulence theory. When we allow $k_x\sim k_y$, however, we obtain $m_x\simeq m_y\simeq4$ in the same inertial range for this weak turbulence model.

Now consider the transport for a mixture of waves and vortex structures. The

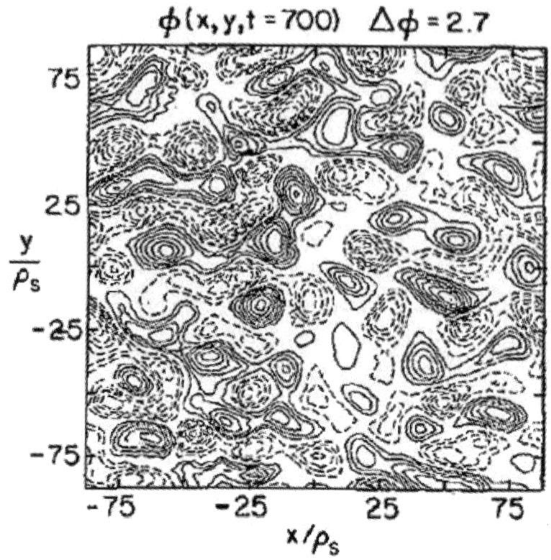

Fig. 8.2 Typical electron drift wave fluctuation spectrum for dissipative Terry-Horton system in Eq. (8.12) with weak growth rate from Landau wave-particle resonance. The dynamical equation may also be termed a Hasegawa-Mima equation containing the electron dissipation that produces the drift wave instabilities [Horton, etal (2010)].

transport from the drift waves as a gas of colliding vortices with number density $n_v = N_v/L_x L_y$ and packing fraction $f_p = N_v \pi r_0^2/L_x L_y = \pi n_v r_0^2 = r_0^2/r_{ij}^2$ where $r_{ij} = (\pi n_v)^{-1/2}$ is the average distance between the vortices. For Fig. 8.2 the vortex line density $n_v = 0.002/\rho_s^2$ and the packing fraction is $f_p = 0.2-0.3$. The vortices are treated as independent except during collisions where they move across the magnetic field with a step on the order of $\Delta x \sim r_0$ and as statistically independent of the wave fluctuation component. With these assumptions we arrive at the diffusion coefficient [Horton (1986)]

$$D_p = f_p^2 \frac{r_0}{r_n}\frac{cT_e}{eB} + \alpha \frac{\overline{\lambda}_x}{r_n}\frac{cT_e}{eB} \tag{8.13}$$

where the first term in Eq. (8.13) arises from the vortex-vortex collisions and the second term from the turbulent wave fluctuations taken at the mixing length limit $e\widetilde{\varphi}/T_e = \alpha^{1/2}\overline{\lambda}_x/r_n$ and with spatial correlation length $\overline{\lambda}_x$. For systems with $f_p \lesssim 1$ with large scale vortices $r_0 \gg \overline{\lambda}_x$ the vortex-vortex induced diffusion can dominate the wave transport. For small f_p the diffusion reduces to the usual drift wave turbulence theory mixing length value. Huld, *et al.* (1991) investigate the nature of coherent structures in basic plasma experiments.

Thus, we find that the vortex contributions to the anomalous transport process, typically neglected in the analysis of magnetic fusion systems, should be considered as potentially effective transport mechanism. The role of vortices in producing transport is well recognized in the corresponding atmospheric and oceanographic systems.

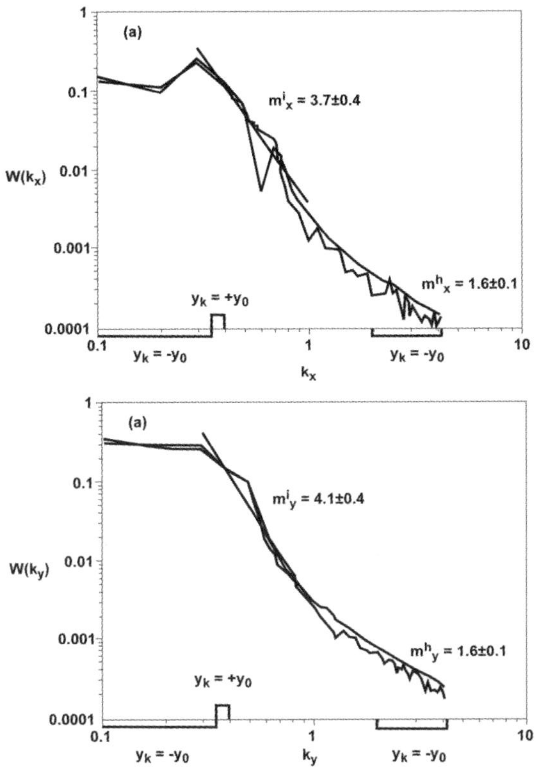

Fig. 8.3 Anisotropic wavenumber spectrum from the Novakoskii model electron drift wave spectrum. In this model the isotropic rings of a growth γ_0 and damping $-\gamma_0$ are added as shown along the k-axis.

8.2.1 *Consequences of sheared flows on the drift wave power spectrum*

Sheared flows strongly influence the three wave coupling interactions by a nonlinear transverse redistribution of the fluctuation energy in k_x, k_y-space. The transverse sheared flow $v_y(x) = (dv_y/dx)(x - x_0)$ defines preferred directions. So the turbulent k-space is intrinsically anisotropic. An initial clump of fluctuation energy released at $t = 0$ from the region (K_x, K_y) will propagate in time to the region $K_x - K_y(dv_y/dx)t$ to a new region as shown in Fig. 8.4 This process can allow an initial small amplitude signal to gain amplitude and under certain conditions, obtain sufficient amplitude to become a self-organized coherent structure as discussed in Chapter 10.

The initial wave packet localized at high k_x is swept by the sheared flow through the region of low-k_x where the amplitude increases sufficiently for nonlinear three-wave coupling to become strong. In some cases, as shown here in Horton, *et al.* (2010), the wave packet becomes a coherent vortex structure.

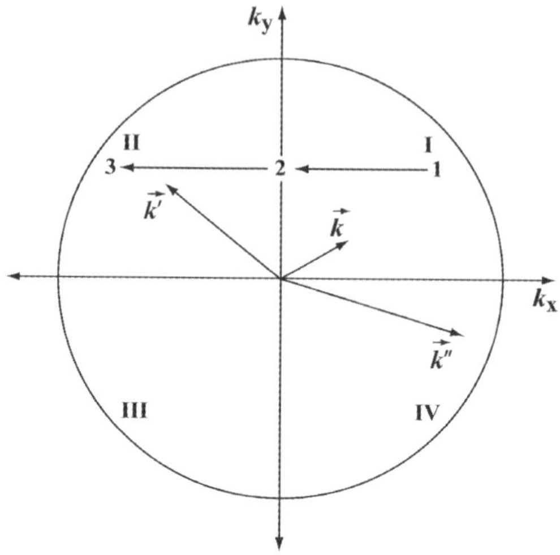

Fig. 8.4 The initial wave packet localized at high k_x at position 1 is swept by the sheared flow through the region of low k_x, region 1, and then moves to position 3. The amplitude increases sufficiently for nonlinear three-wave coupling to become strong in region 1 with $\boldsymbol{k} = \boldsymbol{k}' + \boldsymbol{k}''$ [Kim, *et al.* (1996)].

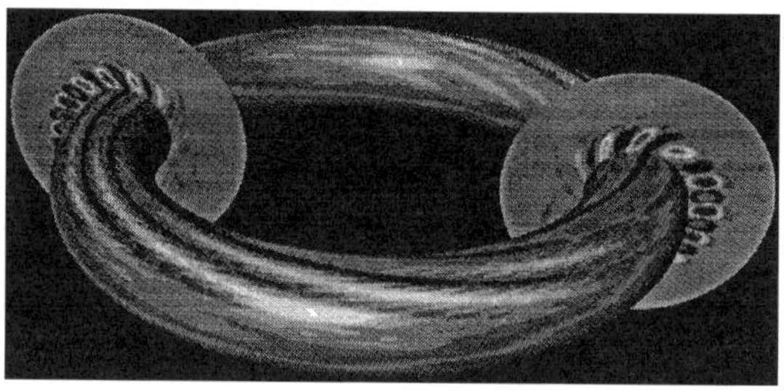

Fig. 8.5 Contours of the potential filaments aligned along the helical magnetic field lines in a simulation of the tokamak [Parker, *et al.* (1996)] driven by the ion temperature gradient.

8.3 Ion Temperature Gradient Induced Transport

In the presence of a strong temperature gradient the toroidal drift waves become dominant. Extensive simulations of this toroidal ITG mode are available in the literature. Here we show in Fig. 8.5 a typical result for the vortices created by the turbulent ITG from Parker, *et al.* (1996). An example with specific poloidal rotation v_θ is shown in Fig. 8.6 for five values of v_θ/c_s.

An example for the helical toroidal system (Heliotron-E) with electromagnetic ITG and MHD modes is developed in Hong, *et al.* (1991). In the top panel of Fig. 8.7 the electrostatic mode for $\beta/\beta_{\text{crit}} \to 0$ is shown and in the bottom panel where $\beta = 0.015$, just below the critical beta for the ideal MHD instability, the complex electrostatic field and the fluctuating magnetic vector potential ψ are shown. The

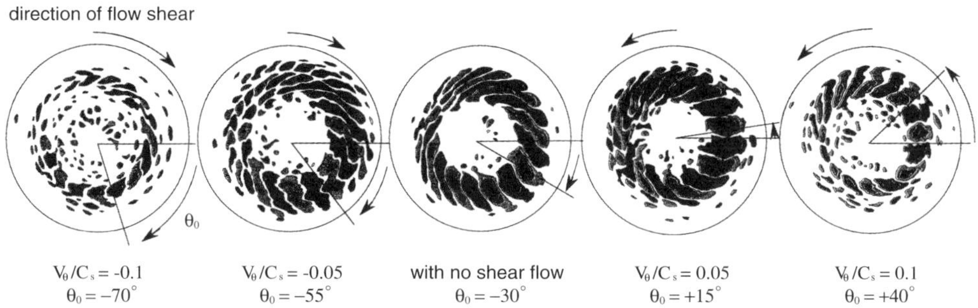

Fig. 8.6 The effect of the sheared poloidal flow $v_E(r)$ on the ITG turbulence. Contours of the toroidal drift-wave eigenvalues in the cross-sectional plane showing the radial scales for $a/\rho_s = 100$. The angle θ_0 of the Bloch-like wave function where $k_r = 0$ is clearly evident [Kim, Kishimoto, *et al.* (1996)].

magnetic fluctuations are given by $\delta \boldsymbol{B}_\perp = (-ik_y\psi_k, \partial\sigma_k/\partial x)$. The peak of the wave functions is at the rotational transform $\iota\!\!\!-$ magnetic surface.

The eigenfunction ψ increases from zero at $\beta_e = 0$ to $\psi_{1,1} \sim \frac{1}{2}\phi_{1,1}$ at $\beta_e = 0.01$ to the MHD polarization with $\psi_{1,1}(r) \simeq \phi_{1,1}(r)$ at $\beta_e = 0.02$. The eigenfunctions for the electrostatic limit and the transitional finite β_c value of $\beta_c = 0.015$ are shown in Fig. 8.7 for $\nu_e = 0$. As the growth rate increases with β_e, the wave functions broaden, approaching the MHD function at sufficiently high growth rates and β_e.

A nonlocal transport may be computed from the perturbed fields ϕ, ψ, $\widetilde{n}$, $v_\|$, and $\widetilde{p}_i$. Here, we limit consideration for the localized resistive-g and η_i modes in the low-β_e limit. Finally, we find that the compressibility effect on the new η_i modes is not important, except near marginal stability for the finite g case, as demonstrated in Yagi, *et al.* (1989). For other low-m η_i mode cases such as $m = 3/n = 2$ and $m = 5/n = 3$, no essential difference appears from the present $m = 1/n = 1$ study.

The small-scale turbulence from drift waves and ion acoustic waves couples to Alfvén waves and creates zonal flows on the large scale. These fluctuations propagate in the ion diamagnetic direction [Mikhailovskii, *et al.* (2006)]. The zonal flows merge with flows driven by the neutral beam heating and other sources of hydrodynamic flows in the plasma. The flows are also generated by the geodesic mode interactions [Zhang and Lin (2010); Kendl and Scott (2005)].

8.4 Nonlinear Three-Mode Interactions and Drift-Wave Turbulence in a Tokamak Edge Plasma

The three-wave coupling is a classic problem in drift wave turbulence [Horton and Hasegawa (1994) and Horton and Ichikawa (1996)]. The source and the sink regions of the turbulence as shown in Fig. 8.3 (the source region) and decay in the inertial range as shown in Fig. 8.3. Parameterizing the numerically computed one-

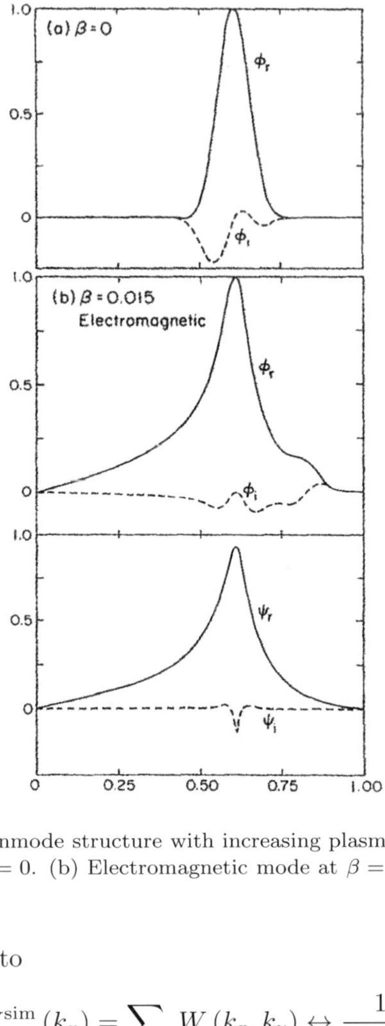

Fig. 8.7 Change in the eigenmode structure with increasing plasma pressure β in the Heliotron. (a) Electrostatic mode at $\beta = 0$. (b) Electromagnetic mode at $\beta = 0.015$ [Hong, *et al.* (1991)] in the Heliotron.

dimensional projections to

$$
\begin{aligned}
W^{\text{sim}}(k_x) &= \sum_{k_y} W(k_x, k_y) \leftrightarrow \frac{1}{|k_x|^{m_x}} \\
W^{\text{sim}}(k_y) &= \sum_{k_y} W(k_x, k_y) \leftrightarrow \frac{1}{|k_y|^{m_y}}
\end{aligned}
\tag{8.14}
$$

the regression fits to the simulation (sim) turbulence data determine the values of $m_x \pm \delta m_x$ and $m_y \pm \delta m_y$. More spectral experiment information is given in Terry (2000).

Exact solutions of three-wave ITG gyrofluid equations are used to compare with the exact kinetic theory three-wave solutions to attempt to verify closure models used in the fluid description. The system is reduced to set of three ODEs that have a structure similar to that of the famous Lorenz attractor problem in fluid dynamics. However, in the plasma case there is the "pump depletion" energy conserving behavior giving rise to periodic oscillations of the energy between the drift waves and the background profile gradient and the $k_y = 0$ mass flows. The reduced slab drift wave model driven by ∇T_e has been extended by Nakata, *et al.* (2010) using these methods. These interactions are described in Chapter 14.

The simulations of the type shown in Fig. 8.3 give a typical form of the drift wave spectrum showing the change in the indices m_x, m_y in the high-(m^h) and low-(m^l) wavenumber regimes as expected from the scaling theory. The exponents change as shown in Fig. 8.3 from $k_2\rho_s \gtrless 1$. In this simple model

$$\begin{aligned} m_x^h &= 1.6 \pm 0.1, \quad m_y^h = 1.6 \pm 0.1, \quad \text{for} \quad k\rho_s > 1 \\ m_x^l &= 3.7 \pm 0.4, \quad m_y^l = 4.1 \pm 0.4, \quad \text{for} \quad k\rho_s < 1. \end{aligned} \tag{8.15}$$

In the steady-state turbulence shown in Fig. 8.2 and Fig. 8.3 the total turbulent energy density and enstrophy are $W = 0.8$ and $U = 0.3$ in units of $(\rho/L_n)^2 n_0 T_e$. The space-time-averaged kurtosis and skewness of ϕ are 2.9 and -0.1, respectively. The system appears to be within the state of weak turbulence for these small values of γ_0.

A related simulation of drift wave turbulence driven by the parallel velocity ($\gamma_{\max} \simeq |du_\parallel/dx|$) shear flow in the ion-acoustic wave equation gives the spectra indices $m_x = 2$ for $0.2 < k_x\rho < 2$ and $m_x \cong 3$ for $2 < k_x\rho < 7$ [Horton and Hasegawa (1994); Dong and Horton (1993)]. Huld (1991) show vortices and vortex transport produced by the KH instability in the edge of low-temperature density plasma.

Even in the isotropic, 2-D Euler limit the question of the value of spectral index remains an active area of research. High resolution simulations of the forced-damped 2-D Euler equation persistently show energy spectral decays faster than the rate $k^{-3}[\ln(k/k_1)]^{-1/3}$ for $k \gg k_1$ for enstropy transfer in space filling, isotropic, homogeneous 2-D turbulence theory [Kraichnan, 1971]. The reason for the large decay indices has been traced to the emergence of long-lived, coherent vortex structures [McWilliams (1984); Legras, *et al.* (1988)]. The high resolution simulations of Legras, *et al.* (1988) show clearly the correlation of larger decay indices ($m \simeq 4$) with the appearance of monopole, dipole, and tripolar vortices. Legras, *et al.* (1988) compare the spectral indices and the associated vortex structures for $(512)^2$ simulations obtained with the three types of drivers (i) constant, single mode $\boldsymbol{k} = (k_l, 0)$, forcing; (ii) stochastic, narrow-band external forcing $\widetilde{F}(k_1 < |\boldsymbol{k}| < k_2)$ taken with news random phases at each time step; and (iii) narrow-band linear growth rate. A large-scale damping rate $-\nu_L\phi(\boldsymbol{k})$ and a hyperviscosity $-\nu_s(k_\perp^2)^p\phi(\boldsymbol{k})$ are used to produce the steady state. The resulting turbulence is characterized by the spectral index m and its vortex content as follows: (i) $m = 3.6$ and a low-packing fraction of relatively large vortices; (ii) $m = 3.5$ and a denser packing of smaller vortices; and (iii) $m = 4.2$ and the prominence of several dipolar and one tripolar vortex. Thus, it appears that for 2-D turbulence the space filling, inverse cascade spectrum of $k^{-3}[\ln(k/k_l)]^{-1/3}$ is an idealization that is not achieved in practice due to the strong, self-organization of the coherent structures, described in Chapter 10, suppressing the turbulent flows. A further demonstration of the suppression of turbulent cascades by coherent vortices in 2-D turbulence is given by McWilliams (1990a,b).

8.5 Inertial Spectral Ranges in 2D and 3D Turbulence

Drift wave simulations have been carried out in models where the growth and damping is suppressed to study the transfer of energy between modes $\boldsymbol{k}$ from the nonlinear interactions. Such regions of $\boldsymbol{k}$ space are called inertial regions stemming from the origin of the nonlinear transfers in the dissipationless neutral fluid equations coming from the inertial terms $\rho\boldsymbol{v}\cdot\nabla\boldsymbol{v}$ alone. Kolmogorov used physical reasoning on the nature of interacting eddies and mathematical invariance scaling arguments to show that for 3D fluid turbulence the spectral energy density $\epsilon(\boldsymbol{k})$ must scale as

$$\epsilon(\boldsymbol{k}) = Ck^{-5/3}\ln\left(\frac{k}{k_0}\right). \tag{8.16}$$

The derivation of this spectrum is straight forward and given in many text on fluid turbulence [Frisch (1995); Lesieur (2008)].

The situation is less straightforward in 2D turbulence owing to the existence of many invariants of the flows in addition to energy in the region of no growth or damping of the fluctuations. Kraichnan (1971) gave a set of definitive arguments to show that there are two inertial ranges in 2D neutral fluid turbulence. The first range is the energy transfer range also of the form of the Kolmogorov spectrum in Eq. (8.16). The second region is one in which there is a cascade of enstrophy rather than energy. The energy has an inverse cascade in 2D turbulence transferring energy from high-k to low-k. A striking example is in the ETG turbulence reported in Fu, *et al.* (2012). This is the range of greatest relevance to 2D fluid turbulence and continues down to the smallest scales where the viscous dissipation finally absorbs the vorticity of the fluctuations. The 2D enstrophy cascade given by Kraichnan (1971) is

$$\epsilon(\boldsymbol{k}) = C\eta^{2/3}k^{-3}\left[\ln\left(\frac{k}{k_1}\right)\right]^{-1/3}. \tag{8.17}$$

Here η is the rate of enstrophy transfer per unit mass. The wavenumber k_1 marks the bottom of the range where enstrophy is pumped into the turbulence. Transfer in the energy and enstrophy inertial ranges is computed according to an almost-Markovian Galilean-invariant turbulence model. Transfer in the two-dimensional energy inertial range is found to be much less local than in three dimensions, with 60% of the transfer coming from wavenumber triads, where the smallest wavenumber is less than one-fifth the middle wavenumber.

Batchelor (1969), Leith (1968), and Kraichnan (1967, 1971) show that two-dimensional neutral fluid turbulence can exhibit two kinds of inertial range: an energy transfer range of the form

$$E(k) = C_E k^{-3} \tag{8.18}$$

and an enstrophy transfer range as in Eq. (8.17)

$$E(k) = C' k^{-5}. \tag{8.19}$$

In these equations, $E(k)$ is the isotropic energy spectrum, normalized so that $\int E(k)dk$ is the average kinetic energy per unit mass. The wealth of inviscid constants of motion in two-dimensional flow makes the universality of C and C_E, more questionable than for the inertial range in three dimensions. If Eq. (8.18) exists, the energy transfer is from higher to lower wavenumbers, while in Eq. (8.18) the enstrophy transfer is from lower to higher wavenumbers. There is zero enstrophy transfer in range lower wavenumber regime and zero energy transfer in the higher wavenumber range.

Worked Problem 8.1: Inertial Spectra

Problem:

Derive the energy spectrum $E(k)$ for (1) 3-D isotropic turbulence in the inertial range $k_0 < k < k_d$ from the constant transfer of a turbulent energy flux $F(k)$ from the large-scale source $k_0 = 1/L_0$ to the small-scale dissipation $k_d = (\varepsilon/\nu^3)^{1/4}$ where ε is the turbulent energy per unit of mass and ν is the viscosity. (2) For 2-D isotropic turbulence, apply the same arguments to the flux of η conserved vorticity to derive the 2-D high-k spectrum.

Solution:

(1) The nonlinear convection rate at scale k is $\nu(k) = kv(k) = k(kE(k))^{1/2}$ (which is also known as the eddy-turnover-rate and the eddy viscosity rate). The flux of energy (per unit mass) is then

$$F(k) = \nu(k)v^2(k) = k^{5/2}\, E^{3/2}(k). \tag{8.20}$$

Taking the energy flux $F(k) =$ constant one finds that

$$E(k) = \frac{F^{2/3}}{k^{5/3}} \qquad \text{for} \qquad k < k_\nu = \left(\frac{\varepsilon}{\nu^3}\right)^{1/4}. \tag{8.21}$$

For $k > k_d$ the viscous dissipation of energy gives $E(k) = (\varepsilon/\nu^3)^2\, k^{-7}$, and the Heisenberg interpolation formula is

$$E(k) = E_0 \left(\frac{k_0}{k}\right)^{5/3} \left(1 + \left(\frac{k}{k_\nu}\right)^{8/3}\right)^{-2}. \tag{8.22}$$

(2) For 2-D turbulence the mean-squared-vorticity is conserved so that the vorticity flux

$$k(kE(k))^{1/2}k^2E(k)k = \eta \tag{8.23}$$

is conserved in the nonlinear transfer to $k = k_D$. This leads to the steeper spectral exponent

$$E(k) = \frac{\eta^{2/3}}{k^3} \qquad \text{for} \qquad k_1 < k < k_\nu \tag{8.24}$$

where k_1 is the region of creation of vorticity. For $k_0 < k < k_1$ there is an inverse cascade of energy that is given by Eq. (8.21).

The subject of spectral distrib ution in turbulence is a rich field of research developed in numerous large works including Monin and Yaglom (1965), Kraichnan (1964), Krommes (1999, 2002), Leslie (1973), Stanisic (1985), and Frisch (1995).

Experiments with various types of electromagnetic wave scattering diagnostics have reported the shape of the turbulent spectrum $E(k_\perp)$ under a variety of conditions. Chapter 2 reviews the early experiments. A recent spectrum with two slopes, of the type described in Eq. (8.14), is reported for Tore Supra by Hennequin, *et al.* (2004).

In magnetic plasma the turbulence is called quasi-2D turbulence. A review article by Horton and Hasegawa (1994) develops the drift wave turbulence problem in some detail. In plasma confinement systems the turbulence is anisotropic, as shown in Fig. 8.6 in a global simulation by Kim, *et al.* and in Chapter 13 Fig. 13.1 in a simulation by Sydora, *et al.* Coherent structures develop as shown by the spiraling radial fingers in the potential contours in these figures. Large values of temperature-to-density gradient parameter defined by $\eta_i = d\ell n\, T_i / d\ell n\, n_i$ are important for H-mode plasma confinement regimes. In these H-mode plasmas there is reduction of the growth rate from a sheared $\boldsymbol{E} \times \boldsymbol{B}$ flow. An example of the turbulence showing this stabilization effect from a modest level of sheared flow in a plasma with a variable level of magnetic shear is shown in Fig. 8.8.

The drift wave turbulence generates a sheared background flow velocity $v_E(x) = d\Phi(x)/B\,dx = d\Phi/d\Psi$ where Ψ is the magnetic flux function, and $\Phi(\Psi, t)$ is the electrostatic potential. The radial electric field $E_r(r,t) = -d\Phi/dx$, which is created by numerous processes, including the turbulence itself through the Reynolds' stress tensor. Figure 8.8 shows how the flow shear dv_E/dx strongly changes the linear growth rate γ of the drift waves. In the high growth rate (solid) curve there is no shear flow in the system, whereas in the lower growth rate (dashed) curve there is a constant shear flow given by $dv_E/dx = 0.05(c_s/L_n)$. The shear flow combines with the magnetic shear S (on the x-axis) to shift the waves off the mode rational surface creating a $\langle k_\parallel \rangle \neq 0$ which breaks the symmetry of turbulence and produces a Reynolds stress transporting flow momentum.

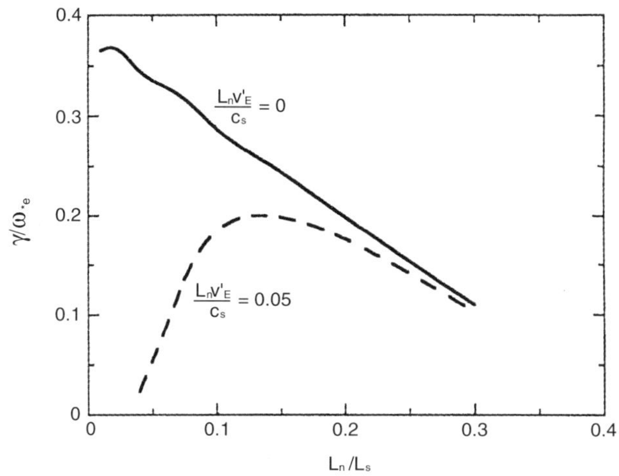

Fig. 8.8 The gyrokinetic growth rate as a function of magnetic shear $S = L_n/L_s$ with and without $\boldsymbol{E} \times \boldsymbol{B}$ shear flow [Hamaguchi and Horton (1993)].

References

Batchelor, G. K. (1969). Computation of the energy spectrum in homogeneous two-dimensional turbulence, *Phys. Fluids* **12**, p. II-233, doi:10.1063/1.1692443, http://link.aip.org/link/doi/10.1063/1.1692443.

Chandrasekhar, S. (1961). *Hydrodynamic and Hydromagnetic Stability* (Oxford University Press), pp. 481-498, ISBN:0-486-64071-X.

Chew, G. F., Goldberger, M. L., and Low. F. E. (1956). *Proceedings of the Royal Society of London. Series A*, Mathematical and Physical Sciences, **236**, 1204, pp. 112-118, http://www.jstor.org/stable/99870.

Dong, J. Q. and Horton, W. (1993). Kinetic quasitoroidal ion temperature gradient instability in the presence of sheared flows, *Phys. Fluids B* **5**, p. 1581 (1993); http://dx.doi.org/10.1063/1.860898.

Frisch, U. (1995). *Turbulence: The Legacy of A. N. Kolmogorov* (Cambridge University Press) ISBN:0521-45103-5.

Fu, X., Horton, W., Sen, A. K., Sokolov, V., Lin, Z., and Xiao, Y. (2012). *Phys. Plasmas* **19**, p. 032303, doi:1063/1.3686143.

Gill, A. E. (1982). *Atmospheric Ocean Dynamics* (Academic Press, London) International Geophysics Series, **30**.

Hamaguchi, S. and Horton, W. (1990). *Phys. Fluids B* **2**, pp. 1833-1851.

Hamaguchi, S. and Horton, W. (1991). *Phys. Fluids B* **4**, 319–328.

Hamaguchi, S. and Horton, W. (1992). *Plasma Phys. Contr. Fusion* **35**, pp. 203-233.

Hasegawa, A. and Mima, K. (1977). *Phys. Rev. Lett.* **39**, p. 205, http://link.aps.org/doi/10.1103/PhysRevLett.39.205.

Hennequin, P., Subot, R., Honoré, C., Hoang, G. T., Garbet, X., Truc, A., Fenzi, C., and Quéméneur, A. (2004). Scaling laws of density fluctuations at high-k on Tore Supra, *Plasma Phys. Control. Fusion* **46**, p. B121, doi:10.1088/0741-3335/46/12B/011.

Hong, B.-G., Horton, W., Hamaguchi, S., Wakatani, M., Yagi, M., and Sugama, H. (1991). Transition from resistive-G to η_i driven turbulence in stellarator systems, doi:10.1063/1.859683, http://link.aip.org/link/doi/10.1063/1.859683

Horton, W., Kim, J.-H., Chagelishvili, G. D., Bowman, J. C., and Lominadze, J. G. (2010). Angular redistribution of nonlinear perturbations: A universal feature of nonuniform flows, *Phys. Rev. E* **81**, p. 066304, doi:10.1103/PhysRevE.81.066304.

Horton, W. and Ichikawa, Y.-H. (1996). *Chaos and Structures in Nonlinear Plasmas* (World Scientific, Singapore) ISBN:81-7764-234-0.

Horton, W. and Hasegawa, A. (1994). Quasi-two-dimensional dynamics of plasmas and fluids, *Chaos* **4**, 2, p. 227.

Horton, W. (1986). Statistical properties and correlation functions for drift waves, *Phys. Fluids* **29**, 1491, doi:10.1063/1.865667, http://link.aip.org/link/doi/10.1063/1.865667.

Huld, T., Nielsen, A. H., Pecseli, H. L., and Rasmussen, J. Juul. (1991). Coherent structures in two-dimensional plasma turbulence, *Phys. Fluids B* **3**, p. 1609, doi:10.1063/1.859680.

Kendl, A. and Scott, B. D. (2005). Shear flow reduction by the geodesic transfer mechanism in tokamak edge turbulence, *Phys. Plasmas* **12**, p. 064506, http://dx.doi.org/10.1063/1.1938147.

Kim, J-Y., Kishimoto, Y., Wakatani, M., and Tajima, T. (1996). Poloidal shear flow effect on toroidal ion temperature gradient mode: A theory and simulation, *Phys. Plasmas* **3**, p. 3689, doi:10.1063/1.871939.

Kraichnan, R. H. (1984). *Phys. Fluids* **1**, pp. 1030-1048.

Kraichnan, R. H. (1971). *J. Fluid Mech.* **47**, pp. 525-535, doi:10.1017/S0022112071001216.

Kraichnan, R. H. (1967). *J. Fluid Mech.* **47**, p. 513-524, http://handle.dtic.mil/100.2/AD653111.

Kraichnan, R. H. (1964). *Phys. Fluids* **1**, pp. 1048-1062.

Krommes, J. A. (2002). Fundamental statistical descriptions of plasma turbulence in magnetic fields, *Phys. Reps.* **360**, pp. 1-352, http://dx.doi.org/10.1016/S0370-1573(01)00066-7.

Krommes, J. A. (1999). *Plasma Phys. Control. Fusion* **41**, pp. A641-A652.

Kulsrud, R. M. (1983). MHD Description of Plasma, *Handbook of Plasma Physics I*, eds. M. N. Rosenbluth and R. Z. Sagdeev, *Basic Plasmas Physics I*, eds. A. A. Galeev and R. N. Sudan (North-Holland Publishing Company).

Legras, B., Santangelo, P., and Benzi, R. (1988). High-resolution numerical experiments for forced two-dimensional turbulence, *Europhys. Lett.* **5**, p. 37, doi:10.1209/0295-5075/5/1/007.

Leith, C. E. (1968). *Phys. Fluids* **11**, p. 671, doi:10.1063/1.1691968.

Lesieur, M. (2008). Turbulence in Fluids, *Series: Fluid Mechanics and its Applications* (Berlin: Springer) ISBN:978-1-6434-0.

Leslie, D. C. (1973). *Theory of Turbulence* (Clarendon Press, Oxford).

McWilliams, J. C. (1984). Emergence of isolated coherent vortices in turbulent flow, *J. Fluid Mech.* **146**, pp. 21-43 (Cambridge University Press) doi:10.1017/S0022112084001750.

McWilliams, J. C. (1990). The vortices of two-dimensional turbulence, *J. Fluid Mech* **219**, pp. 361-385 (Cambridge University Press) doi:10.1017/S0022112090002981.

McWilliams, J. C. (1990). The vortices of geostrophic turbulence, *J. Fluid Mech.* **219**, pp. 387-404 (Cambridge University Press) doi:10.1017/S0022112090002993.

Meiss, J. D. and Horton, W. (1983). Solitary drift waves in the presence of magnetic shear, *Phys. Fluids* **26**, p. 990, doi:10.1063/1.864251, http://link.aip.org/link/doi/10.1063/1.864251

Mikhailovskii, A. B.. Smolyakov, A. I, Kovalishen, E. A., *et al.* (2006). *Phys. Plasmas* **13**,

p. 042507.

Monin, A. S. and Yaglom, A. M. (1965). *Statistical Fluid Mechanics* (MIT Press).

Moreau, R. (1983) Two-Dimensional Turbulence, ed. R. Journal de Mechanique/Journal of Theoretical and Applied Mechanics, supplement, ISBN 2-04-011785-7.

Nakata, T.-H. Watanabe, H. Sugama, H., and Horton, W. (2010). Formation of coherent vortex streets and transport reduction in electron temperature gradient driven turbulence *Phys. Plasmas* **17**, p. 042306, doi:10.1063/1.3356048.

Novakovskii, S. V., Mikhailovskii, A. B., and Onishchenko, D. G. (1988). *Phys. Letts. A* **132**, p. 33.

Parker, S. E., Mynick, H. E. Artun, M., Cummings, J. C., Decyk, V., Kepner, J. V., Lee, W. W., and Tang, W. M. (1996). *Phys. Plasmas* **3**, p. 1959.

Pedlosky, J. (1987). Geophysical Fluid Dynamics, 2nd Ed., *Science*.

Stanisic, M. M. (1985). *The Mathematical Theory of Turbulence*, (Universitext, Springer-Verlag, Berlin, Heidelberg, New York, Tokyo) (paperback), 429 pages, ISBN 0-387-96107-0.

Sugama, H., Watanabe, T.-H., and Horton, W. (2003). Comparison between kinetic and fluid simulations of slab ion temperature gradient driven turbulence, *Phys. Plasmas* **10**, pp. 726-736, `http://dx.doi.org/10.1063/1.1544664`.

Sydora, R. D., Decyk, V., and Dawson, J. M. (1996). Fluctuation-induced heat transport results from a large global 3D toroidal particle simulation model, *Plasma Phys. Control. Fusion* **38**, p. A281, doi:10.1088/0741-3335/38/12A/021.

Terry, P. W. (2000). *Rev. Mod. Phys.* **72**, pp. 109-165, doi:10.1103/RevModPhys.72.109.

Terry, P. W. and Horton, W. (1982). *Phys. Fluids* **25**, p. 491.

Yagi, M., Wakatani, M., Sugama, H., Hong, B-G., and Horton, W. (1989). *J. Phys. Soc. Jpn.* **12**, p. 4265.

Zhang, H. and Lin, Z. (2010). *Phys. Plasmas* **17**, p. 055902.

Chapter 9

Impurity Transport Studies

This chapter analyzes the transport of impurity ions of charge Z and mass m_z in magnetized plasma. In small concentration $\alpha_z = n_z Z^2/n_e \to 0$ the impurities respond as test particles in the drift wave turbulence. Typically, however, the concentration is sufficient to make the impurities an active component in the lowest frequency part of the drift wave spectrum.

Turbulent particle transport experiments in the C-Mod Alcator tokamak give detailed data on the transport of the Boron and Argon impurities. In this chapter we introduce partial differential equations (PDEs) for the multi-component plasmas. The PDEs are solved for the fluctuation fields vector $\boldsymbol{X_k}$ composed of the electric potential φ, hydrogenic density δn_i and the impurity densities δn_z. A typical set of impurities are the ions from the material coating the plasma facing metal walls, which would be Beryllium with $Z = 4$ for JET and Boron with $Z = 5$ for Alcator from the walls and RF antennas along with some metallic ions.

Linear eigenmode analysis shows that there are the usual drift waves with hydrogen, and there are impurity drift waves supported by the impurity density gradients. The impurity drift waves add new degrees of freedom to the plasma allowing for hydrogen flux to be inward or outward. In these new impurity modes the working gas ions behave quasi-adiabatically since the parallel phase velocity is now lower than in the hydrogen drift waves. The interaction between these two types of drift waves makes the transport matrix large and with additional collective modes. Examples are given for the particle fluxes for C-Mod in the L-, H-, ITB- and I-confinement modes.

The probability density of low concentration tracer particles like impurities calculated for a spectrum of drift waves is found to have non-Gaussian or non-normal spatial distributions in owing to the large-scale convection cells in the turbulence. An example for the tracer-charged particles in a spectrum compute from the two PDEs for resistive-g turbulence is given by Del-Castillo-Negrete (2000). The authors show how to analyze the long-tailed PDFs of the test particles with fractional derivatives. The authors argue that such non-diffusive transport is common for tracer particles. Thus, for trace amounts of impurities in a hydrogenic plasma, one may expect to see much faster particle transport than given by diffusion coefficient.

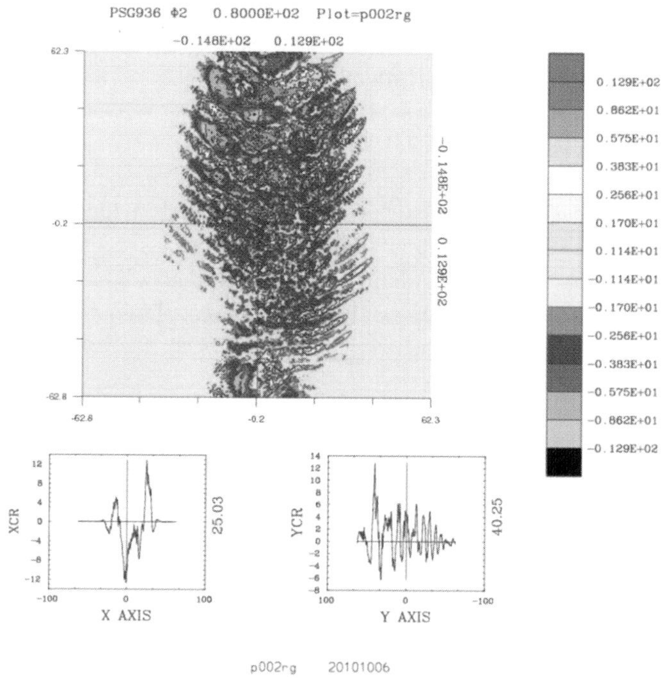

Fig. 9.1 Dispersion and diffusion of a passively convected impurity injected as a localized Gaussian puff at $t = 0$ in plasma with a steady turbulent state of drift wave turbulence. For the passive convection all impurities behave the same from the $\boldsymbol{E} \times \boldsymbol{B}$ motion provided $k_\perp \rho_s \ll 1$. The motion of the center-of-mass of the injected cloud is down the density gradient of the background plasmas however there is large dispersion with long fingers of impurity going toward the high density core plasma on the left. The lower two panels are cuts through the maximum of δn_z at $x = 25, y = 40$ showing the corregation in y and fingers in x.

Thus, the description of the particle data requires both a diffusivity D and a convection velocity V [Horton and Rowan (1994)]. Neoclassical theory gives a base level of D and V from Coulomb collisions in the absence of waves.

In the limit of a small concentration of impurities these charged molecules or ions are passively convected by the drift waves. Figure 9.1 shows a snapshot of the evolution of the impurity concentration for passive convection by the $\boldsymbol{E} \times \boldsymbol{B}$ motion in a drift wave turbulence. The waves produced a faster diffusive transport in the direction of the wave propagation, often called the zonal directions in the terminology taken from geophysical fluid dynamics, and a slower diffusion in the direction of the density gradient called the meridional direction in geophysical fluid dynamics. This transport often deviates some from the normal gaussian diffusion of the particles as revealed by constructing the probability distributions of the particles at different stages of the evolution.

This passive convection is a famous and difficult problem in the dispersal of pollutants in the atmospheric flows and thus has a rich literature associated with the problem. No attempt is made to review this area, but note that the early work was that of Taylor giving an important formula for the average of the particle displacement in homogeneous turbulent flows as

$$R^2(t) = 2\langle V_E^2\rangle \int_0^t d\tau(t-\tau)C_L(\tau) \tag{9.1}$$

where $C_L(\tau)$ is the Lagrangian autocorrelation function for the $\boldsymbol{E}\times\boldsymbol{B}$ velocity of the particle. The function $C_L(\tau)$ generically goes as τ^2 for τ smaller than τ_c the correlation time of the turbulence and linearly as $C_L(\tau) \sim \tau$ for times long compared with the correlation time of the turbulence. This leads to diffusion at the rate

$$D = \langle V_E^2\rangle \tau_L \tag{9.2}$$

for times long compare with the Lagrangian autocorrelations time τ_L.

The problem has been investigated for many different models of drift wave turbulence. In work by Basu, *et al.* (2003) and Hasegawa and Wakatani (1983), the collisional turbulence model is used with a constant $C_\nu = \left\langle k_\parallel^2 v_e^2\right\rangle / \langle\omega_{*e}\nu_e\rangle$. The collisionality parameter C_ν varies between 0.1 for the strongly collisional regime to the value of $C_\nu = 5$ for the weakly collisional regime. The corresponding energy spectra are shown along with the corresponding running diffusion coefficients D_x and D_y for the range of $C_\nu = \{0.1, 0.5, 2.0, 5.0\}$. The dimensionless values of the diffusion drop from about 1.7 for the small $C_\nu = 0.1$ turbulence to the range of $D_x \sim 0.1$ to 0.2 for the large $C_\nu = 2$ to 5 range of the turbulence model which contains many monopole and dipole vortex structures.

The transport across the magnetic field is adequately characterized as diffusivity while the transport in the direction of the wave propagation which is the symmetry direction in the equilibrium (zonal direction for example) is super diffusivity meaning the variance of the spread of the particles in the direction of wave propagation is faster that $2D_yt$. When the motion is examined in detail we see that the impurities are trapped and detrapped in the wave potentials and carried as trapped particles with the wave propagation sufficiently often and far to produce a super-diffusion transport. In the example from Basu, *et al.* (2003) the waves produce a super-diffusion transport with $\langle Y^2\rangle = D_{\text{sup}}t^{1.4}$. Figure 9.1 shows the rapid dispersal of the impurities in the direction of the wave propagation in the y-direction.

9.1 Drift Wave Eigenmodes with Active Impurity Components

Turbulent transport from drift waves is a key problem for fusion physics across all magnetic confinement geometries. To analyze plasmas with impurities three-component fluid equations are used to find the eigenmodes and eigenfrequencies of a nonuniform, magnetized plasma with a four dimensional fluctuation vector $\boldsymbol{X}_k$ composed of fluctuations of the electron density, the working gas ion density, the

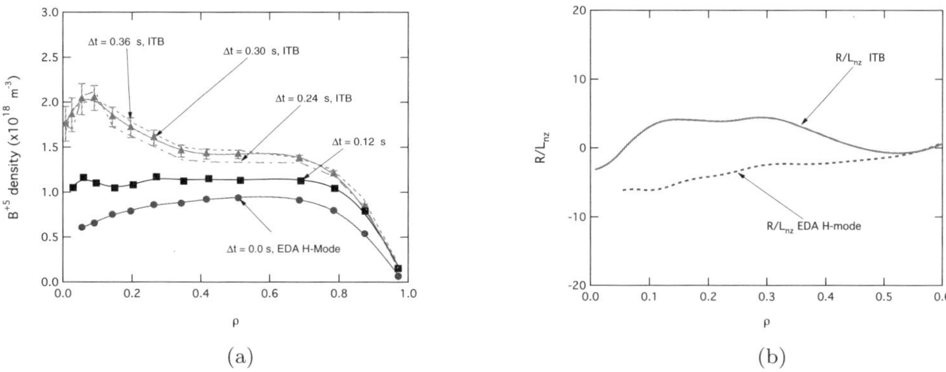

Fig. 9.2 (a) Time evolution of boron density profile in the C-Mod discharge with 3.5 MW ICRF heating, from the H-mode (circles), transition (squares), to ITB (triangles in the error bars); (b) boron density gradient profile in H-mode (dash) and in the ITB (solid) at time $t = 0.30\,\mathrm{s}$ [Rowan, *et al.* (2008)].

impurity density and the electrostatic plasma potential. This structure of the eigenmodes and eigenvectors is shown for two collisionality regimes: (i) the collisional drift waves appropriate for the scrape-off layer (SOL) and the edge plasma in limiter discharges and (ii) the trapped electron mode taken in the limit of a Terry-Horton fluid description for the core plasma. From the eigenmodes and eigenvectors, the out-of-phase part of the density to potential fluctuations are derived which determines the quasilinear particle fluxes as a function of the power spectrum of the plasma potential fluctuations and the gradient parameters characterizing the L-, H- and ITB-confinement modes. Dynamical equations for the nonlinear interaction of the fields expanded in eigenvectors are derived.

First we look at data from C-Mod with boronized walls and with boron injection to obtain a perspective on the wide range of plasma gradients and confinement regimes. Figure 9.2 shows the fully-ionized boron density profile evolution in a C-Mod ITB shot. Multiple types of drift wave modes exist in such plasmas. In the nonlinear state the modes are coupled in a complex manner. For weakly turbulent states the quasilinear particle fluxes for hydrogenic and impurity gases are calculated from the out-of-phase φ and δn_i and δn_z fluctuations. The qualitative changes of the particle fluxes with variation of the sign and strength of the density gradient lengths are compared with C-Mod plasmas in the L-, H-, ITB- and I-mode regimes. The I-mode has its own low-frequency instability under conditions described in Coppi and Zhou (2012).

Nonlinear simulations show a modest change in the amplitude of the turbulence, for a change, from Boron to Argon in the fully-ionized hydrogenic plasma. The change of plasma confinement regimes arises from the balancing of the hydrogen and impurity particle fluxes. The phase shifts in the fluctuations change the fluxes.

A 1-$\frac{1}{2}$ D radial transport code TBD is used with formulas derived from the turbulence runs to examine the effect of the various parts (diffusion and convection) of the quasilinear fluxes.

From gyrokinetic equations one derives the Vlasov-Poisson dispersion relation with the impurities. There are three types of drift-wave instabilities that produce turbulent particle and impurity fluxes:

(1) the drift wave with hydrodynamic ion response giving ITG modes with quasi-adiabatic electrons,

(2) the trapped electron-ITG mode (TEM-ITG) that has trapped electron resonances and FLR-hydrodynamic ions and impurities, and

(3) the impurity drift wave where the impurity has FLR-hydrodynamic response and the hydrogenic working gas has adiabatic response. The impurity mode is strongly unstable when the hydrogenic ions and the impurity ions have reversed (opposite direction) radial gradients and the turbulent flux then transports each ion type down its respective density gradient leading to a thorough mixing of the two ion types. Simulations show turbulence growth from initial plasma states with $\delta n_e = \varphi = 0$ with only a tiny injection of the impurity.

Onset of the impurity mode requires that the impurity density component to effective charge $Z_{\rm eff}$ be of order unity. Often the impurity contribution is specified by the value of $\alpha_z = n_z Z^2/n_e$. For a small impurity contribution $\alpha_z \ll 1$ to the effective charge $Z_{\rm eff}$ the impurity mode is stable.

Owing to the numerous system parameters, the exact stability condition is complicated and must be determined numerically. A DTRANS-Impurity code is available on the IFS (`http://pecos.ph.utexas.edu/~vortex`).

To model the impurity mode turbulence we generalize the multi-fluid models given in Futatani, *et al.* (2010) to have the hydrogenic ion response given by the quasi-adiabatic response

$$\delta n_i = -\frac{n_i e\phi_k}{T_i}\left[1 + i\sqrt{(\pi/2)}\frac{(\omega_k - \omega_{*i})}{(|k_\| v_i| + |\omega_{Di}|)}\right].$$

This hydrogenic response is the ion analog of what is called the "$i\delta_k$" response function used in the Terry-Horton model for TEM turbulence [Terry and Horton (1983)] in Chapter 8.

While some studies suggest that the effect of heavy impurities is more dramatic on the fluctuations and transport than the lighter impurities we have not found a strong difference in the Argon and Boron dynamics in the C-Mod simulations. The difference in the fluxes is explained by the different profiles of the impurities and the spatial distribution of the turbulence.

9.1.1 *Eigenmodes and the quasilinear fluxes with impurities*

The eigenvectors of the impurity drift wave matrix give the polarization of the fluctuations that determine the relative strength and direction of the impurity and hydrogenic and impurity ion transport.

For example, the trapped electron drift eigenvalue problem is given by the matrix

$$\begin{bmatrix} 1 & Z & -1 & -f_p \\ -i\omega & 0 & 0 & i\omega_{*i} - i\left(\omega + i\mu_i k_\perp^2\right) k_\perp^2 \rho_{s,i}^2 \\ 0 & -i\omega & 0 & i\omega_{*z} - i\left(\omega + i\mu_z k_\perp^2\right) k_\perp^2 \rho_{s,z}^2 \\ 0 & 0 & -i\omega + i\omega_{De} + \nu_{\text{eff}} & i\omega_{*e} f_t + i\omega_{De} f_t \end{bmatrix} \begin{bmatrix} \frac{\delta n_i}{n_e} \\ \frac{\delta n_z}{n_e} \\ \frac{\delta n_e}{n_e} \\ \frac{e\phi}{T_e} \end{bmatrix} = \begin{bmatrix} 0 \\ 0 \\ 0 \\ 0 \end{bmatrix} \tag{9.3}$$

where f_p is the passing and f_t the trapped fractions of the electrons. The matrix in Eq. (9.3) is of the form $\omega \boldsymbol{B_k X}_k = \boldsymbol{A X_k}$ where ω is the complex eigenvalue.

Due to the complex dependence of the dynamics on the fluctuation frequency ω the linear equations of matrix by the $\boldsymbol{A}$ and $\boldsymbol{B}$ matrices in Eq. (9.3) lack the symmetries of the $\boldsymbol{A}$ and $\boldsymbol{B}$ matrices in the standard orthogonal form of Hamiltonian systems. Thus we introduce the left $\boldsymbol{y}_i^* = \langle i \mid$ and the right $\boldsymbol{x}_j = \mid j\rangle$ eigenvectors of the auxiliary matrix $\boldsymbol{M} = \boldsymbol{B}^{-1}\boldsymbol{A}$.

The left eigenvectors are taken from

$$\boldsymbol{y}_i^* \boldsymbol{M} = \lambda_i \boldsymbol{y}_i^* \tag{9.4}$$

and the right eigenvectors are taken from

$$\boldsymbol{M}\boldsymbol{x}_j = \lambda_j \boldsymbol{x}_j. \tag{9.5}$$

Then post-multiply Eq. (9.4) by $\boldsymbol{x}_j$ and subtracting from pre-multiply Eq. (9.5) by $\boldsymbol{y}_i^*$ leads to

$$(\lambda_j - \lambda_i)\, \boldsymbol{y}_i^* \boldsymbol{x}_j = 0 \tag{9.6}$$

from which we derived that $\langle i \mid j\rangle = \boldsymbol{y}_i^* \boldsymbol{x}_j = 0$ for nondegenerate $\lambda_i \neq \lambda_j$ eigenvalues. Now we can normalize $\langle i \mid i\rangle = 1$ and proceed as in quantum mechanical perturbation theory.

To get the eigenvectors X_α for each root ω_α of the dispersion relation $D(\omega, k_x, k_y, k_\parallel, \mu) = \det(\boldsymbol{M} - \lambda \boldsymbol{I}) = 0$ where the parameters of the system are listed in the vector μ needs to be solved. We factor out the value of the X_4 component $(e\phi_k/T_e)$ and write the eigenvector as the independent $\boldsymbol{X} = (e\phi_k/T_e)\widehat{\boldsymbol{X}}$. Then the components of the drift wave polarization vector $\widehat{\boldsymbol{X}}$ are

$$\widehat{\boldsymbol{X}} = \begin{bmatrix} \dfrac{\delta n_i}{n_e} / \dfrac{e\phi}{T_e} \\ \dfrac{\delta n_z}{n_e} / \dfrac{e\phi}{T_e} \\ \dfrac{\delta n_e}{n_e} / \dfrac{e\phi}{T_e} \\ 1 \end{bmatrix}. \tag{9.7}$$

Physically, the $\widehat{\boldsymbol{X}}$ vector describes the "polarization" of density fluctuations related to the electrostatic potential fluctuation. An example of the polarization diagram is shown in Fig. 9.3.

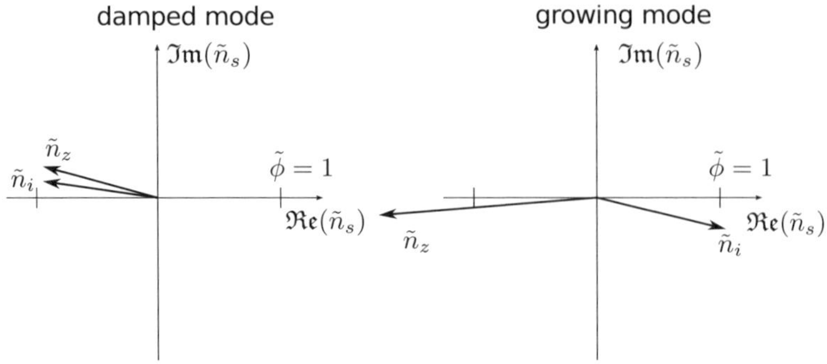

Fig. 9.3 Polarization diagram of eigenmodes of the usual drift wave with impurities.

Now the particle fluxes are given by the polarization vector as

$$\begin{aligned} \Gamma_i &= \mathrm{Re}\sum_k \frac{ik_y\phi_k^*}{B}\delta n_i = -n_e\frac{T_e}{eB}\sum_k \frac{k_y|e\phi_k|^2}{T_e^2}\mathrm{Im}\widehat{X}_1(k) \\ \Gamma_z &= \mathrm{Re}\sum_k \frac{ik_y\phi_k^*}{B}\delta n_z = -n_e\frac{T_e}{eB}\sum_k \frac{k_y|e\phi_k|^2}{T_e^2}\mathrm{Im}\widehat{X}_2(k) \\ \Gamma_e &= \mathrm{Re}\sum_k \frac{ik_y\phi_k^*}{B}\delta n_e = -n_e\frac{T_e}{eB}\sum_k \frac{k_y|e\phi_k|^2}{T_e^2}\mathrm{Im}\widehat{X}_3(k). \end{aligned} \tag{9.8}$$

Analytically we know that formulas that the first row of Eq. (9.3) gives that $\Gamma_e = \Gamma_i + Z\Gamma_z$. Figure 9.4 shows an example of the boron and deuterium flux as in Alcator C-Mod as a function of the boron impurity gradient R/L_{nz}.

The quasilinear formula verified numerically in Eq. (9.8) are valid when there are overlapping resonances in the Hamiltonian motion of the test particles. The condition are given in Horton and Ichikawa (1996). In the nonlinear state the fastest growing modes couple to the damped eigenmodes driving them up to the level required for the power flow from the unstable waves to balance the damping from the stable modes. Figure 9.1 shows an example of the k_y spectrum for fixed k_z and integrated over k_x.

The nonlinear waves will saturate when the finite k_x part of the spectrum grows. The rms values of the density fluctuation is then

$$\left(\frac{\delta n_i}{n_i}\right)_{\mathrm{rms}} \sim Z\left(\frac{\delta n_z}{n_e}\right)_{\mathrm{rms}} \sim \frac{1}{\langle k_x^2 L_n^2\rangle^{0.5}}. \tag{9.9}$$

The saturation level in Eq. (9.9) has the physical meaning that the average gradient of the rms impurity fluctuations $\langle k_x^2\rangle^{1/2}\delta n_{z,\mathrm{rms}}$ reaches the level of the background density gradient.

The spectrum has an isotropic part at higher $k_\perp$ and an anisotropic part from zonal flow and density flattening part at low k_y values. There are coherent vortices that come and go in the turbulence and are strong in the limit of small parallel

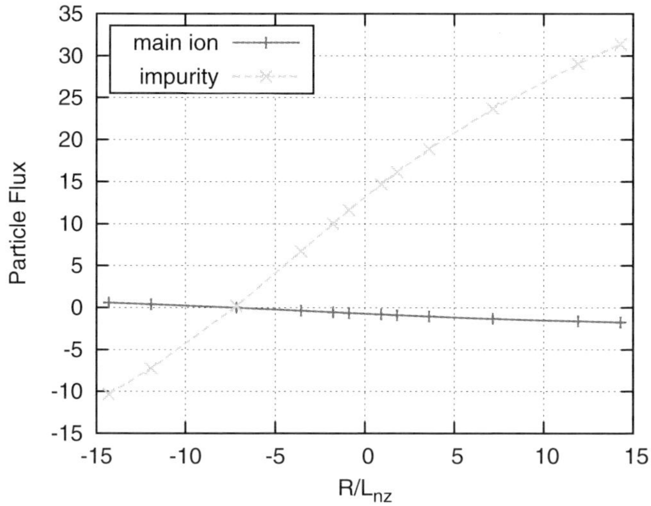

Fig. 9.4 Quasilinear fluxes of the main ion and the impurity, in this case Boron(B^{5+}) as function of the impurity density gradient. Transport coefficients D and V can be extracted from the slope and the intercept of the curve.

diffusivity $\nu_{\parallel} \ll |\omega_k|$. The turbulent wavenumber spectrum that enters Eq. (9.8) maybe modeled as

$$\left|\frac{e\phi_k}{T_e}\right|^2 = \frac{\rho_\tau^2}{L_n^2}\frac{1}{(1+k_\perp^2\rho_s^2)^{n/2}}. \tag{9.10}$$

with the spectral index $n = 3$ to 5 from simulations and scattering experiments.

Figure 9.4 shows the quasilinear flux of ion density and impurity density. The quasilinear fluxes are calculated for $0.1 < k_\perp\rho_s < 1.0$ with Eq. (9.10). In the linear growth rate and the frequency, there is no remarkable difference by changing the impurity species. The quasilinear flux of impurity Γ_z is a function of the impurity species. There is no significant phase shift between electron density and the potential. In other words, the adiabaticity of electron motion is high enough for $\Gamma_e \cong 0$.

The ion density flux is interesting. The direction of the convective part of the flux depends on the phase velocity. There are reversals of the fluxes under certain conditions. Obviously, reversal of a flux of ions is of critical importance. The flux of impurities has a dependence on the species. The higher mass impurity has smaller quasilinear flux.

From the experimentally inferred value of D_z we can infer the fluctuation levels of δn_i, δn_z and $e\phi/T_e$ from the polarization vector. An assumption on the slope of the power spectrum of $\int I(k)dk = \langle |e\phi/T_e|^2\rangle$ is needed to model the spectrum. The dominant $\langle k_y\rho_s\rangle$ is estimated from the integral in Eq. (9.8). A typical drift wave spectrum is

$$I\left(k_{\perp}\right)=\frac{I_1}{\left(k_x^2+k_y^2\right)^{3/2}} \tag{9.11}$$

for $k_x^2 \geq \left\langle k_x^2\right\rangle$ where the turbulent mixing radial correlation length $l_c \approx 1/\left\langle k_x^2\right\rangle^{1/2}$. The associated mixing length amplitude is

$$\left\langle\left(\frac{e\phi}{T_e}\right)^2\right\rangle^{1/2} \sim \frac{l_c}{L_{n_\alpha}} \tag{9.12}$$

when $1/L_{n_\alpha}$ is the dominant driving gradient ($\alpha=i,z,e$) for the growth rate.

The turbulent diffusivity is then

$$\begin{aligned} D_z &= \langle k_y\rho_s\rangle\left\langle\left(\frac{e\phi}{T_e}\right)^2\right\rangle\frac{T_e}{eB} \\ &\approx \frac{\rho_s}{l_c}\left(\frac{l_c}{L_{n_z}}\right)^2\frac{T_e}{eB}=\frac{\rho_s l_c}{L_{n_z}^2}\frac{T_e}{eB} \end{aligned} \tag{9.13}$$

when the impurity gradient $1/L_z$ determines the growth rate. Both magnetic shear $1/L_s$ and the E_r-shear control the radial correlation length l_c.

9.2 The Three-Component Fluid Equations

The Braginskii equations, which describe the evolution of fluid moments, are obtained by taking moments of the Fokker-Planck equations and using a heat flux closure model in the thermal balance equation. Here we use the negligible ion temperatures and the isothermal electron models to close the system of fluid equations [Futatani, *et al.* (2010)].

The equations of the evolution of densities are

$$\frac{\partial n_s}{\partial t}+\nabla\cdot\left(n_s\boldsymbol{v}_s\right)=0 \tag{9.14}$$

where n_s and v_s are the density and the fluid velocity of the species $s=e$, i, and z for electrons, hydrodynamic ions and impurities.

The conservation of charge follows from Eq. (9.14) by computing $\partial_t\rho_{ch}=0$ with $\rho_{ch}=\sum_s e_s n_s$. In the initial evolution following the impurity injection there is a rapid evolution ($\Delta t\sim qR/v_e$) of $\rho_{ch}(x,t)$ to zero from the parallel electron motion taking the system to $\nabla\cdot\boldsymbol{J}=0$ where $\boldsymbol{J}=\sum_s e_s n_s\boldsymbol{v}_s$.

The evolution of fluid velocities is given by

$$n_s m_s\frac{d\boldsymbol{v}_s}{dt}=n_s e_s\left(\boldsymbol{E}+\boldsymbol{v}_s\times\boldsymbol{B}\right)-\nabla p_s-\nabla\cdot\boldsymbol{\Pi}_s \tag{9.15}$$

where p_s and $\boldsymbol{\Pi}_s$ are the pressure and the off-diagonal momentum stress tensor. For low ion temperatures a small ion viscosity is kept in the ion momentum stress tensors.

9.2.1 *Reduction of three-component fluid equations*

An outer product of $\boldsymbol{B}$ to Eq. (9.15) is taken to derive the perpendicular components of drift velocity. The result is

$$\begin{aligned}\boldsymbol{v}_{s\perp} &= \frac{\boldsymbol{E}\times\boldsymbol{B}}{B^2} + \frac{1}{e_s n_s}\frac{\boldsymbol{B}}{B^2}\times\nabla p_s + \frac{m_s}{e_s}\frac{\boldsymbol{B}}{B^2}\times\frac{d\boldsymbol{v}_{s\perp}}{dt} \\ &= \boldsymbol{v}_E + \boldsymbol{v}_* + \boldsymbol{v}_{\rm pol}\end{aligned} \tag{9.16}$$

where the second line of Eq. (9.16) defines the three components of the velocity field; the $\boldsymbol{E}\times\boldsymbol{B}$ drift $\boldsymbol{v}_E = \boldsymbol{E}\times\boldsymbol{B}/B^2$, the diamagnetic drift $\boldsymbol{v}_* = \frac{1}{e_s n_s}\frac{\boldsymbol{B}}{B^2}\times\nabla p_s$ and the polarization drift $\boldsymbol{v}_{\rm pol} = \frac{m_s}{e_s}\frac{\boldsymbol{B}}{B^2}\times\frac{d\boldsymbol{v}_{s\perp}}{dt}$. The electron polarization drift is negligible owing to the proportionality to the electron mass. The polarization drifts for the hydrodynamic and impurity ions reduce to

$$\boldsymbol{v}_{\rm pol,s} = -\frac{m_s}{e_s B^2}\frac{d}{dt}\nabla\phi \tag{9.17}$$

and the polarization current is

$$\boldsymbol{j}_{\rm pol} = -\frac{n_i m_i + n_z m_z}{B^2}\frac{d}{dt}\nabla\phi. \tag{9.18}$$

In the simplest description of drift waves the divergence of the cross field polarization current $\boldsymbol{j}_{\rm pol}$ balances the divergence of the parallel electron currents.

The advection due to the diamagnetic drift is often negligible. For the polarization drift,

$$|v_{\rm pol}| = \frac{m_s}{eB}\frac{d\boldsymbol{v}_E}{dt} \sim \frac{\rho_s\omega}{c_s}v_E \tag{9.19}$$

where $1/\omega$ is a typical time scale of the phenomena. In toroidal fusion plasmas, the typical plasma of core tokamak, has the space-time scales such that $\rho_s \sim 10^{-3}$m, sound speed $c_s \sim 10^5\,{\rm ms}^{-1}$ and the drift wave-ITG turbulence has a frequency of the order $\omega \sim 10^5{\rm s}^{-1}$. Therefore, the order of polarization drift becomes $v_{\rm pol} \sim 10^{-3}v_E$. Therefore, $v_{\rm pol}$ can be neglected compared to other drift velocity, when compared directly with the drift velocity from $\boldsymbol{E}\times\boldsymbol{B}$ and diamagnetic pressure gradient terms v_{*s}. However, the divergence of the plasma current from $\boldsymbol{v}_{\rm pol}$ is comparable to the other plasma current contributions.

The characteristic dispersion scale of the ion gyroradius with the electron temperature in the three component plasma becomes

$$\rho_\tau^2 = \rho_{s,i}^2\frac{n_i}{n_e} + Z^2\rho_{s,z}^2\frac{n_z}{n_e}. \tag{9.20}$$

The definition of effective charge $Z_{\rm eff}$ that determines the resistivity η and the impurity dilution of the hydrogenic fusion fuel component is

$$Z_{\rm eff} = \frac{n_i + Z^2 n_z}{n_i + Z n_z}, \qquad 1 < Z_{\rm eff} < Z. \tag{9.21}$$

For tungsten $W(Z = 74)$, the $Z_{\text{eff}} = 3$ limit corresponds to the small concentration $n_z/n_e = 4 \times 10^{-4}$.

There is an enhancement in the dispersion radius ρ_s^2 from the impurities. The ratio of the impurity density n_z to electron density n_e follows from the effective charge Z_{eff} relative to the impurity charge Z through

$$\frac{n_z}{n_e} = \frac{Z_{\text{eff}} - 1}{Z(Z-1)}. \tag{9.22}$$

For the case $Z_{\text{eff}} = 3$ from fully ionized Beryllium Be^{4+}, the fraction is $n_z/n_e = (3-1)/(4 \times 3) = 1/6$. Therefore, the enhancement term $1 + (A - Z)n_z/n_e$ to ρ_x^2 is $1 + (5-2) \times 1/6 = 1.5$. Here, the upper limit of $Z_{\text{eff}} = 3$ was taken. In a same limit for fully ionized, C^{6+} for $n_z/n_e = (3-1)/(6 \times 5) = 1/15$. Therefore, the term $1 + (A - Z)n_z/n_e$ for fully ionized carbon is $1 + (12 - 6) \times 1/15 = 1.4$. For a partially stripped heavy metal, for example, W^{30+} the concentration is $n_z/n_e = (3-1)/(30 \times 29) = 1/435$. Therefore, the term $1+(A-Z)n_z/n_e$ for tungsten W^{30+} is then $1 + (184 - 74) \times 1/435 = 1.253$.

For the drift frequencies with the impurities, one defines

$$\begin{aligned}
\omega_{*i} &= k_y \frac{T_e}{eB} \frac{1}{n_e} \frac{dn_i}{dx} \\
\omega_{*z} &= k_y \frac{T_e}{ZeB} \frac{1}{n_e} \frac{dn_z}{dx} \\
\omega_{*e} &= -k_y \frac{T_e}{eB} \frac{1}{n_e} \frac{dn_e}{dx} \\
\omega_{De} &= k_y \frac{2T_e}{eBR} \cos\theta.
\end{aligned} \tag{9.23}$$

The charge neutrality condition and Eq. (9.23) give

$$\omega_{*e} = \omega_{*i} + Z^2 \omega_{*z}. \tag{9.24}$$

Other impurity system parameters are

$$\rho_{s,i}^2 = \frac{m_i T_e}{e^2 B^2} \quad \text{and} \quad \rho_{s,z}^2 = \frac{m_z T_e}{Z e^2 B^2}, \quad D_\parallel = \frac{\sigma_\parallel T_e}{n_e e^2} \quad \text{and} \quad \sigma_\parallel = \frac{n_e e^2}{m_e \nu_e}. \tag{9.25}$$

The determinant of the fluctuation matrix Eq. (9.3) gives a cubic polynomial dispersion equation

$$\begin{aligned}
&i k_\perp^2 \rho_T^2 \omega^2 (\omega - \omega_{De}) \\
&\quad - \nu_\parallel \left\{ 1 + k_\perp^2 \rho_T^2 + \left(\mu_i \rho_{s,i}^2 + \mu_z Z \rho_{s,z}^2\right) \frac{k_\perp^4}{\nu_\parallel} + i(Z^2 - Z)\frac{\omega_{*z}}{\nu_\parallel} \right\} \omega^2 \\
&\quad + (\nu_\parallel + i\omega_{De}) \left\{ \omega_{*i} + Z\omega_{*z} - i\left(\mu_i \rho_{s,i}^2 + \mu_z Z \rho_{s,z}^2\right) k_\perp^4 \right\} \omega = 0
\end{aligned} \tag{9.26}$$

where $\nu_\parallel = k_\parallel^2 D_\parallel$ and $\rho_T^2 = \rho_{s,i}^2 + \rho_{s,z}^2 Z$. For all k values, the cubic equation factors into the modes $\omega = 0$ and two finite ω roots of a equivalent quadratic equation.

9.2.2 *Trapped electron mode (inside the SOL)*

Inside the scrape-off layer (SOL) the plasma collisionality is low so the collisional drift wave evolves into the collisional trapped electron mode [Horton (1976)]. In the core plasma the few collisions transform the instability to the collisionless (CTEM) trapped electron mode. The modulation of the toroidal magnetic field strength is modeled as

$$B = B_0\left(1 - \frac{r}{R}\cos\theta\right). \tag{9.27}$$

The small aspect ratio $\epsilon = r/R \ll 1$ is considered here, then the fraction of trapped particles, f_t, can be defined by

$$f_t = \frac{\text{trapped electrons}}{\text{total electrons}}. \tag{9.28}$$

The trapped particle fraction f_t is estimated by

$$f_t = \left(\frac{B_{\max}}{B_{\min}} - 1\right)^{1/2} = \left(\frac{1+\epsilon}{1-\epsilon} - 1\right)^{1/2} \tag{9.29}$$

$$= 1 - f_p \tag{9.30}$$

where f_p is the fraction of passing electrons.

From the electron kinetic equation, we derive the fluctuating δf_e given by

$$\delta f_e = f_M\left\{\frac{e\varphi}{T_e} - \frac{1}{\tau}\int\frac{ds'}{v_\parallel}\varphi(s')\frac{\omega - \omega_{*e}}{\omega - \omega_{De} + i\nu_{\text{eff}}}\right\} \tag{9.31}$$

where the orbit integral is discontinuous with

$$\begin{aligned} \frac{1}{\tau}\int\frac{ds'}{v_\parallel}\varphi &= 0 \qquad \text{(for passing electrons)} \\ &\simeq \varphi \qquad \text{(for trapped electrons).} \end{aligned} \tag{9.32}$$

The passing electron integral in Eq. (9.32) vanishes due to phase mixing of φ averaged over the irrational toroidal field lines. The details of the trapped particle mode analysis are given in Kadomtsev and Pogutse (1970). From Eq. (9.31) and Eq. (9.32) one derives

$$\delta n_p = \int d^3v\,\delta f_p = f_p\frac{n_e e\varphi}{T_e} \tag{9.33}$$

$$\delta n_t = \int d^3v\,\delta f_t = f_t\frac{n_e e\varphi}{T_e}\left[1 - \frac{\omega - \omega_{*e}}{\omega - \omega_{De} + i\nu_{\text{eff}}}\right]. \tag{9.34}$$

There are two regimes of the trapped electron mode: the dissipative (DTEM) and the collisionless (CTEM) contribution. For the DTEM, $\nu_{\text{eff}} \gg |\omega| \sim \omega_{De}$ we have

$$\frac{\delta n_e}{n_e} = \left[1 + if_t\left(\frac{\omega - \omega_*}{\nu_{\text{eff}}}\right)\right]\left(\frac{e\delta\varphi}{T_e}\right).$$

For the CTEM, $\nu_{\text{eff}} \ll \omega_{De}$, the electron density response is

$$\frac{\delta n_e}{n_e} = \left[1 + i\pi f_t \left\langle \delta(\omega - \omega_{De}) \right\rangle (\omega - \omega_{*e})\right] \left(\frac{e\delta\varphi}{T_e}\right)$$

for the collisionless (CTEM) mode.

The equation for the passing electron fluctuation is $\delta n_p = (en_p/T_e)\varphi_k$ where $\delta n_p = n_p - n_{p0}$ is linear and in phase with the potential φ_k. The trapped electron density in the dissipative limit may be modeled with the fluid equation

$$\frac{\partial n_t}{\partial t} + \boldsymbol{v}_E \cdot \nabla n_t = -\nu_{\text{eff}} \left(n_t - n_{t0}\right) \tag{9.35}$$

with the steady-state trapped density given by $n_{t0} = f_t n_e$. The fluctuating trapped electron density δn_t is given by

$$\left(-i\omega + \nu_{\text{eff}}\right) \delta n_t + \frac{-ik_y \phi}{B} \frac{dn_t}{dr} = 0. \tag{9.36}$$

For the transition from the collisional regime to the hot core plasma the dissipation is replaced with $\nu_{\text{eff}} \to c_D|\omega_{De}|$ with constant c_D chosen to get agreement with collisionless gyrokinetic limit. For this system we have the dynamics

$$\begin{aligned}
&\frac{dn_i}{dt} + \mu_i \frac{m_i}{eB^2} \nabla^4 \phi = \nabla \cdot \left(\frac{m_i n_i}{eB^2} \frac{d}{dt} \nabla \phi\right) \\
&\frac{dn_z}{dt} + \mu_z \frac{m_z}{eB^2} \nabla^4 \phi = \nabla \cdot \left(\frac{m_z n_z}{ZeB^2} \frac{d}{dt} \nabla \phi\right) \\
&\frac{d}{dt} \delta n_t = -\nu_{\text{eff}} \delta n_t = n_e f_p \frac{e\varphi}{T_e} + \delta n_z + n_{t0}
\end{aligned} \tag{9.37}$$

The linear fluctuations are given by

$$\begin{aligned}
&-i\omega \delta n_i - \frac{ik_y \phi_k}{B} \frac{dn_i}{dr} = i \frac{k_\perp^2 m_i n_i}{eB^2} \omega \phi_k = i\omega k_\perp^2 \frac{m_i n_i T_e}{e^2 B^2} \left(\frac{e\varphi_k}{T_e}\right) \\
&-i\omega \delta n_z - \frac{ik_y \phi_k}{B} \frac{dn_z}{dr} = i\omega \frac{k_\perp^2 m_z n_z T_e}{e^2 B^2} \left(\frac{e\varphi_k}{T_e}\right) \\
&\delta n_e = \left(n_e f_p + \frac{k_y \frac{dn_t}{dr} \frac{T_e}{e}}{B\left(\omega + i\nu_{\text{eff}}\right)}\right) \left(\frac{e\varphi_k}{T_e}\right) \\
&\delta n_i + Z\delta n_z = \delta n_e + n_e f_p \frac{e\varphi_k}{T_e}.
\end{aligned} \tag{9.38}$$

The system of linear Eq. (9.38) determines the polarization of the fluctuations for the trapped electron modes.

For large collisional detrapping we obtain the approximation for the unstable drift wave given by

$$\omega_k = -\frac{(\omega_{*i} + \omega_{*z} Z)\nu_{\text{eff}}}{-k_\perp^2 \rho_\tau^2 \nu_{\text{eff}} + i\omega_{*e} f_t - i\omega_{*z} Z - f_p \nu_{\text{eff}} - i\omega_{*i} - (\mu_i \rho_{s,i}^2 + \mu_z \rho_{s,z}^2 Z + \mu_e \rho_{s,e}^2) k_\perp^4}$$

$$\gamma_k = -\frac{(k_\perp^2 \rho_\tau^2 + f_p)\omega_k^2 - (\mu_i \rho_{s,i}^2 + \mu_z \rho_{s,z}^2 Z)\nu_{\text{eff}}}{-k_\perp^2 \nu_{\text{eff}} \rho_\tau^2 + i\omega_{*e} f_t - i\omega_{*z} Z - f_p \nu_{\text{eff}} - i\omega_{*i} - (\mu_i \rho_{s,i}^2 + \mu_z \rho_{s,z}^2 + \mu_e \rho_{s,e}^2) k_\perp^4}. \tag{9.39}$$

The electron viscosity is negligible compared to the ion viscosity.

9.2.3 *Spectral expansion of the nonlinear fields for impurity turbulence*

To investigate the nonlinear fluctuations we expand the turbulent fluctuations in terms of the eigenvectors of the linear system. The advantage of the eigenmode expansion over that of Fourier modes is that the linear coupling are taken into account exactly in each eigenvector. The time evolution of the expansion coefficient $C_{k,i}(t)$ is due entirely to the nonlinear interactions.

The vector field $\boldsymbol{X}(t)$ is expressed as the sum over a set of k_x, k_y, k_z modes and for each $\boldsymbol{k}$ vector there is a sum over the three distinct eigenmodes, $j = 1, 2, 3$. The time-dependent expansion coefficients are defined as

$$\boldsymbol{X}(t) = \sum_{j,\boldsymbol{k}} C_{\boldsymbol{k},j}(t) \boldsymbol{X}_j e^{i\boldsymbol{k}\cdot\boldsymbol{x} - i\omega_{\boldsymbol{k},j} t} + \text{c.c.} \tag{9.40}$$

where the dynamics $dC_{\boldsymbol{k},j}/dt$ is due to the nonlinear terms in Eq. (9.40) and Eq. (9.41) and $X_j(\boldsymbol{k})$ is the eigenvector for $j = 1, 2, 3$ eigenmodes.

Substituting into Eq. (9.40) we find

$$\begin{aligned} &\sum_{\boldsymbol{k},j} \left(\omega C_{\boldsymbol{k},j} + i \frac{dC_{\boldsymbol{k},j}}{dt} \right) B X_{\boldsymbol{k},j} e^{i\boldsymbol{k}\cdot\boldsymbol{x} - i\omega_j t} \\ &= \sum_{\boldsymbol{k},j} A(\boldsymbol{k}) C_{\boldsymbol{k},j} X_j + \sum_{\boldsymbol{k}',j'} \sum_{\boldsymbol{k}'',j''} N(X_{\boldsymbol{k}'}, X_{\boldsymbol{k}''}) C_{\boldsymbol{k}',j'} C_{\boldsymbol{k}'',j''}. \end{aligned} \tag{9.41}$$

Now multiplying through Eq. (9.41) by $\boldsymbol{B}^{-1}$ and contracting, or projecting, the equation onto the left eigenvector $\boldsymbol{y}_{l,\boldsymbol{k}''} = \langle l, k'' \mid$ yields

$$\begin{aligned} &\sum i \langle y_{l,\boldsymbol{k}''} | x_{l,\boldsymbol{k}} \rangle \frac{dC_{l,\boldsymbol{k}}}{dt} \\ &= \sum_{m,\boldsymbol{k}_1,n,\boldsymbol{k}_2} \langle y_{l,\boldsymbol{k}''} | N(x_{m,\boldsymbol{k}_1}, x_{n,\boldsymbol{k}_2}) \rangle C_{m,\boldsymbol{k}_1}(t) C_{n,\boldsymbol{k}_2}(t) \exp\{i(\omega_{\boldsymbol{k}} - \omega_{\boldsymbol{k}_1} - \omega_{\boldsymbol{k}_2})t\}. \end{aligned} \tag{9.42}$$

The diagonal terms are normalized such that $\langle y_{l,\boldsymbol{k}''} \mid x_{l,\boldsymbol{k}} \rangle = \delta_{l,\boldsymbol{k}}$. Then we arrive

Table 9.1 Typical ITER Drift Wave Frequencies and Plasma Parameters.

	Core plasma	Edge plasma
Major Radius	$R = 6\,\mathrm{m}$	$R = 6\,\mathrm{m}$
Minor Radius	$a = 2\,\mathrm{m}$	$a = 2\,\mathrm{m}$
Magnetic field	$B = 5\,\mathrm{T}$	$B = 5\,\mathrm{T}$
Safety factor	$q = 2$	$q = 5$
Electron Temperature	$T_e = 10\,\mathrm{KeV}$	$T_e = 1\,\mathrm{KeV}$
Electron Density	$n_e = 10^{20}\,\mathrm{m}^{-3}$	$n_e = 10^{19}\,\mathrm{m}^{-3}$
Impurity Density(Be^{+4})	$n_z/n_e = 1\%$	$n_z/n_e = 3\%$
Parallel Wavenumber	$k_\parallel = 0.5q/R \sim 0.17\,\mathrm{m}^{-1}$	$k_\parallel \sim 0.42\,\mathrm{m}^{-1}$
Electron Thermal Velocity	$v_e = 1.5 \times 10^9\,\mathrm{m/s}$	$v_e = 4.7 \times 10^8\,\mathrm{m/s}$
Sound Speed	$c_s = 3.4 \times 10^7\,\mathrm{m/s}$	$c_s = 1.1 \times 10^7\,\mathrm{m/s}$
Ion Sound Gyroradius	$\rho_s = c_s/\omega_{ci} = 7.3 \times 10^{-3}\,\mathrm{m}$	$\rho_s = 3 \times 10^{-3}\,\mathrm{m}$
Electron Parallel Diffusion	$D_\parallel = \frac{v_e}{k_\parallel} = 8.8 \times 10^9\,\mathrm{m^2s^{-1}}$	$D_\parallel = \frac{v_e}{k_\parallel} = 1.1 \times 10^9\,\mathrm{m^2s^{-1}}$
Electron Collision Rate	$\nu_e = 1.1 \times 10^5\,\mathrm{s}^{-1}$	$\nu_e = 1.6 \times 10^5\,\mathrm{s}^{-1}$
Ion Collision Rate	$\nu_i = 1.6 \times 10^3\,\mathrm{s}^{-1}$	$\nu_i = 2.3 \times 10^3\,\mathrm{s}^{-1}$
Parallel Diffusion Rate	$\nu_\parallel = k_\parallel^2 D_\parallel = 6.6 \times 10^5\,\mathrm{s}^{-1}$	$\nu_e/(a/2R) = 2.2 \times 10^5\,\mathrm{s}^{-1}$
Electron Density Gradient Length	$L_{n_e} = 2.0\,\mathrm{m}$	$L_{n_e} = 5.0\,\mathrm{m}$
Impurity Density Gradient Length	$L_{n_z} = 2.0\,\mathrm{m}$	$L_{n_z} = -2.0\,\mathrm{m}$
Trapped Fraction	$f_t < 0.1$	$f_t = \sqrt{r/R} = \sqrt{a/2R} = 0.41$

at the nonlinear dynamical equation

$$i\frac{dC_{l,\boldsymbol{k}}}{dt} = \sum_{\boldsymbol{k}} \langle y_{l,\boldsymbol{k}} | N\left(x_{m,\boldsymbol{k}_1}, x_{n,\boldsymbol{k}_2}\right)\rangle C_{m,\boldsymbol{k}_1}(t) C_{l,\boldsymbol{k}_2}(t). \tag{9.43}$$

Equation (9.43) is of the form that occurs in quantum mechanics for the phonon-phonon interactions with a summation over $\boldsymbol{k}$-space lattice and a second summation over the states j. In the plasma problem the matrices are not Hermitian. The interactions come in resonant $\boldsymbol{k}$-vector triads but contain the additional complexity that each triad has 3 degrees of freedom for the interaction within each triad.

When the fields are arbitrarily restricted to one vector triad the system becomes integrable. The single triad has a reversible "pump depletion" behavior. As more interacting vector triads are included, the dynamics become chaotic and the fields diverge and mix in their phase space.

9.2.4 *Application to ITER*

Subroutines are available for solving systems of drift wave turbulence equations for the multi-component fusion plasmas (`http://pecos.ph.utexas.edu/~vortex`). The subroutines are fast compared with gyrokinetic simulations and suitable for real-time analysis and control of tokamak discharges.

Table 9.1 shows typical values for the ITER plasma assuming a 3% Beryllium corresponding to a charge fraction $f_Z = Zn_z/n_e = 48$ and a $Z_{\text{eff}} = 1.4$. For the core plasma gradients in the first column we take the H-mode parameters with almost flat electron density and $R/L_{ni} = 3$ balanced by $R/L_{nz} = -5$. In the second column for the edge plasma we take $R/L_{ni} = 10$ and $R/L_{nz} = 20$ and $R/L_{ne} = 12$ for the steep edge gradients. For these model values we estimate a $D_{\text{edge}} = 3\text{m}^2/\text{s}$ and $D_z = 1\,\text{m}^2/\text{s}$ for the core.

9.3 Impurity Transport in High-Density Regimes

High purity plasmas are associated with high plasma density operation. The self-cleaning ability of high-density plasmas has been well documented by Alcator C-Mod [Greenwald, *et al.* (1995)] and the Frascati tokamak FTU [Apicella, *et al.* (1997)] experiments. A scaling law relating plasma purity, radiated power, and machine dimensions has been derived from a number of experiments [Marmar, *et al.* (1982). Based on these experiments, average Z_{eff} values of around 1.2 should be possible in a high-density fusion reactor. The problem of plasma purity is associated with the problem of reducing the power load on the first wall where magnetic sputtering and evaporation produce impurities from the plasma-wall interactions.

One solution to reduce the impurity concentration is to use a cold radiating plasma mantle. The radiating plasma mantle may allow the use of molybdenum tiles as the first wall material. The high density regimes operating with strong screening of the main body of the plasma column from impurities have been observed. Experimental results [Weynants, *et al.* (1999)] indicate the possibility of operating with a radiating mantle able to dissipate up to 90% of the total power lost from the core plasma without energy confinement degradation. Thermal loads have been calculated for an ideal continuous first wall under the assumption that 70% of the input power is radiated, which would reduce the maximum thermal heat flux to the plasma-facing components to $0.5\,\text{MW/m}^2$ or less in compact-high-field tokamaks.

9.4 Trace Impurity Transport Studies in the Texas Experimental Tokamak (TEXT)

The results of impurity transport experiments in the Texas Experimental Tokamak (TEXT) are compared with the predictions of turbulence transport models [Gentle (1981)]. In these early experiments, scandium Sc was injected into the plasma using laser ablation and the time resolved profiles of the Sc ionization stages were measured along with the potential fluctuation profile fom the HIBP. The experiment was simulated using a one-dimensional (1D) radial transport code with the standard density transport flux

$$\Gamma = -D\frac{\partial n}{\partial r} + nV. \tag{9.44}$$

The diffusion coefficient D and convective velocity V parameters were varied until the time dependent 1D simulations reproduce the data. This representation for the

Table 9.2 Ionization Rates $\langle \sigma v \rangle$ for Atoms by Electron Impact [Voronov (1997)].

Atom	1st Ionization Energy	$\langle \sigma v \rangle$ at $E = T_e$
H	13.6 eV	8.7×10^{-9} cm^3/s
He	24.6 eV	5.5×10^{-9} cm^3/s
Li	5.4 eV	3.6×10^{-8} cm^3/s
Be	9.3 eV	2.7×10^{-8} cm^3/s
B	8.3 eV	2.0×10^{-8} cm^3/s
C	11.3 eV	2.1×10^{-8} cm^3/s
N	14.5 eV	1.7×10^{-8} cm^3/s
O	13.6 eV	1.2×10^{-8} cm^3/s

empirical impurity transport is compared with the $\boldsymbol{E} \times \boldsymbol{B}$ turbulent diffusivities and mobility based on the fluctuation data and the measured radial electric field. The agreement is best with the $\boldsymbol{E} \times \boldsymbol{B}$ diffusivity taken in the strong turbulence regime where

$$D = c_D \left(\frac{\widetilde{\phi}}{B_T} \right) \tag{9.45}$$

while the comparison with the weak turbulence diffusivity and the collisional (no fluctuations) diffusivity results in qualitative disagreements. Here $\widetilde{\phi}$ is the local rms value of the electrostatic potential. The percolation theory gives the diffusivity

$$D_{\text{perc}} = \left(\frac{\widetilde{\phi}}{B_T} \right)^{7/10} \left(\frac{\Delta\omega}{k_\perp^2} \right)^{3/10} \tag{9.46}$$

derived by Isichenko, *et al.* (1992).

Predicting the transport of impurities in tokamak confinement systems is of key importance for projections of the operation of fusion systems. In addition, the analysis of the transport of impurity ions injected at sufficiently low levels of concentration that the collective modes responsible for the fluctuations are not materially altered provides important information for testing basic transport theory.

Now we review the results of studies of the impurity transport experiments in which trace amounts of scandium are injected by laser ablation and the observed transport is compared with the $\boldsymbol{E} \times \boldsymbol{B}$ turbulent diffusivities for test particles. For this purpose experiments were performed to provide both impurity profile evolution data and fluctuation data in the same, and in similar discharges. For injection experiments where $n_z Z^2 \ll n_e$ the effect of the impurity on the collective modes is found to be negligible. Thus, we may consider that the impurity scandium ions are being convected by the $\boldsymbol{E} \times \boldsymbol{B}$ fluctuation velocity as test particles.

In complementary research [McWilliams, *et al.* (1990)] applies turbulent-transport-theory-based formulas using measured fluctuations as part of the input data to the transport formula.

New impurity injection experiments in LHD with traces of Vanadium (V), Manganese (Mn), and Cobalt (CO) show that the turbulence and plasma are not appreciably disturbed. These elements are neighbors to the intrinsic impurities of iron Fe and Chromium Cr from the stainless steel vessel walls [Sudo, *et al.* (2012)].

The study of test particle (ion) transport by the turbulent $\boldsymbol{E}\times\boldsymbol{B}$ drifts in electric fields that model drift wave-type fluctuations has been an active area of theoretical research [Horton (1981, 1985); Kleva and Drake (1984); Isichenko, *et al.* (1992); Misguich, *et al.* (1988); Gruzinov, *et al.* (1990); Isichenko and Horton (1991)]. Rather well understood test particle diffusivities from the $\boldsymbol{E}\times\boldsymbol{B}$ drifts are reported in the extensive computer experiments [Horton (1981, 1985); Kleva and Drake (1984); Isichenko, *et al.* (1992)]. The results show that once the fluctuation level exceeds a relatively low level, the diffusion leaves the weak turbulence regime in which

$$D_\perp = \frac{\widetilde{E}_\perp^2\,\tau_c}{B^2} \tag{9.47}$$

and enters the strong turbulence, or renormalized turbulence regime, in which $D_\perp \propto \widetilde{E}_\perp\,\lambda_\perp/B \sim \widetilde{\phi}/B$. At still higher fluctuation amplitudes, the diffusion increases even more slowly with the fluctuating potential. In the percolation theory limit [Isichenko, *et al.* (1992); Gruzinov, *et al.* (1990); Isichenko and Horton (1991)] and enters the strong turbulence, or renormalized turbulence regime, in which $D_\perp \propto \widetilde{E}_\perp\,\lambda_\perp/B \sim \widetilde{\phi}/B$. At still higher fluctuation amplitudes, the diffusion increases even more slowly with the fluctuating potential. In the percolation theory limit [Isichenko, *et al.* (1992); Gruzinov, *et al.* (1990); Isichenko and Horton (1991)] predicting $D_\perp \propto (\widetilde{\phi}/B)^{7/10}(\Delta\omega/k_\perp^2)^{3/10}$ where $\Delta\omega, k_\perp$ are the characteristic frequency band and mean wavenumber of the fluctuations. These predictions are tested numerically in Isichenko, *et al.* (1992) and compared with earlier works [Kleva and Drake (1984); Horton (1985); Misguich, *et al.* (1988); Gruzinov, *et al.* (1990)]. The three regimes of turbulent diffusivity are summarized in Table 9.1 along with the neoclassical Pfirsch-Schlüter regime collisional transport formula.

In addition to the diffusivity the inward radial electric field E_r produces a mobility μ component to the impurity flux that is also proportional to the fluctuation level. In collisional transport theory the mobility μ_z and diffusivity D_z are related by the Einstein relationship

$$D_z = \mu_z\,T_z. \tag{9.48}$$

Due to the radial electric field $E_r = -d\Phi/dr < 0$, this value of the mobility produces a strong inward impurity flux such that the Boltzmann relationship

$$n_z(r) = n_0 \exp\left(-\frac{Ze\Phi(r)}{T_z}\right) \tag{9.49}$$

is an equilibrium state at zero impurity flux $\Gamma_z = 0$. In the turbulent $\boldsymbol{E}\times\boldsymbol{B}$ transport the mobility-to-diffusivity ratio can be much less than the value given by

Table 9.3 Scandium Transport Experiments in TEXT.

$m_z = 45\,\text{amu} \qquad Z = 21$
$Z = 12, 16, 17, 18$
$n_z \leq 2 \times 10^7\,\text{cm}^{-3}$
For $Z = 18$ and $m_z = 45$ rm the experiments have
$\nu_{zi} = 10^4/\text{s}$
$\rho_z = 0.03\,\text{cm} = 300\,\mu\text{m}$
$v_z = 3 \times 10^6\,\text{cm/s}$
The working gas is hydrogen at
$T_i = 300\,\text{eV}$, and impurities are $45\,\text{Sc}^{21}$ $(3d^4\,\text{s}^2)$
and 48 Ti^{22} $(3d^2 4\,\text{s}^2)$.

the Einstein relation. Since the theoretical studies for the effective mobility are not as complete as those for the diffusivity we express here the effective coefficient c_E (with E for Einstein relation) in the mobility relationship $\mu_z = c_E(D_z/T_z)$ that gives the best agreement with the evolution of the impurity density profiles.

The test particle scandium injection experiments in the Texas Experimental Tokamak (TEXT) [Gentle (1981)] can be explained within a factor of 2 by the formulas

$$\Gamma_z = -D_z \frac{dn_z}{dr} + \mu_z n_z Ze E_r \tag{9.50}$$

with $D_z = c_D(\widetilde{\phi}/B_T)$ and $\mu_z = c_\mu(\widetilde{\phi}/B_T T_z)$ with $c_D = 0.28 \pm 0.09$ and $c_\mu = 0.043 \pm 0.007$ for hydrogen as a working gas. For these values of the coefficients the deviation from the Einstein formula is given by $c_E = \mu T_z/D_z = c_\mu/c_D \simeq 0.15$ compared with unity. While with He for the working gas, $c_D = 0.22 \pm 0.09$, $c_\mu = 0.16 \pm 0.07$, $c_E \simeq 0.73$. Thus, for helium as the working gas the mobility is closer to the limiting Einstein value. While the domains of applicability of the turbulent transport formulas are still an active area of theoretical research, these formulas are consistent with the Scandium experiments. Theoretical work and computer experiments performed for the nonlinear $\boldsymbol{E} \times \boldsymbol{B}$ drift motion of test particles in drift wave-type fluctuating electric fields interprets the data.

Analysis of the fluctuation data from the heavy ion beam probe in the Texas Experimental Tokamak (TEXT) with respect to the regime of the $\boldsymbol{E} \times \boldsymbol{B}$ transport indicated that the strong nonlinearity condition $R_E = \Omega_E/\omega_k > 1$ is well satisfied. Here Ω_E is the rotation rate around the maximum and minimum of $\widetilde{\phi}$ which for a fluctuation with $\boldsymbol{k} = (k_x, k_y)$ is approximately given by $\Omega_E = k_x k_y \widetilde{\phi}/B \sim k_\perp^2 D$. The value of the dimensionless nonlinearity parameter R_E is $R_E \simeq 1.5$ to 2.5. The measure R_E is named the Kubo number.

In the regime $R_E > 1$, the nonlinearity determines the correlation time between the ion (guiding center) and the fluctuation. This shortening of the correlation time with increasing amplitude is in contrast with quasilinear theory where the correlation time is independent of the amplitude. Thus, the structure of the transport of trace amounts of impurity elements provides an important comparison with the nonlinear regime of $\boldsymbol{E} \times \boldsymbol{B}$ turbulent transport. Transport experiments in a Q-machine are also explained by the strongly nonlinear turbulent diffusion formula, with over an order of magnitude variation in fluctuation level reported by McWilliams, *et al.* (1990).

The theoretical value of c_D determined from the regression fits to the ensemble averaged diffusivity is a functional of the fluctuation spectrum. Some results for the sensitivity of c_D to changes in modeling of the fluctuation spectrum are shown in Horton, *et al.* (1987) and Kim, *et al.* (1990) for a related problem concerning electron transport. The effect of Coulomb pitch angle scattering along with the turbulent $\boldsymbol{E} \times \boldsymbol{B}$ flow is reported and the neoclassical diffusivity is recovered at a sufficiently low fluctuation level.

9.5 Thermalization of Impurities and the Collisional Fluxes

The laser ablation method creates a burst of scandium ions in a localized radial region. The scandium ions stream over the magnetic surface in time $qR/v_{zi} \simeq 10^{-4}$ s and scatter in velocity space by Coulomb collisions with the working gas ions n_i, m_i, Z_i, T_i. This initial relaxation is given by the kinetic equation

$$\frac{\partial f_z}{\partial t} + \frac{v_\parallel}{qR}\frac{\partial f_z}{\partial \theta} = C_{zi}(f_z) = \frac{\partial}{\partial \boldsymbol{v}} \cdot \left(\frac{\boldsymbol{v} - \boldsymbol{u}_i}{\tau_s} f_z + D_v \frac{\partial f_z}{\partial \boldsymbol{v}} \right) \tag{9.51}$$

where the slowing-down time τ_s from the collisions with the thermal ions is given by

$$\frac{1}{\tau_s} = \frac{16\pi^{1/2}}{3} \frac{n_i\, e^4\, Z^2\, Z_i^2}{m_z\, m_i\, v_i^3} \ell n\, \Lambda. \tag{9.52}$$

In Braginskii (1965) the impurity-ion collision frequency $\nu_{zi} = 1/\tau_s$ occurs with

$$\nu_{zi} = \frac{4(2\pi)^{1/2}}{3} \frac{n_i\, e^4\, Z^2\, Z_i^2\, \ell n\, \Lambda}{m_z\, m_i^{-1/2}\, T_i^{3/2}} \tag{9.53}$$

where Eq. (9.53) is equivalent to Eq. (9.52) with $v_i = (2T_i/m_i)^{1/2}$. For a hydrogen discharge at $n_i = 3 \times 10^{13}\,\mathrm{cm}^{-3}$, $T_i = 300\,\mathrm{ev}$, this time is $\tau_s \simeq 10^{-4}\,\mathrm{s}$ and for the helium discharge the time is approximately twice as long. The velocity diffusion coefficient in Eq. (9.51) is given by $D_v = T_i/(m_z\, \tau_s)$ so that in the absence of driving forces the impurity-ion collisions force $f_z(\boldsymbol{x}, \boldsymbol{v}, t)$ to $f_z = n_z(r,t)\exp[-m_z(\boldsymbol{v} - \boldsymbol{u}_i)^2/2T_i]$ for times long compared with τ_s. Here $\boldsymbol{u}_i(\boldsymbol{x}, t)$ and $T_i(\boldsymbol{x}, t)$ are the center of mass flow velocity and temperature of the working gas.

Note that collisional impurity transport formulas are often presented in terms of the ion-impurity collision frequency $\nu_{iz} = (n_z\, m_z/n_i\, m_i)\nu_{zi}$ and the thermal ion gyroradius $\rho_i = (2m_i\, T_i)^{1/2}/eZ_i\, B_T$ (rather than the gyroradius ρ_z of the impurity ions) which changes the appearance of the formulas.

In the regimes of the TEXT experiments the impurity-ion collision frequency is high enough that the impurity mean free path $v_z \tau_s$ is less than qR so the collisional transport is in the Pfirsch-Schlüter regime. Typical values for these experiments with Sc are $\rho_z(Z = 18) \simeq 0.025\,\text{cm}$, $\nu_{zi} = 1/\tau_s = 10^4\,/\text{s}$ and $v_z \tau_s = 3 \times 10^6\,\text{m/s}\cdot 10^{-4}\,\text{s} = 300\,\text{cm}$. For $q = 3$ the P-S diffusivity is then

$$D_z^{\text{PS}} = (1 + 2q^2)\nu_{zi}\,\rho_z^2 = (1 + 18)(0.025\,\text{cm})^2 10^4\,\text{s}^{-1} \cong 120\,\text{cm}^2/\text{s} \tag{9.54}$$

which is about one order of magnitude smaller than the observed rate of diffusion. Earlier experiments [Marmar, *et al.* (1982); Synakowski, *et al.* (1989); Kim, *et al.* (1990)] attempted to use collisional transport theory to explain impurity transport and met with difficulty. Note that $\nu_{zi}\,\rho_z^2$ is the same for all charge states Z and independent of m_z.

The Alcator impurity study of Marmar, *et al.* (1982) reports that the diffusivities of order $2.6 \times 10^3\,\text{cm}^2/\text{s}$ are required to explain the transport of the silicon ions. Marmar, *et al.* (1982) performed parametric variations for determining the impurity confinement time τ_z related to the constant d diffusivity used for the impurity flux $\Gamma_z = -d_z(\partial n_j/\partial r)$ in a 1D transport analysis of the injection experiments. The empirical impurity confinement $\tau_z = 0.173\,a_L^2/d_z$ derived from d_z and the limiter radius a_L is

$$\tau_I(\text{ms}) = 0.075\,\frac{a_L\, m_{bg}}{q_L}\, R^{3/4}\left(\frac{Z_{\text{eff}}}{Z_{bg}}\right) \tag{9.55}$$

independent of the charge state of the silicon impurity. Here bg is the background working gas. The values of τ_z range from 10 to 40 ms over the range of values available for the limiter radius a_L, the major radius R, the working or background gas charge Z_{bg} and mass m_{bg} and the safety factor q_L at the limiter radius. In contrast to the theoretical model used in the TEXT study, no direct information about the fluctuation spectrum is used in the empirical confinement law.

For a general $\boldsymbol{E}_\perp(\boldsymbol{x}, t)$ field the exact motion of the particles has been studied extensively in theoretical works. The studies show that the initial distribution

$$n_z(\boldsymbol{x}, t) = \sum_{j=1}^{N} \delta(\boldsymbol{x} - \boldsymbol{r}_j(t)) \tag{9.56}$$

evolving with

$$\frac{d\boldsymbol{r}_j}{dt} = \frac{\boldsymbol{E}(\boldsymbol{r}_j, t) \times \boldsymbol{B}}{B^2} \tag{9.57}$$

is well described by an effective diffusion coefficient D_x after the convection over several spatial correlation lengths $\ell_c = 1/\Delta k_\perp$. After this transit the impurity density spreads, in the absence of ionization and recombination, according to

$$\frac{\partial n_z}{\partial t} = \frac{-1}{r}\frac{\partial}{\partial r}(r\,\Gamma_z) \tag{9.58}$$

with

$$\Gamma_z = -D\,\frac{dn_z}{dr} + Ze\,n_z\,\mu\,E_r. \tag{9.59}$$

Now the precise determination of the effective diffusion coefficient D and mobility μ are difficult problems in nonlinear dynamics and depend on the spectral features of the fluctuations [Isichenko, *et al.* (1992); Misguich, *et al.* (1988); Gruzinov, *et al.* (1990); Isichenko and Horton (1991)].

Here we use some of the results from the studies of Horton (1981, 1985), Kleva and Drake (1984), Misguich, *et al.* (1988), and in particular the results of Isichenko, *et al.* (1992) for the nonlinear $\boldsymbol{E}\times\boldsymbol{B}$ transport. No attempt is made here to discuss the nonlinear dynamics theories in these references. The principle results for D are given in Table 9.1. The main point is that for the measured fluctuation spectrum in TEXT the impurity dynamics is highly nonlinear with the decorrelation rate ν determined by the turbulent $\boldsymbol{E}\times\boldsymbol{B}$ motion itself rather than parallel transit motions.

The associated mobility is then

$$\mu = c_E\,\frac{D}{T_z} \tag{9.60}$$

where c_E is a constant between zero and unity. For $c_E = 1$ the mobility is given by the Einstein relation so that the $\Gamma_z = 0$ state corresponds to the Boltzmann impurity density profile $n_z = n_0 \exp(-Ze\,\Phi(r)/T_z)$ when $E_r = -d\Phi/dr$. Some $\boldsymbol{E}\times\boldsymbol{B}$ nonlinear dynamics studies [Aydemir (1990)] indicate that the mobility may fall below this value so that $c_E \ll 1$. Since c_E has not been accurately determined in the literature we leave the coefficient $c_E = c_\mu/c_D$ to be determined by the impurity transport data.

In regard to theoretical value of c_D a study by Isichenko and Horton (1991) of the stochastic $\boldsymbol{E}\times\boldsymbol{B}$ diffusion produced by a six-wave model for the fluctuation spectrum gives the nonlinear regression fit of $D_x = 0.3\widetilde{\phi}^{0.92}$ in Fig. 4 of Isichenko and Horton (1991) for the normalized Reynolds number $R_E = \widetilde{\phi}$ in the range [1,100]. The corresponding Kolmogorov entropy calculated from the exponential rate of separation of neighboring $\boldsymbol{E}\times\boldsymbol{B}$ trajectories is $K = 0.2\widetilde{\phi}^{0.56}$. For the TEXT experiment this implies an e-folding time of $\tau_K = 1/K\,\langle\omega_*\rangle \sim 5\times10^{-5}\,$s at $R_E = \widetilde{\phi} \simeq 1.0$ for the rate of separation of neighboring impurity ion guiding centers. The Kolmogorov entropy defines the exponential rate of separation of neighboring impurity ions.

9.6 Scandium and Titanium Transport with Ionization and Recombination

The transport of Sc was measured in H and He discharges with common external control parameters, $B_T = 2.8\,$T, $I_p = 250\,$kA, $\overline{n}_e = 2\times10^{19}\,\text{m}^{-3}$. In the experiment, a small quantity of Sc was injected into the plasma using laser ablation. The electron density perturbation was less than 1%.

After injection, the impurity trace is transported through the plasma and then out of the plasma through the last closed flux surface. On encountering a surface outside of the plasma, the Sc ion sticks and is not recycled in significant quantity. For measurement of transport, the impurity spatial distributions are as important as the temporal data. Because Sc($m = 45, Z = 21$) must be very close in its transport properties to Ti($m = 48, Z = 22$), an ambient, intrinsic, impurity, the spatial distributions for Ti can be combined with the temporal data for Sc in an appropriate analysis to provide the transport coefficients for either species.

Because the two ions together have significant concentration over the entire plasma and since they peak in different locations, they will adequately describe the spatial structure of Sc and Ti impurities. The analysis that is described below properly accounts for the small (atomic) differences in the two elements, Sc and Ti. In the analysis, the impurity transport of both Sc and Ti are simulated with a fixed background plasma model consisting of measured profiles of T_e and n_e.

A special high-energy (500 KeV) accelerator from RPI is used to measure the mean electrostatic potential $\overline{\phi}(r)$, and the root-mean-square fluctuating potential $\widetilde{\phi}(r)$. The system [Schoch, *et al.* (1986)] is called the heavy ion beam probe.

For each ion species the impurity flux, Γ, is represented with the standard formula

$$\Gamma_z = -D \frac{\partial n_z}{\partial r} + n_z V \tag{9.61}$$

where both D and V may be dependent on radius, but are taken independent of the ion species or ion charge.

The impurity profiles are simulated using the set of impurity transport equations

$$\frac{\partial n_z}{\partial t} + \nabla \cdot \Gamma_z = -n_e n_z (S_z + \alpha_z) + n_e n_{z-1} S_{z-1} + n_e n_{z+1} \alpha_{z+1} \tag{9.62}$$

where the S_z is the ionization rate coefficient for the zth impurity ionization stage and α_z is the recombination rate coefficient. The functions S_z and α_z depend principally on the $T_e(r)$ profile which is well-known from Thomson scattering and electron cyclotron emission (ECE) data.

In the simulation there are two sets of transport equations; one for titanium Ti and one for scandium Sc. The diffusivity $D(r)$ and the convection $V(r)$ are common. The agreement between experiment and simulation is excellent for these two as it is for the intervening stages of ionization which are omitted for clarity of presentation.

The predicted flux is

$$\Gamma_j = -D_j \left(\frac{\partial n_j}{\partial r} + c_E \frac{n_j e_j}{T_j} \frac{\partial \Phi}{\partial r} \right). \tag{9.63}$$

In this simulation the renormalized turbulence theory transport formulas given by

$$D = c_D \frac{\widetilde{\phi}}{B_T} \tag{9.64}$$

$$V_j = c_\mu \frac{\widetilde{\phi}}{B_T} \frac{e_j}{T_j} E_r \tag{9.65}$$

with the constant dimensionless ratio

$$c_E = \frac{c_\mu}{c_D} \tag{9.66}$$

are used.

The design of the divertor in tokamaks can strongly influence the impurity transport. So experimental studies are made with variations of the divertor design. A comparison of the particle flux control from the divertor in the ASDEX Upgrade (AUG) [Gruber, *et al.* (1999)], with that in the JET tokamak is given by the ASDEX team in Angioni, *et al.* (2007). The comparison is aimed at optimizing the design of the divertor for the ITER machine. High spatial resolution transport measurements of trace amounts of the high-Z elements of vandium, manganese and cobalt that neighbor the elements of the stainless steel in the vacuum vessel walls are reported in Sudo, *et al.* (2012). Using the UV emission for the ionization stage and the K_α line radiation for following the motion of the heavy elements in the KeV core plasma the experiments indicate that in LHD there is a transport barrier that keeps the injected trace in the core and prevents the intrinsic iron and chromium elements from the wall from reaching the core.

9.7 Mass Flows and Transport of Impurities in the Tokamak

Once the impurities are released, they flow on the magnetic surface to set upon equilibrium poloidal flow pattern that gives rise to the Pfirsch-Schlüter enhancement factor of $1 + 2q^2$. The flows are fast and of fundamental importance. Here we give the key equations for the establishment of these flows.

Designating the working gas by 1 and the impurity ions by 2 we have

$$e_1 n_1 \left(\boldsymbol{E} + \frac{\boldsymbol{v}_1}{c} \times \boldsymbol{B} \right) - \nabla p_1 - m_1 n_1 \nu_{12} (\boldsymbol{u}_1 - \boldsymbol{u}_2) = 0 \tag{9.67}$$

$$e_2 n_2 \left(\boldsymbol{E} + \frac{\boldsymbol{v}_2}{c} \times \boldsymbol{B} \right) - \nabla p_2 - m_2 n_2 \nu_{21} (\boldsymbol{u}_2 - \boldsymbol{u}_1) = 0 \tag{9.68}$$

with the equal and opposite frictional forces related through $m_1 n_1 \nu_{12} = m_2 n_2 \nu_{21}$. Taking the perpendicular and parallel components we have

$$n_1 \boldsymbol{u}_1 = \frac{n_1 \boldsymbol{B}}{B^2} \times \left[-\boldsymbol{E} + \frac{\nabla p_1}{e_1 n_1} + \frac{m_1 n_1 \nu_{12}}{e_1 n_2} (\boldsymbol{u}_1 - \boldsymbol{u}_2) \right] \tag{9.69}$$

and

$$e_1 n_1 E_\parallel - \nabla_\parallel p_1 = m_1 n_1 \nu_{12}(u_{1\parallel} - u_{2\parallel}) \tag{9.70}$$

with the appropriate interchange of 1 and 2 for the second ion species.

Now the continuity equation $\nabla \cdot n\boldsymbol{u}_i = 0$ in the toroidal magnetic field forces the parallel flow to be established on the time scale qR/v_T such that

$$\frac{B_\theta}{rB}\frac{\partial}{\partial\theta}(n_1 u_{\parallel 1}) = -\nabla \cdot \left[\frac{n_1 \boldsymbol{B}}{B^2} \times \left(-\boldsymbol{E} + \frac{\nabla p_1}{e_1 n_1}\right)\right] \tag{9.71}$$

$$= \frac{2\sin\theta}{BR}\left(n_1 E_r - \frac{1}{e_1}\frac{\partial p_1}{\partial r}\right) \tag{9.72}$$

giving

$$n_1 u_{\parallel 1} = \frac{2r\cos\theta}{B_\theta R}\left(n_1 E_r - \frac{1}{e_1}\frac{\partial p_1}{\partial r}\right). \tag{9.73}$$

Computing $u_{\parallel 1} - u_{\parallel 2}$ from Eq. (9.73) and substituting into Eqs. (9.69) and (9.70) gives the poloidal angle variations on the

$$\frac{1}{Br}\frac{\partial}{\partial\theta}(e_1 n_1 \Phi + p_1) = \frac{m_1 n_1 \nu_{12}}{B_\theta^2 R}(2r\cos\theta)\left(\frac{p_1'}{e_1 n_1} - \frac{p_2'}{e_2 n_2}\right) \tag{9.74}$$

and

$$\frac{1}{Br}\frac{\partial}{\partial\theta}(e_2 n_2 \Phi + p_2) = \frac{m_2 n_2 \nu_{21}}{B_\theta^2 R}(2r\cos\theta)\left(\frac{p_1'}{e_1 n_1} - \frac{p_2'}{e_2 n_2}\right). \tag{9.75}$$

Now the poloidal E_θ and $\partial p_{1,2}/\partial\theta$ forces crossed with the toroidal magnetic field generate the radial convective fluxes given by the radial component of Eq. (9.69). From Eq. (9.69) and Eq. (9.74)-Eq. (9.75) we derive the radial fluxes

$$(n_1\boldsymbol{u}_1)_r = -\frac{B_\phi}{B^2}\left(1 + \frac{2r B^2\cos\theta}{R B_\theta^2}\right)\left(\frac{m_1 n_1 \nu_{12}}{e_1 B}\right)\left(\frac{p_1'}{e_1 n_1} - \frac{p_2'}{e_2 n_2}\right) \tag{9.76}$$

$$(n_2\boldsymbol{u}_2)_r = -\frac{B_\phi}{B^2}\left(1 + \frac{2r B^2\cos\theta}{R B_\theta^2}\right)\left(\frac{m_1 n_1 \nu_{12}}{e_2 B}\right)\left(\frac{p_2'}{e_2 n_2} - \frac{p_1'}{e_1 n_1}\right). \tag{9.77}$$

These formulas show that there is ambipolarity in the ion-impurity flux before making the surface average.

The formulas Eq. (9.76) and Eq. (9.77) also give the full r-θ pattern of convection from E_θ and $\partial p_{1,2}/\partial\theta$ before making the flux surface average. Once the surface average $dS = rd\theta(R_0 + r\cos\theta)d\phi$ is carried out by integrating over dS, the direct collisional random walk flux and the toroidal flow enhancement combine as

$$D_z^{ps} = (1 + 2q^2)\nu_{zi}\,\rho_z^2 \tag{9.78}$$

with $\Gamma_i = 2\pi\int_{-\pi}^{\pi} R(\theta)rd\theta\,(n_i\boldsymbol{u}_i)_r$, where D_z^{ps} is defined such that

$$\Gamma_z = -D_z^{ps}\frac{1}{T_z}\left(\frac{\partial p_z}{\partial r} - \frac{n_z Z}{n_i Z_i}\frac{\partial p_i}{\partial r}\right).$$

For an isothermal plasma the condition $\Gamma_z = 0$ is achieved by the inward pinch velocity v_z when

$$n_z = n_i^{(ZT_i/T_z)} \simeq (n_i)^Z \tag{9.79}$$

when $T_z = T_i$.

Formula Γ_2 is derived by Rutherford (1974) with a generalization to include the thermoelectric effect in the frictional force in Eq. (9.68) which he takes as $\boldsymbol{R}_{iz} = -c_1(\frac{m_i n_i}{\tau_{iz}})(\boldsymbol{v}_i - \boldsymbol{v}_z) - c_2 n_i \nabla_\parallel T_i$ with the values of c_1, c_2 first order rational polynomials in $\alpha = n_z Z^2/n_e$. The Rutherford formula for the impurity particle flux is

$$\Gamma_z = -\frac{n_i \rho_i^2 q^2}{T_i Z^2 n_z \tau_{iz}} \left[(c_1 + c_2^2/c_3) \left(\frac{\partial p_z}{\partial r} - \frac{n_z Z}{n_i} \frac{\partial p_i}{\partial r} \right) - \frac{5}{2} \frac{c_2}{c_3} (n_z Z) \frac{\partial T_i}{\partial r} \right] \tag{9.80}$$

where c_2 and c_3 arise from the parallel ion thermal conductivity

$$q_{i\parallel} = c_2 n_i T_i u_\parallel - c_3 \left(\frac{n_i T_i \tau_{iz}}{m_i} \right) \nabla_\parallel T_i. \tag{9.81}$$

In the trace impurity experiments the terms proportional to n_z/n_i in Eq. (9.80) can be dropped in the Rutherford formula.

Now the question arises as to how this impurity flux depends on the charge state and the mass of the impurity ions. The $\boldsymbol{E} \times \boldsymbol{B}$ turbulent flux is independent of Z and m_z provided $k_\perp \rho_z \ll 1$ and $k_\parallel v_z/\omega \ll 1$. Now using Eq. (9.53) we find that

$$\rho_z^2 \nu_{zi} = \frac{m_z T_z}{e^2 Z^2 B^2} \cdot \frac{n_i e^4 Z_i^2 Z^2 \ell n \Lambda}{m_z m_i^{-1/2} T_i^{3/2}} \tag{9.82}$$

so that both m_z and Z^2 drop out. Thus, all charge states of Sc and Ti etc. have the same D_z^{ps} and are transported at the same rate from the collisions with the working gas ions.

In conclusion we note that the Pfirsch-Schlüter transport is an $\boldsymbol{E} \times \boldsymbol{B}$ convection across the magnetic surfaces that arises from the collisional friction between the different species. Each species has its own mass flow velocity required by the $\nabla \cdot (n_s \boldsymbol{u}_s) = 0$ in the toroidal magnetic geometry.

In 1992 time frame the JET device was equipped with Tritium in the neutral beam injectors to test the production of nuclear reactions in the D-T plasma. Up to 1 g of Tritium was used which as a radioactivity of 10^4 Ci where one Currie 1 Ci= 3.7×10^{10} decays/s which is the radioactivity of one 1 g of radium. The neutral beam power pulse was a box shape with first step to 5MW for 2s and then to 14 MW for 2 s and then back down to 5 MW for 2 s. There was ion cyclotron radio frequency heating in addition to the neutral beam injection heating. The ion temperature rose to approximately 20 KeV and then dropped as the impurity content of the plasma from the blast of the hot plasma on the carbon and beryllium tiles released those atoms into the plasma. The analysis of the discharge with transport codes interpreting the multitude of diagnostics show that the fusion Q

reached 0.4 [discharge 26147]. The ion cyclotron resonance heating accelerates a high-energy ion tail on the distribution functions that was 20 MeV in these JET D-T experiments where the bulk ion temperature was 20 KeV. In Chapter 11 the dispersion relations for the anisotropic, high energy particle distribution function are derived. In these experiments the radiation level reached 1000 Ci and decayed to 50 Ci in 36 hours.

References

Angioni, C., Carraro, L., Dannert, T., Dubuit, N., *et al.* (2007). *Phys. Plasmas* **14**, p. 055905.

Apicella, M. L., *et al.* (1997). *Nucl. Fusion* **37**, p. 381, doi:10.1088/0029-5515/37/3/I12.

Aydemir, A. (1990). *Phys. Fluids B* **2**, p. 2135, doi:10.1063/1.859433.

Basu, R., Naulin, V., and Rasmussen, J. Juul. (2003). Particle diffusion in anisotropic turbulence, *Nonl. Sci. Numer. Sim.* **8**, pp. 477-492, doi:10.1063/1.1578075.

Braginskii, S. I. (1965). *Rev. Plasma Phys.* ed. M. A. Leontovich (Consultants Bureau, New York), **I**, p. 205.

Coppi, B. and Zhou, T. (2012). Interpretation of the I-Regime and transport associated with relevant heavy particle modes, *Phys. Plasmas* **19**, p. 012302, doi:10.1063/1.3671944.

Del-Castillo-Negrete, D. (2000). *Phys. Plasmas* **7**, p. 1702.

Futatani, S., Horton, W., Benkadda, S., Bespamyatnov, I. O., and Rowan, W. L. (2010). *Phys. Plasmas* **17**.

Gentle, K. W. (1981). *Nucl. Technol. Fusion* **1**, p. 479.

Greenwald, M., Boivin, R. L., Bonoli, P., Christensen, C., Fiore, C., Garnier, D., Goetz, J., Golovato, S., Graf, M., Granetz, R., Horne, S., Hsu, T., Hubbard, A., Hutchinson, I., Irby, J., Kurz, C., LaBombard, B., Lipschultz, B., Luke, T., Marmar, E., McCracken, G., Niemczewski, A., O'Shea, P., Porkolab, M., Rice, J., Reardon, J., Schachter, J., Snipes, J., Stek, P., Takase, Y., Terry, J., Umansky, M., Watterson, R., Wolfe, S., Bombarda, F., May, M., and Welch, B. (1995). *Phys. Plasmas* **2**, p. 2308, doi:10.1063/1.871254.

Gruber, O., Bosch, H.-S.,Günter, S., Herrmann, A., Kallenbach, A., *et al.* (1999). Overview of ASDEX Upgrade Results, *Nucl. Fusion* **39**, p. 1321, doi:10.1088/0029-5515/39/9Y/309.

Gruzinov, A. V., Isichenko, M. B., and Kalda, J. (1990). *Zh. Eksp. Teor. Fiz.* **97**, p. 476 [*Sov. Phys. JETP* **70**, p. 263] ISSN:0044-4510.

Hasegawa, A. and Wakatani, M. (1983). *Phys. Rev. Lett.* **50**, p. 682, `http://link.aps.org/doi/10.1103/PhysRevLett.50.682`.

Horton, W. and Ichikawa, Y.-H. (1996). *Chaos and Structures in Nonlinear Plasmas* (World Scientific, Singapore).

Horton, W. and Rowan, W. (1994). *Phys. Plasmas*, p. 901.

Horton, W., Choi, D. I., Yushmanov, P. N., and Parail, V. V. (1987). *Plasma Phys.* **29**, p. 901, doi:10.1088/0741-3335/29/7/008.

Horton, W. (1985). *Plasma Phys. Control. Fusion* **27**, p. 937, doi:10.1088/0741-3335/27/9/001.

Horton, W. (1981). *Plasma Phys.* **23**, p. 1107, doi:10.1088/0032-1028/23/12/002.

Horton, W. (1976). *Phys. Fluids* **19**, p. 711, doi:10.1063/1.861517.

Isichenko, M. B., Horton, W., Kim, D. E., Heo, E. G., and Choi, D.-I. (1992). *Phys. Fluids*

B **4**, p. 3937, doi:10.1063/1.860300.

Isichenko, M. B. and Horton, W. (1991). Scaling laws of stochastic $\boldsymbol{E} \times \boldsymbol{B}$ plasma transport, *Comments Plasma Phys. Control. Fusion* **14**, 5, pp. 249-262, ISSN:0374-2806.

Kadomtsev, B. B. and O. P. Pogutse. (1970). *Phys. Rev. Lett.* **25**, p. 1155, `http://link.aps.org/doi/10.1103/PhysRevLett.25.1155`.

Kim, D. E., Choi, D.-I., Horton, W., Yushmanov, P. N., and Parail, V. V. (1990). *Phys. Fluids B* **2**, p. 547.

Kleva, R. G. and Drake, J. F. (1984). *Phys. Fluids* **27**, p. 1686, doi:10.1063/1.864823.

Marmar, E. S., Rice, J. E., Terry, J. L., and Seguin, F. H. (1982). *Nucl. Fusion* **22**, p. 1567, doi:10.1088/0029-5515/22/12/003.

McWilliams, R., Okuboda, M. K., and Wolf, N. S. (1990). *Phys. Fluids B* **2**, p. 523.

Misguich, J. H., Balescu, R., and Nakach, R. (1988). *Plasma Phys. Control. Nucl. Fusion Res.* Proc. 12th Int. Conf., Nice, 12-19, 3, p. 521, (International Atomic Energy Agency, Vienna, 1989).

Rowan, W. L., Bespamyatnov, Igor O., and Fiore, C. L. (2008). Light impurity transport at an internal transport barrier in Alcator C-Mod, *Nucl. Fusion* **48**, 10, p. 105005, `http://stacks.iop.org/0029-5515/48/i=10/a=105005`.

Rutherford, P. H. (1974). *Phys. Fluids* **17**, p. 1782, doi:10.1063/1.1694975.

Schoch, P. M., Forster, J. C., Jennings, W. C., and Hickok, R. L. (1986). *Rev. Sci. Instr.* **57**, 8, pp. 1825-1827, doi:10.1063/1.1139141, ISSN:0034-6748.

Sudo, S., Tamura, N., Suzuki, C., Muto, S., Funaba, H., and LHD Group (2012). Multiple-tracer TESPEL injection for studying impurity behavior in a magnetically confined plasma, *Nucl. Fusion* **52**, p. 063012, doi 10.1088/0029-5515/52/6/063012.

Synakowski, E. J., Bengtson, R. D., Ouroua, A., Wootton, A. J., and Kim, J. Y. (1989). *Nucl. Fusion* **29**, p. 311.

Terry, P. W. and Horton, W. (1983). *Phys. Fluids* **26**, p. 106.

Voronov, G. S. (1997). *Atomic Data and Nuclear Data Tables* **65**, p. 1.

Weynants, R. R., *et al.* (1999). *Nucl. Fusion* **39**, p. 1637, doi:10.1088/0029-5515/39/11Y/303.

Chapter 10

Coherent Structures in Plasmas

10.1 Kelvin-Helmholtz Instability and Vortices in Magnetized Plasma

Vortices are created in fluids by relative motions of parts of the fluids with different speeds and thus occur naturally in nature as easily seen in rivers and atmospheres. The most efficient way to mix inhomogeneous materials is to create vortices in the material.

When one mixes the various ingredients in a cake recipe the process is quick and easy by using an appropriate spoon in circular motion to create the vortical flow patterns. Mixing machines of many types create homogeneous fluids from the inhomogeneous initial materials. For gases the processes is also widely used to create a homogeneous mixture. One popular car maker advertised for years that their compound vortex combustion chamber gave their engines higher efficiency than that of its competitors. The same situation applies to plasmas. A well-documented plasma example is that from the flow of the solar wind plasma around the Earth's magnetosphere. Both spacecraft data and computer simulations show that Kelvin-Helmholtz instability grows from the subsolar apex of the magnetopause along the flanks forming vortex convection cells which mixes the solar wind plasma with its higher helium concentration with the plasma trapped in the Earth's magnetic field with its higher oxygen concentration.

In this chapter we investigate the nature of the vortices in magnetized, inhomogeneous plasmas. The theoretical structure of the problem and the numerical simulations of the resulting nonlinear partial differential equations are examined. Also shown are the data from controlled experiments measuring precisely the vorticity of the electrically conducting plasma with a probe designed expressly to measure plasma vorticity.

Many of the partial differential equations in plasma physics that describe the nonlinear evolution of the plasma have the vorticity dynamics as their central component. This statement applies to such diverse phenomenon as the magnetized Rayleigh-Taylor instability, the dynamics of magnetic reconnection and the propagation of drift waves. For shear Alfvén waves the two essential partial differential equations are one for the evolution of the plasma vorticity forced by the divergence of the parallel plasma electrical currents and the second equation for the dynamics of the magnetic flux function associate with the bending of the magnetic field lines.

These plasma vortices are quasi-two-dimensional since the ambient magnetic field $B\boldsymbol{e}_z$ makes the spatial variation slow along the magnetic field. Mathematically, this appears more concisely expressed in the Fourier transformed space derived from the 3D Euclidean space. In Fourier space each fluctuation is represented by a wavenumber vector $\boldsymbol{k}$. The slow variation along the field is contained in the energy spectrum of the fluctuations as concentrated in the regions where $k_\perp \gg k_\parallel$ where $k_\parallel = \boldsymbol{b}\cdot\boldsymbol{k}$ and $\boldsymbol{k}_\perp = \boldsymbol{b}\times(\boldsymbol{k}\times\boldsymbol{b})$. Here $\boldsymbol{b}$ is the unit vector along the magnetic field. Often the nonlinear dynamics is most accurately described by computer simulations that advance the fields in time in Fourier transform space. Thus, at the web site `http://orion.ph.utexas.edu/~vortex` one finds a set of computer codes developed at the IFS for describing the nonlinear dynamics of N-fields in $\boldsymbol{k}$-space with adaptive time integrators. The adaptive integrators varying the size Δt of the time stepping to take into account how fast the particular part of the $\boldsymbol{k}$-space is varying. Since the frequency of the motions in fluctuations increases rapidly as the space scale becomes smaller, the use of adaptive time integrators is an important aspect of the nonlinear simulations.

In order to model the Kelvin-Helmholtz instability we consider the electrostatic response of the plasma in a two-fluid model, where the dynamics for the electrostatic potential $\varphi(\boldsymbol{x},t)$ is obtained from the conservation of charge under the assumption of quasineutrality, i.e.,

$$\nabla\cdot\boldsymbol{j} = 0. \tag{10.1}$$

For the 2D model with $\boldsymbol{j} = \boldsymbol{j}_p$, the polarization current, the dynamics reduces as follows, with the nonlinear polarization current

$$\boldsymbol{j}_p = -\frac{m_i n_i}{B^2}\frac{d}{dt}\nabla\varphi = -\frac{m_i n_i}{B^2}\left(\frac{\partial}{\partial t} + \boldsymbol{v}_E\cdot\nabla\right)\nabla\varphi \tag{10.2}$$

where

$$\boldsymbol{v}_E = \frac{\boldsymbol{E}\times\boldsymbol{B}}{B^2} = \frac{1}{B}\widehat{\mathbf{z}}\times\nabla\varphi. \tag{10.3}$$

Substituting Eq. (10.2) into the current closure statement of Eq. (10.1) yields the dynamical equation for the vorticity

$$\left(\frac{\partial}{\partial t} + \vec{v}_E\cdot\nabla\right)\nabla^2\varphi = \frac{\partial}{\partial t}\nabla^2\varphi + [\varphi,\nabla^2\varphi] = 0 \tag{10.4}$$

where we have introduced the Poisson bracket $[f,g]$ between the two 2D functions f, g:

$$[f,g] \equiv \frac{\partial f}{\partial x}\frac{\partial g}{\partial y} - \frac{\partial f}{\partial y}\frac{\partial g}{\partial x}. \tag{10.5}$$

In the vorticity equation the Poisson bracket is arising from the directional derivative of the vorticity along the $\boldsymbol{E} \times \boldsymbol{B}$ flow velocity. Now we consider the shear flows and vorticities in the LAPD experiments.

In the velocity shear layer $v_E(x)$ the density scale length L_{n_e} is calculated from the profile of the mean ion saturation current giving $L_{n_e} \sim 15\,\mathrm{cm}$ which is large enough to justify neglect the density gradients in the shear flow modeling. This scale length is approximately $15\,\rho_i$ where $\rho_i = 3\,\mathrm{mm}$.

The simple KH model in Eq. (10.4) has canceling electron and ion currents from the $\vec{E} \times \vec{B}$ motion. The vorticity equation, Eq. (10.4), is the statement that the divergence of the cross-field current vanishes.

Considering a slab approximation for this plasma we take x as the radial coordinate and y to be the periodic coordinate. Equation (10.4) is linearized around a steady-state poloidal flow $v_y(x)$ like that in the LAPD plasma rotation. By doing so we obtain Rayleigh's eigenvalue equation given by

Rayleigh Equation

$$\varphi_k''(x) - k^2\varphi_k(x) - \frac{kv_y''(x)}{kv_y(x) - \omega}\varphi_k(x) = 0. \tag{10.6}$$

With boundary conditions on $\varphi(x)$ at x_1 and x_2, the Rayleigh equation defines a non-normal eigenvalue problem with non-orthogonal basis vectors. The shear flow problem then has numerous special features given in Antar (2003), Burrell (1997), Kim, *et al.* (2006), and Horton, *et al.* (2010).

10.2 Drift Wave Models for LAPD

During the plasma rotation from the bias pulse, the maximum velocity in the flow corresponds to an electric field $E \approx 1.5\,\mathrm{kV/m}$ which occurs approximately at $r \approx 0.4\,\mathrm{m}$. Taking the magnetic field to be $B = 0.075\,\mathrm{T}$ along the z direction we obtain

$$v_{\max} = \boldsymbol{E} \times \boldsymbol{B} \approx 2 \times 10^4\,\mathrm{m/s}. \tag{10.7}$$

The eigenvalue problem for an idealized triangular flow profile that models the equilibrium electric field profile in Fig. 10.5 is solved and the flow profile defined by

$$v_y(x) = v_{\max} \begin{cases} 0 & x \leq -1 \\ (1 - |x|) & |x| \leq 1 \\ 0 & x \geq 1 \end{cases} \quad . \tag{10.8}$$

Here the velocity is normalized to $v_{\max}$, and all lengths to $a = 0.1\,\mathrm{m}$, corresponding to the half width of the shear layer. The flow in Eq. (10.8) is called a jet with maximum velocity $v_{\max}$. Dimensionless variables are used with $x = \pm 1$ measuring the width of the stream and $v_y = 1$ measuring the maximum velocity $v_{\max}$.

Differentiation with respect to x gives

$$v_y''(x) = \delta(x+1) - 2\delta(x) + \delta(x-1). \tag{10.9}$$

Therefore the eigenfunctions obey the simple equation

$$\varphi_k''(x) - k^2\varphi_k(x) = 0 \tag{10.10}$$

except at the points $x = 0, \pm 1$. The jet flow has a steep, localized vorticity gradient, modeled by Eq. (10.9) giving secondary connection conditions, which are derived by integrating Eq. (10.6) across the singular points. Doing so results in the three continuity conditions

$$\Delta[\varphi_k'(-1)] = -\frac{k}{\omega}\varphi_k(-1) \tag{10.11}$$

$$\Delta[\varphi_k'(0)] = -\frac{2k}{kv_0 - \omega}\varphi_k(0) \tag{10.12}$$

$$\Delta[\varphi_k'(1)] = -\frac{k}{\omega}\varphi_k(1) \tag{10.13}$$

where we have defined the operator

$$\Delta[f(x_0)] \equiv f(x_0^+) - f(x_0^-). \tag{10.14}$$

In practice, the profiles are continuous, but vary rapidly over a small scale δ. The jump conditions are well satisfied if $k\delta \ll 1$; for $k = m/r$ this is equivalent to $m \ll r/\delta$.

Equation (10.6) is invariant under the transformation $x \to -x$, which means that if $\varphi_k(x)$ is a solution to Eq. (10.6) with eigenvalue ω, then $\varphi(-x)$ is also a solution with the same eigenvalue. This fact allows us to find eigenfunctions with a definite parity. Let us first look for odd eigenfunctions, i.e., eigenfunctions such that $\varphi_k(x) = -\varphi_k(-x)$. The general solution to the Rayleigh equation, Eq. (10.6), in this case is of the form

$$\varphi_k^{\text{odd}}(x) = \begin{cases} Ae^{-k|x|}\text{sgn}(x) & |x| > -1 \\ B\cosh(kx)\text{sgn}(x) + C\sinh(kx) & |x| \le 1 \end{cases} \tag{10.15}$$

where we have chosen the solution to decay at $x \to \pm\infty$. By using the matching conditions at the singular points, we obtain the dispersion relation

$$\omega_k = \frac{1}{2}\left(1 - e^{-2k}\right). \tag{10.16}$$

All the odd modes are neutrally stable and are called varicose modes in the neutral fluid literature. These modes, however, are important in the nonlinear analysis and magnetic reconnection in Chapter 15.

Even Parity Modes

Now we proceed to consider the even modes, which satisfy $\varphi_k(x) = \varphi_k(-x)$ which are called sinuous modes in fluids. The general solution to the Rayleigh equation in this case is

$$\varphi_k^{\text{even}}(x) = \begin{cases} Ae^{-k|x|} & |x| > 1 \\ B\cosh(kx) + C\sinh(k|x|) & |x| \leq 1. \end{cases} \tag{10.17}$$

Applying the conditions at the singular point we obtain

$$\omega_k = \frac{1}{4}\left[\left(2k + e^{2k} - 1\right) \pm \sqrt{G(k)}\right] \tag{10.18}$$

where

$$G(k) \equiv 9 - 10e^{-2k} - 12k + e^{-4k} - 4e^{-2k}k + 4k^2. \tag{10.19}$$

The function $G(k)$ is negative for $0 \leq k \leq k_c$, with $k_c \approx 1.833$. Hence, in this range, the eigenvalues are

$$\omega_{k\pm} = \omega_R \pm i\gamma_k \tag{10.20}$$

where

$$\omega_R = \frac{1}{4}\left(2k + e^{2k} - 1\right) \tag{10.21}$$

$$\gamma_k = \frac{1}{4}\sqrt{-G(k)}. \tag{10.22}$$

In this case modes with eigenvalue ω_{k+} are unstable and grow according as to the values in Table 10.1 and Fig. 10.1

$$|\varphi_k(x,t)| = |\varphi_k(x)|\, e^{\gamma_k t}. \tag{10.23}$$

On the other hand we see that for every growing mode there corresponds a damped mode, due to the Hamiltonian structure of the system [Morrison (1998); Balmforth and Morrison (2002)], namely, modes with eigenvalue ω_{k-}. If we include dissipation from viscosity and resistivity, the Hamiltonian symplectic structure is broken. However, dissipation is more important for the high-k modes, while the nonlinearity transfers energy to low-k modes.

The general wave function is of the form

$$\varphi(x,y,t) = \varphi_0(x) + \text{Re}\left[\sum_k e^{\gamma_k t}\varphi_k(x)e^{i(ky-\omega_k t)}\right]. \tag{10.24}$$

Within the range $0 \leq k \leq k_c$ we have the following frequencies and growth rates:

The fastest growing mode corresponds to $m = 7$. Figure 10.2 shows isolines of the total potential

$$\varphi(x,y,t) = \varphi_0(x) + \varphi_k(x,y,t), \tag{10.25}$$

Table 10.1 Linear KH eigenfrequencies for the triangular flow profile model in Eq. (10.26) with frequency and growth rate given for the LAPD experiment.

Mode number m	ω_0 (KHz)	$\gamma\,(ks^{-1})$
1	5	13.6
2	18	33.3
3	36.8	52.8
4	59.7	70.1
5	85.6	84
6	113.5	93.7
7	143	98.4
8	173.6	97.4
9	205	89.1
10	237	69.4
11	273.7	0

from a simulation with the step-up flow profile where

$$\varphi_0(x) = \begin{cases} 0 & x \leq -1 \\ \frac{1}{2}x^2 + x + \frac{1}{2} & -1 \leq x \leq 0 \\ -\frac{1}{2}x^2 + x + \frac{1}{2} & 0 \leq x \leq 1 \\ 1 & x > 1 \end{cases} \tag{10.26}$$

is the basic flow stream function from a vee-shaped electric field.

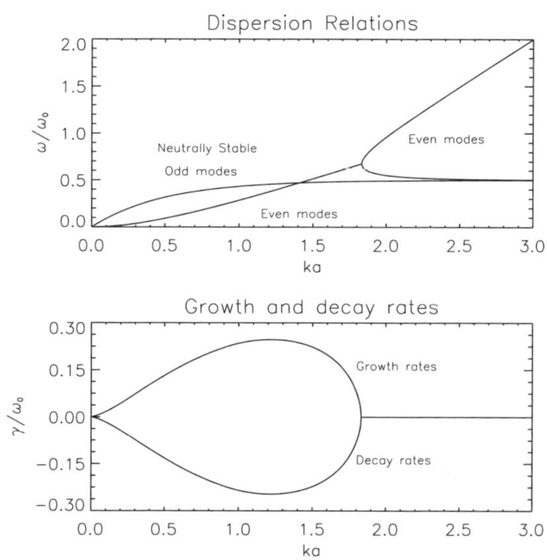

Fig. 10.1 Frequency and growth rate as a function of $k_\theta a$ for the model of the equilibrium radial electric field measured in the plasma. There are two modes: the unstable Kelvin Helmholtz mode and the neutrally stable modes with odd symmetry corresponding to a wavy motion of the jet. The maximum velocity v_0 drops to zero at $\Delta r = \pm a$ and the reference frequency is $\omega_0 = v_0/a$.

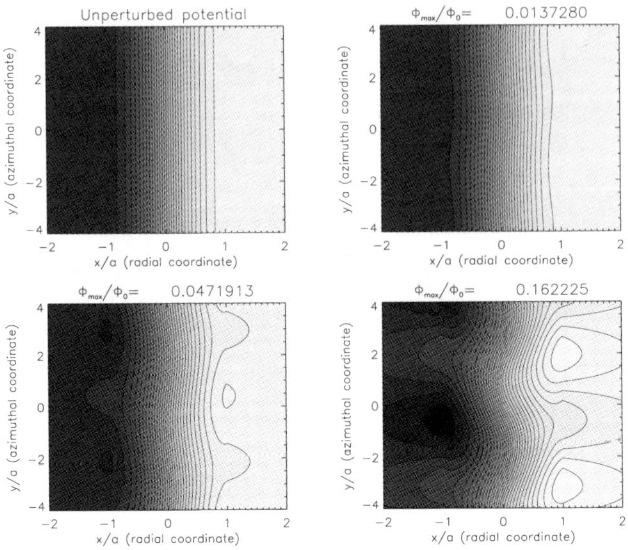

Fig. 10.2 Simulation data for the stream function $\psi = \phi/B$ from the isopotentials of the unstable eigenmode of Eq. (10.6). The last frame in the saturated state shows the alternation of the vortex directions across the jet.

In the turbulent flow layer there is a broad band of fluctuation frequencies that are difficult to associate with the linear eigenmode frequencies shown in Table 10.1. The power spectrum of the floating potential and the vorticity for different time series $\Delta T = N_T \Delta t$ was computed for time spans from $N_T = 256$ to 2048 and the results of the spectra over the M ensembles have been averaged and defined by dividing the stationary section of the bias pulse into M records of length $N_T \Delta t$. The result shows a broadband frequency power spectrum where the power decreases as f^{-3} from $f_1 = 1/N_T \Delta t \sim 1\,\mathrm{kHz}$ to $f_{\max} = \pi/\Delta t = 1.6\,\mathrm{MHz}$ in the case where $N_T = 1024$. Thus, the bulk of the power is at the low frequency end of the spectrum.

The swirling motion of neutral fluids and plasmas is quantified by the vorticity, which can be viewed as the source function for the velocity field, in the same sense that an electric current is the source for a magnetic field. Vorticity is thus a primary physical quantity in equations for fluids and plasma dynamics. For quasi two-dimensional velocity flow fields that occur in geophysical fluid dynamics (GFD) and in magnetized plasmas, it is essential to measure the vorticity field in order to understand quantitatively the dynamics of vortices and other coherent structures in the flow [Maurel and Petitjeans (1999); Horton, *et al.* (2000)].

For shallow water and atmospheric flows of depth H in rotating vessels and on planets with Coriolis frequency f there is a stream function analogous to the plasma stream function $\varphi(x,y,t)/B$. For quasi two-dimensional flows there is a scalar stream function whose Laplacian is the vorticity. In GFD the stream function is $\psi = gH(x,y,t)/f$, where $gH(x,y,t)$ is the gravitational potential energy in the height

H of a column of neutral fluid and f is the Coriolis parameter from the rotation of the planet. In magnetized plasmas the stream function is $\psi = \varphi(x, y, t)/B$, where φ is the electrostatic potential and B is the ambient magnetic field strength.

The fact that the plasma stream function is the electrostatic potential, which can be measured directly by suitable Langmuir probes, gives an unique advantage to the measurement of vorticity in plasmas over the measurement of vorticity in complex fluid dynamics. In neutral fluids, vorticity is constructed from velocity field measurements using Particle Image Velocimetry (PIV) [Wilbert and Gharib (1991)], in which digital images of neutrally buoyant test particles (or seeds) in the fluid are processed with sophisticated computer algorithms to infer the flow velocity, an expensive and somewhat complicated procedure that leads to significant levels of noise.

The coalescence and scattering of monopolar and dipolar drift wave wave vortices are analyzed with a background of drift waves with and without the KdV nonlinearity and compared with fluid Rossby wave vortices and waves [Horton, *et al.* (1990)] using theory and simulations.

A plasma vorticity probe is designed for measuring directly the vorticity of the plasma motion. Fast digital measurements of the field aligned vorticity $\omega(x, y, t) = \nabla^2\varphi/B$ can be measured with this vorticity probe. Here we describe the design of the vorticity probe and its use in a large steady-state cylindrical plasma in the Large Plasma Device (LAPD) [Wilbert and Gharib (1991)]. The experimental situation chosen is one in which the Kelvin-Helmholtz instability is present due to a strongly localized radial electric field at the mid radius of the confinement vessel.

The vorticity probe measures both the DC and AC components of the vorticity with appropriate time filter. The principle of the vorticity probe is to use Langmuir probes in the stencil of the discretized Laplacian operator in the layout as shown in Fig. 10.4.

The experiments are performed in the upgraded Large Plasma Device at UCLA, which is an 18 m long, $\lesssim$ 1 m diameter magnetized plasma column created by a pulsed discharge from a Barium Oxide coated emissive cathode. The typical plasma parameters in LAPD are $n_e \sim 5 \times 10^{12}\,\mathrm{cm}^{-3}$, $T_e \lesssim 15\,\mathrm{eV}$, $T_i \lesssim 1\,\mathrm{eV}$, $B < 0.2\,\mathrm{T}$ with helium, argon, and neon as the working gas ($P_{\rm fill} \sim 5 \times 10^{-5}$Torr). Discharges in LAPD are typically fully ionized, but some residual neutrals are present. A conservative estimate of ion-neutral collisionality is obtained by assuming a neutral density equal to the density before ionization (consistent with the fill pressure at room temperature). This assumption yields a small ion-neutral collision frequency $\nu_{\rm in} = n_0 v_i \sigma_{\rm in} \sim 100/\mathrm{s}$. These collisions produce a very small background viscosity for the plasma that is negligible for the wavenumbers relevant to the fluctuations considered here. A drawing of the apparatus with a schematic of the discharge circuit and rotation biasing circuit is shown in Fig. 10.3.

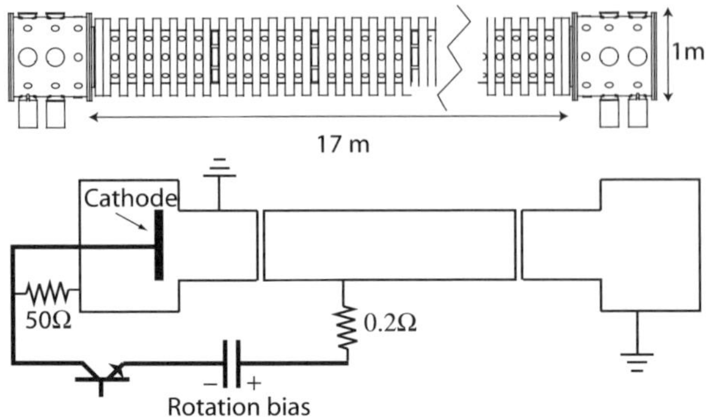

Fig. 10.3 Schematic of the Large Plasma Device (LAPD), including a diagram of the circuit used for plasma column biasing.

10.2.1 *The vorticity probe*

It is well known that the most natural quantity for describing eddies or vortices in neutral fluids and plasmas is the fluid vorticity, defined as

$$\boldsymbol{\omega} = \nabla \times \boldsymbol{v}. \tag{10.27}$$

This gives a local measure of the circulation of the velocity field at every point in the fluid plasma.

In the limit of a uniform magnetic field $B\widehat{\boldsymbol{z}}$ and with an electrostatic field $\boldsymbol{E} = -\nabla\varphi(\boldsymbol{x}, \boldsymbol{t})$, the ions and electrons move across the $\boldsymbol{B}$-field with velocities

$$v_x = -\frac{1}{B}\frac{\partial\varphi}{\partial y}(x, y, t) \tag{10.28}$$

$$v_y = \frac{1}{B}\frac{\partial\varphi}{\partial x}(x, y, z, t). \tag{10.29}$$

The parallel component of the vorticity vector $\boldsymbol{\omega} = \boldsymbol{\nabla} \times \boldsymbol{v}$ is given by

$$\omega = \frac{\partial v_y}{\partial x} - \frac{\partial v_x}{\partial y} = \frac{1}{B}\nabla_\perp^2\varphi(x, y, z, t). \tag{10.30}$$

The vorticity is measured with the vorticity probe shown in Fig. 10.4. From the known spacing $\Delta x = \Delta y = h$ of the four corners of the probe with respect to the center of the probe, we use the sum of the floating potentials of the corner, minus four times the floating potential at the center, $\varphi_1 + \varphi_2 + \varphi_3 + \varphi_4 - 4\varphi_0$ to obtain the vorticity ω.

The vorticity calculated in this way from the floating potential is valid as long as one assumes that the number of primary electrons at probe positions is small and that the temperature is the same at all probe tips. For the measurements in the LAPD more than a few meters, the condition is satisfied. Measurements of temperature profiles show typical temperature scale length of the order of $L_{T_e} \sim 100\,\mathrm{mm}$ which is sufficiently greater than the probe tip distance $h = 5\,\mathrm{mm}$. If this were not the case we would need to measure the temperature profile at each probe tip in order to get the plasma potential φ_p, from the probe floating potential φ_f

$$\varphi_p(eV) = \varphi_f(eV) - \frac{T_e}{2e}\ln\left(2\pi\frac{m_e}{m_i}\right) \tag{10.31}$$

$$= \varphi_f(eV) + 3.5T_e \quad \text{for Helium.} \tag{10.32}$$

The floating potential φ_f of the probe is discussed further in Section 10.2.3.

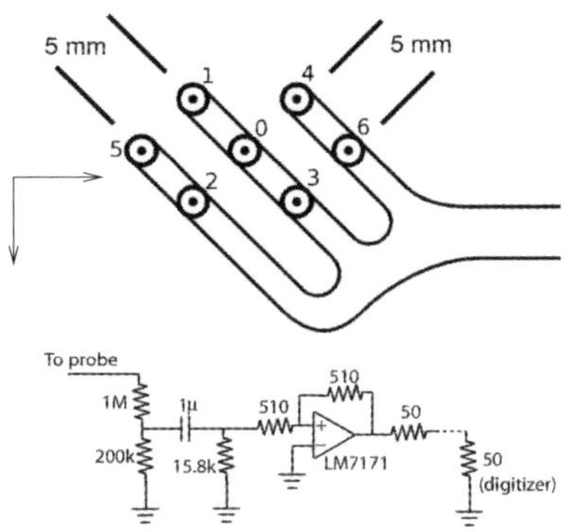

Fig. 10.4 Vorticity probe design. The probe is inserted into the LAPD plasma radially, so that the magnetic field is perpendicular to the surface of the probe tips.

This probe stencil is based on the finite difference approximation to $\nabla^2\varphi$ given by

$$\nabla^2\varphi = \frac{(\varphi_{1,0} + \varphi_{0,1} + \varphi_{-1,0} + \varphi_{0,-1} - 4\varphi_{0,0})}{h^2}. \tag{10.33}$$

The vorticity is then given by dividing by the magnetic field B as in Eq. (10.30). The velocities into and out of the four sides of the square are also of interest. For the left side, for example, the velocity is given by

$$v_x(\text{left}) = \frac{1}{B}\frac{(\varphi_1 - \varphi_2)}{2h}. \tag{10.34}$$

Clearly, the vorticity in Eq. (10.33) vanishes for linearly potential with $\varphi = C + Ax + By$. Only quadratic variations over the probe yield nonzero vorticity.

Just as in numerical simulations, the five-point sample of $\varphi(x, y, z, t)$ yields the true vorticity only for that part of the spectrum $\varphi(k_x, k_y, z, t)$ that has $k_x h < 1$ and $k_y h < 1$. Higher $\boldsymbol{k}$-modes yield an anomalous value for ω, although the value can still be interpreted unless $k_x h \gg 1$. For example, a mode with $k_x h = \pi$ and $k_y h = \pi$ leads to the value $-8\varphi_{0,0}/h^2$ for the right-hand side of Eq. (10.33). For larger $k_x h$ values the vorticity is small from the simulations due to inverse cascade in quasi-2D turbulence.

The rate of convective transport of ω is given by

$$\frac{\partial\omega}{\partial t} + \boldsymbol{v}\cdot\nabla\omega = S_\omega(\boldsymbol{x}, \boldsymbol{t}) \tag{10.35}$$

where the source/sink function S_ω includes various mechanisms that produce and absorb vorticity.

Since the $\boldsymbol{E} \times \boldsymbol{B}$ drift is incompressible for a uniform $\boldsymbol{B}$-field, Eq. (10.35) also has the conservation form

$$\frac{\partial \omega}{\partial t} + \nabla \cdot (\omega \boldsymbol{v}) = S_\omega. \tag{10.36}$$

Averaging using Eq. (10.36) over the symmetry direction y (or θ) yields the transport equation,

$$\frac{\partial}{\partial t} v_y'(x,t) + \frac{\partial}{\partial x}\langle \omega v_x \rangle = \overline{S}_\omega(x,t) \tag{10.37}$$

for the generation of sheared zonal flows $v_y' = d\langle v_y \rangle / dx$.

An important theorem is that the average flux of vorticity $\langle v_x \omega \rangle$ equals the divergence of the Reynolds stress $\langle v_x v_y \rangle$ per unit mass. Both the vorticity flux and the Reynolds stress can be measured by the vorticity probe.

In the LAPD experiment, the differential axial confinement of the electrons at the radius of the cathode controls the source of vorticity S_ω through the biasing of the cathode/anode plasma source assembly with respect to the confinement chamber walls. For tokamaks a similar E_r well structure is formed by the differential confinement of the ions at the edge of the plasma. This E_r well becomes a strong feature after the transition to the H-mode [Burrell (1997)].

The vorticity probe was constructed from seven Tantalum tips, as shown in Fig. 10.4. The tips are cylindrical, 0.02 inch in diameter and 1 mm long. The tips are oriented along the direction of the magnetic field (into the page in Fig. 10.4). The Tantalum tips are held in alumina ceramics, which in turn are held by a stainless steel structure that is slotted in order to minimize perturbation to the plasma. The central five tips are arranged in a diamond pattern, with the outer four tips separated from the central tip by 5 mm. The five inner tips are used to measure the floating potential, which is then used to compute a finite difference estimate of the vorticity, while the two outer tips allow measurement of the ion saturation current, in order to obtain the plasma density and radial particle flux $\langle nE_\theta \rangle / B$, as well as the Reynolds stress $nm_i \langle v_r v_\theta \rangle = (\rho / B^2)\langle E_r E_\theta \rangle$ [Lochte-Holtgreven (1968)].

In order to compute the finite-difference estimate of the vorticity, four times the floating potential on the central tip must be subtracted from the floating potential on the surrounding four tips. During biased rotation experiments in LAPD, the DC floating potential can reach values of order 20 V, substantially larger than the observed fluctuation amplitude ($\sim$ 1 V). The floating potential measurements are therefore performed using AC coupled amplifiers in order to reject the large low-frequency floating potential signal and to maximize the use of the dynamic range of the available digitizers. The amplifiers are constructed using wide band operational amplifiers (National Semiconductor LM7171) with a flat response from 1 KHz to 10 MHz. A schematic of the measurement circuit is shown in Fig. 10.3. Data is acquired using 100 MS/s, 14 bit VME-based digitizers (eight channels per board, four available boards).

The origin of the peaking of the electron temperature peak in the shear flow layer is not well understood and suggests further research studies on the anomalous electron viscous heating and magnetic fluctuations.

10.2.2 *Vorticity probe measurements on the Kelvin-Helmholtz instability*

The Kelvin-Helmholtz instability is excited in the LAPD experiment when biasing the floating anode-cathode plasma source with respect to the chamber walls. This results in a sharply localized radial electric field as shown in Fig. 10.5. This electric field, along with the axial magnetic field, creates a sheared poloidal $\boldsymbol{E} \times \boldsymbol{B}$ jet stream at the edge of the plasma column. This jet stream flow forms as one of the natural self-organized states after the K-H instability in which the vorticity ω as function of the potential φ is a steady-state solution of the vorticity equation

$$\vec{v}_E \cdot \nabla\omega = 0. \tag{10.38}$$

The nonlinear equation, Eq. (10.38), has vortex solutions of the $\omega = F(\varphi - ux)$ moving with speed u in the symmetry direction $y = r\theta$ of the cylinder. Various types of solutions exist with or without embedded vortices. The simplest solution is a jet stream with localized $v_\theta(r)$. In Section 10.2.3 we show a nonlinear simulation that illustrates this feature.

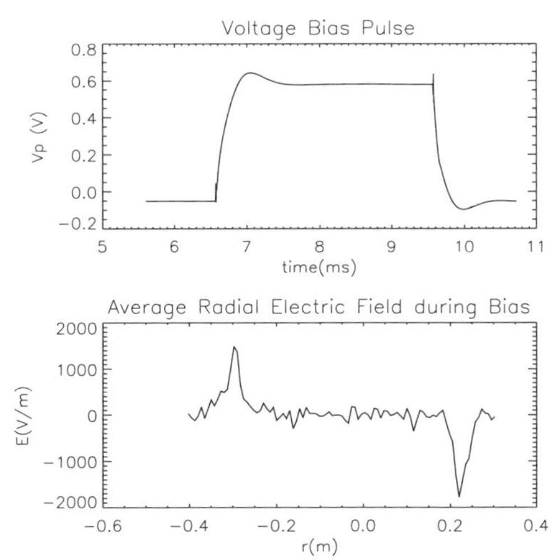

Fig. 10.5 Top: Typical cathode/anode bias voltage pulse with respect to the chamber wall for establishing the rotational jet. Bottom: Radial electric field measured in the stationary section of the bias pulse.

In Fig. 10.6 we show vorticity probability distribution functions or PDFs at representative points of the shear layer. In each plot we combine vorticity data from 25 similar shots, during 2 ms of the bias pulse between $t = 7.5\,\text{ms}$ and $t = 9.5\,\text{ms}$, see Fig. 10.5. This corresponds to a total of 77525 data points divided into 50 equally spaced bins of fluctuating vorticity amplitudes normalized to the standard deviation.

The heavy tails of the vorticity probability distributions functions shown in Fig. 10.6 signify that large values of the vorticity occur frequently. The situation here for vorticity fluctuations driven by the background plasma jet is PDFs shown have heavy tails as do those for the fluctuations $\langle v_x \delta n \rangle$ from the density fluctuations and density fluxes $\langle v_x \delta n \rangle$ from coherent intermittent structures measured in the linear PISCES-A [Antar (2003)] and the ADITYA tokamak machines [Jha, *et al.* (1992)].

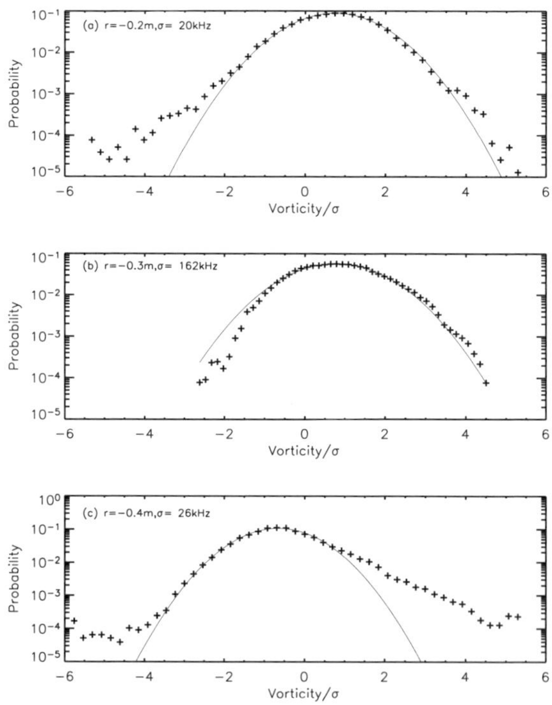

Fig. 10.6 Vorticity PDF for four radial position across the shear flow layer for the left side ($r < 0$) of the plasma column. The sign of the excess vorticity counts over the Gaussian value changes from negative (counterclockwise rotation) to positive (clockwise rotation) in moving across the shear flow layer.

In Fig. 10.7 we show relevant statistical quantities for an ensemble of 25 experimental runs under identical conditions. The measurements of vorticity show a high degree of reproducibility over the experiments. First we consider the degree of correlation of the floating potential difference between the outermost probe pins 5 and 6 as the probe is scanned across the shear flow layer. The spacing between these probes is $d = 2\sqrt{2}h\,\mathrm{mm} = 14\,\mathrm{mm}$. Frame (a) shows that the normalized correlation function $\langle\varphi_5\varphi_6\rangle$ is near unity except in the shear layer where it drops to about 0.7. Thus, we infer that the radial correlation length of the fluctuations drops from greater than 14 mm to about 10 mm in the center of the shear flow layer. Next we consider the vorticity mean values, as the probe is scanned along the radial

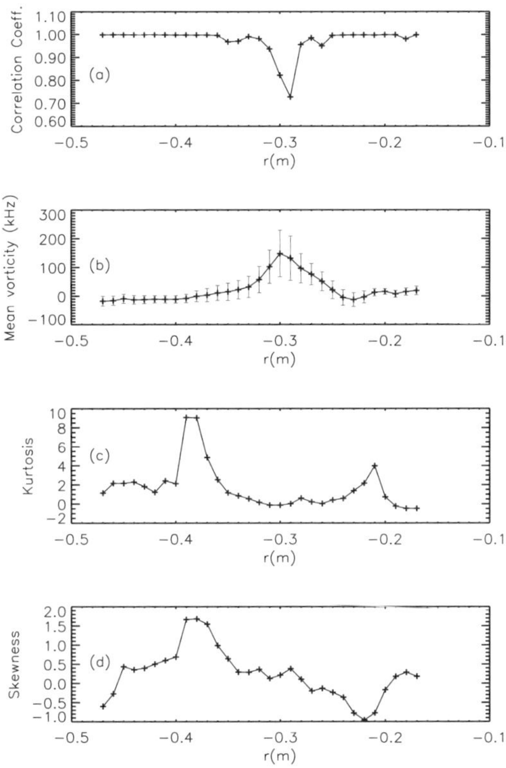

Fig. 10.7 LAPD data analysis with the vorticity probe. Radial variation on the left side of the cylinder for the mean, the kurtosis and skewness [frame (d)]. The antisymmetry of the skewness agrees with the change if vortex rotation direction across the layer and coincides with the larger values of kurtosis in frame (c) that occurs in for a field of vortices.

direction, by averaging vorticity over time at each point in the shear layer and later averaging over the ensemble of the 25 experimental samples. Frame (b) of Fig. 10.7 shows the mean vorticity $\langle\omega\rangle$ as a function of the radius. From this plot we see that the vorticity reaches a maximum value $\omega_{\max} = 150\,\mathrm{kHz}$ which is comparable to $v_{\max}/a \sim 10^4\,(\mathrm{m/s})/0.05\,\mathrm{m} \sim 200\,\mathrm{kHz}$ from the equilibrium radial electric field profile at $B = 0.075\,\mathrm{T}$. The positive core value agrees with sign of chain of vortices expected from the KH instability [Horton, *et al.* (1987)].

Frames (c) and (d) show that outside the core of the jet there is an intermittency of the vorticity probability distribution function consistent with the heavy tails shown in Fig. 10.6. While the error bars on the mean vorticity suggest that it may be consistent to take the vorticity as vanishing outside the core of the jet stream, the small relative bias that is about equal to the width of the error bar is such as to give vortices rotating counter-clockwise outside the stream $r < -0.4\ m$ and clockwise inside the stream $-0.4\ m < r < 0$ on the left side of the cylinder. As will be shown later, the simulations show such a pattern of counter-rotating vortices on the flanks of the jet stream.

There are two types of aliasing errors: (i) the usual single probe sampling error for signal frequencies higher than the Nyquist $f_N = 1/2\delta t$ and (ii) spatial aliasing errors for short wavelength signals. For each pin there are anti-aliasing filters in frequency. This helps reduce anti-aliasing from $k_N = \pi/h$ under the assumption that small scale structures exist at higher frequency. The sample rate (and anti-aliasing filter) is at much higher frequency than the frequency of the observed fluctuations. The probe sampling rate is at few MHz, and the fluctuations are below 5 kHz.

For spatial aliasing problem we consider the effect of the finite probe space h on a single wavelength signal, the ratio of vorticity measured with the probe to the actual value of form factor vorticity is given by

$$R(\vec{k}, h) = 4\frac{\sin^2\left(\frac{k_x h}{2}\right) + \sin^2\left(\frac{k_y h}{2}\right)}{h^2(k_x^2 + k_y^2)}. \qquad (10.39)$$

For $kh < 1$ we have $R \approx 1$. For $k'_x = k_x + 2\pi n$, $k'_y = k_y + 2\pi m$, the probe picks up the same signal for $\vec{k}$ and $\vec{k}'$ which is the aliasing error. However, the probe value is smaller as given by $R(\vec{k}, h)$ at the spatial Nyquist frequency $k_N = \pi/h$ where the ratio function is $R = 4/\pi^2 \approx 0.444$. So, keeping probe data for $k < k_N \approx 628\,\mathrm{m}^{-1}$ largely suppresses the aliasing errors. Using a range of probe sizes h would allow an accurate assessment of the power spectrum in the high-k components.

10.2.3 *Kelvin Helmholtz turbulence with drift waves*

Now we examine the nonlinear simulations of the LAPD shear layer during the rotation bias pulse. A nonlinear pseudo-spectral code is used in Fourier decomposition in space to compute gradients and Poisson brackets in k-space and adaptive fifth/sixth order Runge-Kutta method to advance in time the solution in k-space. Then we transform the advanced solution back to configuration space where the quadratic nonlinear terms are computed.

There are two versions of the code with different boundary conditions available at `http://pecos.ph.utexas.edu/~vortex`. One code is implemented with periodic boundary conditions in both x (radial) and y (poloidal) coordinates [Horton

(1989)]. There are two disadvantages to using the local periodic slab model for the simulation. First, the DC rotation through

$$\langle \omega \rangle = \frac{1}{r}\frac{\partial (r v_\theta)}{\partial r} = \frac{\partial v_\theta}{\partial r} + \frac{v_\theta}{r}.$$

The first gradient component is dominant, for the profile in Perez (2006a) is about 200 kHz while the second rotational frequency Ω, which is symmetric about the peak of the jet, is $\Omega = v_{\theta,\max} r \sim 60\,\text{kHz}$. The second disadvantage is that the radial boundary conditions are not periodic. However, we can still get some qualitatively important results by using the local potential shown in Fig. 10.8.

The top frame of Fig. 10.8 shows the initial state in which opposite vortices are shed from the right and left sides of the jet, so that there is a definite skewness $\langle \omega^3 \rangle$ that changes sign across the layer as measured with the vorticity probe. After some time, the simulation shows that the average value of vorticity along the y-direction of Fig. 10.8 tends to decrease at each x-point due to the alternate counter rotating vortices lining up along the y-direction. In both linear and nonlinear simulations we have taken a slab of the rotating shear layer and therefore we have ignored the fact that the plasma is rotating as a whole, giving rise to a net background positive (clockwise when viewed into the magnetic field vector) vorticity associated to the equilibrium rotation. This rotational vorticity explains the dominance of positive mean values in Fig. 10.8 and a peak at the center of about 150 kHz, where the flow velocity is maximum. Theoretical estimates give about 60 kHz for the background equilibrium vorticity, however, in the present experiment the floating potentials measured with the probe where high pass filtered at about $\sim 1\,\text{kHz}$, so that part of the DC (equilibrium) potential gets through the filter giving rise to steeper gradients and hence higher values for the vorticity which explains why the vorticity maximum is higher than expected.

10.3 Experimental Measurement of Vorticity Dynamics Studies and the Reynolds Stress

The simple analytic model used in simulations takes the plasma density as constant over the shear layer. From the ion saturation current $I_{\text{sat}}(r)$ measured taken through a low pass filter we know just inside the shear flow layer there is a sharp increase of the electron density. In and outside the layer we find that density gradient scale length $L_{ne} \sim 15\,\text{cm}$ which gives a sufficiently low gradient frequency ω_{*e} to neglect the $\delta n_e/n_e$ from $\vec{E} \times \vec{B}$ convection across the density gradient. Thus, the shear flow of the jet reduces the particle flux $\langle n_e v_x \rangle$ across the layer.

The cross-correlation signals between the widest pin spacing in the probe shows that the pin signals are correlated and thus contain useful information rather than noise. The cross correlation drops from near unity to about 70% in the shear flow layer for the outermost pin tips separated by 14 mm which is approximately 20 ion

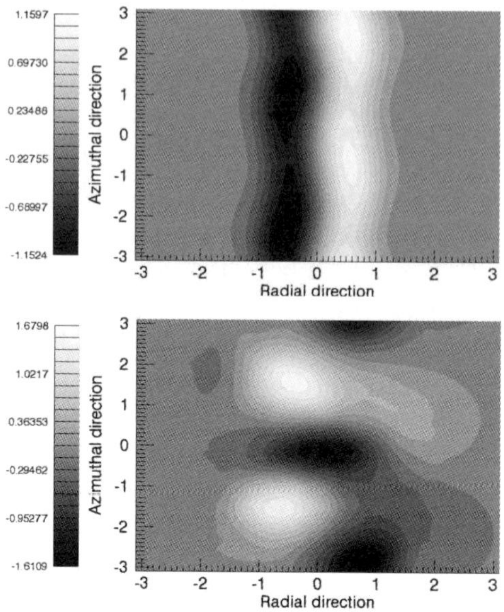

Fig. 10.8 Top panel is the linear stage and bottom panel is the nonlinear stage of the jets. In the nonlinear state, the jet disassembles into positive and negative vortices. Afterwards there is a coalescence of the vortices and a separation of the direction of rotations according to the direction of the jet. In the experiment the momentum is transported away by the divergence of the Reynolds stress tensor.

gyroradii. The mean vorticity shows a large peak of 1.5×10^5 /s at the center and small residual positive and negative values of order $\pm 1 \times 10^4$ /s just outside the shear flow layer. The small residual values are just at the estimate of one standard deviation $\sigma_\omega(r)$ in the spread (variance of ω) about the mean of the vorticity PDF. These residual values of ω are statistically significant and thus we see a confirmation of the odd symmetry of the sign of vorticity shed from the plasma jet as shown in the computer simulations.

A set vorticity of probes with different tip spacing would allow the collection of a relatively complete picture of the vorticity field from $\boldsymbol{E} \times \boldsymbol{B}$ flows in a magnetized plasma. Future research would integrate test particle motion and compare with tagged ions by the laser induced fluorescence technique to investigate the details of the transport across the plasma jet or E_r well.

The probability distribution functions (PDF) of the vorticity were computed across the shear layer, as sampled at $\delta t = 0.3\mu$s over a steady state period of $T = 2$ ms. The PDF shows strong tails in the regimes of maximum vorticity $dE_r/dr\, v'_\theta = -E'_r/B$ near $r = -0.4$ m, where E'_r is a maximum. These results are interpreted as showing the existence of a string of vortices of opposite signs on either side of the $E'_r = 0$ layer. The antisymmetric form of the skewness moment of the vorticity in Fig. 10.7, as well as the local peaks of the kurtosis function, are consistent with a string of oppositely signed vortex chains on either side of the maximum of the $E_r(r)$ profile. This antisymmetric double chain of vortices was also studied with simple pseudo-spectral code simulation of the Kelvin-Helmholtz instability for the experimental system parameters.

Other quantities available from the probe are the ion saturation current, for a measure of the plasma density and particle flux $\Gamma = \langle nE_\theta \rangle / B$ and the Reynolds stress $\Pi_R = nm_i \langle v_r v_\theta \rangle = (\rho/B^2)\langle E_r E_\theta \rangle$. Analysis of these fluxes are given in Perez (2006b), along with a comparison to test particle transport in models for the measured fluctuations.

Plasma has the unique advantage over a neutral fluid in that the stream function for quasi-two dimensional flow is the electric potential, which can be sensitively measured directly by electrical probes with standard Langmuir probe techniques. Vorticity in neutral fluids is inferred from particle image velocimetry of neutrally buoyant test particles in the flow. This measurement is a relatively complex procedure compared with the measurement of the vorticity probe and is subject to high noise level when the vorticity is constructed from the measured velocity field.

The measurement of vorticity is of fundamental importance in theory of turbulent fluids and plasmas. Closure models lead to various relations for the probability distributions and moments of the vorticity. A classical neutral fluid example is the Proudman-Reid formula $d^2\overline{\omega}^2/dt = \frac{1}{3}(\overline{\omega}^2)^2$ discussed in detail in Monin and Yaglom (1965) where this model is compared with that of Millionshchikov and Betchov. Developing the vorticity probe for plasma should provide new insight into plasma turbulence modeling.

The nonlinear stability of the plasma is intimately connected to the formation of vortex structures. Kim, *et al.* (2006) show had the presence of sheared cross-field plasma flow that is linearly stable by the Rayleigh condition can have convective amplification of a initial wave packet disturbance. The sheared flow transforms as a motion $\boldsymbol{k}$-space from the initial region of the wave packet. In the process of $\boldsymbol{k}$-space convection the wave packet amplified and is typically a nonlinear coherent vortex under the conditions given by Kim, *et al.* (2006) and Horton (1989).

In tokamaks, scrape-off layers are formed by applying horizontal magnetic field creating an X-point in the in the poloidal magnetic field structure. The magnetic separatrix then become a boundary between the core plasma and the scrape-off layer plasma. These driven plasmas are found to have large, intermittent coherent structures often called "blobs" expelled from inside the separatrix to the outer scrape-off layer plasma. In the scrape-off layer the plasma streams along the magnetic field line to the carbon coated metal plates placed in a small volume called the divertor chamber. Antar, *et al.* (2001, 2010) have made images of these structures and constructed the probability distributions for these events in the spherical tokamak called MAST standing for Mega Ampere Spherical Tokamak.

Fast imaging cameras with 10μs time resolutions and fast reciprocating probes are used to study the characteristics of the "blobs" ejected from the confinement region. The structures are larger and more frequent on the out-board side where the toroidal magnetic field curvature is such as to accelerate the plasma out of the confinement region. The outboard side is also called the low-field side of the torus and the magnetic separatrix is called the LCFS for last closed flux surface. This basic mechanism is described by the instabilities in Chapter 6 and perhaps the simulation shown in Fig. 8.4 shows one of the mechanims for the formation of the structures inside the separatrix that become accelerated into "blobs" after crossing the separatrix. Antar, *et al.* (2001, 2010) report seeing structures from 5 to 10 cm across the magnetic field. Once in the open magnetic field region outside the separatrix they move with speeds of $10^3\,\mathrm{m/s}$ along the field lines to the strike points on the divertor plates. The accumulative distribution of the blobs makes a pdf (probability distribution function) that has skew (skewness ~ 3) and a flat (kurtosis ~ 20). In contrast, on the high-field inboard side the intermittency is not recorded and the distributions are gaussian. Krasheninnikov (2001) anticipated the likely features of the ejected coherent structures and chose to call them "blobs".

10.4 Electromagnetic Vortices

Drift wave-Alfvénic vortex solutions are known for the electromagnetic structures in the finite pressure magnetized plasma. The generalization follows from the continuity of the plasma currents, which leads to fourth order differential equations for the electromagnetic fields. Special solutions for two types of boundary value problems are known for the fourth-order partial differential equations governing the vortex structures. There are two types of vortex structures which we call the intrinsic electromagnetic drift-Alfvén vortex and the electron scale Alfvén vortex structure.

The electron scale electromagnetic vortices have more complicated radial structures with inner and outer regions than the Rossby-drift-wave vortex. The complex vortex structures guarantee the continuity of the perturbed magnetic field $\delta \boldsymbol{B}_\perp$ and the parallel current $j_\parallel$ on the boundary between the inner and outer regions of the vortex. The allowed regions of propagation speeds and amplitude for these vortices are derived in this chapter.

10.4.1 *Basic considerations of Alfvén-drift wave vortices*

Solitary drift wave vortices are well known and are the subject of Chapters 5 and 6 in Horton and Ichikawa (2000). Such vortex solutions were first given by Larichev, *et al.* (1976) in the course of study of nonlinear Rossby waves in rotating fluids. Solitary vortex solutions which seem ubiquitous in nature have been discussed by Mikhailovskii, *et al.* (1984), Mikhailovskii, *et al.* (1985), Shukla, *et al.* (1985a), Su, *et al.* , (1991, 1992) and Hazeltine, *et al.* (1985). In Shukla, *et al.* (1985b) the drift-Alfvén vortex is analyzed. In a work by Mikhailovskii, *et al.* (1985) two other nonlinear equations of electromagnetic perturbations in magnetized plasma are derived, and the possibility of the existence of solitary vortex solutions for these equations was discussed without presentation of explicit solutions. In work by Hazeltine, *et al.* (1985) a simple electromagnetic generalization of drift-solitary wave for certain nonlinear equations was suggested.

A direct application of the Rossby-type vortex solutions to the electromagnetic Alfvén vortex solutions, as given by Mikhailovskii, *et al.* (1984) and Shukla, *et al.* (1985a), has the problem that the perturbed magnetized field $\delta \boldsymbol{B}_\perp$ and the parallel current $j_\parallel$ are discontinuous on the boundary between the inner and outer regions of the vortex. These discontinuities do not appear in the solution given by Shukla, *et al.* (1985b), but only because that solution is an approximate one. Liu and Horton (1986a,b) show that the nonlinear vortex solutions of the nonlinear equations used by Shukla, *et al.* (1985b) also has the same difficulty. Liu and Horton (1986a,b) derive vortices in rapidly rotating cylindrical plasmas.

In this section electromagnetic vortex solutions that are free of the discontinuities are constructed. These continuous electromagnetic vortex solutions are important for space plasmas where the Cluster mission has found evidence for such structures. They may also be found in laboratory plasmas where $\beta = 2\mu_0\langle p\rangle/B^2$ is finite. The essential steps for deriving the vortex solutions the nonlinear equation are to solve the equations in separated spatial regions requiring localization of the solution in the core and continuity of the fields and their first derivatives on the boundary between the two regions. The condition of the continuity of the fields yields the propagation speed u of the vortex solutions.

The Rossby wave-drift wave vortices are solutions of a second order linear differential equation. The electromagnetic vortex solutions are solutions of fourth order partial differential equations. The electromagnetic solutions have a more complicated structure in the inner region and decay more slowly in the outer region with the continuity of $\delta \boldsymbol{B}_\perp$ and $j_\parallel$ throughout the structure.

The analysis here of the corresponding nonlinear dispersion relations is intrinsically electromagnetic vortices. As the plasma beta vanishes, one of the solutions reduces to the electrostatic vortices given by Meiss and Horton (1983) and Liu and Horton (1986a,b). More general vortex structures are given by Schepp, *et al.* (1994) with FLR terms in the vorticity equation for nonlinear Alfvénic drift wave equations for analysis of coherent Alfvén wave structures from the two-component fluid theory with FLR effects taken into account.

10.4.2 *Special solutions for vortex boundary value problems*

In this section we give the special solutions for the two types of two-dimensional boundary value problems arising from the characteristic fourth order partial differential equations. In the following sections we apply these solutions to the nonlinear equations in magnetized plasma for the corresponding solitary vortex solutions.

Worked Problem: Find the localized solutions of radius a for the Type 1 and Type 2 vortex partial differential equations.

Type 1 drift Alfvénic vortex equation:

Find localized solutions of the fourth-order pdes given by

$$\nabla_\perp^4 \phi(r,\theta) - \alpha_1 \nabla_\perp^2 \phi(r,\theta) = 0, \qquad (r > a) \tag{10.40}$$

$$\nabla_\perp^4 \phi(r,\theta) + \alpha_2 \nabla_\perp^2 \phi(r,\theta) + \alpha_3[\phi(r,\theta) - \alpha_4 r\cos\theta] = 0, \quad (r < a) \tag{10.41}$$

where

$$\nabla_\perp^2 = \frac{1}{r}\frac{\partial}{\partial r}\left(r\frac{\partial}{\partial r}\right) + \frac{1}{r^2}\frac{\partial^2}{\partial\theta^2}$$

and $\alpha_1, \alpha_2, \alpha_3, \alpha_4$ are constants determined by parameters of the large scale drift-Alfvén wave equations. The values of α_n are given in Eqs. (10.81) and (10.82). Find the localized, continuous solution of ϕ along with the conditions on α_i $(i = 1, 2, 3, 4)$.

Type 2 electron inertial scale drift Alfvénic vortices:

Find the localized solutions of the equations

$$\nabla_\perp^4 \phi - \beta_1 \nabla_\perp^2 \phi + \beta_2 \phi = 0 \qquad (r > a) \tag{10.42}$$

$$\nabla_\perp^4 \phi + \beta_3 \nabla_\perp^2 \phi + \beta_4 \phi + \beta_5 r\cos\theta = 0 \qquad (r < a) \tag{10.43}$$

where β_1, β_2, β_3, β_4, and β_5 are constants determined by small space scale electron skin depth vortices of radius a. Find localized, continuous solution of ϕ along with the conditions on β_i $(i = 1, 2, 3, 4, 5)$ which are specified in Eqs. (10.108) and (10.109).

The study is limited to solutions of the form

$$\phi(r,\theta) = \Phi(r)\cos\theta \tag{10.44}$$

for type 1 and type 2 problems with the boundary conditions

$$\phi(r,\theta) \to 0 \text{ as } r \to \infty, \text{ and } \phi \text{ is regular when } \quad r = 0. \tag{10.45}$$

In the following we give in some detail the procedure for solving Problem 1; while for Problem 2, the procedure is similar and we only give the solutions [Liu and Horton (1986a,b).

Substituting Eq. (10.44) into Eq. (10.40) and Eq. (10.41) gives

Outside Vortex Core

$$\left(\frac{d^2}{dr^2} + \frac{1}{r}\frac{d}{dr} - \frac{1}{r^2}\right)\left[\left(\frac{d^2}{dr^2} + \frac{1}{r}\frac{d}{dr} - \frac{1}{r^2} - \alpha_1\right)\Phi(r)\right] = 0 \qquad (r > a) \tag{10.46}$$

and

Inside Vortex

$$\left(\frac{d^2}{dr^2}+\frac{1}{r}\frac{d}{dr}-\frac{1}{r^2}\right)\left[\left(\frac{d^2}{dr^2}+\frac{1}{r}\frac{d}{dr}-\frac{1}{r^2}+\alpha_2\right)\Phi(r)\right]$$
$$+\alpha_3\left[\Phi(r)-\alpha_4 r\right]=0. \qquad (r<a). \tag{10.47}$$

Due to the property of the operator $d^2/dr^2 + d/rdr - 1/r^2$, the general solution of Eq. (10.46) must satisfy the second order equation

$$\left(\frac{d^2}{dr^2}+\frac{1}{r}\frac{d}{dr}-\frac{1}{r^2}-\alpha_1\right)\Phi(r)-f(r)=0 \tag{10.48}$$

where

$$f(r)=Ar+Br^{-1}; \quad A,B= \text{ constants.} \tag{10.49}$$

The solution of Eq. (10.48) which vanishes at infinity is

$$\Phi_{\text{out}}(r)=A_1K_1(\sqrt{\alpha_1}r)+A_2r^{-1} \tag{10.50}$$

where K_1 is the first order McDonald function and A_1, A_2 are integration constants. From the reality requirement of $\sqrt{\alpha_1}$ we have

$$\alpha_1>0. \tag{10.51}$$

The solutions are exponentially localized when $A_2 = 0$ and are extended, or global when $A_2 \neq 0$

Inner Solutions

From the form of Eq. (10.47) and the boundary condition at the origin, the solution of Eq. (10.47) can be any one of the following three expressions, depending on the relations between the coefficients α_2 and α_3 in Eq. (10.41):

$$\Phi_{\text{in}}(r)=A_3J_1(\lambda r)+A_4I_1(\mu r)+\alpha_4 r \tag{10.52a}$$

$$\Phi_{\text{in}}(r)=A_3J_1(\lambda^{(1)}r)+A_4J_1(\lambda^{(2)}r)+\alpha_4 r \tag{10.52b}$$

$$\Phi_{\text{in}}(r)=A_3I_1(\mu_1 r)+A_4I_1(\mu_2 r)+\alpha_4 r \tag{10.52c}$$

where J_1, I_1 are the first order Bessel and modified Bessel functions, respectively, and A_3, A_4 are integration constants. Substituting Eq. (10.52a)-Eq. (10.52c) into Eq. (10.47) yields the following conditions corresponding to the different solutions:

Solution Eq. (10.52a) in which

$$\lambda^2=\frac{1}{2}\left(\sqrt{\alpha_2^2-4\alpha_3}+\alpha_2\right) \tag{10.53a}$$

$$\mu^2=\frac{1}{2}\left(\sqrt{\alpha_2^2-4\alpha_3}-\alpha_2\right) \tag{10.53b}$$

corresponds to

$$\alpha_3 < 0. \tag{10.54}$$

For positive α_3 the solution of Eq. (10.52b) is

$$\left(\lambda^{(1),(2)}\right)^2 = \frac{1}{2}\left(\alpha_2 \pm \sqrt{\alpha_2^2 - 4\alpha_3}\right) \tag{10.55}$$

with the conditions

$$\alpha_2 > 0 \quad \text{and} \quad \alpha_2^2 > 4\alpha_3 > 0. \tag{10.56}$$

Solution of Eq. (10.52c) is

$$\mu_{1,2}^2 = \frac{1}{2}\left(-\alpha_2 \pm \sqrt{\alpha_2^2 - 4\alpha_3}\right) \tag{10.57}$$

with the conditions

$$\alpha_2 < 0 \quad \text{and} \quad \alpha_2^2 > 4\alpha_3 > 0. \tag{10.58}$$

It is easy to show that Eq. (10.52a)-Eq. (10.52c) with corresponding conditions Eq. (10.54), Eq. (10.56), and Eq. (10.58), give inner solutions of Φ which are regular at the origin.

To further determine the solution, we need to impose some matching conditions to the electromagnetic fields Φ_{out} and Φ_{in} on the common boundary $r = a$. This will be done in Section 10.4.3 after we discuss the applications of the solutions.

By analogy, Problem 2 can be solved in a similar way. The solutions of Eq. (10.42) and Eq. (10.43) are found in Liu and Horton (1986b).

As in Problem 1, we leave the determination of the integration constants to the following sections where the continuity conditions for the magnetic field and parallel current are derived.

Comparison between the solutions of Eq. (10.50) and Eq. (10.52a)-Eq. (10.52c) of Problem 1 and solutions and the corresponding equations of Problem 2 shows that the inner solutions for the two problems are quite similar, but the outside solutions are rather different. Problem 1 has a long tail solution while the solution of Problem 2 decreases to zero very rapidly when $r \to \infty$. The solutions of Problem 2 for the electron drift Alfvénic vortices are localized dipole solutions like the Rossby wave vortices [Larichev, *et al.* (1976)] or drift wave vortices [Meiss and Horton (1983)] obtained from solving the corresponding one-potential equivalent linear equation of the nonlinear drift wave equation. The radial structure of the electromagnetic solutions given here are rather different from the drift vortex solutions because their equations are fourth order differential equations while the drift wave solutions are derived from second-order equations.

The outer solutions for both Problems 1 and 2 are uniquely given by Eq. (10.50) while the inner solutions can be either one of the three equations, Eq. (10.52a),

Eq. (10.52b) and Eq. (10.52c) according to the different relations between the coefficients α_i and β_i.

Here we limit the detailed analysis to solution Eq. (10.52a) for Problem 1 and to solution Eq. (10.52a) for Problem 2 with the corresponding conditions of Eq. (10.54). The treatment of the other cases is similar. Generalizations that include the effects of inhomogeneity and shear were discussed in Su, *et al.* (1991, 1992), respectively.

10.4.3 *Drift Alfvén wave vortices*

Consider a plasma which is low $\beta = 8\pi n_o T_e / B_o^2 \ll 1$, inhomogeneous ($n_o = n_o(x)$), isothermal ($T_i, T_e = \text{const.}$) and immersed in a strong, uniform, constant magnetic field $\boldsymbol{B}_o = B_o \widehat{\boldsymbol{z}}$. One can choose the three perturbed fields describing the system to be parallel component of vector potential $A_\parallel(x, y, z, t)$ which describes the magnetic field line perturbation $\delta \boldsymbol{B}_\perp(x, y, z, t)$ perpendicular to $\boldsymbol{B}_o$, the electrostatic potential perturbation $\phi(x, y, z, t)$, and the normalized density perturbation $\widetilde{n}(x, y, z, t) \equiv \delta n(x, y, z, t)/n_o$. Using the two-fluid model, neglecting the electron mass effect and the parallel motion of the ions, one uses the quasineutrality condition with Ampere's law, the continuity equation and the parallel momentum balance equation of the electrons to describe the low-frequency dynamics of the plasma system. The nonlinear dynamics given in the system of three nonlinear pde equations are

$$\left\{\partial_t + \frac{c}{B_o}[\phi,]\right\} \nabla_\perp^2 \phi + \frac{v_A^2}{c} \left\{\partial_z - \frac{1}{B_o}[A_\parallel,]\right\} \nabla_\perp^2 A_\parallel = 0 \qquad (10.59)$$

$$\left\{\partial_t + \frac{c}{B_o}[\phi,]\right\} \widetilde{n} + \frac{c}{4\pi n_o e} \left\{\partial_z - \frac{1}{B_o}[A_\parallel,]\right\} \nabla_\perp^2 A_\parallel + \frac{c\kappa_n}{B_o} \partial_y \phi = 0 \qquad (10.60)$$

$$\frac{1}{c}\partial_t A_\parallel + \left\{\partial_z - \frac{1}{B_o}[A_\parallel,]\right\} \left(\phi - \frac{T_e}{e}\widetilde{n}\right) + \frac{T_e \kappa_n}{e B_o} \partial_y A_\parallel = 0 \qquad (10.61)$$

where $\nabla_\perp^2 \equiv \partial_x^2 + \partial_y^2$ and the directional derivatives $\boldsymbol{v}_E \cdot \nabla$ and $\delta \boldsymbol{B} \cdot \nabla$ give $[f, g] \equiv \partial_x f \partial_y g - \partial_y f \partial_x g$ the Poisson bracket. The Alfvén speed v_A in cgs and mks units is

$$v_A \equiv \sqrt{\frac{B_o^2}{4\pi n_o m_i}}, \quad \text{in cgs Alfvén velocity}$$

$$v_A \equiv \frac{B_0}{(\mu_0 n_0 m_i)^{1/2}} \quad \text{in mks units}$$

for ions of mass m_i in grams and kilograms, respectively. The equilibrium number density of ions and electrons is $n_o(x)$ in cm^{-3} and m^{-3}, respectively. The electron temperature T_e is in electron volts.

The local constant density gradient is defined as $\kappa_n \equiv -d \ln n_o / dx = -1/L_n$.

In Eq. (10.59)-Eq. (10.61) we include the nonlinear terms that arise from the ion polarization nonlinearity, the $\boldsymbol{E} \times \boldsymbol{B}_o$ drift fluid convection, and the coupling of the derivative along the perturbed magnetic field $\boldsymbol{B}\cdot\boldsymbol{\nabla}$ with the parallel electron flow through the $j_\parallel \propto \nabla_\perp^2 A_\parallel$.

Kadomtsev Gauge for Parallel Electric Field

A standard formulation of electromagnetic plasma waves is to introduce a scalar potential function ψ which describes the axial perturbed induction electric field from Faraday's law $E_z(x,y,z,t) = -\partial_z\psi = -\partial A_\parallel/\partial t$. Then we find the relation between induction electric field. The electrostatic electric field and new potential field ψ is that $A_\parallel$ and ϕ, ψ are related by

$$\frac{1}{c}\partial_t A_\parallel = -\partial_z(\phi - \psi). \tag{10.62}$$

The potential ψ was introduced by Kadomtsev to describe the approach to ideal MHD when $\|\psi/\phi\| \to 0$.

Suppose that the perturbation functions ϕ, ψ, and $\widetilde{n}$ are in the form of a traveling helical disturbance

$$\phi(x,y,z,t) = \phi(x,\eta)$$
$$\psi(x,y,z,t) = \psi(x,\eta)$$
$$\widetilde{n}(x,y,z,t) = \widetilde{n}(x,\eta)$$

where coordinate n follows the helical structure is $\eta = y + \alpha z - ut$, with $u, \alpha =$ consts. Substituting Eq. (10.62) and Eq. (10.63) into Eq. (10.59)-Eq. (10.61) yields the reduced vortex equations

$$\widehat{L}_1\nabla_\perp^2\phi - \frac{v_A^2\alpha^2}{u^2}\widehat{L}_2\nabla_\perp^2(\phi - \psi) = 0 \tag{10.63}$$

$$\widehat{L}_1\widetilde{n} - \frac{e}{T_e}\frac{\alpha^2 v_A^2}{u^2}\rho_s^2\widehat{L}_2\nabla_\perp^2(\phi - \psi) - \frac{e}{T_e}\frac{v_{de}}{u}\partial_\eta\phi = 0 \tag{10.64}$$

$$\widehat{L}_2\left(\psi - \frac{T_e}{e}\widehat{n}\right) + \frac{v_{de}}{u}\partial_\eta(\phi - \psi) = 0 \tag{10.65}$$

where the Lie derivative operators are $\widehat{L}_1 = \partial_\eta - c/B_o u[\phi,]$ $\widehat{L}_2 = \partial_\eta - c/B_o u[\phi - \psi,]$. The diamagnetic drift velocity of electrons is displaced $v_{de} \equiv cT_e\kappa_n/eB_o$ and the spatial dispersion scale is $\rho_s^2 \equiv T_e/m_i\omega_{ci}^2$ with the ion gyrofrequency $\omega_{ci} \equiv eB_o/m_i c$. Equation (10.65) has the exact solution

$$\frac{T_e}{e}\widetilde{n} = \psi + \frac{v_{de}}{u}(\phi - \psi). \tag{10.66}$$

Substituting Eq. (10.66) into Eq. (10.64) and eliminating $\widetilde{n}$ gives

$$\widehat{L}_2\left\{\psi - \rho_s^2\frac{\alpha^2 v_A^2}{u(u - v_{de})}\nabla_\perp^2(\phi - \psi)\right\} = 0. \tag{10.67}$$

Using the identity $\widehat{L}_1\psi = \widehat{L}_2\psi$ and substituting Eq. (10.67) into Eq. (10.68) yields

$$\widehat{L}_1 \left\{ \psi - \rho_s^2 \left(1 - \frac{v_{de}}{u}\right)^{-1} \nabla_\perp^2 \phi \right\} = 0. \tag{10.68}$$

By using the property of the Poisson bracket we can write Eq. (10.66) and Eq. (10.67) as

$$\left[\phi - \psi - \frac{B_o u}{c} x,\ \psi - \rho_s^2 \frac{\alpha^2 v_A^2}{u(u - v_{de})} \nabla_\perp^2 (\phi - \psi)\right] = 0 \tag{10.69}$$

and

$$\left[\phi - \frac{B_o u}{c} x,\ \psi - \rho_s^2 \left(1 - \frac{v_{de}}{u}\right)^{-1} \nabla_\perp^2 \phi\right] = 0. \tag{10.70}$$

The general solutions of nonlinear equations Eq. (10.69) and Eq. (10.70) are

$$\psi - \frac{\alpha^2 v_A^2}{u(u - v_{de})} \rho_s^2 \nabla_\perp^2 (\phi - \psi) = f_1 \left(\phi - \psi - \frac{B_o u}{c} x\right) \tag{10.71}$$

$$\psi - \rho_s^2 \left(1 - \frac{v_{de}}{u}\right)^{-1} \nabla_\perp^2 \phi = f_2 \left(\phi - \frac{B_o u}{c} x\right) \tag{10.72}$$

where $f_1(w), f_2(w)$ are arbitrary nonsingular functions of their arguments.

For the construction of the Rossby wave vortices, we consider the case where f_1, f_2 are linear functions of their arguments

$$f_1(w) = C_1 w; \qquad f_2(w) = C_2 w. \tag{10.73}$$

Following the standard procedure [Larichev, *et al.* (1976); Meiss and Horton (1983)], we introduce the polar coordinates defined by $r = \sqrt{x^2 + \eta^2}, \qquad \theta = \tan^{-1} \eta/x$ and look for the solutions of Eq. (10.71) and Eq. (10.72) in two regions divided by a circle $r = a$ in the $r-\theta$-plane: the inner region ($r < a$) and the outer region ($r > a$).

Outer and Inner Boundary Conditions

From the requirement of locality, we impose the asymptotic conditions,

$$\phi \to 0,\ \psi \to 0, \qquad \text{as } r \to \infty \tag{10.74}$$

to the solutions in the outside region. From the requirement of regularity we impose the conditions

$$\phi,\ \psi \text{ finite as } \quad r \to 0 \tag{10.75}$$

in the inner region.

Outer Solutions

Condition Eq. (10.74) implies that for the outer region, after substituting Eq. (10.73) into Eq. (10.71) and Eq. (10.72), C_1 and C_2 must be zero. Therefore, for $r > a$ we have the coupled equations

$$\psi - \left(1 - \frac{v_{de}}{u}\right)^{-1} \left(\frac{\alpha v_A}{u}\right)^2 \rho_s^2 \nabla_\perp^2 (\phi - \psi) = 0 \tag{10.76}$$

$$\psi - \left(1 - \frac{v_{de}}{u}\right)^{-1} \rho_s^2 \nabla_\perp^2 \phi = 0. \tag{10.77}$$

For the inner region, the only physical requirement imposed on the solution is Eq. (10.75), which does not imply an explicit requirement on the choice of integration constants C_1 and C_2.

Basic Inner Solutions

For simplicity consider solutions

$$C_1 = 0, \qquad C_2 \neq 0 \tag{10.78}$$

in the inner region. For this choice Eq. (10.71) and Eq. (10.72) for $r < a$ reduce to

$$\psi - \left(1 - \frac{v_{de}}{u}\right)^{-1} \left(\frac{\alpha v_A}{u}\right)^2 \rho_s^2 \nabla_\perp^2 (\phi - \psi) = 0 \tag{10.79}$$

$$\psi - \left(1 - \frac{v_{de}}{u}\right)^{-1} \rho_s^2 \nabla_\perp^2 \phi = C_2 \left(\phi - \frac{B_o u}{c} r \cos\theta\right). \tag{10.80}$$

Eliminating the ψ Field

Eliminating ψ from Eq. (10.76)-Eq. (10.77) and Eq. (10.79)-Eq. (10.80), we have fourth order equations for ϕ in both the inner and outer regions:

$$\nabla_\perp^4 \phi - \frac{1}{\rho_s^2} \left(1 - \frac{v_{de}}{u}\right) - \left(\frac{u}{\alpha v_A}\right)^2) \nabla_\perp^2 \phi = 0, \qquad (r > a) \tag{10.81}$$

$$\nabla_\perp^4 \phi + \frac{1}{\rho_s^2} \left(1 - \frac{v_{de}}{u}\right) \left[\left(\frac{u}{\alpha v_A}\right)^2 - 1 + C_2\right] \nabla_\perp^2 \phi +$$

$$+ \frac{1}{\rho_s^4} C_2 \left(1 - \frac{v_{de}}{u}\right)^2 \left(\frac{u}{\alpha v_A}\right)^2 \left(\phi - \frac{B_o u}{c} r \cos\theta\right) = 0. \qquad (r < a). \tag{10.82}$$

Equation (10.81) and Eq. (10.82) with boundary conditions Eq. (10.74)-Eq. (10.75) form a boundary value problem which belongs to the type 1 problem that we solved in Section 10.3.2.

Equation (10.52a) with the corresponding condition Eq. (10.54) as the inner solution is used. Substituting the corresponding coefficients of Eq. (10.81) and

Eq. (10.82) into Eq. (10.50) and Eq. (10.52a)-Eq. (10.52b) and using Eq. (10.77) and Eq. (10.80) gives the special solutions of the nonlinear equations Eq. (10.68) and Eq. (10.69) under assumptions Eq. (10.44), Eq. (10.54), Eq. (10.73), and Eq. (10.78).

The solutions of the vortex equations parameterized by C_2 are

$$\phi_{\text{out}} = \left\{A_1 K_1(\nu r) + A_2 \frac{1}{r}\right\} \frac{B_o u}{c} a \cos\theta, \qquad (r > a)$$

$$\phi_{\text{in}} = \left\{A_3 J_1(\lambda r) + A_4 I_1(\mu r) + \frac{r}{a}\right\} \frac{B_o u}{c} a \cos\theta, \; (r < a) \tag{10.83}$$

and

$$\psi_{\text{out}} = A_1 \left[1 - \left(\frac{u}{\alpha v_A}^2\right)\right] K_1(\nu r) \frac{B_o u}{c} a \cos\theta, \qquad (r > a)$$

$$\psi_{\text{in}} = \left\{\left(C_2 - \frac{\rho_s^2 \lambda^2}{1 - v_{de}/u}\right) A_3 J_1(\lambda r) + \left(C_2 + \frac{\rho_s^2 \mu^2}{1 - v_{de}/u}\right) A_4 I_1(\mu r)\right\} \frac{B_o u}{c} \cos\theta$$

$$(r < a) \tag{10.84}$$

where

$$\nu^2 = \frac{1}{\rho_s^2}\left(1 - \frac{v_{de}}{u}\right)\left[1 - \left(\frac{u}{\alpha v_A}\right)^2\right] \tag{10.85}$$

$$\lambda^2 = \frac{1 - v_{de}/u}{2\rho_s^2} \left\{ \left\{ \left[C_2 - 1 + \left(\frac{u}{\alpha v_A}\right)^2\right]^2 - 4\left(\frac{u}{\alpha v_A}\right)^2 \right\}^{\frac{1}{2}} \right\}$$

$$+ \left[C_2 - 1 + \left(\frac{u}{\alpha v_A}\right)^2\right] \Bigg\} \tag{10.86}$$

$$\mu^2 = \frac{1 - v_{de}/u}{2\rho_s^2} \left\{ \left[C_2 - 1 + \left(\frac{u}{\alpha v_A}\right)^2\right]^2 - 4\left(\frac{u}{\alpha v_A}\right)^2 \right\}^{\frac{1}{2}}$$

$$- \left[C_2 - 1 + \left(\frac{u}{\alpha v_A}\right)^2\right] \Bigg\}. \tag{10.87}$$

Equation (10.83)-Eq. (10.84) are localized vortices with dipole structure. Now we still need to determine the integration constants A_1, A_2, A_3, A_4 by matching the solutions in both regions on the border $r = a$. For solutions with continuous density, flow velocity, magnetic field, vorticity and current density, we impose following matching conditions:

$$\phi_{\text{in}} \mid_{r=a} = \phi_{\text{out}} \mid_{r=a} \tag{10.88a}$$

$$\frac{\partial \phi_{\text{in}}}{\partial r} \mid_{r=a} = \frac{\partial \phi_{\text{out}}}{\partial r} \mid_{r=a} \tag{10.88b}$$

$$\nabla_\perp^2 \phi_{\text{in}} \mid_{r=a} = \nabla_\perp^2 \phi_{\text{out}} \mid_{r=a} \tag{10.88c}$$

and

$$\psi_{\text{in}} \mid_{r=a} = \psi_{\text{out}} \mid_{r=a} \tag{10.89a}$$

$$\frac{\partial \psi_{\text{in}}}{\partial r} \mid_{r=a} = \frac{\partial \psi_{\text{out}}}{\partial r} \mid_{r=a} \tag{10.89b}$$

$$\nabla_\perp^2 \psi_{\text{in}} \mid_{r=a} = \nabla_\perp^2 \psi_{\text{out}} \mid_{r=a} . \tag{10.89c}$$

Substituting Eq. (10.83)-Eq. (10.84) into Eq. (10.88a)-Eq. (10.89c) gives

$$A_1 = 2\frac{\lambda^2 + \mu^2}{K_1(\nu a)\Delta} \tag{10.90}$$

$$A_2 = (1 - 2\frac{\lambda^2 + \mu^2}{K_1(\nu a)\Delta})a \tag{10.91}$$

$$A_3 = -2\frac{\nu^2}{J_1(\lambda a)\Delta} \tag{10.92}$$

$$A_4 = 2\frac{\nu^2}{I_1(\mu a)\Delta} \tag{10.93}$$

where

$$\Delta = (\lambda^2 + \mu^2)\left[2 - \frac{\nu a K_2(\nu a)}{K_1(\nu a)}\right] - \nu^2 \left[\frac{\lambda a J_2(\lambda a)}{J_1(\lambda a)} + \frac{\mu a I_2(\mu a)}{I_1(\mu a)}\right] . \tag{10.94}$$

According to the choice of the integration constants C_1, C_2 in Eq. (10.82), from Eq. (10.78) and Eq. (10.79) we can see that the condition (10.57.c) is satisfied automatically after conditions (10.56.a)-(10.57.a) are satisfied.

Nonlinear Vortex Dispersion Relation

The matching conditions Eq. (10.88b) and Eq. (10.89b) give the following relation between the vortex parameters ν, λ, μ

$$(\mu^2 + \lambda^2)\nu\frac{K_2(\nu a)}{K_1(\nu a)} + (\nu^2 - \mu^2)\lambda\frac{J_2(\lambda a)}{J_1(\lambda a)} + (\nu^2 + \lambda^2)\mu\frac{I_2(\mu a)}{I_1(\mu a)} = 0. \tag{10.95}$$

At this stage the special solutions of nonlinear equation Eq. (10.71) and Eq. (10.70) are specified. There are now only three free parameters u, a, α. All other parameters A_i, μ, λ, ν are now determined.

These electromagnetic solitary vortex solutions ϕ and ψ are continuous on the boundary up to their second derivatives, therefore both the perturbed perpendicular magnetic field

$$\delta \boldsymbol{B}_\perp = \nabla_\perp A_\| \times \widehat{\boldsymbol{z}} = -\frac{\alpha c}{u} \widehat{\boldsymbol{z}} \times \nabla_\perp (\phi - \psi)$$

and the parallel plasma current $j_\| = -c/4\pi \nabla_\perp^2 A_\| = -c^2 \alpha / 4\pi u \nabla_\perp^2 (\phi - \psi)$ are continuous across the boundary $r = a$. Since the discontinuity encountered in the Rossby-type vortices is eliminated in these vortices, we call them the intrinsic drift-Alfvén vortices.

The constant parameter u which describes the propagation speed of the vortices is constrained by the nonlinear dispersion relation of the vortices

$$\nu^2 = \left(1 - \frac{v_{de}}{u}\right) \left[1 - \left(\frac{u}{\alpha v_A}\right)^2\right] / \rho_s^2 > 0 \tag{10.96}$$

and is further restricted by the condition

$$C_2 \left(1 - \frac{v_{de}}{u}\right) \left(\frac{u}{\alpha v_A}\right)^2 < 0 \tag{10.97}$$

which arises from the choice of the inner solution form.

At the limit $\kappa_n = 0$ the equations describing drift-Alfvén vortex reduce to the equations describe Alfvén vortex. Accordingly, from Eq. (10.96) the nonlinear dispersion relation for Alfvén vortices becomes

$$\alpha^2 v_A^2 - u^2 > 0. \tag{10.98}$$

Therefore, the vortex structure speed u is limited to the range

$$-\alpha v_A < u < \alpha v_A. \tag{10.99}$$

The further restriction corresponding to Eq. (10.99) becomes that the parameter C_2, defined in Eq. (10.73), must be negative:

$$C_2 < 0 \tag{10.100}$$

when $v_{de} = 0$. For $0 < u < v_{de}$ there are solutions with $c_2 > 0$.

10.4.4 *Electromagnetic electron skin depth vortices and electromagnetic short-wavelength drift vortices*

On the small space scales of the electron skin depth there are electron vortices corresponding to the electron scale electromagnetic turbulence. Mikhailovskii, *et al.* (1985) derived two sets of nonlinear equations for these small scale structures. For the details of the derivation and validity condition of the equations we refer to Mikhailovskii, *et al.* (1985) and Liu and Horton (1986b). These are the natural vortices from unstable ETG modes and inertial Alfvén waves in high-density plasma with $\omega_{pe} > \omega_{ce}$.

In terms of the perturbation fields ϕ and ψ we defined in Eq. (10.62) the first set of equations which describes the electromagnetic electron vortices is

$$\widehat{L}_1\nabla_\perp^2\phi - \left(\frac{\omega_{pe}^2}{c^2} - \frac{\kappa_n\omega_{ce}}{u}\right)\partial_\eta\phi - \left(\frac{\alpha c_{Ae}}{u}\right)^2\widehat{L}_2\nabla_\perp^2(\phi-\psi) = 0 \tag{10.101}$$

$$\widehat{L}_1\left[\nabla_\perp^2(\phi-\psi) + \left(\frac{\omega_{pe}}{c}\right)^2\psi\right] = 0 \tag{10.102}$$

where $\omega_{pe} = (4\pi n_o e^2/m_e)^{1/2}$, is the electron plasma frequency and $\omega_{ce} = eB_o/m_e c$, is the electron gyrofrequency. The electron scale electromagnetic disturbances propagate at the speed given by

$$c_{Ae} \equiv \sqrt{\frac{B_o^2}{4\pi m_e n_o}} = \frac{c\omega_{ce}}{\omega_{pe}} \tag{10.103}$$

the Alfvén velocity calculated by electron mass. Other notations are the same as in Section 10.3.3.

The set of equations which describes the electromagnetic short-wavelength drift vortices is

$$\widehat{L}_1\nabla_\perp^2\phi - \frac{1}{\rho_{ei}^2}\left(1 - \frac{v_{di}}{u}\right)\frac{\partial\phi}{\partial\eta} - \left(\frac{\alpha v_A}{u}\right)^2\widehat{L}_2\nabla_\perp^2(\phi-\psi) = 0 \tag{10.104}$$

$$\widehat{L}_1\left\{\nabla_\perp^2(\phi-\psi) + \left(\frac{\omega_{pe}}{c}\right)\right\}^2\psi\} = 0 \tag{10.105}$$

where

$$v_{di} \equiv -\frac{cT_i\kappa_n}{eB_o} \tag{10.106}$$

is the ion diamagnetic drift velocity and

$$\rho_{ei}^2 = \frac{T_i}{m_e\omega_{ce}^2} \tag{10.107}$$

is the electron gyroradius taken at the ion temperature.

Adiabatic Ion Response

For these small scale vortices in plasma with finite ion temperature, T_i the ion density response, is adiabatic with

$$\widetilde{n}_i = -n_0\left(\frac{e_i\phi}{T_i}\right) \tag{10.108}$$

at these small space scales. The similarity between these two sets of equations is obvious: Eq. (10.102) and Eq. (10.48) are the same; Eq. (10.104) and Eq. (10.109) have the same structures, only the constant coefficients of the second and third terms for them are different. In the following we only present the details of the first set of equations: the solutions to the second being obtained simply by changing the coefficients.

Following the same procedure given in the Section 10.3.4, we can find the equivalent linear equations for Eq. (10.104) and Eq. (10.105) as

$$\nabla_\perp^2(\phi-\psi)+\frac{\omega_{pe}^2}{c^2}\psi - C_1\left(\phi-\frac{B_o u}{c}x\right)=0 \tag{10.109}$$

$$\begin{aligned}\nabla_\perp^2\phi-\left(\frac{\omega_{pe}^2}{c^2}-\frac{\kappa_n\omega_{ce}}{u}\right)\phi+\left(\frac{\alpha c_{Ae}}{u}\right)^2\left(\frac{\omega_{pe}^2}{c^2}-C_1\right)\psi&\\ +C_2\left(\phi-\frac{B_o u}{c}x\right)=0&\end{aligned} \tag{10.110}$$

where C_1, C_2 are integration constants.

In the following subsection we give two types of solitary vortices solutions for Eq. (10.109) and Eq. (10.110) with the localization and regularity conditions Eq. (10.74)-Eq. (10.75).

Consider the drift wave-type of vortex solution of Eq. (10.109)-Eq. (10.110). The linear algebraic relations between the two functions ϕ and ψ in both the inner and outer regions are

$$\psi(r,\theta)=a_1\phi(r,\theta) \qquad (r>a) \tag{10.111}$$

$$\psi(r,\theta)=a_2\phi(r,\theta)+a_3\frac{B_o u}{c}r\cos\theta \qquad (r<a). \tag{10.112}$$

Substituting Eq. (10.111)-Eq. (10.112) into Eq. (10.109)-Eq. (10.110) yields

$$\nabla_\perp^2\phi-k^2\phi=0 \qquad (r>a) \tag{10.113}$$

$$\nabla_\perp^2\phi+p^2\phi+q\frac{B_o u}{c}r\cos\theta \qquad (r<a) \tag{10.114}$$

where a_1 is the real root of quadratic equation

$$a_1^2-\left(1+\frac{c^2}{\omega_{ce}^2}\left(\frac{u}{\alpha v_A}\right)^2\frac{\kappa_n\omega_{ce}}{u}\right)a_1-\left(\frac{u}{\alpha v_{Ae}}\right)^2\left(1-\frac{c^2}{\omega_{ce}^2}\frac{\kappa_n\omega_{ce}}{u}\right)=0 \tag{10.115}$$

and

$$\begin{aligned}k^2=\frac{\omega_{pe}^2}{c^2}\Bigg\{1+\frac{1}{2}\Bigg[\left(\frac{\alpha v_{Ae}}{u}\right)^2-\frac{c^2}{\omega_{pe}^2}\frac{\kappa_n\omega_{ce}}{u}\pm\Bigg(\left(\left(\frac{\alpha v_{Ae}}{u}\right)^2-\frac{c^2}{\omega_{pe}^2}\frac{\kappa_n\omega_{ce}}{u}\right)^2&\\ +4\left(\frac{\alpha v_{Ae}}{u}\right)^2\Bigg]^{1/2}\Bigg\}>0&\end{aligned} \tag{10.116}$$

$$p^2=\frac{\omega_{pe}^2}{c^2}\left\{\left(\frac{u}{\alpha v_{Ae}}\right)^2\frac{a_2+a_3+(1-a_2)\left(1-\frac{c^2}{\omega_{pe}^2}\frac{\kappa_n\omega_{ce}}{u}\right)}{(a_2+a_3)(1-a_2)^2}-1\right\}>0 \tag{10.117}$$

$$q = \frac{\omega_{pe}^2}{c^2}\left\{\frac{a_3+1}{1-a_2} - \left(\frac{u}{\alpha v_{Ae}}\right)^2 \frac{a_2+a_3+(1-a_2)\left(1-\frac{c^2}{\omega_{pe}^2}\frac{\kappa_n\omega_{ce}}{u}\right)}{(a_2+a_3)(1-a_2)^2}\right\}. \tag{10.118}$$

Following the standard procedure from Section 10.3.4, the solution of Eq. (10.115) and Eq. (10.116) is given by

$$\phi_{\text{out}} = A_1 \frac{K_1(kr)}{K_1(ka)} \frac{B_o u}{c} a\cos\theta \qquad (r > a)$$

$$\phi_{\text{in}} = \left\{A_2 \frac{J_1(pr)}{J_1(pa)}\frac{a}{r} - \frac{q}{p^2}\right\} \frac{B_o u}{c} r\cos\theta \qquad (r < a). \tag{10.119}$$

Inserting Eq. (10.119) into Eq. (10.116)-Eq. (10.110) gives the solution of ψ as

$$\psi_{\text{out}} = a_1 A_1 \frac{K_1(kr)}{K_1(ka)} \frac{B_o u}{c} a\cos\theta \qquad (r > a)$$

$$\psi_{\text{in}} = \left\{a_2 A_2 \frac{J_1(pr)}{J_1(pa)}\frac{a}{r} + \left(-a_2\frac{q}{p^2} + a_3\right)\right\} \frac{B_o u}{c} r\cos\theta \ (r < a). \tag{10.120}$$

Matching the inner and outer solutions on the boundary $r = a$ by requiring ϕ to be continuous up to its second derivatives gives

$$A_1 = -\frac{q}{p^2+k^2} \tag{10.121}$$

$$A_2 = \frac{k^2}{p^2+k^2}\frac{q}{p^2} \tag{10.122}$$

and the relation

$$\frac{1}{ka}\frac{K_2(ka)}{K_1(ka)} + \frac{1}{pa}\frac{J_2(pa)}{J_1(pa)} = 0. \tag{10.123}$$

Equation (10.119)-Eq. (10.120) with the coefficients given by Eq. (10.121)-Eq. (10.123) give the usual solitary vortex solutions for Eq. (10.101)-Eq. (10.102), but with five parameters a_2, a_3, u, a, α. In fact a_2, a_3 are determined by u, a, α in the following way: given u, a, α one determines k, p from Eq. (10.109) and Eq. (10.115); then solving Eq. (10.110) and the continuity condition of ψ on the boundary

$$a_1\phi_{\text{out}}\mid_a = a_2\phi_{\text{in}}\mid_a + a_3\frac{B_o u}{c} a\cos\theta. \tag{10.124}$$

Taking consideration of equation Eq. (10.111)-Eq. (10.112), finally determines a_2, a_3.

10.4.5 *Comparisons of the electromagnetic vortices*

The analysis shows that there are two intrinsic drift-Alfvén vortices. Let us compare some of the properties of these vortices. First we observe that changing the coefficients

$$\frac{\omega_{pe}^2}{c^2} - \frac{\kappa_n \omega_{ce}}{u} \to \frac{1}{\rho_{ei}^2}\left(1 - \frac{v_{di}}{u}\right); \quad v_{Ae} \to v_A$$

in Eq. (10.115)-Eq. (10.117). The solutions Eq. (10.111)-Eq. (10.112) gives the intrinsic electromagnetic short-wavelength drift vortices.

(1) Common features

The intrinsic electromagnetic solitary vortices share the basic feature of the Rossby wave vortices: both vortices have localized dipole structure and move in the y-x-plane with constant speed u at an angle $\gamma = \tan^{-1}(1/\alpha)$ with respect to the ambient magnetic field direction without changing their shapes. The localized structures decay to zero as $r \to \infty$. For the electromagnetic electron vortices (and electromagnetic short-wavelength drift vortices) the asymptotic form as $r \to \infty$ for both Rossby and intrinsic vortices is $e^{-\lambda r}/\sqrt{r}$, where for Rossby-type vortices $\lambda = k$, whereas for intrinsic vortices $\lambda = \min\{\lambda_1, \lambda_2\}$ with λ_1, λ_2 given by Eq. (10.116).

Both the drift wave-Alfvén vortices and the skin depth scale vortices, in contrast to the exponentially localized electrostatic drift wave vortices, decay slowly to zero as $\varphi \sim A_\parallel 1/r$ for $r \to \infty$ [Mikhailovskii, *et al.* (1984); Shukla, *et al.* , (1985a,b)].

(2) Differences in Structure

Comparing the drift-Alfvén vortices with the corresponding electrostatic drift wave vortices, we find that the intrinsic vortices have a more complicated radial structure in both inner and outer regions. In the inner region, the amplitude of the intrinsic vortices is a linear combination of both regular and modified Bessel functions J_1 and I_1. In the outer region of the Alfvénic electron scale vortices contain both the first-order McDonald functions. In contrast, the amplitude of the electrostatic vortices is simply $J_1(pr)$ in the inner region and $K_1(kr)$ in the outer region.

(3) Different allowed regions of the vortex propagation speed

A key difference between intrinsic electromagnetic vortices and their corresponding Rossby vortices is that they have different allowed regions of vortex propagation. For electrostatic vortices, the allowed region of vortex propagation is determined by the single nonlinear dispersion relation condition

$$k^2 > 0. \tag{10.125}$$

In contrast, the expressions for Eq. (10.124) for Alfvén and drift-Alfvén vortices, given in works by Mikhailovskii, *et al.* (1984), Shukla, *et al.* (1985a,b), are given by Eq. (10.101) and Eq. (10.98). For electromagnetic electron vortices (or electromagnetic short-wavelength drift vortices), $k^2 > 0$ in Eq. (10.125). Thus, we find that the allowed regions for electromagnetic vortices is narrower than the allowed region for the electrostatic vortices. In conclusion, there are four types of electromagnetic solitary vortices in plasma [Liu and Horton (1986a)]. These vortices are natural self-organized physical states of the electromagnetic collective modes of the magnetized plasma.

References

Antar, G. Y., Assas, S., Bobkov, V., Noterdaeme, J.-M., Wolfrum, E., Herrmann, A., Rohde, V., and ASDEX Upgrade Team. (2010). Convective transport suppression in the scrape-off layer using Ion cyclotron resonance heating on the ASDEX Upgrade Tokamak, *Phys. Rev. Lett.* **105**, p. 165001, doi:10.1103/PhysRevLett.105.165001.

Antar, G. Y., Counsell, G., Ahn, J. W., Yang, Y., Price, M., Tabasso, A., and Kirk, A. (2006). The poloidal distribution of turbulent fluctuations and avaloids in the Mega Ampere Spherical Tokamak (MAST), *Phys. Plasmas* **13**, p. 052508, doi:10.1063/1.2198210.

Antar, G. Y. (2003). On the origin of "intermittency" in the scrape-off layer of linear magnetic confinement devices, *Phys. Plasmas* **10**, 9, p. 3629, doi:10.1063/1.1599855.

Antar, G. Y.,. Krasheninnikov, S. I., Devynck, P., Doerner, R. P., Hollmann, E. M., Boedo, J. A., Luckhardt, S. C., and Conn, R. W. (2001). Experimental evidence of intermittent convection in the edge of magnetic confinement devices, *Phys. Rev. Lett.* **87**, p. 065001, doi:10.1103/PhysRevLett.87.065001.

Antar, G. Y., Devynck, P., Garbet, X., and Luckhardt, S. C. (2001). Turbulence intermittency and burst properties in tokamak scrape-off laye, *Phys. Plasmas* **8**, p. 1612, doi:10.1063/1.1363663.

Balmforth, N. J. and Morrison, P. J. (2002). Hamiltonian Description of Shear Flow, *Large-Scale Atmosphere-Ocean Dynamics II*, eds. J. Norbury and I. Roulstone (Cambridge, Cambridge) pp. 117-142.

Burrell, K. H. (1997). Effects of $\boldsymbol{E} \times \boldsymbol{B}$ velocity shear and magnetic shear on turbulence and transport in magnetic confinement devices *Phys. Plasmas* **4**, 5, p. 1499.

Hazeltine, R. D., Holm, D. D., and Morrison, P. J. (1985). Electromagnetic solitary waves in magnetized plasmas, *J. Plasma Phys.* **34**, pp. 103-114, doi:10.1017/S0022377800002713.

Horton, W., Su, X. N, and Morrison, P. J. (1990). Drift wave vortices and turbulent transfer, *Fizika Plazmy* **16**, Aug. 1990, pp. 969-977. In Russian Translation (ISSN 0367-2921).

Horton, W., Tajima, T., and Kamimura, T. (1987). Kelvin-Helmholtz instability and vortices in magnetized plasma, *Phys. Fluids* **30**, pp. 3485-3495, `http://link.aip.org/link/doi/10.1063/1.866429`.

Horton, W. (1989). Drift wave vortices and anomalous transport, *Phys. Fluids B* **1**, p. 524, `http://dx.doi.org/10.1063/1.859168`.

Horton, W., and Ichikawa, Y. (2000). *Chaos and Structures in Nonlinear Plasmas* (World Scientific).

Horton, W., Kim, J.-H., Chagelishvili, G. D., Bowman, J. C., and Lominadze, J. G. (2010). Angular redistribution of nonlinear perturbations: A universal feature of nonuniform flows, *Phys. Rev. E*, **81**, p. 066304.

Jha, R., Kwa, P. K., Mattoo, S. K., Rao, C. V. S., Saxena, Y. C., and ADITYA Team. (1992). *Phys. Rev. Lett.* **69**, p. 1375.

Kim, J.-H., Perez, J. C., Horton, W., Chagelishvili, G. D., Changishvili, R. G., Lominadze, J. G., and Bowman, J. C. (2006). Self-sustaining vortex perturbations in smooth shear flows, *Phys. Plasmas* **13**, 6, Article No. 062304.

Krasheninnikov, S. I. (2001). On scrape-off layer plasma transport, *Phys. Lett. A* **283**, p. 368, doi:10.1016/S0375-9601(01)00252-3.

Larichev, V. D. and Reznik, G. M. (1976). *Dokl. Akad. Nauk USSR* **231**, p. 1077.

Lochte-Holtgreven, W. (Ed.) (1968) American Vacuum Society Classics, *Plasma Diagnostics*, OSTI Identifier: 4798641.

Liu, J. and Horton, W. (1986a). Electromagnetic solitary vortices in rotating plasma, *Phys. Fluids* **29**, pp. 1828-1835, `http://link.aip.org/link/doi/10.1063/1.865611`.

Liu, J. and Horton, W. (1986b). Electromagnetic vortices in magnetized plasma, *J. Plasma Phys.* **36**, pp. 1-24.

Maurel, A. and Petitjeans, P. (1999). *Vortex Structure and Dynamics* (Workshop, Rouen, France, Springer).

Meiss, J. D. and Horton, W. (1983). *Phys. Fluids* **26**, p. 990, `http://link.aip.org/link/doi/10.1063/1.864251`.

Mikhailovskii, A. B., Aburdzhaniya, G. D., Onishchenko, O. G., and Churikov, A. P. (1984). *Phys. Lett.* **101A**, p. 263.

Mikhailovskii, A. B., Lakhin, V. P., Mikhailovskaya, L. A., and Onishchenko, O. G. (1985). *Sov. Phys. JETP* **59**, p. 1198.

Monin, A. S. and Yaglom, A. M. (1965). *Statistical Fluid Mechanics* (MIT Press).

Morrison, P. J. (1998). Hamiltonian description of the ideal fluid, *Rev. Mod. Phys.* **70**, pp. 467-521.

Perez, J. C., Horton, W., Bengtson, R. D., and Carter, T. (2006). Study of strong cross-field sheared flow with vorticity probe in the large plasma device, *Phys. Plasmas* **13**, p. 057701, `http://link.aip.org/link/doi/10.1063/1.2179423`.

Perez, J. C., Horton, W., Gentle, K., Rowan, W. L., Lee, K., Dahlburg, R. B. (2006). Drift wave instability in the Helimak experiment, *Phys. Plasmas* **13**, 3, Article No. 032101, `http://link.aip.org/link/doi/10.1063/1.2168401`.

Schepp, T. J., Pegoraro, F., Kuvshinov, B. N. (1994). Generalized two-fluid theory of nonlinear magnetic structures, *Phys. Plasmas* **1**, p. 2843, `http://link.aip.org/link/doi/10.1063/1.870523`.

Shukla, P. K., Yu, M. Y., and Varma, R. K. (1985a). Formation of kinetic Alfvén vortices, *Phys. Lett.* **109A**, pp. 322-324, doi:10.1016/0375-9601(85)90183-5.

Shukla, P. K., Yu, M. Y., and Varma, R. K. (1985b). Drift-Alfvén vortices, *Phys. Fluids* **28**, p. 1719, `http://link.aip.org/link/doi/10.1063/1.864964`.

Su, X. N., Horton, W., and Morrison, P. J. (1991). Drift wave vortices in inhomogeneous plasmas, *Phys. Fluids B* **3**, pp. 921-930.

Su, X. N., Horton, W., and Morrison, P. J. (1992). Drift wave vortices in nonuniform plasmas with sheared magnetic fields, *Phys. Fluids B* **4**, pp. 1238-1246.

Wilbert, C. E. and Gharib, M. (1991). *Exp. Fluids* **20**, pp. 181-193.

Chapter 11

Fluctuating Magnetic Fields and Chaotic Orbits

In this chapter we develop the kinetic theory for the response of the plasma to fluctuating magnetic fields. Fluctuating magnetic fields often lead to the direct loss of electrons and electron thermal energy by the motion of the fast electrons along the magnetic field lines. The kinetic theory particle response $\delta f(\boldsymbol{x}, \boldsymbol{v}, t)$ is derived to general electromagnetic field fluctuations.

In Section 11.1-Section 11.2 we first derive the electromagnetic field equations governing the self-consistent dynamics of the magnetized particle orbits and the electric field $\delta \boldsymbol{E}(\boldsymbol{x}, t)$ and fluctuating magnetic field $\delta \boldsymbol{B}(\boldsymbol{x}, t)$. Then in Section 11.3 we derive the ion losses and in Section 11.4 the electron losses from turbulent transport in these fields.

Anisotropic electron phase space densities are required to maintain the plasma current density in the tokamak fusion system. For steady-state fusion plasmas the current maintenance and control is required for stability and performance optimization. Enhanced confinement regimes require specific current profiles. The two methods for driving the current profiles use the injection of multi-megawatts of high frequency radio frequency (RF) waves. The first most thoroughly validated method uses gigahertz lower hybrid waves launched with complex antennas designed to send the waves in the direction opposite to the plasma current. The wave-electron Landau resonances then transfer the wave momentum to the electrons producing the local current density $j_{\|} = -en_e v_{\|e}$. The method is effective and used to produce the record long steady-state discharges over several minutes in time in Tore Supra. These discharges pass gigajoules of energy through the plasma to the walls and the specially designed axisymmetric divertor, giving the only actual steady-state fusion-grade tokamak plasma operation tests in the period of 2000 to 2011. The second method to drive and control the plasma current is the use of electron cyclotron resonance (ECH) waves at over one hundred gigahertz frequency. These waves resonate with the electrons at the upper hybrid resonance layer and drive up the perpendicular electron kinetic energy. They have been used successfully in the DIII-D and TCV tokamaks to heat, drive current, and control the size of the lower order magnetic islands that appear particularly at the $q = 3/2$ and $q = 2$ magnetic surfaces. Both the LHCD and ECCD will be used in ITER to control the plasma.

These high-powered waves drive the electron velocity distributions away from the Maxwellian in the higher than thermal energy region called the superthermal energy distributions. Ion cyclotron resonance (ICRF) and neutral beam injection (NBI) have a similar action on the ion phase space density.

11.1 Kinetic Theory Formulas for the Fluctuating Electromagnetic Fields

The fluctuating phase space density $\delta f(\boldsymbol{x}, \boldsymbol{v}, t)$ in a magnetized plasma is given by the time-history integral specifying the effect of the fluctuation fields on the background distribution function

$$\left(\frac{\partial}{\partial t} + \boldsymbol{v} \cdot \boldsymbol{\nabla} + \frac{q}{m}\boldsymbol{v} \times \boldsymbol{B}_0 \cdot \frac{\partial}{\partial \boldsymbol{v}}\right) \delta f(\boldsymbol{x}, \boldsymbol{v}, t) = -\frac{q}{m}(\delta \boldsymbol{E} + \boldsymbol{v} \times \boldsymbol{\delta B}) \cdot \frac{\partial F(\boldsymbol{v})}{\partial \boldsymbol{v}}. \quad (11.1)$$

The plasma electric current and charge density are given by

$$\delta \boldsymbol{j} = \sum_s q_s \int d^3 v \boldsymbol{v} \delta f_s \quad (11.2)$$

$$\delta \rho = \sum_s q_s \int d^3 v \delta f_s \quad (11.3)$$

where s is the species index of the charged particles. The solution of Eq. (11.1) is

$$\delta f(\boldsymbol{x}, \boldsymbol{v}, t) = -\frac{q}{m} \int_{-\infty}^{t} dt' \left[\delta \boldsymbol{E}(\boldsymbol{x'}, \boldsymbol{v'}, t') + \boldsymbol{v}' \times \delta \boldsymbol{B}(\boldsymbol{x'}, \boldsymbol{v'}, t')\right] \cdot \frac{\partial F(\boldsymbol{v}')}{\partial \boldsymbol{v}'} \quad (11.4)$$

with $\boldsymbol{x}' = \boldsymbol{x}(t'), \boldsymbol{v}' = \boldsymbol{v}(t')$ being the orbits in the absence of the fluctuations specified by the derivatives on the left-hand side of Eq. (11.1).

Let the fluctuating fields vary as $\exp(i\boldsymbol{k} \cdot \boldsymbol{x}' - i\omega t')$. Maxwell's equations give $\delta \boldsymbol{B} = \boldsymbol{k}/\omega \times \delta \boldsymbol{E}$. Eliminating $\delta \boldsymbol{B}$ in Eq. (11.4) gives

$$\begin{aligned} \delta f(\boldsymbol{x}, \boldsymbol{v}, t) &= -\frac{q}{m} \int_{-\infty}^{t} dt' \exp(i\boldsymbol{k} \cdot \boldsymbol{x}' - i\omega t') \left[\delta \boldsymbol{E}(\boldsymbol{k}, \omega) + \boldsymbol{v}' \times \left(\frac{\boldsymbol{k}}{\omega} \times \delta \boldsymbol{E}\right)\right] \\ &\quad \cdot \frac{\partial F(\boldsymbol{v}')}{\partial \boldsymbol{v}'} \\ &= -\frac{q}{m\omega} \delta \boldsymbol{E} \cdot \int_{-\infty}^{t} dt' \exp(i\boldsymbol{k} \cdot \boldsymbol{x}' - i\omega t') \left[\boldsymbol{I}(\omega - \boldsymbol{v}' \cdot \boldsymbol{k}) + \boldsymbol{v}'\boldsymbol{k}\right] \cdot \frac{\partial F(\boldsymbol{v}')}{\partial \boldsymbol{v}'}. \end{aligned} \quad (11.5)$$

The unperturbed gyro-orbits expressed such that $\boldsymbol{x}' = \boldsymbol{x}$ and $\boldsymbol{v}' = \boldsymbol{v}$ at $t' = t$ are given by

$$\tau = t - t' \quad (11.6)$$

$$v_x = v_\perp \cos \phi \quad (11.7)$$

$$v_y = v_\perp \sin\phi \tag{11.8}$$

$$v'_x = v_\perp \cos(\phi + \Omega\tau) \tag{11.9}$$

$$v'_y = v_\perp \sin(\phi + \Omega\tau) \tag{11.10}$$

$$v'_z = v_\parallel \tag{11.11}$$

$$x' = x - \frac{v_\perp}{\Omega}\left[\sin(\phi + \Omega\tau) - \sin(\phi)\right] \tag{11.12}$$

$$y' = y + \frac{v_\perp}{\Omega}\left[\cos(\phi + \Omega\tau) - \cos(\phi)\right] \tag{11.13}$$

$$z' = z - v_\parallel \tau. \tag{11.14}$$

Using the orbits in Eq. (11.6)-Eq. (11.14), the fluctuating distribution function reduces to

$$\delta f = -\frac{q}{m\omega}\delta\boldsymbol{E} \cdot \int_0^\infty d\tau \ \exp\left(i\boldsymbol{k}\cdot\boldsymbol{x'} - i\omega t'\right)\left[\boldsymbol{I}(\omega - \boldsymbol{v'}\cdot\boldsymbol{k}) + \boldsymbol{v'}\boldsymbol{k}\right]\cdot\frac{\partial F(\boldsymbol{v'})}{\partial \boldsymbol{v'}}. \tag{11.15}$$

First we reduce the integrand of Eq. (11.15) to

$$\delta\boldsymbol{E} \ \cdot \ \left[\boldsymbol{I}(\omega - \boldsymbol{v'}\cdot\boldsymbol{k}) + \boldsymbol{v'}\boldsymbol{k}\right]\cdot\frac{\partial F(\boldsymbol{v'})}{\partial \boldsymbol{v'}} \tag{11.16}$$

$$= \delta E_z\left(\omega\frac{\partial F}{\partial v_z} - (k_x v_x + k_y v_y)\left(\frac{\partial F}{\partial v_z} - \frac{v_z}{v_\perp}\frac{\partial F}{\partial v_\perp}\right)\right). \tag{11.17}$$

The exponential phase function $\boldsymbol{k}\cdot\boldsymbol{x'} - \omega t'$ can be rewritten, using $k_x = k_\perp\cos\theta$ and $k_y = k_\perp \sin\theta$ as

$$\boldsymbol{k}\cdot\boldsymbol{x}' - \omega t' = \boldsymbol{k}\cdot\boldsymbol{x} - \omega t - \left\{k_\perp\cos\theta\frac{v_\perp}{\Omega}\left[\sin(\phi+\Omega\tau) - \sin\phi\right]\right\}$$
$$- \left\{k_\perp\sin\theta\frac{v_\perp}{\Omega}\left[\cos(\phi+\Omega\tau) - \cos\phi\right]\right\} - k_z v_\parallel\tau + \omega\tau$$
$$= \boldsymbol{k}\cdot\boldsymbol{x} - \omega t - \frac{k_\perp v_\perp}{\Omega}\left[\sin(\phi - \theta + \Omega\tau) - \sin(\phi-\theta)\right] + (\omega - k_\parallel v_\parallel)\tau. \tag{11.18}$$

Let us define the wave-particle phase function

$$\beta_k(\tau) = -\frac{k_\perp v_\perp}{\Omega}\left[\sin(\phi - \theta + \Omega\tau) - \sin(\phi - \theta)\right] + (\omega - k_\parallel v_\parallel)\tau \tag{11.19}$$

from Eq. (11.18). Then the phase becomes $\exp[i(\boldsymbol{k}\cdot\boldsymbol{x'} - \omega t')] = \exp[i(\boldsymbol{k}\cdot\boldsymbol{x} - \omega t) + i\beta_k(\tau)]$. With $\theta = 0(k_x = k_\perp, \ k_y = 0)$ the fluctuation phase is

$$\beta_k(\tau) = -\frac{k_\perp v_\perp}{\Omega}\left[\sin(\phi+\Omega\tau) - \sin\phi\right] + \left(\omega - k_\parallel v_\parallel\right)\tau. \tag{11.20}$$

The following Bessel function identity is now used

$$\exp(iz\sin\phi) = \sum_{n=-\infty}^{\infty} \exp(in\phi) J_n(z) \tag{11.21}$$

to carry out the τ-integral in Eq. (11.15). Later we will use the identities

$$\int_0^{2\pi} d\phi e^{-iz[\sin(\phi+\Omega\tau)-\sin(\phi)]} \begin{pmatrix} \sin\phi\sin(\phi+\Omega\tau) & 1 \\ \sin\phi\cos(\phi+\Omega\tau) & \sin\phi \\ \cos\phi\sin(\phi+\Omega\tau) & \cos\phi \\ & \\ \cos\phi\cos(\phi+\Omega\tau) & \sin(\phi+\Omega\tau) \end{pmatrix}$$

$$= 2\pi \sum_{n=-\infty}^{\infty} e^{-in\Omega\tau} \begin{pmatrix} (J_n')^2 \\ -\frac{in}{z} J_n J_n' \\ \frac{in}{z} J_n J_n' \\ \frac{n^2}{z^2} J_n^2 \\ J_n^2 \\ -iJ_n J_n' \\ \frac{n}{z} J_n^2 \\ iJ_n J_n' \\ \frac{n}{z} J_n^2 \end{pmatrix}. \tag{11.22}$$

The fluctuating plasma current is then

$$\delta\boldsymbol{j} = \sum_s q_s \int d^3v\boldsymbol{v}\delta f_s = -\sum_s \frac{\epsilon_0\omega_{ps}^2}{n_0\omega} \int_0^\infty \int_{-\infty}^\infty \int_0^{2\pi} v_\perp dv_\perp dv_\| d\phi$$

$$\times\boldsymbol{v}\int_0^\infty d\tau e^{-iz[\sin(\phi+\Omega\tau)-\sin(\phi)]} e^{-i(\omega-k_zv_z)\tau}$$

$$(\delta E_x v_\perp \cos(\phi+\Omega\tau) + \delta E_y v_\perp \sin(\phi+\Omega\tau))$$

$$\times\left(k_z\frac{\partial F}{\partial v_z} + \frac{1}{v_\perp}(\omega - k_z v_z)\frac{\partial F}{\partial v_\perp}\right). \tag{11.23}$$

The wave-particle gradient $\widehat{D}_k F$ is defined by

$$\widehat{D}_k(\boldsymbol{k},\omega,\boldsymbol{v})F = \left(k_z v_\perp \frac{\partial F}{\partial v_z} + (\omega - k_z v_z)\frac{\partial F}{\partial v_\perp}\right). \tag{11.24}$$

Then the components of the electric current are

$$\delta j_x = \sum_s \frac{\epsilon_0 \omega_{ps}^2}{n_0 \omega} \int_0^\infty \int_{-\infty}^\infty \int_0^{2\pi} v_\perp dv_\perp dv_\| d\phi \int_0^\infty d\tau e^{-iz[\sin(\phi+\Omega\tau)-\sin(\phi)]} v_\perp \cos\theta$$

$$(\delta E_x v_\perp \cos(\phi+\Omega\tau) + \delta E_y v_\perp \sin(\phi+\Omega\tau))\, \widehat{D}_k F$$

$$= -\sum_s \frac{\epsilon_0 \omega_{ps}^2}{n_0 \omega} \int_0^\infty \int_{-\infty}^\infty v_\perp dv_\perp dv_\|$$

$$\frac{2\pi i\, \widehat{D}_k F}{\omega - k_z v_z - n\Omega} \sum_{n=-\infty}^\infty \left(\frac{n^2\Omega^2}{k_\perp^2 v_\perp^2} v_\perp J_n^2 \delta E_x + \frac{in\Omega}{k_\perp v_\perp} v_\perp J_n J_n' \delta E_y \right) \tag{11.25}$$

$$\delta j_y = -\sum_s \frac{\epsilon_0 \omega_{ps}^2}{n_0 \omega} \int_0^\infty \int_{-\infty}^\infty \int_0^{2\pi} v_\perp dv_\perp dv_\| d\phi$$

$$\int_0^\infty d\tau e^{-iz[\sin(\phi+\Omega\tau)-\sin(\phi)]} v_\perp \sin\phi\, (\delta E_x v_\perp \cos(\phi+\Omega\tau) + \delta E_y v_\perp \sin(\phi+\Omega\tau))\, \widehat{D}_k F$$

$$= -\sum_s \frac{\epsilon_0 \omega_{ps}^2}{n_0 \omega} \int_0^\infty \int_{-\infty}^\infty v_\perp dv_\perp dv_\|$$

$$\frac{2\pi i\, \widehat{D}_k F}{\omega - k_z v_z - n\Omega} \sum_{n=-\infty}^\infty \left(-\frac{in\Omega}{k_\perp v_\perp} v_\perp J_n J_n' \delta E_x + v_\perp (J_n')^2 \delta E_y \right). \tag{11.26}$$

In matrix form the two cross-field plasma currents are

$$\begin{pmatrix} \delta j_x \\ \delta j_y \end{pmatrix} = \sum_s \sum_{n=-\infty}^\infty \frac{\epsilon_0 \omega_{ps}^2}{n_0 \omega} \int_0^\infty \int_{-\infty}^\infty v_\perp dv_\perp dv_\| \frac{2\pi i\, \widehat{D}_k F}{\omega - k_z v_z - n\Omega}$$

$$\begin{pmatrix} \frac{n^2\Omega^2}{k_\perp^2 v_\perp^2} v_\perp J_n^2 & \frac{in\Omega}{k_\perp v_\perp} v_\perp J_n J_n' \\ -\frac{in\Omega}{k_\perp v_\perp} v_\perp J_n J_n' & v_\perp (J_n')^2 \end{pmatrix} \cdot \begin{pmatrix} \delta E_x \\ \delta E_y \end{pmatrix}. \tag{11.27}$$

With the plasma defined by $\boldsymbol{j} = \sum_s \boldsymbol{\sigma}_s \cdot \delta \boldsymbol{E}$,

$$\boldsymbol{\sigma}_s(\boldsymbol{k},\omega) = \sum_{n=-\infty}^\infty \frac{ie_s^2}{m_s} 2\pi \int_0^\infty \int_{-\infty}^\infty v_\perp dv_\perp dv_\| \frac{\widehat{D}_k F}{\omega - k_z v_z - n\Omega}$$

$$\times \begin{pmatrix} \frac{n^2\Omega^2}{k_\perp^2 v_\perp^2} v_\perp J_n^2 & \frac{in\Omega}{k_\perp v_\perp} v_\perp J_n J_n' \\ -\frac{in\Omega}{k_\perp v_\perp} v_\perp J_n J_n' & v_\perp (J_n')^2 \end{pmatrix} \cdot \begin{pmatrix} \delta E_x \\ \delta E_y \end{pmatrix}. \tag{11.28}$$

The electric conductivity is $\boldsymbol{J} = \boldsymbol{\Sigma} \cdot \boldsymbol{E}$ where $-i\omega\epsilon_0 \mathcal{X} = \boldsymbol{k} \cdot \boldsymbol{\sigma} \cdot \boldsymbol{k}$ where

$$\boldsymbol{\Sigma} = -\sum_s \sum_{n=-\infty}^{\infty} \frac{\omega_{ps}^2}{n_0\omega^2} \int_0^{\infty} \int_{-\infty}^{\infty} 2\pi v_\perp dv_\perp dv_\parallel \frac{i\widehat{D}_k F}{\omega - k_z v_z - n\Omega} \times \begin{pmatrix} \frac{n^2\Omega^2}{k_\perp^2 v_\perp^2} v_\perp J_n^2 & \frac{in\Omega}{k_\perp v_\perp} v_\perp J_n J_n' \\ -\frac{in\Omega}{k_\perp v_\perp} v_\perp J_n J_n' & v_\perp (J_n')^2 \end{pmatrix}. \tag{11.29}$$

The formulas for $\mathcal{X}$ and $\boldsymbol{\sigma}$ are required for computing the fluctuating plasma currents for arbitrary applied electric and magnetic fluctuations. In tokamaks external antennas inside the first plasma facing wall are installed for control of edge localized modes (ELMs) and for launching radio frequency (RF) waves to heat the plasma and to drive the plasma current.

11.2 Requirements for Self-Consistent Fields

Now the electromagnetic fields are made self-consistent with the Maxwell field equations

$$i\boldsymbol{k} \times \boldsymbol{E} = i\omega \boldsymbol{B} \tag{11.30}$$

$$i\boldsymbol{k} \times \boldsymbol{B} = \mu_0 \boldsymbol{J} - \frac{i\omega}{c^2}\boldsymbol{E} \tag{11.31}$$

where the current density $\boldsymbol{J}(\boldsymbol{k}, \omega)$ is given by the kinetic calculation summarized in Eq. (11.29) with $\boldsymbol{J} = \boldsymbol{\sigma} \cdot \boldsymbol{E}$. This substitution gives the linear homogeneous system of equations

$$\boldsymbol{k} \times (\boldsymbol{k} \times \boldsymbol{E}) + i\mu_0\omega \boldsymbol{J} + \frac{\omega^2}{c^2}\boldsymbol{E} = 0. \tag{11.32}$$

In matrix or dyadic form Eq. (11.32) becomes

$$\left(\boldsymbol{kk} - k^2\boldsymbol{I} + i\mu_0\omega\boldsymbol{\sigma} + \frac{\omega^2}{c^2}\boldsymbol{I}\right) \cdot \boldsymbol{E} \tag{11.33}$$

or, equivalently, the matrix

$$\left[\left(k^2 - \frac{\omega^2}{c^2}\right)\boldsymbol{I} - \boldsymbol{kk} - i\omega\mu_0\boldsymbol{\sigma}\right] \cdot \boldsymbol{E} = 0 \tag{11.34}$$

that constrain the fields to be the naturally occurring collective self-consistent fluctuations in the plasma.

For the linear, homogeneous system of Eq. (11.34) to have nontrivial solution with $\boldsymbol{E} \neq 0$, we must have

$$D(\boldsymbol{k}, \omega) = \det\left[\left(k^2 - \frac{\omega^2}{c^2}\right)\boldsymbol{I} - \boldsymbol{kk} - i\omega\mu_0\boldsymbol{\sigma}\right] = 0. \tag{11.35}$$

The complex valued function $D(\boldsymbol{k},\omega)$ vanishes at numerous points in the complex ω-plane. These complex roots give multiple modes with $\omega_\alpha = \omega_\alpha(\boldsymbol{k})$ $\alpha = 1,2,3...n_{\rm modes}$ and their electric field polarization vectors follow from Eq. (11.34) by computing ratios of the $\boldsymbol{E_k}$ components with the roots $\omega = \omega_\alpha(\boldsymbol{k})$ substituted into the matrix.

In the cold plasma limit the conductivity in Eq. (11.29) reduces to

$$\boldsymbol{\sigma} = \sum_s \omega_{ps}^2 i\epsilon_0 \begin{bmatrix} \dfrac{\omega}{\omega^2-\Omega^2} & \dfrac{-i\Omega}{\omega^2-\Omega^2} & 0 \\ \dfrac{i\Omega}{\omega^2-\Omega^2} & \dfrac{\omega}{\omega^2-\Omega^2} & 0 \\ 0 & 0 & \dfrac{1}{\omega} \end{bmatrix}.$$

For weak electron-neutral collisions $1/\omega \to 1/(\omega + i\nu_{en})$ and $\sigma_\parallel = n_e e^2/m_e\nu_{en}$ for $|\omega| < \nu_{en}$.

11.2.1 *Dispersion relation for two-temperature Maxwellian distribution*

Tokamak plasmas are strongly heated with powerful neutral beams (NBI) and radio frequency (RF) antennas. The particle energies above the thermal energy given by the temperature are preferentially heated especially in the RF heating methods designed to drive the plasma current. Thus, the particle distribution functions have anisotropies in their high energy parts. These anisotropies can drive instabilities by transferring energy from the particles to the waves. In one case called "alpha channeling", the antenna system is designed to have the fusion reaction produced alpha particles drive the waves up to significant amplitudes where they may then heat up the thermal ions in the plasma. This mechanism is thought to have the potential of allowing a fast, direct energy transfer from the alpha particles produced by fusion reactions to the hydrogen isotopes of deuterium and tritium used for fuel. In the standard collisional scenario the alpha power goes into heating the electrons which on a long time scale of seconds transfer their energy to the deuterium and tritium.

Let us evaluate the dispersion relation for the bi-Maxwellian distribution function using the plasma dispersion function $Z(\alpha)$ derived in Worked Problem 2.2. The bi-Maxwellian distribution function is given by

$$F\left(v_\perp, v_\parallel\right) = \frac{1}{\sqrt{2\pi}} \frac{1}{v_{T\parallel}} \frac{1}{v_{T\perp}^2} \exp\left(-\frac{v_\perp^2}{2v_{T\perp}^2}\right) \exp\left(-\frac{v_\parallel^2}{2v_{T\parallel}^2}\right) \tag{11.36}$$

The velocity gradients of F are

$$\widehat{D}_k F = k_z v_\perp \frac{\partial F}{\partial v_z} + (\omega - k_z v_z) \frac{\partial F}{\partial v_\perp} \tag{11.37}$$

$$\frac{\partial F}{\partial v_z} = \frac{1}{\sqrt{2\pi}}\frac{1}{v_{T\|}}\frac{1}{v_{T\perp}^2}\frac{-v_z}{v_{T\|}^2}\exp\left(-\frac{v_\perp^2}{2v_{T\perp}^2}\right)\exp\left(-\frac{v_\|^2}{2v_{T\|}^2}\right) \tag{11.38}$$

$$\frac{\partial F}{\partial v_\perp} = \frac{1}{\sqrt{2\pi}}\frac{1}{v_{T\|}}\frac{1}{v_{T\perp}^2}\frac{-v_\perp}{v_{T\|}^2}\exp\left(-\frac{v_\perp^2}{2v_{T\perp}^2}\right)\exp\left(-\frac{v_\|^2}{2v_{T\|}^2}\right). \tag{11.39}$$

The integration over all parallel velocities of the response function gives

$$\frac{1}{\sqrt{2\pi}}\int dv_z \frac{k_z v_z}{\omega - k_z v_z - n\Omega}\frac{1}{v_{T\|}}\frac{1}{v_{T\|}^2}\exp\left(-\frac{v_z^2}{2v_{T\|}^2}\right) \tag{11.40}$$

$$= -\frac{1}{v_{T\|}^2}(1 + \alpha_n Z(\alpha_n)) \tag{11.41}$$

and

$$\frac{1}{\sqrt{2\pi}}\int dv_z \frac{\omega - k_z v_z}{\omega - k_z v_z - n\Omega}\frac{1}{v_{T\|}}\exp\left(-\frac{v_z^2}{2v_{T\|}^2}\right) \tag{11.42}$$

$$= 1 - \frac{n\Omega}{k_z v_{T\|}}Z(\alpha_n) \tag{11.43}$$

where $\alpha_n = (\omega - n\Omega)/\sqrt{2}\,|k_\||v_T$. Here the $Z(\alpha_n)$ is the plasma dispersion function tabulated in Fried and Conte (1961) and introduced in Worked Problem 2.2 of Chapter 2.1. The graphs of Z and Z' are shown in Fig. 2.1 and Fig. 2.2. Note that in the low-temperature limit $|\alpha| \gg 1$ and from Worked Problem 2.2 the cold plasma response is $Z(\alpha) \simeq -1/\alpha$ and thus the integral Eq. (11.43) reduces to the usual cyclotron response $\omega/(\omega - n\Omega)$ for low thermal (cold) particles.

Worked Problem: Conductivities for Maxwell-Boltzmann Plasma

Carry out the integration over the perpendicular velocities in the response functions for the currents in Eq. (11.27) for a Maxwellian distribution $\exp(-mv_\perp^2/2T_\perp)$ of the perpendicular velocities defining $v_\perp^2 = T_\perp/m$ to find the formulas for the kinetic conductivities.

Solution:

The integral over the Maxwell distribution of $J_0^2(k_\perp v_\perp/\Omega) = \exp(-b)I_0(b)$ where $b = k_\perp^2 v_{T\perp}^2/2\Omega^2$. When we obtain for the $\sigma_{x,x}$ conductivity

$$\frac{n^2\Omega^2}{k_\perp^2 v_{T\perp}^2}\exp\left(-\frac{k_\perp^2 v_{T\perp}^2}{\Omega^2}\right)I_n(1 + \alpha_n Z(\alpha_n)) + \frac{n^2\Omega^2}{k_\perp^2 v_{T\perp}^2}\exp\left(-\frac{k_\perp^2 v_{T\perp}^2}{\Omega^2}\right)I_n(1 - n\Omega Z(\alpha_n)) \tag{11.44}$$

and for the upper σ (2×2) matrix elements D_{11}, $D_{12}=D_{21}^{*}$ and D_{22} we derive

$$\begin{aligned}
D_{11} &= \frac{n^2\Omega^2}{k_\perp^2}\exp\left(-\frac{k_\perp^2 v_{T_\perp}^2}{\Omega^2}\right) I_n \\
&\quad\times\left\{\left(\frac{v_{T\perp}^2}{v_{T\|}^2}+1\right)(1+\alpha_n Z(\alpha_n)) - \frac{\omega}{k_\perp v_{T\|}} Z(\alpha_n)\right\} \\
D_{12} &= in\exp\left(-\frac{k_\perp^2 v_{T_\perp}^2}{\Omega^2}\right)(I_n - I_n') \\
&\quad\times\left\{\left(\frac{v_{T\perp}^2}{v_{T\|}^2}+1\right)(1+\alpha_n Z(\alpha_n)) - \frac{\omega}{k_\perp v_{T\|}} Z(\alpha_n)\right\} \qquad (11.45) \\
D_{22} &= \exp\left\{\frac{n^2\Omega^2}{k_\perp^2} v_{T\perp}^2 I_n + 2\frac{k_\perp^2 v_{T_\perp}^2}{\Omega^2}(I_n - I_n')\right\} \\
&\quad\times\left\{\left(1-\frac{v_{T\perp}^2}{v_{T\|}^2}\right)(1+\alpha_n Z(\alpha_n)) - \frac{\omega}{k_\perp v_{T\|}} Z(\alpha_n)\right\}.
\end{aligned}$$

In the limit of low perpendicular temperature, one has $I_{n=1}\to k_\perp^2 v_{T\perp}^2/2\Omega^2$ and $I'_{n=1}\to 1/2$.

11.2.2 *Electromagnetic dispersion relation for cross-field beam injected plasma*

Both strong neutral beam injection NBI and cyclotron resonance heating create strong anisotropies and possibly beams in the plasma. Ion cyclotron waves are observed during perpendicular neutral beam injection in JET [Dendy, *et al.* (1995)].

For the parameterization of a cross-field particle beam with plasma $v_\perp = v_b$ phase space distribution, we take

$$F = g_1(v_\perp)\, g_2(v_z) \qquad (11.46)$$

$$g_2(v_z) = \left(\frac{m}{2\pi T_\|}\right)^{1/2}\exp\left(-\frac{mv_z^2}{2T_\|}\right) = \frac{1}{\sqrt{\pi}\, v_{T\|}}\exp\left(-\frac{v_z^2}{v_{T\|}^2}\right) \qquad (11.47)$$

$$g_1(\overline{v}_\perp) = \frac{v_\perp}{v_r}\frac{1}{V_0^2}\frac{2}{\sqrt{\pi}}\frac{1}{\overline{v}_{T\perp}}\exp\left[-\left(\frac{\overline{v}_\perp - 1}{\sqrt{2}\,\overline{v}_{T\perp}}\right)^2\right] \qquad (11.48)$$

where $\overline{v}_\perp = v_\perp/v_b$. First we consider the low-perpendicular thermal spread limit. From the cyclotron harmonics $n=\pm1$ we derive

$$\int_0^\infty \int_{-\infty}^\infty v_\perp dv_\perp dv_\parallel \frac{2\pi}{\omega - k_z v_z \mp \Omega} \left(v_\perp k_z \frac{\partial F}{\partial v_z} + (\omega - k_z v_z) \frac{\partial F}{\partial v_\perp} \right) \frac{1}{2} v_\perp J_0^2 \tag{11.49}$$

$$= -\frac{1}{v_{T\parallel}^2}(1 + \alpha Z(\alpha))B_{k_\perp} + \left(1 - \frac{n\Omega}{k_z v_{T\parallel}} Z(\alpha)\right) A_{k_\perp} \tag{11.50}$$

where the key stability functions of $k_\perp$ and the beam distribution functions are

$$A_{k_\perp} = \int dv_\perp v_\perp \frac{v_\perp}{2} \frac{\partial g_1(v_\perp)}{\partial v_\perp} J_0^2 \tag{11.51}$$

$$B_{k_\perp} = \int dv_\perp v_\perp \frac{v_\perp}{2} J_0^2 g_1(v_\perp) . \tag{11.52}$$

11.2.3 *Anisotropy-driven growth rates*

The dispersion relation at $k_\perp = 0$ reduces to

$$\omega^2 - k^2c^2 + \sum_s \frac{1}{2}\omega_{ps}^2 \int_{-\infty}^\infty d^3v \frac{1}{\omega - k_z v_z \pm \Omega} v_\perp \left(v_\perp k_z \frac{\partial F}{\partial v_z} + (\omega - k_z v_z) \frac{\partial F}{\partial v_\perp} \right) = 0 \tag{11.53}$$

which gives electromagnetic ion cyclotron waves (EMICs) and whistlers. Let $\omega \to \omega_R + i\gamma$, $\gamma \ll \omega_R(k_\perp, k_z)$, where the subscript R is for the real part of the complex $\omega(k)$-function and $\gamma(k)$ is the imaginary part. Evaluating the resonant particle-wave interactions gives the imaginary terms:

$$2\gamma\omega_R(k_\perp, k_z) + \sum_s \frac{1}{2}\omega_{ps}^2 \left[-\langle v_\perp^2 \rangle \pi \frac{\partial f^{(0)}}{\partial v_z} + \int d^3v v_\perp \times \left(1 \mp \frac{\Omega_s/k_z}{v_z - (\omega \pm \Omega_s)/k_z}\right) \frac{\partial F}{\partial v_\perp} \right] \tag{11.54}$$

$$2\gamma\omega_R(k_\perp, k_z) + \sum_s \frac{1}{2}\omega_{ps}^2 \left[-\langle v_\perp^2 \rangle \pi \frac{\partial f^{(0)}}{\partial v_z} \mp \frac{2\Omega_s}{k_z} f^{(0)}(v_z) \right]_{v_z = (\omega \pm \Omega_s)/k_z} = 0 \tag{11.55}$$

$$\gamma = \frac{\pi}{2\omega_R(k_\perp, k_z)} \sum_s \left[\frac{\langle v_\perp^2 \rangle}{2} \frac{\partial f^{(0)}}{\partial v_z} \pm \frac{\Omega_s}{k_z} f^{(0)}(v_z) \right] . \tag{11.56}$$

If there is a Maxwellian distribution in the parallel velocity v_z, then

$$\frac{\partial f}{\partial v_z} = -\frac{v_z}{v_{T\parallel}^2} f \tag{11.57}$$

The average $v_\perp^2$ is

$$\langle v_\perp^2 \rangle = 2v_{T\perp}^2 . \tag{11.58}$$

The growth rate function $\gamma(k_z)$ from Eq. (11.56) reduces to

$$\begin{aligned}\gamma_{\boldsymbol{k}} &= \frac{\pi}{2\omega_R\left(k_\perp, k_z\right)} \sum_s \left[v_{T\perp}^2 \left(-\frac{\left(\omega_R(k_\perp, k_z) \pm \Omega_j\right)}{k_z} \right) \pm \frac{\Omega_s}{k_z} \right] \\ &\quad \times \frac{1}{\sqrt{2\pi}\, v_{T\|}} \exp\left(-\frac{\left(\omega_R(k_\perp, k_z) \pm \Omega_j\right)^2}{2k_z^2 v_{T\|}^2} \right) \\ &= \frac{\pi}{2\omega_R\left(k_\perp, k_z\right)} \sum_s \left[-\frac{T_\perp}{T_\|} \frac{\left(\omega_R(k_\perp, k_z) \pm \Omega_j\right)}{k_z} \left(1 - \frac{T_\perp}{T_\|} \right) \right] \\ &\quad \times \frac{1}{\sqrt{2\pi}\, v_{T\|}} \exp\left(-\frac{\left(\omega_R(k_\perp, k_z) \pm \Omega_j\right)^2}{2k_z^2 v_{T\|}^2} \right).\end{aligned}$$

For a beam of speed V_0 across the magnetic field, we derive

$$\frac{\partial f}{\partial v_z} = -\frac{v_z}{v_{T\|}^2} f \tag{11.59}$$

$$\left\langle v_\perp^2 \right\rangle = 3v_{T\perp}^2 + V_0^2 \tag{11.60}$$

$$\begin{aligned}\gamma_{\boldsymbol{k}} &= \frac{\pi}{2\omega_R\left(k_\perp, k_z\right)} \sum_s \left[-\frac{3v_{T\perp}^2 + V_0^2}{2v_{T\|}^2} \pm \frac{\Omega_j}{k_z} \left(1 - \frac{3v_{T\perp}^2 + V_0^2}{2v_{T\|}^2} \right) \right] \\ &\quad \times \frac{1}{\sqrt{2\pi}\, v_{T\|}} \exp\left(-\frac{\left(\omega_R(k_\perp, k_z) \pm \Omega_j\right)^2}{2k_z^2 v_{T\|}^2} \right).\end{aligned} \tag{11.61}$$

11.3 Quasilinear Fluctuation Diffusion Tensor

The reaction of the fluctuations on the background phase space distribution is given by

$$\frac{\partial\langle f\rangle}{\partial t} = \frac{e^2}{m^2} \sum_{\boldsymbol{k}} \widehat{D}_{\boldsymbol{k}} \boldsymbol{E}_{\boldsymbol{k}}^* \cdot \operatorname{Im}\left(\chi_{\boldsymbol{k}}\right) \cdot \boldsymbol{E}_{\boldsymbol{k}} \widehat{D}_k \langle f\rangle \tag{11.62}$$

where the operator $\widehat{D}_k$ is defined in Eq. (11.24) and the resonant part of the conductivity tensor $\chi_{\boldsymbol{k}}$ describes the linear wave-particle resonance

$$\omega_{\boldsymbol{k}} - n\Omega = k_\| v_\| \tag{11.63}$$

for all waves $\boldsymbol{k}, \omega_{\boldsymbol{k}}$ and cyclotron harmonics n in the fluctuation spectrum.

11.4 Electron Diffusion from Magnetic Flutter in Tokamaks

11.4.1 *Polarization relations for low-frequency electromagnetic fluctuations*

At moderate plasma pressure, $\beta < 1$, there are general polarization features of the low frequency, drift wave fluctuations that simplify the transport analysis. In contrast, for $\beta \gtrsim 1$, all three electric field components are coupled in a complicated manner; however, the drift-type modes are then unstable only under rather special conditions, such as the $\eta = d\ln T/d\ln = -\frac{1}{2}$ mode found by Mikhailovskii and Fridman (1967).

In moderate pressure plasma the highly elongated fluctuations $k_\perp \gg k_\| $ of the drift wave and tearing mode type decouple from the fast compressional motion with $\omega^2 \ll k^2 v_A^2$. The polarization of the slow convective modes is determined by the total pressure balance across the magnetic field $B\delta B_\|(k) + 4\pi\delta p_\perp(k) = 0$ that is maintained on the scale $k_\perp^{-1}$ during the slow oscillations. From this relation and the formula for $\delta p_\perp(k)$ the magnitude of the magnetic compression follows as

$$\frac{\delta B_\|(k)}{B} = \frac{1}{2}\beta\left[\frac{iekE_y(k)}{k_yT}\right] \cong \frac{1}{2}\beta\left[\frac{e\varphi(k)}{T}\right] \tag{11.64}$$

which is valid for modes rotating in either the ion or the electron diamagnetic direction. The change in the magnetic field strength $\delta B_\|(k)$ ensures that there is no compression of the plasma $\nabla\cdot\boldsymbol{v}_E = 0$ in the convection of the plasma across the finite beta equilibrium with $B^2(r) + 8\pi p_\perp(r) = \text{const.}$

The perpendicular magnetic perturbation $\delta B_\perp(k)$ considerably exceeds $\delta B_\|(k)$ and takes on the magnitude and phase required to short out the parallel electric field, $E_\|(k) = -ik_\|\varphi(k) + i(\omega/c)A_\|(k) \approx 0$. The important case where the inductive part of the parallel electric field is not able to cancel the parallel electric field occurs when the $\boldsymbol{k}\omega$ components are such as to incur in an anomalously low parallel conductivity $\boldsymbol{\sigma}_\|(\boldsymbol{k},\omega)$. The fluctuations for which the parallel conductivity is low are $\omega \sim \omega_{*e}$ and $k_\| v_e \gtrsim \omega$ since from the electron current $\Gamma_\|(k)$,

$$\boldsymbol{\sigma}_\|(k) \cong \frac{im_en_ee^2}{2k_\|^2T_e}(\omega - \omega_{*e})Z'\left(\frac{\omega}{|k_\||v_e}\right). \tag{11.65}$$

For fluctuations in this domain it follows from $\nabla\times\delta\boldsymbol{B}_\perp = 4\pi\delta j_\|/c$ that the perpendicular magnetic perturbation is

$$\frac{\delta B_\perp(k)}{B} \cong i\left(\frac{\beta}{2}\right)^{1/2}\left(\frac{\omega_{*e}}{k_\| v_A}\right)k_\perp\rho_i\left[\frac{e\varphi(k)}{T_e}\right] \tag{11.66}$$

where $e_y(k) \cong -ik_y(\varphi(k)$ is used.

In characterizing the moderate β drift-tearing mode fluctuations it must be observed that the field line bending from $\delta B_\perp(k)$ couples the parallel transport into a radial motion. As an example, we observe that in the absence of line bending the electron temperature fluctuation determined for $\nu_e > k_\parallel v_e$ from the thermal balance equation with $q_{\parallel e} = -\chi_{\parallel e} \nabla_\parallel T_e$ where $\chi_{\parallel e} = n_e T_e / m_e v_e$ yields

$$\frac{\delta T_e(k)}{T_e} = \left(\frac{i\nu_c}{k_\parallel^2 v_e^2} \right) \left[\omega - \omega_{*e} \left(1 - \frac{3}{2}\eta_e \right) \right] \left[\frac{e\varphi(k)}{T_e} \right] = \left[\frac{\delta T_e(k)}{T_e} \right]_{\text{conv}} \tag{11.67}$$

from the balance of convection and parallel diffusion. In the presence of line bending, however, the thermal transport exceeds that due to convection. At the magnetic amplitude $\delta B_\perp(k)$ where this occurs in the parallel inductive field $(\omega/c)A_\parallel(k)$ due to $\delta B_\perp(k) = ik_y A_\parallel(k)$ is such that

$$E_\parallel^{(A)} = \frac{i\omega}{c} A_\parallel(k) \simeq \left(\frac{\omega}{\omega_{*e}} \right) \left[-ik_\parallel \varphi(k) \right].$$

The opposite parity in $k_\parallel$ of $A_\parallel(k)$ and $\varphi(k)$ is a general feature of the low-frequency fluctuations. The relationship of electromagnetic drift modes to magnetohydrodynamic modes is given by Horton, *et al.* (1983).

The polarization relations for the low-frequency drift and tearing mode fluctuations at $\beta < 1$ allow the particle and energy flux formulas to be simplified. First consider the contribution of $\delta B_\parallel(k)$ to the Lorentz force $F_y^j(k, v_\perp, v_\parallel)$. Using Eq. (11.64) to relate $E_y(k)$ and $\delta B_\parallel(k)$ we observe that $F_y^j \simeq E_y(k)[J_0 + i\beta J_1/(k_\perp \rho_j)] \simeq E_y(k) J_0(k_\perp v_\perp/\omega_{cj})$for $\beta < 1$ and arbitrary $k_\perp \rho_j$. Now consider the contribution to F_y^j due to $\delta B_\perp(k)$. For the $\delta B_\perp(k)$ given in Eq. (11.66), it follows that $F_y^j \cong E_y J_0 (1 + v_\parallel/v_A)$ when $\omega_{*e} \sim k_\parallel v_A$. For the plasma pressure $\beta < 1$ the Alfvén speed is intermediate to the ion and electron thermal velocities, so that the magnetic force is small on the ions with $F_y^i \cong J_0 E_y$ and dominant on the electrons $F_y^e \cong (v_\parallel/c) B_x(k)$. Taking into account these simplifications, we have

$$\Gamma_i = \sum_k \left| \frac{cE_y(k)}{B} \right|^2 \int d\boldsymbol{v} \left(\frac{\partial f_i}{\partial x} + \frac{\omega \Omega_i}{k_y} \frac{\partial f_i}{\partial \varepsilon} \right) J_0^2 \left(\frac{k_\perp v_\perp}{\Omega_i} \right) \operatorname{Im} g_k^i(\boldsymbol{v}) \tag{11.68}$$

$$\Gamma_e = \sum_k \left| \frac{B_x(k)}{B} \right|^2 \int d\boldsymbol{v} \left(\frac{\partial f_e}{\partial x} + \frac{\omega \Omega_e}{k_y} \frac{\partial f_e}{\partial \varepsilon} \right) v_\parallel^2 \operatorname{Im} g_k^e(\boldsymbol{v}) \tag{11.69}$$

and the energy flux Q_j is obtained by including the kinetic energy $\frac{1}{2} m_j v^2$ under the velocity integral. For fluctuations with sufficiently low frequencies, $|\omega| \ll k_y \langle v^2 \rangle / 2\Omega_j \langle d \ln f_j / dx \rangle$ the anomalous flux is driven directly by $\partial f_j / \partial x$. The finite $(\omega/k_y)(\partial f / \partial \varepsilon)$ terms give the convection velocity to the fluxes Γ_s.

11.4.2 *Qualitative picture of anomalous transport due to magnetic fluctuations*

Low-frequency fluctuations near a neutral sheet such as in the geomagnetic tail plasma, e.g. [Galeev (1978)], or in the laboratory near the mode rational surfaces of the helical magnetic fields used for confinement, e.g. [Callen (1977)], produce parallel inductive electric fields $E_\parallel = -\partial A_\parallel / c\partial t$ and perpendicular magnetic fields $\delta \boldsymbol{B}_\perp = \nabla A_\parallel \times \widehat{\boldsymbol{z}}$. Such magnetic perturbations lead to anomalous transport. According to the quasilinear formula Eq. (11.69), the electron transport is proportional to $\langle |\delta B_x(k)|^2 \rangle$ and the interaction strength $\mathrm{Im}\, g_k^e(\boldsymbol{v})$. For time dependent magnetic fluctuations Callen describes the process as transport due to "magnetic flutter". Even in the case of static magnetic perturbations with $\langle \delta B_\perp(\boldsymbol{k})|^2 \rangle = \langle \delta B_\perp(\boldsymbol{k})|^2 \rangle \delta(\omega)$, there is an enhanced transport due to the particle motion along the magnetic field line. The Im $g_k^e = -\pi\delta(k_\parallel v_\parallel)$. The anomalous transport produced by the static magnetic field perturbations is described by Stix (1973) as transport due to "magnetic braiding."

For the electron flux in Eq. (11.69) the resonance $-$ Im $g_\parallel^e(v_\parallel) = \pi\delta / k_\parallel v_\parallel)$ gives the simple formula

$$D_M = \frac{\pi v_e}{|k_\parallel|} \frac{\langle \delta B_x^2 \rangle}{B^2}$$

for the electron diffusion from the magnetic fluctuations. Broadband kilohertz magnetic fluctuations with $\langle \delta B_x^e \rangle^{1/2} / B \sim 10^{-5}$ to 10^{-4} are widely observed in tokamaks. The technique is given in Colas, *et al.* (1998) was that the $e\boldsymbol{v}_w \delta B_x$ Lorentz force changes the polarization of the incident electromagnetic wave from $x \leftrightarrow 0$ polarization. This diamagnetic is called cross-polarization scattering.

Stix (1973) points out that a magnetic perturbation $\delta \boldsymbol{B}_\perp(\boldsymbol{x})$ that is resonant with a helical magnetic field produces a chain of magnetic islands of width Δ_k about the resonant surface where $\boldsymbol{B} \cdot \nabla = ik_\parallel B = 0$. Writing $\boldsymbol{B} = B_y(x)\widehat{y} + B_0 \widehat{z}$ for the local magnetic field and $L_s^{-1} = dB_y / Bdx$ for the local shear, the width Δ_k of the magnetic island is given by

$$\Delta_k = \left(\frac{2L_s}{k_y} \frac{\delta B_x}{B_0} \right)^{1/2} . \tag{11.70}$$

When two or more of these chains of islands occur, as shown in Fig. 11.1, the magnetic field becomes ergodic first along the separatrix and then throughout the region between the islands with increasing amplitude of the magnetic perturbation. Owing to their high speed, the electrons follow the braided field lines to produce a transport across the ambient helical field. The magnetic perturbations arise from a variety of reasons such as filamentation of an Ohmic heating current, electromagnetic fields from auxiliary heating methods or collective instabilities in the plasma.

The low-frequency drift waves and tearing modes, which arise from the gradient and magnetic energy density in nonuniform plasmas produce small scale magnetic perturbations. For example, Callen (1977) suggests that the drift wave fluctuations measured by Mazzucato (1976) have sufficient amplitude to produce a substantial anomalous thermal transport due to their magnetic component.

The radial gradient of the large plasma current density $j_\phi(r)$ in a tokamak that is driven either by (i) the ohmic inductive electromotive force, (ii) by radio frequency current drive or (iii) by the intrinsic pressure gradient drive of the bootstrap currents

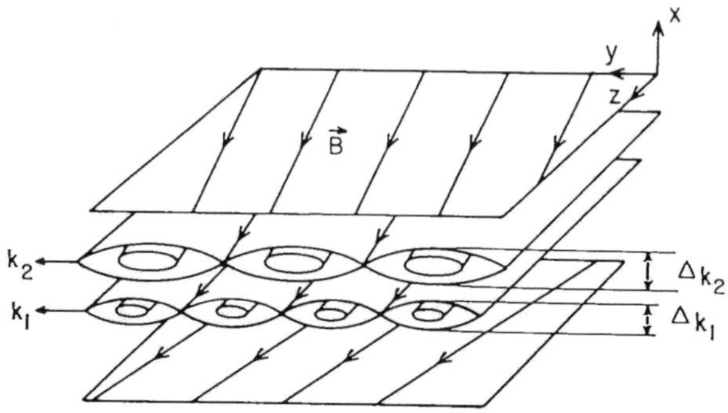

Fig. 11.1 The magnetic island chains by two nearby resonant perturbations $\boldsymbol{k}_1 \cdot \boldsymbol{B} = 0$ and $\boldsymbol{k}_2 \cdot \boldsymbol{B} = 0$ in the sheared magnetic field. The $\delta \boldsymbol{B}_\perp(\boldsymbol{k})$ amplitudes are small enough that overlapping has not yet occurred [Galeev (1978)].

adds a symmetry breaking term to the ballooning interchange mode equation [Zheng and Furukawa (2010)]. The ballooning mode equation is localized about the rational surface in the absence of this current gradient and even symmetry. With the current gradient there is coupling to the odd symmetry mode that evolves into the tearing mode or varicose mode in the nonlinear state [Muragalia, *et al.* (2010); Ishizawa and Nakajima (2010)]. The instability is thus a source of the magnetic turbulence giving the anomalous transport of electrons. Chapter 15 analyzes this form of turbulence.

For finite beta drift wave turbulence, the case emphasized by Callen (1977), $\delta B_x(x)$ vanishes at the resonant surface, rises to a maximum of δB_{xA} at $k_\| v_A \simeq \omega_{*e}$, and decreases at $\delta B_{xA}(x_A/x)$ for $k_\| v_A > \omega_{*e}$. Here, x_A is the distance from the resonant surface $k_\| = 0$ to the region where $k_\| v_A = k_y v_A x_A / L_s = \omega_{*e}$ which is $x_A = \rho_s \beta^{1/2}(L_s/r_n)$. For a drift wave island width $\Delta_k^{\rm dw} > x_A$, one finds that

$$\Delta_k^{\rm dw} \simeq \left(\frac{L_s x_A}{k_y} \frac{\delta B_{xA}}{B_0} \right)^{1/3}$$

and the distance L_k along the magnetic field to circumnavigate the magnetic island $L_k = L_s/k_y \Delta_k \gg L_s$. Since L_k is greater than L_s the drift wave frequency ω_k resonates with the electron transit frequency around the island $\omega_k \cong v_e/L_k$. Including the buildup of an ambipolar potential $\Phi(x)$ to equalize the ion and electron loss rates, Callen (1977) estimates that the electron diffusion and ambipolar potential are given by

$$\Gamma_e \simeq -\frac{n_e}{2} \sum_k \Delta_k^2 (\nu_e + \gamma_k) \left[\frac{d \ln n_e}{dx} - \frac{e}{T_e} \frac{d\Phi}{dx} - \frac{\omega e B}{c k_y T_e} \right] \simeq 0,$$

$$\frac{e}{T_e} \frac{d\Phi}{dx} = \left(1 - \frac{\omega_{\overline{k}}}{\omega_{*e}} \right) \frac{d \ln n_e}{dx} \approx \frac{d \ln n_e}{dx} \tag{11.71}$$

giving electrostatic confinement of the electrons with radially outward $E_r > 0$ electric field. The electron thermal transport $Q_e = -n_e \chi_e (dT_e/dx)$ is estimated by

$$\chi_e = \frac{3}{4} \sum_k \Delta_k^2 \left(\nu_e + \gamma_k\right) \tag{11.72}$$

where ν_e is the electron collision frequency and $1/\gamma_k$ is the correlation time or the lifetime of the magnetic island Δ_k. Equation (11.72) is a nonlinear generalization of the quasilinear formula in which the island width

$$\Delta_k \propto \begin{cases} \left[\dfrac{\delta B_{xA}}{B}\right]^{1/3} & \text{for drift waves} \\ \\ \left[\dfrac{\delta B_x(0)}{B}\right]^{1/2} & \text{for tearing modes} \end{cases} \tag{11.73}$$

replaces the linear radial excursion $v_\parallel B_x(k)/B\Delta(\omega - \boldsymbol{k}\cdot\boldsymbol{v}) = [B_x(k)/B]v_\parallel \tau_c$ that occurs in the quasilinear formula in Eq. (11.69).

The magnitude of the anomalous transport produced by magnetic braiding depends sensitively on the correlation time of the electron with the magnetic perturbation $\delta\boldsymbol{B}_\perp$, as well as the obvious amplitude dependence. Rechester and Rosenbluth (1978) and daSilva, *et al.* (2001) introduce the method of stochastic trajectories to obtain formulas that describe the finite correlation time τ_c or length L_c along the field line.

First the statistical properties of the magnetic field in the presence of the stochastic perturbations $\delta\boldsymbol{B}_\perp = \sum_k \delta\boldsymbol{B}_k \exp\left(i\boldsymbol{k}\cdot\boldsymbol{x}\right)$ are established. Rechester and Rosenbluth introduce the stochasticity parameter

$$s = \frac{1}{2}\frac{(\Delta_{k_1} + \Delta_{k_2})}{|(r_{k_1} - r_{k_2})|} \tag{11.74}$$

that compares the total width of two magnetic islands with their radial separation. When this parameter exceeds unity the magnetic surfaces are destroyed between the chains of islands. In this region where the field lines wander ergodically, a small circle of radius l_0 is mapped by following the field line into a complex, elongated shape that has the same area as the initial circle owing to the constraint $\nabla\cdot\boldsymbol{B} = 0$. The stretching of a typical side of this area is given by

$$I(z) = l_0 \exp\left(\frac{z}{L_c}\right) \tag{11.75}$$

where $L_c = \pi R/\ln(\pi s/2)$ which is called the stochastic instability of trajectories. To conserve area the width $w(z)$ of the area must decrease according to $w(z) = l_0 \exp\left(-z/L_c\right)$. When the elongation $l(z)$ exceeds the cross-field correlation distance δ different regions separated by δ evolve independently. This condition, $l(z) = \delta = 1/\Delta k_\perp$, defines the correlation length

$$L_{c0} = L_c \ln\left(\frac{\delta}{l_0}\right) \tag{11.76}$$

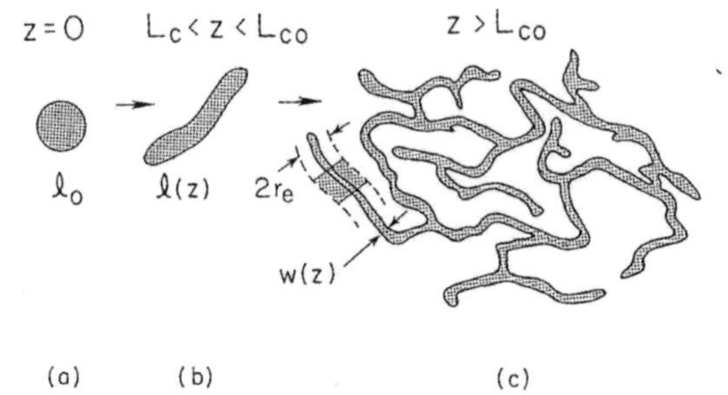

Fig. 11.2 In the stochastic regime of strong island overlapping $s \gg 1$, the small circle in (a) is mapped by the magnetic field to (b) showing the stochastic instability of trajectories. When the filament length in (b) exceeds the cross-field correlation distance, the mapping evolves as shown in (c) [Rechester and Rosenbluth (1978)].

for the area. For $L \gg L_{c0}$ the magnetic mapping is a random walk with the average squared displacement of the connected area, shown in Fig. 11.2, described by

$$\langle \Delta r^2 \rangle = 2D_F L \tag{11.77}$$

where

$$D_F = \pi \sum_k \left| \frac{B_x(\boldsymbol{k})}{B} \right|^2 \delta(k_\parallel) \tag{11.78}$$

is the field (F) diffusion coefficient.

Now consider the motion of a small patch of electrons whose guiding centers follow the magnetic field lines. The small patch of electrons of radius r_e move along the field mapping into the complicated thin filaments of width $r_e \exp(-z/L_c)$ until they suffer a collision at the average distance $z = \lambda_e = v_e/\nu_e$. At the time of the collision the initial patch of electrons has spread the average squared perpendicular distance $\langle \Delta r^2 \rangle = 2D_F \lambda_e$ provided $\lambda_e > L_{c0}$. After the collision the thin filaments expand to r_e, and subsequently the new subsections of radius r_e repeat the process. These random steps give rise to the thermal diffusion

$$\chi_e = \frac{1}{2} \langle \Delta r^2 \rangle \nu_e = v_e D_F \tag{11.79}$$

valid for $\lambda_e \gg L_c \sim 1/\Delta k$. Rechester and Rosenbluth (1978) note that this transport applies to the electron thermal energy since the particle transport must be limited by an ambipolar potential.

These stochastic magnetic fields are created with external magnetic coils on the vessel walls to produce what is called an ergodic magnetic limiter in some tokamaks. The stochastic magnetic fields occur automatically in LHD owing to the broken toroidal symmetry from the $l = 2/n = 10$ external coils.

In the collisional regime $\lambda_e < \lambda_c$ the electrons change direction along the field line many times before reaching uncorrelated regions. During the parallel diffusion

$\Delta z^2 = \chi_{\| e} t$ the patch of electrons expands across the magnetic field at the rate $(\chi_{\perp e} t)^{1/2}$ by collisional diffusion while the field line mapping contracts the width of the patch according to $\exp(-z/L_c)$. The balance between these processes occurs when the width is of order $L_c(\chi_{\perp e}/\chi_{\| e})^{1/2}$. A small area of this radius becomes uncorrelated when $L_c(\chi_{\perp e}/\chi_{\| e})^{1/2} \exp(-z/L_c) \approx 1/\Delta k_\perp$ which yields the collisional correlation length

$$L_{c\delta} = L_c \ln\left[\frac{\delta}{L_c}\left(\frac{\chi_{\| e}}{\chi_{\perp e}}\right)^{1/2}\right]. \tag{11.80}$$

The associated time for parallel diffusion $t_\delta = L_{c\delta}^2/\chi_{\| e}$ is the time during which a particle may be thought of as orbiting along a single field line before collisionally scattered to a new uncorrelated field line. The random walk of the electron patch proceeds with the time-step t_δ and mean radial step size $\langle \Delta r^2 \rangle = D_F L_{c\delta}$. For this collisional regime the thermal transport is

$$\chi_e = \frac{\langle \Delta r^2 \rangle}{t_\delta} = D_F\left(\frac{\chi_{\| e}}{L_{c\delta}}\right) \tag{11.81}$$

which is smaller than the collisionless χ_e in Eq. (11.79) by the factor λ_e/L_c. To within the logarithmic factor, the transport in Eq. (11.81) is simply $\chi_e = \chi_{\| e}\langle B_x^2/B^2 \rangle$ which follows from the fluid description of the transport due to magnetic turbulence as shown in the next section.

Rechester and Rosenbluth (1978) suggest that finite beta drift waves and tearing modes with fluctuation scales $k_\| R \sim 1$ and $k_\perp \rho_i \sim 1$ have sufficient amplitude in confinement experiments to be well into the regime of strong stochasticity where the field diffusion is $D_F \cong R\langle \delta B_x^2 \rangle / B^2$ and χ_e follows from Eq. (11.79) for $\lambda_e > L_c$ and Eq. (11.81) for $\lambda_e > L_c$.

11.4.3 *Electron thermal transport from magnetic fluctuations in the fluid approximation*

Kadomtsev and Pogutse (1978) address the problem of calculating the anomalous electron transport from stochastic magnetic perturbations within the MHD approximation for thermal conduction. The basic transport law is taken as

$$\nabla \cdot \boldsymbol{q} = 0 \tag{11.82}$$

with the collisional formula for the local thermal flux

$$\boldsymbol{q} = -\chi_\| \boldsymbol{h}(\boldsymbol{h} \cdot \nabla T_e) - \chi_\perp \nabla T_e \tag{11.83}$$

where $\chi_\|$ exceeds $\chi_\perp$ be several orders of magnitude. The collisional values of approximately $\chi_\| = nT_e/m_e\nu_e$ are and $\chi_\perp = nT_e\nu_e/m_e\omega_{ce}^2$ and give the ratio $\chi_\perp/\chi_\| = (\nu_e/\omega_{ce})^2 = \gamma^2$ where γ is assumed small. The vector $\boldsymbol{h}$ along the magnetic field is written as the sum

$$\boldsymbol{h} = \widehat{\boldsymbol{e}}_z + \boldsymbol{b} \tag{11.84}$$

of the constant vector in the direction of the ambient magnetic field and small stochastic vector $\boldsymbol{b} = \delta \boldsymbol{B}_\perp / B$. It is seen from Eq. (11.83) that for $\langle b^2 \rangle \chi_\parallel \sim \chi_\perp$ the contribution from the parallel transport is comparable to that from the cross-field transport.

The stochastic magnetic field $\boldsymbol{b}(\boldsymbol{r}_\perp z)$ has $\nabla_\perp \cdot \boldsymbol{b} = 0$ since $\boldsymbol{B}_0 = \text{const}$. The stochastic field $\boldsymbol{b}$ is characterized by (i) the small root-mean-square value $b_0 = \langle b^2 \rangle^{1/2}$, (ii) isotropic properties in the plane perpendicular to the ambient magnetic field, (iii) the cross-field correlation distance δ and (iv) the parallel field correlation distance L_c. In this and subsequent sections the logarithmic factors that distinguish L_c, L_{c0} and $L_{c\delta}$ are disregarded [Krommes, *et al.* (1983)].

The equation for the transverse coordinates $\boldsymbol{r}_\perp$ of a point following the field lines is

$$\boldsymbol{r}_\perp(z) = \int_0^z \boldsymbol{b}\left[r(z'), dz'\right] dz'. \tag{11.85}$$

In the limit of small b_0 the quasilinear approximation for $\langle r_\perp^2(z) \rangle$ becomes at large z

$$\langle r_\perp^2 \rangle = 4 D_F z \tag{11.86}$$

where

$$D_F = \frac{1}{4} \int_{-\infty}^{+\infty} \langle \boldsymbol{b}(0,z) \boldsymbol{b}(0,0) \rangle dz. \tag{11.87}$$

The parallel correlation length L_c is defined by

$$D_F = \frac{1}{4} b_0^2 L_c \tag{11.88}$$

with which the quasilinear diffusion may be written as $\langle r_\perp^2 \rangle = b_0^2 L_c z$ for $z > L_0$. The quasilinear approximation of neglecting $\boldsymbol{r}(z)$ in Eq. (11.87) is valid when $\langle r_\perp^2 \rangle^{1/2} \simeq b_0 L_c \leq \delta$ where δ is the cross-field correlation distance. Thus, the amplitude parameter R, defines the quasilinear regime by $R < 1$.

To determine the field diffusion D_F in the regime $R > 1$, Kadomtsev and Pogutse (1978) use the statistical theory of plasma turbulence [Dupree (1967)]. With $N(\boldsymbol{r}, z)$ defined as the conserved density of magnetic field lines, the magnetic transport equation is

$$\frac{\partial N}{\partial z} + \boldsymbol{b} \cdot \nabla N(\boldsymbol{r}_\perp, z) = 0. \tag{11.89}$$

Splitting the density into its mean value $\overline{N}$ and its fluctuation N', the well-known coupled turbulence equations

$$\frac{\partial \overline{N}}{\partial z} = -\nabla \cdot \langle \boldsymbol{b} N' \rangle = D_F \nabla_\perp^2 \overline{N}, \qquad \frac{\partial N'}{\partial z} - D_F \nabla_\perp^2 N' = -\boldsymbol{b} \cdot \nabla N' \tag{11.90}$$

are used to obtain

$$D_F = \frac{1}{2} \int \frac{b^2(k)}{i k_z + k_\perp^2 D_F} dk \tag{11.91}$$

where the factor 1/2 results from the angle average of the isotropic spectrum. Here, the Fourier spectrum of the isotropic magnetic correlation function is

$$b^2(\boldsymbol{k}) = \frac{1}{(2\pi)^3} \int d\boldsymbol{r} \langle \boldsymbol{b}(0)\boldsymbol{b}(\boldsymbol{r})\rangle \exp(-i\boldsymbol{k}\cdot\boldsymbol{r}).$$

The quasilinear formula is recovered from Eq. (11.91) when $\Delta k_z > k_\perp^2 D_F$ where $D_F = (\pi/2)\int d\boldsymbol{k} b^2(\boldsymbol{k})\delta(k_z) = b_0^2 L_c$. For large amplitudes $R = b_0 L_c/\delta > 1$ the cross-field decorrelation dominates, $k_\perp^2 D_F > \Delta k_z$, and Eq. (11.91) reduces to

$$D_F^2 = \frac{1}{2}\int \frac{b^2(\boldsymbol{k})}{k_\perp^2} d\boldsymbol{k} \cong b_0^2\delta^2. \tag{11.92}$$

Thus, the field diffusion varies according to

$$D_F = \begin{cases} b_0^2 L_c \text{ for } R = \dfrac{b_0 L_c}{\delta} < 1, \\ b_0\delta \quad \text{for } R = \dfrac{b_0 L_c}{\delta} > 1. \end{cases} \tag{11.93}$$

Returning to the solution of Eq. (11.82) and Eq. (11.83), the temperature field is split into its mean T_e and fluctuating components T_e' where linearization gives

$$\left(k_z^2 + \gamma^2 k_\perp^2\right)\delta T_e(\boldsymbol{k}) = ik_z b_z(\boldsymbol{k})\frac{d\overline{T}_e}{dx}. \tag{11.94}$$

The mean value of the radial component of the heat flux is

$$q_e = -\chi_\parallel\langle b_x^2\rangle\frac{d\overline{T}_e}{dx} - \chi_\parallel\left\langle b_x\frac{dT_e'}{dx}\right\rangle = -\frac{\chi_\parallel}{2}\int\frac{\gamma^2 k_\perp^2 b^2(k)dk}{k_z^2+\gamma^2k_\perp^2}\frac{d\overline{T}_e}{dx}. \tag{11.95}$$

When we restrict consideration to $b_0 > \gamma = \nu_e^e/\omega_{ce}$ and demand that $R < 1$, it follows that for $\Delta k_z > \gamma k_\perp$ the quasilinear approximation for the effective transport resulting from Eq. (11.95) is

$$\chi_e = \left(\chi_\parallel\chi_\perp\right)^{1/2}\left(\frac{b_0^2 L_c}{\delta}\right) = \left(\frac{cT_e}{eB}\right)\left(\frac{b_0^2 L_c}{\delta}\right) \tag{11.96}$$

where $(\chi_\parallel\chi_\perp)^{1/2} = D_B = cT_e/eB$ and $b_0 L_c \le \delta$ is the limit of the quasilinear regime.

The rms fluctuation level $b_0 = \left\langle\delta B_\perp^2/B^2\right\rangle^{1/2}$ in a tokamak ranges from 10^{-5} to 10^{-4} which is well above ν_e/ω_{ce}.

In the limit $\chi_\perp \to 0$ formula Eq. (11.96) breaks down because of the neglect of higher-order terms in $\boldsymbol{b}$. Kadomtsev and Pogutse (1978) give a theory for the nonlinear regime defined by $b_0 > \delta/R$. They show that in this regime the thermal conductivity is

$$\chi_e = \chi_\parallel\left(\frac{b_0^2}{\delta}\right)^2 \qquad \text{for} \quad R = \frac{b_0 L_c}{\delta_0} < 1. \tag{11.97}$$

In both parameter ranges of the ratio $\chi_\parallel/\chi_\perp$ the anomalous thermal conductivity is related to the field diffusion by

$$\chi_e = D_f \overline{v} \tag{11.98}$$

where $\overline{v} = \chi_\parallel / L^*$ with L^* the characteristic length over which the temperature perturbation is transported. Equation (11.98) with $D_F = b_0^2 L_c$ agrees with Eq. (11.79) and Eq. (11.81) to within logarithmic factors of order unity and extends the results for χ_e outside the quasilinear regime.

In a plasma with high electrical conductivity it is natural to consider that $\delta B_\perp$ arises from displacements with $\xi \sim \delta$ which bend the field according to $\delta B \approx B_0 \delta / L_c$, according to Kadomtsev and Pogutse (1978). Such plasma displacements put the magnetic turbulence in the $R \sim 1$ regime where χ_e is only weakly dependent on $\gamma = (\chi_\perp/\chi_\parallel)^{1/2}$. The value of χ_e is

$$\chi_e = \begin{cases} \chi_\parallel \left(\dfrac{\delta^2}{L_c^2} \right) & \text{for } \lambda_e < L_c, \\ v_e \left(\dfrac{\delta^2}{L_c} \right) & \text{for } \lambda_e > L_c \end{cases} \tag{11.99}$$

from formulas Eq. (11.97) and Eq. (11.98).

For the cross-field correlation distance δ [Kadomtsev and Pogutse (1978)] propose that for low-frequency fluctuations of the drift mode type, the natural scale for the destruction and reconnection of the magnetic field lines is the collisionless skin depth $\delta = c/\omega_{pe}$. The argument is that at these low frequencies the magnetic surfaces are frozen into the electrons on this scale. With this assumption the formula Eq. (11.99) for the anomalous electron thermal conductivity becomes

$$\chi_e = \frac{c^2}{\omega_{pe}^2} \frac{v_e}{L_c} \quad \text{for} \quad \nu_e < \frac{v_e}{L_c} \tag{11.100}$$

where the parallel correlation length L_c is now the effective length of the system. In a toroidal confinement device with major radius R and inverse rotational transform or safety factor q, the effective length is $L_c = qR$. In this case the formula agrees with the empirical formula proposed by Ohkawa (1978) to explain the parametric dependence observed in toroidal confinement experiments.

11.4.4 *Kinetic theory of transport due to magnetic fluctuations*

More information about the character of the anomalous electron thermal conductivity due to magnetic turbulence is obtained from a kinetic description of the fluctuating particle distribution. In the analysis of the transport due to tearing modes, Galeev (1978) shows that the net particle and thermal flux across the ambient magnetic field is limited by the randomization or stochastization of the particle-fluctuation phase relation. For example, the cross-field electron diffusion

$\langle \Delta x^2 \rangle = \chi_e \tau$ in the sheared magnetic field results in the decorrelation in time τ of the parallel phase relationship according to

$$\left\langle \exp\left(i \int^z k_\parallel dz' \right) \right\rangle \cong \exp\left(-\tfrac{1}{6} k_\parallel'^2 v_\parallel^2 \chi_e \tau^3 \right)$$

where $k_\parallel' = \partial k_\parallel / \partial x = k_y / L_s$.

To describe both the turbulent phase stochastization and the collisional diffusion from electron-ion ν_{ei} collisions and electron-electron ν_{ee} collisions, Galeev (1978) introduces the renormalized kinetic equation with a Batnagar-Gross-Krook collision operator that conserves particles and parallel momentum. From this kinetic equation the fluctuating parallel electron current $j_\parallel(k)$ is calculated to determine the radial transport due to $\langle \delta B_x(k)^2 \rangle$. In the quasistatic, renormalized quasilinear regime defined by

$$\omega_k < \nu_{\text{eff}} < \nu_{ei} \tag{11.101}$$

the formula for the anomalous thermal conductivity reduces to

$$\chi_e = \chi_{\parallel e} \sum_k \left| \frac{\delta B_x(\boldsymbol{k})}{B} \right|^2 \frac{\nu_{\text{eff}}}{k_\parallel^2 \chi_{\parallel e} + \nu_{\text{eff}}} \tag{11.102}$$

where

$$\nu = \lambda_e^2 \left(k_\parallel' \right)^2 (\chi_{\perp e} + \chi_e) \tag{11.103}$$

with $\chi_{\parallel e} = \lambda_e^2 \nu_{ei}$ and $\chi_{\perp e} = \rho_e^2 \nu_{ei}$ being the classical parallel and perpendicular electron thermal conductivities. Here, $\lambda_e = v_e / \nu_{ei}$ is the electron mean-free-path and $\rho_e = v_e / \omega_{ce}$ the electron gyroradius.

The $k_\parallel$ width of the correlation function is $\Delta k_\parallel = (\nu_{\text{eff}} / \chi_{\parallel e})^{1/2}$ which defines the characteristic scale length L^* for stochastization of the electron-fluctuation phase

$$L_e^* = \frac{1}{\Delta k_\parallel} \lambda_e \left(\frac{\nu_{ei}}{\nu_{\text{eff}}} \right)^{1/2} . \tag{11.104}$$

When the $\delta B_x(k)$ spectrum is broad compared with $\Delta k_\parallel$ the conductivity formula reduces to

$$\chi_e = v_e \frac{\lambda_e}{L_e^*} D_F \tag{11.105}$$

where the magnetic field diffusion coefficient D_F is given in Eq. (11.78). At the collisionless limit of condition Eq. (11.101) where the mean-free-path exceeds the phase stochastization length the electrons diffuse at the same rate as the magnetic field

$$\chi_e = v_e D_F \tag{11.106}$$

as discussed in the previous subsection. For $\lambda_e < L_e^*$ the anomalous transport is reduced by λ_e / L_e^* but does not go directly to the collisional regime of Rechester and Rosenbluth (1978) given in Eq. (11.81). Between the collisional and collisionless regime there is a plateau or semicollisional regime where the mean-free-path is long compared to the shear length $L_s \approx L_0$, but shorter than the phase stochastization length L_e^*. In this amplitude regime where χ_e exceeds $\chi_{\perp e}$ the length L_e^* varies inversely with $\chi_e^{1/2}$ according to Eq. (11.103) and Eq. (11.104). Expressing L_e^* in terms of χ_e and solving Eq. (11.102) yields the plateau thermal conductivity

$$\chi_e = \chi_{\parallel e} b_0^4 \overline{k}_y^{-2} \lambda_e^2 \qquad (11.107)$$

where b_0 is defined in the previous subsection and $\overline{k}_y$ is a mean wavenumber in the spectrum. The plateau or semicollisional formula Eq. (11.107) is valid for b_0 constrained by $\chi_e > \chi_{\perp e}$ and $\nu_{\text{eff}} < \nu_{ei}$, or equivalently $L_e^* > \lambda_e$, which defines the domain

$$\frac{L_s^{1/2}}{\overline{k}_y \lambda_e^{3/2}} > b_0 > \left(\frac{\rho_e}{\overline{k}_y \lambda_e^2} \right)^{1/2} . \qquad (11.108)$$

For b_0 less than the lower limit in Eq. (11.108), χ_e equals $\chi_{\perp e}$ plus a small increment from the magnetic turbulence. For higher fluctuation levels the mean-free-path exceeds the stochastization length and the collisionless formula Eq. (11.106) applies. At still larger amplitudes $b_0 > \delta / L_s$ the thermal conductivity becomes $\chi_e = v_e b_0 \delta_x$.

The limited region of radial correlation in the magnetic turbulence also influences the form of the anomalous thermal flux. Here, we define $\delta_x = l/\overline{k}_x$ as the radial scale distance over which the magnetic fluctuations are correlated. There are two effects that arise from the finite radial correlation distance. First, there is a further loss of the phase coherence due to the perpendicular diffusion over δ_x, which is described by $\nu_{\text{eff}} \to \nu_{\text{eff}} + k^2 \chi_e$; and secondly, there is an additional decorrelation from the parallel motion of the electron along the radially tilted field line which is described by the nonlinear $\boldsymbol{k} \cdot \boldsymbol{B} = k_{\parallel 0} B + k_x \delta B_x$. With regard to the first effect, a comparison of $k_x^2 \chi_e$ with ν_{eff} in Eq. (11.103) shows that the radial decorrelation dominates when

$$\frac{L_s}{\lambda_e} > \frac{\overline{k}_y}{k_x} = \overline{k}_y \delta_x \qquad (11.109)$$

From the second effect it is adequate to take into account that the mean value of the stochastic $k_\parallel$ is given by

$$\left\langle k_\parallel^2 \right\rangle = k_\parallel^2 + k_x^2 b_0^2 \qquad (11.110)$$

from which it is evident that the nonlinear broadening of the parallel phase dominates when

$$b_0 > \frac{\max(k_\parallel)}{k_x} = \frac{\delta_x}{L_s} . \qquad (11.111)$$

Taking these two effects into account generalizes correlation function in Eq. (11.102) so that the quasistatic thermal conductivity becomes

$$\chi_e = \chi_{\| e} \sum_{\boldsymbol{k}} \left| \frac{B_x(\boldsymbol{k})}{B} \right|^2 \frac{\nu_{\text{eff}} + k_x^2 \chi_e}{\left(k_\|^2 + k_x^2 b_0^2\right) \chi_{\| e} + \nu_{\text{eff}} + k_x^2 \chi_e}. \tag{11.112}$$

The approximation of replacing $k_\|^2$ the average nonlinear $\left\langle k_\|^2 \right\rangle$ given by Eq. (11.110) appears adequate to recover the essential results of the nonlinear physics.

First, it is observed from Eq. (11.112) that in the quasilinear regime, $b_0 < \delta_x / L_s$, that the scale length for stochastization of the electron phase is shorter by $1/L_e^* = \Delta k_\| = [(\nu_{\text{eff}} + k_x^2 \chi_e)/\chi_{\| e}]^{1/2} \cong k_x (\chi_e / \chi_{\| e})^{1/2}$ when condition Eq. (11.109) is satisfied. For $\lambda_e < L_e^*$ the plateau formula that follows from L_e^* is

$$\chi_e = \chi_{\| e} \left(\frac{b_0^4 L_s^2}{\delta_x^2} \right). \tag{11.113}$$

which agrees with Eq. (11.97) from Kadomtsev and Pogutse (1978) and connects the thermal conductivity to Eq. (11.107) when condition Eq. (11.109) is reversed. For the anomalous conductivity χ_e to exceed the collisional conductivity $\chi_\perp$, it is necessary that $b_0 > (\rho_e \delta_x / \lambda_e L_s)^{1/2}$.

In the nonlinear regime $b_0 > \delta_x / L_s$ the stochastization of the parallel phase is unimportant according to Eq. (11.112). Clearly, there remain two subdomains depending on the magnitude of the nonlinear broadening $k_x^2 b_0^2 \chi_{\| e}$. In the lower amplitude subdomain clearly Eq. (11.112) reduces simply to

$$\chi_e = \chi_{\| e} b_0^2 \tag{11.114}$$

where now the condition for applicability becomes $\delta_x / L_x < b_0 < \delta / \lambda_e$. When the nonlinear parallel decorrelation is substantial, the only consistent solution of Eq. (11.112) has $k_x^2 b_0^2 \chi_{\| e} \sim k_x^2 \chi_e$ and consequently $\chi_e \cong \chi_{\| e} b_0^2$, valid when the mean-free-path is less than the parallel correlation length. The parallel correlation length is given by $1/L_\| \sim k_x b_0 \sim b_0 / \delta_x$. Thus, the well-known self-similar form $\chi_e \cong \chi_{\| e} b_0^2$ remains valid until $b_0 > \delta_x / \lambda_e$ whereupon

$$\chi_e = v_e D_F \cong v_e b_0 \delta_x \tag{11.115}$$

gives the nonlinear, collisionless thermal transport.

The self-similar result $\chi_e = \chi_{\| e} \langle \delta B_x^2 / B^2 \rangle$ is universal in the sense that the formula applies for both strong collisions and high-level magnetic fluctuations. In the regime of strong collisions the result is obtained by Rechester and Rosenbluth (1978). As shown by the kinetic analysis in this section, there are plateau or semi-collisional regimes of practical importance where the transport follows different laws.

Additional studies clarifying the assumptions [Rose (1982)] and relationships [Krommes, *et al.* (1983)] between the different regimes of anomalous magnetic transport have been developed. Closely-related formulas for χ_e from electromagnetic fluctuations are derived in Zhang and Mahajan (1988), Hong and Horton (1989), and Horton, *et al.* (1987).

In conclusion, there are different mechanisms for anomalous transport depending on the regime of the plasma. At the present, there are no simple rules for determining which of the formulas presented here govern transport in a specific plasma. The method now used is to evaluate the stability to the various drift waves to determine the dominant type of magnetic turbulent fluctuations. Transport codes like the Multimode Code of Bateman and Kinsey carry out this analysis and use transport coefficients derived for the dominant instability. This method has been successful in explaining many to tokamak transport features for a variety of machines and in different confinement modes. The method is exploredin some detail in Chapter 18.

References

Callen, J. D. (1977). Drift-wave turbulence effects on magnetic structure and plasma transport in tokamaks, *Phys. Rev. Lett.* **39**, pp. 1540-1543, `http://link.aps.org/doi/10.1103/PhysRevLett.39.1540`.

Colas, L., Zou, X. L., Paume, M., Chareau, J., Guiziou, L., Hoang, G. T., Michelot, Y., Grésillon, D. (1998). Internal magnetic fluctuations and electron heat transport in the Tora Supra Tokamak: Observation by cross-polarization scattering, *Nuclear Fusion* **38**, 6, pp. 903-918 .

da Silva, D. C., Caldas, I. L., and Viana, R. L. (2001). Field-line diffusion and loss in a tokamak with an ergoci magnetic limiter, *Phys. Plasmas* **8**.

Dendy, R. O., Hastie, R. J., McClements, K. G., and Martin, T. J. (1995). A model for ideal $m = 1$ internal kink stabilization by minority ion cyclotron resonant heating, *Phys. Plasmas* **2**, p. 1623, `http://dx.doi.org/10.1063/1.871457`.

Dupree, T. H. (1967). Nonlinear theory of drift-wave turbulence and enhanced diffusion, *Phys. Fluids* **10**, p. 1049, doi:10.1063/1.1762220.

Fried, B. D. and Conte, S. S. (1961). *The Plasma Dispersion Function* (New York: Academic Press).

Galeev, A. A. (1978). Stationary magnetic field line reconnection in a tearing-unstable plasma slab, *Phys. Fluids* **21**, p. 1353, doi:10.1063/1.862376.

Hong, B. G. and Horton, W. (1989). Anomalous electron thermal conduction from magnetic turbulence, (Brief Communications) *Phys. Fluids B* **1**, 12, p. 2527-2530.

Horton, W., Choi, D. I., and Hong, B. G. (1983) *Phys. Fluids* **26**, p. 1461, doi:10.1063/1.864316.

Horton, W., Choi, D. I., Yushmanov, P. N., and Parail, V. V. (1987). Electron diffusion in tokamaks due to electromagnetic fluctuations, *Plasma Phys.* **29**, p. 901, doi:10.1088/0741-3335/29/7/008.

Ishizawa, A. and Nakajima, N. (2010). Turbulence driven magnetic reconnection causing long-wavelength magnetic islands, *Phys. Plasmas* **17**, p. 072308, doi:10.1063/1.1388176, `http://link.aip.org/link/doi/10.1063/1.1388176`.

Kadomtsev, B. B. and Pogutse, O. P. (1978). *Proc. 7th Int. Conf. on Plasma Phys. Control. Nucl. Fusion Research* (International Atomic Energy Agency, Vienna), **I**, p. 649.

Krommes, J. A., Oberman, C., and Kleva, R. G. (1983). *J. Plasma Phys.* **30**, p. 11, doi:10.1017/S0022377800000982.

Mazzucato, E. (1976). *Phys. Rev. Lett.* **36**, p. 792, `http://link.aps.org/doi/10.1103/PhysRevLett.36.792`.

Mikhailovskii, A. B. and Fridman, A. M. (1967). *Zh. Tekh. Fiz.* **37**, p. 1782.

Muragalia, *et al.* (2010) *Nucl. Fusion.*

Ohkawa, T. (1978). *Phys. Lett. A* **67**, p. 35.

Rechester, A. B. and Rosenbluth, M. N. (1978), Electron heat transport in a tokamak with destroyed magnetic surfaces, *Phys. Rev. Lett.* **40**, pp. 38-41, doi:10.1103/PhysRevLett.40.38.

Rose, H. A. (1982). Test-particle transport in stochastic magnetic fields: A fluid representation, *Phys. Rev. Lett.* **48**, pp. 260-263, doi:10.1103/PhysRevLett.48.260

Stix, T. H. (1973). Magnetic Braiding in a Toroidal Plasma, *Phys. Rev. Lett.* **30**, pp. 833-835, doi:10.1103/PhysRevLett.30.833.

Zhang, Y. Z. and Mahajan, S. M. (1988). Renormalized perturbation theory: Vlasov-Poisson system, weak turbulence limit, and gyrokinetics, *Phys. Fluids* **31**, p. 2894, `http://dx.doi.org/10.1063/1.866998`.S

Zheng, L. J. and Furukawa, M. (2010). Current-interchange tearing modes: Conversion of interchange-type modes to tearing modes, *Phys. Plasmas* **17**, p. 052508, doi:10.1063/1.3425857.

Chapter 12

Toroidal Confinement Systems

12.1 Toroidal System with High Temperature Plasmas

Most of this book is concerned with the turbulent transport in the tokamak confinement system which has set the records for the highest achieved fusion power in the laboratory. Looking ahead to achieving a steady-state fusion power system with full external control of the plasma conditions, one is led to search for a toroidal magnetic confinement system without the large internal plasma current that must be maintained in the tokamak by an external driver.

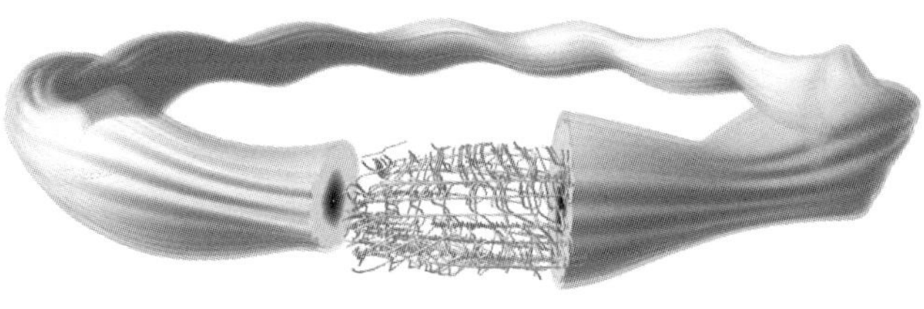

Fig. 12.1 Simulation of the Large Helical Device plasma with helical period periodicity $\ell = 2$ and toroidal periodicity $n = 10$ during an unstable motion when the plasma pressure exceeds that of ideal MHD stability. The motion is driven by the curvature of the toroidal system with line-bending magnetic field tension as the chief stabilizing force. The treads shown in the core of the cut-away section are drawn from a 3D rendering of the shear mass flows generated by the Reynolds stresses from the MHD usable motion. Owing to the $\ell = 2, n = 10$ variations of the equilibrium there is extensive mode coupling from the $\cos\theta$ ballooning interchange motion in the MHD unstable motion [Miura and Nakajima (2010)].

The alternative to providing the confining poloidal magnetic field by the large plasma current I_p is to design external coils that create the helical magnetic field along with the toroidal magnetic field. From a sketch of these two magnetic field structures one is led to consider helical coils around the torus. The Large Helical Device is a superconducting toroidal magnetic confinement machine designed from experience with the earlier series of Heliotron machines built in Kyoto University in the 1980-1990s. The physics of Heliotron E, the last machine in the series before construction of the LHD, is thoroughly documented in the book, *Stellarator and Heliotron Devices* by Wakatani (1998).

The stability analysis of Fu, *et al.* (2000) shows that both magnetic curvature and the Pfirsch-Schlüter current contribute to the pressure limit of the MHD stability of the helical/stellarator device. They report that the system is stable without a close conducting wall up to dimensionless plasma pressure $\beta = 0.05$.

Helical toroidal systems have numerous advantages over the tokamak for the design of a steady state power reactor as discussed in Chapter 9.6 of Wakatani (1998). The external control of both the toroidal and poloidal magnetic fields is a large advantage in view of the unpredictable current disruption events know to occur in tokamaks operating near their peak performance. The helical system has a natural magnetic divertor with a large surface area from the stochasticity of the magnetic field lines readily understood from the bending of the straight $R = \infty$ helical coils with $\ell = 2$ and $n = 10$ with its regular magnetic separatrix into the toroidal shape which adds the standard $B_{\rm tor} = B_0/[1+r/R\cos(\theta)]$ toroidal field as a perturbation $B_n = B_n(r/a)^2\cos(2\theta - 10\phi)$ producing a wide edge region of stochastic magnetic field lines. This combination of the helical and toroidal magnetic field produces what is called an ergodic limiter that has been studied in several tokamaks including TEXT, JFT-2M and Tore Supra. The ergodic divertor lowers the power delivered to the plasma facing components. In his Chapters 8 and 9 Wakatani describes in some detail the favorable features of a a steady-state fusion reactor based on the design of the like the Large Helical Device at NIFS.

Gyrokinetic simulations of the drift waves in the Wendelstein 7-X stellarator are given in Xanthopoulos and Jenko (2007). For LHD the simulations are in Nunami, *et al.* (2012).

The six rings of poloidal field coils shift the plasma to a smaller major radius in the left column of Fig. 12.2 and to a larger major radius in the right column. For the inward shift there is a magnetic hill with $t(0) < 0.5$ and for the outward shift there is a magnetic well with $t(0) > 0.5$ in the right column of Fig. 12.2. For the outward shift plasma there are substantial density "ears" in the plasma $n_e(p)$ profile from a transport barrier as shown in Fig. 12.2b.

12.2 Helical Toroidal Systems: LHD, Heliotrons, and Stellarators

The torodial heliotrons like the Large Helical Device shown in Fig. 12.2 are relatively simple in design with respect to the current carrying coil design. The helical toroidal machines now have a large database. The transport is rather difficult to compute due to the nonaxisymmetry of the combined helical ripples $\epsilon_h\cos(2\theta - 10\phi)$ and smaller toroidal magnetic field strength modulation as $\epsilon_t\cos(\theta)$. The broken symmetry leads to the nonambipolar rate of ion and electron transport creating two

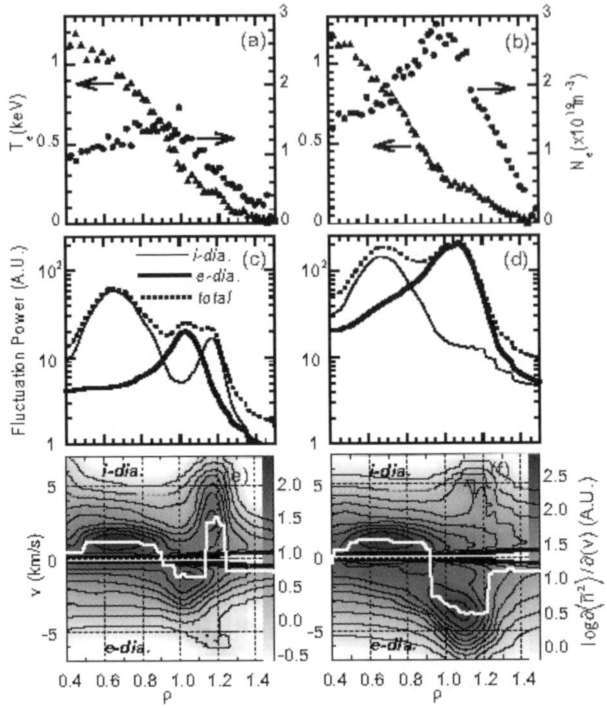

Fig. 12.2 The results of CO_2 laser PCI imaging for an L-H transition in the LHD plasma from Tanaka, *et al.* (2008). The left column is data at $t = 2.5$ s before the transition and the right column is at $t = 2.7$ s after the transition. (a)-(b) show the YAG laser incoherent Thomson scattering derived electron density and temperature profiles. (c) and (d) show the profiles of fluctuation power. (e) and (f) show fluctuation components are poloidally dominated and Doppler shifted by $E_r \times B_t$ poloidal rotation.

distinct modes of operation. The tokamak-like regime is called the ion-root where the ions are lost faster than the electrons and there is an inward radial electric field at the stable ion point of matched ion and electron radial transport. This regime occurs in hot-ion plasmas created by either NBI or ICRF as the dominant plasma heating process. Ion dominant heating occurs with positive ion based NBI with acceleration voltage 40 KV and ICRF heating with large contents of minority ions. In ICRF heated plasma, E_r field was not measured yet since they do not inject NBI which is necessary for CXRS measurements. Also, with ion heating case using positive ion based NBI, E_r is sometimes positive in edge region ($\rho > 0.7$) [Tanaka, *et al.* (2010); Ida, *et al.* (2001)]. With strong ion heating at high density where energy equipartition from electron to ion is fast, the E_r is negative.

Figure 12.2 shows a snapshot for the fluctuation spectrum across the entire cross-section of the plasma for wavenumbers in the range of the drift wave fluctuations in the L-mode at $t = 2.5$ s on the left and after a transition to the H-mode at $t = 2.7$ on the right for a Large Helical Device plasma. The top-left and top-right figure

Table 12.1 Plasma Parameters Achieved in Highlighted Plasmas in LHD.

$T_i(0)$(KeV)	$T_e(0)$(KeV)	$n_e(0)(10^{19}\ \mathrm{m}^{-3})$	$\beta(\%)$	$\tau_E(s)$	P_{abs} (MW)	$B(T)$
0.22	0.22	125	1.43	0.039	21.4	2.51
-	0.43	2.3	5.1	0.040	11.9	0.425
5.6	3.8	1.6	0.80	0.046	19.1	2.9
-	18.9	0.24	0.19	0.043	3.7	2.71

shows the electron density and temperature profiles. The middle left and right figures show the intensity of the electron density fluctuations in the wavenumber range $k = 10\,\mathrm{cm}^{-1}$ before and after the L-H transition. The lower row of figures gives the radial variation of the dominant power resolved by the direction of the phase velocity. The rotation in the ion-diagmagnetic velocity arises from a positive radial electric field and rotation in the electron-diamagnetic direction arises from the negative radial electric field. Thus, the transition to the H-mode is consistent with a substantially increased inward (negative) pointing radial electric field. This is also typical for tokamak plasmas that have stronger negative radial electric fields and are required for higher confinement. From this data one sees the ability of the phase contrast imaging method to give time-resolved fluctuation levels over large spatial regions [Tanaka, *et al.* (2008)].

The second stable root from the condition that the total ion radial loss rate balance the electron loss rate occurs in hot electron plasmas driven by strong electron cyclotron resonance heating ECRH power. This regime is called the electron root and the radial electric field points outward or $E_r > 0$. In this state the plasma rotates in the ion diamagnetic direction with respect to the laboratory frame. The PCI infrared laser imaging system is able to see both these rotation regimes and confirms the general description of the radial electric field and the plasma rotation [Tanaka, *et al.* (2008)]. In addition the LHD has a heavy ion beam probe HIBP that directly measures the local plasma electric potential in the plasma. Both regimes have their advantages. In Table 12.1 the record parameters for plasma confinement in terms of plasma temperatures and densities and beta achieved in different regimes are shown as reported by the Yamada for the LHD group at the 23 IAEA Fusion Conference in 2010. LHD has heavey ion beam probe HIBP that directly measures the local plasma electric potential in the plasma and the CXRS diagnostic. The radial electric field is inferred by a combination of the HIBP and the CXRS signals.

An example of the turbulent transport in the LHD device for the hot ion mode is shown in Fig. 12.3 from simulations by Nunami, *et al.* (2011). In this simulation the ion temperature gradient drives the drift wave turbulence and the electrons may be taken with the simple adiabatic response. The figure shows a snapshot in time of the nonlinear state of turbulence with shaded contours of the electric potential turbulence for annular volume sectors centered on flux surfaces locations at $r/a = 0.46$, 0.65, and 0.83. The ambient background plasma profiles are taken

Fig. 12.3 Contours of the electric potential turbulence for a simulation of the ion temperature gradient driven instability in the Large Helical Device. The simulations from the GKV-X code are performed with simple adiabatic electron response for the electron dynamics at the three different radial flux surfaces locations at $r/a = 0.46$, 0.65, and 0.83 in the non-axisymmetric helical torus. The ambient background plasma profiles are taken from codes for the magnetic configuration reconstructed from the experimental data for the high ion temperature discharge.

from codes for the magnetic configuration with reconstructed from the experimental data for the high ion temperature discharge. The non-axisymmetric geometry is complex and requires the use of the GKV-X code described further in Chapter 19 for describing the drift wave turbulence in the Large Helical Device.

The continuum gyrokinetic code of Watanabe, *et al.* (2007) computes the ion temperature gradient driven drift wave turbulence with the adiabatic $\delta n_e = en_0\delta\varphi/T_e$ response for plasma profiles taken from experimental data. A typical result for the electrostatic potential is shown in Fig. 12.3 and on the book cover.

12.3 Neoclassical Dynamics and Transport in Toroidal Systems

Chapter 7.3.4 develops the resistive-viscous partial differential equations for the plasma flows and associated rotation with convective transport in Eqs. (7.58)-(7.53). In Eq. (7.76) the parallel resistivity $\eta_\|$ produces convective cells in the toroidal equilibrium with a $m = 1$ or $\omega s\theta$ and $\sin\theta$ structure as a result of the guiding-center drifts in the toroidal curvature of the axisymmetric toroidal system. The $\boldsymbol{E} \times \boldsymbol{B}$-convection of plasma from these

$$\frac{\nu_\alpha}{(\Delta B/B)} < \frac{v_\alpha}{L_c}\left(\frac{\Delta B}{B}\right)^{1/2} \tag{12.1}$$

become the dominant loss mechanism from the Coulomb collisional scattering rate ν_α giving rotations of the particle velocity vector v over 90° in the time

$$\tau_\alpha = \frac{1}{\nu_\alpha} = \frac{4\pi\epsilon_0^2 m_\alpha^2 \mathrm{v}_\alpha^3}{n_\beta e_\alpha^2 e_\beta^2 \ell n\,\Delta} \tag{12.2}$$

from collisions with particles n_β of charge e_β.

There are various definitions of the collisional scattering rate $\nu_{\alpha/\beta}$ depending on the calculation being performed. The definition given in Eq. (12.2) is from the Appendix of Helander and Sigmar (2002). See the NRL formulary at `http://nrlformulary.gov` for another commonly used definition of $\nu_{\alpha/\beta}$.

In the banana regime the locus of the guiding-center motion projected onto a poloidal cross-section of the torus traces out the shape of a banana. Other projections of the guiding-center orbits on the cross-section can give figures resembling potatoes. Clearly, these large orbital excursions create a fast loss of particles when the Coulomb scattering on collective modes of drift waves is taken into account. The Coulomb scattering loss rate is intrinsic to the system and is the subject of the books on tokamaks and helical toroidal confinement systems [Helander and Sigmar (2002); Wakatani (1998)]. The potential-parallel current systems given in Eq. (7.76) is called the Pfirsch-Schlüter (PS) regime of plasma transport. As the temperature of the plasma increases this transport mechanism makes a transition from the PS-regime to first a "plateau" regime in which the dissipation from the Landau-particle resonance at

Plateau Regime

$$k_\parallel \mathrm{v}_\parallel + \omega_D = 0 \tag{12.3}$$

replaces the resistivity $\eta_\parallel$. Further increase in the temperature in the plasma enters the "banana" regime in which the trapped particles occur. The collective drift-wave instabilities associated with these large trapped particle orbits are called trapped particle modes. There is a trapped electron mode

TEM

with frequency

$$\omega_{\boldsymbol{k}}(\mathrm{TEM}) \cong \frac{\epsilon^{1/2}\omega_{*e} I_0(b) e^{-b}}{1+\frac{T_e}{T_i}\left(1-I_0(b)e^{-b}\right)} < \frac{v_e}{L_c}\left(\frac{\Delta B}{B}\right)^{1/2} \tag{12.4}$$

and a lower-frequency trapped ion mode

TIM

$$\omega_{\boldsymbol{k}}(\mathrm{TIM}) = \frac{\epsilon^{1/2}\omega_{*i} F\left(k_\perp^2 \rho_b^2\right)}{1+T_i/T_e} < \frac{v_i}{L_c}\left(\frac{\Delta B}{B}\right)^{1/2} \tag{12.5}$$

where $\Delta B/B$ is the depth of the magnetic well over connection length L_c that traps the guiding-center motion.

These trapped particle instabilities and transport mechanisms are universal in that they occur in tokamaks, toroidal helical systems, stellarators and mirror machines. The transport regimes are different for the banana regime and the plateau regime.

12.4 Large Helical System Fields and Transport

The structure for the fluctuations in the toroidal geometry is a more complicated problem than first appears. While the Fourier decomposition in the double periodicity in the two toroidal angles ϕ and θ is a natural choice for satisfying the single valued condition on the toroidal surface ψ, such a decomposition as $\delta n_e \cos(m\theta + n\phi - \omega t)$ does not describe the alignment of the fluctuation with the magnetic flux tube away from the rational surfaces. The rational surfaces are of special importance but have zero measure in terms of their density. Thus we need to introduce representations that allow the conflicting requirements of weak variation along the magnetic field lines while maintaining single valued functions in the toroidal magnetic geometry. The methods used are called the ballooning representation and in simulation codes in Chapter 19 the flux tube representation, it does describe well the long, thin flux tubes defined by the helical magnetic fields $\boldsymbol{B} = (B_\theta, B_\phi)$. The fast motion and waves parallel to the helical magnetic field is the dominant effect in determining the mode structure and long, thin fluctuation filaments.

12.4.1 *Fluctuations in the ballooning mode representation*

For fluctuations to be unstable in both MHD models and for the full kinetic theory dynamics the fluctuating potential φ must satisfy, to the lowest order, the magnetic differential equation

$$\boldsymbol{B} \cdot \nabla \varphi = 0 \tag{12.6}$$

for field lines that have an irrational value of the winding number or twist function

$$q(\psi) = \frac{rB_\phi}{RB_\theta} \neq \text{rational number.} \tag{12.7}$$

Then the condition Eq. (12.6) and the condition of double periodicity are in conflict for the fluctuations simply expanded in the double Fourier series $\exp(im\theta - in\phi)$ since the operator in the magnetic differential equation then gives

$$\boldsymbol{B} \cdot \nabla \varphi = ik_\parallel \varphi \neq 0. \tag{12.8}$$

The resolution of this difficulty is given by the ballooning mode representation developed in Connor, *et al.* (1978). The idea behind the method is to extend the flux tubes to infinity and use a new coordinate η on the infinite domain to describe the variation along the field line. The functions are Fourier decomposed in $\exp(im\eta)$ modes on the infinite domain $\eta = [-\infty, = \infty]$. These Fourier coefficients $\varphi_m = \int d\eta \exp(im\eta)\varphi(\eta)$ are then used to construct the periodic function in the poloidal angle θ through the Fourier summation

$$\varphi(\theta, \psi) = \sum_m \varphi_m \exp(-im\theta). \tag{12.9}$$

This decomposition is performed for each toroidal mode number n with the nonlinear fields then being the summation over all the ballooning modes with $\sum_n \exp(in\phi)\varphi(\theta, \psi)$.

The details of the theoretical construction are given [Connor, *et al.* (1978); Connor, *et al.* (1991); Connor, *et al.* (1991); Hastie, *et al.* (1979)] and in the expanded article in the *Proceedings of the Royal Society*. The reasoning is intricate and the calculations long and complicated. A small expansion parameter is introduced to solve the equations in the order: (i) the ballooning eigenmode structure for variation of the fluctuation along the flux tube given by the function $\varphi(\eta)$ on the infinite domain of the coordinate η. This is called the ballooning mode eigenfunction and typically extends well beyond the spatial domain of 2π of the poloidal coordinate. The eigenvalue for this 1D eigenmode equation is then a key function called $\omega_0(\psi, \theta_0, n)$; (ii) the first-order correction to the lowest order equation is derived; (iii) the second order correction to the eigenvalue equation is derived. The effective potential of the second-order equation then determines the radial mode structure. This equation has a bounded solution when the variation of the lowest order eigenvalue $\omega_0(\theta_0)$ with respect to the variable θ_0 vanishes. With this condition satisfied there is then a rather standard second-order radial Storm-Liouville eigenvalue problem to solve for the radial mode structure in the flux coordinate ψ.

There are only a limited number of cases where this full ballooning procedure has been carried out. In the case of MHD without flow the self-adjointness of the operators in the formula for δW gives considerable simplification and the effective radial potential for the radial eigenvalue problem can be derived from the radial variation of the ω^2(ψ radial coordinate). This MHD ω^2 is extremely important in itself since it defines what is a stable magnetic confinement system. So this class of problems has been solved extensively for a wide range of axisymmetric toroidal machines.

12.4.2 *Drift waves in the ballooning representation*

For the case of drift wave fluctuations the operators in the linear partial differential equation are not self-adjoint and there is no useful variational form of the associated eigenvalue problem. Here the ballooning mode theory procedure following the Connor, *et al.* model is worked out by Choi and Horton (1980). The procedure is essentially the same as described above. The theory is interesting and the results show that often the drift waves in the torus are strongly localized to the outer equatorial plane and in other cases to the top and bottom of the torus.

Energetically, we can understand these localizations by the destabilizing effect of the magnetic curvature vector $\boldsymbol{\kappa} = \boldsymbol{b} \cdot \boldsymbol{\nabla} \boldsymbol{b}$. At the outboard mid-plane or equatorial plane there is a large effective acceleration of the plasma given by $g_{\text{eff}} = 2v_{th}^2/R_c$ which produces a local ballooning interchange driving mechanism. For two strong a localization the stabilizing role of the ion acoustic waves over comes to the destabilization from the g_{eff}. At the top and bottom of the torus where the poloidal angle $\theta = \pm\pi/2$ the effective g_{eff} acceleration is horizontal and thus tangential to the flux surface at $\theta = \pm\pi/2$. This also creates and interchange release of energy, but the wave function must be peaked in these regions just before the values $\theta = \pm\pi/2$ to use this effective force to move plasma down the ∇B.

In Fig. 12.4 the typical drift wave ballooning mode structure is shown and in Fig. 12.5 the corresponding eigenvalue shown versus the strength of the toroidicity parameter $\epsilon = r/R$. Note the sudden decrease in the damping rate of the eigenvalue for the toroidicity induced mode for $\epsilon \gtrsim 0.05$ for the symmetric eigenmode labeled with the filled circles in Fig. 12.5 [Chen and Cheng (1980)].

The ballooning mode representation gives the method of calculating the parallel wavenumber $k_\parallel$ for general flux surfaces with irrational magnetic winding numbers $q(\psi)$. The parallel wavenumber $k_\parallel$ has a spectrum due to the summation over the integer m values of the transform of the infinite domain ballooning mode. These means that the modes can now be radial extended structures with the wave functions in the linear mode being a coherent sum of localized ballooning modes. This behavior gives rise to a rich spectrum seen experimentally most clear in the Alfvén modes in the tokamak. Since the Alfvén wave frequency is $\omega^2 = k_\parallel^2 v_A^2$ the overlapping and coupling of the different poloidal harmonics is most critical. This result is that new modes called the Alfvén gap mode modes occur as discrete modes in the gaps between the $k_\parallel$ lines drawn for different poloidal mode numbers.

A similar overlapping of the linear poloidal modes occurs for the drift waves where again the value of $k_\parallel$ is critically important for the mode structure and growth rates. In Fig. 12.4 the variation of the resonant ion region for the ITG modes in the torus are shown as the value of $k_\parallel$ changes from near zero in part (a) at the top of the figure, to an intermediate value in frame (b) and to the larger value in frame (c). The nature of the mode and its growth rate is changing from the toroidal ITG in the frame (a) to the slab-like ITG mode in frame (c). The details of the how the ITG eigenmode in the ballooning mode representation is given in Rewoldt, *et al.* (1982), Rewoldt and Tang (1983), Dong, *et al.* (1992), and Horton, *et al.* (1979).

The simulation codes use what is called the flux tube representation that is a numerical version of the infinite domain coordinate η defined in Eq. (12.8). The development of this numerically flux tube representation that is described in Cowley and Sudan (1997) and Beer and Hammett (1996). The work is rather technical so we

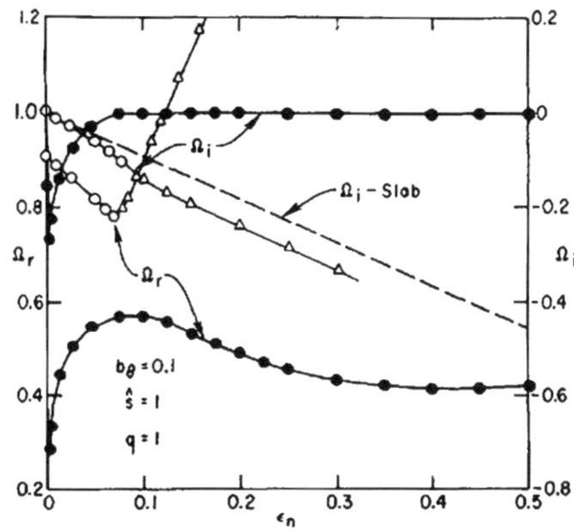

Fig. 12.4 Plot of eigenmode frequencies Ω vs. ϵ_n for $b_\theta = 0.1$, $q = 1$, and $\widehat{s} = 1$. $\circ$, $\bullet$ and Δ correspond, respectively, to the $n = 0$ slab-like, $n = 0$ and $n > 0$ toroidicity-induced eigenmodes [Chen and Cheng (1980)].

leave this discussion here and refer the reader to these works and reviews by Brizard and Hahm (2007) for the details of how the flux tube method is implemented in the simulation and gyrokinetic equations. The method used by Rewoldt is rather simpler consisting of a set of harmonic oscillator functions as given in Chapter 3 on each rational surface up to the maximum number of rational surfaces kept the simulations.

The relationship between the kinetically-modified MHD ballooning mode and the finite-β ITG mode is explored in Hong, *et al.* (1989) for tokamaks and for a stellarator in Hong, *et al.* (1991) and Yagi, *et al.* (1989). Boozer, *et al.* (1990) compare the properties of stellarators and tokamaks.

12.5 Toroidal Alfvén Eigenmodes

Fast neutral beam injection that exceeds the stability limit of the internal $m = 1$ modes gives rise to nonlinear relaxation oscillations called "fishbones" due to the character of the times series from the poloidal magnetic field probes. The bursts of the $m = 1$ oscillations ejects a fraction of the injected energetic ions and limits their density and pressure. A simple description of the dynamics consistent with the data from Heidbrink, *et al.* (1986) and McGuire, *et al.* (1983), is given by Coppi and Porcelli (1986).

Alpha particle driven toroidal Alfvén eigenmodes (TAEs) were first observed in deuterium-tritium plasmas in TFTR [Nazikian, *et al.* (1997)]. The modes were in the core of the plasma with amplitudes $\delta B/B \sim 10^{-5}$ with toroidal mode numbers $n = 2-4$ in a low magnetic shear regime. The dimensionless alpha particle pressure $\beta_\alpha \sim 2 \times 10^{-4}$ below the valued expected for onset of the modes from theory. A new interpretation of the alpha particle driven instabilities more consistent with the data from the TFTR experiments is given in Nazikian, *et al.* (2003). The Alfvén waves in the tokamak are coupled in their poloidal angle-θ variation due to the $B_\phi(\theta) = B_0/(1 + r/R \cos\theta)$ variation of the toroidal magnetic field. The coupling results in the formation discrete eigenmodes where the frequencies $\omega_\pm(m,n)$ of the uncoupled modes derived in the $\epsilon = r/R \to 0$ limit intersect. The frequency of the gap eigenmode is derived by setting the frequencies to Alfvén wave continuum modes $\omega_1^2 = k_\parallel^2(m,n)v_A^2$ and $\omega_2^2 = k_\parallel^2(m+1,n)v_A^2$ equal to each other in the limit $\epsilon \to 0$. The finite ϵ-coupling of the two continuum ω_1, ω_2 is degenerate at the radial position $r = r_{\rm TAE}$.

The presence of the toroidal Alfvén eigenmodes, or TAE modes, is analogous to the discrete bound electron hole-modes in the semi-conductors in the band-gaps of solid state physics. Whereas the Alfvén wave continuum modes have strong electron Landau damping, the discrete Alfvén gap modes are weakly damped and easily destabilized by energetic electron or ion Landau wave-particle resonances. The Alfvén gap modes have been extensively developed since the fusion plasma has a high energy population of alpha (He^4 helium) particles that destabilize the Alfvén gap modes. Laboratory simulation experiments by Wong, *et al.* (1991) with high energy neutral beam produced hydrogen nuclei show that the nonlinear TAE modes have a wide variety of frequency spectra of Alfvén waves, and that the waves produce a fast loss of the resonant ions through turbulent transport processes [Berk, *et al.* (1997); Todo, *et al.* (2010); Gorelenkov, *et al.* (2007); Fu and Van Dam (1989); Breizman and Sharapov (1995)]. In an homogeneous magnetized plasma, linear ideal-MHD arguments show the existence of a shear-Alfvén wave of frequency ω_A with the dispersion relation

$$\omega_A^2 = k_\parallel^2 v_A^2 \tag{12.10}$$

where v_A is the local Alfvén velocity and $k_\parallel$ is the wavenumber in the direction of the equilibrium magnetic field $\boldsymbol{B}_0$.

Let us consider axisymmetric toroidal plasmas. In the cylindrical limit, the periodicities of the system require that there exists two integers, a toroidal mode number n and a poloidal mode number m, such that

$$k_\parallel = \frac{n - m/q(r)}{R_0}, \tag{12.11}$$

where R_0 is the distance from the symmetry axis of the tokamak to the magnetic axis. In a non-homogeneous plasma in a sheared magnetic field, both $k_\parallel$ and v_A are functions of τ, where τ is the local radial flux coordinate. The simple dispersion relation Eq. (12.10) is still valid in this configuration and it is called the Alfvén continuum. Since phase velocity is a function of radius, a wave packet with finite radial extent would suffer from phase-mixing, the so-called continuum damping. Except for energetic particle modes, resonant drive-by fast particles is not enough to overcome this damping. However, a toroidal coupling of two successive poloidal modes m and $m+1$ breaks up the continuous spectrum. This is illustrated in

Fig. 12.4, which shows the Alfvén continuum for $n = 1$, $m = 2$, and $n = 1$ and $m = 3$, in cylindrical geometry, where two poloidal continuum are decoupled, and in toroidal geometry, with a two-mode coupling model. The latter is obtained with equilibrium plasma parameters corresponding to JT-60U shot E32359 at $t = 4.2\,\mathrm{s}$ assuming concentric circular magnetic flux surfaces, retaining toroidicity effects in the first order in inverse aspect ratio. Though we show only the $\omega > 0$ half-plane, the continuum is symmetric with respect to ω. Coupled modes are (n, m) and $(-n, -m-1)$ for $\omega > 0$, and $(n, m+1)$ and $(-n, -m)$ for $\omega < 0$. The gap is centered at a radius τ_A such that $q(\tau_A) = (m + 1/2)/n$, where the two continuous spectra would cross in the absence of coupling, and where $|k_{\|}| = 1/2qR_0$. The resulting discrete eigenmode is a TAE, at a frequency $\omega_A = v_A/2qR_0$.

For a deuterium plasma with typical magnetic field $B_0 \sim 1\,\mathrm{T}$ and density $n_i \sim 10^{20}\,\mathrm{m}^{-3}$, the Alfvénic energy is $E_A \equiv m_i v_A^2/2 \sim 10\,\mathrm{KeV}$, which is in the range of passing particles induced by NBI. For ITER parameters, $E_A \sim 1\,\mathrm{MeV}$, which is in the range of passing α-particles born from fusion reactions. In both cases, TAEs can be driven unstable by resonance with energetic particles. For far-passing particles, the resonance condition is $\Omega = \omega_A$, where

$$\Omega = n\omega_\zeta + l\omega_\theta, \tag{12.12}$$

where $\omega_\zeta = v_{\|}/R_0$ and $\omega_\theta = v_{\|}/qR_0$ are frequencies of toroidal motion and poloidal motion, respectively, and $l = -m$ for co-passing particles, $l = m$ for counter-passing particles [TS98]. Since we analyze TAEs driven by co-injected ions, we can simplify following discussions by considering only co-passing particles. Then, the resonance condition is

$$\omega_A - n\frac{v_{\|}}{R_0} + m\frac{v_{\|}}{qR_0} = 0. \tag{12.13}$$

In JET the TAE modes are routinely observed and used to interpret the $q(r,t)$ profiles in the evolving plasma. An early example is given by Fasoli, *et al.* (1995) for the measurement of the damping of toroidicity-induced Alfvén eigenmodes.

An important low-order TAE mode example, as shown in Fig. 12.4 occurs for the coupling of the $m = 1$ and $m = 2$ forward and backward $\omega - (m+1, n, r)$ Alfvén wave continuum modes. Writing the condition for

$$\omega_+ = k_{\|}(m,n,r)v_A \quad \text{and} \quad \omega_- = -k_{\|}(m,n,r)v_A \tag{12.14}$$

to be degenerate $\omega_+ = \omega_-$ gives

$$\left(\frac{n}{R} - \frac{m}{qR}\right)v_A = -\left(\frac{n}{R} - \frac{m+1}{qR}\right)v_A \tag{12.15}$$

which reduces to

$$\frac{2n}{R} = (m + 1 + m)/qR \tag{12.16}$$

giving the resonant radius for the gap mode of

$$q\left(r_0, m, n\right) = \frac{m + 1/2}{n}. \tag{12.17}$$

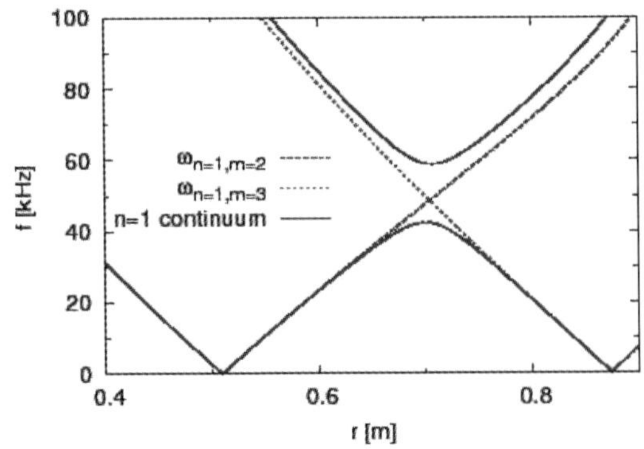

Fig. 12.5 Alfvén continuum for $n = 1$ with (solid line) and without (dashed line) coupling between $m = 2$ and $m = 3$ poloidal modes. Note that the discrepancy, relatively far from resonance, between the upper branch of coupled continuum and the uncoupled $m = 2$ branch, is accounted by terms of second-order in the inverse aspect ratio.

Extended radial TAE-modes occur for $m = 1, n = 1$ coupled to $m = 2, n = 1$ modes at $q(r_0, m, n) = m + (1/2)/n$ as shown in Fig. 12.5.

Nonlinear simulations of the energetic particle destabilization and evolution of these TAE modes is described in Todo, *et al.* (2010). The saturation level is shown to require both nonlinaer particle dynamics and nonlinear MHD dynamics when the amplitudes reach high levels with $\delta B/B \sim 10^{-2}$. There is a zonal glow and nonlinear coupling to high n-modes up to $n \sim 10$.

Other low-frequency MHD modes where the high energy injected ions create instabilities are called energetic particles modes or EP-modes for short. One of the more interesting and still incompletely understood of the energetic particle modes that is of concern as to how it might appear in the burning plasma driven by the fusion products of alpha particles is the "fish bone" mode. The name is derived from the plot of the fluctuating magnetic field versus time and has the structure resembling the skeleton with backbone of a large-flat bonita type of fish. Recent progress in the research of fishbones is given in Okabayashi, *et al.* (2011). The waves are driven by a population of high energy ions resonanting with the low frequency modes through the bounce averaged gradient-B drift frequency with the transit motion frequency. The polarization of the magnetic field fluctuations indicate the mode has a compressional magnetic component in addition to the line-bending shear Alfvén wave component. The radial gradient of the fast ion distribution function is important. The mode ejects some fraction of the fast ions from the plasma and repeats in rapid succession, creating the "fish-bone" signature on the magnetic field probes. In addition, the new results on DIII-D show that when the mode occurs sufficiently off-axis the magnetic activity destabilizes or drives the resistive-wall mode.Experiments showing the importance of the toroidal Alfvén eigenmodes were carried out on TFTR by Wong, *et al.* (1991).

References

Beer, M. A. and Hammett, G. W. (1996). Toroidal gyrofluid equations for simulations of tokamak turbulence, *Phys. Plasmas* **3**, 11, pp. 4046-4064, doi:10.1063/1.871538, `http://link.aip.org/link/?PHP/3/4046/1`.

Berk, H. L, Breizman, B. N., and Petviashvili, N. V. (1997). Spontaneous hole-clump pair creation in weakly unstable plasmas, *Phys. Lett. A* **234**, 3, pp. 213-218, doi:10.1016/S0375-9601(97)00523-9.

Boozer, A. H., Baldwin, D. E., Horton, W., Glasser, A. H., Krommes, J. A., Neilson, G. H., Shaing, K-C., Sadowski, W. L., and Weitzner, H. (1990). Altnerate transport, *Phys. Fluids B* **2**, 12, pp. 2870-2879, `http://dx.doi.org/10.1063/1.859357`.

Breizman, B. N. and Sharapov, S. E. (1995). *Plasma Phys. Control. Fusion* **37**, p. 1057.

Brizard, A. J. and Hahm, T. S. (2007). Foundations of nonlinear gyrokinetic theory, *Rev. Mod. Phys.* **79**, pp. 421-468, doi:10.1103/RevModPhys.79.421.

Chen, L. and Cheng, C. Z. (1980). Drift-wave eigenmodes in toroidal plasmas, *Phys. Fluids* **23**, p. 2242, doi:10.1063/1.862907.

Choi, D.-I. and Horton, W. (1980). *Phys. Fluids* **23**, pp. 356-365, doi:10.1063/1.862980.

Connor, J. W., Hastie, R. J., and Taylor, J. B. (1991a). Resonant magnetohydrodynamic modes with toroidal coupling. Part I: Tearing modes, *Phys. Fluids B* **3**, p. 1532, `http://dx.doi.org/10.1063/1.859724`.

Connor, J. W., Hastie, R. J., and Taylor, J. B. (1991b). Resonant magnetohydrodynamic modes with toroidal coupling. Part II: Ballooning-twisting modes, *Phys. Fluids B* **3**, p. 1539, `http://dx.doi.org/10.1063/1.859993`.

Connor, J. W., Hastie, R. J., and Taylor, J. B. (1978). *Phys. Rev. Lett.* **40**, p. 396, doi:10.1103/PhysRevLett.40.396.

Coppi, B. and Porcelli, F. (1986). Theoretical model of fishbone oscillations in magnetically-confined plasmas, *Phys. Rev. Lett.* **57**, p. 2272.

Cowley, S. C. and Sudan, R. N. (1997). Turbulence and intermittency in plasmas, *Phys. Reports* **282**, 1-4, p. 3-4, ISSN:0370-1573, doi:10.1016/S0370-1573(97)88184-7, `http://www.sciencedirect.com/science/article/pii/S0370157397881847`.

Dong, J. Q., Horton, W., and Kim, J-Y. (1992). Toroidal kinetic ITG mode study in high-temperature plasmas, *Phys. Fluids B* **4**, pp. 1867-1876.

Fasoli, A., Borba, D., Bosia, G., Campbell, D. J., Dobbing, J. A., Gormezano, C., Jacquinot, J., Lavanchy, P., Lister, J. B., Marmillod, P., Moret, J.-M., Santagiustina, A., and Sharapov, S. E. (1995). *Phys. Rev. Lett.* **75**, 4, pp. 645-648.

Fu, G. Y., Ku, L. P., Cooper, W. A., *et al.* (2000). Magnetohydrodynamics stability of compact stellarators, *Phys. Plasmas* **7**, pp. 1809-2000.

Fu, G. Y. and Van Dam, J. W. (1989). *Phys. Fluids B* **1**, p. 1949.

Gorelenkov, N. N., Berk, H. L., Fredrickson, E., Sharapov, S. E., and JET EFDA Contributors. (2007). Predictions and observations of low-shear beta-induced shear Alfvén-acoustic eigenmodes in toroidal plasmas, *Phys. Letts. A* **370**, 1, pp. 70-77, doi:10.1016/j.physleta.2007.05.113.

Hastie, R. J., Hesketh, K. W., and Taylor, J. B. (1979). Shear damping of two-dimensional drift waves in a large-aspect-ratio tokamak, *Nucl. Fusion* **19**, p. 1223, doi:10.1088/0029-5515/19/9/006.

Helander, P. and Sigmar, D. J. (2002). *Collisional Transport in Magnetized Plasmas* (Cambridge Monographs on Plasma Physics).

Heidbrink, W. W., *et al.* (1986). *Phys. Rev. Lett.* **57**, p. 835.

Hong, B.-G., Horton, W., Hamaguchi, S., Wakatani, M., Yagi, M., and Sugama, H. (1991). Transition from resistive-G to η_i driven turbulence in stellarator systems,

Phys. Fluids B **3**, 7, p. 1638, doi:10.1063/1.859683, `http://link.aip.org/link/doi/10.1063/1.859683`.

Hong, B.-G., Horton, W., and Choi, D-I. (1989). Pressure gradient-driven modes in finite beta toroidal plasmas, *Plasma Phys. Control. Fusion* **31**, p. 1291, doi:10.1088/0741-3335/31/8/006.

Horton, W., Jr., Choi, Duk-In, and Estes, R. D. (1979). Diamagnetic frequency profile effects on toroidal drift waves, *Phys. Fluids* **22**, pp. 519-521.

Ida, K., Funaba, H., Kado, S., Narihara, K., Tanaka, K., Takeiri, Y., Nakamura, Y., Ohyabu, N., Yamazaki, K., Yokoyama, M., Murakami, S., Ashikawa, N., deVries, P. C., Emoto, M., Goto, M., Idei, H., Ikeda, K., Inagaki, S., Inoue, N., Isobe, M., Itoh, K., Kaneko, O., Kawahata, K., Khlopenkov, K., Komori, A., Kubo, S., Kumazawa, R., Liang, Y., Masuzaki, S., Minami, T., Miyazawa, J., Morisaki, T., Morita, S., Mutoh, T., Muto, S., Nagayama, Y., Nakanishi, H., Nishimura, K., Noda, N., Notake, T., Kobuchi, T., Ohdachi, S., Ohkubo, K., Oka, Y., Osakabe, M., Ozaki, T., Pavlichenko, R. O., Peterson, B. J., Sagara, A., Saito, K., Sakakibara, S., Sakamoto, R., Sanuki, H., Sasao, H., Sasao, M., Sato, K., Sato, M., Seki, T., Shimozuma, T., Shoji, M., Suzuki, H., Sudo, S., Tamura, N., Toi, K., Tokuzawa, T., Torii, Y., Tsumori, K., Yamamoto, T., Yamada, H., Yamada, I., Yamaguchi, S., Yamamoto, S., Yoshimura, Watanabe, K. Y., Watari, T., Hamada, Y., Motojima, O., and Fujiwara, M. (2001). Reduction of ion thermal diffusivity associated with the transition of the radial electric field in neutral-beam-heated plasmas in the Large Helical Device, *Phys. Rev. Lett.* **86**, p. 23, doi:10.1103/PhysRevLett.86.5297.

McGuire, K., *et al.* (1983). *Phys Rev. Lett.* **50**, p. 891.

Miura, H. and Nakajima, N. (2010). Influences of ballooning modes with moderate wavenumber on MHD equilibrium in LHD, *Nucl. Fusion* **50**, p. 054006, doi:10.1088/0029-5515/50/5/054006.

Nazikian, R., Kramer, G. J., Cheng, C. Z., and Gorelenkov, N. N., *et al.* (2003). *Phys Rev. Lett.* **91**, pp. 125003-1, doi:10.1103/PhysRevLett.91.125003.

Nazikian, R., Fu, G. Y., Batha, S. H., *et al.* (1997). Alpha particle driven toroidal Alfvén eigenmode in the tokamak fusion test reactor, *Phys. Rev. Lett.* **78**, p. 2976.

Nunami, M., Watanabe, T-H., Sugama, H., and Tanaka, K. (2012). Gyrokinetic turbulent transport simulation of a high ion temperature plasma in large helical device experiment, *Phys. Plasmas* **19**, p. 042504, `http://dx.doi.org/10.1063/1.4704568`.

Nunami, M., Watanabe, T-H., Sugama, H., and Tanaka, K. (2011). Linear gyrokinetic analyses of ITG modes and zonal flows in LHD with high ion temperature, *Plasma Fusion Res.* **6**, p. 1403001, doi:10.1585/pfr.6.1403001.

Okabayahsi, M., Matsunaga, G., deGrassie, J. S., Heidbrink, W. W., In, Y., Liu, Y. Q., Reimerdes, H., Solomon, W. M., Strait, E. J., Takechi, M., Asakura, N., Budny, R. V., Jackson, G. L., Hanson, J. M., La Haye, R. J., Lanctot, M. J., Manickam, J., Shinohara, K., and Zhu, Y. B. (2011). Off-axis fishbone-like instability and excitation of resistive wall modes in JT-60U and DIII-D, **18**, p. 056112, doi:10.1063/1.3575159.

Rewoldt, G., Tang, W. M., and Chance, M. S. (1982). Electromagnetic kinetic toroidal eigenmodes for general magnetohydrodynamic equilibria, *Phys. Fluids* **25**, pp. 480-490.

Rewoldt, G. and Tang, W. M., (1983). Beam-ion and alpha-particle effects on microinstabilities in tokamaks, *Phys. Fluids* **26**, pp. 3619-3623.

Tanaka, K., Michael, C. A., Vyacheslavov, L. N., Sanin, A. L., Kawahata, K., Akiyama, T., Tokuzawa, T., and Okajima, S. (2008). Two-dimensional phase contrast imaging for local turbulence measurements in large helical device (invited), *Rev. Sci. Instrum.* **79**, p. 10E702.

Tanaka, K., Michael, C., Vyacheslavov, L., Funaba, H., Yokoyama, M., Ida, K., Yoshinuma, M., Nagaoka, K., Murakami, S., Wakasa, A., Ido, T., Shimizu, A., Nishiura, M., Takeiri, Y., Kaneko, O., Tsumori, K., Ikeda, K., Osakabe, M., Kawahata, K., and LHD Experiment Group. (2010). Turbulence Response in the High Ti Discharge of the LHD, *Plasma Fus. Res.* **5**, p. S2053, doi:10.1585/pfr.5.S2053.

Todo, Y., Berk, H. L., and Breizman, B. N. (2010). Nonlinear magnetohydrodynamic effects on Alfvén eigenmode evolution and zonal flow genertation, *Nucl. Fusion* **50**, pp. 084016-084025, doi 10.1088/0029-5515/50/8/084016.

Wakatani, M. (1998). *Stellarator and Heliotron Devices, Oxford University Press*, ISBN:0-19-507831-4.

Watanabe, T.-H., Sugama, H., and Ferrando-Margalet, S. (2007). Gyrokinetic simulation of zonal flows and ion temperature gradient turbulence in helical systems, *Nucl. Fusion* **47**, p. 1383, doi:10.1088/0029-5515/47/9/041.

Wong, K. L., Fonck, R. J., Paul, S. F., Roberts, D. R., Fredrickson, E. D., Nazikian, R., Park, H. K., Bell, M., Bretz, N. L., Budny, R., Cohen, S., Hammett, G. W., Jobes, F. C., Meade, D. M., Medley, S. S., Mueller, D., Nagayama, Y., Owens, D. K., and Synakowski, E. J. (1991). Excitation of toroidal Alfvén eigenmodes in TFTR, *Phys. Rev. Lett.* **66**, pp. 1874-1877, doi:10.1103/PhysRevLett.66.1874.

Xanthopoulos, P. and Jenko, F. (2007). Gyrokinetic analysis of linear microinstabilities for the stellarator Wendelstein 7-X, *Phys. Plasmas* **14**, 042501, doi:10.1063/1.2714328.

Yagi, M., Wakatani, M., Sugama, H., B. G. Hong, B. G., and Horton, W. (1989). Ion-temperature-gradient driven drift mode coupled to the resistive interchange mode in a heliotron/torsatron, *J. Phys. Soc. Jpn.* **58**, 12, pp. 4265-4268.

Chapter 13

Temperature Gradient Driven Instabilities

13.1 Ion Temperature Gradient Instabilities

Magnetized plasmas with a significant temperature gradient across the magnetic field are a common situation owing to the thermal insulating properties of the magnetic field. Such temperature gradients occur in the magnetosphere, the solar corona and with particularly high gradients in the laboratory experiments aimed at magnetic confinement for nuclear fusion. In the large tokamaks experiments reviewed in Chapter 1 we see that the core ion temperatures have reached 40 KeV in the TFTR device (Sec. 1.7.2), and in the JT60-U (sec. 1.7.3) machine with $R/a = 3.1\,\mathrm{m}/0.7\,\mathrm{m}$. Thus, we know that the magnetic field can provide strong thermal insulation withstanding gradients greater than 20 KeV/m under ideal conditions of the closed irrational magnetic flux tubes. Table 13.1 gives the parameters for a well-known JT60-U shot that has a core ion temperature of 38 KeV and core electron temperature of 12 KeV. The ion (deuterium) quantities T_i and v_ϕ are inferred from spectroscopic measurement of carbon ions.

The whole concept of magnetic confinement for thermonuclear fusion of plasmas depends on this thermal insulating principle. From the deuterium-tritium reactivity function shown in Fig. 1.1 one finds that the core ion plasma temperatures of order $10^7 - 10^8$ K are required for significant fusion nuclear reactivity. This high temperature plasma must be partially insulated from the vessel walls which must

Table 13.1 JT60-U High-β_p Experiment 17110 with Internal Transport Barrier: parameters in Phase I of shot.

R/a	$3.1\,\mathrm{m}/0.7\,\mathrm{m}$
B_ϕ	$4.4\,\mathrm{T}$
I_p	$2\,\mathrm{MA}$
P_{NBI}	$27\,\mathrm{M}$
$n_D(0)$	$4.1 \times 10^{19}\,\mathrm{m}^{-3}$
$T_i(0)/T_e(0)$	$38\,\mathrm{KeV}/12\,\mathrm{KeV}$
$n_D \tau_E T_i$	$1.1 \times 10^{21}\,\mathrm{m}^{-3} \cdot s \cdot \mathrm{KeV}$
$v_\phi(0)$	$-100\,\mathrm{km/s}$

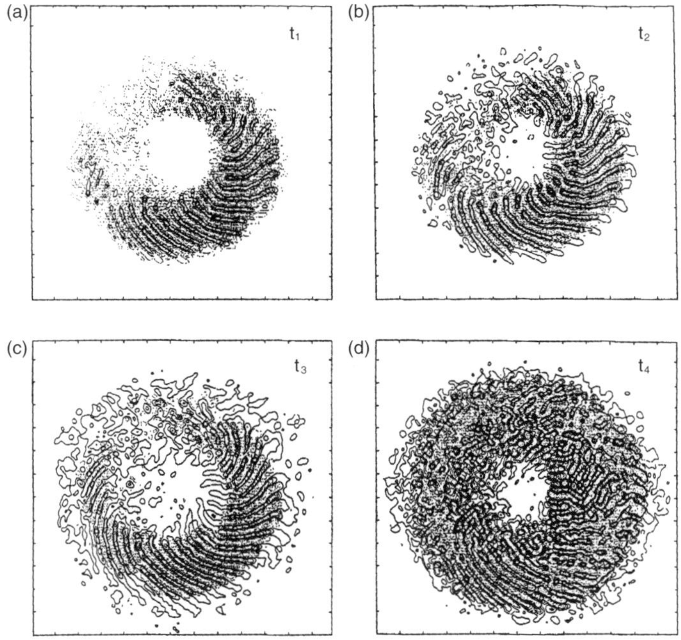

Fig. 13.1 Evolution of the isopotential contours for the ITG turbulence for $a/\rho_i = 28$, $L_T/R = 0.2$, $T_i = T_e$ with hyperbolic tangent profiles for the temperature. The density profile is flat [Sydora, *et al.* (1996)].

remain below 5000 K to retain their integrity. Thus, steep temperature gradients of order 20 KeV/m must be stably maintained in the thermonuclear fusion device. The associated ion thermal diffusivity χ_i is then of order or less than $1\,\mathrm{m}^2/\mathrm{s}$ in devices with minor radius a few meters.

Such magnetic thermal insulation is consistent with the results from the latest generation of tokamak and spherical tokamak experiments under favorable conditions. The thermal flux across the magnetic surfaces is large $q_i = -n_i\chi_i dT_i/dr$ reaching $0.1\,\mathrm{MW/m}^2$ and requiring large heating rates of many MW to maintain the plasma against turbulent losses. For the alpha particle decay products from the fusion reactions to provide the power to maintain the ion temperature, the core electron temperature T_e must also be sufficiently high – in the 10 KeV range. The next chapter, Chapter 14, is devoted to instabilities driven by the electron temperature gradient and the associated turbulent transport of electron thermal energy across the magnetic flux tubes. This electron thermal transport problem is more difficult to control in the magnetic confinement experiments than the ion thermal turbulent transport as emphasized by Kadomtsev (1992).

In both laboratory plasmas and in astrophysical plasma where a strong local heating takes place there is clearly the potential source of plasma instability. There can be MHD (single conducting fluid in Chapter 6) instabilities controlled by the constraints of frozen-in magnetic flux and the adiabatic equation of state. These fast instabilities limit the plasma pressure gradients with respect to the confining magnetic pressure. These fast, single fluid instabilities have been eliminated in the magnetic confinement devices proposed for nuclear fusion. Here, we consider that the plasma is stable to ideal MHD motions and study the remaining smaller scale drift wave instabilities. The drift waves have dispersive frequencies and finite parallel electric fields which means the associated waves and vortices have resonant wave-particle interactions. Figure 13.1 shows the growth of the ion temperature gradient driven drift wave vortices starting on the outside of the torus in frame (a) and spreading throughout the torus in frames (b), (c), and (d).

The drift wave dispersion relation in a collisionless plasma does not assume a local thermal Maxwell-Boltzmann distribution function. In the high temperatures of fusion and geospace plasmas the particle distributions in energy typically have high energy tails and different effective temperatures across and parallel to the magnetic field. In low beta $\beta = p/(B^2/2\mu_0)$ plasmas the ion temperature gradient instability, known as the ITG mode, has nearly an electrostatic polarization for the coupled drift waves and the ion acoustic waves that gives rise to the collective instability. In higher pressure plasmas with $\beta > m_e/m_i$ the ITG mode couples with the shear Alfvén wave becoming a dispersive oscillation with an electromagnetic polarization. As the plasma pressure increases the inductive electric field from the time changing δB_x cancels part of the electrostatic component of the parallel electric field. This cancelation reduces the energy transfer rate $\langle j_\parallel E_\parallel \rangle$ and reduces the growth rate of the instability. There are numerous studies of these weakly electromagnetic ITG modes including Rewoldt, *et al.* (1998), Hong, *et al.* (1989a, 1989b) and Kim, *et al.* (1993) that detail how the ITG modes change with increasing plasma beta. At certain intermediate values of plasma beta both the ITG and the kinetic ballooning mode of the MHD polarization are present with different mode frequencies and growth rates.

We begin with a thermodynamic picture of the energy source for the turbulence from the temperature gradient. Then we consider the kinetic theory in the electrostatic limit of the ion temperature gradient driven stability. In Section 13.3 we consider the stability analysis for the long-tailed ion energy distribution models characteristic of some neutral beam heated plasmas and astrophysical plasmas. For example, the dense plasma in the wake of the solar wind behind magnetized planets are typically parameterized by a power laws distribution that is called the "kappa" distribution [Kivelson and Russell (1995)] with power law energy E distributions parameterized as $f_\kappa(E) \simeq 1/E^\kappa$. The energy spectrum index parameter κ for heavy-tailed distributions varies from $\kappa = 2$ or 3 energy distributions with every large high energy fluxes suitable for modeling shock accelerated protons to large values of kappa where the bulk of the distribution is close to the Maxwell-Boltzmann distribution. The κ-distribution is used in the space physics [Kivelson and Russell (1995)] for the hot geotail plasma and the solar corona [Aschwanden (2005)] for the proton distributions. In both these astrophysical plasma the time for Coulomb collisions exceeds the dynamical times scales by orders of magnitudes.

13.2 Mechanism of the Ion Temperature Gradient Instabilities

For the Maxwellian phase space distributions the electrostatic ITG (ion temperature gradient) instability starts when the ratio of the ion temperature gradient $d\ln T_i/dr$ exceeds the density gradient $d\ln n_i/dr$ by a number that depends on the compressibility of the ion gas. This critical gradient is best understood by considering the maximum work that can be extracted from the plasma temperature profile with a Carnot cycle operating between the high and low temperature regions. The "engine" for the cycle is the drift wave convection between the high temperature and low temperature regions and the working gas is the ion component of the plasma with an adiabatic gas constant $\Gamma = (d+2)/d$ where $d =$ number of degrees of freedom active in the dynamics of the ion gas. The situation is shown in Figs. 3.3 and 13.1 where the radially-extended vortex cells, tilted to follow the twist of the magnetic field lines, cycle the hot interior plasma to the lower temperature plasma [Sydora, *et al.* (1996)].

One uses the Carnot cycle to calculate the maximum energy $W_{\rm ITG}$ that can be extracted per cycle from the drift waves. The plasma is adiabatically compressed and expanded between the high and low ion temperature regions so that the adiabatic gas constant controls the stability condition. We find that the condition for a net energy release from the temperature gradient is given by

$$\eta_i > \eta_{\rm crit} = \frac{2}{3} \tag{13.1}$$

for the usual $d = 3$ degrees of freedom gas equation of state.

This η parameter is widely used in the plasma literature and defined for a charge species s by

$$\eta_s = \frac{d\ln T_s}{d\ln n_s}. \tag{13.2}$$

In the early plasma literature the instabilities driven by ion temperature gradient were often called the η_i modes [Horton and Varma (1972)].

The temperature gradient instability condition follows from a calculation of the work done $W = Q_1 - Q_2$ from the thermal energies Q taken from the hot ion region at $T_i(x)$ and adiabatically expanding the gas to the lower the temperature to the lower temperature regions at $T_i(x+\Delta x)$. Then the convection recompresses the gas to return to the high temperature region at $T_i(x)$. Subtracting the work done in the compression minus the work gained in the expansion and adding the work done by the gas in the two isothermal strokes gives the net energy W released for one cycle. The energy W is proportional to the differences $\Delta T/T(\Delta T/T - 2/3\Delta n/n)$.

For the general value of the adiabatic gas constant the factor 2/3 is replaced by $\Gamma - 1$. Thus for the slab model of the ITG threshold, where there is only the parallel compressible motion, there is a one-degree-of-freedom compression and the threshold is raised to the value of $\Gamma - 1 = 2$. In toroidal systems there are three-degrees-of-freedom and the threshold starts at 2/3. The corresponding changes in the wave-particle resonances are shown in Fig. 13.2.

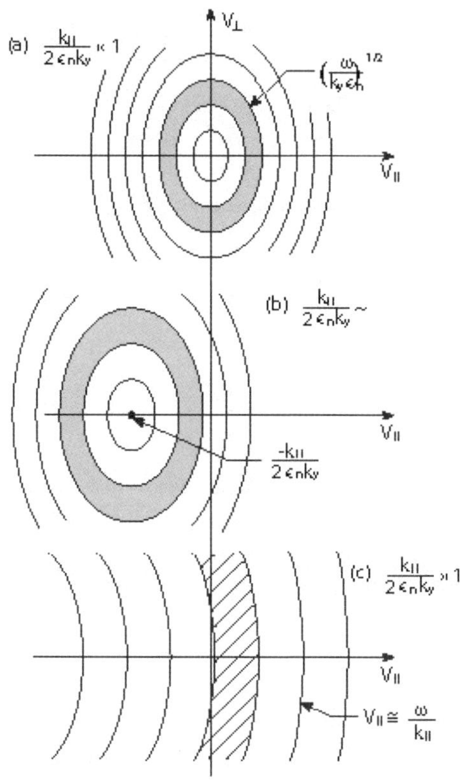

Fig. 13.2 The band of resonant ion velocities for a given drift-wave fluctuation $(\omega, k_y, k_\parallel)$. (a) the ideal delta function resonance curve $\omega_k = \omega_D$, (b) the shifted contours of resonant $\omega_k - k_\parallel v_\parallel = \omega_D$, and (c) the slab Landau resonance $\omega_k = k_\parallel v_\parallel$.

These threshold formulas are known to agree with the results from Nyquist diagrams for the stability condition of the electrostatic drift waves when other complications such radiative damping are absent.

Here some intermediate steps in the Carnot cycle construction are given for clarity. The different temperatures $T_i(x + \Delta x)$ and $T_i(x)$ are the driving force of the convection. In the ideal isentropic limit the work W done from the release of energy from the hot $T_2(x)$ plasma to reservoir $T_i(x+\Delta x)$ to the lower temperature $T_1(x + \Delta x)$. Using conservation of energy (first law of thermodynamics) the $W =$

$Q_2 - Q_1$ and $Q_2 = T_2 \Delta S_2$ and $Q_1 = -T_1 \Delta S_1$, with $\Delta S_1 + \Delta S_2 = 0$ gives the maximum work W that can be extracted as per

$$W = N\left(T_2 - T_1\right) \Delta \underbrace{\Delta \ell n\left(T^{3/2}/N\right)}_{\Delta S_2} = N \Delta T \left(\frac{3}{2}\frac{\Delta T}{T} - \frac{\Delta N}{N}\right) \tag{13.3}$$

from the gradients of $T(x)$ and $N(x)$.

Thus, in the absence of Landau damping and other non-ideal loss processes such as wave radiation from the convection region in the calculation, the maximum energy per cycle that the drift wave can extract from the distribution satisfies the condition in Eq. (13.3). We will return to the generalization of Eq. (13.3) to non-Maxwellian $f_0(H,x)$ distributions later.

The complete dispersion relation from solving the eigenvalue problem in the toroidal geometry has been developed into a computer code by Rewoldt. The fastest growing eigenmode from this numerical solution with the ITG and TEM physics for typical discharge parameters in TFTR is shown in Fig. 13.3 from Rewoldt, *et al.* (1998). One sees in Fig. 13.3(a) that for the collisionality of $\nu_{*e} = 1$ the mode onsets with a well defined ion temperature gradient of $d \ln T_i / d \ln n_i = 1$ and rotates in the plasma rest frame in the ion diamagnetic direction. For higher temperatures where there is a substantial fraction of trapped electrons Fig. 13.3(b) shows that the mode is unstable even with no ion temperature gradient $d \ln T_i / d \ln n_i = 0$ from the eigenvalue code. This instability is the trapped electron mode (TEM) and has growth rate with $\gamma_k / \omega_{*e} = 0.25$ with even for no ion temperature gradient.

In the next section, Chapter 13.3, we show in detail how the electrons respond and destabilize the drift wave in this weakly collisional regime where those particles with large pitch angles are trapped by the mirror force in the low-B field region on the outside of the torus. These types of modes exist in all toroidal confinement devices and produce drift wave instabilities that are universal to toroidal confinement geometries. The growth rate in Fig. 13.2a for $d \ln T_i / d \ln n_i = 0$ is from the mirror trapped electrons resonantly interacting with the drift wave.

The change in the unstable region for $k_\perp L_n$ versus $k_y \rho_i$ in a stability diagram with varying values of $\eta_\perp$ and $\eta_\parallel$ is worked out in Hojo (1995). The result shows that at low values $\eta_\perp$, the critical condition is a value of $\eta_\parallel$ equal to 3 as given by the thermodynamics of the one-degree of freedom gas. The change in the unstable region for $k_\perp L_n$ versus $k_y \rho_i$ in a stability diagram with varying values of $\eta_\perp$ and $\eta_\parallel$ is worked out in Hojo (1995). The result shows that at low $\eta_\parallel$ the critical condition is value of $\eta_\perp > 1$ to 2 consistent with the thermodynamics.

For the critical gradients over all k modes by Hojo (1995) explains how plasma with high T_i/T_e can remain stable to the ITG modes for $\eta_\perp$ moderately large. The analysis is done specifically for the high T_i/T_e ratios relevant to the tandem mirror plasma produced in GAMMA 10 which contains a 5 KeV perpendicular ion temperature plasma in a relatively small radius cylindrical plasma. So one might expect a strong ITG instability. This analysis and related analysis by Kim (1992)

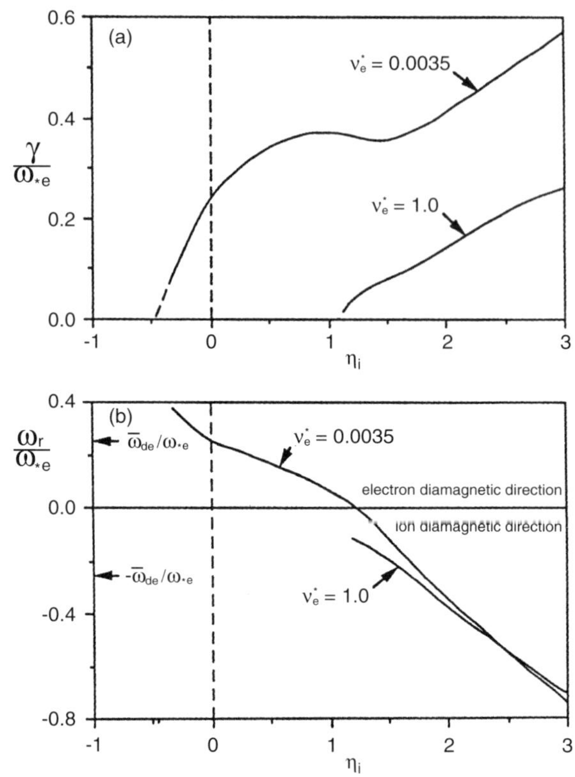

Fig. 13.3 The normalized (a) growth rate and (b) frequency of the complex kinetic toroidal eigenvalue problem. At high collisionality only the ITG mode appears for $\eta_i > 1.2$ while at low collisionality the trapped electron mode appears for $\eta_i < 1.5$ [Rewoldt, *et al.* (1998)].

shows why these systems are stable to ITG modes [Chen and Migliuolo (1992)]. A similar stability to ITG at high T_i/T_e ratios was reported in the TFTR supershots [Horton, *et al.* (1992)], although the analysis for the stability are rather different for the tokamak.

The kinetic stability analysis of the temperature gradient driven modes is determined by kinetic plasma response functions $P(\omega, k)$, $Q(\omega, k)$, $R(\omega, k)$ which are integrals of the gradients of the phase space density function f weighted by the Bessel function $J_0^2, v_\perp J_1 J_0$ and $v_\perp^2 J_1^2$ where $J_n = J_n(k_\perp v \sin\alpha/\Omega)$ with α the pitch angle and $v = (2H_0/m)^{1/2}$ the particle speed. We use the symbol H_0 for the kinetic energy of the ion of mass m and charge e. To start with we will take $\sin\alpha$ as a parameter by using $\lambda = k_\perp v \sin\alpha/\Omega$ as the dimensionless variable for the mean or characteristic speed v_0 in the $f(H_0, x)$ ion distributions.

The simplest, and most studied response function $P(\omega/k_y, \lambda)$ describes the electrostatic response of the plasma. For electrostatic modes the modes are given by dispersion relation

$$D_{\mathrm{ESk},\omega} = 1 + \frac{T_e}{T_i}[1 - P(k,\omega)] = 0. \tag{13.4}$$

The function $P(\omega/k_y, \lambda)$ may be thought of as the generalization of the plasma dispersion function $Z(\omega/k_{\|} v_{th})$ when the grad-B and curvature drifts dominate the parallel phase velocity resonance.

The $P(\omega/k_y, \lambda)$ function is defined by

$$P(\omega/k_y,\lambda) = -\int \frac{dH_0 H_0^{1/2}\left(\omega \dfrac{\partial f}{\partial H_0} + \dfrac{k_y}{eB_z}\dfrac{\partial f}{\partial x}\right) J_0^2}{\omega - k_y \overline{v}_D H_0 + i0^+}. \tag{13.5}$$

We may define an effective temperature T (or $k_B T$) from the mean ion energy in the distribution function $f(H_0, x)$ such that T becomes the thermodynamic temperature when $f(H_0,x) \Rightarrow f_{\mathrm{Max-Boltzmann}}$.

For a Maxwellian $f_{\mathrm{Max-Boltzmann}} = N\exp(-H_0/T)/(2\pi T/m)^{3/2}$ the P-function is thoroughly studied in the literature [Similon, *et al.* (1984)].

The $\overline{v}_D$ in equation Eq. (13.5) is the sum of the grad-B and curvature drift frequencies bounce-averaged and taken at some representative pitch angle (typically the variation with the pitch angle α is weak). The transition from the toroidal response function P to the slab model response function Z is given in Kim and Horton (1991).

The gradient $\partial/\partial x$ is with respect to the flux function ψ with $B_z dx = d\psi$ in the magnetic flux coordinate (per unit of dy where ∇y is the bi-orthogonal direction to the magnetic field vector and the radial gradient direction).

We may first approximate the spatial gradient as

$$\frac{\partial f}{\partial x} \equiv \frac{1}{L_n} f(H_0) \tag{13.6}$$

defining the length L_n. This local approximation is adequate for the first stage of analysis as seen comparing with case of a Maxwellian where

$$\frac{\partial f_M}{\partial x} = \left[\frac{1}{M}\frac{\partial N}{\partial x} + \frac{1}{T}\frac{\partial T}{\partial x}\left(\frac{H_0}{T} - \frac{3}{2}\right)\right] f_M \equiv \frac{f_M}{L_n}\left[1 + \eta\left(\frac{H_0}{T} - \frac{3}{2}\right)\right] \tag{13.7}$$

the energy dependence of $\eta(H_0/T - 3/2)$ arises describing that higher energy ions have a shorter gradient scale length $L_n(H_0/T \gg 1) \ll L_T$ than that of the density itself. The instabilities from this effect are often called η_i-modes due to the historical definition of η as

$$\eta_i = \frac{d_x \ell n\, T_i}{d_x \ell n\, N}. \tag{13.8}$$

For the Maxwellian distribution the electrostatic instability starts at

$$\eta_i > \eta_{\mathrm{crit}} = \frac{2}{3}. \tag{13.9}$$

This critical ratio of the temperature gradient instability threshold condition is understood by constructing a Carnot cycle between the different temperatures $T_i(x+\Delta x)$ and $T_i(x)$ and asking for the ideal isentropic limit of the work W done from the release of energy from the hot reservoir as explained with Eq. (13.3).

The analysis also shows that the case of inverted temperature and density gradients is strongly unstable. This regime of negative $\eta_{\rm crit}$ occurs when the pressure gradient is negligible across a region with a temperature and density gradient. The same situation occurs in the electron temperature gradient driven modes where the negative η_e instability has been demonstrated in basic laboratory experiments. The situation is common in between two different plasmas as across transition layers as the plasma pause and the chromosphere-corona boundary. Using conservation of energy (first law of thermodynamics) the $W = Q_2 - Q_1$ and $Q_2 = T_2\Delta S_2$ and $Q_1 = T_1 \Delta S_1$, with $\Delta S_1 + \Delta S_2 = 0$ gives the maximum work W that can be extracted by the turbulence per convection cycle as given in Eq. (13.3). The formula leads to the maximum turbulence energy density formula

$$\frac{W^{\rm turb}(t)}{nT} = \frac{3}{2}\frac{\Delta T}{T}\left(\frac{\Delta T}{T} - \frac{2}{3}\frac{\Delta n}{n}\right) \tag{13.10}$$

that can be a useful real-time diagnostic for feedback control of the auxiliary power $P_{\rm aux}(t)$.

Thus, without Landau damping and other loss processes we have a bound, valid nonlinearly, on the maximum turbulent energy W that can be extracted from the distributions. We will return to the generalization of Eq. (13.10) for non-Maxwellian $f_0(H,x)$ distributions in Chapter 14.7.

13.2.1 *Nyquist analysis*

In the kinetic stability analysis, condition Eq. (13.3) results from the Nyquist analysis of the P-function. Namely, we must find the $\omega_{\rm crit}$ at which Im $P(\omega) = 0$ and then evaluate the Re P at $\omega = \omega_{\rm crit}$.

For real ω the resonant ion contribution is fixed for each value of ω/k_y through

$$\text{Im } P = \pi T \int_0^\infty \delta\left(\omega - k_y \overline{v}_D H_0\right) J_0^2 \left(\omega \frac{\partial f_0}{\partial H} + \frac{k_y}{e}\frac{\partial f_0}{\partial B_x}\right) H_0^{1/2} dH_0. \tag{13.11}$$

Only phase velocities ω/k_y in the same direction as the pitch-angle averaged-drift velocity contribute to Im P through the resonant ions of energy $H_0 = \omega/k_y\overline{v}_D$. If we use the approximation in Eq. (13.6) then

$$\text{Im } P = \pi T \left(\frac{\omega}{k_y v_D}\right)^{1/2} \frac{J_0^2}{|k_y v_D|}\left(\omega\frac{\partial f_0}{\partial H} + \frac{k_y f_0}{eBL_n}\right)\Bigg|_{H_0=\omega/k_y\overline{v}_D}. \tag{13.12}$$

The critical character frequency label by point c in Fig. 13.4 is

$$\omega_* = -\frac{k_y}{eL_n}\frac{f_0(H)}{(\partial f_0/\partial H)} \tag{13.13}$$

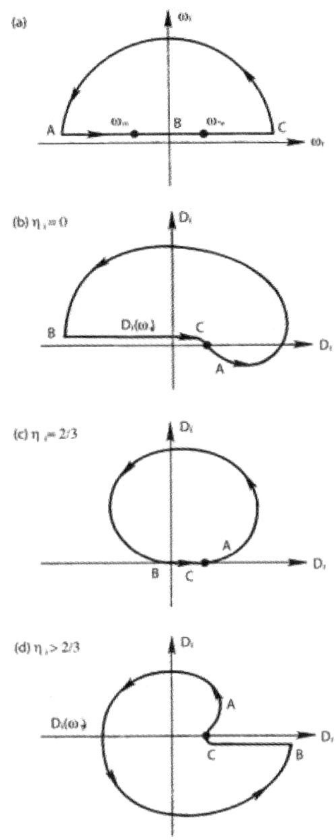

Fig. 13.4 The Nyquist diagram for the ion temperature gradient driven instability. Frame (a) is the contour enclosing the unstable upper-half plane that is mapped by the dispersion function in frames (b), (c) and (d) for three values of the stability parameter η_i. For the first case in (b) there is no temperature gradient and the mapped contour does not encircle the origin, so there is no unstable root. In frame (c) with $\eta_i > 2/3$ there is a marginally stable root. For frame (c) with $\eta_i > 2/3$ the D-function contour encircles the origin proving that there is an unstable root within the A-B-C contour in frame (a) of the complex omega plane.

as a function of H. The wave-ion drift resonance condition that

$$\omega = k_y \overline{v}_D H_0 = \omega_* \tag{13.14}$$

which is the critical frequency for the point (labeled as in Fig. 13.4) of crossing the imaginary axis in the complex P-plane. The change in the sign of the imaginary part Im $P(\omega)$ at this critical frequency reverses the flow of energy between the waves and the resonant ions.

Now that the critical frequency is determined, we evaluate the Re P, which now is the principal value integral with a well-defined value for $\omega = \omega_{\rm crit}$ to complete the Nyquist diagram in Fig. 13.4.

In the simplest case of electrostatic ITG modes the electron density response is adiabatic with respect to the fluctuating potential $e\widetilde{\phi}$ giving

$$\widetilde{n}_e = \frac{Ne\widetilde{\phi}}{T_e}. \tag{13.15}$$

The ion kinetic density perturbation is

$$\widetilde{n}_i = \frac{-Ne\widetilde{\phi}}{T_i}[1 - P(k,\omega)] \tag{13.16}$$

where $\widetilde{\phi}$ is the fluctuation electrostatic potential ($\boldsymbol{E} = -\nabla\widetilde{\phi}$) and T_i is a characteristic ion energy for the distribution f_0. Thus, the quasi-neutrality condition yields the dispersion relation

$$D_{k\omega}^{\mathrm{ES}} = 1 + \frac{T_e}{T_i}[1 - P(k,\omega)] = 0 \tag{13.17}$$

for the ITG-ES waves that largely control turbulent ion thermal transport in toroidal magnetic confinement systems.

13.2.2 *Full 3 × 3 electromagnetic ion temperature gradient modes*

The wave fluctuations satisfy the matrix equation $\boldsymbol{A} \cdot \boldsymbol{X} = 0$ defined by

$$\begin{bmatrix} a & b & c \\ b & d & e \\ c & e & f \end{bmatrix} \begin{bmatrix} \phi \\ \psi \\ \dfrac{\delta B_{\|}}{B} \end{bmatrix} = 0 \tag{13.18}$$

where the six complex response functions are

$$\begin{aligned} a &= -1 + \frac{T_e}{T_i}(P-1) \\ b &= 1 - \frac{\omega_{*e}}{\omega} \\ c &= Q \\ d &= \frac{k_\perp^2\rho_s^2\omega_A^2}{\omega^2} - \left(1 - \frac{\omega_{*e}}{\omega}\right) + \frac{\omega_{De}}{\omega}\left(1 - \frac{\omega_{*p_e}}{\omega}\right) \\ e &= -\left(1 - \frac{\omega_{*p_e}}{\omega}\right) \\ f &= \frac{2}{\beta_e} + \frac{T_i}{T_e}R. \end{aligned} \tag{13.19}$$

The polarization of the mode is given by the ratio of the components of the vector $X^T = (\phi, \psi, \delta B_{\|}/B)$.

The physics contained in Eq. (13.18) is the following: The first row of the matrix equation is the condition of quasi-neutrality $\delta\widetilde{n}_e = \delta\widetilde{n}_i$. The second row is from Ampére's law, $\nabla_\perp^2 A_{\|} = -\mu_0\delta j_{\|}$, where $k_\perp^2\widehat{\omega}_A^2 = -\partial_s v_A^2 k_\perp^2(s)\partial_s$ is the line-bending operator. The third row is from the "radial" ($\nabla\Psi$) component of Ampére's law.

Here the three ion kinetic response functions for ($|\omega| \gg k_\parallel v_i$) are given by P, Q, and R with analytic continuations from the upper-half of the complex ω plane

$$P = \left\langle \frac{\omega - \omega_{*i}(\epsilon)}{\omega - \omega_{Di}} J_0^2 \right\rangle, \tag{13.20}$$

$$Q = \left\langle \frac{\omega - \omega_{*i}(\epsilon)}{\omega - \omega_{Di}} \left(\frac{m_i}{b_i T_i}\right)^{1/2} v_\perp J_0 J_1 \right\rangle, \tag{13.21}$$

$$R = \left\langle \frac{\omega - \omega_{*i}(\epsilon)}{\omega - \omega_{Di}} \cdot \frac{m_i v_\perp^2}{b_i T_i} J_1^2 \right\rangle. \tag{13.22}$$

Here $\omega_{*i}(\epsilon) = \omega_{*i}[1 + \eta_i(\epsilon - 3/2)]$ with $\eta_i = d\ell n\, T_i / d\ell n\, n_i$, and likewise of ω_{*e} with $\eta_e = d\ell n\, T_e / d\ell n\, n_e$. The quantities $\omega_{*pj} = \omega_{*j}(1 + \eta_j)$, where p stands for the pressure gradient and j denotes the species. Similon, *et al.* (1984) describe the calculation and properties of the analytic P, Q, R functions.

13.2.3 *Limiting cases of the electromagnetic ITG dispersion relation*

Let us consider various limits of $D(\omega) = \det|A| = 0$ from Eq. (13.18). For a low-β plasma, $f \gg 1$ and the determinant D of Eq. (13.18) is

$$D = \left(ad - b^2\right) f \simeq 0. \tag{13.23}$$

For this system the MHD modes have $a \simeq -b$ and $d \cong -b$. Equation (13.14) for $f \neq 0$ gives the kinetically modified MHD modes $\omega^2 - \omega\omega_{*pi} - k_\parallel^2 v_A^2 + \gamma_{\rm mhd}^2 = 0$ with small $E_\parallel$. There are electron drift modes with finite $E_\parallel$ given by

$$\omega^2 \left(1 + k_\perp^2 \rho_s^2\right) - \omega\omega_{*e} - k_\parallel^2 v_A^2 k_\perp^2 \rho_s^2 = 0. \tag{13.24}$$

The $\delta B_\parallel \neq 0$ compressional mode has $f(\omega) \simeq 0$ and is stable until the mirror mode instability condition $\beta(p_\perp/p_\parallel - 1) > C_m \approx 1$ is satisfied. The mirror mode is discussed in Chapter 6.2

13.2.4 *Ion temperature gradient and trapped electron transport*

Now we consider the general case of the electron response $\widetilde{n}_e$ to the potential fluctuations $\widetilde{\phi}_k(\omega)$. The full response is called the "non-adiabatic" electron response.

The stability and the transport produced by the ion temperature gradient (ITG) and the trapped electron (TEM) modes are now coupled. An analytical model [Zhu (1999, 2000)] as well as the numerical Weiland model [Guo and Weiland (1997); Jarmen, *et al.* (1987); Nordman, *et al.* (1990); Weiland (2000)] will be used as tools to calculate the local stability of these complex modes. Other tools for the TEM-ITG analysis are an analytical model [Zhu (1999, 2000)] and the fluid-based Weiland model [Guo and Weiland (1997); Jarmen, *et al.* (1987); Nordman, *et al.* (1990); Hammett, *et al.* (1993); Weiland (2000)].

13.3 Analytical TEM-ITG Drift-Wave Model

Without the ion-wave resonances, the dispersion relation $D(\boldsymbol{k},\omega,\boldsymbol{P})$ with temperature and density gradient driven drift-waves ($\omega \sim \omega_* = k_y T_e/eBL_n$) is given [Zhu (1999) ,(2000)] by

$$D(\boldsymbol{k},\omega,\boldsymbol{P}) = 1 + \frac{T_e}{T_i} - \left(\frac{T_e}{T_i} + \frac{\omega_*}{\omega}\right)\Gamma_0 - \eta_i \frac{\omega_*}{\omega} b_i(\Gamma_1 - \Gamma_0)$$
$$- \int_0^\infty 2\pi v^2 dv F_e^M(v) \int_{-1}^{+1} d\mu h_e(v,\mu,\omega) = 0 \tag{13.25}$$

where the perturbed electron phase-space distribution fraction is $\widetilde{f} = (e\phi/T_e) F_e^M(1-h_e)$ with $F_e^M(v)$ the Maxwell distribution for the electron velocity. Moreover, $\eta_{i,e} = d\log T_{i,e}/d\log n_{i,e}$ and the usual finite ion gyro-radius Bessel function $\Gamma_n(b_i) = e^{-b_i} I_n(b_i)$ where $b_i = k_y^2\rho_i^2$. The non-adiabatic part of the electron distribution $h_e(v,\mu)$ satisfies the kinetic equation

$$\nu(v)\frac{\partial}{\partial\mu}\left(1-\mu^2\right)\frac{\partial h_e}{\partial\mu} + i\left[\omega - \omega_{De}\frac{v^2}{v_e^2} - k_\parallel v\mu\right] h_e$$
$$= i\left[\omega - \omega_*\left[1 + \eta_e\left(\frac{v^2}{v_e^2} - \frac{3}{2}\right)\right]\right] \tag{13.26}$$

where $\nu(v) = \nu_e(v_e/v)^3$. $\omega_{De} \simeq \omega_*(2L_n/R) = 2w_*\epsilon_n$, $\eta_e = d\log T_e/d\log n_e \neq 0$, $v_e = (2T_e/m_e)^{1/2}$, and $\mu = v_\parallel/v$.

The region of resonant wave-particle interactions is shown as the shaded elliptical region in Fig. 13.2 in frame (a). As the wave develops a positive parallel wavenumber $k_\parallel$ the resonance shifts to the left as shown in frame (b) and eventually becomes the usual Landau resonance as shown in frame (c) when the $k_\parallel v_\parallel \gg \omega_D$.

The model has the relevant electron temperature gradients and the density gradient as the primary driving forces for instability in this strongly heated plasma with no toroidal momentum input. The theoretical studies cited in Chapter 14 show that the ETG is the key instability for producing a strong electron thermal turbulent transport without large particle transport. The instability is insensitive to the details of magnetic shear when the q-profile is monotonically increasing and the shear parameter $s < 1$. The case of hollow or inverted q-profiles that occur in electron transport barrier are explained in Horton, *et al.* (2005) as discussed in Chapters 14 and 18.

The dispersion function $D(\boldsymbol{k},\omega,\boldsymbol{P}_d)$ depends on the $d = 8$ dimensional system parameter vector $\boldsymbol{P}$, which in dimensionless form is

$$\boldsymbol{P}_q = \left(\eta_i, \eta_e, \epsilon_n = \frac{L_n}{R}, \epsilon = \frac{r}{R}, q, \widehat{s} = \frac{d\ln q}{d\ln r}, \nu_* = \frac{\nu_e}{\epsilon\omega_{be}}, \tau = \frac{T_e}{T_i}\right), \tag{13.27}$$

where $\omega_{be}^T = v_e\epsilon^{1/2}/L_c$ is the effective bounce frequency, $\epsilon = r/R$, $L_c = qR$ is the connection length. The reciprocal of the collisionality ν_* is the number of electron bounces before an effective ($\Delta\theta = \epsilon^{1/2}$) detrapping collision. Thus, there is a large dimensionless set of the response functions, at least $2^8 = 256$ for high and low values of the eight parameters.

13.3.1 *Nonadiabatic electron response functions*

In the classic ITG instability the electron response is taken as adiabatic and produces the usual drift wave fluid dispersion relation in the ITG stable regime. In reality in the toroidal plasma, however, the drift wave in this regime becomes unstable if there is a resonance with the electron motion. The most relevant electron-wave resonance at these frequencies is the trapped electron mode. In this instability the wave frequency is resonant with the frequency of the bounce motion of the electrons that are trapped between the regions of max B on the in-board side, or at the smallest major radius $R_{\min}$ of the torus. For typical aspect ratio tokamaks there is of order 30-50% of the electron population executing these trapped particle orbits. These orbits are the ones that give the "banana" shape figure for the guiding center projected on to the poloidal cross-sectional plane. In the spherical tokamaks like NSTX and MAST the trapped electron traction can be the dominant set of electron orbits due the large mirror ration $R_M = B_{\max}/B_{\min}$ on the outer magnetic surfaces. The suprathermal and thus less collisional electrons, are the first to be trapped by R_M. As the temperature increases more and more of the electron population becomes collisionless and trapped. Thus the trapped electron instability has a strong dependence on the collisionality parameter $\nu_{\text{eff}} = \nu_e/\omega_{be}$.

The perturbed, fluctuating, electron distribution functions δf_e is conveniently described with a decomposition in Legendre polynomials in the transport calculations of electrical conductivity from collisions. These basis functions diagonalize the second order PDE of the dominant part of the collision operator.

To calculate the nonadiabatic part of the fluctuating electron distribution function we proceed as follows. We write

$$h_e(v, \mu, \boldsymbol{k}, \omega) = \Sigma_n h_{e,n}(v) P_n(\mu). \tag{13.28}$$

The zeroth order, $n = 0$, is isotropic with no the pitch-angle dependence. In general there are three velocity, or energy, ranges for the electrons for a given wave ω, k. We divide the electron energy ranges as follows depending on the collision frequency and the wave parameters:

(1) Low electron energies: the parallel electron motion is collision-dominated for,

$$\frac{v}{v_e} \leq \left(\frac{\nu_e}{k_\| v_e}\right)^{1/4} \rightarrow h_{e_0}(v) = \frac{\omega - \omega_* \left[1 + \eta_e \left(v^2/v_e^2 - 3/2\right)\right]}{\omega + i\left(k_\|^2 v_e^2/6\nu_e\right)\left(v/v_e\right)^5}. \tag{13.29}$$

The factor of six multiplying the collision frequency ν_e in these response function arises from the eigenvalues $L(L+1)$ of the Lorentz pitch angle scattering operator in Eq. (13.26) where both the $L = 1$ and $L = 2$ eigenmodes arises in the fluctuating distribution function of the electrons.

(2) Intermediate electron energies: the electron Landau resonance is formed:

$$\left(\frac{\nu_e}{k_\parallel v_e}\right)^{1/4} < \frac{v}{v_e} \le \left(\frac{\nu_e}{k_\parallel v_e \epsilon^{3/2}}\right)^{1/4}$$

$$h_{e_0}(v) = \frac{-i\pi}{2} \frac{\omega - \omega_* \left[1 + \eta_e \left(\frac{v^2}{v_e^2} - \frac{3}{2}\right)\right]}{|k_\parallel| v_e \left(\frac{v}{v_e}\right)}. \tag{13.30}$$

(3) High electron energies: the electrons divide into passing and trapped populations. The trapped electron response is bounce-averaged, such as $\overline{\phi} = (1/\tau) \oint ds \phi(s)/v_\parallel$.

$$\frac{v}{v_e} \ge \left(\frac{\nu_e}{k_\parallel v_0 \epsilon^{3/2}}\right)^{1/4} \to h_{e_0}(v) = \left(\frac{\epsilon}{2}\right)^{1/2} \frac{\omega - \omega_* \left[1 + \eta_e (v^2/v_e^2 - 3/2)\right]}{\omega - \omega_{De} v^2/v_e^2 + i(\nu_e/\epsilon)(v_e/v)^3} \tag{13.31}$$

where the passing electron contribution averages to zero.

Substituting Eq. (13.29)-Eq. (13.31) into Eq. (13.25) yields the dispersion relation

$$D_0(\boldsymbol{k}, \omega, \boldsymbol{P}) = 1 + \frac{T_e}{T_i} - \left(\frac{T_e}{T_i} + \frac{\omega_*}{\omega}\right)\Gamma_0 - \eta_i \frac{\omega_*}{\omega} b\left(\Gamma_1 - \Gamma_0\right) - \sum_m^3 \left(\Delta h_m^0 + \eta_e h_m^1\right) = 0 \tag{13.32}$$

where $\omega = \omega_0 + i\gamma$, $\Delta = 1 - \omega/\omega_*$ and the dimensionless energy integral Green's functions are

$$h_1^n = -\frac{2}{\pi^{1/2}} \int_0^{t_1} \frac{dt t^{1/2} e^{-t} (t - 3/2)^n}{1 - \Delta + i(t/t_0)^{5/2}} \tag{13.33}$$

$$h_2^n = i\pi^{1/2} \frac{\omega_*}{|k_\parallel| v_e} \int_{t_1}^{t_2} dt e^{-t} \left(t - \frac{3}{2}\right)^n \tag{13.34}$$

$$h_3^n = -\left(\frac{\epsilon}{2}\right)^{1/2} \frac{2}{\pi^{1/2}} \int_{t_2}^{\infty} dt t^{1/2} \frac{e^{-t}(t - 3/2)^n}{1 - \Delta - \epsilon_n t + i(t_3/t)^{3/2}}, \tag{13.35}$$

with

$$t_0 = \left(\frac{6\nu_e \omega_*}{k_\parallel^2 v_e^2}\right)^{2/5}, \; t_1 = c_1 \left(\frac{\nu_e}{k_\parallel v_e}\right)^{1/2}, \; t_2 = c_2 \left(\frac{\nu_e}{k_\parallel v_e \epsilon^{3/2}}\right)^{1/2}, \; t_3 = \left(\frac{\nu_e}{\epsilon \omega_*}\right)^{2/3}. \tag{13.36}$$

h_1^n, h_2^n and h_3^n correspond to the collisional drift wave, the collisionless drifts wave and the trapped electron mode, respectively.

The electron energy dependence of the dissipative part of the fluctuating phase space density that gives the growth-damping contributions to the growth rate is shown in Fig. 13.5. The solid line gives the response when there is an electron tem-

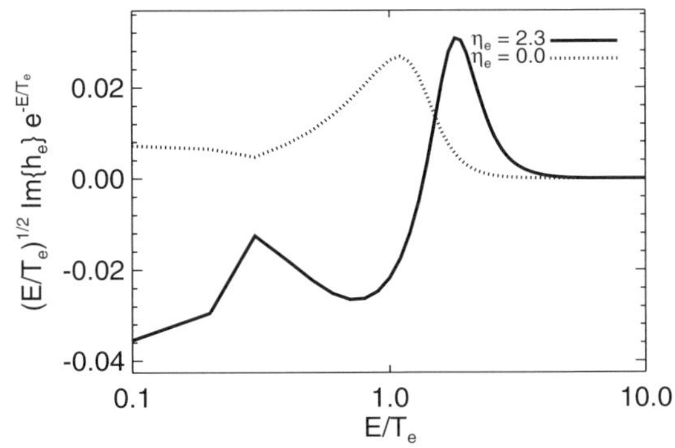

Fig. 13.5 Comparison of the nonadiabatic TEM electron response with and without the electron temperature gradient. The plasma parameters are from a TCV discharge discussed in detail in Chapter 14. The case $\eta_e = 2.3$ (solid) has both inward at low electron energy and outward particle transport at high electron energies. Whereas the reference case with $\eta_e = 0$ (dotted) has only outward particle transport. The discontinuous slope at $E/T_e \sim 0.5$ arises from the transitions between collisionless drift wave and trapped electron responses.

perature gradient of 2.3 times the density gradient and the dashed line when there is no electron temperature gradient. For the case with the temperature gradient ones sees the change in the direction of the energy flow from the electrons to the wave at a energy of $E/T_e \sim 2$ where the phase space gradient of the electron distribution has a steeper unstable gradient owing to the temperature gradient. At subthermal energies the phase space density varies as $n_e/T_e^{3/2}$ and thus is inverted and stable for the $\eta_e = 2.3$ regime.

With the assumption of $\gamma/\omega_0 \ll 1$ and $h_e \ll 1$, the mode frequency ω_0 and growth rate γ are

$$\omega_0(k, \boldsymbol{P}) = \omega_* \frac{[\Gamma_0 + \eta_i b_i(\Gamma_1 - \Gamma_0)]}{1 + (T_e/T_i)(1 - \Gamma_0)} \tag{13.37}$$

and

$$\frac{\gamma\omega_*}{\omega_0^2}\left[\Gamma_0 + \eta_i b_i(\Gamma_1 - \Gamma_0)\right] = \Delta \sum_{m=1}^{3} \text{Im}\ h_m^0 + \eta_e \sum_{m=1}^{3} \text{Im}\ h_m^1 \tag{13.38}$$

where

$$\Delta = 1 - \frac{\omega_0}{\omega_*}. \tag{13.39}$$

Im h_m^n for even n are positive definite and determine the electron particle diffusivity D and the heat diffusivity χ_e. For odd n, Im h_m^n are indefinite in sign. Also, the contribution of Δh_m^0 and $\eta_e h_m^1$ to the growth rates determines which is the more critical parameter between the density gradient and the temperature gradient.

In Chapter 14 we see that the electron temperature gradient also drives drift wave unstable at smaller space-time scales. These higher k, ω turbulence is called the ETG mode and as with the ITG mode occurs both in cylindrical and toroidal plasmas under conditions with $\eta_e > \eta_{\text{crit}}$. The frequency and growth rate now scale with v_e/L_{T_e} where $v_e = \sqrt{(T_e/m_e)}$ and the electron gyroradius ρ_e or $\rho_{e,i}$ where the later gyroradius is the analog of the ρ_s of the ITG modes that has the mixed measure of electron temperature and the ion mass. Here the $\rho_{e,i} = (m_e T_i)^{1/2}/eB$ so that there is an interchange of the roles of electrons and ions in-going from the ITG to the ETG instability. A important difference is that owing to the small scales of the ETG the quasineutral approximation that is well satisfied for the ITG drift waves is marginal and often not satisfied for the ETG modes. The condition for the ETG to be quasineutral is that $k_\perp \lambda_{De} \ll 1$ which requires that $\omega_{pe} \gg \omega_{ce}$.

For comparison of the two temperature gradient driven instabilities we show in Fig. 13.6 the results of the ETG modes derived in the next chapter plotted along with the TEM results for a TCV discharge driven strongly with third harmonic electron cyclotron heating in the core plasma.

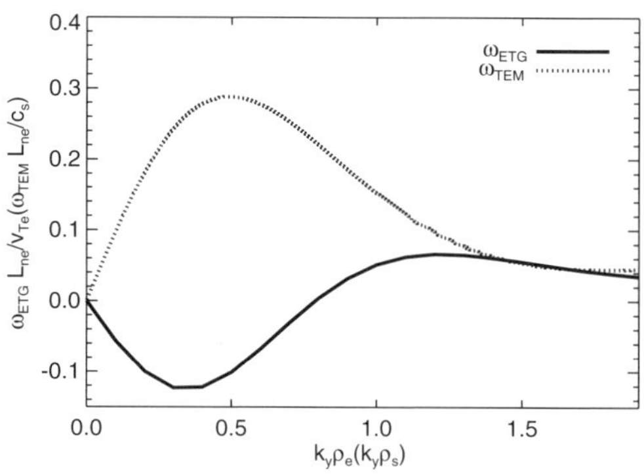

Fig. 13.6 The ETG (solid) and TEM (dotted) mode frequencies with increasing $k_y\rho_s(k_y\rho_e)$ at $r/a = 0.7$ and $t = 1.0\,\text{s}$ for TCV discharge 29892 in Table 14.1. The frequencies are respectively normalized by v_{T_e}/L_n and c_s/L_n. The high frequency -small scale ETG mode, analyzed in Chapter 14, has many features similar to the TEM-ITG as shown here.

Carefully note the change in the scaling of the frequencies and the wavenumbers in Fig. 13.6 and reversal of the direction of rotation of the waves. The longer wave length modes of the TEM may cause a larger transport when present with sufficient amplitude however, the smaller scale ETG modes grow much faster and so the TEM modes live in a bath of the ETG modes. The ETG modes have an inverse cascade of their energy to wave lengths comparable to the ion gyroradius where they can feed the lower frequency drift waves. Both forms of turbulence exist in plasmas with steep electron temperature gradients however the smaller scale modes are not sensitive to the details of the geometry such as the fraction of trapped electrons and are well above the usual electron collision frequencies. Thus, the universal nature of the anomalous electron transport across all magnetic confinement geometries is likely due to the small scale ETG modes. When the trapped electron mode at low collisionality with significant trapped electrons is present then the associated transport may mask that due to the ETG turbulence.

The role of zonal flows in controlling the level of transport from the trapped electron mode is explored in Ernst, *et al.* (2009). The work shows that the zonal flows can reduce the level of the electron thermal transport and the fluctuation level in general for the both the trapped electron modes and the modes driven by the ion temperature gradients. The method of investigation uses and compares the results of the particle-in-cell gyrokinetic codes and the continuum codes. These types of simulations are discussed in detail in Chapter 19 on simulations methods.

More generally, the works [Nevins, *et al.* (2005, 2006)] characterize the role of the interaction of the microturbulence and the sheared flows.

13.4 Internal Transport Barriers for ITG/TEM Models

The turbulent transport from the ion temperature gradient driven drift waves and the trapped electron modes transport the ion momentum flow in the plasma fluid in addition to the transport of thermal energy and particles. The first theories and computations of the flux of momentum from the turbulent $\boldsymbol{E} \times \boldsymbol{B}$ drift wave fluctuations were for cylindrical plasmas [Connor and Wilson (2000)]. In the neutral fluid literature the calculations of the momentum flux from turbulence is an long standing research area and the turbulent flux is called the Reynolds stress tensor given by $\rho_m \langle v_i v_j \rangle$ where ρ_m is the mass density and the components of the fluctuating velocity are v_i. The early calculations of Hamaguchi and Horton (1992) for the ITG turbulence included the calculation of the momentum stress tensor and showed that this flux can generate a mean sheared background flow.

A clear indication of the formation of a diffusion barrier is shown by integrating ensembles of test particles in a specified spectrum of drift waves existing on top of the background plasma with flow shear or reversed magnetic field. This method gives a direct picture of how the transport is reduced by the sheared fields. The reversed magnetic field breaks the Chirikov overlap condition so that the stochastic diffusion across the minimum $q_{\min}$ surface is not possible. There are trajectories that cross the $q_{\min}$ surface but these are better described by long-flight orbits than diffusion. Some works giving the details of the orbits in realistic fluctuation fields are Horton, *et al.* (1998), Evstatiev, *et al.* (2003), and Batista, *et al.* (2006).

Perpendicular shear flows are responsible for both symmetry breaking and suppression of turbulence, resulting in a shearing rate at which there is a maximum contribution to the momentum transport. The $\boldsymbol{E} \times \boldsymbol{B}$ momentum transport is shown to be quenched by increasing flow shear more strongly than the standard linear quench rule for turbulent heat diffusivity.

The analogous fluid turbulence calculations and experiment are given by Howard and Krishnamurti (1986). In their experiment in a Bernard convection in water, the low thermal gradient driving the Bernard convection cells analogous to the drift waves in the plasma interact with one another to create a mean circulation in the box. A simple mathematical model with a few key k_x, k_y modes was derived by Howard and used to explain the bifurcation with increasing temperature gradient to a state of turbulence with mean sheared flows.

Here show the plasma physics analog of this fluid experiment in which an increasing ion temperature gradient drives turbulence which in turn drives a mean ion $\boldsymbol{E} \times \boldsymbol{B}$ flow that reacts back on the turbulence. Since the reaction of the sheared flow on the ITG turbulence is stabilizing this situation has been characterized as a "predator-prey" situation. Here the sheared flow feeds on the turbulence and at some point when sufficiently strong sheared flows are present the ITG turbulence as the prey is reduced. This process can go through cycles as often seen in the simulations and in the dithering of plasmas between the two states of low-mode L and high-mode H of confinement [Diamond and Kim (1991)].

To understand and make a mathematical model of this problem we consider the dynamics of the ITG driven drift wave turbulent energy $W(t)$ in the toroidal magnetic geometry. The standard representation for the toroidal axisymmetric magnetic field in the torus is

$$\boldsymbol{B} = I(\psi)\nabla\phi + \nabla\phi \times \nabla\psi \tag{13.40}$$

where the toroidal angle is ϕ and the poloidal magnetic flux function is ψ and the toroidal field is $I(\psi)/R$ where $\nabla\phi$ is the unit vector in the toroidal direction divided by R the distance from the axis of symmetry to the spatial point x, y, z under consideration for the $\boldsymbol{x}$ field.

From the analysis of the ITG turbulence in this chapter we can write that the thermal ion flux q_i and the turbulent wave energy density $W(r,t)$ are given by

$$q_i = -n_i \chi_i(W)\left(\frac{dT_i}{dr}\right) \tag{13.41}$$

$$\gamma(W) = \gamma^{\ell} - \gamma^{n\ell} W \tag{13.42}$$

$$W = \sum_k \omega_k (\partial\epsilon_k/\partial\omega_k)\frac{|E_k|^2}{8\pi n_e T_e} \simeq \frac{1}{2}\sum_k \frac{e^2|\phi_k|^2}{T_e^2} \tag{13.43}$$

$$\frac{dW}{dt} = \left[2\gamma_0\left(\frac{R}{L_{T_i}} - \left(\frac{R}{L_{T_i}}\right)_{\mathrm{crit}}\right) - 2\gamma^{n\ell} W\right] W \tag{13.44}$$

$$\frac{3}{2}\frac{\partial}{\partial t}(n_i T_i) = \frac{1}{r}\frac{\partial}{\partial r} r\left(n_i \chi_0 W \frac{\partial T_i}{\partial r}\right) + P_E(r,t) \tag{13.45}$$

$$P_E' = \frac{\partial}{\partial r}\left(\frac{2P_E(r)}{3n_i(r)}\right). \tag{13.46}$$

The radial gradient of the power deposited per ion units of KeV/m per ion persecond is given in Eq. (13.46).

The ion temperature profile is determined by the combination of the injected heating power $P_E(r,t)$, turbulent profile, and the turbulent flux $q_i(W, dT_i/dr)$ defined in Eq. (13.41)-Eq. (13.45). Clearly the profile of the injected heating power $P_E(r,t)$ is important in determining the temperature gradient. In this simple model the spatial dependence of the heating power is reduced to the gradient defined in Eq. (13.46).

The transport problem is now reduced to a system of spatially local ODEs for the turbulence level dW/dt and the temperature gradient dT_i/dr. To this we add the equation for the momentum transport. Already at this simple stage there is interesting dynamics of the evolution of the turbulence and temperature gradient. There is a stable fixed point the system evolves to given by the formulas Eq. (13.47)-Eq. (13.48). The energy confinement time is then determined by the formula in Eq. (13.49).

$$W^*(P_E') = \left(\frac{\gamma^0}{2\gamma^{n\ell}}\right)\left[\sqrt{\left(\frac{T_i'}{T_i}\right)_c^2 + \frac{6\Delta^2\gamma^{n\ell}P_E'}{\gamma^0\chi_0 T_i}} - \left(\frac{T_i'}{T_i}\right)_c\right] \tag{13.47}$$

$$\frac{T_i'(P_E')}{T_i} = \frac{1}{2}\left[\sqrt{\left(\frac{T_i'}{T_i}\right)_c^2 + \frac{6\Delta^2\gamma^{n\ell}P_E'}{\gamma^0\chi_0 T_i}} - \left(\frac{T_i'}{T_i}\right)_c\right]. \tag{13.48}$$

The energy confinement time varies as

$$\tau_E = a\left(\frac{\gamma^{n\ell}}{\chi_0\gamma^0}\right)^{1/2}\left(W_*^{-1/2}\right). \tag{13.49}$$

The flow velocity is

$$\boldsymbol{u} = \frac{\boldsymbol{B}}{B_0}\left(u_\parallel + \frac{B_T}{B_p}u_\perp\right) - \frac{u_\perp B}{R_0 B_p}R^2\nabla\phi.$$

We break the flow velocity into poloidal and toroidal components given by

$$u_p = \left(\frac{B_t}{B}\right)u_\perp - \left(\frac{B_p}{B}\right)u_\parallel \tag{13.50}$$

$$u_t = \left(\frac{B_t}{B}\right)u_\parallel + \left(\frac{B_p}{B}\right)u_\perp. \tag{13.51}$$

The critical value of the flow shear divided by magnetic shear is

$$\left(\frac{L_s u_\perp'}{c_s}\right)_c = \gamma_{\text{crit}} = 2\sqrt{\frac{T_i}{T_e}(\eta_i - \eta_{ic})}. \tag{13.52}$$

For the JT60-U data we find the balance in Eq. (13.52) is

$$\left(\frac{L_s u_\perp'}{c_s}\right)_c = 2\sqrt{\frac{4\,\text{KeV}}{2\,\text{KeV}}(2-1)} = 2.8 \approx 3. \tag{13.53}$$

The local gradients of the momentum injection profiles are

$$P_t' = \frac{\partial}{\partial r}\left(\frac{P_t}{m_i n_i}\right) \qquad \text{and} \qquad P_p' = \frac{\partial}{\partial r}\left(\frac{P_p}{m_i n_i}\right) \tag{13.54}$$

and strongly influence the amount of flow shear. The nonlinear dynamics of the heating and flow shear are given by the four coupled ODEs:

$$\frac{dW}{dt} = 2\gamma_0\left[(\mu - \mu_c) + c_1 F_\parallel - c_2 F_\perp - \gamma^{n\ell} W\right] W \tag{13.55}$$

$$\frac{d\mu}{dt} = -\chi_x W \mu + P_E' \tag{13.56}$$

$$\frac{dF_\perp}{dt} = -\nu^{\mathrm{nc}} F_\perp + \bar{c}_2 W F_\perp + \sqrt{F_\perp}\, P_\perp' \tag{13.57}$$

$$\frac{dF_\parallel}{dt} = -\bar{c}_1 W F_\parallel + \sqrt{F_\parallel}\, P_\parallel'. \tag{13.58}$$

A typical model for the NBI power deposition is given by

$$P_{bi}(r) = \frac{P_T n\sigma}{(2\pi r)(2\pi r)\sin\theta}\left[e^{-(a-r)n\sigma/\sin\theta} + e^{-(r+a)n\sigma/\sin\theta}\right] \tag{13.59}$$

where θ is the angle of injection with tangential injection having $\theta_{\min} \simeq (a/R)^{1/2}$. The condition of depositing the power in the core plasma puts a narrow range on $a/\lambda_{\text{eff}} = an\sigma(E_0)/\sin\theta$ so that we can estimate $a/\lambda_{\text{eff}} \approx 2$ in calculations with Eq. (13.58). Using this estimate for $\lambda_{\text{eff}} \simeq a/2$ yields the coefficient in Eq. (13.59):

$$\frac{P_{\text{tot}}}{(2\pi R)(\pi r a)} \simeq \frac{20 \times 10^6\,\text{Watts}}{2 \times 10^7\,\text{cm}^3} \simeq 1\,\text{MW/m}^3.$$

The ion heating rate at $n = 3 \times 10^{13}\,\text{cm}^{-3}$ is the

$$\frac{2P_{bi}}{3n_i} = 150\,\text{KeV/s} \tag{13.60}$$

For the TCV 29892 discharge we calculated the drift-waves and TEM modes at the radial position $r/a = 0.7$ at $t = 1.0\,\text{s}$, $(T_e, T_i) = (1.30, 0.76)\,\text{KeV}$, $n_e = 5.74 \cdot 10^{19}\,\text{m}^{-3}$, $B_T = 1.09\,\text{T}$, $\eta_e = 2.3$ and $\widehat{s} = 0.65$. The relative plasma pressure is $\beta_e = 2\mu_0 p_e/B_T^2 = 0.025$ so that the $\omega \ll k_\parallel v_{\text{A}}$ and the modes are approximately electrostatic. Figure 13.7 shows the comparison of the growth rates between the case $\eta_e = 2.3$ (solid) and the reference case $\eta_e = 0$ (dotted) where in the absence of an electron temperature gradient where the density gradient drives turbulence. The comparison shows that the electron temperature gradient η_e leads to the distinct transition around $\nu_* = 0.45$ between a low growth rate collisional drift wave mode and the high growth rate trapped electron mode. At ν_* lower than 0.45, where the data point (+), $\nu_* = 0.06$, corresponds, the electron temperature gradient destabilizes the trapped electron mode by a factor 2 over the density driven TEM mode.

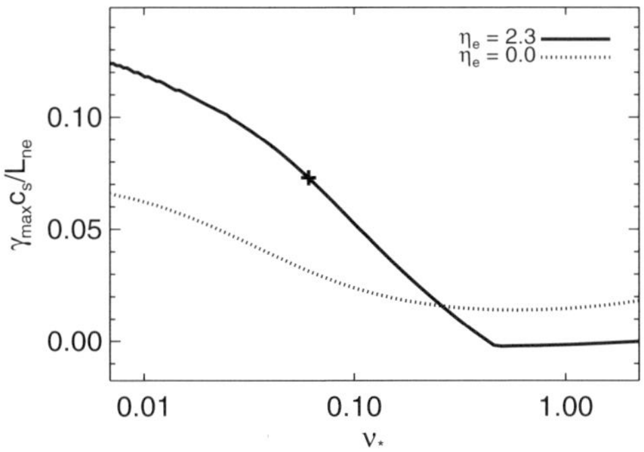

Fig. 13.7 The comparison of the maximum TEM growth rate versus the collisionality ν_* between the $\eta_e = 2.3$ (solid data) and $\eta_e = 0$ (dotted). The variation of collisionality ν_* is obtained by varying $T_e = 200 \sim 3000\,\text{eV}$ and with the fixed temperature ratio T_e/T_i.

As far as the magnetic shear is concerned, the shear dynamics principally plays a role in determining the parallel wavelength $k_{\|}L_n = k_y\rho_s\widehat{s}^{1/2}$. Therefore collisional and collisionless drift waves are affected by the magnetic shear through collisional dissipation and Landau damping. So, the magnetic shear dynamics has little influence over the trapped electron mode. The electron response functions for $\eta_e = 2.3$ and 0 with $\nu_* = 0.06$ of the experiment are compared in Fig. 13.8 and Fig. 13.9. The wave/guiding-center drift resonances occur at $E/T_e \sim \omega_0/\omega_{De}$. With the temperature driven case $\eta_e = 2.3$, the resonance shifts from $E/T_e = 1.1$ to 1.9 and it is clearly seen that the temperature gradient term $\eta_e(E/T_e - 3/2)$ stabilizes the lower energy contribution $(E/T_e < 3/2)$ to the TEM modes and destabilizes the higher energy contribution $(E/T_e > 3/2)$ to the TEM modes.

For all time slices and for all the radial profiles, the mode frequency, growth rate and electron response functions are calculated. The mode frequencies and the maximum growth rates for each radial point are shown at each time slice in Fig. 13.8.

At $r/a = 0.45 - 0.55$, there is a transition between drift waves and trapped electron modes. The drift waves propagate in the ion drift wave direction with low frequencies $\omega_0 \ll 0.1c_s/L_{ne}$ contrary to the TEM modes with higher frequencies $\omega_0 \sim 0.1c_s/L_{ne}$ at $r/a \gtrsim 0.5$, which propagate in the electron drift wave direction.

For each time slice, the drift wave modes are weakly destabilized. Beyond the transition radial position $r/a = 0.5$, the trapped electron mode is η_e-dominated and the temperature gradient stabilizes the drift wave mode. Towards the edge, the density and temperature gradients contribute to the instability with comparable magnitudes.

The imaginary parts of the electron response function are shown in Fig. 13.9. A positive value of the product of the response function and the sign of the mode frequency ω_0 implies the destabilizing contribution of the trapped electrons. The sign of the functions in Fig. 13.9 indicates destabilizing contribution by positive values of the response function.

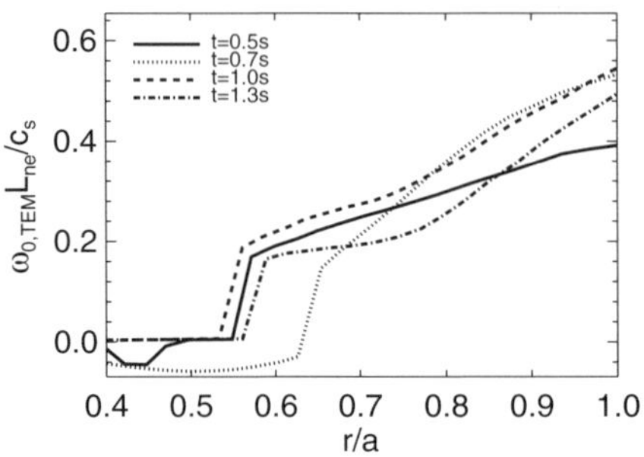

(a) TEM frequencies

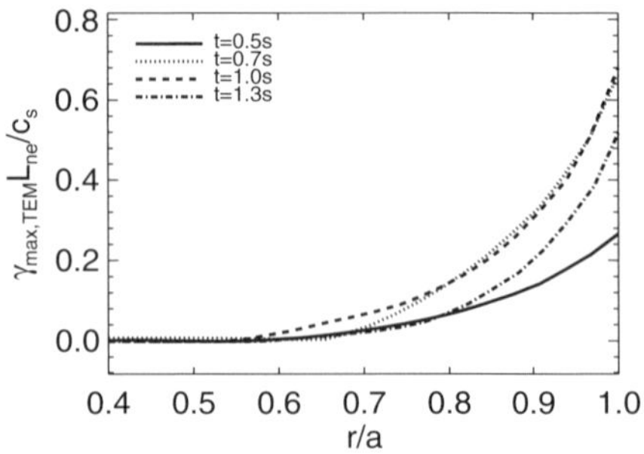

(b) Maximum TEM growth rates

Fig. 13.8 The TEM mode frequencies and the maximum growth rates of TCV 29892 at $t = 0.5$ (solid), 0.7 (dotted), 1.0 (dashed) and 1.3 s (dashed-dot). The maximum growth in panel (b) rate is obtained among the different k_y values.

At $r/a = 0.5$ as in Fig. 13.9(a)-(b), for $t = 0.5$, 1.0, 1.3 s, the collisionless and collisional drift waves are stabilized by the electron temperature gradient, while for $t = 0.7$ s, the TEM/ITG mode propagates in the ion diamagnetic drift direction, $\omega_0 < 0$, due to the large ion temperature gradient $\eta_i = 3$ and the electron temperature gradient drives the TEM instability for the lower energy $E/T_e < 3/2$ and drift wave instability. The ETG modes have the broad energy range of the resonances due to electron Landau damping $\omega_0 - \omega_{De}(E/T_e) - k_\parallel v_\parallel = 0$ and strong resonances due to the lower collisional frequencies $\nu_e L_{n_e}/v_{T_e} \ll \nu_e L_{n_e}/c_s$ compared to the TEM modes.

At $r/a = 0.7$ as in Fig. 13.9(c)-(d), for all the times, the electron temperature drives the TEM/ITG instability at the higher energy $E/T_e > 3/2$ and stabilizes at the lower energy. The ETG modes in panel Fig. 13.9(d) are destabilized by the temperature gradient at the lower electron energy. The high collisional frequency at $t = 0.5$ s weakens the resonance and the ETG instability. The stronger electron resonances at later times are consistent with the power balance transport [Erba, *et al.* (1999)].

13.5 The Weiland ITG/TEM Transport Model

The Weiland model [Jarmén, *et al.* (1987); Nordman, *et al.* (1990, 2005); Guo and Weiland (1997); Weiland (2000); Weiland, *et al.* (2005); Weiland and Holod (2005)] is often used to predict the radial global transport. In this section the model is used as a local stability tool to analyze the instabilities present in the plasma and their thresholds, and to calculate the local turbulent transport. The version of the Weiland model used here includes electromagnetic effects, collisions, impurity and magnetic shear effects but neglects varying correlation length [Weiland, *et al.* (2005);

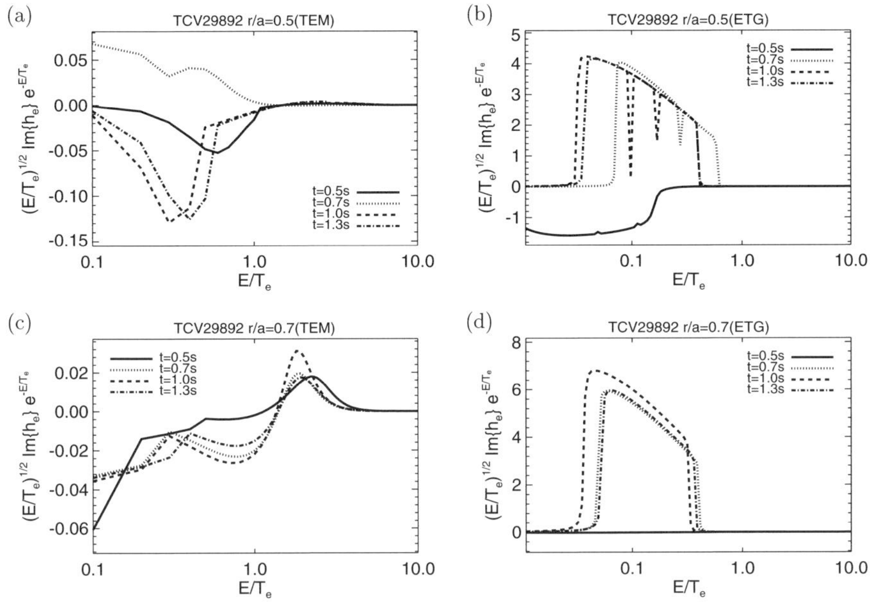

Fig. 13.9 The imaginary part of electron response functions at $r/a = 0.5$ (top panel) and $r/a = 0.7$ (bottom panel). The left and right panels are for TEM, Eq. (13.29)–Eq. (13.30), and ETG, Eq. (13.31) respectively.

Weiland and Holod (2005)] and toroidal and poloidal rotation effects [Weiland and Nordman (2006); Eriksson, *et al.* (2007)]. The two latter effects are included in a more refined version of the model.

In principle, the TEM (ITG) mode mainly drives electron (ion) thermal transport but the Weiland model (W). With a gyro-Bohm type mixing-length formula [Weiland and Nordman (1988)], the thermal diffusivities are

$$\chi^{\mathrm{W}}_{e,i} = \frac{R/L_{Te,i} - R/L_T^{\mathrm{crit}}(\widehat{w}_{\mathrm{re},i}, L_n)}{R/L_{Te,i}} \times \left\{ f_{e,i} \frac{m_i^{1/2}}{k_\theta \rho_s e^2 B^2 L_{ne}} T_e^{3/2}(eV) \sum_{g>0} \frac{\widehat{g}^3}{\left(\widehat{w}_{\mathrm{r}} \mp F_{e,i}\right)^2 + \widehat{g}^2} \right\} \tag{13.61}$$

where

$$F_{e,i} = \frac{10}{3} \frac{L_{ne}}{R} \frac{T_{e,i}}{T_e}. \tag{13.62}$$

Above e and i denote electron and ion, respectively. Here $f_e = f_t$, is the trapped electron fraction and $f_i = 1$. The mode frequency, $\widehat{w}_{\mathrm{r}}$, and growth rate, $\widehat{g}$, are normalized to the diamagnetic drift frequency of the electrons ($w_* = K_\theta \rho_s c_s / L_{ne}$). The model assumes that the radial correlation length is of the same order as the poloidal one, i.e. $k_r^2 \approx k_\theta^2$. A heat pinch arises in Eq. (13.61) from the threshold function $p(\widehat{\omega}_{\mathrm{re},i}, L_n) = R/L_T^{\mathrm{crit}}$. The part of Eq. (13.61) within the curly brackets gives the mixing length diffusivity.

In Eq. (13.61), the sum over g is over all unstable modes. Hence, TEM modes can drive ion transport and ITG modes the electron transport depending on the relative sizes of the resonance factor, $(\widehat{\omega}_{\mathrm{r}} \mp F_{e,i})$, and the growth rate. In the case of ion transport, the resonance factor will decrease with F_i if both R/L_{ne} and T_e/T_i are much larger than unity. In such a case, a significant TEM drive of χ_i can be expected [Asp, *et al.* (2007)].

Worked Problem 13.5.1

Find the 2 × 2 matrix eigenvalue problem for polarization and stability of the compressional drift waves.

The perturbative motions induced by ϕ and ψ produce a compressional change in the magnetic field $\delta B_{\|}$ given by

$$\frac{\delta B_{\|}}{B} = -\frac{1}{f}(c\phi + e\psi) \tag{13.63}$$

where properly $1/f$ is the inverse of a complicated integral operator f involving bounce averages over the particle trajectories. Here f is taken as a local quantity. The compressional change in the magnetic field $\delta B_{\|}$ is dictated by Amperé's law from the third row of the matrix equation Eq. (13.18) with the plasma currents flowing across the magnetic field lines in the $\widehat{x} \propto \nabla\Psi$ direction.

Substituting Eq. (13.63) into the first and second rows of the full matrix equation Eq. (13.18) gives the reduced 2 × 2 symmetric matrix

$$\begin{bmatrix} a - \dfrac{c^2}{f} & b - \dfrac{ce}{f} \\ b - \dfrac{ce}{f} & d - \dfrac{e^2}{f} \end{bmatrix} \begin{bmatrix} \phi \\ \psi \end{bmatrix} = 0. \tag{13.64}$$

The dispersion relation given by the determinant is

$$D_k(\omega, \mu) = ad - b^2 - \frac{1}{f}\left(c^2 d - 2bce + e^2 a\right) = 0. \tag{13.65}$$

The compressional MHD limit is $a \simeq -b \simeq d$, which reduces Eq. (13.65) to

$$ad - b^2 - \frac{d}{f}(c + e)^2 = 0. \tag{13.66}$$

To find analytically the kinetic ballooning interchange drift mode and connect that mode with the variational formulas, we take $\eta_e = 0$ and note $a \simeq -b \simeq d \simeq e$ for the dominant terms owing to the small $E_{\|}$. Then the full determinant D factors as

$$ad - b^2 + \frac{\tau(1 - \omega_*/\omega)[Q - (1 - \omega_*/\omega)]^2}{\frac{2}{\beta_e} + \frac{T_i}{T_e}R} = 0 \tag{13.67}$$

as shown by Eq. (13.66). Here $\omega_{*e} = \omega_*$ for $\eta_e = 0$. The last term in Eq. (13.73) gives the kinetic compressional response. For the near-to-MHD regime, the response function ϕ in Eq. (13.13) reduces to

$$Q - 1 + \frac{\omega_*}{\omega} \simeq -\frac{\omega_{*pi} - \omega_*}{\omega} + i\Delta_Q = -\frac{\omega_{*p}}{\omega} + i\Delta_Q, \tag{13.68}$$

with ω_{*p} having the total pressure gradient and a wave-particle resonant part $i\Delta_Q$ from wave-particle resonance. A reasonable approximation for the compressional response function R in the region $\omega\omega_{Di} > 0$ is

$$R \cong c_0 \frac{[\omega - \omega_{*i}(1 + 2\eta_i)]}{\omega - \omega_{Di} + ic_1|\omega_{Di}|} \tag{13.69}$$

where c_0 and c_1 are positive fitting coefficients of order $c_0 \simeq 1$ and $c_1 \simeq 0.1$. Here $\Delta_Q = \operatorname{Im} Q(\omega)$ and $\Delta_R = \operatorname{Im} R(\omega)$ are real numbers giving the resonant ion contributions.

The resonant modes in the high-β region have

$$\omega = \omega_0 + i\gamma_k \simeq \omega_{*i}(1 + \eta_i) + i\gamma_k \tag{13.70}$$

which is an ion diamagnetic drift wave. Here ω_0 and γ_k are real valued functions of k, P. A Taylor series expansion of the dominant terms in Eq. (13.67) gives

$$\begin{aligned} ad - b^2 &\simeq \left(\frac{\omega_*}{\omega} - 1\right)\left[\frac{\omega_A^2}{\omega_0^2} b - b\left(1 - \frac{\omega_{*pi}}{\omega_0}\right) - \frac{\omega_*\omega_D}{\omega_0^2}\right] \\ &+ \left(\frac{\omega_*}{\omega_0} - 1\right)\frac{i\gamma_k}{\omega_0}\left[\frac{-2\omega_A^2}{\omega_0^2} b - \frac{b\omega_{*pi}}{\omega_0} + \frac{2\omega_*\omega_D}{\omega_0^2}\right] \end{aligned} \tag{13.71}$$

for $\omega = \omega_0 + i\gamma_k$ with $|\gamma_k| \ll \omega_0$. Thus the growth rate γ_k is determined by

$$\frac{i\gamma_k}{\omega_0}\left[\frac{2\omega_A^2}{\omega_0^2} + \frac{\omega_{*pi}}{\omega_0}\right] b + \frac{T_e}{T_i}\frac{(\omega_{*p}/\omega_0 - i\Delta_Q)^2}{c_0[1 - \frac{\omega_{*i}(1+2\eta_i)}{\omega_0} - i\Delta_R]} = 0 \tag{13.72}$$

for $\Delta_Q \sim \Delta_R \ll 1$. Since $\omega_{*p}/\omega_0 \gtrsim 1$, the significant resonant contribution comes from $i\Delta_R$ in the denominator. Thus we obtain the growth rate formula

$$\frac{\gamma_k}{\omega_0}\left(\frac{2\omega_A^2}{\omega_{*p}^2} + 1\right) b = -\frac{T_e}{T_i}\frac{(\omega_*/\omega_0)^2\Delta_R}{c_0\left[\left(\frac{\eta_i}{1+\eta_i}\right)^2 + \Delta_R^2\right]} \tag{13.73}$$

with $\gamma_k > 0$ for $\Delta_R \equiv \operatorname{Im}[R(\omega_0)] < 0$.

The results of applying the Weiland model to the TCV discharge 29892 are shown in Fig. 13.10. The trend of the diffusivities at $r/a = 0.7$ agrees with the trend of the experimental heat fluxes shown in Fig. 13.9 up to $t = 1.0\,\text{s}$. At $t = 1.3\,\text{s}$ the MHD activity reduces the gradients in the outer part of the plasma, which results in a marked drop in the diffusivities at $r/a = 0.7$. This drop was also seen in the nonlinear ETG simulations for the same reason.

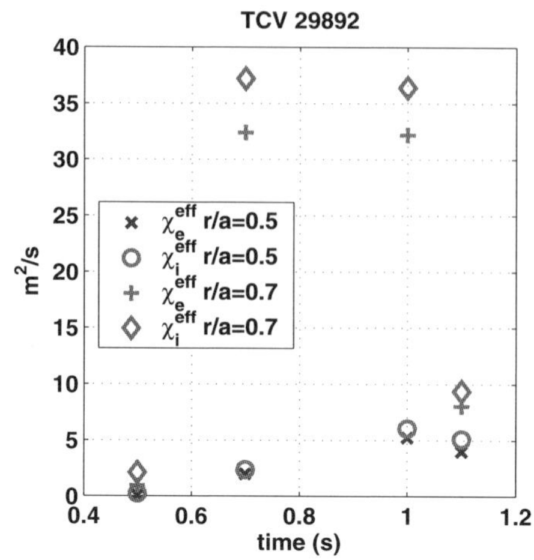

Fig. 13.10 Effective electron and ion diffusivities calculated by the Weiland model at $r/a = 0.5$ and 0.7 [Asp, *et al.* (2008)].

The diffusivities in Fig. 13.10 for $r/a = 0.5$ follows the same trend as for $r/a = 0.7$ with the exceptions that between $t = 1.0\,\mathrm{s}$ and $1.3\,\mathrm{s}$ the diffusion only decreases by 20% and that the diffusivities at $0.7\,\mathrm{s}$ are not of the same magnitude as at $1.0\,\mathrm{s}$.

During the ohmic phase at $t = 0.5\,\mathrm{s}$, the transport is much lower, as the ohmic heating of 0.49 MW is much smaller than the total heating with ECRH, $P_{\mathrm{tot}} \sim 1.2\,\mathrm{MW}$. This ohmic phase is also the only phase in which the ITG instability is about twice as big as the TEM. As a consequence, the ITG modes drive most of the ion transport and the TEM most of the electron. In the following phases the TEM dominates both the ion and electron transport.

The TEM electron heat diffusivities, however, are higher than those from the power-balance calculations. In the next chapter on the ETG turbulence we show that the ETG turbulence gives closer agreement with the power-balance diffusivity. In Chapter 14 we introduce a metric called the Average Relative Variance to measure the degree to which a model and data agree.

An error analysis shows that the diffusivities can be reduced by, on average, $\sim 75\%$ for the ions and $\sim 65\%$ for the electrons by decreasing all gradients (R/L_{ne}, R/L_{Te} and R/L_{Ti}) by 30%, which is within the error bars of the ion gradients. This would give diffusivities of similar magnitude to the experiment in inner part of the plasma, whereas further out they would still be overestimated. The local version of the Weiland model hence gives an inadequate quantitative picture of the heat transport. However, the TEM dominance remains even when decreasing the gradients and hence the qualitative analysis remains valid. The local version can therefore give a good idea of which modes are present and how the transport behaves qualitatively.

When the magnetic shear is weak the hydrodynamics electron response is stronger and the simulations shows a "bursty" time behavior of the heat flux. In this regime there are long periods with low levels of transport when the simulation is dominated by the nonlinearly excited velocity shear. These periods end when the shear flow drops sufficiently with a new bursts of high level transport characterized by large numbers of modes interacting. In sheared-slab geometry, the only physics limiting the $(k_y = 0, k_z = 0)$ mode is collisionality. However, in the toroidal plasma with a high fraction of trapped electrons the limit is more complex due to the trapped electrons which are hydrodynamic in their response interacting with the passing electrons that are near-to-adiabatic in their response.

Dorland and Hammett (1993) show a way to implement periodic radial boundary conditions in a sheared magnetic field to overcome the difficulty of edge effects where there is an absence of rational surfaces with Dirichlet boundary conditions. Periodic boundary conditions allow the heat flux leaving the box on the cold side to re-enter the simulation domain on the warm side. This recycling represents adequately the steady injection from heat sources on the high temperature side of the box and a sink on the low temperature side. The benefit of this method over simply zeroing out the $(k_y = 0, k_z = 0)$ modes is that these modes are important in the nonlinear dynamics with respect to the generation and decay of the sheared flows and this mechanics is lost if these modes are completely removed from the simulation.

References

Aschwanden, M. (2005). *Physics of the solar corona: An introduction with problems and solutions* (Springer-Verlag) ISBN:3-540-30765-6.

Asp, E., Weiland, J., Garbet, X., Parail, V., Strand, P., and the JET EFDA contributors. (2007). *Plasma Phys. Control. Fusion* **49**, p. 1221.

Asp, E., Kim, J.-H., Horton, W., Porte, L., Alberti, S., Karpushov, A., Martin, Y., Sauter, O., Turri, G., and the TCV TEAM. (2008). Electron thermal transport analysis in tokamak configuration variable, *AIP*, doi:10.1063/1.2965828.

Batista, A. M., Calderas, I. L., Lopes, S. R., Viana, R. L., Horton, W., and Morrison, P. J. (2006). Nonlinear three-mode interaction and drift-wave turbulence in a tokamak edge plasma, *Phys. Plasmas* **13**, 4, p. 042510.

Chen, J., Sen, A. K., and Migliuolo, S. (1992). *Phys Fluids B* **4**, 3, pp. 512-520, ISSN:0899-8221.

Connor, J. W. and Wilson, H. R. (2000). A review of theories of the L-H transition, *Plasma Physics Controlled Fusion* **42**, p. R1-R75.

Diamond, P. H., and Kim, Y-B. (1991) *Phys. Fluids B* **3**, p. 1626.

Dorland, W. and Hammett, G. W. (1993). *Phys. Fluids B* **5**, pp. 812-835.

Erba, M., Horton, W., and Ottaviani, M. (1999). Predictive tests of ITG based models of tokamak heat transport on ITER database discharges, *Nucl. Fusion* **39**, p. 495, doi:10.1088/0029-5515/39/4/307.

Eriksson, A., Nordman, H., Strand, P., Weiland, J., Tala, T., Asp, E., Corrigan, G., Giroud, C., de Greef, M., and Jenkins, I. (2007). *Plasma Phys. Control. Fusion* **49**, p. 1931.

Ernst, D. R., Lang, J., Nevins, W. M., Hoffman, M., Chen, Y., Dorland, W., and Parker, S. (2009). Role of zonal flows in trapped electron mode turbulence through nonlinear gyrokinetic particle and continuum simulation, *Phys. Plasmas* **16**, p. 055906.

Evstatiev, E. G., Horton, W., and Morrison, P. J. (2003). Multi-wave model for plasma-wave interaction, *Phys. Plasmas* **10**, 10, p. 4090.

Guo, S. and Weiland, J. (1997). *Nucl. Fusion* **37**, pp. 1095-1107, doi:10.1088/0029-5515/37/8/I05.

Hamaguchi, S. and Horton, W. (1992). *Phys. Fluids B* **4**, p. 319.

Hammett, G. W., Beer, M. A., Dorland, W., Cowley, S. C. and Smith, S. A. (1993). *Plasma Phys. Control. Fusion* **35** p. 973, doi:10.1088/0741-3335/35/8/006.

Hojo, H. (1995). *Jpn. J. Phys.* **34**, pp. 6501-6502.

Hong, B. G., Horton, W., and Choi, D-I. (1989). Pressure gradient-driven modes in finite beta toroidal plasmas, *Plasmas Phys. Control Fusion* **31**, pp. 1291-1303, doi:10.1088/0741-3335/31/8/006.

Hong, B. G., Horton, W., and Choi, D-I. (1989). Drift-Alfvén kinetic stability theory in the ballooning mode approximation, *Phys Fluids B* **1**, p. 2527.

Horton, W., Wong, H. V., Morrison, P. J., Wurm, A., Kim, J. H., Perez, J. C., Pratt, J., Hoang, G. T., LeBlanc, B. P., and Ball, R. (2005). Temperature gradient driven electron transport in NSTX and Tore Supra, *Nucl. Fusion* **45**, pp. 976-985, doi:10.1088/0029-5515/45/8/025.

Horton, W., Park, H-B., Kwon, J-M., Strozzi, D., Morrison, P. J., and Choi, D-I. (1998). Drift wave test particle transport in reversed shear profile, *Phys. Plasmas* **5**, pp. 3910-3917.

Horton, W., Lindberg, D., Kim, J. Y., Dong, J. Q., Hammett, G. W., Scott, S. D., Zarnstorff, M. C., and Hamaguchi, S (1992). Ion-temperature-gradient-driven transport in a density modification experiment on the Tokamak Fusion Test Reactor, *Phys. Fluids B* **4**, pp. 953-967, `http://dx.doi.org/10.1063/1.860112`.

Horton, W., Estes, R. D., and Biskamp, D. (1980). *Plasma Phys.* **22**, p. 663.

Horton, C. W., Jr. and Varma, R. K. (1972). Electrostatic stability theory of tokamaks from two-component fluid equations, *Phys. Fluids* **15**, pp. 620-631.

Howard, L. N. and Krishnamurti, R. (1986). *J. Fluid Mech.* **170**, pp. 385-410.

Jarmén, A., Andersson, P., and Weiland, J. (1987). *Nucl. Fusion* **27**, p. 941, doi:10.1088/0029-5515/27/6/006.

Kadomtsev, B. B. (1992). *Tokamak Plasma: A Complex Physical System* (Institute of Physics Pub., Bristol, UK and Philadelphia) ISBN: 0750302348.

Kim, J.-Y. and Horton, W. (1991). Transition from toroidal to slab temperature gradient driven modes, *Phys. Fluids B* **3**, pp. 1167-1170.

Kim, J.-Y. (1992). *Phys. Fluids B* **512**.

Kim, J.-Y., Horton, W., and Dong, J. Q. (1993). Electromagnetic effect on the toroidal ion temperature gradient mode, *Phys. Fluids B* **5**, p. 4030.

Kivelson, M. G. and Russell, C. T. (1995). *Introduction to Space Physics* (Cambridge Univ. Press) ISBN:0-521-45714-9.

Nevins, W. M., Candy, J., Cowley, S., Dannert, T., Dimits, A. M., Dorland, W., Estrada-Mila, C., Hammett, G.W., Jenko, F., Pueschel, M. J., and Shumaker, D. E. (2006). Characterizing electron temperature gradient turbulence via numerical simulation, *Phys. Plasmas* **13**, p. 122306.

Nevins, W.M., Hammett, G.W., Dimits, A. M., Dorland, W., and Shumaker, D. E. (2005). Discrete particle noise in particle-in-cell simulations of plasma microturbulence, *Phys. Plasmas* **12**, p. 122305.

Nordman, H., Weiland, J., and Jarmén, A. (1990). *Nucl. Fusion* **30**, p. 983.

Nordman, H., Strand, P., Eriksson, A., and Weiland, J. (2005). *Plasma Phys. Control. Fusion* **47**, p. L11.

Rewoldt, G., Beer, M. A., Chance, M. S., Hahm, T. S., Lin, Z. H., and Tang, W. M. (1998). Sheared rotation effects on kinetic stability in enhanced confinement tokamak plasmas, and nonlinear dynamics of fluctuations and flows in axisymmetric plasmas, *Phys. Plasmas* **5**, pp. 1815-1821, `http://link.aip.org/link/doi/10.1063/1.872851`.

Similon, P., Sedlak, J. E., Stotler, D., Berk, H. L., Horton, W., and Choi, D-I. (1984). Guiding-center dispersion function, *J. Comp. Phys.* **54**, 2, pp. 260-277, doi:10.1016/0021-9991(84)90118-9.

Sydora, R. D., Decyk, V., and Dawson, J. M. (1996). Fluctuation-induced heat transport results from a large global 3D toroidal particle simulation model, *Plasma Phys. Control. Fusion* **38**, p. A281, doi:10.1088/0741-3335/38/12A/021.

Weiland, J. and Nordman, H. (2006). *3rd EPS Conf. Plasma Phys.* **301**, p. 2.186 (Rome: ECA).

Weiland, J. and Nordman, H. (1988). *Theory of Fusion Plasmas: Proceedings of the Joint R.K. Wakerling* (Editrice Compositori for Societa Italiana di fisica, Bologna) former key: WeiNor88.

Weiland, J. (2000). *Collective Modes in Inhomogeneous Plasma* (IOP Publishing Ltd.).

Weiland, J., Asp, E., Garbet, X., Mantica, P., Parail, V., Thomas, P., Suttrop, W., Tala, T., and the EFDA-JET Contributors. (2005). *Plasma Phys. Control. Fusion* **47**, p. 441.

Weiland, J. and Holod, I. (2005). *Phys. Plasmas* **12**, p. 012505.

Zhu, P., Bateman, G., Kritz, A. H., and Horton, W. (2000). Predictive transport simulations of internal transport barriers using the Multi-Mode model, *AIP* **7**, pp. 2898-2908, `http://ojps.aip.org/pop/popcr.jsp`.

Zhu, P., Horton, W., and. Sugama, H. (1999). *Phys. Plasmas* **6**, p. 2503, doi:10.1063/1.873522.

Chapter 14

Electron Temperature Gradient Driven Turbulence

14.1 Electron Transport and the Critical Temperature Gradient

The problem of turbulent electron thermal losses has plagued all types of magnetic confinement experiments from the beginning of the of the fusion program. In the first reports of the Princeton Stellarator program led by Lyman Spitzer in the 1960s the electron temperature would not rise above a few hundred electron volts with intense heating. The confinement in these early experiments was characterized by the number of Bohm confinement times that could be achieve. The definition of the Bohm diffusivity is $T_e/16\,\mathrm{eB}$ where the coefficient of $1/16$ has no particular theoretical significance, but instead is historical from the early work of Bohm. Now, it is more conventional to define the Bohm D_B and related gyroBohm D_{gB} diffusivities by

$$D_B = \frac{k_B T_e}{eB} \qquad \text{and} \qquad D_{gB} = \frac{\rho_s k_B T_e}{aeB} \tag{14.1}$$

where the scale length a is to be associated with the scale length of the gradient driving the turbulence which for ETG would be the L_{T_e} scale length of the electron temperature gradient. In mks units for the charge and magnetic field and electron volts for the temperature, the Boltzmann constant is dropped. For example, $T_e = 1\,\mathrm{KeV}$ plasma in $B = 1\,T = 10^4\,\mathrm{G}$ magnetic field and minor radius $a = 1\,\mathrm{m}$ the formula for $D_B = 1000\,\mathrm{V}/1\ T = 10^3\,\mathrm{m}^2/\mathrm{s}$ and $\rho_s = 3\,\mathrm{mm}$ so the gyroBohm diffusivity is $D_{gB} = 3\,\mathrm{m}^2/\mathrm{s}$.

There is experimental evidence for both the Bohm and gyroBohm scaling in the magnetic confinement databases. Sometimes the transport is characterized as making a transition from one form to the other. When good confinement is obtained the numerical factor $\chi_e = c_B D_B$ in the Bohm diffusivity is given as a measure of the success of the confinement system. The toroidal quadrupoles and octupoles discussed in Chapter 5.4 were the first devices to obtain the low values of $c_B \lesssim 1/100$.

Scaling with the gyroBohm formula is easier to understand in terms of drift wave turbulence and brings in the scale length ρ_s of the turbulence driving the transport in a natural way. The formula for D_{gB} follows simply from using the drift wave wavelengths for the mixing lengths and the growth rate time scale for the decorrelation time scale. Thus, the numerical simulations discussed in Chapters 13-19 almost exclusively express their results in the units of D_{gB}. For Bohm turbulent diffusivity the mixing length increases from ρ_s to $(\rho_s L_T)^{1/2}$ and the correlations time remains of order L_T/c_s.

The electron thermal diffusivity, χ_e is reported in the first high electron temperature magnetic confinement experiments greatly exceeded the collisional transport value and this situation remains true today. Thus, it seems clear that the electron thermal transport is due to the drift wave turbulence over a wide range of conditions. Kadomtsev, in his 1992 monograph, devotes a chapter to the universality of the large, or anomalous value of the electron thermal diffusivity χ_e. Kadomtsev argues that intrinsic electromagnetic drift wave processes are responsible for this universal anomalous thermal transport.

Rebut and Lallia (1988) put forth a formula based on general arguments about electromagnetic fluctuations in magnetic confinement systems. The resulting formula was used extensively for a period to interpret the electron confinement data from JET.

The theoretical models for the anomalous electron transport based on drift waves are now almost exclusively used to interpret the electron transport data. There remain some disagreements on the particular type of drift wave fluctuations that dominant in producing the turbulent thermal diffusivity. In terms of broad categories, the two types of fluctuations that are found in theory and simulations are (i) the larger scale drift wave fluctuations on the scale of the ion inertial scale length ρ_s derived in the previous Chapter 13 and (ii) the smaller scale, faster growing gyroradius electron scale drift wave turbulence. There are cases where the power balance studies compared with databases favor one or the other type of instabilities. Since the turbulence is on two different space-time scales separated by the $(m_i/m_e)^{1/2}$ both types of turbulence may play a role in confinement experiments producing the measured anomalous electron thermal losses.

Electron energy confinement analysis of the Tore Supra, TCV, NSTX and MAST plasmas supports the conclusion that the ETG model is able to explain a wide range of anomalous electron transport data. Detailed thermal transport analysis supports the drift waves thermal loss models in ASDEX [Ryter, *et al.* (2001a); Ryter, *et al.* (2001b)], Tore Supra [Hoang, *et al.* (2003)], and the Frascati Tokamak Upgrade [Jacchia, *et al.* (2002)].

This chapter concentrates on the dynamics and transport from the small spatial scale-fast electron temperature gradient driven turbulence. The reader may observe that some of the ETG physics is similar to that of that of the ITG modes in Chapter 13 with the role of the electrons and ions interchanged.

The general scheme of the electron temperature gradient driven turbulence is shown in Fig. 14.1. The figure shows the small scale ETG is related to the larger scale ITG turbulence of Chapter 13. In the ETG turbulence the parallel wave

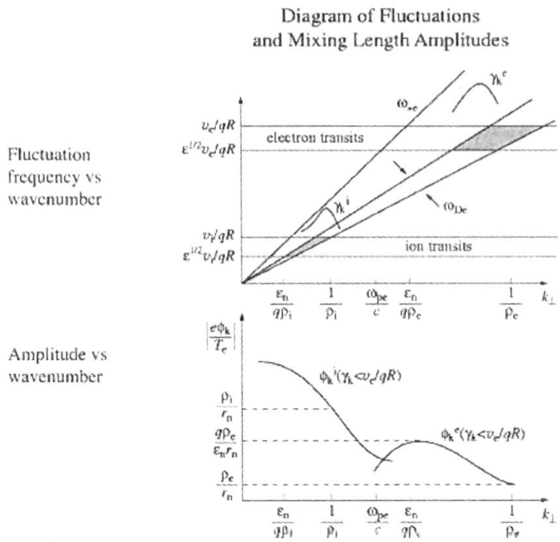

Fig. 14.1 A map for the distribution of the fluctuations and wave amplitudes ranging from the ion transit frequency through the ITG-TEM regime of Chapter 13 to the high-frequency, small-scale regime of the ETG turbulence. The upper part (a) gives the frequency versus wavenumber region of the turbulence and particle transit and bounce motions. The lower part (b) gives the amplitudes for the mixing length level of saturation, where the estimate is for the regime with $L_T \sim L_n < a \ll R$ for the typical well-confined tokamak plasma [adopted from Horton, *et al.* (1988)].

phase velocity $\omega/k_\parallel$ is higher than and comparable to the electron thermal velocity in contrast to the lower phase velocity of the ITG-TEM turbulence. This makes the ETG turbulence less sensitive to the details of geometry than for the lower frequency ITG-TEM modes. The instability has a strong inverse cascade, as shown theoretically and in computer simulations, which transfers fluctuation energy to the ITG instability scale. Thus, wave energy from nonlinear interactions is transferred from the ETG to the ITG-TEM regime. ETG spectral simulations are typically in boxes of order $3\rho_i \times 3\rho_i$ so that the pile-up of low-k energy can be absorbed into the dynamics of the ITG modes at the box-size limit of the k-space. The formulas are for a tokamak geometry with parameters $\epsilon_n = L_n/R$ and magnetic shear $s \sim 1$. Here $\rho_i = v_i/\omega_{ci}$ is the thermal ion gyroradius and $\rho_e = v_e/\omega_{ce}$ is the electron thermal gyroradius. This range of scales and frequencies is typically two orders of magnitude and dictates the use of different diagnostic methods to measure the turbulence.

Owing to the faster parallel phase velocity the quiver velocity of the electrons along the magnetic field from the parallel electric field and pressure gradient is relatively high for a given wave amplitude. This builds up the wave parallel electric current giving a significant magnetic perturbation even at relative low electron plasma beta or thermal-to-magnetic pressure ratio $\beta_e = 2(\omega_{pe}v_e/\omega_{ce}c)^2 = 8\pi n_e T_e/B^2$. One introduces the parallel vector potential $A_\parallel$ to describe the electromagnetic parts of the ETG turbulence. The parallel vector potential is solved for from the parallel electron current by

$$\nabla^2 A_\parallel = -\mu_0 \delta j_\parallel. \tag{14.2}$$

A typical simulation for NSTX plasma is shown in Fig. 14.2 where contours of isopotentials are shown for the three fluctuating fields $\phi(x, y, t)$, $A_\parallel(x, y, t)$ and $\delta T_e(x, y, t)$ in the saturated state. The figure shows that there are different dominant scales for the three fields. The electron temperature fluctuations have the smallest spatial scale lengths of order ρ_e, the electrostatic potential has the intermediate scale length size of order $\sqrt{\rho_e L_{T_e}}$ to $q\rho_e$ and the vector potential field has the scale length of order c/ω_{pe}. The $\boldsymbol{E} \times \boldsymbol{B}$ flow velocities and the $\boldsymbol{\delta B}_\perp$ vectors are given by the local tangents to the isolines of the electric potential and the vector potential.

The directional derivatives along these vector fields are given by the Poisson bracket operators designated by ϕ and $A_\parallel$ respectively. As seen in Chapter 10 these convective nonlinearities create vortex structures in the fields, as is evident in Fig. 14.2 in the snapshot in time for this simulation for ETG turbulence.

The electromagnetic component of the turbulence introduces a magnetic flutter described by Callen (1977) and Horton (1984). The space scale of the magnetic flutter is the electron collisionless skin depth $\delta_e = c/\omega_{pe}$. The relationship of the skin depth scale to the electrostatic scale of the inverse cascaded electrostatic potential is important to determine the parameter scaling of the electron thermal diffusivity. In Fig. 14.1 the regime is shown with the skin depth larger than the electrostatic scale length $l_{es} = q\rho_e/\epsilon_n = qR\rho_e/L_n$. First one considers that $L_n \sim L_{T_e}$ typical of L-modes and I-modes. Then one considers H modes $L_n > L_{T_e}$ where the scale length becomes L_n of order R which must be small compared with the temperature gradient scale length L_{T_e} for a confined hot plasma. As the dimensionless electron pressure increases to exceed $\beta_{e,\mathrm{crit}}$ then the skin depth becomes smaller than the electrons scale length l_{es}. Now the mixing length is controlled by the electromagnetic skin depth and the theoretical turbulent diffusivity, χ_e, is reduced compared to the extrapolated high-T_e values given in the electrostatic turbulence. In the electromagnetic regime has

$$\beta_e > \beta_{e,\mathrm{crit}} = \frac{2m_e}{m_i} \tag{14.3}$$

and the mixing length, l_{em}, is now

$$l_{em} = \delta_e = \frac{c}{\omega_{pe}}. \tag{14.4}$$

The electromagnetic transport formulas for the electron thermal diffusivities with this transition are used in Horton, *et al.* (2002, 2003) and are given for $\beta_e > \beta_{\mathrm{crit}}$

$$\chi_e^{em} = C_{em} q^\nu \frac{c^2}{\omega_{pe}^2} \frac{v_e}{(L_{T_e} R)^{1/2}} \left(\frac{R}{L_{T_e}} - \frac{R}{L_c} \right) \tag{14.5}$$

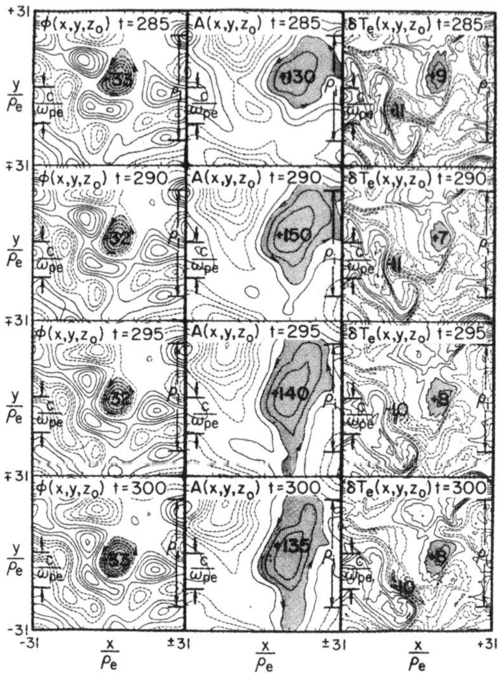

Fig. 14.2 ETG turbulence simulations used to interpret the electron thermal losses in NSTX [Kaye, *et al.* (2007)]. Contours of the electrostatic potential in the nonlinear saturated state showing the results of the inverse cascade and the formation of radially extended streamers.

and for $\beta_c < \beta_{\text{crit}}$ by

$$\chi_e^{es} = C_e^{es} q^\nu \left(\frac{R}{L_{T_e}}\right)^{3/2} \left(\frac{\rho_e^2 v_e}{L_{T_e}}\right) \left(\frac{R}{L_{T_e}} - \frac{R}{L_c}\right). \tag{14.6}$$

The magnetic shear and q dependence of the growth rate are used in choosing the exponents ν in the thermal flux. The exponent ν varies between 1 and 2. The critical gradient $(\nabla T_e)_c$ is derived from the marginal stability analysis of the dispersion relation. The presence of a critical gradient in the thermal flux expressions gives the flux formula the form of a diffusivity plus a radial transport velocity $V_r(r,t)$ where the relationship with the critical gradient is

$$V_r = \chi_e \frac{(\nabla T_e)_c}{T_e} = -\frac{\chi_e}{L_{T_{e,c}}} \tag{14.7}$$

giving an inward thermal pinch velocity with $L_{T_{e,c}} V_r/\chi_e \sim 1$. The following power balance studies will yield $V_r = \chi_e^{pb}/L_{Te,c} \geq 10\,\text{m/s}$. The gyroBohm estimate for χ_e is $\chi_e/L_{Te} = c_s(\rho_s/L_{Te})^2$. The velocity V_r is inward giving rise to the off-diagonal heat pinch effect. The sheared slab regime has

$$V_r = -1.9|s|\chi_e/qR\,(1 + Z_{\text{eff}} T_e/T_i) \tag{14.8}$$

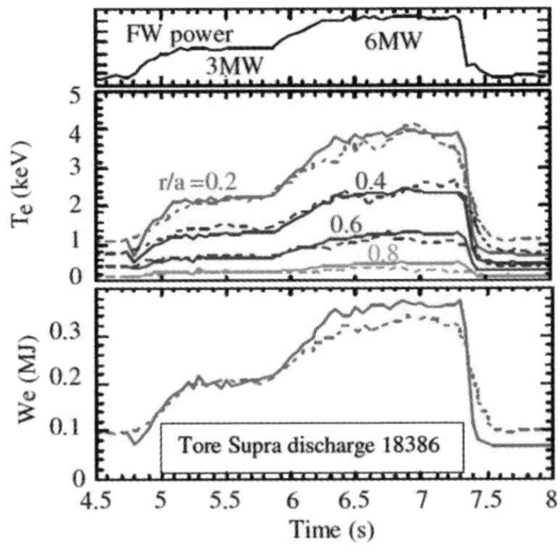

Fig. 14.3 Predictive simulation of an existing Tore Supra FWEH discharge [#18368, $I_p = 0.6\,\text{MA}$, $B = 2.2\,\text{T}$, $n_e(0) = 4 \times 10^{19}\,\text{m}^{-3}$], top panel with the fast wave heating power rising from 3 to 6 MW. Middle panel shows the time evolution of electron temperature (T_e) at various radii. The botom panel shows the rise of the total electron energy context at (W_e).

from Eq. (14.7). In the absence of magnetic shear the compression stabilization rather than electron Landau damping determines the transport velocity V_r with $(\nabla T_e)_c$. Peaked density profiles with small L_{ne} give an inward particle pinch velocity V_n. Pinch terms are equivalent to off-diagonal transport terms and are a standard feature of turbulent transport.

The formulas Eq. (14.6) are able to interpret the Tore Supra database since in the plasma profile the plasma β_e is above the β_{crit} in the core region and below the β_{crit} in the outer part of the profile.

The transport formulas were used in the CRONOS transport code along with the WAVE code for the plasma heating profile from the Fast Wave Ion Cyclotron heating [Moroz and Colestock (1991)] used to drive the Tore Supra plasma. The results showed a good predictive performance as shown in Fig. 14.3 where the FWICRH heating was first held at 3 MW and then ramped up to 6 MW during a long $\Delta t = 2\,\text{s}$ pulse .

An important difference in the large-scale ITG turbulence and the small scale ETG turbulence is that the electron parallel fluctuating currents $j_\parallel(k, \omega)$ are relatively larger in terms of their parallel acceleration of the electron fluid. This means that the electron inertial is important in determining the phase of the accelerated electrons. Thus, the scale length $\delta_e = c/\omega_{pe}$ is an important dispersion scale length in addition to the ion inertial scale length $\rho_s = c_s/\omega_{ci}$ dominant in Chapter 13. As the plasma pressure increases these dispersion scale lengths become equal when the dimensionless electron pressure $\beta_e = \beta_{e,\text{crit}} = 2m_e/m_i$ given in Eq. (14.3). At this $\beta_{e,\text{crit}}$ pressure value the electron thermal pressure, the Alfvén wave velocity is equal to the electron thermal velocity, so it is natural that the waves and fluctuations become electromagnetic. When the cross field correlation length for the electron transport is taken as the electron skin depth $\delta_e = c/\omega_{pe}$ then some of the features of the empirical scaling laws introduced by Rebut, *et al.* (1991) and Kadomtsev (1992) are present. The first principles understanding of this anomalous electron transport, however, comes from the electron temperature gradient driven drift waves and their nonlinear dynamics. Experiments in the Tore Supra device over a wide range of electron heating powers (ohmic heating to 8 MW of RF Fast Wave heating power) confirmed the ability of the ETG transport models to systematically explain the anomalous electron transport rates.

The dynamics of the large-scale structures in the toroidal ETG turbulence has been explained with theory and simulations. Simulations [Li and Kishimoto (2004)] have shown how the radially extended streamers emerge from the nonlinear interactions in the ETG turbulence.

The simulations using ETG formulas reproduced Tore Supra FWEH experiments and CLM experiments [Wei, *et al.* (2010)]. The most direct evidence of fluctuations in the tokamak follows from the electromagnetic scattering of microwaves off the ETG driven electron density fluctuations. The techniques are those described in Chapter 2 extended to smaller space scales by Mazzucato (2009, 2010) on NSTX. Earlier studies of the power balance on a large database of RF heated discharges in Tore Supra by Horton, *et al.* (2000, 2003), and Hoang, *et al.* (2001, 2003) showed that theory reproduces the $T_e(r,t)$ profiles measured with Thomson laser scattering for plasma heating powers from Ohmic to 8 MW of radio frequency heating power where the core temperature increased from about 1 KeV to 3 KeV. Both the spatial profile and the time dependence with a step-up in heating power are well reproduced. Similar good agreement with the profiles of $T_e(r,t)$ were found in the NSTX experiment with high-resolution Thomson scattering diagnostics in Kaye, *et al.* (2007). In the NSTX and MAST experiments the ion thermal transport flux remains close to the neoclassical collisional level while the electron transport is larger and requires the ETG level turbulent transport for predicting the measured electron temperature profiles from the deposited power. A typical example of the comparison of the ETG thermal flux calculated from simulations compared with the power balance flux for a fast wave heated NSTX discharge [Kaye, *et al.* (2007); Stutman, *et al.* (2009)] is shown in Fig. 14.4.

Measurements with coherent scattering of 280 GHz microwaves in NSTX with a five-channel heterodyne receiver showed that the presence of the ETG fluctuations at $k_\perp \rho_e \sim 0.2$-0.4 with amplitudes strongly correlated with the strength of the electron temperature gradient. The plasma beta is in the range of 3% to 6% so that the plasma density fluctuation producing the scattering has $k_\perp \delta_e \sim 2$. The electron temperature gradient varied from $R/L_{T_e} = 4$ to 8 in the experiments which produced and order of magnitude variation in the spectral density of the scattered microwave power.

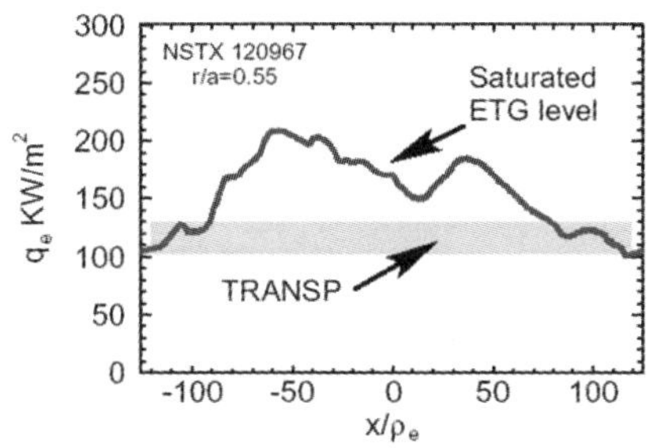

Fig. 14.4 Comparison of the power balance electron thermal flux computed for the NSTX parameters from the TRANSP data analysis code for discharge 120967 compared with the turbulent thermal flux computed from a local psuedospectral gyrofluid simulation. The system of partial differential equations used in the turbulence simulation are given Eqs. 14.21-14.25.

14.2 Electron Temperature Gradient Transport

In the high-$\boldsymbol{k}$ regime of ETG turbulence, the thermodynamics of the electron gas explains the direct drive of the turbulence from the electron temperature gradient $\boldsymbol{\nabla} T_e$. Thermodynamics shows that there is a critical temperature gradient $\boldsymbol{\nabla} T_e \equiv -T_e/L_{Te}^{\mathrm{crit}}$ for the ETG turbulence to extract energy from the temperature difference, $\Delta T = T_1 - T_2$, over the correlation length l_c.

The Carnot cycle shown in Fig. 3.3 gives the upper bound on the released turbulence energy $W \leq W_c = \Delta T \Delta S_e$, where ΔS_e is the change in the electron entropy density along the isothermal side T_1 in the core and T_2 in the lower temperature zone [Horton, *et al.* (2004)] of the nonlinear convective heat engine. The simulations of the associated partial differential equations for the electron gas directly confirm the onset of the turbulence above the critical gradient and the presence of elongated convection cells. As the gradient rises above the critical value, stronger nonlinear features appear in the fluctuations. In addition to thermal energy the fluctuations produce a transport of parallel electron momentum and magnetic flux that also develop into strong coherent structures for the parallel electron current and the magnetic flux function. The transport of magnetic flux results in the turbulence developing into a form of the small-scale tearing mode turbulence. Large-scale tearing modes are described in Chapter 15 as driven by the gradients of the background current density profile driving mechanism, but on a smaller spatial scale. Even in the absence of the dj/dr driven tearing modes the nonlinear ETG modes at finite electron pressures produce a small-scale, microtearing magnetic turbulence. This ∇T_e driven magnetic turbulence twists and shears the magnetic field lines which then reconnect releasing magnetic energy. This transforms a fraction of poloidal magnetic energy into the parallel electron parallel acceleration and into electron thermal energy.

Figure 14.2 shows the extended radial structures that convect the plasma in such a way as to take thermal energy from the high-temperature region T_1 to the lower

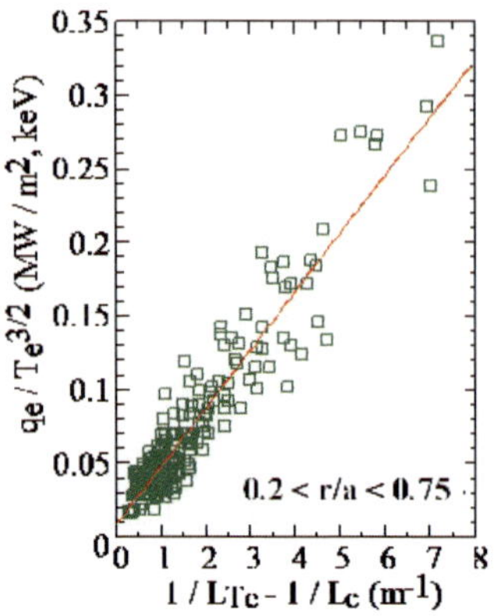

(a)

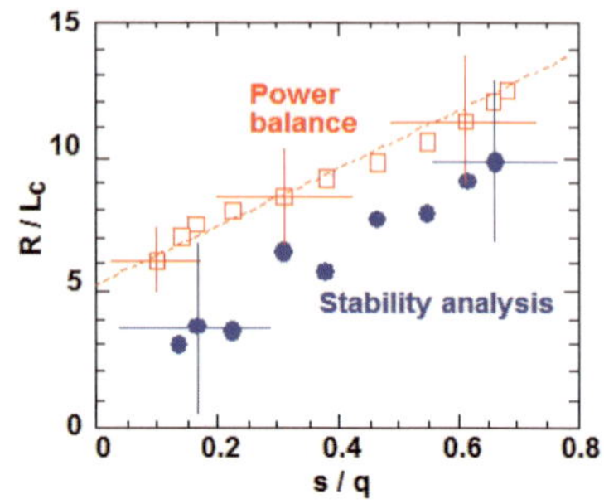

(b)

Fig. 14.5 Reduction of the large Tore Supra database for the scaling of the electron thermal flux q_e versus the temperature gradient for discharges steady state, hot electron discharges driven by fast wave electron heating (FWEH) . The core deposited heating power ranges from 0.5 MW to 7 MW. The scaling with $T_e^{3/2}$ agrees with Eq. (14.5) and the increase of the critical gradient scale with magnetic shear s/q agrees with Eq. (14.8) [Hoang, *et al.* (2001, 2003)].

temperature region T_2 across the convection cell. One can place an upper bound on the energy available to do work on the turbulence and convective cells from the Carnot cycle theorem of thermodynamics.

On the adiabatic sides of the Carnot cycle, a gas constant, $\Gamma = (d+2)/d$, describes the electron gas, where $d = 1, 2, 3$ for the number of degrees of freedom active in the ETG dynamics. Kinetic theory guides the choice of d with the slab and toroidal modes having $d = 1$ and $d = 3$, respectively. The drift-wave vortex or streamer gives the convective $\boldsymbol{E} \times \boldsymbol{B}$ motion between (n_1, T_1) and (n_2, T_2) producing the Carnot cycle for the convection period. From the Carnot-cycle calculation, the change of entropy density is given by,

$$\Delta S_e = n_e k_B \left(\frac{3}{2} \frac{\Delta T_e}{T_e} - \frac{\Delta n_e}{n_e} \right) \tag{14.9}$$

for $d = 3$ where k_B is the Boltzmann constant. Using the correlation length ℓ_c and the gradients $T_1 - T_2 = -\ell_c \mathrm{d}T_e/\mathrm{d}r$ and $n_1 - n_2 = -\ell_c \mathrm{d}n_e/\mathrm{d}r$ gives the maximum released electron energy density W_e released by the cycle as

$$W_e = \frac{3}{2} n_e T_e \ell_c^2 \frac{\mathrm{d}T_e}{\mathrm{d}r} \left[\frac{\mathrm{d} \ln T_e}{\mathrm{d}r} - \frac{2}{3} \frac{\mathrm{d} \ln n_e}{\mathrm{d}r} \right]. \tag{14.10}$$

In this ideal limit without dissipation in the system, the critical gradient expressed in terms of the major radius, R, of the torus is

$$\frac{R}{L_{Te}^{\mathrm{crit}}} = \frac{2}{d} \frac{R}{L_{ne}} \tag{14.11}$$

as follows from $W_e > 0$. Here d is the number of degrees of freedom in the electron dynamics. This critical gradient value R/L_{Te}^{crit} is expressed in terms of the temperature-to-density gradient ratio $\eta_e = L_n/L_T$ is well known from the Nyquist stability analysis of both ITG and ETG. Adding wave dissipation induced by magnetic shear, $R/L_s = s/q$, increases the critical temperature gradient [Horton, *et al.* (2004)] in proportion to $|s|/q$. The corresponding linear contribution is given in Hahm and Tang (1989) eigenmode calculation for the offset in the critical gradient from magnetic shear.

14.2.1 *Two-space scales for electron transport*

The role of the two space scales in the electron thermal heat flux, $q_e(r,t) = -n_e \chi_e \mathrm{d}T_e/\mathrm{d}\delta r$, is made clear from the kinetic theory formula for q_e. Kinetic theory gives the total energy flux as

$$Q_e = \frac{3}{2} T_e \Gamma_e + q_e \tag{14.12}$$

where Γ_e is the particle flux and q_e thermal conduction flux.

For a background plasma with local Maxwellian electrons parameterized by n_e, T_e and the energy density $\mathrm{g}(\epsilon)\mathrm{d}\epsilon$ for $\epsilon = m_e v^2/2T_e$, the cross field electron thermal flux is given by

$$q_e = n_e T_e \left[\sum_{q,\Omega} \frac{q_y |\overline{\Phi}_q|^2}{BT_e} f_t \int_0^\infty \mathrm{d}\epsilon \left(\epsilon - \frac{3}{2}\right) \mathrm{g}(\epsilon)\, \mathrm{Im}\left\{h(\epsilon, q, \omega)\right\} \right. \tag{14.13}$$

$$\left. + \sum_{k,\omega} \frac{k_y |\Phi_k|^2}{BT_e} \int_0^\infty \mathrm{d}\epsilon \left(\epsilon - \frac{3}{2}\right) \mathrm{g}(\epsilon)\, \mathrm{Im}\{h(\epsilon, k, \omega)\} \right] \tag{14.14}$$

where $\int_0^\infty \mathrm{d}\epsilon \mathrm{g}(\epsilon) = 1$, $\overline{\Phi}_q$ is the bounce-averaged potential fluctuation, f_t is the fraction of trapped electrons, $h(\epsilon, k, \omega)$ is the non-adiabatic electron phase-space-density response function and $\mathrm{Im}\{\cdot\}$ is the imaginary part. The notation is used that k and ω are the small-scale, high-frequency part of the fluctuation spectrum while Ω, q are for the large-scale, low-frequency part of the fluctuation spectrum.

In the thermal-flux formula, the low frequency large-scale contribution arises only from the trapped electrons, since the bounce-averaged potential $\overline{\Phi}_q(r) = 0$ for passing electrons (except for r exactly on a rational surface). The small-scale response Φ_k, involves resonance with both passing and trapped electrons through the response function,

$$h_k(\epsilon, \omega) = \frac{\left\{\omega - \omega_{*e}\left[1 + \eta_e\left(\epsilon - \frac{3}{2}\right)\right]\right\} J_o^2\left(\frac{k_\perp v_\perp}{\omega_{ce}}\right) f_M}{\omega - \epsilon\omega_{De} - k_\| v_\| + i\delta^+} \tag{14.15}$$

where the wave frequency, $\omega \gg \Omega_q$ and is given approximately by

$$\omega_{\boldsymbol{k}} = -\frac{\omega_{*e}\left[I_o(b_e)\mathrm{e}^{-b_e} + \eta_e b_e(\mathrm{I}_1(b_e) - I_o(b_e))\mathrm{e}^{-b_e}\right]}{Z_{\text{eff}} T_e/T_i + 1 - I_o(b_e)\mathrm{e}^{-b_e} + k^2\lambda_{\text{De}}^2} \tag{14.16}$$

where $b_e = (k_y \rho_e)^2$. The formula Eq. (14.15) gives the nonadiabatic contribution to the fluctuating electron distribution function through $\delta f_{k,\omega} = [1 - h_k(\epsilon,\omega)][e\phi_k/T_e]$. Equation (14.16) extends up to frequencies that are a fraction of $v_e/L_{Te} \leq 10^7$/s. Thus, thermal and supra-thermal electrons with large and small pitch angles carry the anomalous electron heat flux in the ETG turbulence.

For example, Eq. (14.16) yields the frequency in the case of TCV 29892 at $r/a = 0.7$. For the η_e values in this discharge, the ETG mode changes direction of rotation at $k_y \rho_e \simeq 0.3 - 0.8$ and this is where the linear ETG growth rate reaches a maximum. The transport flux is dominated by longer wavelength turbulence generated by the mode coupling nonlinearities as shown in the next subsection.

14.2.2 *Nonadiabatic ion response*

The effect of nonadiabatic ions occurs in the transitional wavenumber space scale region where $k_\perp \rho_i \sim$1-5. The wave response is associated with the part of the frequency spectrum shown in Fig. 13.6 where the wave propagates in the ion diamagnetic direction. In this low k part of the ETG spectrum the wave resonates with the guiding center drift motion of the ions with a resonant energy ϵ determined by

$$\omega_{k_y} = \omega_{Di}\epsilon + k_\parallel v_i \epsilon^{1/2}$$

where $\epsilon = m_i v^2/(2T_i)$ and α is the pitch angle. At $B = B_{\min}$ the pitch angle is defined by $\cos\alpha = v_\parallel/v$. The guiding center drift dominates the resonance for the conditions present in TCV.

In the low-$k_y\rho_e$ part of the spectrum in Fig. 13.6 the waves are resonant with ion guiding center drifts. The resonant ions have energy $\epsilon_r = E_{\rm res}/T_i = \omega_{k_y}/\omega_{Di} \lesssim 1$ in the low-k_y part of the spectrum and no resonance when the direction of rotation changes to that of the electrons. These resonant ions add the dissipative response to the ion density fluctuation through the nonadiabatic ion response,

$$\delta n_i = -\frac{Ze\phi}{T_i} n_{i0} \left\langle 1 - \frac{\omega_k - \omega_{*i}(\epsilon)}{\omega_k - \omega_{Di}\epsilon + i0^+} J_0^2 \right\rangle . \tag{14.17}$$

From Eq. (14.17) one calculates by making an expansion for $k_\perp \rho_i > 1$ in γ_k/ω_k, that the contribution of the resonant ions to the growth rate is

$$\gamma_i(k) = \frac{\pi^{1/2}\omega_k(\omega_{*i} - \omega_k)}{|\omega_{Di}|(1+b_i)^{1/2}} . \tag{14.18}$$

Equation (14.18) applies only for rotation of the mode in the ion diamagnetic direction as shown in Fig. 13.6 at low $k_y\rho_e$. Since the upper bound on this growth rate is given by v_e/L_{T_e} the contribution is small compared with the maximum ETG growth rate from the shorter wavelength part of the spectrum where the mode rotates in the electron direction. In many works on ETG turbulence the ions are taken as adiabatic. At the same time this ion contribution gives a small particle flux

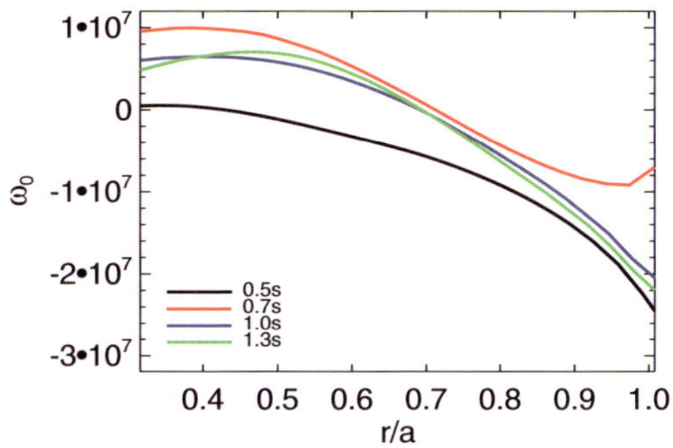

Fig. 14.6 The ETG mode frequencies along the radial positions r/a for $k_y\rho_e = 1.0$ of TCV 29892. In the core region the frequency increases monotonically in time with $t = 0.5, 0.7, 1.0$ from Eq. (14.16) until $t = 1.3$ where a magnetic island has formed.

due to the difference in the phase between the density fluctuation and the potential fluctuation from Eq. (14.17).

14.3 Electron Thermal Transport in TCV

The Tokamak Configuration Variable (TCV) [Tonetti, *et al.* (1991)] is a plasma with high-power density (up to $8\,\mathrm{MW/m^3}$) deposited in the core of the plasma by Electron Cyclotron Resonance Heating (ECRH) at significant plasma densities ($\leq 7\times 10^{19}\,\mathrm{m^{-3}}$) with a strong turbulent electron thermal transport χ_e.

The discharge chosen for analysis is distinguished by having four distinct high-confinement mode (H-mode) phases. An ohmic H-mode with type III edge localized modes (ELMs), which turns into a type I ELMy H-mode when the ECRH is switched on. The ELMs then vanish, which gives rise to a long quasi-stationary ELM-free H-mode. This ELM-free phase can be divided into two parts, one without magnetohydrodynamics (MHD oscillations) and one with the oscillations. The MHD mode in the latter case causes the plasma energy confinement to drop by $\sim 15\%$. For all four phases both large-scale trapped electron (TEM) and ion temperature gradient (ITG) modes and small-scale electron temperature gradient (ETG) modes are analyzed and found to co-exist. One finds, however, that the analytical TEM formulas have difficulty in predicting both the magnitude and the radial profile of the electron thermal flux. Collisionality governs the drive of the TEM, which for the discharge in question implies the instability can be driven by either (1) the temperature or (2) density gradient. The TEM response function is calculated to be relatively small and to have sharp resonances in its energy dependence defined by the condition $\omega_k = \overline{\omega}_{De}$. The ETG turbulence, on the other hand, is driven solely by the electron temperature gradient. Both trapped and passing electrons add to the ETG instability and turbulent thermal flux.

Average Relative Variance

For a quantitative comparison of the results of the TEM and ETG modes and also with the Weiland model, a dimensionless error measure, the so-called average relative variance (ARV) is introduced. According to this ARV metric method the ETG model explains 70% of the variation in the electron heat diffusivity whereas the predictive capabilities of the TEM-ITG models are poor with an ARV less than 50%. These results for TCV anomalous transport support the conclusion that the ETG model is able to explain a wide range of anomalous electron transport data, in addition to earlier evidence from ASDEX (Ryter (2001a,b), Tore Supra [Hoang, *et al.* (2003)], and the Frascati Tokamak Upgrade [Jacchia, *et al.* (2002)]).

The Tokamak Configuration Variable (TCV) [Tonetti, *et al.* (1991)] is a compact tokamak with high-power electron cyclotron resonance heating (ECRH) that is ideally suited for studies of turbulent electron thermal transport. In the configuration of TCV analyzed here, the plasma energy confinement $H_{98(2,y)} > 1$. The interesting feature of this particular discharge, TCV 29892, is that it contains no less than four different high-confinement mode (H-mode) phases whereof two are quasi-stationary without edge localized modes (ELMs) [Porte, *et al.* (2006)]. The differences between this quasi-stationary regime and other regimes obtained at DIII-D [Lohr, *et al.* (1988)] and Alcator C-Mod [Fielding, *et al.* (1996)], are that on TCV the confinement is achieved with pure electron heating at low-q_{95} (< 3), with no auxiliary ion heating, no active fueling and no cryo-pumping. This shot is an ITER-relevant plasma, as the plasma has a hybrid scenario-like flat q-profile with low magnetic shear.

The TCV shot 29892 begins with a typical ohmic H-mode with type III ELMs. When the ECRH is turned on at $t = 0.7\,\mathrm{s}$, the plasma rapidly switches to a type I ELMy H-mode. When the ELMs disappear, a quasi-stationary ELM-free H-mode emerges. After a single ELM event at $t = 0.9\,\mathrm{s}$, the quasi-stationary ELM-free H-mode is recovered but the residual magnetohydrodynamic (MHD) activity keeps the energy confinement lower than in the previous phase. The suppression of the ELMs in this case coincides with the onset of the ECRH power modulation. The transition has also been observed to arise spontaneously [Porte, *et al.* (2006)]. The transition to an ELM-free H-mode is an interesting topic in itself.

This discharge exhibits a higher density ($\sim 6 \cdot 10^{19}\ \mathrm{m}^{-3}$) than what is commonly found with ECRH heated tokamaks. The use of the third harmonic (X3) heating system instead of the second harmonic system allows for strong electron heating at plasma density up to approximately $11 \cdot 10^{19}\ \mathrm{m}^{-3}$. This in turn also permits considerable ion heating due to the faster thermal equilibration between ions and electrons increasing with higher density. Unfortunately there is no charge exchange recombination spectroscopy (CXRS) measurements during this shot, that would provide profiles of carbon ion temperature, density and rotation. Such data is available for a sister shot.

The core electron power deposition in this discharge varies through the four phases with an average of 1-2 MW/m^3 and has a maximum of 8 MW/m^3 on-axis. Thus, the power deposition is strongly peaked on-axis. With a core electron density, $n_e \sim 6 \cdot 10^{19}\,m^{-3}$ this gives a heating rate of 800 KeV/s per electron. The electron temperature is measured by Thomson scattering with 25 cords every 25 ms and the ECRH power deposition profile is calculated with the TORAY-GA ray-tracing code [Matsuda (1989)].

The electron heating is focused in the core plasma which gives rise to an accurately defined flow of thermal energy through the electron channel from the core plasma ($1 < T_e < 3$ KeV) to the edge plasma. This large thermal flow yields a precise estimate of the radial electron heat flux $q_e(r)$ and the electron heat diffusivity χ_e as there is accurate electron temperature profile data (5% error). The profiles are stationary for more than five energy confinement times, τ_E, and therefore the power balance χ_e is accurately determined. The greatest uncertainty in q_e and χ_e arrives from the ion temperature, T_i, and the uncertainties of the measurements of the plasma gradients. The error analysis performed shows that allowing T_i to vary between $0.33\,T_e$ to T_e only yields a 10% variation in the electron flux q_e in the ECRH heated phases. For the ohmic phase the error may be considerably larger.

The TCV deuterium plasma analyzed has electron plasma beta $\beta_e > 2\%$ at the core such that $\beta_e m_i/m_e \simeq 73$, where m_i and m_e are the masses of ions and electrons. The TCV 29892 plasma has a toroidal magnetic field $B \simeq 1.27$ T, plasma current $I_p \simeq 400$ kA and Alfvén velocity $v_A \ll v_e = (T_e/m_e)^{1/2}$. This implies that we have an electromagnetic regime where the Alfvén wave falls in the low parallel velocity $v_\parallel < v_e$ part of the electron energy distribution where electron Landau damping is significant. Hence, wave damping is important and the coupling to drift waves is strong. Due to the steep electron temperature gradient and the electron plasma pressure we expect the electromagnetic electron temperature gradient (ETG) drift-wave turbulence to be an important transport mechanism. The plasma also enters the trapped electron (TEM) and ion temperature gradient (ITG) modes in the electron energy spectrum.

The tools for the TEM-ITG analysis are an analytical model [Zhu, *et al.* (1999); Zhu, *et al.* (2000)] and the fluid-based Weiland model [Guo and Weiland (1997); Jarmen, *et al.* (1987); Nordman, *et al.* (1990); Weiland (2000)]. For the ETG analysis we use a code based on the three-coupled gyrofluid differential equations, the linear modes of which describes well the ETG instability [Horton, *et al.* (2005); Kim, *et al.* (2006)]. This ETG code has proven itself by accurately modeling a large database of electron thermal transport from ASDEX [Ryter, *et al.* (2001a); Ryter, *et al.* (2001b)], Tore Supra [Hoang, *et al.* (2003)] and the Frascati Tokamak Upgrade [Jacchia, *et al.* (2002)].

To obtain a quantitative measure of how well the theoretical models explain the observed heat flux, the metric called the average relative variance (ARV) [Weigend, *et al.* (1990); Detman and Vassiliadis (1997)]) is introduced. The ARV metric gives a dimensionless error measure for inter-model comparisons. The ARV increases quadratically with the magnitude of the model χ_e so it increases with a mismatch in the model's relative error with respect to the database χ_e. The model ETG thermal flux is also in good agreement with the power balance as shown in a figure for the thermal diffusivities in Section 14.2. The ARV parameters for the analytical TEM model and the local Weiland model show that these models fail to predict quantitatively the thermal transport. The ETG model, on the other hand, can, according to this metric, explain 70% of the variation of electron heat diffusivity.

14.4 ECRH Driven Discharges

The TCV discharge 29892 analyzed in this chapter is described by Porte, *et al.* (2006) in the conference proceedings of the 21st IAEA Fusion Energy Conference [Porte, *et al.* (2006)]. This section gives the basic features of the shot needed to understand the underlying heat transport and energy confinement. TCV is a tokamak with a high-power density electron cyclotron resonance heating (ECRH) system in Table 14.1. In shot 29892, the third-harmonic X-mode (X3) heating system was

Table 14.1 TCV machine and discharge 29892 parameters. Radial values are given at $r/a = 0.7$ in the Ohmic H-mode and the ELM-free H-mode.

Fixed parameters		
Plasma major radius, R (m)	0.89	
Plasma minor radius, a (m)	0.22	
Plasma elongation	1.75	
Plasma triangularity	0.51	
Plasma volume (m^3)	1.47	
Plasma current, I_p (kA)	400	
Toroidal field at the magnetic axis (T)	1.3	
Time dependent parameters	$t = 0.5$ s	$t = 1.0$ s
Total heating, P_{tot} (MW)	0.49	1.13
Global energy confinement time, τ_E (ms)	37	32
H-factor $\text{H}_{98(2,y)}$	0.72	1.1
Safety factor, q	1.29	1.30
Safety factor, q_{95}	2.18	2.16
Magnetic shear, s	0.80	0.65
Density gradient length, L_{ne} (m)	0.15	0.21
Temperature gradient length, L_{Te} m)	0.12	0.09
Electron temperature, T_e (KeV)	0.57	1.3
Electron diamagnetic drift velocity, v_{de} (km/s)	2.0	4.5
Electron magnetic drift velocity, v_{De} (km/s)	0.5	1.3
Ion Larmor radius at T_e, ρ_s (mm)	2.7	4.1
Power-balance diffusivity, χ^{PB} (m^2/s)	0.28	0.83
Gyro-Bohm diffusivity, $D_{gB} = \rho_s v_{de}$ (m^2/s)	5.4	18.5
Drift wave frequency, $\omega = k_y v_{de}$ (kHz) at $k_y \rho_s = 0.3$	341	353
Electron collisionality, ν_*[1]	0.24	0.06
ω_{pe}/ω_{ce}	1.8	1.9

used to enable significant electron heating at the plasma density $n_e \sim 6 \cdot 10^{19}$ m^{-3} [Porte, *et al.* (2006)]. The electron heating gave rise to a significant ion heating $P(e \to i)$ through thermal equilibration, $P_{e \to i} = 0.2 \pm 0.1$ MW. Ion dynamics may directly or indirectly influence the electron transport. Ion heating from the electrons is also of fundamental importance for ITER and in nuclear fusion producing machines in which the ion heating is from the alpha particles through the electrons. Electron-to-ion heat transfer is vital to sustain the fusion reactions. In addition, this discharge features an ITER hybrid scenario q profile with low magnetic shear.

X3 (third harmonic extraordinary mode polarized electron cyclotron wave) heating is applied at $t = 0.6$ s to a target plasma consisting of an ohmic ELMy H-mode. The X3 ECH power is injected vertically from the top of the machine with the resonating surfaces close to the plasma axis which makes the heating peak in the inner core of the plasma as shown in Fig. 14.7. Moreover, the high temperature ($T_e \approx 1\,4$KeV) caused by the high-energy confinement time, ($\tau_E \approx 37$ ms), of the target plasma helps to increase the X3 absorption. The X3 heating does not produce a significant population of supra-thermal electrons that would have influenced the transport analysis. The ion temperature profile used for the calculation of the thermal exchange power density in Fig. 14.7 was taken from a sister shot where CXRS data was available. To validate the $T_i(r, t)$ profiles they were cross-checked with Neutral Particle Analyzer (NPA) [Karpushov, *et al.* (2006)] measurements of the energy spectra of the deuterium ions escaping the plasma.

The electron heat flux is proportional to the difference between the integrated electron power deposition P_e and the integrated electron-ion thermal equilibration $e \to i$, energy transfer power from electrons to ions through collisions. During the ohmic phase at $t = 0.5$ s, the total thermal equilibration is comparable to the total

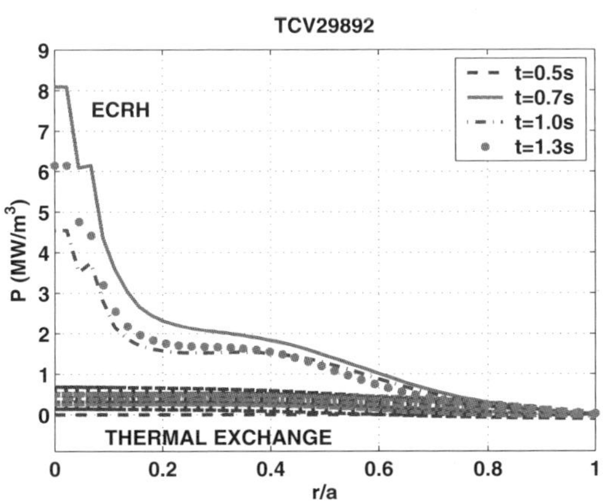

Fig. 14.7 Deposition of ECRH in TCV 29892 calculated by TORAY-GA [Matsuda (1989)] and the thermal equilibration power to the ions. The latter shows the range of values expected assuming a 5% error in n_e and T_e and assuming a 20% error in T_i.

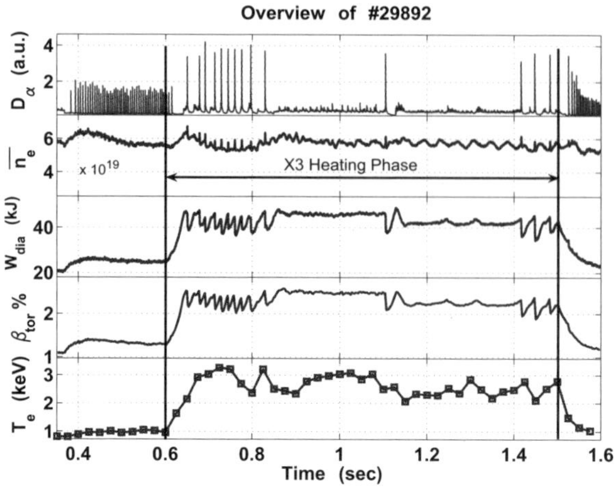

Fig. 14.8 Temporal evolution of TCV29892. From top to bottom are shown D_a emission, line-averaged density, total stored energy measured by the diamagnetic loop, toroidal β and on-axis electron temperature. At $t < 0.6$ s, standard TCV H-mode with Type I ELMs. Full ECRH power for $0.6 \leq t < 0.8$ s and Type I ELMy H-mode. Modulated ECRH power for $t \geq 0.8$ s and quasi-stationary ELM-free H-mode. Single-ELM event at $t = 1.1$ s triggers MHD activity which reduces the confinement as shown by the decrease in W_{dia} at the D_α-spike at $t = 1.1$ s.

power deposition $P_e = P_{\text{oh}} \sim 0.4$ MW. For the ohmic regime, a 10% error in the ion temperature produces a 20% error in the electron heat flux. In the ECRH X3-heated phases of the discharge, a 10% error in the ion temperature leads to approximately a 10% variation in the electron heat flux.

The time history of the discharge TCV2989 is shown in Fig. 14.8 with five panels for plasma-neutral ionization $D_\alpha(t)$, mean density $\overline{n}_e(t)$, stored plasma energy $W_{\text{dia}}(t)$, plasma beta and electron temperature.

The D_α line is from electron transitions from the $n = 3$ to the $n = 2$ quantum (Bohr) levels of the deuterium atom. The signal is used to monitor the interaction of the hot plasma with neutral gas component that surrounds the plasma in edge and scrape-off layer. When there is a burst of brightness in the alpha line radiation there is a excursion or a plume of the hot plasma into the outer layer of the confinement chamber. Thus, the edge localized modes called ELMs give rise to bursts of the red $\lambda = 656$ nm radiation visible light from the plasma. For hydrogen working gas the line radiation is from the neutral hydrogen atom and for the deuterium working gas the line is from the neutral deuterium atoms. Owing to the higher nuclear mass of deuterium the D_α line has a slight shorter wavelength of $\lambda = 656$ nm than the hydrogen H_α line.

The transport analysis is carried out in each of the four H-mode phases shown in Fig. 14.8 and Fig. 14.9. The first phase (< 0.6 s) is a standard ohmic H-mode with Type III ELMs. At $t = 0.6$ s the ECRH X3 heating is switched on at full power and the plasma enters into a type I ELMy H-mode. The X3 heating is at

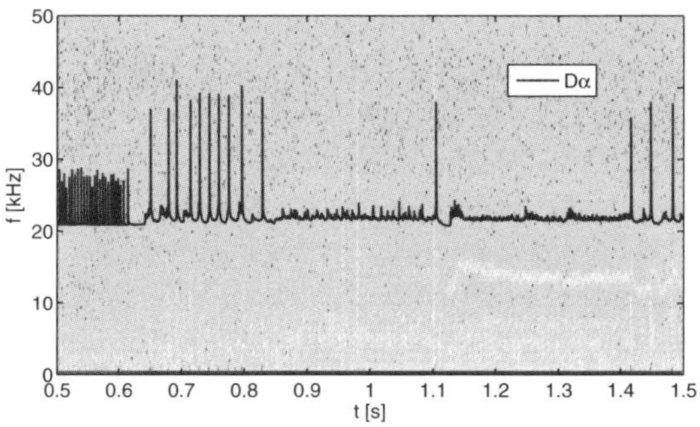

Fig. 14.9 Spectral temporal evolution of TCV29892 also including the D_a emission in red. The $m/n = 3/2$ mode with a frequency of 15 kHz emerges after the single ELM event at 1.1 s.

full power (~ 1.35 MW coupled to the plasma) until 0.8 s when the power from one gyrotron is fully modulated at a frequency of 127Hz with a duty cycle of 50%. This modulation results in an average heating power reduction of approximately 17%. Although, in this case, there was a strong correlation between X3 power modulation and the onset of the ELM-free H-mode, power modulation is not necessary [Porte, *et al.* (2006)]. The ELM-free H-mode regime has been achieved without modulation. In contrast to other machines like DIII-D [Lohr, *et al.* (1988)] and Alcator C-Mod [Fielding, *et al.* (1996)] which have also produced ELM-free H-modes, the TCV tokamak accomplishes this with no direct ion heating, no active fueling and no cryo-pumping.

The ELM-free nature of the discharge is seen from the steady D_a signal in the top frame of Fig. 14.8 and Fig. 14.9 while the high-energy confinement is shown in the measurement of total energy content obtained from a diamagnetic loop, $W_{\rm dia}$ in the third frame of Fig. 14.8. The constant $W_{\rm dia}$ and line averaged density (second frame) show that this phase is quasi-stationary. The phase ends with a single ELM event at 1.1 s at which time a $m/n = 3/2$ magnetic island is triggered by a core $m = 1$ sawtooth oscillation as seen in Fig. 14.9. This MHD mode has a frequency of 15 kHz and a magnetic island width of about 5 cm [Turri, *et al.* (2008)]. A magnetic island of this width located at a normalized minor radius around $r/a = 0.6$ gives rise to a decrease of total stored energy of about 25% according to the model of Chang and Callen (1990, 1992). The observed energy drop in Fig. 14.8 third frame, is closer to 15%. There is therefore no doubt that it is the destabilization of the $m/n = 3/2$ MHD mode that causes the drop in confinement in this fourth and last H-mode phase of the discharge. Note that although affected by this MHD mode, this fourth phase is a quasi-stationary ELM-free H-mode. Moreover, $m/n = 1/1$ modes associated with sawteeth activity and intermittent $m/n = 2/1$ are present throughout the discharge [Turri, *et al.* (2008)].

Simulations of ELMs are performed with MHD codes like JOREK [Huysmans and Czarny (2007)]. Using models of the H-mode plasma profiles with sharp gradients in the pressure profile and the toroidal current profiles produce bursts of complex interchange modes called the peeling-ballooning mode that releases up to 10% of stored plasma energy in a short burst. These events are monitored by the light emission from the ionization-recombination lines of the neutral deuterium atoms in the edge plasma. Examples of the ELMs are shown in Fig. 14.9.

For axisymmetric tokamak geometry without external toroidal momentum input the radial electric field E_r is arbitrary within the ordering of $|v_{E_r}| \sim \rho_i v_i/a$. Only, small deviations from axisymmetry such as the toroidal field ripple lead to a particular E_r. The poloidal rotation velocity is of the order

$$v_\theta = C_{E_r} \frac{T_e}{eBL_{ne}} \tag{14.19}$$

with the numerical coefficient C_{E_r} bounded by 2. The standard observed tokamak core rotation is in the electron diamagnetic direction from $E_r < 0$. The Heavy Ion Beam Probe on TEXT (HIPB) [Hallock, *et al.* (1994)] measured this negative $E_r < 0$ directly. This magnitude of E_r is consistent with a collisional loss of ions in the edge or scrape off layer plasma. In outer edge region the sign of E_r flips to hold the electrons in resulting in a narrow shear poloidal flow shear layer at the edge of the plasma. Bengtson, *et al.* verified this picture with probes and the HIBP in the Ohmic heated L-mode TEXT tokamak.

In the single null divertor configurations the inward E_r shear is enhanced by having the grad-B and curvature drifts directed toward the separatrix divertor chamber. This breaks up-down symmetry and results in a lower threshold power for the transition from the L-mode to the H-mode presumably due to the stronger radial shear in the $E_r(r)$ scrape-off layer field.

This upper limit to E_r is used to analytically and numerically estimate the effect on the growth rates for ITG and ETG turbulence [Hamaguchi and Horton (1991); Sugama and Horton (1995)]). The effect is rather complicated since at first order there is a radial shift of the wave functions proportional to the E_r shear divided by the magnetic shear. The shear-flow effect is significant on weakly growing ITG modes and generally weak on the ETG modes. From the electron temperature T_e profile and the system parameters in Table 14.1, the range of the shearing frequency is $\mathrm{d}v_{E_r}/\mathrm{d}r = 5 \sim 8 \times 10^4\ \mathrm{s}^{-1}$. This shearing rate may stabilize some of the slower growing TEM modes given in Section 13.2.4 but is too low to stabilize the turbulence considered here. For ETG this level of sheared rotation is negligible due to the fact that the electrons are interacting with much higher frequency fluctuations.

14.5 Electron Temperature Gradient Turbulence Modeling

The electron temperature gradient drift wave frequency $\omega_{0,\mathrm{ETG}}$ is given in Eq. (14.16) arising from the $\boldsymbol{E} \times \boldsymbol{B}$ convection of the electron density in the presence of adiabatic ions $\delta n_i = -eZ_i n_i \phi/T_i$. The frequency of the mode is shown in

Fig. 14.6 for $k_y\rho_e = 1.0$. The smaller electron temperature gradient toward the edge gives the change of the wave direction from electron diamagnetic to ion diamagnetic drift direction.

A simple gyrokinetic simulation for the ETG was carried out with by Jenko and Dorland (2002) using the codes they developed called GENE and GS2. The results were that the ETG thermal flux for usual tokamak parameters is large and universal in nature. The comparison with the slab ETG simulation is given and the torodial curvature-driven flux is about 30 to 40 times larger than the slab ETG thermal flux for their particular set of parameters. The slab ETG simulation with the GTC code are given by Fu, *et al.* (2012) and show a validation of the model with the results measured in a steady-state hydrogen cylindrical plasma in the Columbia Linear Machine by Wei, *et al.* (2010). The simulations required large computing power owing to the small scale of the ETG simulation and the experiments require special microelectronic detectors for the megaHertz frequencies and 100 micron wavelengths. The GENE simulations used teraflop computers with billions of effective particles as discussed in Chapter 19. In this chapter we analyze the turbulence using gyrofluid equations for the electron dynamics with an electromagnetic field $A_\parallel$ to generate the associated magnetic turbulence. The gyrokinetic simulations use several simplifications including adiabatic ion response and electrostatic fields with a limited number of parameter variations.

Gyrofluid models of the ETG mode were developed to extend the convection fluid model for the study of drift wave turbulence [Horton, *et al.* (1988)], Li and Kishimoto (2004, 2005), [Holland and Diamond (2002)]. Here we describe the finite electron Larmor radius (gyrofluid) fluid model [Horton, *et al.* (1988); Horton and Tajima (1988); Horton, *et al.* (1990)]. The model was used for the analysis of the Tore Supra discharges [Horton, *et al.* (2004); Horton, *et al.* (2005)] and upgraded [Kim, *et al.* (2006)] with electron Landau damping effect [Hammett and Perkins (1990)] for NSTX. The model is constructed in a local toroidal geometry where the magnetic curvature and grad-$\boldsymbol{B}$ drifts are kept. The magnetic shear is low and positive in these discharges which allows the approximation of a constant $k_\parallel = |s|/qR$ in the analysis. This constant $k_\parallel$ approximation is acceptable when the growth rate is not near marginal stability. The higher-order finite Larmor radius (FLR) dynamics is modeled by cross-field viscosities that absorb the energy cascaded to high-k fluctuation components. The kinetic Landau damping physics is represented by the Hammett and Perkins closure model [Hammett and Perkins (1990)], $q_\parallel = -\sqrt{8/\pi}\, k_\parallel v_e \delta T_e(k)/|k_\parallel|$, for the fast parallel electron thermal dynamics.

The equations of the system can be described as

$$\widehat{\boldsymbol{T}}\Phi = \boldsymbol{L}\Phi + \widehat{\boldsymbol{N}}(\Phi, \Phi), \tag{14.20}$$

where $\widehat{\boldsymbol{T}}$ is the first-order time derivative operator, the $\widehat{\boldsymbol{L}}$ and $\widehat{\boldsymbol{N}}$ operators are the linear spatial operator and $\widehat{\boldsymbol{N}}$ the nonlinear spatial operator, respectively.

The radial and poloidal position (x, y), the toroidal position z and time t are normalized by the electron gyroradius ρ_e, electron density gradient length L_n and L_n/v_{Te}. The 3-component field vector $\boldsymbol{\Phi} = (\phi, A_\parallel, \delta T_e)$ consists of the electrostatic fluctuation ϕ, the parallel magnetic potential fluctuation $A_\parallel$ and the electron temperature fluctuation δT_e, all of which are normalized by T_e/e, $\beta_e T_e/eV_{Te}$ and T_e with the additional factor ρ_e/L_n.

The time derivative operators are by

$$\widehat{\boldsymbol{T}}_{11} = \left(1 - \nabla_\perp^2\right)\frac{\partial}{\partial t},$$

$$\widehat{\boldsymbol{T}}_{22} = \left(\frac{\beta_e}{2} - \nabla_\perp^2\right)\frac{\partial}{\partial t},$$

$$\widehat{\boldsymbol{T}}_{33} = \frac{\partial}{\partial t} \quad \text{and} \quad \widehat{\boldsymbol{T}}_{ij} = 0 \quad \text{where} \quad i \neq j. \tag{14.21}$$

The linear spatial derivative operators are given by

$$\widehat{\boldsymbol{L}}_{11} = \left(1 - 2\epsilon_n + (1+\eta_e)\nabla_\perp^2\right)\frac{\partial}{\partial y} - \mu\nabla^4, \widehat{\boldsymbol{L}}_{12} = \nabla_\perp^2\frac{\partial}{\partial z}, \ \widehat{\boldsymbol{L}}_{13} = 2\epsilon_n\frac{\partial}{\partial y} \tag{14.22}$$

$$\widehat{\boldsymbol{L}}_{21} = -\frac{\partial}{\partial z}, \ \widehat{\boldsymbol{L}}_{22} = -\frac{\beta_e}{2}(1+\eta_e)\frac{\partial}{\partial y} + \frac{\eta}{\mu_0}\nabla_\perp^2, \ \widehat{\boldsymbol{L}}_{23} = \frac{\partial}{\partial z} \tag{14.23}$$

$$\widehat{\boldsymbol{L}}_{31} = -\left(\eta_e - 4\epsilon_n(\Gamma - 1)\right)\frac{\partial}{\partial y}, \ \widehat{\boldsymbol{L}}_{32} = -(\Gamma-1)\nabla_\perp^2\frac{\partial}{\partial z} \tag{14.24}$$

$$\widehat{\boldsymbol{L}}_{33} = -2\epsilon_n(\Gamma-1)\frac{\partial}{\partial y} + \chi_\perp\nabla_\perp^2 + \frac{\partial^2}{\partial z^2} + (\Gamma-1)\sqrt{\frac{8}{\pi}}\,|k_\parallel| \tag{14.25}$$

where electron plasma beta $\beta_e = 2\mu_0 p_e/B^2$, the ratio of electron and ion temperature $\tau = T_e/T_i$ and the ratio of electron density gradient length L_n and magnetic field gradient R, $\epsilon_n = L_n/R$. Background viscosity μ, resistive diffusivity η/μ_0 and heat diffusivity $\chi_\perp$ are included.

The linear operator matrix $\widehat{\boldsymbol{L}}_{ij}$ is a function of five-dimensional parameter vector $\{\mu^5\} = \{\eta_e, \epsilon_n, T_e/T_i, \beta_e, \Gamma\}$ and weakly on the three dissipation coefficients $\{\mu, \eta/\mu_0, \chi_\perp\}$.

The nonlinear derivative terms in the dynamics are given by

$$\widehat{\boldsymbol{N}}_1 = \widehat{\boldsymbol{N}}_{111} + \widehat{\boldsymbol{N}}_{122} = -\left[\phi, \nabla_\perp^2\phi\right] + \frac{\beta_e}{2}\left[A_\parallel, \nabla_\perp^2 A_\parallel\right]$$

$$\widehat{\boldsymbol{N}}_2 = \widehat{\boldsymbol{N}}_{212} + \widehat{\boldsymbol{N}}_{221} + \widehat{\boldsymbol{N}}_{223} = -\left[\phi, \nabla_\perp^2 A_\parallel\right] - \frac{\beta_e}{2}\left[A_\parallel, \phi - \delta T_e\right]$$

$$\widehat{\boldsymbol{N}}_3 = \widehat{\boldsymbol{N}}_{313} + \widehat{\boldsymbol{N}}_{322} = [\phi, \delta T_e] - (\Gamma-1)\frac{\beta_e}{2}\left[A_\parallel, \nabla_\perp^2 A_\parallel\right]$$

where the Poisson bracket $[f, g] = (\partial f/\partial x)(\partial g/\partial y) - (\partial f/\partial y)(\partial g/\partial x)$.

The energy transfer equations without viscosity, resistivity and diffusion derived from the dynamics are

$$\frac{\partial W_E}{\partial t} = \left\langle \phi \widehat{\boldsymbol{L}}_{11}\phi \right\rangle + \left\langle \phi \widehat{\boldsymbol{L}}_{12} A_\parallel \right\rangle + \left\langle \phi \widehat{\boldsymbol{L}}_{13} \delta T_e \right\rangle - \left\langle \phi \widehat{\boldsymbol{N}}_{122} \right\rangle$$

$$= \left\langle \phi \nabla_\perp^2 \frac{\partial A_\parallel}{\partial z} \right\rangle + 2\epsilon_n \left\langle \phi \frac{\partial \delta T_e}{\partial y} \right\rangle - \frac{\beta_e}{2} \left\langle \phi \left[A_\parallel, \nabla_\perp^2 A_\parallel \right] \right\rangle$$

$$\frac{\partial W_B}{\partial t} = -\left\langle \nabla_\perp^2 A_\parallel \widehat{\boldsymbol{L}}_{21} \phi \right\rangle - \left\langle \nabla_\perp^2 A_\parallel \widehat{\boldsymbol{L}}_{22} A_\parallel \right\rangle - \left\langle \nabla_\perp^2 A_\parallel \widehat{\boldsymbol{L}}_{23} \delta T_e \right\rangle$$

$$+ \left\langle \nabla_\perp^2 A_\parallel \left(\widehat{\boldsymbol{N}}_{221} + \widehat{\boldsymbol{N}}_{223} \right) \right\rangle$$

$$= -\left\langle \phi \nabla_\perp^2 \frac{\partial A_\parallel}{\partial z} \right\rangle + \left\langle \delta T_e \nabla_\perp^2 \frac{\partial A_\parallel}{\partial z} \right\rangle + \frac{\beta_e}{2} \left\langle \phi \left[A_\parallel, \nabla^2 A_\parallel \right] \right\rangle$$

$$- \frac{\beta_e}{2} \left\langle \delta T_e \left[A_\parallel, \nabla^2 A_\parallel \right] \right\rangle$$

$$\frac{\partial W_{\delta T_e}}{\partial t} = \frac{1}{\Gamma - 1} \left(\left\langle \delta T_e \widehat{\boldsymbol{L}}_{31} \phi \right\rangle + \left\langle \delta T_e \widehat{\boldsymbol{L}}_{32} A_\parallel \right\rangle + \left\langle \delta T_e \widehat{\boldsymbol{L}}_{33} \delta T_e \right\rangle - \left\langle \delta T_e \widehat{\boldsymbol{N}}_{322} \right\rangle \right)$$

$$= \left(\frac{\eta_e}{\Gamma - 1} - 4\epsilon_n \right) \left\langle \phi \frac{\partial \delta T_e}{\partial y} \right\rangle - \left\langle \delta T_e \nabla_\perp^2 \frac{\partial A_\parallel}{\partial z} \right\rangle + \frac{\beta_e}{2} \left\langle \delta T_e \left[A_\parallel, \nabla_\perp^2 A_\} \right] \right\rangle$$

where the three components of the total turbulent energy are

$$W_E = \frac{1}{2} \left\langle \tau |\phi|^2 + |\boldsymbol{\nabla} p \phi|^2 \right\rangle$$

$$W_B = \frac{1}{2} \left\langle \frac{\beta_e}{2} |\boldsymbol{\nabla}_\perp A_\parallel|^2 + |\nabla_\perp^2 A_\parallel|^2 \right\rangle$$

$$W_{\delta T_e} = \frac{1}{2(\Gamma - 1)} \left\langle |\delta T_e|^2 \right\rangle$$

and the integration $\langle \cdot \rangle$ is done over $dxdydz$. The total energy evolves as

$$\frac{\partial}{\partial t} \left(W_E + W_B + W_{\delta T_e} \right) = -\left(\frac{\eta_e}{\Gamma - 1} - 4\epsilon_n \right) \left\langle \delta T_e \frac{\partial \phi}{\partial y} \right\rangle - \left\langle \sqrt{\frac{8}{\pi}} |k_\parallel| |\delta T_e|^2 \right\rangle \quad (14.26)$$

which in terms of the turbulent thermal flux

$$q_e = \langle v_{Ex} \delta T_e \rangle = -\left\langle \frac{\partial \phi}{\partial y} \delta T_e \right\rangle$$

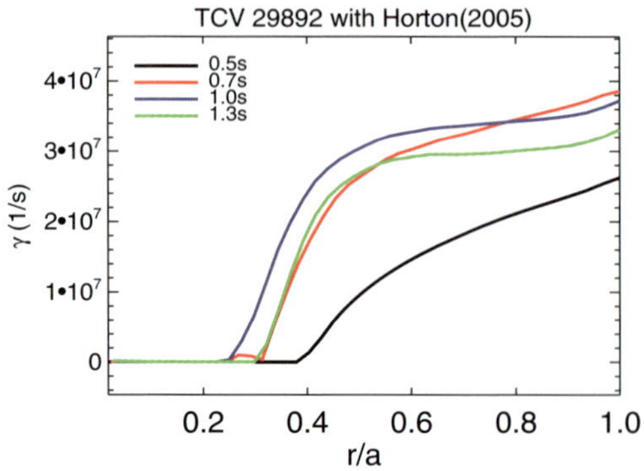

Fig. 14.10 The ETG mode growth rates used to calculate the $\chi_{\rm ETG}$ in the four regimes of the TCV discharge 29892.

becomes

Energy Balance Theorem

$$\frac{d}{dt}(W_E + W_B + W_{\delta T_e}) = q_e\left(\frac{\eta_e}{\Gamma - 1} - 4\epsilon_n\right) - \left(\frac{8}{\pi}\right)^{1/2}\left\langle |k_\parallel|\,|\delta T_e(k)|^2\right\rangle. \quad (14.27)$$

Thus, the heat flux times electron temperature gradient beyond the compressional threshold drives up the turbulent fluctuations. Thus, the $\boldsymbol{E}\times\boldsymbol{B}$-induced turbulent energy flux $q_e = -\langle \delta T_e \partial\phi/\partial y\rangle$, determines the total energy evolution according to the theorem in Eq. (14.21).

The linear stability analysis of the gyrofluid model shows the large increase of the growth rates from $t = 0.5$ s to 0.7 s Fig. 14.10 for the TCV discharge. In the inner region $r/a < 0.3$ the mode is almost linearly stable. The strong ETG instabilities at the later times are consistent with the behavior of the heat diffusivities derived from the power balance analysis with the deposited ECH power and the accurately measured electron temperature profile.

The nonlinear simulations in Fig. 14.11(a) for $r/a = 0.7$ for the TCV transport data show that the electron heat fluxes is overestimated by a factor of two. By increasing the damping in the pdes, the turbulence level can be brought down to the experimental level. The simulations predict the sharp increase of the electron heat fluxes observed between the ohmic phase and the later ECRH phases and comparable heat fluxes during the three $t = 0.7$ s, 1.0 s, 1.3 s ECRH phases. The error bars in Fig. 14.11 are the standard deviation at the nonlinear stage of the simulation. In the nonlinear stage of the simulation, the vortices and radially extended streamer structures with size of $10 - 20\rho_e$ are produced as shown in Fig. 14.11(b). Thus, the turbulent energy cascades from the source region of scale ρ_e to the ion scale $(m_i/m_e)^{1/2} \sim 40\rho_e$ through the formation of vortex merging and streamer formation.

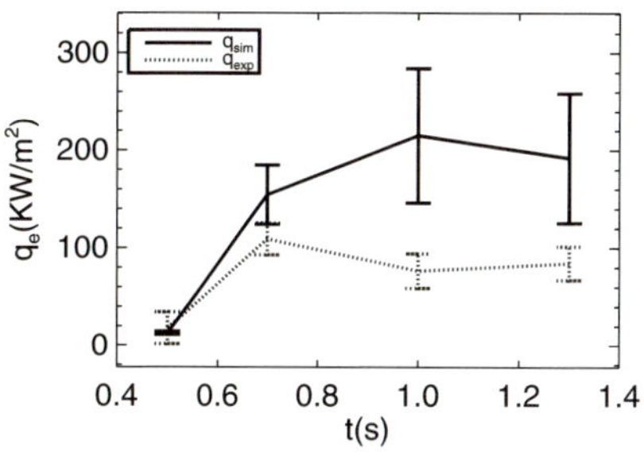

(a) Electron heat fluxes.

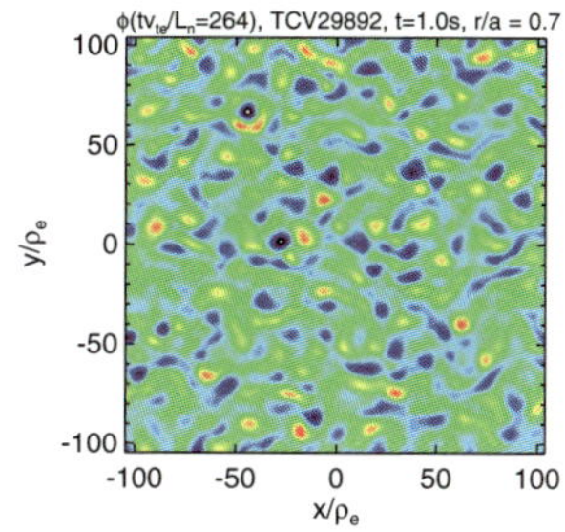

(b) Electrostatic potential.

Fig. 14.11 Electron heat fluxes (a) from nonlinear ETG fluid simulations. The solid and dashed line represent the simulations and the experiment. The contours of electrostatic potential $\phi(x, y)$ (b) from the nonlinear simulation at TCV29892 $r/a = 0.7$ and $t = 1.0\,\mathrm{s}$ where $\rho_e = 75\mu\,\mathrm{m}$ and the box corresponds to $1.5\,\mathrm{cm} \times 1.5\,\mathrm{cm}$.

Gyrokinetic simulations of the ETG turbulence show similar results as the gyrofluid simulations described here, but require extensive computer simulation resources [Nakata, *et al.* (2010)]. An example giving similar comparisons of the power balance χ_e with the simulations from a gyrokinetic code are given in Joiner, *et al.* (2006).

14.6 Validation Analysis of the Electron Transport Modeling

Since there has historically been a multitude of models for the anomalous electron thermal transport problem a validation method is needed to rank the success of the various models. Here we briefly describe the use of one metric commonly used in weather systems to compare the TEM-ITG model and the ETG model for the discharge analyzed in detail in this chapter. There will be a need for a community wide method of comparing models in the future as the number of models and the associated number of simulations codes is continually expanding.

Validation Analysis of the ETG Model

When the nonlinear simulations are too heavy numerically to give the heat transport over the whole minor radius, one may use the theoretical scaling law to estimate the ETG-driven transport in the discharge. The gyroBohm heat diffusivity with length scale $q\rho_e$ and time scale L_{T_e}/v_{T_e} [Horton, *et al.* (2004)] is

$$\chi_e = \alpha_{\mathrm{ETG}}\; q^2 \rho_e^2 v_{T_e} \left(\frac{1}{L_{T_e}} - \frac{1}{L_{T_e}^{\mathrm{crit}}} \right) \tag{14.28}$$

where q is the safety factor and $\alpha_{\rm ETG}$ is a scaling coefficient obtained from comparison with experimental data or with more rigorous theoretical calculations.

In the region $r/a > 0.4$, we calculated χ_e in Eq. (14.28) using the inverse of the critical temperature gradient length,

$$\frac{1}{L_{T_e}^{\rm crit}} = \left(1 + Z_{\rm eff}\frac{T_e}{T_i}\right)\left(\frac{1.33 + 1.9|s|}{q}\right)/R \tag{14.29}$$

[Horton, *et al.* (2004)]. The critical temperature length turns out to be five to ten times larger than the electron temperature gradient length, L_{T_e} at $r/a > 0.4$ derived from the measure electron temperature profiles.

The average relative variance (ARV) [Weigend, *et al.* (1990); Detman and Vassiliadis (1997)] is the degree to which the theoretical models for χ_e explain the power balance data $\chi^{\rm pb}$ and is given in terms of the data and the model by

$$\mathrm{ARV}_j = \frac{\sum_i \left(\chi_{e,i} - \chi_i^{\rm pb}\right)^2}{\sum_i \left(\chi_i^{\rm pb} - \left\langle \chi_i^{\rm pb}\right\rangle\right)^2} \tag{14.30}$$

with the spatial and temporal indices i and j, respectively. For TCV 29892, the power balance diffusivities χ_i^{pb} are accurately known from the plasma data and the theoretical heat diffusivities $\chi_{e,i}$. Here $\left\langle \chi_i^{\rm pb}\right\rangle$ is the average power balance diffusivity over all radial points for each temporal index j. For each temporal slice with index j, there is a total of 41 points whereof there are 30 radial points for $r/a > 0.4$.

The ARV is a widely-used metric measure for how well a model's prediction follows the behavior of experimental data. When the averages of $\chi_{e,i}$ and $\chi_i^{\rm pb}$ are comparable, ARV < 1 indicates that a model behaves in the same way as the experiment as shown in Fig. 14.12 where $\chi_e^{\rm pb}$ increases as $\chi_e^{\rm ETG}$. A smaller ARV means better agreement.

The model with $\alpha_{\rm ETG} \sim 0.8$ shows the best agreement beyond $r/a = 0.4$ in Fig. 14.12. The gyroBohm ETG heat diffusivity formula with a free parameter $\alpha_{\rm ETG}$, Eq. (14.28) explains not only the electron heat diffusivity increase with the H-mode transition but also gives ARV ~ 0.3 for $t = 0.7\,$s and $1.0\,$s. For $t = 0.5\,$s, the flat experimental heat flux profile gives a small denominator in Eq. (14.30) producing the large ARV. With $\alpha_{\rm ETG} \sim 1$ and the assumption that the dominant time scale is $\omega \sim v_{T_e}/L_{T_e}$, we can estimate the correlation length $l_c \sim 2\pi q \rho_e \sim 10\rho_e$, consistent with the nonlinear simulation result.

14.7 LHCD Driven Discharges and Anisotropic Electron Phase-Space Distribution Functions

Lower hybrid current drive suggested by Fisch (1978) and demonstrated on PLT tokamak [Bernabei, *et al.* (1982)] is the most robust and efficient method of driving

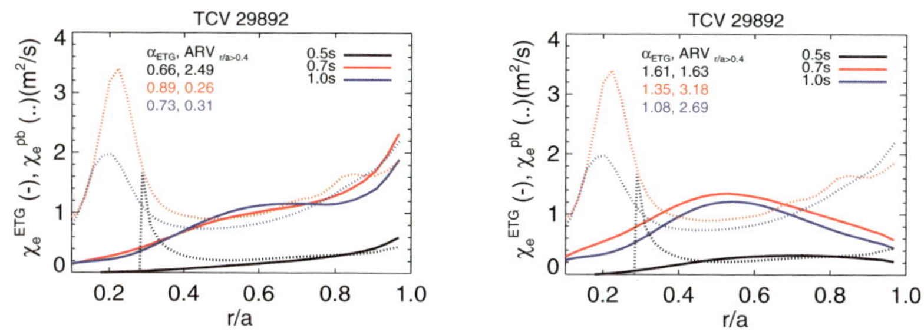

(a) (b)

Fig. 14.12 The comparison between the electron gyroBohm heat diffusivity, χ_e in Eq. (14.28), with ETG time and spatial scales (the lines with $*$) and the power balance heat diffusivity χ_e^{pb} (no mark) for each time slice $t = 0.5$ (solid), 0.7 (dotted) and 1.0 (dashed) are shown.

the tokamak current with external radio frequency waves in steady-state tokamak operation [Litaudon (2010)]. The method has been tested in Tore Supra for discharges at megawatts of RF power over periods of minutes [Hoang, *et al.* (2003)].

The Tore Supra power deposition code is ALCYON and the transport code is CRONOS, given in Artaud, *et al.* (2010). Simulations have been performed with CRONOS using ALCYON for the FW power deposition calculation.

Intense electron heating from electron cyclotron resonant waves (ECRH) and from lower hybrid waves (LHRH and LHCD) distort the electron phase space distribution functions producing significant changes in the unstable parameter space for the ETG turbulence. Heavy parallel velocity tails from LHCD and close to runaway [Knoepfel and Spong (1979)], some times called slide away, electron distribution functions have much lower rates of electron Landau damping and thus higher rates of radial gradient driven growth rates for ETG turbulence.

The electron cyclotron resonance heating in TCV is shown through tomographic reconstruction of the ECE emission on the high-field side of the plasma to create a two-component energy distribution with the high-energy part with components having an effective temperature of 10 to 50 KeV. The high-energy part is essential in creating an increased absorption of the second 2X and third 3X harmonic heating with the extraordinary wave (X) polarization. The extraordinary wave has the wave electric field pointing perpendicular to the magnetic field, which then gives much stronger coupling to the plasma through mode conversions to electrostatic waves than the ordinary wave with the wave electric field parallel to the toroidal magnetic field [Blanchard, *et al.* (2002)].

The electron distributions functions computed from LHCD wave codes produce skewed parallel velocity distributions as opposed to drifting Maxwellians. These skewed distributions have large parallel thermal fluxes $q_{\|}$ with steep radial gradients.

A velocity distribution with a large parallel heat flux $q_{\|}(r)$ has a velocity skewness that makes the distribution intrinsically different from that of a shifted, bi-

Maxwellian. Thus, it is important to compute the kinetic Vlasov response functions for these skewed distribution functions. Methods for driving an electron drift velocity to maintain a steady-state tokamak plasma current I_p, will typically create a large parallel thermal flux $q_{\|}(r)$. The best known method of driving the plasma current is with microwaves focused by the antennas to carry the required unidirection momentum to be absorbed by the electrons. This involves the use of high-power levels of microwaves that become lower hybrid plasma waves. This method of current drive is designated as LHCD for lower hybrid current drive. Contours of the constant levels of a the electron distribution functions computed for the simulations of $P_{\rm LHCD} = 8\,{\rm MW}$ apply to the Tore Supra plasma using the CIEL2F3 simulation code. All the electric current $j_{\|} = -en_e u_{\|}$ and parallel thermal flux $q_{\|}$ is carried by the passing electrons. The plasma response function for the LHCD distribution is defined as $P_{\rm LHCD}$.

In Table 14.2 the drift wave stability parameters for the LHCD driven plasmas are given. The dimensionless parameter $\alpha_q = q_{\|}/n_e T_e v_e$ defined by the generalized drift wave driving term is given. For example, the LHCD distribution of f_e and $q_{\|}$ have a large gradient of $q_{\|}(r)$, giving $\frac{L_{Te}}{L_{q\|}} \gtrsim 5$. The combined ray tracing/Fokker-Planck code DELPHINE of Peysson [Peysson, *et al.* (2011); Saoutic, *et al.* (1994)] is used to compute the electron phase space distribution functions.

The $P^{\rm LHCD}(\omega, k_{\|}, k_{\perp})$ dispersion function reduces to the well-known $Z(\omega/k_{\|}v_T)I_0 e^{-b}$ response function when the electron distribution function is replaced by a non-relativistic bi-Maxwellian with $v_T = (T_{\|}/m_e)^{1/2}$ and $b = k_{\perp}^2 T_{\perp}/m\omega_c^2$. The complex relativistic electrostatic response function is computed as

$$P(\omega, k_{\|}, k_{\perp}) = -\int_0^{\infty}\int_{-\infty}^{+\infty} \frac{\left(\frac{\omega}{v_{\|}}\frac{\partial f}{\partial p_{\|}} + \frac{k_y}{eB}\frac{\partial f}{\partial x}\right) J_0^2\left(\frac{k_{\perp}p_{\perp}}{eB}\right) dp_{\|}dp_{\perp}^2}{\omega - k_{\|}v_{\|}\omega_D + i\varepsilon} \tag{14.31}$$

with the Landau initial value rule given by $\varepsilon \to 0^+$ for real ω. For ${\rm Im}(\omega) < 0$ the function $P(\omega)$ is the analytic continuation of the polarization function $P(\omega)$ defined in Eq. (14.31). The phase distribution function $f(p_{\|}, p_{\perp}, x)$ has relativistic electrons with $p/mc = \gamma > 1$.

14.7.1 *Comparison of T_e and $q_{\|e}$ gradients*

The ratios of the gradients of T_e and $q_{\|}$ to the growth rate are

$$\frac{dT_e}{dx} : \frac{k_{\|}}{\omega + i\varepsilon}\frac{1}{n_{0e}}\frac{dq_{\|}}{dx}. \tag{14.32}$$

Thus, the comparison involves the gradient scale length

$$\frac{1}{L_{q_{\|}}} = \frac{1}{q_{\|}}\frac{dq_{\|}}{dx} \gg \frac{1}{L_{T_e}} \tag{14.33}$$

and the magnitude of the heat flux parameter

$$\alpha_{q_\parallel} = \frac{q_\parallel}{n_e T_e v_e} \equiv \frac{u_{\parallel q}}{v_e} \tag{14.34}$$

of the parallel thermal flux. For the LHCD electron distributions we have that $u_{\parallel q} \gtrsim u_{\parallel d} \cong j_\parallel / n_e e$ where $j_\parallel = I_p/\pi\Delta^2$ is the driven plasma current density required to confine the tokamak with current I_p over region $\Delta < a$.

The dimensionless α_q parameter measures the skewness of the $v_\parallel$-distribution. For example, the LHCD distribution f_e which has

$$\frac{L_{T_e}}{L_{q_\parallel}} \gtrsim 5. \tag{14.35}$$

The combined ray tracing/Fokker-Planck code DELPHINE of Imbeaux and Peysson (2005) is used to compute the driven electron distributions $f(v_\parallel, v_\perp, x)$.

14.7.2 *Kinetic dispersion relation for LHCD plasma*

The electrostatic modes are given by

$$D_k = k^2 + \sum_i \frac{n_i e_i^2}{\varepsilon_0 T_i} + \frac{n_e e^2}{\varepsilon_0 T_{\parallel e}} \left[1 + \left(\frac{T_{\parallel e}}{T_{\perp e}} - 1 \right) (1 - \Gamma_0(b_e)) - P_e^{\text{LHCD}} \right] \tag{14.36}$$

where $b_e = k_\perp^2 T_{e\perp} / m_e \omega_{ce}^2$ and

$$P_e^{\text{LHCD}}(\omega, k_\parallel, k_\perp)$$
$$= -T_{e\parallel} \int \frac{\frac{\partial f_e}{\partial E_\parallel} J_0^2 d^3p \left[\omega - \frac{k_y T_{e\parallel}}{eBLn_e} \left(1 + \eta_{e\parallel} \left(\frac{\epsilon_\parallel}{T_\parallel} - \frac{1}{2} \right) \right) + \eta_{e\perp} \left(\frac{\epsilon_\perp}{T_\perp} - 1 \right) \right]}{\omega - k_\parallel v_\parallel - k_y \widehat{v}_D (\epsilon_\parallel + \epsilon_\perp/2)}. \tag{14.37}$$

In the limit of a two-temperature Maxwellian, this P function reduces to the standard toroidal $P_e^{\max}$[Kim, *et al.* (1992)] in Eq. (13.5).

14.7.3 *Hydrodynamic-FLR limit of P^{LHCD}-function*

To gain insight into the form of the unstable modes driven by the LHCD distribution function we take the hydrodynamic or fluid limit of the integrals. Thus, we obtain the dispersion relation for the LHCD plasma

$$\begin{aligned} D_{f\ell}^{\text{LHCD}} &= k^2 + \sum_i \frac{n_i e_i^2}{\varepsilon_0 T_i} + \frac{ne^2}{\varepsilon_0 T_{\parallel e}} \left[1 + \left(\frac{T_\parallel}{T_\perp} - 1 \right) k_\perp^2 \rho_{\perp e}^2 - 1 + \frac{\omega_{*ne}}{\omega} \right] \\ &\quad - \left(1 - \frac{\omega_{*\pi e}}{\omega} \right) \frac{k_\parallel^2 T_\parallel}{m\omega^2} - \frac{k_y v_{De}}{\omega} \end{aligned} \tag{14.38}$$

which has the form

$$= A + \frac{B}{\omega} + \frac{C}{\omega^2} + \frac{D}{\omega^3} \tag{14.39}$$

Slab Modes $k_y q \rho_{e\|} < 1$

The slab regime ETG modes are given by $k_\| v_{e\|} \sim (v_{e\|}/qR)(1+|s|) > k_y v_{De}$ with the growth rate given by $\gamma_{\rm slab} = |k_\||v_{e\|}(L_{ne}/L_{Te} - \eta_{e,\rm crit})^{1/2}$.

Toroidal ETG Modes $k_y q \rho_{e\|} > 1$

The toroidal regime modes are given by

$$\gamma_{\rm toroidal} = \frac{v_{e\|}}{R}\left(\frac{R}{L_{T_e}} - \frac{R}{L_{T_{e,\rm crit}}}\right)^{1/2} \frac{|k_y|\rho_{e\|}}{\sqrt{1+k_y^2\rho_{e\|}^2(1+\omega_{ce}^2/\omega_{pe}^2)}}.$$

$$-\frac{k_\|^2 v_{e\|}^2}{\omega^2}\left(1 - \frac{\omega_{*e}(1+\eta_{e\|})}{\omega^2}\right) + \frac{2k_y^2\epsilon_n}{\omega^2}\left[1+\eta_{\|e} + \frac{T_e}{T_\|}(1+\eta_{\perp e})\right]. \tag{14.40}$$

The maximum χ_e arises from longest wavelength with $\lambda_\perp \lesssim \rho_i$

$$\chi_e \le \rho_i^2 \frac{v_{\|e}}{R}\left(\frac{R}{L_{T_e}} - \frac{R}{L_{T_{r,\rm crit}}}\right)^{1/2}.$$

There are two useful limiting forms of the LHCD-ETG waves from the fluid expansion of P. For small $k_\| L_{T_e}$ and $\overline{\omega} = \omega - k_\| u_\|$ we find from the fluid equations

$$\overline{\omega}\left(\overline{\omega} - \omega_{*i}\right) = \frac{k_\| k_y T_i}{eBZ_{\rm eff}}\frac{du_\|}{dr} + \frac{k_\|^2 T_i}{m_e Z_{\rm eff}}\left[1 - \frac{k_y T_e}{\overline{\omega} eBn_e}\frac{dn_e}{dr} - \frac{k_y T_e}{eB}\frac{1}{\overline{\omega} - k_\|^2 v_e^2/(\overline{\omega}+i\nu)}\right.$$
$$\left.\times\left[\left(\frac{1}{T_e}\frac{dT_e}{dr} - \frac{k_\| dq_\|/dr}{(\nu - i\omega)p_e}\right)\right]\right]. \tag{14.41}$$

The drift wave frequency ω_{*i} in Eq. (14.41) arises from the $\boldsymbol{E}\times\boldsymbol{B}$ convection of the electron density in the presence of the adiabatic ions $\widetilde{n}_i = -eZ_i n_i\phi/T_i$. Due to the electron temperature gradient η_e, for frequency formula is approximately

$$\omega(k) = \frac{-\omega_{*e}S(b,\eta_{e\perp})}{\frac{Z_{\rm eff}T_e}{T_i} + k_\perp^2\rho_e^2 + k^2\lambda_{De}^2}$$

with $S(b,\eta_e) = e^{-b}I_0 - \eta_{e\perp}be^{-b}(I_0 - I_1) \simeq 1 - k_\perp^2\rho_e^2(1+\eta_{e\perp})$.

For larger $k_\| L_{T_e} \gtrsim 1$ the waves are given by

$$\omega^2 = \frac{k_\|^2 T_i}{Z_{\rm eff} m_e}\left[1 - \frac{k_y T_e}{eB\omega}\left[\frac{1}{T_e}\frac{dT_e}{dr} + \frac{k_\|}{\omega p_e}\frac{dq_\|}{dr}\right]\right]. \tag{14.42}$$

Often the temperature gradient is subdominant to the gradient of the parallel heat flux and the instability becomes

$$\omega_k = \omega_{*i} + i\left(\frac{k_\|^2 T_i}{2m_e}\right)^{1/3}\left|\frac{k_y T_{\|e}}{eBL_{T_{e\|}}}\frac{dq_\|}{dx}\right|^{1/3}.$$

Table 14.2 Parameters for LHCD Driven Plasmas.

Drift Wave Parameters in LHCD Driven Plasmas

$$\omega_{*e} = \frac{k_y T_e}{q_e B n_e}\frac{dn_e}{dr} = \frac{k_y T_e}{eBL_{ne}} \text{ and } \omega_{*eT_e} = \eta_{e\|}\omega_{*e} = -\frac{k_y}{eB}\frac{dT_{e\|}}{dr} = \frac{k_y T_{e\|}}{eBL_{T_e}},$$

where $\eta_{e_\|} = \frac{\partial_r \ell n T_{e\|}}{\partial_r \ell n n_e} = \frac{L_{n_e}}{L_{T_e}}$ and $\eta_{e\perp} = \frac{\partial_r \ell n T_{e\perp}}{\partial \ell n n_e}$

The dangerous instabilities have: $\omega_k \le |\omega_{*e}| \le \frac{v_e}{L_{T_{e\|}}} \simeq (3-5)k_\| v_e$

and thus $k_\| L_{T_{e\|}} \le 1/3$ and $k_\perp \rho_e \lesssim 1$.

The nonuniform parallel thermal flux introduces the electron drift-gradient term

$\frac{k_y\phi}{B}\frac{dT_e}{dx} + \frac{k_\| k_y \phi}{n_e(\omega+i\varepsilon)B}\frac{dq_\|}{dx}$ where $\boldsymbol{E} = -\nabla\phi$.

Comparison of T_e and $q_{\|e}$ gradient terms $\frac{dT_e}{dx} : \frac{k_\|}{\omega+i\varepsilon}\frac{1}{n_{0e}}\frac{dq_\|}{dx}$

Thus, the comparison involves the scale length $\frac{1}{L_{q\|}} = \frac{1}{q_\|}\frac{dq_\|}{dx} \gg \frac{1}{L_{Te}}$

and the magnitude of the heat flux parameter $\alpha_q = \frac{q_\|}{n_e T_e v_e} \equiv \frac{u_{\|q}}{v_e}$

LHCD electron distributions have:

$u_{\|q} \gtrsim j_\|/n_e e$ where $j_\| = I_p/\pi\Delta^2$

is the driven plasma current density required for confinement.

The dimensionless α_q parameter measures the skewness of the $v_\|$-distribution.

These heat-flux gradient driven modes have

$$\omega_k = \left(\frac{k_\|^2 T_i}{Z_{\text{eff}} m_e}\right)^{1/4} |k_\| u_\| q\omega_{*i}|^{1/4} \propto k_\|^{3/4} k_y^{1/4} u_\|^{1/4} v_e^{1/2}. \tag{14.43}$$

The LHCD driven modes have a mean parallel wavenumber $\overline{k}_\|$ so the eigenmodes are shifted off the rational magnetic surfaces. Thus, the modes form magnetic islands in their nonlinear state when the plasma electron beta is appreciable and contribute to the radial transport of $j_\|$ and $q_\|$.

14.7.4 *Analytic quasilinear RF velocity diffusivity*

The intense LH-waves produce the parallel velocity diffusion coefficient $D_\|^{\text{LH}}$ that is modeled by a box function with

$$D_\|^{\text{LH}}(v_\|) = \widehat{D}\nu_e v_e^2 \qquad \text{for} \qquad v_1 \lesssim v_\| \lesssim v_2 \tag{14.44}$$

with $D_\|$ vanishing outside the interval $[v_1, v_2]$. For CIEL and other LHCD systems the upper parallel velocity $v_\|$ may be extended to $v_2 \lesssim c/2 = 1.5\times 10^8$ m/s$\sim 13v_e$. The dimension measure of the LH power is given by Stix parameter for RF power density

$$\widehat{D} = \frac{P^{\text{RF}}}{n_e T_e \nu_e} \tag{14.45}$$

where ν_e is the Coulomb collision frequency for electron collisions that restore the distribution to the local Maxwellian.

In the region $[v_1, v_2]$ a useful analytic model is obtained by taking $\widehat{D} \gg 1$ so that the collisional effects are negligible and the absorbed RF power forces $\partial f/\partial v_\| = 0$ in this $v_\|$-interval. This model is referred to as the plateau LHCD model. In reality the electron Coulomb collisions will always be non-negligible due to the formation of boundary layers at the transition across the $v_\| = v_1$ and $v_\| = v_2$ layers.

The electron distribution function, in the absence of the mirror force and toroidal velocity drifts, is

$$\frac{\partial f}{\partial t} = \frac{\partial}{\partial v_\|}\left[D_\|(v_\|)\frac{\partial f}{\partial v_\|}\right] + \frac{\nu_e}{v^2}\frac{\partial}{\partial v}\left[v^2 g(v)\left(fv + \frac{T_e}{m_e}\frac{\partial f}{\partial v}\right)\right]$$
$$+ \nu_{ei} Z_{\text{eff}}\left(\frac{v_e}{v}\right)^3 \frac{\partial}{\partial \mu}\left[(1-\mu^2)\frac{\partial f}{\partial \mu}\right] \tag{14.46}$$

where ν_e is the electron collision frequency, $g(v)$ is the coefficient of drag (which we take as unity in the following discussion), ν_{ei} is the electron-ion (working gas) collision frequency and $Z_{\text{eff}} = \Sigma_I Z_I^2 n_I/n_e$ gives the enhancement of the electron pitch-angle scattering rate from impurities I over that in the pure hydrogenic plasma for which $Z_{\text{eff}} = 1$.

14.7.5 *High-power LHCD plateau model* F_e

In the limit that $D_\|/\nu_e v_e^2 \gg 1$ in the interval (v_1, v_2), we may obtain the solution of thorough geometrical construction. The key point is that one may map the values of $f(v_\| = v_1, v_\perp) \to f(v_\| = v_2, v_\perp)$ since $\partial f/v_\| = 0$ in this interval. The coordinates of the locus of points $(v_2, v_\perp)$ parameterized by the pitch-angle α is $v_2 = v\cos\alpha$ and $v_\perp = v\sin\alpha = v_2\tan\alpha$. Thus, the value of f along the high velocity resonant plane $v_\| = v_2$ is given by $f(v_2, v_\perp) = f(v_2, v_2\tan\alpha)$ with $0 \le \alpha \le \pi/2$ parameterizing the position along the $v_\| = v_2$ surface.

Now the value of $f(v_2, v_\perp)$ is taken as the same as the value of $f(v_1, v_\perp)$ in this high-power limit due to $\partial f/\partial v_\| = 0$. So, on the surface $v_\| = v_1$ one has $v_\perp = v\sin\alpha$ and $v_\| = v\cos\alpha = v_1$. For parallel velocities below $v_\| = v_1$ the distribution is Maxwellian since this is the solution of $Cf = 0$ in the low-velocity region up to v_1 and for all $\pi/2 < \alpha \le \pi$ (co-current moving electrons). Thus, f is known through the region $[v_1, v_2]$ and on the surface $v_\| = v_2$. Pitch-angle collisions dominate for $v_\| > v_1$.

Typical values of the temperatures in the core plasma are $T_\| \sim 750\,\text{KeV}$ to $1\,\text{MeV}$, and $T_\perp \sim 150 - 300\,\text{KeV}$ with lower energy Maxwellian with temperature of 50-100 KeV. The analytic model [Stevens, *et al.* (1985)] used in the literature is:

$$f_{3T} = \begin{bmatrix} C_n \exp\left(-\dfrac{p_\perp^2}{2T_\perp} - \dfrac{p_\|^2}{2T_{\|F}}\right) & \text{for} \quad p_\| > 0 \\[2ex] C_n \exp\left(-\dfrac{p_\perp^2}{2T_\perp} - \dfrac{p_\|^2}{2T_{\|B}}\right) & \text{for} \quad p_\| < 0 \end{bmatrix}$$

and $f_{3T} = 0$ for $E(p) = mc^2\gamma > E^* \sim 2\,\mathrm{MeV}$. The temperatures are determined by the photon spectrum from the Bremsstrahlung spectroscopy

The Nyquist stability analysis gives the marginal stability condition

$$-\frac{1}{T_{e\|}}\frac{dT_{e\|}}{dr} > \left[\frac{1}{L_{ne}} + \frac{1+2|s|/q}{R}\right] \times \left[1 + \frac{Z_{\text{eff}}T_{e\|}}{T_i} + \left(\frac{T_{\|e}}{T_{\perp e}} - 1\right)(1-\Gamma_0(b_e)) + k_\perp^2\lambda_{De_\|}^2\right]. \tag{14.47}$$

In conclusion the LHCD can stimulate the (a) fast-growing toroidal modes $T_\| > T_{\perp i}$ and (b) fast-growing slab-like modes $k_\perp\rho_{e\|} \sim 1$. Secondly, one finds that (c) short wavelengths $b_e = k_\perp^2\rho_{e\perp}^2 > 1$ are stabilized by $T_{\|e} > T$ and finally (d) that the effect of $Z_{\text{eff}}T_{e\|}/T_i$ is important in controlling the threshold.

Formulas for the anomalous thermal flux associated with the $\boldsymbol{E}\times\boldsymbol{B}$ motion and the magnetic $\delta B_\perp$ fluctuations for LHCD driven tokamaks remain an unsolved problem. The turbulence for the c/ω_{pe} scale electromagnetic turbulence is typical of the ∇T_e-driven short wavelength drift modes. The result is to be used in the formulas in Chapter 11 for the anomalous electron thermal energy transport due to $\langle\delta B_x^2\rangle$.

The design for LHCD for ITER is given in Hoang, *et al.* (2009). Lower hybrid current drive produce record-breaking steady-state tokamak discharges in Tore Supra (Saoutic, *et al.* (1994) and Alcator-C (Bonoli, *et al.* , 1988). In divertor tokamaks, there is a turbulent plasma outside the last closed flux surface that produces refraction and scattering of the LHCD waves (Peysson, *et al.* 2011).

Worked Problem 2×2:

Problem:

Find the 2×2 matrix equations for compressional drift waves.
Use the kinetic matrix formulation in Chapter 13.2.2.

Solution:

The perturbative motions induced by ϕ and ψ produce a compressional change in the magnetic field $\delta B_\|$ given by

$$\frac{\delta B_\|}{B} = -\frac{1}{f}(c\phi + e\psi) \tag{14.48}$$

where properly $1/f$ is the inverse of a complicated integral operator f involving bounce averages over the particle trajectories. Here the diagonal compressional mode dispersion function f is taken as a local quantity defined in Eq. (13.18). The compressional change in the magnetic field $\delta B_\|$ is dictated by Ampere's law from the third row of the matrix equation (13.17) with the plasma currents flowing across the magnetic field lines in the $\widehat{x} \propto \nabla\Psi$ direction.

Substituting Eq. (14.48) into the first and second rows of the full matrix equation (13.17) gives the reduced 2×2 symmetric matrix

$$\begin{bmatrix} a - \dfrac{c^2}{f} & b - \dfrac{ce}{f} \\ b - \dfrac{ce}{f} & d - \dfrac{e^2}{f} \end{bmatrix} \begin{bmatrix} \phi \\ \psi \end{bmatrix} = 0. \tag{14.49}$$

The dispersion relation given by the determinant is

$$D_k(\omega, \mu) = ad - b^2 - \frac{1}{f}\left(c^2 d - 2bce + e^2 a\right) = 0. \tag{14.50}$$

The compressional kinetic-MHD limit is $a \simeq -b \simeq d$, which reduces Eq. (14.50) to

$$ad - b^2 - \frac{d}{f}(c+e)^2 = 0. \tag{14.51}$$

To find analytically how the kinetic ballooning interchange drift mode connects to the compressional mode one takes $\eta_e = 0$ and note that $a \simeq -b \simeq d \simeq e$ for the dominant terms owing to the small $E_{\|}$. Then the full determinant D in Eq. (14.50) factors as

$$ad - b^2 + \frac{\tau(1 - \omega_*/\omega)[Q - (1 - \omega_*/\omega)]^2}{\frac{2}{\beta_e} + \frac{T_i}{T_e} R} = 0 \tag{14.52}$$

follows from Eq. (14.51). Here $\omega_{*e} = \omega_*$ for $\eta_e = 0$. The last term in Eq. (14.52) gives the kinetic compressional response. For the near-to-MHD regime, the response function $Q(k, \omega)$ in Eq. (14.52) reduces to

$$Q - 1 + \frac{\omega_*}{\omega} \simeq -\frac{\omega_{*pi} - \omega_*}{\omega} + i\Delta_Q = -\frac{\omega_{*p}}{\omega} + i\Delta_Q \tag{14.53}$$

with ω_{*p} having the total pressure gradient and a wave-particle resonant part $i\Delta_Q$ from wave-particle Landau resonance at $\omega = \omega_{De}$. An adequate approximation for the compressional response function R in the region $\omega\omega_{Di} > 0$ is

$$R \cong c_0 \frac{[\omega - \omega_{*i}(1 + 2\eta_i)]}{\omega - \omega_{Di} + ic_1|\omega_{Di}|} \tag{14.54}$$

where c_0 and c_1 are positive fitting coefficients of order $c_0 \simeq 1$ and $c_1 \simeq 0.1$. Let us define the resonant contributions by $\Delta_Q = \operatorname{Im} Q(\omega)$ and $\Delta_R = \operatorname{Im} R(\omega)$ as real numbers giving the resonant ion contributions.

The resonant modes in the high-β region have

$$\omega = \omega_0 + i\gamma_k \simeq \omega_{*i}(1 + \eta_i) + i\gamma_k \tag{14.55}$$

which is an ion diamagnetic drift wave. Here ω_0 and γ_k are real valued functions of k and the system parameters $\{\mu\}$. A Taylor series expansion of the dominant terms in Eq. (14.52) gives

$$ad - b^2 \simeq \left(\frac{\omega_*}{\omega_0} - 1\right)\left[\frac{\omega_A^2}{\omega_0^2} b - b\left(1 - \frac{\omega_{*pi}}{\omega_0}\right) - \frac{\omega_* \omega_D}{\omega_0^2}\right]$$

$$+ \left(\frac{\omega_*}{\omega_0} - 1\right)\frac{i\gamma_k}{\omega_0}\left[\frac{-2\omega_A^2}{\omega_0^2} b - \frac{b\omega_{*pi}}{\omega_0} + \frac{2\omega_*\omega_D}{\omega_0^2}\right] \tag{14.56}$$

for $\omega = \omega_0 + i\gamma_k$ with $|\gamma_k| \ll \omega_0$. Thus the growth rate γ_k is determined by

$$\frac{i\gamma_k}{\omega_0}\left[\frac{2\omega_A^2}{\omega_0^2} + \frac{\omega_{*pi}}{\omega_0}\right] b + \frac{T_e}{T_i}\frac{(\omega_{*p}/\omega_0 - i\Delta_Q)^2}{c_0[1 - \frac{\omega_{*i}(1+2\eta_i)}{\omega_0} - i\Delta_R]} = 0 \tag{14.57}$$

for $\Delta_Q \sim \Delta_R \ll 1$. Since $\omega_{*p}/\omega_0 \gtrsim 1$, the significant resonant contribution comes from $i\Delta_R$ in the denominator. Thus one obtains the growth rate formula

$$\frac{\gamma_k}{\omega_0}\left(\frac{2\omega_A^2}{\omega_{*p}^2} + 1\right) b = -\frac{T_e}{T_i}\frac{(\omega_*/\omega_0)^2 \Delta_R}{c_0\left[\left(\frac{\eta_i}{1+\eta_i}\right)^2 + \Delta_R^2\right]} \tag{14.58}$$

with $\gamma_k > 0$ for $\Delta_R \equiv \mathrm{Im}\,[R(\omega_0)] < 0$.

References

Artaud, J. F., Basiuk, V., Imbeaux, F., Schneider, M., Garcia, J., Giruzzi, G., Huynh, P., Aniel, T., Albajar, F., Ané, J. M., Bécoulet, A., Bourdelle, C., Casati, A., Colas, L., Decker, J., Dumont, R., Eriksson, L. G., Garbet, X., Guirlet, R., Hertout, P., Hoang, G. T., Houlberg, W., Huysmans, G., Joffrin, E., Kim, S. H., Kčhl, F., Lister, J., Litaudon, X., Maget, P., Masset, R., Pégourié, B., Peysson, Y., Thomas, P., Tsitrone, E., and Turco, F. (2010). The CRONOS suite of codes for integrated tokamak modeling, *Nucl. Fusion* **50**, p. 043001, doi:10.1088/0029-5515/50/4/043001.

Bernabei, S., Daughney, C., Efthimion, P., Hooke, W., Hosea, J., Jobes, F., Martin, A., Mazzucato, E., Meservey, E., Motley, R., Stevens, J., Von Goeler, S., and Wilson, R. (1982). Lower-hybrid current drive in the PLT tokamak, *Phys. Rev. Lett.* **49**, pp. 1255-1258, `http://link.aps.org/doi/10.1103/PhysRevLett.49.1255`.

Blanchard, P., Alberti, S., Coda, S., Weisen, H., Nikkola, P., and Klimanov, I. (2002). High-field side measurements of non-thermal electron cyclotron emission on TCV plasmas with ECH and ECCD, *Plasma Phys. Control. Fusion* **44**, p. 2231, doi:10.1088/0741-3335/44/10/310.

Bonoli, P. T., Porkolab, M., Takase, Y., and Knowlton, S. F. (1988). Numerical modeling of lower-hybrid RF heating and current drive experiments in the Alcator-C tokamak, *Nucl. Fusion* **28**, p. 991, doi:10.1088/0029-5515/28/6/004.

Chang, Z. and Callen, J. D. (1992). Unified fluid/kinetic description of plasma microinstabilities. Part II: Applications, *Phys. Fluids B* **4**, pp. 1182-1192.

Chang, Z. and Callen, J. D. (1990). Global energy confinement degradation due to macroscopic phenomena in tokamaks, *Nucl. Fusion* **30**, pp. 219-233, doi:10.1088/0029-5515/30/2/003.

Callen, J. D. (1977). Drift-wave turbulence effects on magnetic structure and plasma transport in tokamaks, *Phys. Rev. Lett.* **39**, pp. 1540-1543, `http://link.aps.org/doi/10.1103/PhysRevLett.39.1540`.

Detman, T. and Vassiliadis, D. (1997). Review of techniques for magnetic storm forecasting, *Geophysical Monograph, American Geophysical Union* **98**, pp. 253-266, ISSN:0065-8448.

Fielding, S. J., Ashall, J. D., Carolan, P. G., Colton, A., Gates, D., Hugill, J., Morriss, A. W., Valovic, M., and the COMPASS-D and ECRH Teams. (1996). *Plasma Phys. Control. Fusion* **38**, p. 1091, doi:10.1088/0741-3335/38/8/002.

Fisch, N. (1978). *Phys. Rev. Lett.* **49**, p. 1255.

Fu, X., Horton, W., Xiao, Y., Lin, Z., Sen, A. K., and Sokolov, V. (2012). *Phys. Plasmas* **19**, pp. 032303-032310, `http://dx.doi.org/10.1063/1.3686148`.

Goniche, M., Sharma, P. K., Basiuk, V., Baranov, Y., Castaldo, C., Cesario, R., Decker, J., Delpech, L., Ekedahl, A., Hillairet, J., Kirov, K., Mazon, D., Oosako, T., Peysson, Y., and Prou, M. (2011). Lower hybrid current drive efficiency at high density on Tore Supra, *AIP Conf. Proc. 1406*, pp. 407-410, doi:http://dx.doi.org/10.1063/1.3665003.

Guo, S. and Weiland, J. (1997). *Nucl. Fusion* **37**, pp. 1095-1107, doi:10.1088/0029-5515/37/8/105.

Hallock, G. A., Hickok, R. L., and Hornady, R. S. (1994). The TMX heavy ion beam probe, *Plasma Science, IEEE Transactions on Plasma Science* **22**, 4, pp. 341-349, ISSN:0093-3813, doi:10.1109/27.310639.

Hahm, T. S. and Tang, W. M. (1989). Properties of ion temperature gradient drift instabilities in H-mode plasmas, *Phys. Fluids B* **1**, p. 1185, `http://dx.doi.org/10.1063/1.859197`.

Hamaguchi, S. and Horton, W. (1991). *Phys. Fluids B* **4**, p. 319.

Hammett, G. W. and Perkins, F. W. (1990). *Phys. Rev. Lett.* **64**, pp. 019-3022, doi:10.1103/PhysRevLett.64.3019.

Hoang, G. T., Bécoulet, A., Jacquinot, J., Artaud, J. F., Bae, Y. S., Beaumont, B., Belo, J. H., Berger-By, G., Bizarro, João, P. S., Bonoli, P., Cho, M. H., Decker, J., Delpech, L., Ekedahl, A., Garcia, J., Giruzzi, G., Goniche, M., Gormezano, C., Guilhem, D., Hillairet, J., Imbeaux, F., *et al.* (2009). A lower hybrid current drive system for ITER, *Nucl. Fusion* **49**, p. 075001, doi:10.1088/0029-5515/49/7/075001.

Hoang, G. T., Horton, W. Bourdelle, C., Hu, B., Garbet, X., and Ottaviani, M. (2003). Analysis of the critical electron temperature gradient in Tore Supra, *Phys. Plasmas* **10**, pp. 405-412, doi:10.1063/1.1534113.

Hoang, G. T., Bourdelle, C., Garbet, X., Giruzzi, G., Aniel, T., and Ottaviani, M., Horton, W., Zhu, P., and Budney, R. V. (2001). Experimental determination of critical threshold in electron transport on tore supra, Phys. *Rev. Lett.* **87**, p. 125001, `http://link.aps.org/doi/10.1103/PhysRevLett.87.125001`, doi:10.1103/PhysRevLett.87.125001.

Holland, C. and Diamond, P. H. (2002). *Phys. Plasmas* **9**, p. 3857, doi:10.1063/1.1496761.

Horton, W., Wong, H. V., Morrison, P. J. Wurm, A., Kim, J.-H., Perez, J. C., Pratt, J., Hoang, G. T., LeBlanc, B. P., and Ball, R. (2005). *Nucl. Fusion* **45**, pp. 976-985, doi:10.1088/0029-5515/45/8/025.

Horton, W., Hoang, G. T., Bourdelle, C., Garbet, X., Ottaviani, M., and Colas, L. (2004). *Phys. Plasmas* **11**, 5, p. 2600, doi:10.1063/1.69076.

Horton, W., Hu, B., Dong, J.-Q., and Zhu, P. (2003). Turbulent electron thermal transport in tokamaks, *New J. Phys.* **5**, p. 14, doi:10.1088/1367-2630/5/1/314.

Horton, W., Zhu, P., Hoang, G. T., Aniel, T., Ottaviani, M., and Garbet, X. (2000). *Phys. Plasmas* **7**, pp. 1494-1511, doi:10.1063/1.873969.

Horton, W., Hong, B.-G., Tajima, T., and Bekki, N. (1990). *Comm. Plasma Phys. Control. Fusion* **13**, p. 207.

Horton, W., Hong, B.-G., and Tang, W. M. (1988). *Phys. Fluids* **31**, p. 2971-2983.

Horton, W. and Tajima, T. (1988). *J. Geophys. Res.* **93**, pp. 2741-2748, doi:1029/JA093iA04p02741.

Horton, W., Bekki, N., Berk, H. L., Hong, B.-G., LeBrun, M. J., Mahajan, S. M., Tajima, T., and Zhang, Y. Z. (1988) *12th Intl. Conf. Plasma Phys. Control. Nucl. Fusion Res.* (IAEA, London) CN-50/D-4-3.

Horton, W. (1984). *Handbook of Plasma Physics, II*, M. N. Rosenbluth and R. Z. Sagdeev ed., pp. 384-402, (North-Holland, Amsterdam).

Huysmans, G. T. A. and Czarny, O. (2007). MHD stability in X-point geometry: Simulation of ELMs, *Nucl. Fusion* **47**, pp. 659-666, doi:10.1088/0029-5515/47/7/016, `stacks.iop.org/NF/47/659`.

Imbeaux, F. and Peysson, Y. (2005). Ray-tracing and Fokker-Planck modelling of the effect of plasma current on the propagation and absorption of lower hybrid waves, *Plasma Phys. Control. Fusion* **47**, p. 2041, doi:10.1088/0741-3335/47/11/012.

Jacchia, A., Luca, F. D., Cirant, S., Sozzi, C., Bracco, G., Bruschi, A., Buratti, P., Podda, S., and Tudisco, O. (2002). *Nucl. Fusion* **42**, p. 1116, doi:10.1088/0029-5515/42/9/310.

Jarmen, A., Andersson, P., and Weiland, J. (1987). *Nucl. Fusion* **27**, p. 941, ISSN:0029-5515.

Jenko, F. and Dorland, W. (2002). Prediction of significant tokamak turbulence at electron gyroradius scales, *Phys. Rev. Lett.* **89**, p. 225001, `http://link.aps.org/doi/10.1103/PhysRevLett.89.225001`.

Joiner, N., Applegate, D., Cowley, S. C., Dorland, W., and Roach, C. M. (2006). *Plasma Phys. Control. Fusion* **48**, pp. 685-697, doi:10:1088/0741-3335/48/5/013.

Kadomtsev, B. B. (1992). *Tokamak Plasma: A Complex Physical System* (Institute of Physics Pub., Bristol, UK and Philadelphia) ISBN: 0750302348.

Karpushov, A. N., Duval, B. P., Schlatter, C., Afanasyev, V. I., and Chernyshev, F. V. (2006). *Rev. Sci. Instrum.* **77**, p. 033503, doi:10.1063/1.2185151.

Kaye, S. M., Levinton, F. M, Stutman, D., Tritz, K., Yuh, H., Bell, M. G., Bell, R. E., Domier, C. W., Gates, D., Horton, W., Kim, J.-Y. LeBlanc, B. P., Luhman, Jr., N. C., Maingi, R., Mazzucato, E., Menard, J. E., Mikkelsen, D., Mueller, D., Park, H., Rewoldt, G., Sabbagh, S. A., Smith, D. R., and Wang, W. (2007). *Nucl. Fusion* **47**, p. 499, doi:10.1088/0029-5515/47/7/001.

Kim, J.-H., Perez, J. C., Horton, W., Chagelishvili, G. D., Changishvili, R. G., Lominadze, J. G., and Bowman, J. C. (2006). Self-sustaining vortex perturbations in smooth shear flows, *Phys. Plasmas* **13**, 6, p. 062304.

Kim, J-Y., Horton, W., Choi, D-I., Migliuolo, S., and Coppi, B. (1992). *Phys. Fluids B*, p. 152.

Knoepfel, H. and. Spong, D. A. (1979). Runaway electrons in toroidal discharges, *Nucl. Fusion* **19**, p. 785.

Li, J. and Kishimoto, Y. (2004). *Phys. Plasmas* **11**, pp. 1493-1511, doi:10.1063/1.1669397.

Li, J. Q., Kishimoto, Y., Miyato, N., Matsumoto, T., and Dong, J. Q. (2005). *Nucl. Fusion* **45**, 11, pp. 1293-1302, doi:10.1088/0029-5515/45/11/010.

Litaudon. X. (2010). Real-time control of advanced scenarios for steady-state tokamak operation, *Fusion Sci Tech.* **59**, 3, p. 469.

Lohr, J., Stallard, B. W., Prater, R., Snider,R. T., Burrell, K. H., Groebner, R. J., Hill, D. N., Matsuda, K., Moeller, C. P., Petrie, T. W., St. John, H., and Taylor, T. S. (1988). *Phys. Rev. Lett.* **60**, pp. 2630-2633, doi:10.1103/PhysRevLett.60.2630.

Matsuda, K. (1989). *IEEE Transactions on Plasma Science* **17**, p. 6.

Mazzucato, E. (2010). Study of turbulent fluctuations driven by the electron temperature gradient in the National Spherical Torus Experiment, *Nucl. Fusion* **50**, p. 029801, `http://iopscience.iop.org/0029-5515/50/2/029801`.

Mazzucato, E., Bell, R. E., Ethier, S., Hosea, J. C., Kaye, S. M., LeBlanc, B. P., Lee, W. W., Ryan, P. M., Smith, D. R., Wang, W. X., Wilson, J. R., and Yuh, H. (2009). *Nucl. Fusion* **4**, p. 5, doi:10.1088/0029-5515/49/5/055001.

Mazzucato, E., Smith, D. R., Bell, R. E., Kaye, S. M., Hosea, J. C., LeBlanc, B. P., Wilson, J. R., Ryan, P. M., Domier, C. W., Luhmann, Jr., N. C., Yuh, H., Lee, W., and Park, H. (2008). *Phys. Rev. Lett.* **101**, pp. 075001-1, doi:10.1103/PhysRevLett.101.075001.

Moroz, P. E. and Colestock, P. L. (1991). Electron absorption of fast waves in global wave calculations, *Plasma Phys. Control. Fusion* **33**, 5, pp. 417-431.

Nakata, M., Watanabe, T.-H., Sugama, H., and Horton, W. (2010). Formation of coherent vortex streets and transport reduction in electron temperature gradient driven turbulence, *Phys. Plasmas* **17**, p. 042306, `http://dx.doi.org/10.1063/1.3356048`.

Nordman, H., Weiland, J., and Jarmén, A. (1990). *Nucl. Fusion* **30**, p. 983, doi:10.1088/0029-5515/30/6/001.

Peysson, Y., Decker, J., Morini, L., and Coda, S. (2011). RF current drive and plasma fluctuations, *Plasma Phys. Control. Fusion* **53**, p. 124028, doi:10.1088/0741-3335/53/12/124028.

Porte, L., Coda, S., Alberti, S., Arnoux, G., Blanchard, P., Bortolon, A., Fasoli, A., *et al.* (2006). *21st International Atomic Energy Agency Fusion Energy Conf.* (Vienna) EX/P6-20.

Rebut, P. H., Watkins, M. L., Gambier, D. J., and Boucher, D. (1991). A program toward a fusion reactor, *Phys. Fluids B* **6**, 3, pp. 2209-229.

Rebut, P. H. and Lallia, P. P. (1988). *Plasma Phys. Control. Nucl. Fusion.*

Ryter, F., Imbeaux, F., Leuterer, F., Fahrbach, H.-U., Suttrop, W., and ASDEX Upgrade Team. (2001a). *Phys. Rev. Lett.* **86**, pp. 5498-5501, doi:1-/11-3/PhysRevLett.86.5498.

Ryter, F., Leuterer, F., Pereverzev, G., Fahrbach, H.-U., Stober, J., Suttrop, W., and ASDEX Upgrade Team. (2001b). *Phys. Rev. Lett.* **86**, pp. 2325-2328, doi:10.1103/PhysRevLett.96.2325.

Saoutic, B., Beaumont, B., Becoulet, A., Bizarro, J. P., Fraboulet, D., Garbet, X., Goniche, M., Guiziou, L., Hoang, G. T., Hutter, T., Joffrin, E., Kuus, H., Litaudon, X., Mollard, P., Moreau, D., Nguyen, F., Pecquet, A. L., Peysson, Y., Rey, G., van Houtte, D., and Zabiego, M. (1994). High-power ICRF and LHCD experiments on Tore Supra, *AIP* **289**, pp. 24-31, `http://link.aip.org/link/?APC/289/24/1`.

Stevens, J., Von Goeler, S., Bernabei, S., Bitter, M., Chu, T. K., Efthimion, P., Fisch, N., Hooke, W., Hosea, J., Jobes, F., Karney, C., Meservey, E., Motley, R., and Taylor, G. (1985). Modeling of the electron distribution based on Bremsstrahlung emission during lower-hybrid current drive on PLT, *Nucl. Fusion* **25**, p. 1529, doi:10.1088/0029-5515/25/11/002.

Stutman, D., Delgado-Aparicio, L., Gorelenkov, N., Finkentha, M., Mazzucato, E., and Tritz, K. (2009). *Phys. Rev. Lett.* **102**, p. 115002.

Sugama, H. and Horton, W. (1995). *Plasma Phys. Control. Fusion* **37**, p. 345, doi:10.1088/0741-3335/37/3/012.

Tonetti, G., Heym, A., Hofmann, F., *et al.* (1991). *Proceedings of the 16th Symposium on Fusion Technology* (North-Holland, London, U.K.) R. Hemsworth, ed., p. 587.

Turri, G., Sauter, O., Porte, L., Alberti, S., Asp, E., Goodman, T. P., Martin, Y. R., Udintsev, V., and Zucca, C. (2008). *J. Phys.* **123**, p. 1, doi:10.1088/1742-

6596/123/1/012038.

Wei, X., Sokolov, V., and Sen, A. K. (2010). Experimental production and identification of electron temperature gradient mode, *Phys. Plasmas* **17**, 4, p. 042108.

Weigend, A. S., Huberman, B. A., and Rumelhart, D. E. (1990). *Int. J. Neural Systems* **1**, pp. 193-209, doi:10.1142/SO129065790000102.

Weiland, J. (2000). *Collective modes in inhomogeneous plasma* (IOP Publishing Ltd.).

Zhu, P., Bateman, G., Kritz, A. H., and Horton, W. (2000). Predictive transport simulations of internal transport barriers using the Multi-Mode model, *AIP* **7**, pp. 2898-2908, `http://ojps.aip.org/pop/popcr.jsp`.

Zhu, P., Horton, W., and. Sugama, H. (1999). *Phys. Plasmas* **6**, p. 2503, doi:10.1063/1.873522.

Chapter 15

Magnetic Reconnection Instabilities

15.1 Introduction

Tearing instabilities are spontaneous magnetic reconnection that may occur in sheared magnetic configuration, such as a current sheet and a rational surface in a magnetic confinement. The conventional theory shows that the instability is driven by the current density gradient of a current sheet and is caused by resistivity [Furth, *et al.* (1963)]. The simplest model of magnetic reconnection instability is based on resistive magnetohydrodynamics (MHD), which consists of the vorticity equation and the Ohm's law,

$$\frac{\partial U}{\partial t} + \mathbf{v} \cdot \nabla U = \mathbf{B} \cdot \nabla J \qquad \frac{\partial \psi}{\partial t} = -\mathbf{B} \cdot \nabla \phi + \eta J \tag{15.1}$$

where $U = \nabla^2 \phi$ and $J = \nabla^2 \psi$ are vorticity and current density normal to the two-dimensional plane (along to the guide magnetic field) in two-dimensional MHD (reduced MHD). The parameter η stands for plasma resistivity which violates frozen of magnetic field line into plasma. Assuming the growth time of instability is much slower than Alfvén time, one finds that most of the system is governed by the equation $\mathbf{B} \cdot \nabla \mathbf{J} = 0$ except the region where the equation is singular $\mathbf{B} \cdot \nabla = 0$. It is clearly shown by the linearized form of the equation, $\mathbf{B_0} \cdot \nabla \nabla^2 \psi_1 + \mathbf{B_1} \cdot \nabla \nabla^2 \psi_0 = 0$ called external equation [See Eq. (4.16) in Biskamp (2000)], where $f = f_0 + f_1$. The first term, that is the highest derivative term, vanishes at $\mathbf{B_0} \cdot \nabla = 0$ where it is called neutral sheet or rational surface in toroidal plasma. At the singular surface one needs to solve the resistive layer equation Eq. (15.1), that is obtained by assuming $\partial/\partial x \gg \partial/\partial y$. The asymptotic matching between solutions by the external equation and and the resistive layer equation gives us the dispersion relation of reconnection instability called tearing instability. When a current sheet is narrow it is unstable for long-wavelength perturbation, and magnetic islands appear. The theory shows that the stability parameter of tearing instability Δ' determines the instability threshold [Furth, *et al.* (1963)] and thus the formation of magnetic

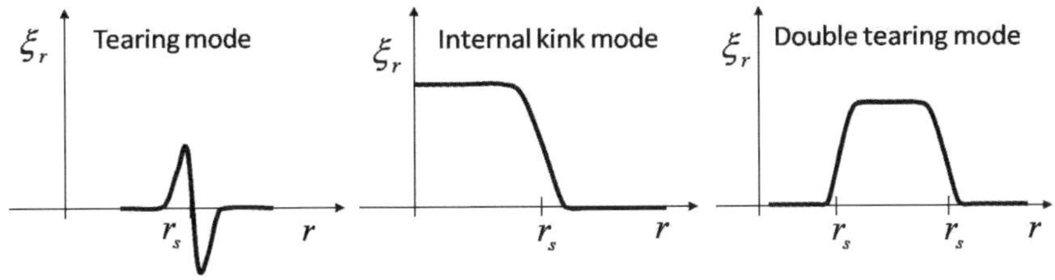

Fig. 15.1 Radial plasma displacement of three reconnection instabilities, tearing instability, internal kink instability, and double tearing instability.

islands. The stability parameter of tearing instability

$$\Delta' = \frac{L}{2}\left[\frac{d\ln\psi_1}{dx}\right]_{0-}^{0+} \tag{15.2}$$

is obtained by solving the external equation.

The tearing mode growth rate is a hybrid of the Alfvén wave time scale $\tau_A = L/v_A$ and slow resistive time scale $\tau_R = \mu_0 L^2/\eta$, where L is the space scale of the current sheet, η is the collisional diffusion of magnetic flux and $v_A = B_0/\sqrt{4\pi m_i n_0}$ is the Alfvén wave velocity. The associated dimensionless parameter $S = \mu_0 v_A L/\eta$ is the Lundquist number. A typical time scale for the tearing instability is $\tau_A^{2/5}\tau_R^{3/5}$. This strong dependence on the resistivity is similar to the time scale of reconnection described by the Sweet-Parker model, which predicts the scaling of $\tau_A^{1/2}\tau_R^{1/2}$.

There are three types of reconnection instabilities; tearing instability, internal kink instability, and double tearing instability. The plasma displacement $\xi_r = -\psi_1/(d\psi_0/dr)$ in cylindrical geometry (r,θ) is shown in Fig. 15.1. The displacement of tearing instability shows inflow to the reconnection point. The displacement of internal kink instability shows rigid body displacement inside r_s where the rational surface is located. The displacement of double tearing instability shows rigid body displacement between the two rational surfaces. The displacement of latter two are similar at the rational surface and they cause faster reconnection than tearing instability, that is scaled $\tau_A^{2/3}\tau_R^{1/3}$.

Fast magnetic reconnection, which is independent, or weakly depends, on the value of the resistivity, occurs when the current sheet thins to the ion inertial length δ_i or the ion Larmor radius ρ_i. Thin sheets require the Hall region around the reconnection point, $\mathbf{E} = -\mathbf{v}\times\mathbf{B} + \mathbf{J}\times\mathbf{B}/en$. In the Hall region, electrons are magnetized and moves with magnetic field lines, $\mathbf{E} = -\mathbf{v_e}\times\mathbf{B}$, where $\mathbf{v_e} = \mathbf{v_i} - \mathbf{J}/en$. Figure 15.1 shows that z-component of this equation $E_z = -\mathbf{v_e}\times\mathbf{B}|_z$ is valid within the ion skin depth d_i or the ion Larmor radius l_{mi} except the vicinity of reconnection point $y = 0$. On the other hand, ions are not tied to magnetic field at all in the Hall region, and they do not need to go through small scale electron layer, resulting in fast reconnection. The motion of electrons is decoupled from ions and it causes the Hall current in the region. Turbulence can arise around a thin current sheet, and its fluctuations can produce fast magnetic reconnection. The turbulence

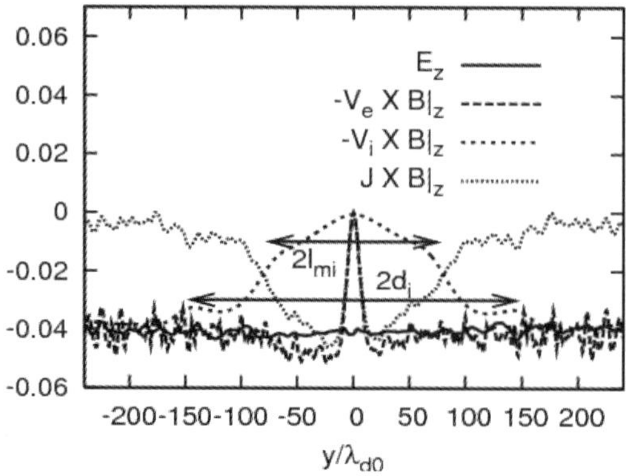

Fig. 15.2 The spacial profile of electric field E_z, $-\mathbf{v_e} \times \mathbf{B}_z$, $-\mathbf{v_i} \times \mathbf{B}_z$, and Hall term $-\mathbf{J} \times \mathbf{B}_z$, where d_i, l_{mi}, and λ_{d0} are the ion skin depth, local ion Larmor radius, and Debye length. [Ishizawa and Horiuchi (2005)].

may accelerate magnetic reconnection and change the threshold for magnetic island formation.

This chapter is mainly divided into three parts, corresponding to the three instabilities. In each section, effects of turbulence and/or two-fluid effects, responsible for collisionless re- connection, on the reconnecting instability are described. Section 15.2 describes nonlinear dynamics of internal kink instability with two-fluid effects. Section 15.3 describes effects of drift-wave turbulence on tearing instability. Section 15.4 describes two-fluid effects on tearing instability. Section 15.5 describes effects of drift-wave turbulence on double-tearing instability. Additional cases, such as tearing instabilities with the combined effects of resistivity, viscosity, and shear flow, are not treated here, but can be found in detail in Chen and Morrison (1990a,b) and Ofman, *et al.* (1991).

Worked Problem:

(a) Derive the linearized equation that governs the exterior perturbed flux function $\delta\psi = \psi_{ky}(x)\cos(k_y x)$ from $\mathbf{B}\cdot\nabla\mathbf{J} = 0$ for a general current sheet $J(x)$.
(b) Show that for a Harris sheet current profile $J(x) = J_0\,\mathrm{sech}^2(x/L_s)$ the tearing mode stability parameter is $\Delta' = 2/L_s(1/(k_y L_s) - k_y L_s)$.

Solution:

(a) The linearized form of $\boldsymbol{B}\cdot\boldsymbol{\nabla}\boldsymbol{J} = 0$ is

$$B_y(x)\left(\partial_x^2 - k_y^2\right)\psi_{ky}(x) - J'(x)\psi_{ky}(x) = 0 \tag{15.3}$$

which has a singularity where the line bending stabilizing term (the first term) vanishes, $B_y(x) = \partial\psi_0/\partial x = 0$. The second term is the kink term, which drives tearing instability, where $J'(x) = d^3\psi_0/dx^3$ represents current density gradient.

To resolve the $\psi_{ky}(x)$ functional at the singular the effect of magnetic flux diffusion $D_m = \eta/\mu_0(\text{mks}) = c^2\eta/4\pi(\text{cgs})$ must be included in a resistive plasma and the effect of electron inertia $(m_e/e)du_{e\|}/dt = -(m_e/ne^2)dJ_\|/dt)$ and parallel electron pressure $\nabla_\| p_e$ in the collisionless plasma.

(b) For the Harris sheet

$$B_y(x) = B_0 \tanh\left(\frac{x}{L_s}\right) \quad \text{and} \quad J(x) = (J_0)\,\text{sech}^2\left(\frac{x}{L_s}\right)$$

the eigenvalue problem in Eq. (15.3) becomes

$$B_0 \tanh\left(\frac{x}{L_s}\right)\left(\frac{d^2}{dx^2} - k_y^2 + \frac{2}{L_s^2}\,\text{sech}^2\left(\frac{x}{L_s}\right)\right)\psi_{k_y}(x) = 0 \tag{15.4}$$

with boundary conditions, in the simplest case with no external forcing, of $\psi_{k_y}(x \to \pm\infty) = 0$ and finite $\psi_{k_y}(0)$, which is allowed by non-ideal MHD effects such as the resistivity or electron inertia. By substitution one finds that $\psi_{k_y}(x) = \exp -k_y|x|(1 + 1/k_y L_s \tanh(|x|/L_s))$. Substituting this solution into Eq. (15.2) gives the stability parameter. For a long wavelength perturbation $k_y/L_s < 1$, the parameter is positive $\Delta' > 0$, and thus tearing mode is unstable.

Worked Problem:

Show that tearing mode is unstable when the stability parameter Δ' is positive and its growth rate depends on the resistivity as $\eta^{3/5}$ by following (a) and (b).

(a) Derive the linearized equation that governs the resistive layer (inner layer) equation for tearing modes by assuming that the equilibrium magnetic field is approximated to $B_y = B_0 x/L_s$, $\partial/\partial x \gg \partial/\partial y$, and the growth rate is small (constant-ψ approximation).

(b) By using the asymptotic matching condition

$$\Delta' = \frac{1}{\psi_1}\int J_1 dx. \tag{15.5}$$

Solution:

(a) The linearized form of Eq. 15.1 is

$$\frac{\partial}{\partial t}\frac{\partial^2 \phi_1}{\partial x^2} = x\frac{\partial}{\partial y}\frac{\partial^2 \psi_1}{\partial x^2}, \qquad \frac{\partial \psi_1}{\partial t} = x\frac{\partial \phi_1}{\partial y} + \eta\frac{\partial^2 \psi_1}{\partial x^2}. \tag{15.6}$$

By assuming the solution has a form $f_1 = \widehat{f}_1(x)\exp(st + iky)$,

$$s^2\frac{d^2\widehat{\xi}_1}{dx^2} = xk^2\frac{d^2\widehat{\psi}_1}{dx^2}, \qquad s\widehat{\psi}_1 = -sx\widehat{\xi}_1 + \eta\frac{d^2\widehat{\psi}_1}{dx^2}, \tag{15.7}$$

where $\widehat{\xi} = -ik\widehat{\phi}$ is the radial plasma displacement. For small growth rate $s \ll 1$, we can expand the magnetic flux and the displacement as $\widehat{\psi}_1 = \widehat{\psi}_1^{(0)} + \widehat{\psi}_1^{(1)}$ and $\widehat{\xi}_1 = \widehat{\xi}_1^{(1)}$. In the 0-th order the first equation becomes $d^2\widehat{\psi}_1^{(0)}/dx^2 = 0$, so that

$\widehat{\psi}_1^{(0)} = \Psi$ is constant because of tearing parity $\psi(-x) = \psi(x)$. In the first order the equations become

$$s^2 \frac{d^2\widehat{\xi}_1^{(1)}}{dx^2} = xk^2 \frac{d^2\widehat{\psi}_1^{(1)}}{dx^2}, \qquad s\Psi = -sx\widehat{\xi}_1^{(1)} + \eta \frac{d^2\widehat{\psi}_1^{(1)}}{dx^2}, \tag{15.8}$$

and they are combined into an equation, which is the resistive layer equation for tearing modes,

$$\left(\frac{d^2}{dz^2} - z^2\right)\chi = z, \tag{15.9}$$

where $\chi = \widehat{\xi}_1^{(1)}/(\alpha\Psi)$, $z = \alpha x$, and $\alpha = (\eta s/k^2)^{-1/4}$.

(b) Since $J = \frac{\partial^2\psi}{\partial x^2}$ in the layer, the matching condition is

$$\Delta' = \frac{1}{\Psi}\int_{-\infty}^{\infty} \frac{1}{x}\frac{d^2\widehat{\psi}_1^{(1)}}{dx^2} dx = C\frac{s^2}{k^2}\left(\frac{k^2}{\eta s}\right)^{3/4}, \qquad C = \int_{-\infty}^{\infty} \frac{1}{z}\frac{d^2\chi}{dz^2} dz. \tag{15.10}$$

The growth rate is $s = \Delta'^{4/5}\eta^{3/5}k^{2/5}C^{-4/5}$, and thus tearing mode is unstable only when $\Delta' > 0$, and its growth rate depends on the resistivity as $\eta^{3/5}$ because the resistivity makes magnetic reconnection.

The regime where $\Delta' \leq 0$ is able to develop magnetic islands in the nonlinear finite beta regime from the growth of magnetic interchange instabilities [La Haye and Sauter (1998); La Haye (2006); Ishizawa and Nakajima (2007); Ishizawa and Nakajima (2010); Fitzpatrick, *et al.* (2006); Waelbroeck, *et al.* (2009)].

15.1.1 *Nonlinear dynamics of the sawtooth events*

In high-temperature tokamak plasmas, when the safety factor $q(r) = rB_\phi/RB_\theta = 1$ in the core, there is a marginally stable $m = 1$ MHD mode or oscillation with a singular layer at the $m/n = 1/1$ rational surface. Inside this singular surface there is a solid-body displacement of the interior plasma that pushes the magnetic at the position of the maximum of the radial displacement $\xi(r,\theta,\phi)$ against the exterior flux. The condition for forced reconnection is created and one finds that the adding a small amount of resistivity to the plasma allows the rapid growth of the $m = 1$ mode in the core of the plasma with minor distortion of the plasma outside the $m = 1$ plasma. The growth rate for this internal kink mode with the finite resistivity is fast occurring as $\gamma = (\eta/\mu_0 r_1^2)^{(1/3)}(v_A/r_1)^{(2/3)}$ where the radius r_1 is the position of the $q(r_1) = 1$ resonant surface.

It was expected that this internal tearing mode would continue to grow at this linear resistive Alfvén hybrid power of 1/3 of the resistive diffusion rate and the 2/3 power of the Alfvén time. However, Aydemir (1992) showed that reconnection accelerates in the nonlinear stage when the two component fluid equations are used and the Ohm's law includes the parallel electron pressure gradient and the electron inertia. The discovery of the is accelerated nonlinear of the $m = 1$ mode has been an important step in understanding the dynamics of the internal kink dynamics that creates the sawtooth oscillations in the tokamak.

Early analytical work on the sawtooth did not anticipate this accelerated growth and conjectured the mode would saturate as the $m \geq 2$ in what is a saturated magnetic island. Kadomtsev (1975) realized that the current sheet at the rational surface would develop with a rapid reconnection if resistive effects were included and derived widely-used reconnection model. The method of calculating the growth of the magnetic islands was found by Rutherford (1973). Magnetic islands at the low order rational surfaces $q = 3/2, 2$ are commonly observed as shown in Chapter 14 on the ECH driven TCV plasma.

The different topology of the $m = 1$ MHD disturbance as a rigid displacement of the core plasma inside the $q = 1$ surface and a relatively undisturbed exterior plasma changes the dynamics and creates the topology for the fast reconnection.

Figure 15.3 shows the nonlinear growth rate computed from the two-component fluid model by Aydemir (1991, 1992). The phenomenon is robust and reproduced now in a number of simulations including high resolution Adaptive Mesh Simulations by Bhattacharjee and Ma. Here we briefly give the equations used by Aydemir for his important discovery of an accelerated rate of growth, sometimes called explosive growth, of the reconnection dynamics. The nonlinear simulations are shown in Fig. 15.3, Fig. 15.4 and Fig. 15.5 for the nonlinear growth and crash of the sawtooth oscillation.

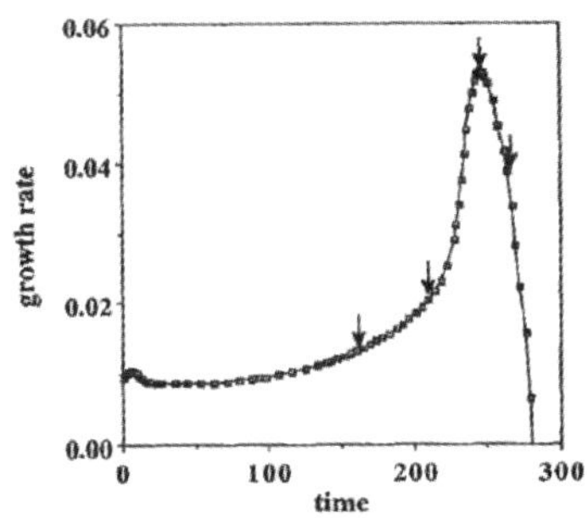

Fig. 15.3 The growth rate as a function of time for an $m = 1$ mode that starts in the semicollisional regime and exhibits a dramatic increase in its growth rate, becoming collisionless during its nonlinear evolution. Time and the growth rate are normalized to the poloidal Alfvén time. The arrows point to approximate points in time where the flux function and current density contours are shown in subsequent figures [Aydemir (1991, 1992)].

The simplified model of the core of a tokamak that includes the effects of the long-mean-free-path electron dynamics and support of a parallel electric field by the electron pressure gradient as in drift waves follows from the reduce two component fluid equations given Chapter 4. Following Chapter 10 we introduce the parallel vector potential and the electrostatic potential to describe the coupled drift-Alfvén wave dynamics. The electron inertial term which introduces the collisionless skin depth $\delta_e = c/\omega_{pe}$ is important in the parallel component of the electron fluid acceleration equations since the dynamics becomes fast and develops small space scales. The interchange driving terms from the toroidal curvature are weak contributors to the dynamics and are omitted in the following equations. The coupling to the parallel ion motion is retained but is also not a key player in this fast reconnection dynamics.

As in the nonlinear drift and Alfvén wave equations the divergence of the polarization current from the ions introduces the cross-field plasma vorticity $U = \nabla^2\phi$ and the Alfvén wave dynamics is described by the parallel component of the vector potential $A_\parallel = -\psi$ where the magnetic flux function ψ is introduced such that the parallel electric current is given by $J = \nabla^2\psi$. The electron pressure is $P_e = p$, and the parallel ion flow velocity v. The evolution of the four fields is given by

$$\frac{\partial U}{\partial t} + [\phi, U] + \nabla_\parallel J = 0 \tag{15.11}$$

$$\frac{\partial \psi}{\partial t} + \nabla_\parallel[\phi - \delta p] = \eta J + \delta_e^2 \left\{ \frac{\partial J}{\partial t} + [\phi, J] \right\} \tag{15.12}$$

$$\frac{\partial p}{\partial t} + [\phi, p] + \beta\nabla_\parallel[v + 2\delta J] = 0 \tag{15.13}$$

$$\frac{\partial v}{\partial t} + \frac{1}{2}\nabla_\parallel p + [\phi - \delta\tau p, v] = 0. \tag{15.14}$$

In this model there are four dynamical fields given by

$$\phi, \quad J = \nabla_\perp^2\psi, \quad U = \nabla_\perp^2\phi, \ v \tag{15.15}$$

and the system parameters are

$$\beta = \frac{nkT_e}{B_T^2/2\mu_o} \quad \delta_i = \frac{c\omega_{pi}}{a}, \quad \delta_s = \frac{c/\omega_{pe}}{a} = \left(\frac{m_e}{m_i}\right)^{1/2} \delta_i. \tag{15.16}$$

The variables have been normalized as follows: $t \to t/\tau_A, \boldsymbol{r} \to \boldsymbol{r}/a, \eta = \tau_A/\tau_R$, where $\tau_A = a/v_A$, $\tau_R = \mu_o a^2/\eta_o$, and $v_A = B_{po}/\sqrt{\rho_o\mu_o}$. Here τ_A and τ_R are the poloidal Alfvén time and the resistive diffusion time defined in terms of the minor radius a, a characteristic poloidal field strength B_{po}, and plasma electrical resistivity η_o. See Hazeltine *et al.* (1985, 1987) for more details.

The four brackets in Eq. (15.11)-Eq. (15.14) are defined by $[\phi, U] = \widehat{\zeta} \cdot \nabla_\perp\phi \times \nabla_\perp U$, where $\widehat{\zeta}$ is a unit vector in the toroidal direction, and $\nabla_\perp$ is the 2D gradient in the plane perpendicular to the magnetic field. The parallel gradient operator is defined as $\nabla_\parallel J = \partial J/\partial\zeta + [J, \psi]$ for any scalar J. The parameter δ_i, in combination with the electron β, is related to the ion Larmor radius, as it can be shown that $\tau\delta_i^2\beta = (\rho_i)^2$. Similarly, the ion Larmor radius based on electron temperature, ρ_s, is defined by $(\rho_s)^2 = \delta_i^2\beta$. Both ρ_i and ρ_s are normalized to the minor radius. Finally, δ_e is the collisionless skin depth.

When the finite-electron-inertia terms $m_e dJ/dt$ are present, it is useful to rewrite the parallel Ohm's law in terms with an auxiliary variable $\chi \equiv \psi - \delta_s^2 J$.

In terms of χ the electron acceleration equation along the reconnecting magnetic field becomes

$$\frac{\partial \chi}{\partial t} + [\phi, \chi] + \frac{\partial}{\partial \zeta}(\phi) = \delta \nabla_{\|} p + \eta J + \mu_e \nabla_{\perp}^2 \chi. \tag{15.17}$$

The last term on the right-hand side of Eq. (15.17) introduces an electron viscosity μ_e that controls the build-up of the smallest scale fluctuations in the simulations. The effective resistivity is $\eta_{\text{effec}} = \eta + \mu_e$ where μ_e is called the hyper-resistivity with $\eta_{\text{hyper}} = \mu_e \delta_s^2$. The effects of hyper-resistivity are examined in Aydemir (1990). The simulation shown here has $\mu_e \ll \eta$.

The equations are related to models from those of Drake and Lee's semicollisional regime [Drake and Lee (1977)]. Observe that the intrinsic Coulomb resistivity is $\eta_0 = m_e \nu_{ei}/e^2 n_e$, where ν_{ei} is the electron-ion collision frequency. The definition of the resistive layer width is $x_\eta^2 = \eta_0/\omega$. A semi-collisional regime is defined by $\delta_s < x_\eta < \delta$. The collisionless regime is defined by $x_\eta < \delta_s$.

The nonlinear increase in the growth rate of the $m = 1$ mode shown in Fig. 15.3 is observed in both semi-collisional regime, where the electron inertia terms in the parallel Ohm's law can be ignored ($x_\eta \gg \delta_s$), and in the collisionless regime where the electron inertia dominates over the collisional effects ($x_\eta \ll \delta_s$).

The ion-sound terms, which come in through Eq. (15.14), are retained, although their effect was shown to be relatively unimportant [Aydemir (1991)].

The four partial differential equations, Eq. (15.11)-Eq. (15.14), are solved in a cylindrical geometry, with periodically identified ends. The variables are Fourier-expanded in poloidal and toroidal directions; pseudo-spectral techniques are used [Orszag (1977)], keeping up to 256 modes. In the radial direction, a nonuniform grid with over 600 grid-points is used. In order to reduce the oscillatory behavior of fields near regions with sharp gradients, a third-order upwind-biased differencing is used for the convective derivatives.

The nonlinear growth rate shown in Fig. 15.3 is defined from the growth of the kinetic energy E_K in the fields with $\gamma = (1/2) d \log E_K / d \log t$.

The parameters used in Fig. 15.3 are $\delta = 0.11, \delta_s = 5 \times 10^{-3}, \beta = 5 \times 10^{-3}$, and $\eta = \mu_0 \eta / r_1 v_A = 1.0 \times 10^{-6}$, which gave a linear growth rate of $\gamma_{\tau_A} = 1 \times 10^{-2}$. The dramatic increase in the growth rate of the mode as the island grows is not observed in a resistive similar resistive-viscous MHD simulation. The new feature of the large nonlinear increase in the growth rate is from the generalized equation for the parallel acceleration of the electrons often called the generalized Ohm's law.

The mode starts linearly in the semi-collisional regime ($\delta_s < x_\eta < \delta_i$) and becomes collisionless during the fastest part of the reconnection. The collisionless physics, provided by the electron inertia terms in the Ohm's law, however, is not essential to this accelerated growth; the same behavior is observed in a purely semi-collisional run where we set the collisionless skin depth, δ_s, to zero. Internal consistency, of course, requires that the electron inertia terms be kept for modes growing on this rapid time scale, as the classical resistive layer width becomes less than the collisionless skin depth.

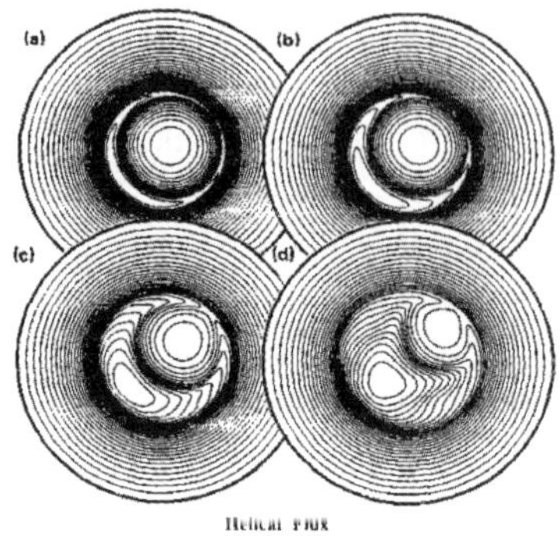

Fig. 15.4 The contours of the helical flux function at various points during the nonlinear evolution of the semicollisional $m = 1$ mode, the growth rate of which is shown in Fig. 15.3.

Further differences between the semicollisional/collisionless mode and its purely resistive counterpart can be seen also in the geometry of the island and the current sheet that develops around the separatrix. Figure 15.4 shows the helical flux contours associated with the nonlinear run of Fig. 15.3, at four times during the nonlinear evolution of the island, pointed out with the arrows in Fig. 15.3.

The early nonlinear behavior is similar to that of a purely resistive island; the reconnection layer is poloidally extended and forms more of a "Y-point" [Waelbroeck (1989); Biskamp (1991); Zakharov and Rogers (1992)] than the classical X-point of an $m \geq 2$-island. However, as the island grows further, a well-defined X-point emerges, widening the outflow region of the reconnection layer. The accelerated

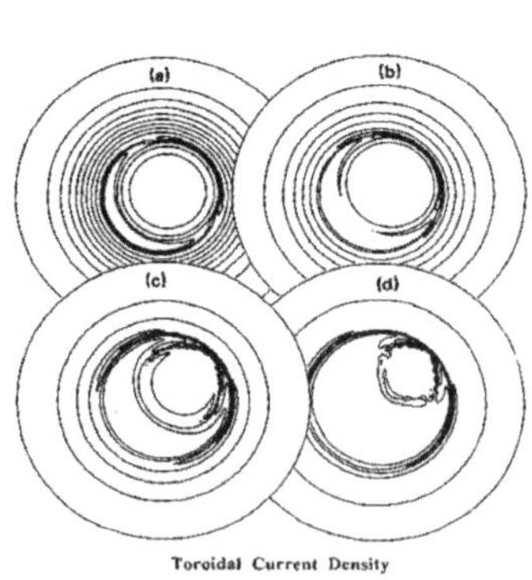

Fig. 15.5 The current density contours for the semicollisional $m = 1$ mode, showing the change in the geometry of the current sheet from a flat ribbon (Y-layer) to a well-defined X-point.

growth (faster reconnection) can be attributed to this change in the geometry of the layer, as it is certainly easier to remove the reconnected flux with a "wider nozzle". Figure 15.5 shows the contours of the current density, at the same times as the flux contours of Fig. 15.4. Again the change in the geometry is clear; the extended current sheet of the 'Y-layer' of early times breaks up and forms an "X-layer", clearly following the contours of the island separatrix. The current sheet that forms at the inner separatrix of the island shrinks and disappears as reconnection approaches completion; the sheet along the outer separatrix lengthens and eventually extends all the way around the circle, as predicted by Kadomtsev (1975).

This novel behavior exhibited by the semicollisional/collisionless $m = 1$ mode can be traced to the coupling between the electron pressure gradient term in the Ohm's law, Eq. (15.12), and the parallel divergence of the electron velocity $(v+2\delta J)$ in Eq. (15.13); linearly, these terms give rise to an adiabatic electron response in the long mean-free-path regime. The exact nonlinear mechanism that gives rise to an increase in the growth rate, however, is not clear at this point. Kleva (1992) have also seen a similar accelerated growth during the reconnection of two flux bundles. Dagazian (1979) has found that the convective nonlinearity $\mathbf{v} \cdot \nabla \mathbf{v}$ can lead to an Alfvénic, but algebraic, growth for kink modes; however, convection does not seem to be responsible for the enhanced growth rate observed in this work.

The nonlinear evolution of the $m = 1$ mode in high temperature plasmas produces a strong nonlinear acceleration of the linear magnetic reconnection rate. Both the electron inertia and electron pressure gradient through ρ_s support the parallel electric field $E_{\|}$ and are important aspects of the nonlinear dynamics of the sawteeth oscillations in tokamaks, as shown in detail in Bhattacharjee, *et al.* (2005).

In the magnetic flux topology of the reconnection dynamics at the $q = 1$ is one where the characteristic "Y" shape flux lines form as seen in many resistive MHD simulations. Concomitant with this change in the geometry of the reconnection layer, the growth rate of the reconnection process increases, resulting in a fast (explosive) reconnection of the remaining magnetic flux. This dynamics explains the observed growth of the sawtooth signals in the soft X-ray data and the subsequent rapid crash times.

Because of the importance of controlling the sawtooth oscillations in tokamaks, a real-time control system with based on this relatively simple simulation model may provide a useful tool for feedback and control of the core sawtooth activity in tokamaks.

15.2 Effects of Drift Wave Turbulence on Magnetic Reconnection

Here, magnetic reconnection caused by turbulence in a current sheet is studied by means of numerical simulations of fluid equations [Ishizawa and Nakajima (2010)]. It is shown that drift wave turbulence causes magnetic reconnection and produces coherent magnetic islands in a current sheet, even when the sheet is so thick that there is no spontaneous magnetic reconnection, i.e. tearing modes are stable, $\Delta' < 0$. Thus, turbulence modifies the threshold of magnetic island formation predicted by the conventional theory. The coherent magnetic islands are formed by the merging of small-scale magnetic islands produced by magnetic reconnection driven by turbulent fluctuations. That is consistent with the fact that the islands are caused by the energy transfer from small-scale turbulence in the Fourier space. In spite of the fact that the turbulence is driven by a short-wavelength instability due to temperature gradient, the length of the magnetic island is the same order as the system size. Even if one use MHD model, similar coherent magnetic islands are formed due to turbulence driven interchange instability which is a generalized Rayleigh-Taylor instabilities [Ishizawa and Nakajima (2010)].

Stronger turbulence causes wider magnetic islands, this suggests that the turbulence can trigger neoclassical tearing modes (NTMs). NTM is one of primary MHD activities that limit the plasma pressure and are nonlinear instabilities. They require finite size magnetic islands, called seed islands, for overcoming a threshold of destabilization. The typical width of seed island is evaluated to be several times as large as the ion Larmor radius by using experimental data. Thus, such turbulence-driven magnetic island can be the seed island for NTM. It was found that the magnetic islands propagate in the electron diamagnetic direction. The propagation of the island is also important for NTM destabilization because of polarization current effects. The appearance of magnetic islands as neoclassical tearing modes was reported in ASDEX Upgrade. The magnetic signal showed that there was a substantial disturbance at the $q = 3/2$ rational surface that was identified as a NTM using a generalized Rutherford equation. The NTMs often form in ELMy H-mode plasmas as shown for the TCV discharge analyzed in Chapter 14. The control of NTM is a key current topic in fusion research.

The derivation of the generalized Rutherford equation for the growth and saturation of the magnetic islands is given in Fitzpatrick and Aydemir (1996) and Sauter, *et al.* (1997). There is a strong dependence on the beta-poloidal of the plasma as with the associated ELMs. Both topics are under development. The role of the drift wave turbulence on the NTM growth and propagation is given in Waelbroeck, *et al.* (2009). The stability of the $m = 1$ internal kink mode can be partially controlled by the resistivity of a close fitting metallic shell surrounding the plasma as shown by Fitzpatrick and Aydemir (1996).

15.2.1 *Two-component fluid simulations*

Effects of turbulence driven by a typical drift-wave instability, which is the ion temperature gradient instability, on magnetic reconnection is studied by means of numerical simulations of a two-dimensional reduced set of two-fluid equations [Ishizawa and Nakajima (2007)]. The model consists of the vorticity equation, the electron density equation, the parallel velocity equation, the generalized Ohm's law, and the ion temperature equation,

$$\left(\frac{\partial}{\partial t} + \boldsymbol{v}\cdot\nabla\right)\nabla^2\phi = -\boldsymbol{B}\cdot\nabla J - \delta_i\beta\nabla\cdot(\boldsymbol{v}_{di}\cdot\nabla\nabla\phi) + \mu\nabla^4\phi \tag{15.18}$$

$$\left(\frac{\partial}{\partial t} + \boldsymbol{v}\cdot\nabla\right) n = -\boldsymbol{B}\cdot\nabla v_{e\|} + \mu\nabla^2 n \tag{15.19}$$

$$\left(\frac{\partial}{\partial t} + \boldsymbol{v}\cdot\nabla\right) v_{\|} = -\beta\boldsymbol{B}\cdot\nabla p + \mu\nabla^2 v_{\|} \tag{15.20}$$

$$\frac{\partial\psi}{\partial t} = -\boldsymbol{B}\cdot\nabla\phi + \delta_i\beta\boldsymbol{B}\cdot\nabla p_e + \eta J \tag{15.21}$$

$$\left(\frac{\partial}{\partial t} + \boldsymbol{v}\cdot\nabla\right) T_i = -(\Gamma - 1)\left(\boldsymbol{B}\cdot\nabla v_{\|} + \kappa_L T_i\right) + \mu\nabla^2 T_i \tag{15.22}$$

where $\boldsymbol{v}\cdot\nabla f = [\phi, f]$, $\boldsymbol{B}\cdot\nabla f = -[\psi, f]$, $\boldsymbol{v}_{di}\cdot\nabla\nabla f = -[\nabla f, p_i]$, are written in terms of the Poisson bracket $[f,g] = \partial_x f\partial_y g - \partial_x g\partial_y f$, where $\nabla = (\partial_x, \partial_y)$, and $\boldsymbol{B} = B_0\boldsymbol{e}_z - \boldsymbol{e}_z\times\nabla\psi$, $\boldsymbol{v} = \boldsymbol{e}_z\times\nabla\phi$, $\boldsymbol{v}_{di} = \boldsymbol{e}_z\times\nabla p_i$ are the magnetic field, the $E\times B$ flow velocity, and the ion diamagnetic velocity respectively. In these equations ϕ, ψ, n, $J = -J_z = \nabla^2\psi$, $v_{\|}$, $v_{e\|} = v_{\|} + \delta_i J$, T_i, T_e, $p_i = nT_i$, $p_e = nT_e$, $p = p_i + p_e$, are the electrostatic potential, the magnetic flux function, the electron density, the negative of current density, the parallel ion velocity, the parallel electron velocity, the ion temperature, the electron temperature, the ion pressure, the electron pressure, and the total pressure, respectively.

Magnetic field configuration in (x, y)-plane is the Harris current sheet, $d\psi_{eq}(x)/dx = \tanh(x/L_s)$. The magnetic shear length L_s of the Harris sheet is associated with the gradient of current density and controls the strength of spontaneous magnetic reconnection called tearing instability, and small L_s, i.e. a thin current sheet, causes strong instability. The uniform current density model $d\psi_{eq}(x)/dx = x/L_s$, which represents an infinitely thick current sheet, is also used. This profile has the neutral sheet at $x = 0$ and is totally stable against tearing instability. The stability parameter of tearing instability $\Delta' = L/2[d\ln\widetilde{\psi}/dx]_{0-}^{0+}$ is obtained by solving the linearized form of $\boldsymbol{B}\cdot\nabla J = \boldsymbol{B}\cdot\nabla\nabla^2\psi = 0$. An equilibrium is stable (unstable) against tearing instability when Δ' is negative (positive).

15.2.2 *Magnetic islands caused by turbulence*

Figure 15.6 shows map of electrostatic potential and equicontours of magnetic flux for the Harris sheet with $L_s/L = 0.3125$, $\Delta' = -0.76$, and $\eta_i = 3.5$. The stability parameter Δ' is calculated for the longest wavelength mode, $k = 1$, where the wavenumber is given by $k_y = \pi k/L$. At $t = 264$, narrow magnetic islands appear at the neutral sheet, and electrostatic potential profile exhibits turbulent fluctuations. At $t = 1188$, magnetic islands become large, and zonal flow, which is uniform flow

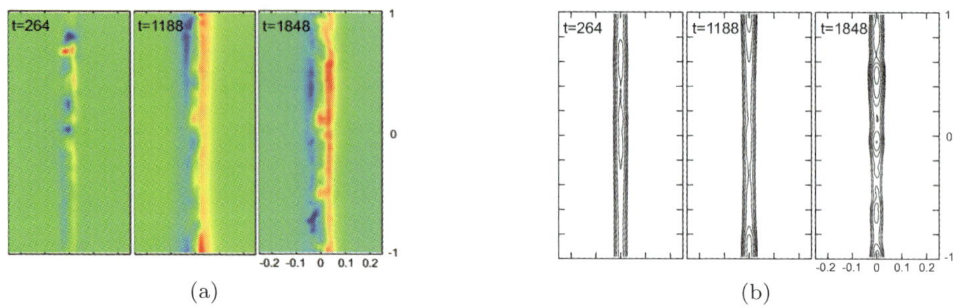

Fig. 15.6 Map of electrostatic potential and equicontours of magnetic flux that indicates long-wavelength magnetic islands, even if there is no spontaneous magnetic reconnection, $\Delta' = -0.76$, where time is normalized by Alfvén transit time and x and y are normalized by the box size [Ishizawa and Nakajima (2010)].

along the y-axis, appears. At the quasisteady-state, $t = 1848$, magnetic islands clearly appear on the neutral sheet $x = 0$. The length of the magnetic island is much longer than the ion Larmor radius and is the same order as the system size, and the width is several times the ion Larmor radius $\rho_i/L = 1/80$. The shape of the magnetic island is deformed and is different from that of tearing instability because of turbulent fluctuations of the magnetic field.

Figure 15.7 shows the magnetic island width as a function of the stability parameter of tearing instability Δ' for the case $\eta_i = 3.5$, where the width is time-averaged in the quasi-steady state. The magnetic island width is several times the ion Larmor radius, even if the current sheet is stable against tearing instability, $\Delta' < 0$. Hence, turbulence modifies the island formation threshold $\Delta' = 0$ given by the conventional theory.

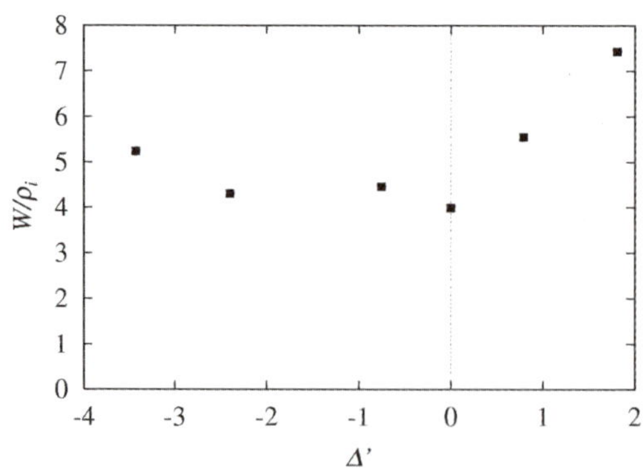

Fig. 15.7 Magnetic island width W as a function of the stability parameter of tearing instability Δ' [Ishizawa and Nakajima (2010)].

Figure 15.8 shows time trace of the reconnected flux that is the largest value of magnetic flux on the neutral sheet, $x = 0$, for each temperature gradient parameter

η_i that indicates the strength of the instability. When turbulence is strong, $\eta_i = 4.5$, the reconnected flux grows rapidly, and then it attains a large value and a quasi-steady state is formed. On the other hand, when the instability is weak, $\eta_i = 2.5$, the reconnected flux grows slowly, and it attains a small value after which it gets saturated. Thus, stronger turbulence causes faster reconnection and wider magnetic islands.

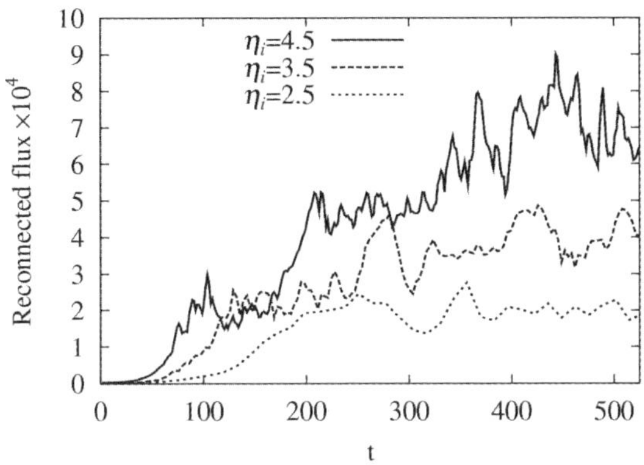

Fig. 15.8 Time trace of the reconnected flux for each temperature gradient parameter η_i for the case $\Delta' = -3.4$ [Ishizawa and Nakajima (2010)].

15.2.3 *Formation process of magnetic islands*

The process of magnetic island formation is examined. Figure 15.9 shows the time evolution of magnetic flux on the neutral sheet, $x = 0$, for the Harris sheet with $\Delta' = -2.4$, where red region represents positive flux, and blue negative flux, that represents the core parts of magnetic islands around O-points. Initially, the patterns show wave structure characterized by the wavelength of ion temperature gradient instability. After $t \approx 200$, large red and blue regions appear, and this means that long-wavelength magnetic islands are formed. Figure 15.10 shows the time evolution of the positions of X-points and O-points for the Harris sheet with $\Delta' = -2.4$.

The red and black points represent X-points and O-points, respectively. Initially, there are many X- and O-points, and then some merge and disappear. In this process small-scale magnetic reconnection may play a role. After the large red and blue regions appear in Fig. 15.9, sometimes X-points and O-points appear and disappear, because the turbulent perturbation causes random magnetic reconnection. Figure 15.9 and Fig. 15.10 also show the propagation of magnetic islands. Initially, they propagate in the negative y-direction, which is the ion diamagnetic direction, because the ion temperature gradient instability propagates in the ion diamagnetic direction. Then, after $t = 200$, the magnetic islands propagate in the positive y-direction, which is the electron diamagnetic direction.

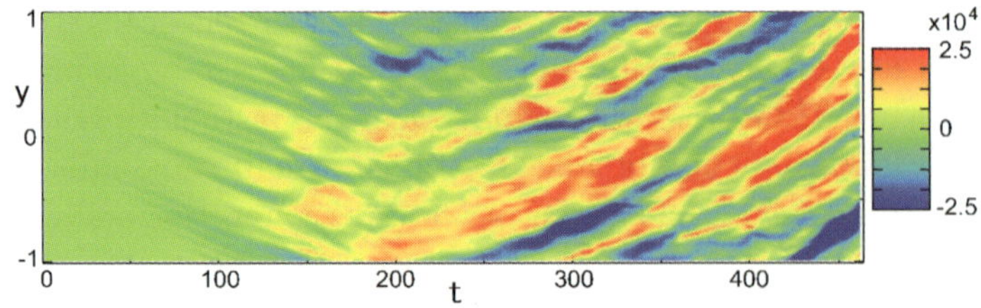

Fig. 15.9 Time evolution of magnetic flux on the neutral sheet, $x = 0$, for the case $\eta_i = 3.5$ and $\Delta' = -2.4$. Blue region represents the core parts of magnetic islands around O-points [Ishizawa and Nakajima (2010)].

The nonlinear dynamics of rotating low-m mode tearing modes with external resonant magnetic perturbations is developed by Fitzpatrick (1993, 1995), and Smolyakov, *et al.* (1995). The analysis shows that the magnetic island rotation slows down as they grow in amplitude. In the presence of a symmetry breaking external magnetic perturbation the magnetic islands become locked to the external perturbation. This can be made a locked mode or a rotating island, depending on whether this a fixed machine associated asymmetry or an applied rotating magnetic perturbation, called RMPs for short, designed to mitigate the build-up of large ELMs (edge localized modes) for example.

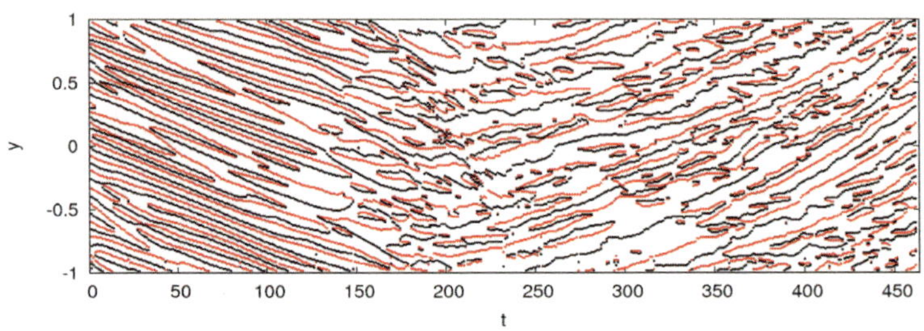

Fig. 15.10 Time traces of X-points and O-points, for the case $\eta_i = 3.5$ and $\Delta' = -2.4$. Black points are O-points and red points are X-points [Ishizawa and Nakajima (2010)].

15.2.4 *Mixing of the eigenmode parities*

When Eq. (15.18)-Eq. (15.22) are linearized by dividing field quantities into equilibrium part and perturbed part $F(x, y) = f_0(x) + f(x, y)$ and $f_0(x)$ has symmetry, the linearized equation has symmetry. The symmetry yields two parities: one is tearing parity, $\psi(-x) = \psi(x)$, $\phi(-x) = -\phi(x)$, $v(-x) = v(x)$, $n(-x) = -n(x)$, $T_i(-x) = -T_i(x)$, the other is interchange parity, $\psi(-x) = -\psi(x)$, $\phi(-x) = \phi(x)$, $v(-x) = -v(x)$, $n(-x) = n(x)$, $T_i(-x) = T_i(x)$. Here, the process how drift wave turbulence which has interchange parity produces magnetic islands which has tearing parity is discussed [Ishizawa and Nakajima (2007)].

The mechanism is forced magnetic reconnection as described below. Consider, for simplicity, resistive interchange mode in slab plasma governed by reduced MHD equations instead of ITG as shown in Fig. 15.11(a). The magnetic flux ψ of interchange mode is odd function of x, ψ_o, in linear evolution, while the potential is even function of x, ϕ_e, and thus the interchange mode does not cause the magnetic flux of tearing mode that is finite at $x = 0$ and even function of x, ψ_e, as shown in Fig. 15.11(a), where $f_e(-x) = f_e(x)$ and $f_o(-x) = -f_o(x)$. Then, in the nonlinear evolution, the interchange mode produces magnetic flux ψ_e, which is even function of x, and potential ϕ_o, which is odd function of x, through the nonlinear terms of reduced MHD equations,

$$\frac{\partial \nabla_\perp^2 \phi_o}{\partial t} = -[\phi_e, \nabla_\perp^2 \phi_e] - [\phi_o, \nabla_\perp^2 \phi_o] + [\psi_e, \nabla_\perp^2 \psi_e] + [\psi_o, \nabla_\perp^2 \psi_o] \quad (15.23)$$

$$\frac{\partial \psi_e}{\partial t} = [\psi_o, \phi_e] + [\psi_e, \phi_o]. \quad (15.24)$$

This is described in Fig. 15.11(b). These ψ_e and ϕ_o force magnetic flux to reconnect at $x = 0$, and then we have a finite amplitude of magnetic flux at $x = 0$ as shown in Fig. 15.11(c). The finite magnetic flux at $x = 0$ means that magnetic reconnection takes place.

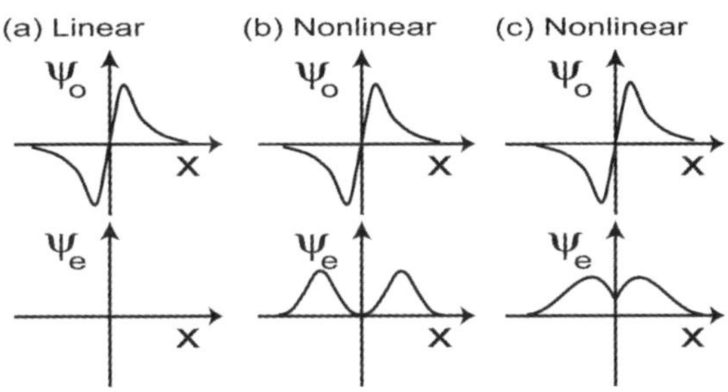

Fig. 15.11 Profile of magnetic flux showing the process turbulence which has interchange parity causes magnetic islands which has tearing parity, where ψ_o is odd part and ψ_e is even part; (a) linear evolution of interchange instability, (b) nonlinear evolution of interchange instability, (c) forced reconnection occurs and producing magnetic islands [Ishizawa and Nakajima (2007)].

Interchange-type modes exchange not only thermal and magnetic energies between flux tubes but also current. In a plasma with a current (or resistivity) gradient, such an interchange can create a current sheet at a mode resonance surface and result in the excitation of current-interchange tearing modes. To reveal this effect, it just needs to include the current (or resistivity) gradient effect in magnetic diffusion equation. This inclusion leads to the coupling of even (tearing) and odd (interchange) modes.

An adaptive eigenfunction independent solution-kinetic (AEGIS-K) shooting code has been developed. AEGIS-K is based on a fully kinetic (nonhybrid) and self-consistent (nonperturbative) description. It includes wave-particle resonances, shear Alfvén continuum damping, trapped particle effects, and parallel electric effects. AEGIS-K has been applied to study the kinetic and rotation stabilization of resistive wall modes. It is shown that the resistive wall modes in ITER configuration can be stabilized by subsonic plasma rotation.

15.3 Equilibrium and MHD Equations

Consider the standard sheared slab configuration,

$$\boldsymbol{B}_0(x) = B_{0y}(x)\widehat{\boldsymbol{y}} + B_{0z}(x)\widehat{\boldsymbol{z}} \tag{15.25}$$

where $B_{0y}(x)$ equals zero at $x = \pm x_s$. The plasma sheet is of length a in the x-direction, has a current in $j_z = (c/4\pi)(dB_{0y}/dx)$ in the z-direction, and its equilibrium flow velocity $\boldsymbol{v}_0 = 0$. The stability of this initial configuration will be examined with respect to two-dimensional, incompressible perturbations. For this restricted class of perturbations, the vector fields are expressible in terms of the two scalar potentials: the flux function $\psi(x, y, t)$, and the stream function $\phi(x, y, t)$,

$$\boldsymbol{B}_\perp = \nabla\psi \times \widehat{\boldsymbol{z}} \tag{15.26}$$

and

$$\boldsymbol{v}_\perp = \nabla\phi \times \widehat{\boldsymbol{z}}. \tag{15.27}$$

15.3.1 *Electron viscosity or hyper-resistivity*

With electron viscosity, the ohm's law becomes

$$\boldsymbol{E} = \eta\boldsymbol{j} - \frac{1}{c}\boldsymbol{V} \times \boldsymbol{B} - \frac{m_e\mu_e}{n_e e^2}\nabla^2\boldsymbol{j}. \tag{15.28}$$

The z-component of the curl of Eq. (15.28) as

$$\frac{\partial\psi}{\partial t} = -\boldsymbol{v}\cdot\nabla\psi + \frac{c^2}{4\pi}\eta\nabla^2\psi - \frac{m_e\mu_e c^2}{4\pi n_e e^2}\nabla^4\psi \tag{15.29}$$

after using Eq. (15.26) and Faraday's law. The z-component of plasma vorticity equation may be written as

$$\frac{\partial}{\partial t}(\nabla^2\phi) = -(\boldsymbol{V}\cdot\boldsymbol{\nabla})\nabla^2\phi + \frac{1}{4\pi\rho}\left[\nabla(\nabla^2\psi)\times\nabla\psi\right]\cdot\widehat{\boldsymbol{z}} \tag{15.30}$$

where ρ is the mass density of the plasma. The close set of Eq. (15.29) and Eq. (15.30) is our starting point for the stability analysis.

15.3.2 *Linear eigenmodes*

Assuming all perturbations in the form $f \sim f(x)\exp(ik_y y + \gamma t)$, we get the linearized version of Eq. (15.29) and Eq. (15.30):

$$\gamma\psi_1 = V_x B_{0y}(x) + \frac{\eta c^2}{4\pi}\left(\frac{\partial^2\psi_1}{\partial x^2} - k_y^2\psi_1\right) - \frac{\mu_e c^2}{\omega_{pe}^2}\left(\frac{\partial^4\psi_1}{\partial x^4} + k_y^4\psi_1\right) \tag{15.31}$$

and

$$\rho\gamma\left(\frac{\partial^2\phi}{\partial x^2} - k_y^2\phi\right) = -\frac{i}{4\pi}k_y B_{0y}''(x)\psi_1 + \frac{ik_y}{4\pi}B_{0y}(x)\left(\frac{\partial^2\psi_1}{\partial x^2} - k_y^2\psi_1\right) \tag{15.32}$$

where η, ρ and μ_e have been taken as constants and the double prime denotes second derivative with respect to x.

Normalizing all lengths to a, time to τ_h, and the magnetic field to some standard measure B_0, Eq. (15.31) and Eq. (15.32) convert to the following dimensionless from

$$\gamma\tau_h\psi_1 = \gamma\tau_h\xi B_{0y}(x) + \frac{1}{S}\left(\frac{\partial^2\psi_1}{\partial x^2} - \alpha^2\psi_1\right) - \frac{1}{R}\left(\frac{\partial^4\psi_1}{\partial x^4} + \alpha^4\psi_1\right) \tag{15.33}$$

and

$$(\gamma\tau_h)^2\left(\frac{\partial^2\xi}{\partial x^2} - \alpha^2\xi\right) = \alpha^2 B_{0y}''(x)\psi_1 - \alpha^2 B_{0y}(x)\left(\frac{\partial^2\psi_1}{\partial x^2} - \alpha^2\psi_1\right) \tag{15.34}$$

where $\alpha = k_y a, \xi = ik_y\phi/\gamma a, S = \tau_r/\tau_h, R = \tau_v/\tau_h$. It is easy to note from Eq. (15.29) that $\tau_v = a^4\omega_{pe}^2/\mu_e c^2$ is the viscosity diffusion time for a magnetic field over a plasma sheet of width a.

15.4 Double Tearing Mode in Plasmas with Anomalous Electron Viscosity

The linear behavior of the double tearing mode in plasmas with a phenomenological anomalous electron viscosity is investigated within the framework of magnetohydrodynamic (MHD) theory. In the large Reynolds number $R = \tau_v/\tau_h$ (τ_v and τ_h are, respectively, the viscosity penetration time of the magnetic field and the Alfvén time for a plasma sheet of width a) limit, the growth rate is found to scale as $R^{-1/5}$ if the two resonant surfaces, at $x = \pm x_s$, are close enough to satisfy $x_s/a \ll (k_y a)^{-11/15}R^{-1/15}$. For larger separation between the resonant surfaces, the growth rate transits to a $R^{-1/3}$ scaling. The transition occurs at $x_s/a \sim (k_y a)^{-11/15}R^{-1/15}$. The $R^{-1/5}$ is shown to be closely correlated with the violation of the constant-ψ approximation. The nonlinear velocity perturbations associated with the unstable double tearing mode are estimated to saturate at a level high enough to serve as a trigger for the formation of transport barriers observed in advanced tokamaks.

One of the characteristics for an advanced tokamak (AT) operation is a nonmonotonic safety factor q-profile, providing a region with negative magnetic shear

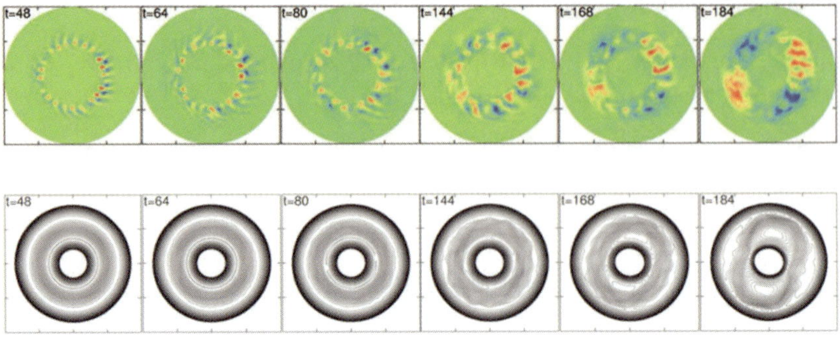

Fig. 15.12 a) Time evolution of electrostatic potential; b) Equicontours of magnetic flux showing the appearance of magnetic islands of double tearing mode [Ishizawa and Nakajima (2007)].

[Lao, *et al.* (1996); Koide, *et al.* (1994)]. Such a configuration is prone to the excitation of a double tearing mode (DTM), an instability peculiar to a plasma with multiple resonant surfaces ($\boldsymbol{k} \cdot \boldsymbol{B} = 0$). With magnetic islands on sufficiently close adjacent resonant surfaces interacting and effectively enhancing each other, the double tearing mode turns out to be a much stronger instability than the standard (i.e., constant-ψ) tearing mode because of large Δ'. Numerical simulations demonstrate that a macro-magnetohydrodynamic (macro-MHD) mode is excited as a result of multi-scale interaction in a quasi-steady equilibrium formed by a balance between micro-turbulence and zonal flow based on a reduced-two-fluid model. This simulation of a macro-MHD mode, a double tearing mode, is accomplished in a reversed shear equilibrium which includes zonal flow and turbulence due to kinetic ballooning modes. In the quasi-steady equilibrium, a macro-scale fluctuation which has the same helicity as the double tearing mode is a part of the turbulence. After a certain period of time, the macro-MHD mode begins to grow. It effectively utilizes free energy of the equilibrium current density gradient and is destabilized by a positive feedback loop between zonal flow suppression and magnetic island growth. Thus, once the macro-MHD appears from the quasi-equilibrium, it continues to grow steadily. This simulation is more comparable with experimental observations of growing macro-MHD activity than earlier MHD simulations starting from linear macro-instabilities in a static equilibrium [Ishizawa and Nakajima (2007)].

On the other hand the AT mode of operation is considered to be highly desirable because of the realization of high performance regimes in which the so-called internal transport barriers (ITBs) are formed with the simultaneous appearance of highly sheared localized poloidal flows. In this section we will attempt to show a possible causal relationship between the DTM and the experimentally-observed shear flows.

The double tearing mode, driven by plasma resistivity, has been studied by several authors [Pritchett, *et al.* (1980); Mahajan and Hazeltine (1982); Goodall and Wesson (1984); Kleva, *et al.* (1989); Ofman (1992); Persson and Dewar (1994); Yu (1996)]. The linear growth rate of the mode goes as $S^{-1/3}$ when the separation of the resonant surfaces ($x = \pm x_s$) is sufficiently small, i.e. x_s satisfies the inequality

$$\frac{x_s}{a} < (k_y a)^{-7/9} S^{-1/9}. \tag{15.35}$$

When the separation of the resonant surfaces is large, i.e., the inequality in Eq. (15.35) is reversed, the islands do not interact strongly, and the growth rate of the double tearing mode scales as $S^{-3/5}$ as it is for the standard tearing mode [Pritchett, *et al.* (1980)]. Here, $S = \tau_r/\tau_h$ is the magnetic Reynolds number with $\tau_r = 4\pi a^2/c^2\eta$ and $\tau_h = a/v_A$ being the resistive diffusion time and the poloidal Alfvén time of a plasma column of scale width a, respectively; c is the speed of light, η is the plasma resistivity, v_A is the poloidal Alfvén velocity. The island growth of the double tearing mode in early nonlinear stage is analyzed by Yu (1996).

In contrast to extensive studies on tearing modes driven by plasma resistivity, investigations on tearing modes driven by plasma viscosity are rather scant. Experiments indicate, however, that anomalous viscosity may be present in tokamak plasmas due to electromagnetic turbulence [Kadomtsev (1992); Scott, *et al.* (1989)]. Bootstrap drive of neoclassical tearing modes in the presence of anomalous viscosity is performed recently by Konovalov, *et al.* (2002). Tearing modes driven by anomalous electron viscosity due to braiding magnetic field lines have been considered as possible candidates responsible for disruptive instability [Rosenbluth and Rutherford (1981); Kaw, *et al.* (1979)]. It is found that the growth rates of $m \geq 2$ and $m = 1$ electron viscosity tearing modes scale as $\gamma \sim R^{-1/3}$ and $\gamma \sim R^{-1/5}$, respectively. Here, $R = \tau_v/\tau_h$ is the fluid dynamic Reynolds number, while $\tau_v = 4\pi a^4 n_e e^2/c^2\mu_e m_e = \omega_{pe}^2 a^4/c^2\mu_e$ is the viscosity diffusion time of plasma current over a sheet of width a, μ_e is the electron viscosity diffusion coefficient. In strong electromagnetic drift-wave turbulence the coefficient μ_e may be comparable to the electron thermal diffusivity which can be a few m^2/s. The electron viscosity tearing mode was, later, studied by Aydemir as a possible candidate for experimentally observed fast sawtooth crashes [Aydemir (1990)]. It was found that the electron viscosity driven modes had higher growth rates than the corresponding resistivity driven modes in present fusion devices if the electron viscosity had a value comparable to that of the anomalous electron thermal diffusivity. The viscosity considered here is expected to contribute to the parallel electron motion equation and to cause plasma current penetration across equilibrium magnetic flux surfaces.

The subject of this section is a detailed linear analysis of the double tearing mode driven by anomalous electron viscosity. In Section 15.4, the governing MHD equations, including resistivity and viscosity, are presented and an approximate analytic dispersion relation for the double tearing mode [following the methods of Pritchett, *et al.* (1980)] is derived and discussed in Section 15.5. In Section 15.6 we turn to a well developed computer code to study the instability in detail; the eigenvalues and the structure of the eigenfunctions for a variety of parameters are obtained. Emphasis is placed on the detailed analysis of the perturbed poloidal velocity profiles. The conclusions and discussion are presented in Section 15.7. The possible correlation of the non-linear DTM driven by electron viscosity with the dynamics of ITB formation is also emphasized.

15.5 Two-Space-Scale Analysis and Dispersion Relation

Equation (15.34) indicates that the dissipation terms (the second and third terms on the right-hand side) are important only in a narrow layer around $x = \pm x_s$ where the magnetic field B_{y0} equals zero. In the regions between the rational surfaces ($|x| < x_s$) and outside the two surfaces ($|x| > x_s$), the effects of the dissipation terms are negligible. Therefore, we adapt the two-space-scale analysis usually used for the resistive tearing mode studies. Ideal magnetohydrodynamics are employed in the outer regions, i.e., the equations obtained when $S \to \infty$ and $R \to \infty$ in Eq. (15.34). The solutions obtained in these two regions (exterior solutions) must join smoothly with the dissipative (interior) solutions obtained from Eq. (15.34) and valid near each resonant surface.

The ideal MHD is first solved substituting

$$\psi_1(x) = B_{0y}(x)\xi(x) \tag{15.36}$$

a consequence of Eq. (15.34), into Eq. (15.35), we find

$$\frac{d}{dx}\left\{\left[(\gamma\tau_h)^2 + (\alpha B_{0y}(x))^2\right]\frac{d\xi}{dx}\right\} = \alpha^2\left[(\alpha B_{0y}(x))^2 + (\gamma\tau_h)^2\right]\xi. \tag{15.37}$$

This equation governs the behavior of the double kink mode in a slab. The inertial terms ($\gamma\tau_h$) are negligible in comparison with the αB_{0y} terms in the ideal MHD regions and Eq. (15.37) reduces to,

$$\frac{d}{dx}\left[(\alpha B_{0y}(x))^2\right]\frac{d\xi}{dx} = \alpha^4 B_{0y}^2(x)\xi \tag{15.38}$$

which may be solved in terms of a power series expansion in the small parameter $\alpha^2 x_s^2$: $\xi = \xi_0 + \xi_1 + \ldots\ldots$ To the lowest order, Eq. (15.38) reduces to

$$\frac{d}{dx}\left[(\alpha B_{0y}(x))^2\right]\frac{d\xi_0}{dx} = 0. \tag{15.39}$$

Because the displacement ξ_0 is symmetric about the $x = 0$ surface, and $|x| = x_s$ are the singularity surfaces, the solution for Eq. (15.39) may be written as Coppi, *et al.* (1976)

$$\xi_0(x) = \xi_\infty = \begin{cases} \text{const}, & |x| < x_s \\ 0, & |x| > x_s. \end{cases} \tag{15.40}$$

The first order solution is given by

$$\frac{1}{\xi_\infty}\frac{d\xi_1}{dx} = \begin{cases} \left(\dfrac{\alpha}{B_{0y}(x)}\right)^2 \displaystyle\int_0^x B_{0y}^2(x')dx', & |x| < x_s \\ \left(\dfrac{\alpha}{B_{0y}(x)}\right)^2 \displaystyle\int_0^{x_s} B_{0y}^2(x')dx', & |x| > x_s. \end{cases} \tag{15.41}$$

Near $x = x_s$, we expand $B_{0y}(x)$ as a Taylor series and neglect the two terms on the right-hand side of Eq. (15.37).

$$\xi = \frac{1}{2}\xi_\infty \left\{1 - \frac{1}{\pi}\arctan\left[\alpha B'_{0y}(x - x_s)/\gamma\tau_h\right]\right\} \tag{15.42}$$

where $B'_{0y} = B'_{0y}(x_s)$.

15.5.1 *Stable double kink modes*

The growth rate of the double kink mode, now, is found by equating $d\xi/dx$ obtained from Eq. (15.42) in the limit $\alpha B'_{0y}(x - x_s)/\gamma\tau_h \to -\infty$ with $d\xi_1/dx$ obtained from Eq. (15.41) in the limit $x \to x_s^-$,

$$\gamma_h\tau_h = -\frac{\pi\alpha^3}{B'_{0y}}\int_0^{x_s} B^2_{0y}(x')dx'. \tag{15.43}$$

Thus we find from Eq. (15.43) that with pure MHD driving energy alone, the double kink mode is stable in slab geometry approaching marginal stability in the limit $\alpha x_s \to 0$ [Pritchett, *et al.* (1980)].

15.5.2 *Growth from resistivity and electron viscosity*

Dissipation (resistivity and viscosity), however, may provide a mechanism for the marginally stable double tearing mode to go unstable. Even with dissipation, the outer region (away from the resonant surfaces $x = \pm x_s$) solutions, Eq. (15.40) and Eq. (15.41), remain unchanged while the inner region is now described by Eq. (15.34) and Eq. (15.35)

$$(\gamma\tau_h)^2\,\xi'' = -\alpha^2 B'_{0y}(x - x_s)\psi_1'' \tag{15.44}$$

$$\gamma\tau_h\psi_1 = \gamma\tau_h B'_{0y}(x - x_s)\xi + \frac{1}{S}\frac{\partial^2\psi_1}{\partial x^2} - \frac{1}{R}\frac{\partial^4\psi_1}{\partial x^4} \tag{15.45}$$

where the double prime denotes second derivative with respect to x, and α^2 has been neglected in comparison with d^2/dx^2.

Under the transformations $x - x_s \to x$, $\psi_1/B'_{0y} \to \psi_1, \xi \to -\xi$, Eq. (15.44) and Eq. (15.45) become

$$\xi'' = \frac{x}{\lambda^2}\psi_1'' \tag{15.46}$$

$$\psi_1 = -x\xi + \frac{\varepsilon}{\lambda}\psi_1'' - \frac{\sigma}{\lambda}\psi_1^{(4)} \tag{15.47}$$

where $\lambda = \gamma\tau_h/\alpha B'_{0y}$, $\varepsilon = 1/S\alpha B'_{oy}, \sigma = 1/R\alpha B'_{0y}$.

These inner region equations have to be solved with the boundary condition that the solution matches the ideal MHD solution as $|x| \to \infty$. Equation (15.46) and Eq. (15.47) may be converted into a single sixth order differential equation for ξ,

$$-\lambda\sigma\left[\frac{24}{x^5}\xi'' - \frac{24}{x^4}\xi^{(3)} + \frac{12}{x^3}\xi^{(4)} - \frac{4}{x^2}\xi^{(5)} + \frac{1}{x}\xi^{(6)}\right]$$

$$+\lambda\varepsilon\left[\frac{2}{x^3}\xi'' - \frac{2}{x^2}\xi^{(3)} + \frac{1}{x}\xi^{(4)}\right] - 2\xi' - \left(x + \frac{\lambda^2}{x}\right)\xi'' = 0 \tag{15.48}$$

where $\xi^{(n)}$ indicates the nth derivative with respect to x. This equation has six independent solutions. It is not difficult to see that the asymptotic form of one of these solutions, $\xi = \text{const}/x$ with $d\xi/dx = \text{const}/x^2$, matches the outer solution perfectly. Let us write this solution as

$$\xi = \frac{1}{2}\xi_\infty + \xi_{\text{odd}}(x) \tag{15.49}$$

and the outer solution, Eq. (15.49), as

$$\frac{1}{\xi_\infty}\frac{d\xi}{dx} = -\frac{\lambda_h}{\pi}\frac{1}{x^2} \tag{15.50}$$

where

$$\lambda_h = -\frac{\pi\alpha^2}{B_{0y}'^2}\int_0^{x_s} B_{0y}'^2(x')dx' \tag{15.51}$$

and $\xi_{\text{odd}}(x)$ in Eq. (15.49) has to satisfy the boundary condition

$$\frac{x^2}{2}\frac{d}{dx}(\ln\xi_{\text{odd}}) = \begin{cases} -\frac{1}{\pi}\lambda_h, & \text{for } x \to -\infty: \\ \frac{1}{\pi}\lambda_h, & \text{for } x \to \infty \end{cases} \tag{15.52}$$

since ξ_{odd} is an odd function of x.

For a given λ_h, σ and ε (representing viscosity and resistivity), Eq. (15.46) and Eq. (15.47) may be solved numerically (with appropriate conditions for $\xi_{\text{odd}}(0) = 0$ and ψ_1) to determine the dependence of the effective eigenvalue λ on λ_h, i.e., the dispersion relation of the mode. The present work, however, is limited to the double tearing mode driven by electron viscosity. Approximate analytical work is presented below.

If resistivity is neglected ($\varepsilon = 0$) Eq. (15.46) and Eq. (15.47) reduce to

$$\xi'' = \frac{x}{\lambda^2}\psi_1'' \tag{15.53}$$

$$\psi_1 = -x\xi - \frac{\sigma}{\lambda}\psi_1^{(4)}. \tag{15.54}$$

Following Coppi, *et al.* (1976), we introduce the function

$$\chi(x) = x\psi_1' - \psi_1 = \lambda^2\frac{d\xi}{dx} + \chi_\infty \tag{15.55}$$

in terms of which, Eq. (15.54) and Eq. (15.55) may be combined into one equation,

$$\sigma\lambda\left[\frac{d^4X}{dx^4}-\frac{4}{x}\frac{d^3X}{dx^3}+\frac{8}{x^2}\frac{d^2X}{dx^2}-\frac{8}{x^3}\frac{dX}{dx}\right]+(\lambda^2+x^2)X=x^2X_\infty \tag{15.56}$$

where the constant X_∞ is to be determined from the asymptotic behavior of the solution.

Manipulating Eq. (15.55), we derive

$$\xi=-\frac{\psi_1}{x}-\frac{\sigma}{\lambda x}\frac{d^4\psi}{dx^4}=-\frac{1}{x}X(x)+\int_x^\infty\frac{1}{x'}\frac{dX}{dx'}dx'-\frac{\sigma}{\lambda x}\left[\frac{2}{x^3}\frac{dX}{dx}-\frac{2}{x^2}\frac{d^2X}{dx^2}+\frac{1}{x}\frac{d^3X}{dx^3}\right] \tag{15.57}$$

and

$$\xi_\infty\simeq\int_{-\infty}^\infty\frac{1}{x}\frac{dX}{dx}dx\simeq 2\int_0^\infty\frac{1}{x}\frac{dX}{dx}dx \tag{15.58}$$

since $\xi\to\xi_\infty$ when $x\to-\infty$. In addition, we know that $d\xi/dx\to\text{const}/x^2$ when $x\to-\infty$. Defining

$$\frac{d\xi}{dx}\,|_{x\to-\infty}=\frac{X_\infty}{x^2} \tag{15.59}$$

and comparing it with Eq. (15.53), we deduce

$$X_\infty=\frac{1}{\pi}\lambda_h\xi_\infty\simeq\frac{2\lambda_h}{\pi}\int_0^\infty\frac{dX}{dx}\frac{dx}{x}. \tag{15.60}$$

Equation (15.61) may be considered as the boundary condition for the solution of Eq. (15.57).

To obtain approximate solution for Eq. (15.57), we notice that in the ideal MHD limit, $\sigma\to 0$, Eq. (15.57) yields

$$X=\frac{x^2}{\lambda^2+x^2}X_\infty. \tag{15.61}$$

Substituting this solution into the boundary condition, Eq. (15.61), we may easily show that $\lambda=\lambda_h$ i.e., the perturbation is an ideal MHD mode as it was supposed to be.

When the ideal MHD mode is marginally stable, $\lambda_h=0$, $X_\infty=0$ and Eq. (15.57) becomes

$$\sigma\lambda\left[\frac{d^4X}{dx^4}-\frac{4}{x}\frac{d^3X}{dx^3}+\frac{8}{x^2}\frac{d^2X}{dx^2}-\frac{8}{x^3}\frac{dX}{dx}\right]+(\lambda^2+x^2)X=0. \tag{15.62}$$

To order x^3, Eq. (15.63) allows the solution

$$X=A\exp\left[-\frac{x^2}{2^{6/5}\sigma^{2/5}}\right],\qquad \lambda=\frac{\sigma^{1/5}}{2^{2/5}} \tag{15.63}$$

where the constant coefficient A is easily obtained from the boundary condition, Eq. (15.61); the final solution is

$$X=\frac{\xi_\infty\sigma^{1/5}}{\sqrt{\pi}2^{2/5}}\exp\left[-x^2/2^{6/5}\sigma^{2/5}\right]. \tag{15.64}$$

Equation (15.55) tells us that in the case of ideal MHD marginal stability, the growth rate of the tearing mode driven by electron viscosity is

$$\gamma = \frac{(\alpha B'_{0y})^{4/5}}{2^{2/5}\tau_h^{4/5}\tau_v^{1/5}}. \tag{15.65}$$

It is the same as that for the $m = 1$ tearing mode driven by electron viscosity [Rosenbluth and Rutherford (1981)].

Now, we look for a more general solution of Eq. (15.57). Introducing $\zeta = x^2/\lambda^{1/3}\sigma^{1/3}$, $\widehat{\lambda} = \lambda/\sigma^{1/5}$, we transfer Eq. (15.57) into

$$\zeta^2\frac{d^4\chi}{d\zeta^4} + \zeta\frac{d^3\chi}{d\zeta^3} - \frac{1}{4}\frac{d^2\chi}{d\zeta^2} + \frac{1}{16}\left(\widehat{\lambda}^{5/3} + \zeta\right)\chi = \zeta\chi_\infty. \tag{15.66}$$

An approximate solution for this equation can be found. First, in comparison with Eq. (15.64), we neglect the term with ζ^2. Then, we assume that the equation has a solution of the form

$$\chi = A\left\{1 + B\int_0^1 t^l(1+t)^m \exp\left[-\frac{\zeta}{\alpha'}\left(\frac{1-t}{1+t}\right)\right]dt\right\} \tag{15.67}$$

where the constants A, B, l, m, and α' are readily determined (by direct substitution) to be

$$A = \chi_\infty, \quad B = -\frac{\widehat{\lambda}^{5/3}}{2^{m+1}\alpha'}, \quad l = \frac{\widehat{\lambda}^{5/3}}{3\alpha'} - \frac{13}{12}, \quad m = \frac{\widehat{\lambda}^{5/3}}{3\alpha'} + \frac{5}{4}, \quad \alpha' = 2^{4/3}. \tag{15.68}$$

Invoking Eq. (15.61), we derive the dispersion relation,

$$\begin{aligned}\lambda = \lambda_h &\left\{\frac{\widehat{\lambda}^{5/2}}{16\cdot 2^{(\frac{\widehat{\lambda}^{5/3}}{3\alpha'}+\frac{1}{4})}} \cdot \frac{\Gamma(\frac{\widehat{\lambda}^{5/3}}{6\sqrt[3]{2}} - \frac{1}{12})}{\Gamma(\frac{\widehat{\lambda}^{5/3}}{6\sqrt[3]{2}} + \frac{17}{12})}\right.\\ &\left.\cdot F\left(-\frac{\widehat{\lambda}^{5/3}}{6\sqrt[3]{2}} - \frac{3}{4}\frac{\widehat{\lambda}^{5/3}}{6\sqrt[3]{2}} - \frac{1}{12}\frac{\widehat{\lambda}^{5/3}}{6\sqrt[3]{2}} + \frac{17}{12}, -1\right)\right\}\end{aligned} \tag{15.69}$$

where F is the hypergeometric function [Oberhettinger (1968)],

$$F(a,b,c,z) = \frac{\Gamma(c)}{\Gamma(a)\Gamma(b)}\sum_{n=0}^{\infty}\frac{\Gamma(a+n)\Gamma(b+n)}{\Gamma(c+n)}\frac{z^n}{n!}. \tag{15.70}$$

It is easy to verify that the function F in Eq. (15.70) converges absolutely. The numerical solution of the dispersion relation, Eq. (15.69), is displayed as a plot of $\widehat{\lambda}$ versus $\widehat{\lambda}_h$ in Fig. 15.13. The approximate analytical results are in good agreement with the numerical results obtained for the $m = 1$ mode when $\widehat{\lambda}$ is not too large [Rosenbluth and Rutherford (1981)]. The general behavior tends to be similar though one notices clear differences for large $\widehat{\lambda}$. Small $\widehat{\lambda}$ corresponds to $\widehat{\lambda}_h \lesssim 0$, i.e., the ideal MHD double kink mode is stable or marginally stable. It is precisely in these two cases that the study of the double tearing mode driven by electron

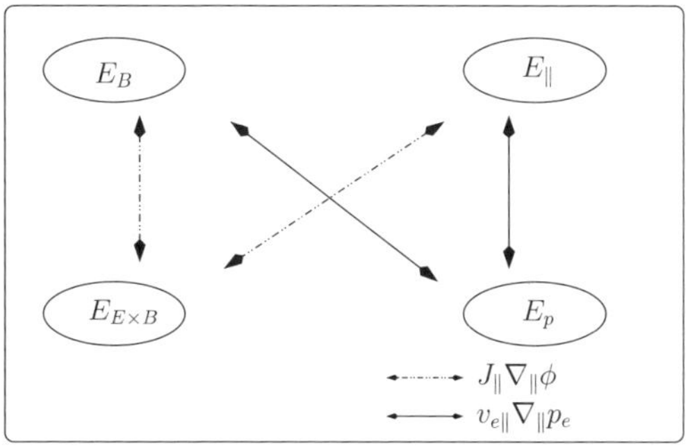

Fig. 15.13 Diagram of the energy transfer among magnetic energy E_B, perpendicular kinetic energy $E_{E\times B}$, parallel kinetic energy $E_{\|}$ and electron thermal energy E_p. The nonlinear transfer powers are from $J_{\|}\nabla_{\|}\phi$ with $J_{\|} = -en_e u_{\|e}$ and $J_{\|}\nabla_{\|}p_e$ where p_e is the electron thermal pressure.

viscosity is important; the approximate analytic solution given above is, therefore, appropriate for the study of the electron viscosity double tearing mode.

The dispersion relation, Eq. (15.69), may be further approximated as

$$\lambda = \lambda_h \left\{ \frac{\widehat{\lambda}^{5/2}}{16 \cdot \sqrt[4]{2}(1 + \frac{\widehat{\lambda}^{5/3}}{6\sqrt[3]{2}}\ln 2)} \left[1 + \frac{(\frac{\widehat{\lambda}^{5/3}}{6\sqrt[3]{2}} + \frac{3}{4})(\frac{\widehat{\lambda}^{5/3}}{6\sqrt[3]{2}} - \frac{1}{12})}{\frac{\widehat{\lambda}^{5/3}}{6\sqrt[3]{2}} + \frac{17}{12}} \right] \times \frac{\Gamma(\frac{\widehat{\lambda}^{5/3}}{6\sqrt[3]{2}} - \frac{1}{12})}{\Gamma(\frac{\widehat{\lambda}^{5/3}}{6\sqrt[3]{2}} + \frac{17}{12})} \right\}. \tag{15.71}$$

Two special limits are easy to understand:

1) if $\lambda_h = 0$ then $\widehat{\lambda} = 1/2^{2/5}$, i.e., $\lambda = \sigma^{1/5}/2^{2/5}$ which is exactly the same as Eq. (15.64);

2) if $\widehat{\lambda}_h < 0$ and $|\widehat{\lambda}_h| \gg 1$, then $\widehat{\lambda} \ll 1$ according to Fig. 15.13 and we have from Eq. (15.71) that

$$\lambda = \text{const}\frac{\sigma^{1/3}}{\lambda_h^{2/3}}. \tag{15.72}$$

Now, we discuss the requirements that will let the viscosity-driven double tearing mode to have a growth rate given in Eq. (15.64), and then to transit to that given by Eq. (15.72). On double differentiation with respect to x, Eq. (15.54) becomes $\xi^{(4)} = \frac{x}{\lambda^2}\psi_1^{(4)} + \frac{2}{\lambda^2}\psi_1^{(3)}$. Substituting $\psi_1^{(4)}$ from Eq. (15.55) into this equation gives us $\xi^{(4)} = x/\lambda\sigma[-\psi_1 + x\xi] + 2/\lambda^2\psi^{(3)}$. Comparing the ξ terms in the equation, we get an estimate for the thickness of the dissipation layer:

$$\Delta \simeq (\lambda\sigma)^{1/6} \simeq [\gamma\tau_h/(\alpha B'_{0y})^2 R]^{1/6}. \tag{15.73}$$

It has to be pointed out that the dispersion relation given in Fig. 15.13 is valid only for $\Delta \ll x_s$; otherwise there are no separable inner and outer regions. it is also essential to remember that the solution in the outer region is valid only when $\alpha x_s < 1$.

Remembering

$$\widehat{\lambda} = \frac{\lambda}{\sigma^{1/5}} = \gamma\tau_h(R/\alpha^4 B_{0y}'^4)^{1/3} \tag{15.74}$$

$$\widehat{\lambda}_h = \frac{\lambda_h}{\sigma^{1/5}} = \gamma_h\tau_h(R/\alpha^4 B_{0y}'^4)^{1/3} \tag{15.75}$$

and the discussion following Eq. (18) (leading to $|\widehat{\lambda}_h| \ll 1$ for $\alpha x_s \ll 1$), we get $\widehat{\lambda} = 1/2^{2/5}$ and the growth rate

$$\gamma\tau_h \simeq 0.8\left(\frac{\alpha^4 B_{0y}'^4}{R}\right)^{1/5}. \tag{15.76}$$

which scales as $R^{-1/5}$.

The conditions $|\hat{\lambda}_h| \ll 1$ and $\Delta \ll x_s$ impose (assuming $B_{0y}' = 1$)

$$\left(\frac{\alpha}{R}\right)^{1/5} \ll \alpha x_s \ll \left(\frac{\alpha^4}{R}\right)^{1/15}. \tag{15.77}$$

If x_s is sufficiently large, then $|\widehat{\lambda}_h| \gg 1$ and $\widehat{\lambda} \ll 1$, and we have

$$\gamma\tau_h = \frac{\text{const.}}{(\gamma_h\tau_h)^{2/3}}\left(\frac{\alpha^4 B_{0y}'^4}{R}\right)^{1/3}. \tag{15.78}$$

with the growth rate scaling as $R^{-1/3}$.

The conditions $|\widehat{\lambda}_h| \gg 1$ and $\alpha x_s \ll 1$ imply

$$\left(\frac{\alpha^4}{R}\right)^{1/15} \ll \alpha x_s \ll 1. \tag{15.79}$$

Therefore, it is explicit that the transition from the $R^{-1/5}$ to $R^{-1/3}$ occurs roughly at $\alpha x_s \sim (\alpha^4/R)^{1/15}$.

The current penetration time due to viscosity over the tearing layer may be estimated as

$$\tau_\Delta \simeq \frac{\Delta^4\omega_{pe}^2}{\mu_e c^2} = \Delta^4\tau_h R \tag{15.80}$$

where Δ is given by Eq. (15.73). This results in

$$\gamma\tau_\Delta = \gamma\left(\frac{\gamma\tau_h}{\alpha^2 B_{0y}'^2 R}\right)^{2/3}\tau_h R = \widehat{\lambda}^{5/3}. \tag{15.81}$$

If the ideal MHD mode is marginally stable, then $\widehat{\lambda} \simeq 1$ from Fig. 15.13. The current penetration time is comparable with the tearing mode growing time and the constant-ψ approximation is not valid. The mode grows according to Eq. (15.76).

On the other hand, if the ideal MHD mode is stable, then $\widehat{\lambda} \ll 1$. The tearing mode growth time is much longer than the current penetration time. In this case, ψ_1 may approximately be considered as a constant over the tearing layer. The growth rate is given by Eq. (15.78) in this regime.

15.6 Numerical Eigenmodes of Resistive Viscous Tearing Modes

The same configuration as used in Pritchett, *et al.* (1980) can be employed in

$$B_{0y}(x) = 1 - (1 + b_c)\text{sech}(\zeta x) \tag{15.82}$$

where

$$\zeta x_s = \text{sech}^{-1}\left[\frac{1}{(1 + b_c)}\right]. \tag{15.83}$$

Equation (15.83) has the properties that $B_{0y}(\pm x_s) = 0$, $B_{0y}(0) = -B_c$, and $B_{0y} \to 1$ as $x \to \pm\infty$. The constant B_c is chosen so as to make $B'_{0y}(x_s) = \pi/2$ [Pritchett, *et al.* (1980)]. No specifications for $B_{0z}(x)$ and $P_0(x)$ are given here since incompressible equations are used. An assumption is made that the resistive and the electron viscosity diffusion times are much longer than the double tearing mode growing time and the equilibrium is static. The resistivity and the viscosity are both assumed to be constant. Equation (15.34) and Eq. (15.35) are solved as an eigenvalue problem using a shooting code. Only symmetric modes are considered for which $d\xi/dx|_{x=0} = d^3\psi/dx^3|_{x=0} = 0$ are the required boundary conditions. In addition, the normalization $\psi_1(0) = 1$ is applied. At the outer boundary $x = \pm x_w$, $\psi_1 = 10^{-5}$ and $\xi = d\xi/dx = 0$ are employed.

With the grid number fixed at 200, the results presented below were checked to be approximately independent of x_w. The growth rate changes less than 0.1% for a 50% change of x_w. The code was benchmarked with the results on the resistive double tearing mode in Pritchett, *et al.* (1980) first by taking a sufficiently large R value such as 10^{10} for $S = 10^6$ and comparing numerical results with Fig. 4 of Pritchett, *et al.* (1980). Then the electron viscosity was introduced, and the related double tearing modes were studied by decreasing R and increasing S. Shown in Fig. 15.14 is the normalized growth rate $\gamma\tau_h$ as a function of the wavenumber α for $x_s = 0.25$ (a) and 0.7 (b).

The lines, from the top to the bottom, correspond to $R = 10^5,\ 10^6,\ 10^7$, and 10^8 in Fig. 15.14(a), and to $R = 10^4, 10^5, 10^6, 10^7$, and 10^8 in Fig. 15.14(b), respectively. The growth rate is checked to be independent of S up to $s = 10^7$. The electron viscosity double tearing mode always dominates in the parameter regime studied here for $R \lesssim S$. The growth rate increases with α in both cases whereas it increases with α for $x_s = 0.25$ and has a maximum around $\alpha \sim 0.35$ for $x_s = 0.7$ in the resistive double tearing case.

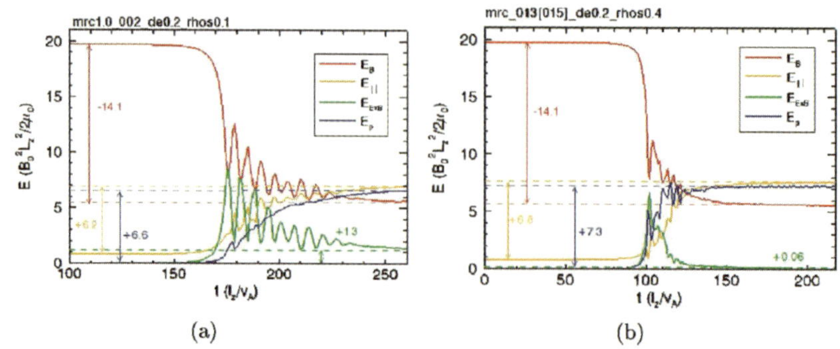

Fig. 15.14 Time evolution of magnetic energy E_B (red), parallel electron flow energy $E_{\|}$ (orange), perpendicular $\boldsymbol{E} \times \boldsymbol{B}$ energy (green) and electron thermal energy E_p for small and large ρ_s values. (a) $\widehat{d}_e = 0.2$, $\widehat{\rho}_s = 0.1$ and (b) $\widehat{d}_e = 0.2$, $\widehat{\rho}_s = 0.4$ with initial mode $(k_x, k_z) = 2\pi(4/L_x, 1/L_z) = 2\pi(0.5, 4)/L_z$. Approximately 70% of initial magnetic energy is released in the cases (a) and (b).

Shown in Fig. 15.15 is the growth rate as a function of the fluid dynamic Reynolds number R for $x_s = 0.25$(the short line) and 0.7(the longer line); the wavenumber is $\alpha = 0.75$. The mode growth rate follows the $R^{-1/3}$ scaling for $x_s = 0.25$ while there is a transition from $R^{-1/3}$ to $R^{-1/5}$ at $R \sim 10^6$ for $x_s = 0.7$. This is in agreement with the analytic results given in Section 15.2. The eigenfunctions $\psi_1(x)$, $\xi(x)$ and the perturbations of the parallel current density j_{iz} and the poloidal velocity V_y are given in Fig. 15.12 and Fig. 15.13 for $x_s = 0.25$, 0.7 and 0.1, respectively. The other parameters for Fig. 15.12(5) are $\alpha = 0.25(0.5)$, $S = 9.4 \times 10^5(10^8)$ and $R = 10^6(10^8)$. It is clearly shown that the changes of ψ_1 across the tearing layers are significant in Fig. 15.14(a); the constant-ψ approximation breaks down in this case.

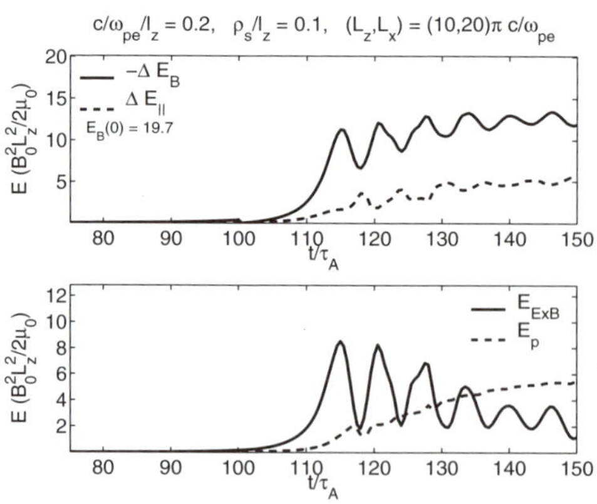

Fig. 15.15 The impulsive release of stored magnetic energy $-\Delta E_B$ from collisionless reconnection into the parallel out-flow energy of the electrons $\Delta E_{\|}$ in top figure in time of 30 Alfvén periods. Lower panel shows the transient increase and subsequent decrease in the $E \times B$ flow energy and the steady increase in the electron thermal energy E_p.

In addition, the perturbation of the parallel current density is spread over the entire region between the two rational surfaces, and peaks at the center Fig. 15.14(b). On the other hand, the perturbation of the flux ψ_1 does not change much across the tearing layers in Fig. 15.14(a); the constant-ψ approximation is valid in this case. In addition, the perturbation of the parallel current density sharply peaks at the two rational surfaces and is about one half the peak current density at the center Fig. 15.14(b).

This is similar to the binary branch studied by Mahajan and Hazeltine (1982) for resistive tearing modes in configurations with a parabolic safety factor q. The V_y profile of Fig. 15.14 has two strong peaks highly localized at the two rational surfaces, and is essentially zero in the between region.

15.7 Magnetic Reconnection Instabilities

Two distinct kinds of tearing modes driven by finite plasma resistivity in configurations with non-monotonic safety factor profile were identified and analyzes by Pritchett, *et al.* (1980), and by Mahajan and Hazeltine (1982). Two analogous kinds of tearing modes but driven by anomalous electron viscosity in the same equilibrium magnetic configuration [Pritchett, *et al.* (1980)], are studied in this paper in this section. One of the principal aims of this investigation is the search for an engine for the creation of transport barriers. It is found that the growth rates of the modes scale as $R^{-1/5}$ and $R^{-1/3}$, respectively. Strictly speaking it is only the first kind that merits the name – double tearing mode – because it straddles the two flux surfaces that are close enough. The second kind is just a combination of two regular standard tearing mode stationed at each of the well-separated rational surfaces. In the double tearing mode the perturbations centered at each rational surface interact with, and enhance each other causing a marked increase in the growth rate. For the regular tearing case, the two rational surfaces are wide apart, and the perturbations centered at each rational surface do not interact but merely develop independently. The structures of the modes are also significantly different. In the double tearing case, the perturbed magnetic flux changes significantly across the tearing layers while it is essentially a constant in the regular tearing case. The perturbed parallel current and poloidal velocity both are non-zero over the whole region between the two rational surfaces and extend far beyond in the former case. Whereas, they peak at the two rational surfaces and are rather localized in the late case. The transition from one to the other occurs at $x_s/a|_{\text{crit}} \sim (k_y a)^{-11/15} R^{-1/15}$. The small R exponent $1/15 \simeq 0.067$ means that the current profile and the poloidal mode number $k_y = m/r_{\text{min}}$ determine the critical value of x_s. The growth time of the double tearing mode is comparable with current penetration time over the tearing layer. In contrast, the growth time is longer than the current penetration time for the regular tearing mode.

Mahajan and Yoshida (2000) have shown that in a collisionless two-fluid model, a combination of the Hall term and the fluid nonlinearity can lead to the formation of a self-organized singular layer that displays the essential observational features of the thin shear layer associated with the high-confinement (H-mode) tokamak discharges [Mahajan and Yoshida (2000)]. The necessary condition for the formation of such a layer is a poloidal velocity of the order of poloidal Mach number unity. A similar situation, i.e., the existence of a narrow layer with a strongly sheared flow, exists for tokamak discharges with internal transport barriers (ITB). The mechanisms for the creation of such flows is far from understood. An important clue does come from the experimental observations that the ITB's are often formed around low mode number rational surfaces [Joffrin, *et al.* (2002)] (and the references therein). Coupled with the fact that the magnetic energy released in the reconnection process following the development of a tearing mode can drive large flows, Mahajan, *et al.* suggest that the tearing mode studied in this work may provide a possible trigger for the formation of the singular layers or ITBs. No attempt is made to describe the detailed dynamics here but make a rough nonlinear estimate for the amplitude of the saturated poloidal velocity is given.

Balancing shearing frequency in the first and viscous dissipation rate in the third term on the right-hand side of Eq. (6), we find $k_y V_y \sim k_\perp^4 \mu_e c^2/\omega_{pe}^2 = v_A(k_\perp^4 a^4/R\alpha$ where v_A is the poloidal Alfvén velocity and $k_\perp^2 = k_x^2 + k_y^2 \simeq k_x^2$. With $k_x \sim \Delta^{-1} \simeq (\lambda\sigma)^{-1/6} \simeq [\gamma\tau_h/(\alpha B_{0y}')^2 R]^{-1/6}$ and $\gamma\tau_h \sim \lambda \sim R^{-1/5}$, it is easy to estimate that the saturated poloidal shearing velocity compared with the thermal velocity is $V_y/v_T \sim R^{-1/5}$. For $R \sim 10^5$, this is the just the magnitude required [Mahajan and Yoshida (2000)] for strong sheared flows in the layer. It is worthwhile to point out that the resistivity driven double tearing modes will generate similar levels of saturated poloidal flows for rather low $S \sim 10^3$ making them unlikely to be the sources for ITB creation. The viscosity driven double tearing mode, on the other hand, is able to drive up the required level of poloidal flows at the large S values of the experiments.

Not a lot of time has been spent on the underlying physical mechanisms that may lead to the anomalous electron viscosity. However, the braiding magnetic field lines [Aydemir (1990)] created by higher mode number MHD instabilities is one possible mechanism. Another possible source could be the unstable electromagnetic micro-modes characteristic to high β (= plasma pressure/magnetic pressure) plasmas [Dong, *et al.* (1999); Gao, *et al.* (2001)]. Since the double tearing modes driven by the electron viscosity may play an important role in the processes of flow creation, it would be worth our while to thoroughly examine the origin of anomalous μ_e.

15.8 Turbulent Impulsive Magnetic Energy Release from Electron Scale Reconnection

Magnetic reconnection may occur as bursts of nonlinear plasma dynamics on the electron collisionless skin length scale $d_e = c/\omega_{pe}$ during which a large fraction of the magnetic energy is converted to plasma thermal energy and plasma flow energy. An example of such a bursty energy release event is given with a simple set of electron Hall equations. The energization mechanism is the cross-field compression of the electron gas between interacting magnetic islands. The electron energization appears to be consistent with the rapid electron energy flux changes measured by the Cluster spacecraft crossing thin current sheets at $-17\,R_E$ in the geotail. The analysis is relevant to planned spacecraft missions for measuring electron scale magnetic reconnection events in the geomagnetic tail.

Magnetic reconnection modeling has a long and complicated history [Coppi, *et al.* (1966); Kadomtsev (1975)]. In the domain of collisionless magnetic reconnection, there are two space/time scales of the dynamics. One is the ion tearing mode dynamics where the nonlinear structure and islands are large compared to ion gyroradius ρ_i and the ion inertial scale length $\rho_s = (m_i T_e)^{1/2}/eB$, and the second is the formation of small scale structures and islands on the scale of electron [Drake and Lee (1977)] skin length $d_e = c/\omega_{pe}$ and up to the scale of the ion inertial scale $\rho_s \leq c/\omega_{pe}$. The time scale for the electron event is corresponding shorter than that for the ion events, and they can have a large power transfer from the magnetic energy releases $\Delta E_B < 0$.

The short duration and large numbers of such events make them an ideal candidate to explain the fast, sometimes called explosive, energy release required by geomagnetic substorms. The time scale for the collisionless ion tearing modes is generally too long for the fast dynamics $\Delta t \ll 100L/v_A < 10\,\text{min}$ observed in geomagnetic substorms.

For geotail reconnection we use the Geocentric Solar Magnetospheric Coordinate System, called the GSM system, with x-axis pointing toward the sun, y-axis from dawn to dusk perpendicular to x-axis and the Earth's magnetic dipole axis. In general, the planet's magnetic dipole axis has a tilt angle θ with respect to the mutually perpendicular z-axis.

Here we investigate a relatively simple nonlinear electron scale magnetic reconnection model based on two nonlinear partial differential equations for the magnetic flux $\psi(x, z, t)$ and the electrostatic potential $\phi(x, z, t)$ that describe in the final state a release of magnetic energy, $-\Delta E_B$, of order 50% available from the reversed field components, $\pm B_{x0}$ in a volume of $\Omega = L_x L_y L_z \sim 2000\, d_e^2 L_y$.

The energy flow from the stored magnetic field energy to the kinetic energy of the ions and electrons and then to the thermal energy of the electrons is described in Fig. 15.13. The degree to which energy is conserved in the dynamics is used to verify the accuracy of the simulations.

During the reconfiguration period there is a interval of time, in the initial evolution of the magnetic island, during which the $\boldsymbol{E} \times \boldsymbol{B}$ flow energy dominates the parallel flow energy and the electron thermal energy. Subsequently the power transfer to the electron pressure $\delta P_e = \delta n T_e$ is dominant.

Indeed, in the early nonlinear phase, the magnetic island associated to the instability undergoes an explosive growth, which confirms the theoretical predictions of Ottaviani and Porcelli (1993) and Bhattacharjee, *et al.* (2005). In this first phase, most of the power transfer goes to the $\boldsymbol{E} \times \boldsymbol{B}$ flow (around 70% of $|\Delta E_B|$).

However, when islands on neighboring resonant surfaces start to interact, the system becomes turbulent. The $\boldsymbol{E} \times \boldsymbol{B}$ energy decreases initially in favor of the parallel electron flows energy, $K_{\|} = 1/2\, n_e m_e u_{\|}^2$, and finally of the electron thermal energy, $\Delta W_e = 3/2\, \Delta p_e \Omega$. As the energy unit scale in the theory we use the initial reversed field energy per unit length along the ambient current, $E_{B0} = B_{x0}^2 L_x L_z/(2\mu_0)$. In the given numerical example in Fig. 15.14a, the energy release in the turbulent state, $-\Delta E_B$, is approximatively 70% of the initial reversed field energy E_{B0}. Of this released magnetic energy, approximately 44% goes to electron thermal energy, 47% to parallel electron flow kinetic energy, $K_{\|}$, and 9% to $\boldsymbol{E} \times \boldsymbol{B}$ flow kinetic energy. The key pathways for the power transfer are shown in Fig. 15.13.

In the model, there are four nonlinear terms and two key kinds of power transfer terms, $J_{\|} \nabla_{\|} \phi$ and $v_{e\|} \nabla_{\|} p_e$ as shown in Fig. 15.13. A third nonlinear transfer, the $\boldsymbol{E} \times \boldsymbol{B}$ convection of the plasma vorticity, $U = \nabla_{\perp}^2 \phi / B_0$, derived from $\vec{v}_E = \vec{E} \times \vec{B}/B_0^2$ with $\vec{E} = -\nabla\phi$ and $B_0 \widehat{\mathbf{y}}$ the out-of-plane guide field, arising from the ambient dawn-to-dusk magnetic field in the geomagnetic tail. The vorticity nonlinearity, $\vec{v}_E \cdot \nabla U$, is well known and recently measured by Perez, *et al.* (2006) in the Large Plasma Device (LAPD). Its effect is to couple energy throughout the cross-field flow wavenumber spectrum, $|\phi_{\vec{k}}(t)|^2$, which appears in the configuration space as vortex merging. Thus, this vorticity nonlinearity has only an indirect effect in the dynamics of the reconnection which accounts for its being dropped in the model of Ottaviani and Porcelli (1993).

15.8.1 *Electron scale magnetic reconnection model*

The latest developments on the electron-scale magnetic reconnection (MR) modes are given in Ottaviani and Porcelli (1993) and Bhattacharjee, *et al.* (2005) and Horton, *et al.* (2007). The impulsive reconnection model of Bhattacharjee has the field-line-breaking mechanism of electron inertial given by finite $d_e = c/\omega_{pe}$. Therefore, it is necessary to use the generalized ohm's Law

$$\vec{E} + \vec{v} \times \vec{B} = \frac{c^2 \mu_0}{\omega_{pe}^2} \frac{\mathrm{d}\vec{J}}{\mathrm{d}t} - \frac{\nabla p_e}{ne} + \frac{\vec{J} \times \vec{B}}{ne} + \eta \vec{J} \tag{15.84}$$

where $\vec{E}$ is the electric field, $\vec{B}$ is the magnetic field, $\vec{v}$ is the plasma flow velocity, c is the speed of light, $\vec{J}$ is the current density, p_e is the electron pressure, ω_{pe} is the electron plasma frequency, n is the electron density, e is the magnitude of the electron charge.

In the simulations shown here, we take the limit of zero plasma resistivity, η, and check energy conservation to five digits. High-order time integrators techniques allow to run simulations for finite times with zero resistivity.

The simplest electron tearing mode dynamics occurs when plasma flow is almost incompressible. This occurs when there is a guide field B_y. Although the pressure of the guide field is not thought to be important to the real event, it is included in our treatment to make a simple justification of dropping the dynamics of the electron pressure equation.

The small flow compression comes from the polarization drift of the ions and is given by $\partial U/\partial t$ where $U = \nabla_\perp^2 \phi / B_y$, with $\vec{E} = -\partial\psi/\partial t - \nabla\phi$ and $\vec{B} = \nabla\psi \times \widehat{\mathbf{y}} + B_y \widehat{\mathbf{y}}$. The vorticity, U, grows to feed plasma into the reconnection layer. The compressional Alfvén wave drops out of the dynamics in this regime owing to its higher frequency and different polarization.

The plasma is described in an electron-Hall fluid limit by the following two field equations

$$\frac{\partial F}{\partial t} + [\phi, F] = \widehat{\rho}_s^2 [U, \psi] \tag{15.85}$$

$$\frac{\partial U}{\partial t} + [\phi, U] = [J, \psi] \tag{15.86}$$

where $J = -\nabla_\perp^2 \psi$, $F = \psi + \widehat{d}_e^2 J$, $U = \nabla_\perp^2 \phi$, and the Poisson bracket is defined by $[\phi, F] = \widehat{\mathbf{y}} \cdot \nabla\phi \times \nabla F$. All quantities are dimensionless (this will be fixed with the proper dimensions). Time is normalized by Alfvén time scale $\tau_A = l_z / v_A$ where the Alfvén velocity $v_A = \sqrt{B^2/\mu_0 \rho}$ and $l_z = L_z / 2\pi$ is the scale over which the ambient current density varies. The dynamics depends on two dimensionless parameters $\widehat{d}_e$ and $\widehat{\rho}_s$ where the electron skin depth $\widehat{d}_e = (c/\omega_{pe})/l_z$ and the ion sound gyroradius $\widehat{\rho}_s = (c_s/\omega_{ci})/l_z$.

In the region of a uniform $B_x = \partial\psi/\partial z$ field, the linearized dynamics from Eq. (15.85) and Eq. (15.86) gives $\omega F_k - k_x B_x \phi_k = \rho_s^2 k_x B_x U_k$ and $\omega U_k = -k_x B_x J_k$. Therefore, we obtain

$$\omega^2 = \frac{k_x^2 v_A^2 (1 + k_\perp^2 \widehat{\rho}_s^2)}{1 + k_\perp^2 \widehat{d}_e^2} \tag{15.87}$$

which is the well-known kinetic Alfvén wave dispersion relation. The properties of these waves in a uniform plasma have recently been studied experimentally by Vincena, *et al.* (2004) in the LAPD. Here we follow the formulation of Bhattacharjee, *et al.* (2005) and Ottaviani and Porcelli (1993), but note there is an important error in the magnetic flux function equation in Bhattacharjee, *et al.* (2005). That is corrected here in Eq. (15.85). The error in the $\rho_s^2[U, \psi]$-term would completely eliminate the energy conservation derived here. The same model has been used in Cafaro, *et al.* (1998) and Grasso, *et al.* (2001) where the formation of fine structures and the saturation of a single magnetic island are investigated. In particular, Grasso, *et al.* (2001) present a numerical study of the energetics of an isolated island by using an Harris pinch equilibrium. A more complete Hamiltonian model, which necessarily conserves energy, is given in Tassi, *et al.* (2008).

Owing to the absence of boundaries in the central plasma sheet and the non-monotonic nature of the cross-tail current density $J_y(z)$, periodic box boundary conditions, as those in Bhattacharjee, *et al.* (2005), are more appropriate in our investigation.

The computation has been done with

$$\psi = \Sigma_{n,m}\psi_{m,n}(t)\exp(inz + im\epsilon x) \tag{15.88}$$

$$\phi = \Sigma_{n,m}\phi_{m,n}(t)\exp(inz + im\epsilon x) \tag{15.89}$$

where n, m are integers, $\epsilon = L_z/L_x = l_z/l_z$ is an aspect ratio of the system. The initial profile are the unstable current sheet $\psi_{0,1}$ with small perturbations $\psi_{1,4}$ and $\phi_{1,4}$, that is,

$$\psi(0) = \cos z + \delta\psi\cos(4z)\cos(\epsilon x)$$

$$\phi(0) = \delta\phi\sin(4z)\sin(\epsilon x).$$

For the example shown here, $\psi_{1,4} = \phi_{1,4} = 0.0001$ are 10 e-folding to reach the first maximum of $-E_B$ in Fig. 15.14. Other initial conditions ($n = 2$) with a spectrum of small modes leads to the same conclusions.

The simulation box is $L_z \times L_x = 10\pi\, c/\omega_{pe} \times 20\pi\, c/\omega_{pe}$ and the unit of energy is $E_0 = B_0^2 L_z^2/2\mu_o$, which for 10 nT reversed field over a $l_z = 5\, c/\omega_{pe} = 50$ km is $2.5 \times 10^6 \mathrm{J}/R_E$.

Equations (2) and (3) are numerically solved in a double periodic box, with three resonant surfaces at $z = 0$ and $z = \pm\pi$. The periodic boundary conditions used in our numerical analysis are justified by the fact that we are interested in the investigation of the dynamics of interacting microscale magnetic islands. The use of periodic boundary conditions is the standard model in condensed matter physics for eliminating the influence of surface effects in bulk matter calculations. Similarly, the physics of the boundary conditions at the lobe plasma and the magnetopause can be eliminated by assuming periodicity at the edge of the numerical box when its size is large compared with c/ω_{pe}, but small compared to the dimensions of the current sheet.

Note that the theoretical models of Ottaviani and Porcelli (1993) and Bhattacharjee, *et al.* (2005) apply only to the early nonlinear phase of our numerical calculation, when the magnetic islands do not interact with each other. This is due to the fact that both models rely on the standard Rutherford nonlinear growth theory for single islands [Rutherford (1973)]. In particular, the existence of a dominant harmonic in the nonlinear magnetic flux is not a valid *Ansatz* for the interacting islands arising in the late phase of our simulation. For this reason, the numerical integration in Bhattacharjee, *et al.* (2005) is stopped when the width of the central magnetic island becomes comparable to the system size. The simulations presented here are continued farther in time, roughly twice as long as those in Bhattacharjee, *et al.* (2005).

The electron reconnection model conserves the total energy of electrons and the magnetic field. The energy component formulas and the conservation law are given by

$$\frac{\mathrm{d}E_{\mathrm{tot}}}{\mathrm{d}t} = \frac{\mathrm{d}}{\mathrm{d}t}\left(E_B + E_{\|} + E_{\boldsymbol{E}\times\boldsymbol{B}} + E_p\right) = \frac{1}{2}\frac{\mathrm{d}}{\mathrm{d}t}\int dxdz\Big[(\nabla\psi)^2 + d_e^2(\nabla^2\psi)^2 + (\nabla\phi)^2 + \left(\rho_s^2(\nabla^2\phi)\right)^2\Big] = 0 \tag{15.90}$$

where E_B is magnetic energy, $E_{\|}$ is electron parallel flow energy, $E_{\boldsymbol{E}\times\boldsymbol{B}}$ is the electron perpendicular flow energy and E_p is the electron thermal energy. As shown in Fig. 15.13, there are four transfers between the different energy components. The magnetic energy E_B and the electron parallel flow energy $E_{|}$ are transferred back and forth to the thermal energy E_p by thermal interaction $\langle v_{e\|}\nabla_{\|}p_e\rangle$. And the energies, E_B and $E_{\|}$, are transferred back and forth to the perpendicular flow energy $E_{\boldsymbol{E}\times\boldsymbol{B}}$ by the electric interaction $\langle j_{\|}\nabla_{\|}\phi\rangle$.

The time history of the energy relevant to the ϕ, U, ψ and J fields are shown in Fig. 15.14. During the saturation process, almost 70% of the magnetic energy is released and about 45-50% of the released energy is transferred to the electron thermal energy. In Fig. 15.14a, we see the released magnetic energy ΔE_B of 14.1 E_0 and the increase of the parallel flow kinetic energy in the electrons is 6.2 E_0. In the bottom panel the perpendicular flow energy increases first to 8 E_0 and then drops to 1.3 E_0 while the electron thermal energy continually increases to 6.6 E_0. Energy is conserved by the 14.1 units of released magnetic energy going to $6.2 + 6.6 + 1.3$ (E_0 unit) units of parallel flow, perpendicular flow and thermal plasma energies. In terms of fractional changes, this is approximately a 70% decrease of magnetic energy transformed into 9% $\boldsymbol{E}\times\boldsymbol{B}$ flow, 44% parallel flow and 47% electron thermal energy. The space and time scales are $10\,c/\omega_{pe}$ and $60L_z/v_A$. In the central plasma sheet, these scales are typically 100 km and 1 min.

In Fig. 15.14b, the case of $\widehat{d}_e = 0.2$, $\widehat{\rho}_s = 0.4$ shows the energy release 70% of the magnetic energy and the transfer of 40%, 0.4% and 52% to the parallel flow energy, the $\boldsymbol{E}\times\boldsymbol{B}$ energy and the thermal energy. The twice larger $\widehat{\rho}_s$ corresponds to a four times large $\nabla_{\|}p_e$-term in ohm's law which makes the transition faster as shown in Fig. 15.13 and increases the final electron thermal energy from 6.6 to 7.3 E_0.

When the Eq. (15.85) and Eq. (15.86) are run as written with no dissipation, the magnetic energy drops sharply over a period of 100 to 150 Alfvén time and the high-k spectrum increases until subscale energy dissipation is required or the simulation is stopped in time. For a convergence test the simulations are then repeated with hyper-resistivity $\eta_4(\nabla^2)^4F$ and hyperviscosity $\mu_4(\nabla^2)^4U$ added to the right-hand side. Now wavenumber spectra are compared. The wavenumber spectra in the range $k_\perp = [1, 20]/L_z$ are essentially the same. For $k_\perp$ greater or equal to $30/L_z$, there is a rollover and a rapid drop off for the runs with high-k damping.

The background electron temperature gradient effect with two parallel heat closure models is being investigated. The first one is an electrostatic closure model. The parallel heat flux convected by $\boldsymbol{E} \times \boldsymbol{B}$ velocity without the consideration of the magnetic fluctuation is zero. This model needs a large but finite parallel heat diffusivity $\chi_{e\|}$. Thermal balance reduces to

$$n\vec{v} \cdot \nabla T_e + \nabla_{\|} q_{e\|} = 0 \tag{15.91}$$

with $q_{e\|} = -n\chi_{e\|}\nabla_{\|}T_e$ where Sugama, *et al.* (2001) gives a model for $\chi_{e\|}$. The electron temperature fluctuation is

$$\frac{\delta T_e}{T_e} = -i\frac{\omega_{*Te}}{\chi_{e\|}k_{\|}^2}\frac{e\phi}{T_e} \tag{15.92}$$

where $\omega_{*Te} = k_x T_e/eB_y L_{Te}$. The second model is an electromagnetic closure model. The parallel thermal diffusivity is taken as infinitely large and the parallel heat flux $q_{e\|}$ is bounded from above. This means the magnetic fluctuation is governed by

$$\nabla_{\|}T_e = \frac{\partial \delta T_e}{\partial y} + \frac{\delta \vec{B}}{B_y} \cdot \nabla T_e = 0 \tag{15.93}$$

reducing to

$$\frac{\delta T_e}{T_e} = \frac{\omega_{*Te}}{ck_y}\frac{ce\psi}{T_e}. \tag{15.94}$$

At the small values of $\omega_{*Te}/\chi_{e\|}k_y^2$ and ω_{*Te}/ck_y, we did not observe the significant effect and we need to investigate further. Drake and Lee (1977) reported that the electron temperature gradient does not cause the instability in collisionless regime.

The electron reconnection scale produces fast, small scale releases of stored, reversed field magnetic energy. In laboratory experiments, the calculations are typically performed for a few, well separated magnetic islands, as observed from soft X-ray emissions in tokamaks. However, in the present work we did not stop the simulations in order to consider only the isolated island evolution, but continued them into the interesting regime of strong island interactions. The pseudo-spectral simulations with high order time integration can resolve the strong nonlinear state down to the scale where microturbulence from lower hybrid drift waves would be expected. The current densities reported in the Cluster spacecraft geotail data have peak values with drift velocities of order twice the ion thermal velocity.

The dynamics exhibits a long exponential growth, ending with a nonlinear, faster than linear exponential growth pulse. Finally, saturation is reached into a turbulent state. Typically, 70% of the initial magnetic energy, which is the difference of the magnetic energy in two different magnetic field geometry, is redistributed into the sum of the parallel and perpendicular flow energies and the electron thermal pressure by the reconnection pulse.

The model developed here is missing the kinetic physics of the wave-particle interactions. Thus, the simulation alone is not able to capture the beam-like features of the ion and electron velocity distribution, reported by spacecraft particle detectors. Some of these features maybe captured by adding ensembles of test particles to be integrated with the evolving reconnection electromagnetic fields. An example of this technique can be found in Horton and Tajima (1991), where the strong ion energization is shown for a simple analytic model of the ion-tearing mode.

The physical nature of the ion tearing mode is far from clear. A common formulation of ohm's law (ion Hall MHD) for the ion tearing mode is $\mathbf{E} = -\mathbf{v} \times \mathbf{B} + \mathbf{J} \times \mathbf{B}/ne$ which keeps $E_{\parallel} = \mathbf{E} \cdot \mathbf{b} = 0$ so that the magnetic flux is not reconnected [Biskamp (2000)] (p. 16). In the case where the $\mathbf{B} \cdot \nabla \cdot \mathbf{P}_i$ is large, magnetic reconnection does occur. An example of this type of ion reconnection is analyzed in Horton and Tajima (1991). This is a type of chaotic ion viscous dissipation that frees the magnetic field from the assumption of the frozen-in theorem.

Some kinetic theory physics, such as that related to a finite ion gyroradius, can be added to the partial differential equations. Perhaps the most important kinetic effect to be added to the fluid model is the electron temperature fluctuation, δT_e. This is driven by the $\widetilde{\mathbf{v}}_E \cdot \nabla T_e$-convection in the nonuniform background T_e field. This addition has been briefly investigated. The background temperature gradient introduces the drift-tearing mode physics, thought to be an important instability for electron thermal transport in high-β laboratory confinement experiments. The simplest version of this modification is to generalize the right-hand side of Eq. (15.85) to $\widehat{\rho}_s^2[U + \eta_e \partial_x \phi, \psi]$, where η_e measures the nonuniformity of the ambient electron temperature profile.

The electron reconnection event similar to that modeled here may have been observed with a substorm of August 24, 2003 during a period where the Cluster spacecrafts crossed a thin current sheet $L_j \leq 2c/\omega_{pi} \approx 1000$ km and measured electron acceleration and tailward high speed flows. The event was found in the Cluster 2003 tail season data by Nakamura, *et al.* (2006). The current sheet is comparable to the ion collisionless skin depth in thickness as the four space craft with a spacing of 200 km flew through the geotail. The event was found by selecting data with high current densities. The current density for this event reached $40\,\mathrm{nA/m^2}$ which is about ten times the nominal value for the center of the current sheet. The spacecrafts were at approximately $(-17, -3.8, 3.3)\,\mathrm{R_E}$ in the geotail. The electron energy flux and pitch angle distribution are observed to increase from $0.5 - 0.8\,\mathrm{KeV}$ up to 5 KeV in crossing a turbulent reconnection zone. Electron acceleration reported during this reconnection event is suggestive of the electron reconnection dynamics analyzed in this present work.

A second example, Wygant, *et al.* (2005) of a microscale magnetic reconnection event, is described by a 15 minutes Cluster repeated crossings of a moving plasma sheet at distance $x = -16\,\mathrm{R_E}$ in the midnight sector. This event was measured during the geomagnetic storm of October 1, 2001. The electric and magnetic fields are measured every 50 ms and the particle distribution every 8 s. In the 15-minute interval, the signature of the c/ω_{pe}-scale reconnection dynamics are seen. The density is $0.1\,\mathrm{cm^{-3}}$, giving $c/\omega_{pe} = 20\,\mathrm{km}$. The measured magnetic field reverses from $B_x = \pm 15\,\mathrm{nT}$ over a current sheet of thickness $\sim 5\,c/\omega_{pe}$ in 4 s, from the motion of the spacecraft through the current sheet. The guide field, B_y, is approximately 5 nT. The plasma parameters upstream ($\beta \ll 1$), in the layer ($\beta \geq 100$), and downstream appear to be consistent with the magnetic reconnection model described in this work.

When both the electron temperature gradient and the current density gradient are retained in the nonlinear dynamics of small scale ETG and microtearing modes we find the following equations:

$$\left(1-\nabla_\perp^2\right)\frac{\partial\phi}{\partial t} = \left\{\left[1-2\epsilon_n+(1+\eta_e)\nabla_\perp^2\right]\frac{\partial}{\partial y}-\mu\nabla_\perp^4\right\}\phi \tag{15.95}$$

$$+\left(\nabla_\parallel\nabla_\perp^2-j_\parallel'\frac{\partial}{\partial y}\right)A+2\epsilon_n\frac{\partial\delta T_e}{\partial y}+\left[\phi,\nabla^2\phi\right] \tag{15.96}$$

$$\left(\nabla_\perp^2-\frac{\beta}{2}\right)\frac{\partial A}{\partial t} = \left(2\nabla_\parallel-j_\parallel'\frac{\partial}{\partial y}\right)\phi+\left[\frac{\beta}{2}(1+\eta_e)\frac{\partial}{\partial y}-\frac{\eta}{\mu_0}\nabla_\perp^2\right]A \tag{15.97}$$

$$-\nabla_\parallel\delta T_e-\left[\phi,\nabla_\perp^2 A\right] \tag{15.98}$$

$$\frac{\partial\delta T_e}{\partial t} = -\left[\eta_e-4\epsilon_n(\Gamma-1)\right]\frac{\partial\phi}{\partial y}-(\Gamma-1)\left(\nabla_\parallel\nabla_\perp^2-j_\parallel'\frac{\partial}{\partial y}\right)A \tag{15.99}$$

$$-\left[2\epsilon_n(2\Gamma-1)\frac{\partial}{\partial y}-\chi_\perp\nabla_\perp^2-\chi_\parallel\left(\nabla_\parallel\right)^2\right]\delta T_e-[\phi,\delta T_e]. \tag{15.100}$$

15.9 Rotating Magnetic Island and Locked Modes

As shown in the TCV data in Fig. 14.9 low order magnetic islands arise spontaneous in the tokamak discharges. Their onset if often associate and correlated with the sawteeth oscillations in the core electron temperature. In the absence of a symmetry breaking magnetic perturbation from the external coils these magnetic islands propagate in the toroidal direction at speeds associated with the electron diamagnetic drift velocity relative to the rest frame of the plasma. The islands, by definition, have a even magnetic perturbation δB_r at the magnetic surface that creates the inside-island closed helical flux surfaces of half-width $w = 2\sqrt{(L_s\psi(x_{m,n})/B_0)}$. Here the magnetic shear length $L_s = qR/s$ with s/q evaluated at the rational surface specified by $q = m/n$. Clearly, in weak magnetic shear confinement modes the magnetic island width can be substantial. In the TCV example in Chapter 14 the profiles are in a normal ELM-free H-mode state so the shear is not weak and the magnetic island grows to be a to be several centimeters in width as inferred from the Thomson scattering diagnostics.

The theory frame work for the formation of the magnetic islands and the locked modes is that of the tearing mode calculations and the existence of a symmetry breaking external magnetic field dB_{ext} [Smolyakov, *et al.* (1995)] develop the two-component fluid equations for the rate of rotation for the island in the presence of the external magnetic field. The island rotation is determined by the projection of the momentum balance equation on to the toroidal direction with the parallel component of the momentum stress tensor evaluated as a function of the viscosity coefficients acting on the ion velocity components v_θ and v_ϕ. The calculations are complicate so the not repeated here.

From the nonlinear equations projected on to the single helicity magnetic island

$$\Psi[r, \theta, \phi(t), t] = \psi(r, t) \cos[m\theta - \ell\phi(t)]. \tag{15.101}$$

A pair of nonlinear equations for the rate of change of the toroidal rotation velocity are derived that with the amplitude of the external perturbation being the control parameter. From these equations the critical value for the external magnetic perturbation for the locking of the magnetic island to the external perturbation is derived.

There are several mechanisms controlling the rotation of these magnetic island so there is observed a variety of behaviors [Fitzpatrick and Hender (1991); Fitzpatrick, *et al.* (1993); Nave and Wesson (1986)]. The comparison of theory and experiments on magnetic islands is given in Waelbroeck and Fitzpatrick (1997), Zohm, *et al.* (2001), and Gunter, *et al.* (1998).

The separatrix region of the magnetic island develops sharp plasma gradients and a strong radial electric field shown in the figures from analytic theory and electrostatic gyrokinetic simulations in Siccinio, *et al.* (2011). The 3/2-magnetic island shown in Chapter 14 for a TCV ECRH discharge is unbiquitous feature of high-temperature discharges as reviewed by Zhom, *et al.* (2001) for ASDEX.

References

Aydemir, A. Y. (1992). Nonlinear studies of $m = 1$ modes in high-temperature plasmas, *Phys. Fluids B* **4**, p. 3469, doi:10.1063/1.860355, `http://link.aip.org/link/doi/10.1063/1.860355`.

Aydemir, A. Y. (1991). Linear studies of $m = 1$ modes in high-temperature plasmas with a four-field model, *Phys. Fluids B* **3**, 11, pp. 3025-3032, doi:10.1063/1.859780, `http://link.aip.org/link/?PFB/3/3025/1`.

Aydemir, A. Y. (1990). Magnetohydrodynamic modes driven by anomalous electron viscosity and their role in fast sawtooth crashes, *Phys. Fluids B* **2**, 9, pp. 2135-2142, doi:10.1063/1.859433, `http://link.aip.org/link/?PFB/2/2135/1`.

Bhattacharjee, A., Germaschewski, K., and Ng, C. S. (2005). Current singularities: Drivers of impulsive reconnection, *Phys. Plasmas* **12**, p. 042305, doi:10.1063/1.1872893, `http://link.aip.org/link/doi/10.1063/1.1872893`.

Biskamp, D. (2000). *Magnetic Reconnection in Plasmas* (Cambridge University Press) `http://books.google.com/books?id=YkzgcUuLC8oC&lpg=PR13&ots=Z3BtXVwmVE&dq=Biskamp`.

Biskamp, D. (1991). Algebraic nonlinear growth of the resistive kink instability, *Phys.*

Fluids B **3**, 12, pp. 3353-3356, doi:10.1063/1.859766, http://link.aip.org/link/?PFB/3/3353/1.

Cafaro, E., Grasso, D., Pegoraro, F., Porcelli, F., and Saluzzi, A. (1998). Invariants and geometric structures in nonlinear Hamiltonian magnetic reconnection, *Phys. Rev. Lett.* **80**, p. 4430, http://link.aps.org/doi/10.1103/PhysRevLett.80.4430.

Chen, X. L. and Morrison, P. J. (1990). Resistive tearing instability with equilibrium shear flow, *Phys. Fluids B* **2**, pp. 495-507 .

Chen, X. L. and Morrison, P. J. (1990). The effect of viscosity on the resistive tearing mode with the presence of shear flow, *Phys. Fluids B* **2**, pp. 2575-2580.

Coppi, B., Laval, G., and Pellat, R. (1966). Dynamics of the geomagnetic tail, *Phys. Rev. Lett.* **16**, pp. 1207-1210, http://link.aps.org/doi/10.1103/PhysRevLett.16.1207.

Coppi, B., Galvao, R., Pellat, R., Rosenbluth, M. N., and Rutherford, P. H. (1976). *Fiz. Plazmy* **2**, p. 961 [(1976). *Sov. J. Plasma Phys.* **2**, p. 533].

Dagazian, R. Y. (1979). Effect of convection on magnetohydrodynamic kink stability, *Phys. Fluids* **22**, 10, pp. 1952-1957, doi:10.1063/1.862484, http://link.aip.org/link/?PFL/22/1952/1.

Dong, J.-Q., Chen, L., and Zonca, F. (1999). Study of kinetic shear Alfvén modes driven by ion temperature gradient in tokamak plasmas, *Nucl. Fusion* **39**, p. 1041, doi:10.1088/0029-5515/39/8/308.

Drake, J. F. and Lee, Y. C. (1977). Kinetic theory of tearing instabilities, *Phys. Fluids* **20**, 8, pp. 1341-1353, doi:10.1063/1.862017, http://link.aip.org/link/?PFL/20/1341/1.

Fitzpatrick, R., Waelbroeck, F. L., and Militello, F. (2006). *Phys. Plasmas* p. 122507.

Fitzpatrick, R. and Aydemir, A. Y. (1996). Stabilization of the resistive shell mode in tokamaks, *Nucl. Fusion* **36**, p. 11, doi:10.1088/0029-5515/36/1/I02.

Fitzpatrick, R. (1995). Helical temperature perturbations associated with tearing modes in tokamak plasmas, *Phys Plasmas* **2**, p. 825.

Fitzpatrick, R. (1993). Interaction of tearing modes with external structures in cylindrical geometry (plasma), *Nucl. Fusion* **33**, p. 1049, doi:10.1088/0029-5515/33/7/I08.

Fitzpatrick, R. and Hender, T. C. (1991). The interaction of resonant magnetic perturbations with rotating plasmas, *Phys. Fluids B* **3**, p. 644, doi/10.1063/1.859863, http://link.aip.org/link/doi/10.1063/1.859863.

Furth, H. P., Killeen, J., and Rosenbluth, M. N. (1963). Finite-resistivity instabilities of a sheet pinch, *Phys. Fluids* **6**, p. 459, doi:10.1063/1.1706761, http://link.aip.org/link/.

Gao, Z., Dong, J.-Q., Liu, G. J., and Ying, C. T. (2001). Study of electromagnetic instabilities driven by ion temperature gradient and parallel sheared flows in high-β plasmas, *Phys. Plasmas* **8**, p. 4080, doi/10.1063/1.1388176, http://link.aip.org/link/doi/10.1063/1.1388176.

Goodall, D. H. J. and Wesson, J. A. (1984). Cine observations of MHD instabilities in a tokamak, *Plasma Phys. Control. Fusion* **26**, p. 789, doi:10.1088/0741-3335/26/6/002.

Grasso, D., Califano, F., Pegoraro, F., and Porcelli, F. (2001). Phase mixing and island saturation in Hamiltonian reconnection, *Phys. Rev. Lett.* **86**, p. 5051, http://link.aps.org/doi/10.1103/PhysRevLett.86.5051.

Günter, S., Gude, A., Maraschek, M., Sesnic, S., Zohm, H., and ASDEX Upgrade Team. (1998). β scaling for the onset of neoclassical tearing modes at ASDEX Upgrade, *Nucl. Fusion* **38**, p. 1431, doi:10.1088/0029-5515/38/10/102.

Hazeltine, R. D., Kotschenreuther, M., and Morrison, P. J. (1985). A four-field model for tokamak plasma dynamics, *Phys. Fluids* **28**, pp. 2466-2477.

Hazeltine, R. D., Hsu, C. T., and Morrison, P. J. (1987). Hamiltonian four-field model for nonlinear tokamak dynamics, *Phys. Fluids* **30**, pp. 3204-3211.

Horton, W. and Tajima, T. (1991). Collisionless conductivity and stochastic heating of the plasma sheet in the geomagnetic tail, *J. Geophys. Res.* **96**, pp. 15811-15829, doi:10.1029/91JA01014.

Ishizawa, A. and Horiuchi, R. (2005). *Phys. Rev. Lett.* **95**, p. 0450003, `http://link.aps.org/doi/10.1103/PhysRevLett.95.045003`.

Ishizawa, A. and Nakajima, N. (2010). Turbulence driven magnetic reconnection causing long-wavelength magnetic islands, *Phys. Plasmas* **17**, p. 072308, doi:10.1063/1.1388176, `http://link.aip.org/link/doi/10.1063/1.1388176`.

Ishizawa, A. and Nakajima, N. (2007). Excitation of macro-magnetohydrodynamic mode due to multi-scale interaction in a quasi-steady equilibrium formed by a balance between micro-turbulence and zonal flow, *Phys. Plasmas* **14**, p. 040702, doi:10.163/1.271669, `http://link.aip.org/link/doi/10.1063/1.2716669`.

Joffrin, E., Challis, C. D., Conway, G. D., *et al.* (2002). Internal transport barrier triggering by rational flux surfaces in tokamaks, *International Atomic Energy Agency 19*th *Fusion Energy Conf.* (Lyon, France), submitted to *Nucl. Fusion.*

Kadomtsev, B. B. (1992). *Tokamak Plasma: A Complex Physical System* (Institute of Physics, Bristol).

Kadomtsev, B. (1975). *Sov. J. Plasma Phys.* **1**, p. 389.

Kaw, P. K., Valeo, E. J., and Rutherford, P. H. (1979). Tearing modes in a plasma with magnetic braiding, *Phys. Rev. Lett.* **43**, pp. 1398-1401, `http://link.aps.org/doi/10.1103/PhysRevLett.43.1398`.

Kleva, R. G. (1992). *Phys. Fluids B* **4**, p. 218, doi:10.1063/1.860436.

Kleva, R. G., Drake, J. F., and Denton, R. E. (1989). Magnetohydrodynamic simulations of density-limit disruptions in tokamaks, *Comm. Plasma Phys. Control. Fusion* **13**, p. 63, ISSN:0374-2806.

Koide, Y., Kikuchi, M., Mori, M., Tsuji, S., Ishida, S., *et al.* (1994). Internal transport barrier on $q = 3$ surface and poloidal plasma spin-up in JT-60U high-β_p discharges, *Phys. Rev. Lett.* **72**, pp. 3662-3665, `http://link.aps.org/doi/10.1103/PhysRevLett.72.3662`.

Konovalov, S. V., Mikhailovskii, A. B., Shirokov, M. S., and Tsypin, V. S. (2002). Transport threshold model of neoclassical tearing modes in the presence of anomalous perpendicular viscosity, *Phys. Plasmas* **9**, p. 4596, doi:10.1063/1.1509455, `http://link.aip.org/link/doi/10.1063/1.1509455`.

La Haye, R. J. (2006). *Phys. Plasmas* p. 055501.

La Haye, R. J. and Sauter, O. (1998). *Nuclear Fusion* **987**.

Lao, L. L., Burrell, K. H., Casper, T. S., Chan, V. S., Chu, M. S., DeBoo, J. C., Doyle, E. J., Durst, R. D., Forest, Greenfield, C. M., Groebner, R. J., Hinton, F. L., Kawano, Y., Lazarus, E. A., Lin-Liu, Y. R., Mauel, M. E., Meyer, W. H., Miller, R. L., Navratil, G. A., Osborne, T. H., Peng, Q., Rettig, C. L., Rewoldt, G., Rhodes, T. L., Rice, B. W., Schissel, D. P., Stallard, B. W., Strait, E. J., Tang, W. M., Taylor, T. S., Turnbull, A. D., and Waltz, R. E. (1996). Rotational and magnetic shear stabilization of magnetohydrodynamic modes and turbulence in DIII-D high performance discharges, *Phys. Plasmas* **3**, p. 1951, doi:10.1063/1.871991, `http://link.aip.org/link/doi/10.1063/1.871991`.

Mahajan, S. M. and Yoshida, Z. (2000). A collisionless self-organizing model for the high-confinement (H-mode) boundary layer, *Phys. Plasmas* **7**, p. 635, doi:10.1063/1.873850, `http://link.aip.org/link/doi/10.1063/1.873850`.

Mahajan, S. M. and Hazeltine, R. D. (1982). Linear theory of electromagnetic instabilities in plasmas with a hollow q-profile, *Nucl. Fusion* **22**, p. 1191, doi:10.1088/0029-5515/22/9/005.

Nakamura, R., Baumjohann, W., Asano, Y., Runov, A., Balogh, A., Owen, C. J., Fazakerley, A. N., Fujimoto, M., Klecker, B., and Reme, H. (2006). *J. Geophys. Res.* **111**, p. A11206.

Nave, M. F. F. and Wesson, J. A. (1986). *Nucl. Fusion* **18**, p. 617.

Oberhettinger, F. (1968). *Handbook of Mathematical Functions* ed. M. Abramowitz and I. A. Stegun (Dover, New York), p. 555.

Ofman, L. (1992). Double tearing instability with shear flow, *Phys. Fluids B* **4**, p. 2751, doi:10.1063/1.860148, `http://link.aip.org/link/doi/10.1063/1.860148`.

Ofman, L., Chen, X. L., Morrison, P. J., and Steinolfson, R. S. (1991). Resistive tearing mode instability with shear flow and viscosity, *Phys. Fluids B* **3**, pp. 1364-1373.

Orszag, S. A. (1977). *Numerical analysis of spectral methods: theory and applications.*

Ottaviani, M. and Porcelli, F. (1993). Nonlinear collisionless magnetic reconnection, *Phys. Rev. Lett.* **71**, pp. 3802-3805, `http://link.aps.org/doi/10.1103/PhysRevLett.71.3802`.

Perez, J. C., Horton, W., Gentle, K., Rowan, W. L., Lee, K., and Dahlburg, Russell B. (2006). Drift wave instability in the Helimak experiment, *Phys. Plasmas* **13**, p. 032101, doi:10.1063/1.2168401, `http://link.aip.org/link/doi/10.1063/1.2168401`.

Pritchett, P. L., Lee, Y. C., and Drake, J. F. (1980). Linear analysis of the double-tearing mode, *Phys. Fluids* **23**, p. 1368, doi:10.1063/1.863151, `http://link.aip.org/link/doi/10.1063/1.863151`.

Persson, M. and Dewar, R. L. (1994). Nonlinear destabilization of linearly stable tearing modes with multiple rational surfaces. *Phys. Plasmas* **1**, p. 1256, doi:10.1063/1.870723, `http://link.aip.org/link/doi/10.1063/1.870723`.

Rosenbluth, M. N. and Rutherford, P. H. (1981). *Fusion* ed. E. Teller, (Academic Press, New York) **1**, Part A, p. 31.

Rutherford, P. H. (1973). *Phys. Fluids* **16**, p. 11, 1903-1909, doi:10.1063/1.1694232, `http://link.aip.org/link/doi/10.1063/1.1694232`.

Sauter, O., La Haye, R. J., Chang, Z., Gates, D. A., Kamada, Y., Zohm, H., Bondeson, A., Boucher, D., Callen, J. D., Chu, M. S. (1997). Beta limits in long-pulse tokamak discharges, *Phys. Plasmas* **4**, 2, pp. 1654-1664

Scott, S. D., *et al.* (1989). *Plasma Phys. Control. Nucl. Fusion* (International Atomic Energy Agency, Vienna) **1**, p. 655.

Siccinio, M., Poli, E., Casson, F. J., Hornsby, W. A., and Peeters, A. G. (2011). Gyrokinetic determination of the electrostatic potential of rotating magnetic islands in tokamaks, *Phys. Plasmas* **18**, p. 122506, doi:10.1063/1.3671964.

Smolyakov, A. I., Hirose, A., Lazzaro, E., Re, G. B., and Callen, J. D. (1995). Rotating nonlinear magnetic islands in a tokamak, *Phys. Plasmas* **2**, pp. 1581-1598.

Sugama, H., Watanabe, T.-H., and Horton, W. (2001). Collisionless kinetic-fluid closure and its application to the three-mode ion temperature gradient driven system, *Phys. Fluids* **8**, pp. 2617-2628, doi:0.1063/1.1367319, `http://link.aip.org/link/doi/10.1063/1.1367319`.

Tassi, E., Morrison, P. J., Waelbroeck, F. L., and Grasso, D. (2008). Hamiltonian formulation and analysis of a collisionless fluid reconnection model, *Plasma Phys. Control. Fusion* **50**, p. 085014.

Vincena, S., Gekelman, W., and Maggs, J. (2004). Shear Alfvén wave perpendicular propagation from the kinetic to the inertial regime, *Phys. Rev. Lett.* **93**, p. 105003, http://link.aps.org/doi/10.1103/PhysRevLett.93.105003.

Waelbroeck, F. L., Militello, F., Fitzpatrick, R., and Horton, W. (2009). *Plasma Phys. Control. Fusion* p. 015015.

Waelbroeck, F. L. and Fitzpatrick, R. (1997). Rotation and locking of magnetic islands, *Phys. Rev. Lett.* **78**, p. 1703-1706, http://link.aps.org/doi/10.1103/PhysRevLett.78.1703.

Waelbroeck, F. L. (1989). Current sheets and nonlinear growth of the $m = 1$ kink-tearing mode, *Phys. Fluids B* **1**, 12, pp. 2372-2380, doi:10.1063/1.859172, http://link.aip.org/link/?PFB/1/2372/1.

Wygant, J. R., Cattell, C. A., Lysak, R., Song, Y., Dombeck, J., McFadden, J., Mozer, F. S., Carlson, C. W., Parks, G., Lucek, E. A., Balogh, A., Andre, M., Reme, H., Hesse, M., and Mouikis, C. (2005). Cluster observations of an intense normal component of the electric field at a thin reconnecting current sheet in the tail and its role in the shock-like acceleration of the ion fluid into the separatrix region, *J. Geophys. Res.* **110**, p. A09206, doi:10.1029/2004JA010708.

Yu, Q. (1996). Nonlinear growth of the double tearing mode, *Phys. Plasmas* **3**, p. 2898, doi:10.1063/1.871649, http://link.aip.org/link/doi/10.1063/1.871649.

Zakharov, L. and Rogers, B. (1992). Two-fluid magnetohydrodynamic description of the internal kink mode in tokamaks, *Phys. Fluids B* **4**, 10, pp. 3285-3301, doi:10.1063/1.860384, http://link.aip.org/link/?PFB/4/3285/1.

Chapter 16

Ignition Physics in Multiscale Plasma Turbulence

16.1 Ignition and Confinement Simulations for Compact High-Field Tokamaks

Tokamak machines designed to reach ignition during a transient pulse driven by the transformer action of the primary voltage applied with coils in a central solenoid have a long history. Such tokamaks, however, are yet to be built on the scale required to test the ignition hypothesis. Tokamaks designs in this class include the Compact Ignition Torus CIT, Fusion Ignition Research Experiment FIRE and IGNITOR. The compact high-field IGNITOR design is the most extensively developed design. In this chapter we show the performance projections for the IGNITOR machine made with the multiscale transport codes called JETTO and the MMM model. The parameters and characteristics of that typical high-field tokamak design are given in this chapter. Currently, there are plans to built a version of this compact high-field tokamak machine in Troisk under an agreement between Italy and Russia. The general description of IGNITOR is found at `http://www.frascati.enea.it/ignitor/`.

The basic plan for this type of tokamak is to use the natural ohmic heating to the maximum extent that technology will allow. As shown in Chapter 15 the energy confinement time scaling law is most favorable for the intrinsic plasma heating produced by Coulomb collisions relaxing the transformer driven plasma current I_p driven by the toroidally symmetric voltage V_{loop}. When auxiliary power P_{aux} from neutral beams or radio frequency antennas is applied the energy confinement time degrades, as detailed first in the Kaye-Goldston scaling law in PLT. The confinement time is found to decrease with applied P_{aux} as $\tau_E \sim 1/(P_{\text{aux}}^{1/2})$. The reason for this degradation, as explained in Chapters 13-15, is that as the auxiliary power P_{aux} increases, the level of the drift wave turbulent transport rapidly increases.

The issue with using the natural ohmic heating of ηj^2 to reach the temperature required for ignition, which is at least 10 KeV, is that the plasma resistivity η decreases with increasing temperature and thus the plasma temperature reaches a plateau for a given geometry and toroidal magnetic field strength. To reach the ignition temperature with ohmic heating alone, the plasma current density j must be increased to high levels. For a given major radius of the torus, the limit on the current density is from the strength of the toroidal magnetic field $B_{\rm tor}$ that can be applied. Thus, the pathway to an ohmic ignition experiment requires the use of the highest magnetic fields that can be maintained over relatively small plasma volumes. Volumes of a few cubic meters with fields of the level of $B_{\rm tor} \sim 10\,{\rm T}$ are sufficient.

Since superconducting wires make a spontaneous transition to normal conductivity for magnetic fields above a critical magnetic field strength, the high field ignition experiment must use normal conductivity copper wire for the toroidal field coil magnets. The ohmic dissipation in these copper coils then gives engineering limits to the maximum field strength in machines of spatial scale of a meter.

The fact that one uses the transformer action to drive the toroidal electric field and associated loop voltage $V_{\rm loop}$, the limits the length of time $\Delta t_{\rm pulse}$ the ohmic heating is applied to the plasma. For example, the design values that have been developed for IGNITOR and CIT would limit the ohmic heating pulses to the order $\Delta t_{\rm pulse} \sim 10\,{\rm s}$. The pulse limit is determined by the amount of magnetic flux swing $\delta\psi_{\rm primary}$ that can be produced in the primary external transformer coil.

Central Solenoid and Plasma Current Densities

The magnetic flux-swing in the primary is produced by currents in core winding around what is called the central solenoid (CS) in a tokamak. The compact ignition experiments have flux swings $\delta\psi_{\rm primary} = \int E_{\rm loop} dt = V_{\rm loop} \sim 10 - 20\,{\rm Volt.sec}$. For comparison the flux swing in the central solenoid of ITER is 100 V.s.

In addition the plasma current density is limited by the minor radius of the machine a and the total plasma current by the maximum toroidal magnetic field and the aspect ratio through the condition in Chapter 5 for the onset of the kink instability. Recall the driving force of the kink is the radial gradient of the toroidal current density. For MHD kink stability the maximum current density is set by $j_{\rm max} = B_{\rm tor}/qR\mu_0$. The safety factor q is limited to $q > 1$ for kink stability as confirmed in tokamaks and in another class of machines called the RFPs, for the Reversed Field Pinch. These toroidal RFP machines operate in the kink unstable regime with $q \leqq 0.5$ and have high levels of magnetic fluctuations [Escande, *et al.* (2000)].

From the above discussions on the required magnetic flux swing in volt-sec, the limit on the current density, and the Lawson condition in Chapter 1 one is lead to the design of a machine with a small major radius and the strongest practical toroidal magnetic field that can be maintained for the time periods of order a few seconds to reach ignition. The time interval available is limited by the volt seconds of the central solenoid. When these limits are applied in transport codes with D-T plasmas, one achieves the ignition temperature and the Lawson condition as shown in the transport simulations in this chapter. Achieving ignition depends on the details of the thermal diffusivities and the evolution of the plasma profiles. Thus sophisticated transport models and simulations are used in these studies. The plasma density is limited both by the ballooning pressure gradient limit and by a radiative collapse density limit known as the Murakami limit or the Greenwald limit for the maximum confined plasma density. The fact that the simulations show ignition under suitable conditions has encouraged numerous designs for the compact ignition experiment to be put forward.

Some compact high-field ignition designs include a modest heating boost from radio frequency RF power, usually at the ion cyclotron frequency, to control and push the plasma into thermonuclear ignition regime more rapidly at the end of the pulse. A design for such an ignition experiment was debated against the current ITER design in a high level panel meeting at the headquarters of the Commissaire Energie Atomic CEA in 1999 in Paris. The CEA panel chose the standard large tokamak design over the compact high-field design for two main reasons [Laval (2001)]. The dominant reason was that it is not clear how the intrinsically pulsed ignition experiment would be transformed into a steady state electric power source. The second more pragmatic reason for the ITER choice was that the magnetic confinement database is very extensive for the standard large tokamak and rather limited for the compact high-field tokamak. The database for the compact high field approach comes from the Alcator experiments with $B_{\max} \leqq 6\,\mathrm{T}$ and the Frascati Tokamak FTU. The Alcator experiments hold the record for the highest values of the Lawson measure of a few times $n_e \tau_E \sim 10^{13}\,\mathrm{s/cm^3}$. The temperature in these record $n_e \tau_E$, discharges was about 1 KeV for both the ions and electrons in deuterium plasmas. The Frascati Tokamak FTU has supporting experiments with pulsed high-field discharges maintained by continuous high speed pellet injections for maintaining the high densities.

There was considerable excitement in the tokamak community over these Alcator results in the 1980s. The Alcator results lead to several proposals to the US Department of Energy from the tokamak community including the University of Texas at Austin. The Princeton Plasma Physics Laboratory proposed the CIT design and the earlier IGNITOR proposal from the team of Coppi and Brussard was refined.

The proposers reasoned that it is not clear what limits the Lawson parameter in the fusion tokamak experiments designed to achieve thermonuclear ignition in a deuterium-tritium plasma. The proposed experiments were well suited to reach ignition for a short time interval of a second or two. The results would bring a much clearer understanding to the ignition problem. The plan was to achieve the burning plasma for about ten energy confinement times. The energy confinement time would be about 100 ms under the extrapolation of the Alcator database and consistent with scaling from other tokamak data bases. Plasma ignition is defined as the interval where fusion alpha products heat the plasma sufficiently to compensate for the plasma energy losses. Computer simulations, described in this Chapter, show that a pulse of α particle power up to 10-20 MW is produced for a period over a few seconds. Crucial issues are the production of peaked density profiles on a time-scale of several energy confinement times, the control of the plasma current penetration for the optimization of ohmic heating and the avoidance of sawtooth oscillations. In some simulations, the programmed application of a 10-20 MW Ion Cyclotron Radio frequency heating and the operation of a high-speed pellet injectors are considered to provide more real-time control of the plasma dynamics.

The designs considered are the IGNITOR [Coppi (1991, 1992)], the FIRE [Meade, *et al.* (2001)] and the later compact ignition toroidal experiment dubbed CIT [Greenwald, *et al.* (1995); Meade, *et al.* (1990)].

Simulations with the TRANSP transport code with full alpha particle physics [Budny (2002)] calibrated with JET and TFTR data shows comparisons of the ignition dynamics in IGNITOR, FIRE and ITER-FEAT. The Budny work includes extensive tables of parameters for the five machines.

For the purpose of this chapter it is sufficient to consider the simulations for the IGNITOR tokamak machine with the parameters shown in Table 16.1. The relevant information that would be gained from such an experiment can be summarized as follows:

(i) Improved understanding of plasma turbulence and transport processes, by advanced exploration of high-plasma-density, high-magnetic-field regimes never accessed before. More specifically, a compact, high-field ignition experiment would provide insight into the ohmic confinement scaling laws and self-organized plasma profiles at multi-KeV electron temperatures. The experiments would provide new information on the conditions required for the spontaneous generation of plasma rotation through anomalous angular momentum transport. Plasma rotation plays an important role in determining the level of turbulent fluctuations controlling the radial transport processes in plasmas.

(ii) Alpha particle physics issues would be answered. The confinement of the fusion alpha particle product in fusion reactors may be limited by collective electromagnetic modes driven by the fusion alphas phase space gradients. The large JET tokamak has a wide range of the high energy deuteron and alpha particle driven instabilities [Nave, *et al.* (1995)]. In strong NBI heated tokamaks an outstanding alpha physics issue is when do the high energy ions create MHD "fishbone oscillations" and Toroidal Alfvén Eigenmodes (TAE) that seriously degrade the quality of confinement [Coppi and Porcelli (1986)]. An outstanding physics issue is whether alpha particle heating causes the same degree of confinement degradation as found from auxiliary NBI plasma heating. Purely ohmic heated plasma has the highest energy confinement laws at lower levels of auxiliary heating power.

(iii) The experiment would enable the control of a fusion-burning plasma over physically significant time scales. With a few second flat top in the ten meg-ampere current range, the compact high-field experiments would provide a long enough pulse length to thoroughly examine alpha particle physics and thermal transients

Table 16.1 Parameters of IGNITOR for a typical inductive operation scenarios. The first column is from Coppi, *et al.* (1999). The second column is the cross-over point (COP) where $P_\alpha \simeq P_{\rm OH}$ at $t = 4.8$ s and the third column (BP) is at $t \simeq 6$ s in the burning plasma.

Parameter	Coppi, *et al.* (1999)	COP	BP
R/a(m/m)	1.32/0.47	1.32/0.46	1.32/0.46
B_T(T)	13	13	13
I_p(MA)	11	12	12
κ_x/δ_x	1.83/0.4	1.83/0.4	1.83/0.4
q_a	3.5	3.5	3.5
β_p	0.26	0.11	0.16
β_T%	1.2	1.6	2.3
β_N	0.67	0.71	0.99
$\langle n_e \rangle$ (10^{20}/m^3)	5	8.9	9.2
$\overline{n}_e$	6.2	8.9	9.6
$n_e(0)$	9.5	8.3	9.7
$n_e/n_{\rm GW}$	0.4	0.5	0.5
$T_e(0)$(KeV)	11.5	12.0	18.9
$T_i(0)$(KeV)	10.5	10.8	15.5
$P_{\rm OH}$(MW)	11	9.8	7.1
$P_{\rm ICRH}$(MW)	0	0	0
$P_{\rm NBI}$(MW)	0	0	0
P_α(MW)	19	10	31
$P_{\rm BREMS}$(MW)	4	12	17
dW/dt(MW)	11	4.8	4.4
W_{th}(MJ)	12	15	20
$Q(Q^*)$	8.6(∞)	5.1 (10)	21.8 (57)
$Z_{\rm eff}$	1.2	1.1	1.1
$\tau_E(s)$	0.62	0.73	0.52
$n\tau_E$(10^{20} m^{-3} s)	6.0	6.5	4.8
$H_{97\rm L}$	1.3	1.4	1.4

in a DT burning plasma. The alpha particle slowing down time is short owing to the high plasma density and relatively low plasma temperature compared with that in conventional large volume-low-field tokamak plasmas.

(iv) The high-field experiment would give first indications about a possible development path to a tritium-poor, reduced neutron production of fusion power. Since the lifetime for the metal walls under the intense neutron fluxes in the DT reactor are unknown, exploring the alternatives without tritium are needed. Ignition is intrinsically a time-evolving event. Thus a discharge of 10 seconds maybe sufficient to answer many questions if the fusion products thermalize in less than one second. In these demonstration experiments, the burning plasma state would last a few confinement times (2-5$\tau_E$with $\tau_E \sim 0.5$ s) and many alpha particle slowing-down times (50-100$\tau_{\rm sd}$), which is adequate, from the physics point of view, to explore the burning fusion plasma state. An example of a density versus temperature phase diagram of the transition to ignition from an ohmically heated plasma with a sufficient continuous fueling rate is shown in Fig. 16.1. The dynamical trajectories of different paths of this ohmically-heated plasma follow from integrating a set of

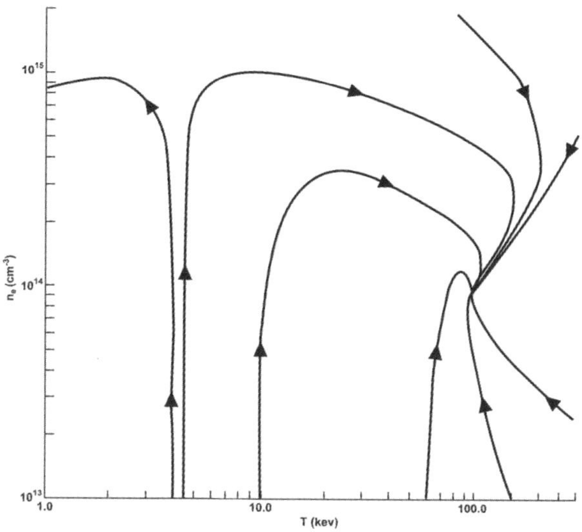

Fig. 16.1 Schematic from integrating a pair of density and temperature equations with ohmic heating, Bohm losses and a suitable constant fueling rate. The $d = 2$ low-order fusion model does not include magnetic field and thus does not include the rapid losses that set in when a beta limit is exceeded.

ordinary differential equations for the density with a source and Bohm loss rate and the energy equation with ohmic heating, Bohm loss rate and Bremmstrahlung radiation. The system has a bifurcation at the ignition temperature of approximately 4 KeV and the dynamics leads to the stable fixed point between 10 and 100 KeV. The simple example shown does not include the magnetic field so that to achieve MHD stability along these curves would take an unrealistically large magnetic field of over 15 T owing to the plasma pressure being about a megapascal at the fixed point. For a pressure limited loss rate the curves would come down into a lower operating temperature. Numerous other practical considerations would need to be included for a realistic scenario.

The concept of ignition is defined [Miyamoto (1989)] as the transition to a new plasma state in which the heating power due to the nuclear fusion alpha particles compensates for all forms of power losses (due to anomalous transport and radiation). It is worthwhile to discuss this concept quantitatively with simulation codes, especially in view of dynamics of the transition given in the Miyamoto definition. Now we show some examples of the passage to a burning plasma.

Consider the power balance equation,

$$\frac{dW}{dt} = P_{\text{ohm}} + P_\alpha + P_{\text{aux}} - P_{\text{loss}} \tag{16.1}$$

where W is the plasma energy content, P_{ohm} is the ohmic power, P_α is the alpha particle heating power, P_{aux} is the auxiliary heating power, and P_{loss} is the loss power, including radiation losses. In the ignited state, $P_\alpha = P_{\text{loss}}$ and the auxiliary power, $P_{\rm aux}$, may be switched off. Thus, ignition is an overheated state with $dW/dt = P_{\rm ohm} > 0$. One may introduce a parameter

$$Q^* = \frac{P_{\rm fus}}{(P_{\rm loss} - P_\alpha)} \tag{16.2}$$

where the fusion power is given as $P_{\rm fus} = (17.6\,{\rm MeV}/3.5\,{\rm MeV})P_\alpha \simeq 5P_\alpha$ for a D-T reacting plasma. With this definition, $Q^* = \infty$ at ignition. Alternatively, using the power balance relation in Eq. (16.1), one may rewrite Eq. (16.2) as

$$Q^* = \frac{P_{\rm fus}}{(P_{in} - \frac{dW}{dt})} \tag{16.3}$$

where $P_{in} = P_{\rm ohm} + P_{\rm aux}$.

A key objective of high-field experiment is to provide multi-megawatts of ohmic power without the symmetry-breaking from an auxiliary power $P_{\rm aux}$.

The fusion power gain parameter Q^* is the measure of fusion power to the input power, taking into account the transient loading/unloading of the plasma energy $W(t)$. One can see that Q^* becomes equal to the widely-used thermonuclear gain parameter, $Q = P_{\rm fus}/P_{in}$, when $dW/dt = 0$. Thus Q^* is the natural extension of Q under non-steady-state operation.

During transient regimes, the difference between Q and Q^* becomes significant and should be kept in mind. For instance, for the simulation example in the first column labeled Coppi of Table 16.1 of an ignited discharge in IGNITOR, one finds $Q^* = \infty$, while $Q = 8.6$.

Table 16.1 gives the parameters of simulation from the IGNITOR Team in the first column. Values at two times called the cross-over point (second column labeled COP) and a burning plasma (third column labeled BP) from the first simulation given in Hu, *et al.* (2003).

A record value of $n_0\tau_E \sim 1 \times 10^{20}\,{\rm m}^{-3}\,{\rm s}$ was reached in ALCATOR C experiments with high peaked density, $n_0 \approx 2 \times 10^{21} m^{-3}$ and a confinement time of about $\tau_E = 0.05\,{\rm s}$. Similar record values of the Lawson parameter were subsequently achieved in high density ohmic discharges in TFTR [Meade, *et al.* (1990)] and FTU [Frigione (2000)].

Experience from ALCATOR C-MOD [Greenwald, *et al.* (1995)] and from FTU [Frigione (2000)] at Frascati indicates that the worst-case discharges in these machines have a energy confinement times that follows the ITER89P L-mode scaling law. A more favorable confinement law called the neo-Alcator and improved ohmic confinement (IOC) scaling describe discharges at slightly lower densities. Regimes of improved confinement at high plasma density have been observed. Ohmic H-modes have been observed in limiter as well as divertor configurations. Enhanced confinement in L-mode operation, such as the Improved ohmic Confinement (IOC) regime observed in ASDEX-U [Bessenrodt-Weberpals, *et al.* (1991)], has now been observed in Alcator C [Greenwald, *et al.* (1984)] at high plasma densities with peaked density profiles. Frigione (2000) reports pellet-assisted ohmic discharges with very good confinement obtained in FTU, with a line-average density $\overline{n}_e \approx 4 \times 10^{20}\,\mathrm{m}^{-3}$ and $\tau_E \approx 0.1\,\mathrm{s}$.

As compared with the record Alcator C experiments, a value of τ_E higher by about a factor of about ten is needed to reach the target value $n_0\tau_E \geq 4\times 10^{20}\,\mathrm{m}^{-3}\,\mathrm{s}$. This factor is only about five if the comparison is made against the recent FTU results.

High-field, high current operation in existing experiments are limited. As a result, the available database is limited compared with the low-field, high $P_{\mathrm{aux}}/P_{\mathrm{ohm}}$ experimental database. A new high-field high-density database is needed. The new database would widen the margin of confidence for extrapolations to high-field burning plasma experiments.

16.2 Turbulent Transport in Ignition Experiments

The plasma physics pioneer, Kadomtsev (1992) stated in regard to understanding the turbulent transport in his monograph (*Tokamak Plasma: A Complex Physical System*) that

> *In the beginning of tokamak research there was the hope that experiments would allow us to determine empirical expressions for the relevant transport coefficients, which would then be explained theoretically.*

The text goes on to explain that up to that time there was no clear explanation, sufficient for predictions, of the turbulent transport. The author feels that the situation has changed now with the advances of the ITG and ETG turbulence theories given in Chapters 13-15. Now with these theories and simulations in hand we are in a position to make complete simulations for the tokamak with turbulent transport. The simulation and methods are explained in more detail in Chapters 18 and 19.

Fusion experiments from the 1990s to 2000 have made it evident that the self-consistent coupling of the turbulence with the plasma profiles plays the crucial role in determining the effective plasma transport and power losses. Thus the plasma fusion Q_{fus} is determined by the drift wave turbulent transport. The strength of this turbulence dependents on the growth rates derived from the profiles of the density and electron and ion temperature profiles along with profiles of the safety factor $q(r,t)$. Thus the tokamak dynamics is like a "weather system" with feedback loops within this complex driven dynamical system. There are bifurcations that arise analogous to those well known in turbulent neutral fluids and atmospheres. As we have already listed in Chapters 13-15, there are a variety of plasma confinements states. In some cases similar system-control parameters result in discharges that take up rather different confinement states. A well-known example occurred for the matched discharges in TFTR, in which one ultimately deviates from the other through the bifurcation to a new state known as Enhanced Reversed Shear (ERS) confinement [Mazzucato, *et al.* (1996)]. These issues of the bifurcations, or sharp transitions, between different plasma confinement states are of fundamental importance to plasma science and to achieving a controlled fusion reaction.

The confinement issue for new fusion devices will be addressed in the future with a wider variety of methodologies including those of real-time control from engineering control theory. The different perspectives and methodologies are required to achieve the confinement required for a practical energy source from nuclear fusion. In the following sections of this chapter, we examine the projections based on multiscale physics in the transport simulation codes named JETTO and MMM for multi-mode modeling.

16.2.1 *Tokamak confinement time based on the L-mode database*

The L-mode database with approximately 3000 discharge entries is summarized by the ITER-97 L-mode formula [Kaye and ITER Group (1997)]:

$$\tau_E^{\mathrm{L97}} = 0.023\kappa^{0.64} R^{1.83} A^{0.06} B_T^{0.03} \overline{n}_e^{0.4} m_{\mathrm{eff}}^{0.2} I_p^{0.96} P_{\mathrm{loss}}^{-0.73}, \tag{16.4}$$

where $P_{\mathrm{loss}} = P_{in} + P_\alpha - dW/dt$ is the loss power, which includes radiation losses, and $P_{in} = P_{\mathrm{OH}} + P_{\mathrm{aux}}$. At ignition, $dW/dt = P_{in}$ and $P_{\mathrm{loss}} = P_\alpha$. For the reference case of Table 16.1, $P_{\mathrm{loss}} = 19\,\mathrm{MW}$ at ignition. Taking the remainder of the parameters from Table 16.1 gives $\tau_E^{97\mathrm{L}} = 0.47\,\mathrm{s}$; thus to obtain the required confinement time of $\tau_E \approx 0.62\,\mathrm{s}$, an enhancement H-factor of 1.3 would be required to reach ignition. Similarly if the older L-mode scaling expression ITER89-P scaling is used, then $\tau_{E89\mathrm{L}} = 0.4\,\mathrm{s}$ and an H factor of 1.55 is required for ignition.

The L97 scaling law has been compared with confinement in the superconducting steady state tokamak Tore Supra [Horton, *et al.* (2000)]. The Tore Supra database has 50 discharges with Fast Wave ICRH that deposits wave power into the electrons, $P_{\mathrm{ICRH}} = P_0 \exp(-r/L_p)$, in the core plasma with $L_p \approx a/5$. Thus, the fast wave ICRH heating experiments are a guide to the simulation of the alpha power heating to the electrons. Tore Supra operates routinely in L-mode and exhibits various levels of enhancement over the ITER-97 L-mode scaling law as a function of density profile peaking. Tore Supra electron heating experiments provide a database for predicting the electron thermal transport in fusion machines. The best Tore Supra discharges have an confinement time enhancement factor of $H = 1.4$ to 1.7 with respect to the ITER-97-L mode formula, which gives a conservative calculation of τ_E.

A key scientific issue that the high-field experiments would address is whether the ohmic power at multi-megawatt levels plays the same role as P_{aux} power in the confinement scaling laws as assumed in the above formulas for the confinement time. The same type of question arises for the alpha heating power P_{alpha} since this would become the dominant form of heating in any burning plasma.

16.2.2 *Improved confinement regimes with peaked density profiles: LOC, SOC and IOC regimes*

Since the high-field tokamak proponents may reach ignition with ohmic heating alone, we start our discussion with the consideration of ohmic confinement modes. The LOC (linear ohmic confinement) mode is a regime of ohmic confinement with a linear increase in the energy confinement time with increasing plasma density. This is also called the neo-ALCATOR scaling regime [Coppi and Mazzucato (1979)]. The LOC regime corresponds to the best confinement mode. With increasing plasma density above a critical density, the LOC regime makes a transition either into a saturated ohmic confinement (SOC) mode or into the L-mode with Goldston confinement scaling. The critical density at which the transition between LOC and SOC regimes occurs is the so-called Shimomura density [Shimomura (1985)], whose expression is

$$n_S \approx \left(\frac{A_i}{2}\right)^{1/2} \frac{B_T}{Rq_\psi} \tag{16.5}$$

where A_i is the mean atomic mass number, B_T is expressed in Tesla, R in meters and n_S in $10^{20}\,\mathrm{m}^{-3}$. The density limit Eq. (16.5) can be re-expressed in terms of plasma current density limit with $\mu_0 j_p = B_p/a$ and $Rq_\psi = B_T a/B_p$ to give the Greenwald limit.

Experiments on ASDEX and on Alcator C MOD have reached a new regime called the IOC that has an internal transport barrier and extents the good confinement of H-modes to higher plasma densities. One method of achieving this new regime is to control the L-H power threshold. Theoretically, this is understood as keeping the ratio of the temperature gradient to the density gradient parameterized in Chapters 3 to 14 by the ratios called "eta-i" and "eta-e" defined by $\eta_i = d\ln T_i/dr,\ d\ln n_i/dr$ and $\eta_e = d\ln T_e/dr,\ d\ln n_e/dr$ sufficiently close to unity. Large values of these dimensionless parameters from flat density profiles lead to strong turbulent mixing from the ITG and ETG instabilities as discussed in detail in Chapters 13 and 14 [Horton (1999); Horton, *et al.* (2000)].

Here, we summarize the improved confinement regimes that can be reached with peaked density profiles. These regimes are listed as follows:

(i) The improved ohmic confinement (IOC) mode, initially discovered in ASDEX [Bessenrodt-Weberpals, *et al.* (1991)].

(ii) The radiative improved (RI) mode, discovered in TEXTOR [Weynants, *et al.* (1999)].

(iii) The P-mode, which is a mode of improved ohmic confinement both in ohmic and in auxiliary heated discharges, first obtained with the help of pellet injection in ALCATOR-C [Greenwald, *et al.* (1984)].

(iv) The supershot, or S-mode, experimentally discovered in TFTR with central neutral beam injection (NBI) and strongly peaked density distributions [Meade, *et al.* (1990)].

(v) Fast Wave ICRH on Tore Supra [Hoang, *et al.* (2000)] and [Horton, *et al.* (2000)].

In the experimental regimes listed above, energy confinement enhancement factors up to 2-3 over that for L-mode have been obtained. Clearly, the relevant question is whether the compact high-field experiments can access these regimes.

The key feature common to these improved confinement regimes is the realization of peaked density profiles. Different injection methods and considerable tokamak operator skills have led to these enhanced confinement regimes. In ASDEX, peaked densities and the IOC mode were obtained after appropriate wall conditioning and decreased gas puffing. In TEXTOR, the transition from the L- to the RI-mode was obtained with impurity seeding. In this way, a strongly radiating layer was established at the edge, with a corresponding decrease of the edge temperature and a steepening of the density gradient deeper inside the plasma. In ALCATOR-C, pellet deposition in the plasma core resulted in peaked density profiles and the establishment of the P-mode. In TFTR, peaked densities were obtained in supershots with central NBI deposition.

Recently, a spontaneous density profile peaking called the I-mode has been produced in Alcator C-MOD H-mode plasmas before the H- to I-mode transition when the additional ICRH resonant layer was placed in the the high field side. Density peaking has also been observed in the same device during ohmic H-modes when the toroidal field was ramped down. These phenomena are not fully understood and are under experimental investigation, with the objectives of extending the duration of the improved energy confinement performance phase and simultaneously increasing the temperature and density peaking.

Experiments on FTU in Frascati [Frigione (2000); Apicella, *et al.* (1997)] with high fields and pellet injections have been successfully carried out. Estimated confinement times of about 90-100 msec have been reached with a magnetic field $B \approx 7\,\mathrm{T}$, a line-average density $\overline{n}_e \approx 4 \times 10^{20}\,\mathrm{m}^{-3}$ and a current $I_p \approx 0.8\,\mathrm{MA}$. The improvement in confinement lasts for more than one confinement time and is recovered after the injection of a second fueling pellet.

From a theoretical viewpoint, peaked density profiles are known to have a beneficial effect on plasma confinement through the quenching of the ion temperature gradient (ITG) driven turbulence. Peaked density profiles reduce the two-dimensionless temperature gradient profile parameters, η_i and η_e, that represent driving terms for the instability of ion and electron temperature gradient modes as shown in Chapters 13 and 14. For flatter density profiles, the ITG stability condition is that L_{T_i}/R exceed a critical value, which is typically not compatible with the overall required temperature difference between the edge and core plasmas. In the 1980s, several theoretical investigations of ITG modes supported the conclusion that the improved confinement in Alcator C pellet fueling experiments [Greenwald, *et al.* (1984)] was due to the simultaneous drop of the η_i parameter defined in Chapter 13. Numerous other machines have shown discharges with improved confinement from density peaking. For example, Tokar, *et al.* (1999) predicts a suppression of the ion thermal flux due to ITG turbulence going from L-mode to the RI-mode in TEXTOR [Samm, *et al.* (1993); Winter, *et al.* (1993)].

The electron temperature gradient driven turbulence has a similar structure to the ITG but operates on the scale of the electron dynamics rather than that of the ions. There are two theoretical forms of the anomalous electron thermal diffusivity that are depressed by high density: the dissipative trapped electron turbulent diffusivity and the short wavelength electromagnetic diffusivity with mixing length proportional to the collisionless skin depth [Horton (1999); Horton, *et al.* (2000)].

For the high-field tokamak parameters given in Table 16.1, ignition would be reached with plasma densities well below the Greenwald density limit. This is a distinct advantage over conventional low field tokamak designs. Density peaking appears crucial to gain access to improved confinement regimes. Strong density peaking in IGNITOR depends on the existence of an inward particle pinch. There are two theoretical models used to explain the particle pinch that occurs in tokamaks. The classical mechanism, known as the Ware pinch: an off-diagonal transport coefficient that is conjugate through the Onsager symmetries [Helander and Sigmar (2002)] to the experimentally verified bootstrap current transport coefficient. An inward convection is required for transport modeling to be consistent with the measured density profiles. The density pinch effect, however, is also modeled through drift wave turbulence driven by the temperature gradients, where again symmetries dictate an off-diagonal transport matrix producing a turbulent inward particle transport [Coppi and Spight (1978); Futatani, *et al.* (2010)].

A cold radiating mantle from the injection of neutral Argon gas, for example, may be advantageous for the formation of peaked density profiles. Indeed, the "R" in the acronym RI-mode refers to the fact that this radiating mantle mode discovered in TEXTOR. The RI-mode was found when a cold plasma mantle concept was being established as a means to solve the reactor thermal exhaust problem [Messiaen, *et al.* (1996)]. Modeling the density profile with the RITM code showed that an essential ingredient for peaking of the density profile is the action of the radiative mantle on the anomalous inward pinch velocity v_{in}, which is taken to have the form $v_{in} = 1/(2T_e)(dT_e/dr)D$, where $D(r)$ is the radial particle diffusivity. This formula for v_{in} may be justified by arguments of profile consistency and more generally justified as a fundamental off-diagonal contribution of the transport matrix for fluctuation-driven particle transport [Shaing (1988)]. This pinch velocity was also found to be essential in explaining the SOC-IOC transition in ASDEX [Becker (1990)]. Direct observation of an inward particle pinch is given by spectroscopic measures during silicon injection in the TEXT device and in Alcator C-MOD. The data is interpreted with a simple diffusion-pinch velocity model of the Einstein type for drift wave turbulence in Horton and Rowan (1994).

16.2.3 *Dynamics of reversed magnetic shear confinement in fast current tramp discharges*

Transient effects can be exploited in tokamaks to produce states with reversed magnetic shear. In ohmically-heated plasmas, a hollow current density $j_\phi(r,t)$ naturally produces a reversed magnetic shear profile $q(r,t)$ during the penetration of the driving toroidal electric field from the central solenoid. This regime is called the current-ramp where the plasma current is rising with $dI_p(t)/dt > 0$. In this regime the plasma current increases from induction loop voltage $2\pi \mathrm{RE}_\phi$ by the rising magnetic flux in the central solenoid. Skin-layers of current driven by rising flux $2\pi\, \mathrm{RE}_\phi$ in the central solenoid are added to the outer surface of the plasma column. The plasma current penetration time based on neoclassical plasma resistivity is slow and can be controlled with programmable currents in the central solenoid. As a consequence, non-monotonic q-profiles form during the current ramp and remain for a significant fraction of the current flat top. Since ignition is expected to be achieved near the end of the current ramp, high-field ohmic tokamaks are well suited for a Reversed Shear (RS) mode of operation [Mazzucato, *et al.* (1996)] of the type observed in JET and in TFTR, among other devices. The PEP mode [Hugon, *et al.* (1992)], which is the JET variant of the RS mode, was obtained with central ICRH and pellet injection. For these modes of operation, energy confinement enhancement factors of H $\approx$ 2-3 were achieved. The physics of confinement enhancement for a non-monotonic q-profile clearly has basic differences with that in a monotonic q-profile. The reversed magnetic shear is one factor in achieving reduced turbulence. Sheared plasma rotation and/or peaked profiles also play important roles in controlling turbulent transport. Thus, although non-monotonic q-profiles occur spontaneously in ohmic heated tokamaks, the detailed shape of the current ramp maybe controlled with the assistance of pellet injection and ICRH heating so as to enhance the RS confinement regimes.

16.2.4 *High-confinement H-modes*

High-confinement H-modes of operation quickly occur in the presence of a magnetic X-point in the poloidal magnetic field structure and heating power levels above a threshold value called P_{LH}. The threshold power P_{LH} is low for single null divertor (SND) geometry with the ion ∇B-drift towards the X-point and the divertor chamber.

The L-H power threshold formula used by the ITER-FEAT physics expert group is

$$P_{\mathrm{LH}}^{\mathrm{IPB99(5)}} = 3.24 \overline{n}_e^{0.62} B^{0.75} R^{0.98} a^{0.81} m_{\mathrm{eff}}^{-1}. \tag{16.6}$$

If an H-mode is accessed, the ITER ELMy H-mode scaling law predicts a confinement time according to the formula

$$\tau_E^{\mathrm{IPB98}} = 0.0562 \kappa^{0.78} R^{1.97} A^{-0.58} I_p^{0.93} B_T^{0.15} \overline{n}_e^{0.41} m_{\mathrm{eff}}^{0.19} P_{\mathrm{loss}}^{-0.69}. \tag{16.7}$$

For the IGNITOR parameters in Table 16.1, these formulae give a power threshold

$$P_{\rm LH}^{\rm IPB98}(\text{IGNITOR}) \approx 19\ \text{MW}. \tag{16.8}$$

This threshold power level would be exceeded by a combination of ohmic and ICRH heating with the resulting confinement time at ignition given by

$$\tau_E^{\rm IPB98} = 0.98\,\text{s} \tag{16.9}$$

from Eq. (16.7). The parameters for the examples in Eq. (16.8) and Eq. (16.9) are from Table 16.1. With this longer confinement time $\tau_E = 0.98\,\text{s}$ the plasma would readily reach ignition.

The poloidal field system is capable of producing an X-point within the vacuum vessel and the RF power would be sufficient for a transition to H-mode confinement. However, the compact, high-field device has a first wall that is not designed to withstand the alpha power flux. The radial heat fluxes would need to be taken to divertor plates with an X-point magnetic flux configuration.

H-mode quality plasmas have sometimes been obtained with the magnetic X-point just outside the vacuum vessel. Indeed, H-modes have been observed in limiter configurations [Weisen, *et al.* (1996)]. There is evidence from Alcator C-MOD of an enhanced L-mode with the magnetic X-point on the vacuum vessel wall [Greenwald, *et al.* (1995)]. In this case, the plasma is prevented from entering the H-mode by reversing the toroidal magnetic field so as to set the ∇B drift away from the single null divertor. For power thresholds well in excess of those that would be required for an L-H transition (if the ∇B drift were in the favorable direction), confinement enhancement factors of 1.2-1.4 relative to L-mode have been obtained [Connor and Wilson (2000)]. In all these cases, a more favorable, more uniform, heat flux distribution to the first wall may result. Thus, improved confinement assisted by real-time X-point control operation is a promising future direction for exploring burning plasmas.

16.2.5 *Turbulent heat diffusivity in the JETTO and MMM transport codes*

Let us examine the question of thermal confinement from the point of view of the heat diffusivity value needed for ignition, relative to models that are known to work well in standard tokamak experiments.

JET plasmas were described by the JETTO model that has either the drift wave models or the empirical Taroni-Bohm [Taroni, *et al.* (1994)] model $\chi_e^{\rm TB}$ for the transport formulas given as follows:

Taroni-Bohm Thermal Transport Coefficient

$$\chi_e^{\rm TB} = 0.33q^2\left(\frac{-a\,\nabla p_e}{p_e}\right)\frac{T_e}{B}. \tag{16.10}$$

ITG formulas in the Multi-Mode Model (MMM) give the major contribution to the total anomalous thermal diffusivity. The fluid model [Nordman (1992, 1993)] is implemented for that part of the Multi-Mode model for the transport by ITG mode. The ITG mode growth rates numerically computed from the Weiland model (Chapter 13.5) are then used to estimate the fluctuation level with the quasilinear approximation. The anomalous transport coefficients, which are functions of the fluctuation level, are thus obtained for ITG modes in the Multi-Mode model. Similar models for resistive-ballooning and kinetic-ballooning modes are included in the MMM model.

The multimode model is described and applied to JET and DIII-D in Chapter 18.

The electrostatic (ES) turbulent electron transport gives

Electrostatic Turbulence

$$\chi_e^{\mathrm{es}} = C_e^{\mathrm{es}} q^2 \left(\frac{R}{L_{T_e}}\right)^{3/2} \left(\frac{\rho_e^2 v_e}{T_e}\right) (|\nabla T_e - (\nabla T_e)_c|)\,, \tag{16.11}$$

where

$$(\nabla T_e)_c \simeq 1.88 \left(\frac{|s| T_e}{qR}\right) \left(1 + Z_{\mathrm{eff}} \frac{T_e}{T_i}\right) \tag{16.12}$$

and $C_e^{\mathrm{es}} \simeq .05$.

The electromagnetic (EM) turbulent electron diffusivity, based on electromagnetic turbulence driven by ETG and collisionless electron skin depth [Kadomtsev (1992)] gives

Electromagnetic Turbulence

$$\chi_e^{\mathrm{em}} = C_e^{\mathrm{em}} \frac{qc^2}{\omega_{pe}^2 \left(L_{T_e} R\right)^{1/2}} (|\nabla T_e - (\nabla T_e)_c|)\,, \tag{16.13}$$

where $C_e^{\mathrm{em}} \simeq 2C_e^{\mathrm{es}}$. The transitional formulas from the ES to the EM regime are given in Hoang, *et al.* (2000b). The transitions give $\chi_e = \min(\chi_e^{es}, \chi_e^{em})$.

Formulae Eq. (16.11)-Eq. (16.13), with $C_e^{\mathrm{es}} = 0.05$ and $C_e^{\mathrm{em}} = 0.1$, have been validated and show good predictive/interpretative performance for the electron power balance in Tore Supra [Hoang, *et al.* (2000); Horton, *et al.* (2000)] and NSTX [Kaye and ITER Group (1997)]. The models are discussed in Chapter 14.

16.3 Simulations with the JETTO and MMM95 Transport Codes

The range of predictions for the IGNITOR parameters in Table 16.1 performance may be assessed by comparing the results of the empirical transport models of Taroni-Bohm and the mixed Bohm-GyroBohm [Erba, *et al.* (1998)] contained in JETTO with the multi-mode model MMM95 [Bateman, *et al.* (1998)].

In these codes the pedestal temperature may be supplied to avoid the complex dynamics of ELMs: for the reference pedestal temperatures we use 2 KeV. More optimistic values are reported for the FIRE simulations with the pedestal $T_e = 3.0$ to 6.0 KeV in a work [Samm, *et al.* (1993)] that compares the fusion performance dependence on the edge temperatures for ITER, AIRES-RS and FIRE.

The simplest, most useful turbulent electron heat diffusivity model is a thresholdless parameterization defined as Taroni-Bohm

$$\chi_e^{\mathrm{TB}} = \alpha_B \frac{q^2(r)a}{B_T}\left(-\frac{dT_e}{dr}\right) \tag{16.14}$$

where the constant α_B is calibrated with certain JET discharges and validated on other JET discharges. The model has good predictive power for JET, TS and JT60U. The theoretical basis from particle-in-cell simulations is given in Kishimoto, *et al.* (1996) and Furnish, *et al.* (1999) for Bohm scaling. The power balance of $P_L = q_r^{\mathrm{TB}} S = P_{\mathrm{aux}}$ with the heat flux q_r and surface area $S = 4\pi^2 Rr$ leads to a global τ_E scaling exponents close to the ITER-97L formula [Kaye and ITER Group (1997)].

The multiple mode transport model (MMM) consistently gives one of the best performances for the ITER profile database. The transport model contains descriptions of the entire standard forms of plasma turbulence including resistive-ballooning, TEM, CTEM, and ITG. To include E_r-shear and a reversed q-profile, the code was modified [Zhu, *et al.* (2000)]. The MMM95-version is calibrated with circular TFTR discharges and tends to be strongly L-mode in character and thus is overly pessimistic for JET, JT60U, and DIII-D. Thus, we expect these two models (JETTO and MMM) to bracket the performance [Nordman (1992, 1993)].

The highest performance discharges are obtained with the mixed Bohm/gyroBohm model which is used in the JETTO model first developed to interpret the JET confinement data. Figure 16.2 shows the time traces from JETTO simulations for several key parameters in an ohmic ignition discharge simulation for the IGNITOR parameters. The time interval where $P_\alpha(t)$ raises above $P_{\mathrm{loss}} = W/\tau_E$ is $t_{ig} = 4\text{-}5\,\mathrm{s}$ shown in Fig. 16.2, comparing a slow and fast current ramp discharge.

16.3.1 *Sawtooth Modeling*

Figure 16.2 shows how the fast current ramp allows the plasma to reach the cross-over time defined by when the alpha particle heating dominates the ohmic heating rate earlier than in the slow current ramp discharge. The top figure shows the alpha heating power and the power loss rate for the slow current ramp, and the lower figure the alpha heating power for the fast current ramp discharge. For the fast current ramp the alpha particle heating dominates shortly after the time $t = 4\,\mathrm{s}$.

This is defined as the cross-over time where one may refer to the plasma as self-heated from the nuclear fusion reactions. This improvement from the fast current ramp is largely due to the movement of the q_{min} surface formation to mid-radius.

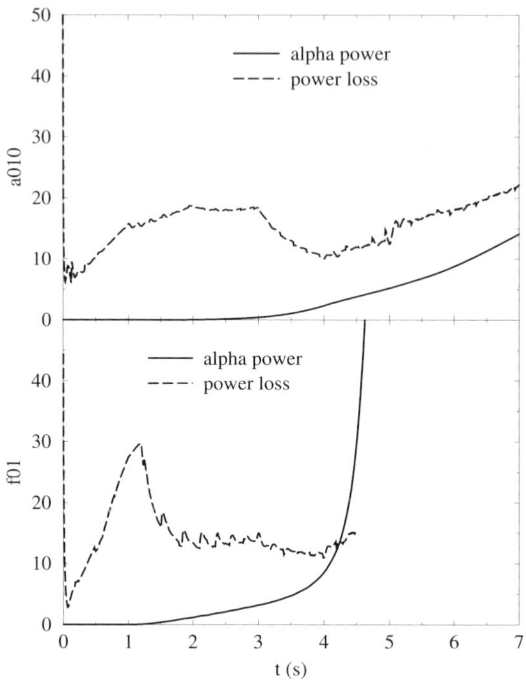

Fig. 16.2 The time evolutions of the alpha power $P_\alpha(t)$ heating the plasma and the power losses from turbulent thermal diffusivity for the compact high-field tokamak with parameters in Table 16.1. The top panel shows the experiment with a slow current ramp and the lower panel the experiment with a fast current ramp where the cross-over point $P_a(t*) > P_{\rm loss}(t_*)$ occurs at $t_* \simeq 4.25\,{\rm s}$ [Hu, *et al.* (2003)].

This $q_{\rm min}$ surface is where the sawtoothing from the internal kink-tearing modes sets in which is now displaced away from the core peaked density and temperature region where the nuclear reactivity is peaked. This spatial displacement and early time cross-over allows the transition to ignition to occur within the ten second pulse of the high field pulsed tokamak discharge.

The simulation's magnetic reconnection from the sawteeth periods and amplitudes at the $q(r,t) = 1$ surface are calculated using the modeling of Porcelli, *et al.* (1996).

The model gives $Q^*_{\rm min} \simeq Q^*_{\rm MMM95} \simeq 5\times$ (5 MW)/(11 MW ohmic)$\simeq$ 2.3 showing for the MMM model that auxiliary heating is necessary to achieve a successful ignition experiment gives this transport model.

16.3.2 *Sawtooth effects*

When the sawtooth is allowed in the simulations, the JETTO simulation model shows that the increase of α heating power and the decrease of the ohmic heating power stop at $t = 4\,{\rm s}$, the time at which the first sawteeth event is triggered. After

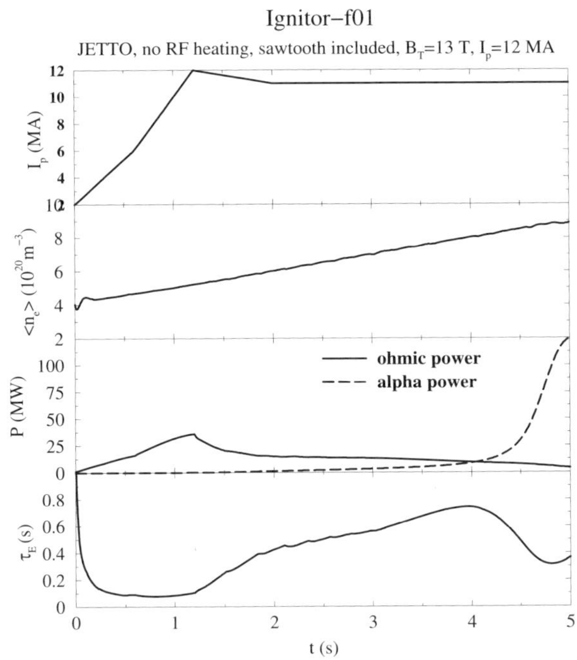

Fig. 16.3 Time history of a) plasma current $I_{\rm p}$, b) volume-average density $\langle n_{\rm e}\rangle$, c) Ohmic power $P_{\rm oh}$ (solid line), α power P_α (dashed line), d) confinement time $\tau_{\rm E}$ in the reference simulation run #ignif01 with peaked density profile and fast current ramp in Fig. 16.2.

that time, the ohmic heating power stays at a constant level while the α heating power remains at a low level ($\sim 3\,$MW), preventing the occurrence of ignition. The corresponding movements of the position of $q_{\rm min} = 1$ surface is shown in Fig. 16.4b. The plasma pressure remains at or below the critical pressure $\beta_{p,c} \simeq 0.13$ [Porcelli, *et al.* (1996)].

16.3.3 *RF power scan*

The effect of auxiliary RF heating on the performance of IGNITOR is evaluated using the JETTO model and the MMM95 models with 10 MW and 20 MW of ion cyclotron RF heating powers applied from $t = 1.0\,$s to $t = 9.0\,$s. The simulation using JETTO model shows that as the RF power increases, the α heating power also increases, but the magnitude of sawtooth oscillation becomes larger at the same time. The fusion gain Q is around 2.2 as the plasma enters steady state beyond 4 s. Hence the overall improvement in Q from heating is not significant. For the simulation using MMM95 model, the RF heating power is shown to greatly increase the α heating power until the sawtooth oscillation is triggered. Both Q^* and Q oscillate about the average value of 5 during the steady state. In simulations with both models, the RF heating enhances the magnitude of sawtooth oscillation.

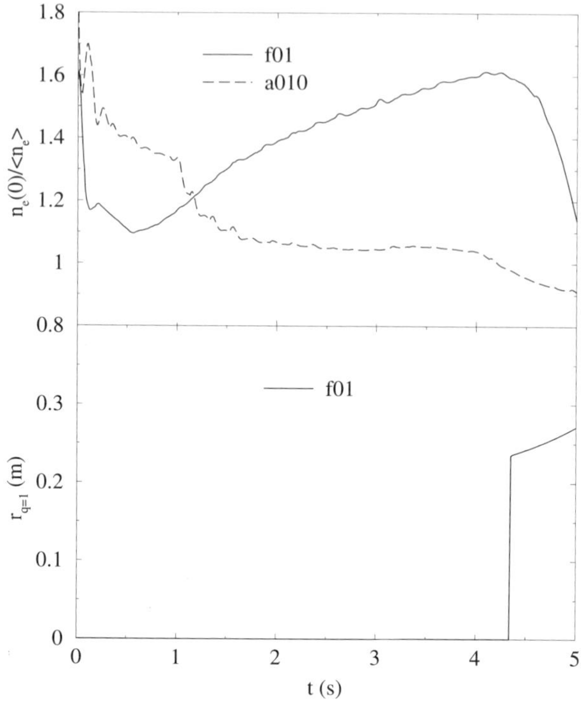

Fig. 16.4 The top frame compares the electron density peaking for the slow and the fast current ramps. The lower frame shows the time evaluation of the $q(r,t) = 1$ surface.

The position of $q = 1$ surface as the function of time for the 20 MW RF heating power cases, one finds the first major sawtooth occurs at $t \sim 3$ s for the JETTO model while the first major sawtooth is not triggered until $t \sim 6$ s for the MMM95 model. In this case, the growth of the $q = 1$ radius is delayed in MMM95 model until $t = 6$ s and then rapidly catches up with that of the JETTO model. The radius $q(r_1) = 1$ saturates at $r_1/\overline{a} = 26\text{cm}/64\text{cm} = 0.4$ and the plasma pressure exceeds the theoretical estimate of $\beta_{p,c} \simeq 0.1 - 0.2$ for stability to the ideal internal kink mode [Bussac, *et al.* (1975); Lütjens, *et al.* (1992)].

The simulations show that (1) control of the sawtoothing and (2) the density profile are critical elements for achieving the goal of ignition. While these two problems have been identified earlier the simulations make quantitative the degree of the problem. By comparing predictions of an empirical-based transport model with a theory-based model, the problem of extrapolating from the present, limited database in terms of high-field/high-density operations to a prediction of ignition is made clear.

The MMM2000 code includes the E_r and reversed magnetic shear turbulent suppression models. The code gives reasonable agreement with two OS-JET discharges and good agreement with two DIII-D central shear discharges [Huysmans,

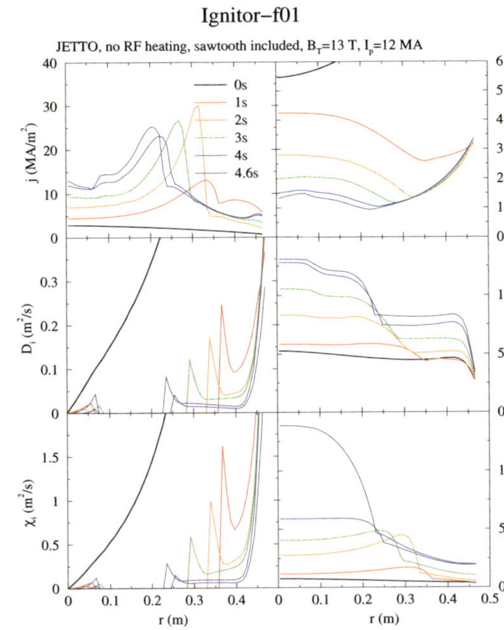

Fig. 16.5 Radial profiles of a) plasma current density j_z, b) safety factor q, c) particle diffusivity D_i, d) electron density n_e, e) ion thermal conductivity χ_i, and f) ion temperature T_i, at six time slices $t = 0, 1, 2, 3, 4$ and s in the reference simulation run #ignif01.

et al. (1997)]. The MMM2000 can yield higher performance than the L-mode confinement of MMM95. In contrast, the empirical formulas of JETTO and the IGNITOR-relevant database taken from the ITER global database give predictions for ignition to occur [Hu, *et al.* (2003)].

Since the high-field compact tokamaks are designed to reach ignition at relatively low plasma temperatures, the projected alpha particle pressure is relatively low. In particular the alpha pressure is lower than the threshold for the excitation of Toroidal Alfvén Eigenmodes (TAE). IGNITOR is designed to operate at such low poloidal beta β_p that neoclassical tearing modes, called NTMs, would only be very weak if even present. The NTM growth rate would be too slow compared to the energy confinement time and thus not influence the approach to ignition. To the extent that a burning plasma regime is achieved and sustained for a time of the order of the skin diffusion time, neoclassical tearing modes might be observed. Thus, neoclassical tearing modes seem to be a minor issue for the high-field compact ignition experiments.

For the fast current ramp shown in Fig. 16.5 there is well-defined off axis minimum in the $q(r)$ profile rise to the improved energy confinement of the RS (reversed shear) regime shown in Fig. 16.4 (top right). Recall that for $q(r) = q_{\min} + 0.5(r - r_m)^2 d^2q/dr^2$ there is a wide gap in the density of rational surfaces so that the drift modes no longer have a significant radial overlapping [Beklemishev and Horton (1992)]. In Figure 16.5 the usual slow current ramp-up in curve (a010), the density peaking or gradient continually forms as the plasma heats and grows in density and temperature. In the case of the fast current ramp the density gradient measured by the peaking ratio increases to 1.6 at $t = 4\,\mathrm{s}$ and then drops. The peaked density profile reduces the parameters driving the ion temperature gradient and the electron temperature gradient modes of Chapters 13 and 14. For the fast current ramp-up Fig. 16.2 predicts ignition.

References

Apicella, M. L., *et al.* (1997). *Nucl. Fusion* **37**, p. 81, doi:10.1088/0029-5515/37/3/I12.

Bateman, G., Kritz, A. H., Kinsey, J. E., Redd, A. J., and Weil, J. (1998). *Phys. Plasmas* **5**, p. 1793, doi:10.1063/1.872848.

Becker, G. (1990). Analysis of energy and particle transport and density profile peaking in the improved ohmic confinement regime, *Nucl. Fusion* **30**, p. 2285, ISSN:0029-5515.

Beklemishev, A. D. and Horton, W. (1992). *Phys. Fluids B* **4**, p. 200.

Bessenrodt-Weberpals, M., McCormick, K., Söldner, F. X., *et al.* (1991). The multiple facets of ohmic confinement in ASDEX, *Nucl. Fusion* **31**, p. 155, ISSN:0029-5515.

Budny, R. V. (2002). Fusion alpha parameters in tokamaks with high DT fusion rates, *Nucl. Fusion* **42**, pp. 1383-1393, PII: S0029-5515(02)55646-5, `http://stacks.iop.org/nf/42/1382stacks.iop.org/NF/42/1382`.

Bussac, M. N., Pellat, R., Edery, D., and Soule, J. L. (1975). Internal kink modes in toroidal plasmas with circular cross-sections, *Phys. Rev. Lett.* **35**, p. 1638, doi:10.1103/PhysRevLett.35.1638.

Connor, J. W. and Wilson, H. R. (2000). A review of theories of the L-H transition, *Plasma Phys. Control. Fusion* **42**, p. R1, doi:10.1088/0741-3335/42/1/201.

Coppi, B. and Spight, C. (1978). *Phys. Rev. Lett.* **41**, p. 551, doi:10.1103/PhysRevLett.41.551.

Coppi, B. and Mazzucato, E. (1979). *Phys. Letts. A* **71**, p. 337, doi:10.1016/0375-9601(79)90071-9.

Coppi, B. and Porcelli, F. (1986). Theoretical model of fishbone oscillations in magnetically confined plasmas, *Phys. Rev. Lett.* **57**, p. 2275, doi:10.1103/PhysRevLett.57.2272.

Coppi, B. and Porcelli, F. (1988). *Fusion Technol.* **13**, p. 447.

Coppi, B., *et al.* (1999). MIT (RLE) Report PTP 99/06.

Coppi, B., Nassi, M., and Sugiyama, L. E. (1992). *Physica Scripta* **45**, p. 112, doi:10.1088/0031-8949/45/2/010.

Erba, M., *et al.* (1998). *Nucl. Fusion* **38**, p. 1013, doi:10.1088/0029-5515/38/7/305.

Escande, D. F., Martin, P., Ortolani, S., Buffa, A., Franz, P., Marrelli, L., Martines, E., Spizzo, G., Cappello, S., Murari, A., Pasqualotto, R., and Zanca, P. (2000). Quasi-single-helicity reversed-field-pinch plasmas, *Phys. Rev. Lett.* **85**, pp. 1662-1665, `http://link.aps.org/doi/10.1103/PhysRevLett.85.1662`.

Frigione, D. (2000). *FTU, High density performance* (International Atomic Energy Agency, Sorrento, Italy).

Furnish, G., Horton, W., Kishimoto, Y., LeBrun, M. J., and Tajima, T. (1999). *Phys. Plasmas* **6**, p. 1227, doi:10.1063/1.873366.

Futatani, S., Horton, W., Benkadda, S., Bespamyatnov, I. O., and Rowan, W. L. (2010). *Phys. Plasmas* **17**.

Greenwald, M., Gwinn, D., Milora, S., Parker, J., Parker, R., Wolfe, S., Besen, M., Camacho, F., Fairfax, S., Fiore, C., Foord, M., Gandy, R., Gomez, C., Granetz, R., LaBombard, B., Lipschultz, B., Lloyd, B., Marmar, E., McCool, S., Pappas, D., Petrasso, R., Pribyl, P., Rice, J., Schuresko, D., Takase, Y., Terry, J., and Watterson, R. (1984). Energy Confinement of High-Density Pellet-Fueled Plasmas in the Alcator C Tokamak, *Phys. Rev. Lett.* **53**, pp. 352-355, doi:10.1103/PhysRevLett.53.352.

Greenwald, M., *et al.* (1995). *Phys. Plasmas* **2**, p. 2308.

Helander, P. and Sigmar, D. J. (2002). *Collisional Transport in Magnetized Plasmas* (Cambridge Monographs on Plasma Physics).

Hoang, G. T., Bourdelle, C., Garbet, X., Aniel, T., Giruzzi, G., Ottaviani, M., Horton, W., Zhu, P., and Budny, R. V. (2000). *Electron transport and improved confinement on Tore Supra* (IAEA, Sorrento, Italy).

Hoang, G. T., Bourdelle, C., and Garbet, X. (2000). *Phys. Rev. Lett.* **84**, p. 4593.

Horton, W. and Rowan, W. (1994). Impurity transport studies in the Texas Experimental Tokamak, *Phys. Plasmas* **1**, p. 901, doi:10.1063/1.870749.

Horton, W. (1999). Drift wave transport, *Rev. Mod. Phys.* **71**, p. 735, doi:10.1103/RevModPhys.71.735.

Horton, W., Zhu, P., Hoang, G., Aniel, T., Ottaviani, M., and Garbet, X. (2000). Electron transport in Tore Supra with fast wave electron heating, *Phys. Plasmas* **7**, pp. 1494-1510, doi:10.1063/1.873969.

Hu, B., Horton, W., Zhu, P., and Porcelli, F. (2003). Density profile control with current ramping in a transport simulation of IGNITOR, *Phys. Plasmas* **10**, p. 1015, doi:10.1063/1.1555623.

Hugon, M., van Milligen, B. Ph., Smeulders, P., *et al.* (1992). Shear reversal and MHD activity during pellet enhanced performance pulses in JET, *Nucl. Fusion* **32**, p. 33, doi:10.1088/0029-5515/32/1/I04.

Huysmans, G. T. A., *et al.* (1997). *Proc. 24th EPS Conf. on Control. Fusion Plasma Phys.* (Berchtesgaden, June; Part IV) p. 1857.

Kadomtsev, B. B. (1992). *Tokamak Plasma: A Complex Physical System* (IOP, Bristol).

Kamada, Y. (1996). Characteristics of and issues regarding combined H-modes, *Plasma Phys. Control. Fusion* **38**, pp. 1173-1188.

Kaye, S. M. and the ITER Confinement Database Working Group. (1997). *Nucl. Fusion* **37**, p. 1303.

Kishimoto, Y., Tajima, T., Horton, W., LeBrun, M. J., and Kim, J.-Y. (1996). *Phys. Plasmas* **3**, pp. 1289-1307.

Laval, G. (2001). Thermonuclear Tokamak Panel Report, CEA, Paris, France.

Lütjens, H., Bondeson, A., and Vlad, G. (1992). Ideal MHD stability of internal kinks in circular and shaped tokamaks, *Nucl. Fusion* **35**, p. 1625, doi:10.1088/0029-5515/32/9/I10.

Mazzucato, E., Batha, S. H., Beer, M., Bell, M., *et al.* (1996). Turbulent fluctuations in TFTR configurations with reversed magnetic shear, *Phys. Rev. Lett.* **77**, p. 3145, doi:10.1103/PhysRevLett.77.3145.

Meade, D. M., *et al.* (2001). Fusion Ignition Research Experiment (FIRE), *Fusion Tech.* **39**, p. 336.

Meade, D. M., *et al.* (1990). *13th Int. Conf. on Plasma Phys. Control. Fusion Research* (Washington, D.C.) **1**, p. 9.

Messiaen, A. M., Ongena, J., Samm, U., *et al.* (1996). Transport and improved confinement in high power edge radiation cooling experiments on TEXTOR, *Nucl. Fusion* **36**, p. 39, doi:10.1088/0029-5515/36/1/I03.

Miyamoto, K. (1989). *Plasma Physics for Nuclear Fusion* (MIT Press, Cambridge, Mass) pp. 8-9.

Nave, M. F. F., Ali-Arshad, S., Alper, B., Balet, B., De Blank, H. J., Borba, D., Challis, C. D., Von Hellermann, M. G., Hender, T. C., Huysmans, G. T. A., Kerner, W., Kramer, G. J., Porcelli, F., O'Rourke, J., Porte, L., Sadler, G. J., Smeulders, P., Sips, A. C. C., Stubberfield, P. M., Stork, D., Reichle, R., Wessom, J. A., and Zwingmann, W. (1995). MHD activity in JET hot ion H-mode discharges, *Nucl. Fusion* **35**, p. 409, doi:10.1088/0029-5515/35/4/I04.

Nordman, H. and Weiland, J. (1992). Concept of marginal stability and recent experimental results from the TFTR tokamak, *Nucl. Fusion* **32**, p. 1653, doi:10.1088/0029-5515/32/9/I14.

Nordman, H. and Weiland, J. (1993). Comments on Ion-temperature-gradient-driven transport in a density modification experiment on the Tokamak Fusion Test Reactor, *Phys. Fluids B* **5**, p. 1032, doi:10.1063/1.860507. [(1992). *Phys. Fluids B* **4**, p. 953].

Porcelli, F., Boucher, D., and Rosenbluth, M. N. (1996). Model for the sawtooth period and amplitude, *Plasma Phys. Control. Fusion* **38**, p. 2163, doi:10.1088/0741-3335/38/12/010.

Samm, U., Bertschinger, G., Bogen, P., Hey, J. D., Hintz, E., Konen, L., Lie, Y. T., Pospieszczyk, A., Rusbuldt, D., Schorn, R. P., Schweer, B., Tokar, M., and Unterberg, B. (1993). Radiative edges under control by impurity fluxes, *Plasma Phys. Control. Fusion* **35**, p. B167, doi:10.1088/0741-3335/35/SB/013.

Shaing, K. C. (1988). Neoclassical quasilinear transport theory of fluctuations in toroidal plasmas, *Phys. Fluids* **31**, p. 2249, doi:10.1063/1.866626.

Shimomura, Y. (1985). JAERI Report JAERI-M 87-080.

Taroni, A., Erba, M., Tibone, F., and Springmann, E. (1994). Global and local energy confinement properties of simple transport coefficients of the Bohm type, *Plasma Phys. Control. Fusion* **36**, p. 1629, doi:10.1088/0741-3335/36/10/003.

Tokar, M. Z., Jaspers, R., Weynants, R. R., Koslowski, H. R., Krämer-Flecken, A., Messiaen, A. M., Ongena, J., and Unterberg, B. (1999). Evidence of suppression of ITG-instability in the radiatively improved mode in TEXTOR-94, *Plasma Phys. Control. Fusion* **41**, p. L9, doi:10.1088/0741-3335/41/3/001.

Weisen, H., Hofmann, F., Dutch, M. J., *et al.* (1996). Ohmic H-modes in the TCV tokamak, *Plasma Phys. Control. Fusion* **38**, p. 1137, doi:10.1088/0741-3335/38/8/006.

Weynants, R. R., Messiaen, A. M., Ongena, J., Unterberg, B., Bonheure, G., Dumortier, P., Jaspers, R., Koch, R., Koslowski, H. R., Krämer-Flecken, A., Mank, G., Rapp, J., Tokar, M. Z., Van Wassenhove, G., Biel, W., Brix, M., Durodié, F., Esser, G., Finken, K. H., Fuchs, G., Giesen, B., Hobirk, J., Hüttemann, P., Lehnen, M., Lyssoivan, A., Mertens, P., Pospieszczyk, A., Samm, U., Sauer, M., Schweer, B., Uhlemann, R., Van Oost, G., Vandenplas, P. E., Vervier, M., Philipps, V., Waidmann, G., and Wolf, G. H. (1999). Overview of radiative improved mode results on TEXTOR-94, *Nucl. Fusion* **39**, p. 1637, doi:10.1088/0029-5515/39/11Y/303.

Winter, J., *et al.* (1993). *Phys. Rev. Lett.* **71**, p. 1549, doi:10.1103/PhysRevLett.71.1549.

Zhu, P., Bateman, G., Kritz, A. H., and Horton, W. (2000). *Phys. Plasmas* **7**, p. 2898, doi:10.1063/1.874140.

Chapter 17

Rotational Instabilities

17.1 Rotation in Cylinders

Instabilities driven by rotation and sheared rotational flows occur in the MHD description of plasmas and are a thoroughly developed subject in the context of space, solar and astrophysical plasmas and laboratory plasmas. In this chapter we restrict attention to the drift wave regime and ask when is rotation a stabilizing or destabilizing influence on the drift waves.

One finds in the literature many varied results for this question: the answer depends on the particular rotational profiles with respect to the density and temperature profiles and the magnetic topology. Experimental results are equally varied. To use sheared flows as a principle for reducing transport, one must specify, in considerable detail, the plasma profiles and the level of the rotational shear. Indicative of this difficulty is that Waltz and others often use a "rule of thumb" to assess the change in the growth rate from the effect of flow shear rather than first principle eigenmode calculations [Waltz, *et al.* (1994)]. Chapter 5 shows examples in the simplest regime of a cylindrical plasma with an externally applied radial electric field in a well diagnosed laboratory plasma. In those experiments, simulation and theory, two regimes are found: one with a suppression of the radial transport by what is defined as a shear flow transport "barrier" in the fusion literature and one of destabilization through the Kelvin-Helmholtz form of the plasma instability in a stronger sheared flow regime. The problem is also difficult to simulate due to the extreme twisting up of the convective structures in the sheared flow. Following the lead of the neutral fluid turbulence community, one must also consider the initial value problem and examine the transient growths of initial disturbances that can bootstrap their way into nonlinear structures.

Analyzing the sheared flow through the modal approach, commonly used by the neutral fluid turbulence community, reveals that the linear operators describing sheared flow are non-normal and that the non-normality results in transient growth and mode coupling. In cases where the nonlinear feedback is positive, this linear transient growth can produce self-sustaining vortices that affect the drift wave turbulent transport [Kim, *et al.* (2006)]. The momentum in the rotational flows is also transported by drift waves through the turbulent $\boldsymbol{E} \times \boldsymbol{B}$ convection. Thus the turbulent plasma transport problem becomes generalized to include the turbulent transport of the momentum. In hydrodynamics the turbulent flux of momentum is given the name of Reynolds stress tensor. When the turbulence is driven by temperature gradients then the Reynolds stress tensor can create a net circulation of fluid flows.

In experiments at low Prandtl number $\mathrm{Pr} = \mu/\chi$ of the Benard convection, Krishnamurti and Howard (1981) showed the onset of a large vortex of sheared flow once the thermally driven transport exceeded a critical level. Low-order models of the plasma physics equivalent of this situation have been developed in a number of magnetic confinement theories including Diamond, *et al.* (1997), Horton, *et al.* (1996), Hu and Horton (1997), Sugama and Horton (1998), Kishimoto, *et al.* (2000), and Uzawa, *et al.* (2009). In magnetic confinement in tokamaks the interest has been driven by the discovery of the transition from the L-mode to the H-mode of plasma confinement when the heating power exceeds a critical level. The first L-H transitions were observed by Wagner, *et al.* (1982) in ASDEX and then subsequently in most major tokamak facilities. These transitions are consistent with a drop in the turbulence level and the onset of strongly sheared rotation in the edge region of the plasma. In Chapter 13.4 these types of transitions are called edge transport barriers. Transport barriers in core plasma are analyzed in Chapter 18 for the JET and DIII-D plasmas.

In plasmas the rotation is associated with the $\boldsymbol{E} \times \boldsymbol{B}$ rotation from a radial electric field across the confining magnetic field. For straight, or nearly straight, cylindrical geometry, as in the LAPD in Chapter 3, the Tandem Mirror central cells, as in Hooper, *et al.* (1982) and Cho, *et al.* (2005), and in the Field Reversed Configuration the rotational modes are often unstable in themselves and thus add to the wave-induced transport processes. An analysis of these linear geometries is given since the theoretical formulations are much clearer and simpler. Chapter 3.3 explores the situation in cylindrical geometry. The situation in the toroidal plasma is more complicated since there are two components of the flow across the magnetic field. The toroidal rotation is in a direction of rotational symmetry and is arbitrary until the heating and transport processes themselves are included in the transport equations. The largest toroidal rotations are up to a substantial fraction of the sound speed, that is with speeds of order a 100 km/s, which are common in the neutral beam driven plasmas. For beam (NBI) driven plasmas the degree of imbalance between the two rotational directions is often varied which is described as balance or unbalanced injection.

In tokamaks the plasma current defines a preferred direction around the torus. Thus, the rotation is described as co-rotation in the direction the current and counter rotation in the opposite direction. One can show that for up-down symmetric toroidal flux surfaces there is a symmetry in the transport of angular momentum in the co- and counter direction. But, in single null divertor machines like JET and

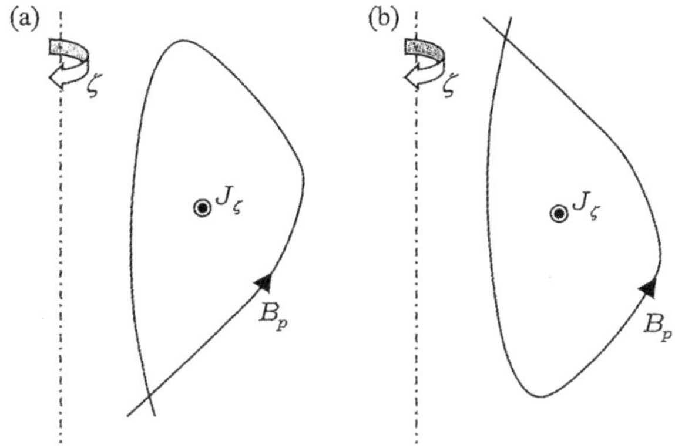

Fig. 17.1 Example of single null divertor configuration with broken top-down symmetry [Parra and Catto (2010)].

ITER, the top-to-bottom symmetry is broken with the consequence there is net radial flux of toroidal angular momentum a torque force at zero rotation speed driving the plasma into rotation spontaneously. One sees the preferred direction from the cross product of the vector to the lower divertor with vector of the plasma current. Examples of such injection experiments are given in Chapters 13 and 15 for TFTR, JET and JT60-U. More recently, the spherical tokamaks with small aspect ratios of NSTX and MAST have reported rapidly rotating toroidal plasmas from unbalanced NBI injection.

The mechanisms of nonlinear interaction of external helical fields [Lazzaro, *et al.* (2002)] with a rotating plasma are investigated analyzing the results of recent systematic experiments on the Joint European Torus (JET) [Gibson and the JET Team (1998)]. The equilibrium neoclassical transport equations for these plasma is found in Sugama and Horton (1998), Hirshman and Sigmar (1997), Helander and Sigmar (2005), and Newton and Helander (2006), and the symmetry arguments are given in Peeters and Angioni (2005) and Parra and Catto (2010). There is a generalization of the Grad-Shafranov equilibrium equation that gives a large class of equilibrium solutions to the rotating ideal MHD plasmas [Goedbloed, *et al.* (2010)] and Sugama and Horton (1997 1998). In the torus the population of mirror trapped ions in the outside weak magnetic field regions gives rise to a frictional drag on the plasma rotation in the poloidal direction. This trapped ion population friction results in a low poloidal drift speed than can often be approximated as vanishing with respect to the high speed toroidal rotation. The equation for the residual poloidal rotation rate is specified by the ion temperature gradient and the ion-ion collisionality regime of the plasma as derived in Hinton and Hazeltine (1976), Helander and Sigmar (2005), Hirshman and Sigmar (1997), Balescu (1988), and Sugama and Horton (1997, 1998).

The generalization of the Grad-Shafranov equilibria required when there is strongly peaked neutral beam injection [Cooper, *et al.* (1980)], strong LHCD [Hoang, *et al.* (2009)] heating, or a strong runaway electron component to the plasma [Knoepfel and Spong (1979)] are complex as found in these references.

17.2 Rotation in Cylindrical Plasma

For a uniform axial magnetic field $\boldsymbol{B} = B\widehat{\boldsymbol{e}}_z$ and a radial electric field $E_r(r)$ the plasma rotates at the angular frequency $\Omega(r) = -E_r/rB$ with respect to the laboratory frame in which the electric field is specified. In the frame moving with the plasma at this rotational velocity $\Omega(r)$ the electric field vanishes and there is a centrifugal force radially outward given by $F_r(r) = n_i m_i\, v_\theta^2/r$ and there is the Coriolis force $\boldsymbol{F}_{\rm cor} = -n_i m_i 2\boldsymbol{v} \times \boldsymbol{\Omega}$. The centrifugal force adds to the effective gravity term introduced in Chapter 4 for describing the effect of the curvature of the magnetic field. So we expect and find instability of the interchange type analyzed in Chapter 4 when the rotational velocity is large and density profile is monotonically decreasing. When the profile is hollow, however, large rotation velocities are stabilizing as shown in experiments in Novosibirsk and University of Maryland.

Now we consider the derivation of the drift wave equations in the presence of the radial electric field. First we consider electrostatic dynamics in a low-dimensionless pressure, $\beta \ll 1$, plasma. Taken into account are the cross-field currents driven by the electrostatic potential through the divergence of the collisionless ion momentum stress tensor [Rosenbluth, *et al.* (1962); Rosenbluth and Simon (1965)]. The eigenmode equation for the rotating plasma with an effective radially outward acceleration $g(r)$ from magnetic curvature and the centrifugal force is

$$\nabla \cdot \left[n \left(1 - \frac{\omega_{*i}(r)}{\omega} \right) \nabla_\perp \varphi \right] + \left[\frac{1}{r}\frac{d}{dr}\left(\frac{n\omega_{*i}(r)}{\omega} \right) + \frac{m^2 g}{r^2\omega^2}\frac{dn}{dr} \right] \varphi = 0. \tag{17.1}$$

The eigenvalues of the higher wavenumber modes $m \gg 1$ are found by using the local approximation to reduce Eq. (17.1) to the local algebraic dispersion relation

$$k_\perp^2 \omega\, (\omega - \omega_{*i}) - k_\theta^2\, \frac{g}{n}\frac{dn}{dr} = 0 \tag{17.2}$$

with $k_\theta \gg k_r$.

In the absence of the FLR term ω_{*i} the dispersion relation is the same as that for internal gravity waves in hydrodynamics. Stably-stratified layers in hydrodynamics have the buoyancy frequency $N^2 = -g(d\ln n/dr) = g/L_n$ [Gill (1982)].

From Eq. (17.2) one sees that the ion diamagnetic frequency ω_{*i}^2 is stabilizing against the interchange term from $g d\ln n(r)/dr$. Thus, from the condition of a negative discriminant of the quadratic equation for ω, one finds that the instabilities are limited to mode numbers below a critical mode number $m < m_{\rm crit}$ determined by ω_{*i} relative to the growth rate at $\omega_{*i} = 0$.

To go beyond the local approximation the eigenvalue problem must be solved. The problem is generally numerical and requires the use of a "shooting code" that repeatedly integrates the second-order differential equation from the $r = 0$ to $r = r_{\text{edge}}$ with appropriate boundary conditions at both end points [Gladd and Horton (1973)]. The exception occurs for a rigidly rotating plasma with $\Omega = \text{const}$. For rigid rotation, there is an analytic solution of Eq. (17.1).

17.2.1 *Solid body rotation eigenmodes*

For the case of a rigidly rotating plasma with $\Omega = \text{const}$, the electric field is given by

$$E_r = -r\Omega B \tag{17.3}$$

with a plasma potential Φ given by

$$\Phi(r) = \frac{1}{2}r^2\Omega B \tag{17.4}$$

that is a quadratic potential well or hill, depending on the direction of rotation. For example, core ohmic tokamak plasmas have $E_r < 0$ with rotation in the electron diamagnetic direction.

For the rigid rotation potential, the Boltzmann phase space density for the ions is a Gaussian density profile

$$n(r) = n_0 \exp\left(-\frac{r^2}{a^2}\right). \tag{17.5}$$

The electrons adjust to have charge neutrality $n_e = n_i$ with $a^2 = k_B T_i/eB\Omega$ for $\Omega > 0$. In mirror machines the electrons rapidly leave the ends of the machines so that $E_r > 0$ and the rotation is in the ion diamagnetic direction.

For this equilibrium with solid body rotation the differential Eq. (17.1) reduces to a form that can be solved with confluent hypergeometric function $F(a, c, x)$ or the Whittaker function $W_{\mu,\kappa}(x)$. The details are given in Rognlien (1973) and Mikhailovskii (1992). The eigenvalues and eigenfunctions for solutions of Eq. (17.1) for rigid body rotation equilibrium are given by a transformation to the Whittaker functions

$$\begin{gathered}\varphi = \frac{e^{\frac{r^2}{2a^2}}}{r}\, y\left(\frac{r^2}{a^2}\right) = \frac{e^{x/2}}{x^{1/2}}\, y(x) \\ y = C_1 W_{\frac{(1+\nu)}{2},\frac{m}{2}}(x) + C_2 W_{\frac{(1+\nu)}{2},\frac{-m}{2}}(x)\end{gathered} \tag{17.6}$$

following Rognlien (1973). The boundary condition gives the eigenvalues

$$\nu = m + 2n \qquad n = 0, 1, 2, \ldots \sim \frac{1}{\Gamma\left(\frac{1}{2} + \frac{m}{2} - \frac{1+\nu}{2}\right)} \longrightarrow 0.$$

Using these eigenvalues, the dispersion relation becomes

$$\omega^2 - \omega_{*i}\omega\left(1 - \frac{1}{\nu}\right) + \frac{m^2 g_0}{a(\nu)} = 0 \tag{17.7}$$

where we define g_0 by $g = r/a\, g_0$, $g_0 = aV_T^2/L_t^2$. The distance between the minimum and maximum of the magnetic field is characterized as L_t and the plasma radius is a.

The roots of Eq. (17.7) are stable for

FLR Stability Condition

$$\frac{m-1}{\sqrt{m}} > \frac{a^2}{L_t \rho_i} \sim \text{of order unity.} \tag{17.8}$$

This FLR stabilization condition is a famous method for stabilizing mirror machines and field-reversed confinement systems, as shown in Eq. (17.8), however, the $m = 1$ mode behaves differently from the $m = 2, 3$ and higher-order azimuthal modes. The $m = 1$ modes typically appears in a nonlinear state of a "wobbling" of the plasma column mode in the central cell of a long mirror machine. This topology of the $m = 1$ mode is a displacement of the core plasma. The $m = 1$ mode appears in the simulations and laboratory as coherent displacement of the entire plasma rotating at the mean rotation rate Ω of the plasma. As the rotation rate increases above a critical value the plasma accelerates into the wall.

Here we take the plasma wall as a perfect conductor giving Dirichlet boundary condition $E_\theta(r = b) = 0$. Fitzpatrick, *et al.* (1993) and others consider resistive walls which are more realistic but give rise to resistive wall modes.

17.2.2 *Rotational instabilities in central cells of mirror machines and cylindrical plasmas*

From the analysis in the previous Section 17.2.1 we derive the eigenfunctions and eigenvalues for typical laboratory magnetic confinement plasmas. The clearest data for fusion grade plasma comes from the central cells of Tandem Mirror Machines which are hot mirror confined plasmas. Examples are from GAMMA 10 at the University of Tsukuba, the TMX at LLNL from 1970-1980 and the FRC experiments at University of Washington and LASL in the 1980-2000 period.

For an axisymmetric tandem mirror the plasma is stabilized by having sufficient plasma pressure in the end-cells to have the total line integral of the pressure gradient times the magnetic radius of curvature throughout the length of the machine to have the MHD stable value. The condition on how much plasma pressure and injected power is required to satisfy this condition is given in Berk and Pratt (2011).

Figure 17.2 shows the lowest radial eigenmodes for the azimuthal mode numbers $m = 1$, 2 and 3 in a collisionless plasma with a gaussian density profile and constant temperatures. The profile of the annular rotation frequency $\Omega(r)$ is first taken as constant and then consider taken as monotonically decreasing from a core value of $\Omega(0)$ to an edge value of $\Omega(r = b)$.

In Fig. 17.2 there is a conducting wall at $r = b$ at which $\phi_m(r = b) = 0$ since otherwise the wall would have an tangential electric field $E_\theta = \mathrm{Re}[(-im/b)\varphi(b)]$ contrary to its being a perfect conductor. If the wall is thin and the resistivity of the wall is finite, then we must modify this boundary condition as in Fitzpatrick, *et al.* (1993) to be that of a resistive wall. The skin time of the wall compared to the growth/damping rate of the plasma wave then becomes a key parameter in the problem. Here we give the results for the perfectly conducting wall meaning the skin time is very short compared with the growth time.

Figure 17.2 shows the eigenfunctions and Fig. 17.3 shows the frequencies $\omega(m, b/a)$ and growth rates $\gamma(m, b/a)$ for the lowest azimuthal mode numbers in a conducting wall cylinder with a constant temperature gaussian density hot plasma. The $m = 1$ wobble mode has a qualitatively different behavior from the $m = 2$ and 3 modes. The $m = 2$ and 3 modes are unstable with the wall well away from the plasma, while the $m = 1$ mode is a stable oscillation in this case. When the wall is moved closer to the plasma by decreasing the wall radius b divided by the radius a of the plasma profile, the $m = 2$ and 3 modes are stabilized, while the $m = 1$ mode is destabilized. This behavior is confirmed in the laboratory experiments of Ilic, *et al.* (1973) in FRC and other plasma experiments.

The $m = 1$ mode is a displacement of the plasma towards the conducting wall with zero frequency in the rest frame of the plasma. Thus we know from electrostatics that there is an induced image charge in the conducting wall. There is then an attractive force between the charges in the conductor and those in the plasma that give rise to the motion of the plasma toward the wall. The effect saturates at a certain relatively low level of the displacement and so a new nonlinear state of

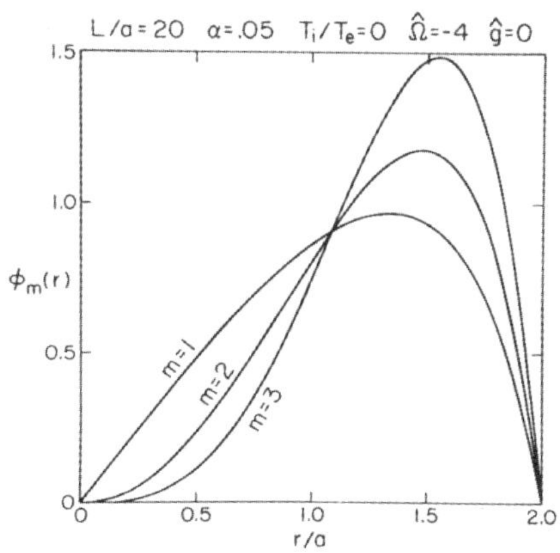

Fig. 17.2 Radial eigenfunctions for the rigidly rotating plasma with hot ions. The density profile is gaussian so the electron diamagnetic frequency is constant in radius. In this example the $\Omega = -4$ means the rotation is from four times the electron diamagnetic frequency and in the ion diamagnetic direction. The choice of positive E_r from a positive plasma potential with $\Phi(r = 0) \sim 4T_e$ is characteristic of open-ended systems and Tandem Mirror plasmas. In open systems electrons rapidly leave on the open magnetic field lines until the plasma potential becomes sufficiently positive to slow their loss rate to balance that of ions. This typically requires $\Delta\phi = \phi(0) \sim 4T_e$.

the rotating plasma is created. This type of nonlinear rotating displaced state was routinely observed in the central cells of the TMX experiment at LLNL [Hooper, *et al.* (1982)] and in the FRC experiments. The dangerous rotational instabilities with respect to magnetic fusion experiments then starts with the $m = 2$ instability. Many experiments end in the crash created by the fast-growing $m = 2$ instability.

Figure 17.3 gives an overview of the instabilities for the central cell of the tandem mirror. The analysis shows that at higher mode numbers the growth rate comes back to zero. This effect is due to the large ion gyroradius effect producing a stabilizing charge separation from the electrons. The effect is contained in the dispersion relation through the B^2 term in the discriminant of the quadratic equation. The B-term arises from the frequency ω_{*i} in Eq. (17.7). This key stabilizing effect is called FLR stabilization for the finite-Larmor-radius stabilization from charge separation. The effect is contained in Chapter 2 in the Bessel functions I_0 and I_1 in, for example, Eqs. (2.10)-(2.20).

17.2.3 *Shear Alfvén wave stabilization of the rotational instabilities*

The eigenvalue equation for mode number m written out in detail is

$$\begin{aligned}\frac{1}{r}\frac{d}{dr}&\left[rn\left(\omega-\omega_E\right)\left(\omega-\omega_E-\omega_{*i}\right)\frac{d\varphi}{dr}\right]\\&+\left[\left(m^2-1\right)\left(\omega-\omega_E\right)\left(\omega-\omega_E-\omega_{*i}\right)n+\left(r\Omega^2+g(r)\right)\frac{dn}{dr}\right]\varphi=0.\end{aligned} \tag{17.9}$$

This form of the eigenmode equation makes clear that the m=1 mode behaves differently from the m=2 and 3 modes as shown in Fig. 17.3.

Now, the radial plasma displacement ξ, derived within the approximation of $\boldsymbol{E}+\boldsymbol{v}\times\boldsymbol{B}=0$ motion bends the magnetic field lines which launches shear Alfvén waves with $\omega^2 = k_\parallel^2 v_A^2$. The dispersion relation with Alfvén waves for $a \gg c/\omega_{pi}$ becomes

$$\omega\left(\omega-\omega_{*i}\right)-k_\parallel^2 v_A^2+\frac{g_{\text{eff}}}{r_p}\frac{k_\theta^2}{k_\perp^2}=0 \tag{17.10}$$

where $g_{\text{eff}} = r\Omega^2 + g(r)$; where the $g(r)$ is from the curvature of the magnetic field lines. In the region close to the axis ($r = 0$) of a cylindrically symmetric field $\boldsymbol{B} = B_r(r,z)\boldsymbol{e}_r + B_z(r,z)\boldsymbol{e}_z$ one derives that $g(r) = rC_s^2/L_t^2$. Thus, both the magnetic curvature of the field lines and the centrifugal force from the rotation combine to form the magnetic form of the Rayleigh-Taylor instability in the plasma column.

The instability now requires that the wave function be sufficiently flute-like that

$$\frac{g}{r_p}\frac{k_\theta^2}{k_\perp^2} > k_\parallel^2 v_A^2 + \frac{\pi}{4}\omega_*^2 \tag{17.11}$$

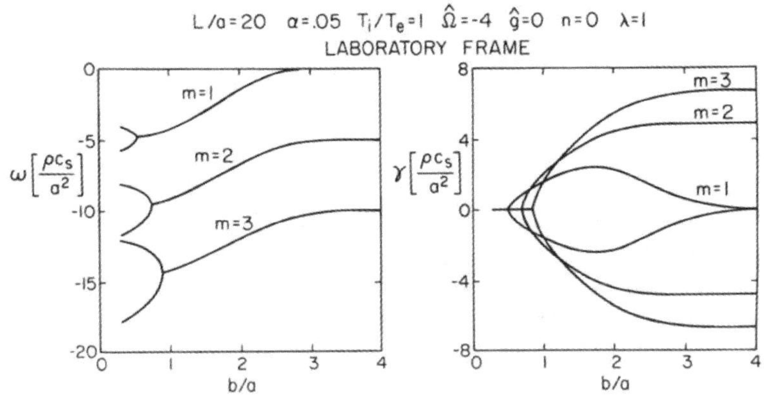

Fig. 17.3 For the case shown in Fig. 17.2 except the ion temperature is taken equal to the electron temperature and the distance b to the conducting wall relative to the plasma radius a of the plasma is taken as the variable parameter. The left panel shows how the two stable frequencies at sufficiently small b/a bifurcate into the complex plane as the wall is moved outward. The first mode to go unstable is the $m = 1$ mode whose frequency subsequently drops to zero as the wall is move far out from the plasma. This is due to a rigid displacement being a symmetry of the equilibrium in the absence of the wall. Next the $m = 2$ and $m = 3$ modes are destabilized and their growth rates shown in the right panel increase monotonically to the "no wall" values. By using coils with the $m = 2$ and $m = 3$ symmetries on the walls a stabilizing feedback control system can be designed to control these instabilities.

is satisfied for the $m \geqq 2$. The boundary conditions at the ends of the flux tubes determine when this line bending stabilizes the rotational instabilities. The condition of a vanishing plasma displacement at the ends of the flux tube is called the "line tying" boundary condition and can eliminate the flute modes that are most strongly unstable. The boundary conditions at the ends of actual machines, or flux tubes in the magnetosphere and solar corona, are complicated as given in Berk, *et al.* (1983) and for the ambipolar trap by Dimov (2005).

The details of the analysis of the eigenvalue analysis with the line-bending stabilization is given by Janssen (1982). The stability diagram of $k_{\|} v_A/\omega_{*i}$ versus Ω_E/ω_{*i} is given for the $m = 2$ mode and lowest-order radial eigenmode. A nonlinear perturbation calculation for small deviations into the unstable domain leads to a nonlinear oscillator equation that shows both shear angular velocity profile and flattened density profile developed in the nonlinear saturated state.

Hooper, *et al.* (1982) report that there are nondisruptive $m = 1$ oscillations at 7 kHz and $m = 0$ oscillations at 13 kHz in the TMX experiments. The $m = 0$ mode is an ion acoustic mode with a phase shift of 180° degrees at the two end cells. The $m = 1$ mode characteristics change with changes in the density profile and are marginally stable drift mode given in Eq. (17.6)-Eq. (17.8). This $m = 1$ mode is a nonlinear wobble of core plasma. The eigenmode is shown in Fig. 17.3 and the dependence of the frequency on the conducting all radius relative the density profile radius is shown in Fig. 17.2.

Table 17.1 Summary of stability condition for rotating plasmas.

Summary of Rotational Stability		
Rotation	$g \rightarrow$	$g + v_\theta^2/r$
	over L_t	over length L_c
FLR stabilization	$\frac{1}{4}\omega_{*i}^2 > \frac{g}{r_p}$	
line bending stabilization	$\frac{g}{r_p} < k_\parallel^2 v_A^2$	
parallel wavenumber $k_\parallel L_c = \pi$	$\beta < \pi \frac{r_p R_c}{L_c^2}$	
Beta stability limit	$\frac{1}{\beta} \geqq \frac{1}{\beta_{\text{balloon}}} + \frac{1}{\beta_{\text{rot}}}$	
	$\beta_{\text{rot}} = \pi \frac{r_p r}{L_c^2} \frac{a^2}{\rho^2} - \kappa \left(\frac{ra}{L_c \rho}\right)^2$	

A summary of the key features of the rotational instabilities is given in Table 17.1.

17.2.4 *Wave energies and the variational principle for determining plasma stability*

Here we calculate the energy W released from the plasma by the displacement ξ_n. One derives the energy δW from the line bending displacement

$$\frac{\delta B_x}{B} = k_\parallel \xi_n \tag{17.12}$$

and the release of the plasma thermal energy

$$\delta W = \frac{1}{2} \int \left[\xi_n^2 \left(-g \frac{d\rho_m}{dr} \right) + \frac{\delta B_\perp^2}{4\pi} \right]. \tag{17.13}$$

The line-bending condition in Eq. (17.11) is now clear as the condition that the total potential energy δW of the disturbed plasma increases. For flute modes $k_\parallel = 0$ the system is unstable.

A summary of the stability conditions is given in Table 17.1. Goedbloed, *et al.* (2010) give a detailed analysis of the shear rotational flow stability in the MHD framework in pages 48-123. Frieman and Rotenberg (1960) develop the Lagrangian MHD formulation of the rotational instability.

17.2.5 *Rotation in tokamaks: transport and stability*

For tokamaks with up-down symmetry, there are symmetries in the transport equation for the angular momentum of the plasma. The up-down reflection symmetry allows one to conclude that in the absence of rotation the radial transport flux of angular momentum vanishes. Thus, there needs to be a seed or external driver in such up-down symmetric systems for the angular momentum flux to develop. This is not the case in the single null divertor configuration which has a strong-weak magnetic field asymmetry in the vertical direction. In this case angular momentum flux can occur spontaneously with the heating of the plasma. An explicit quasilinear calculation is given in Peeters and Angioni (2005, 2007) and a general neoclassical calculation in Parra and Catto (2010).

Rotational effects are particularly strong in the neutral beam heated spherical tokamaks NSTX and MAST [Field, *et al.* (2009)]. On MAST high resolution Doppler shift carbon impurity poloidal flow velocities are measured. The technique is called Doppler CXRS spectroscopy from charge exchange recombination radiation lines yielded from $C^{+5}(n = 8 \rightarrow 7)$ transitions at 529 nm wavelengths.

Neoclassical transport processes of electrons and ions are investigated in detail for toroidally rotating axisymmetric plasmas with large flow velocities on the order of the ion thermal speed. The Onsager relations for the flow-dependent neoclassical transport coefficients are derived from the symmetry properties of the drift kinetic equation with the self-adjoint collision operator. The complete neoclassical transport matrix with the Onsager symmetry is obtained for the rotating plasma consisting of electrons and single-species ions in the Pfirsch-Schlüter and banana regimes [Sugama and Horton (1996, 1997, 1998)]. It is found that the inward banana fluxes of particles and toroidal momentum are driven by the parallel electric field, which are phenomena coupled through the Onsager symmetric off-diagonal coefficients to the parallel currents caused by the radial thermodynamic forces conjugate to the inward fluxes, respectively.

In MAST the neoclassical rotation velocities are shown to compare well with the Doppler CXRS data in a sawtooth-free L-mode discharge with 800 kA plasma current [Field, *et al.* (2009)]. The poloidal velocities for the C^{+6} impurity ions of order 4 km/s are shown to be comparable with the values from the NCLASS [Newton and Helander (2006)]. The plasma is in the small ν_* regime for a strong collisional friction between trapped and passing ions.

17.2.6 *Improved confinement modes*

Improved confinement modes of tokamak plasmas such as high-confinement modes (H-modes) [ASDEX Team (1989)] and reversed shear configurations [Levinton, *et al.* (1995)] are attracting considerable attention as promising means for achieving controlled fusion. Such a reduction of the transport level is generally considered as caused by the large radial electric field shear (or sheared flow). In the Japan Atomic Energy Research Institute Tokamak-60 Upgrade (JT-60U) [Ninomiya and the JT-60 Team (1992)], the internal transport barrier (ITB) with the steep ion temperature gradient is formed in the region where the gradient of the toroidal flow is steepest [Koide, *et al.* (1996)]. In rotating plasmas with the large flow velocities on the order of the ion thermal speed v_{Ti}, the toroidal flow shear influences the transport of particles, heat, and momentum as an additional thermodynamic force, although, in conventional neoclassical theories [Hinton and Hazeltine (1976); Hirshman and Sigmar (1997); Balescu (1988)] the flow velocities are assumed to be on the order of δv_{Ti} and the direct effects of the flow shear on the transport do not appear in the lowest order. Here $\delta \equiv \rho_i/L$ is the drift ordering parameter, ρ_i the ion thermal gyroradius, and L the equilibrium scale length. It is important to derive the transport equations including the flow shear effects at the same order as particle and thermal transport for understanding the ITB physics. Neoclassical ion transport equations for rotating plasmas were obtained by Hinton and Wong (1985) and by Catto, *et al.* (1987). However, neoclassical electron fluxes are also required for a comprehensive description of transport processes. For example, the neoclassical parallel (bootstrap) current is associated with the parallel electron viscosity and is necessary for determining the equilibrium configuration self-consistently. Here we give full transport equations for neoclassical electron and ion fluxes in the rotating plasma with the toroidal flow velocity on the order of the ion thermal speed.

For axisymmetric toroidal systems the magnetic field is given by

$$\boldsymbol{B} = I(\Psi)\nabla\zeta + \nabla\zeta \times \nabla\Psi \tag{17.14}$$

where ζ is the toroidal angle, Ψ represents the poloidal flux, and $I(\Psi) = RB_T$. In the axisymmetric systems, the poloidal flow decays in a few transit or collision times and the lowest-order flow velocity $\boldsymbol{V}_0$ is in the toroidal direction [Hinton and Wong (1985)]:

$$\boldsymbol{V}_0 = V_0\widehat{\zeta} \quad V_0 = RV^{\zeta} = -R\Phi_0'(\Psi) \tag{17.15}$$

where $\Phi_0(\Psi)$ denotes the lowest-order electrostatic potential in δ (which corresponds to Φ_{-1} in the paper by Hinton and Wong (1985) and $\boldsymbol{E}_0 \equiv -\nabla\Phi_0'\nabla\Psi$. The toroidal angular velocity $V^{\zeta} = -\Phi_0'$ is directly given by the radial electric field and is a flux-surface quantity. The superscript ζ on V^{ζ} denotes the contravariant component of the toroidal velocity which is the angular velocity in radians per second.

For particle species a with the mass m_a and the charge e_a, the phase space variables $(\boldsymbol{x}, \varepsilon, \mu, \xi)$ are defined in terms of the spatial coordinates $\boldsymbol{x}$ in the laboratory frame and the velocity $\boldsymbol{v}' \equiv \boldsymbol{v} - \boldsymbol{V}_0$ in the moving frame [Hinton and Wong (1985); Sugama and Horton (1997)] as

$$\begin{aligned} \boldsymbol{x}' = \boldsymbol{x}, \quad \epsilon = \frac{1}{2}m_a(v')^2 + e_a\widetilde{\Phi}_1 - \frac{1}{2}m_a V_0^2, \quad \mu = \frac{m_a(v_\perp')^2}{2B} \\ \frac{\boldsymbol{v}_\perp'}{v_\perp'} = \boldsymbol{e}_1\cos\xi + \boldsymbol{e}_2\sin\xi. \end{aligned} \tag{17.16}$$

Here $(\boldsymbol{e}_1, \boldsymbol{e}_2, \boldsymbol{b} \equiv \boldsymbol{B}/B)$ are unit vectors which form a right-handed orthogonal system at each point, $\boldsymbol{v}' = v'_{\|}\boldsymbol{b} + \boldsymbol{v}'_{\perp}$, $v'_{\|} = \boldsymbol{v}' \cdot \boldsymbol{b}$, and

$$\xi_a \equiv e_a \widetilde{\Phi}_1 - \frac{1}{2} m_a V_0^2 \tag{17.17}$$

where $\widetilde{\Phi}_1 \equiv \Phi_1 - \langle \Phi_1 \rangle [= \mathcal{O}(\delta)]$ is the poloidal angle-dependent part of the electrostatic potential. The magnetic flux surface average is denoted by $\langle \cdot \rangle$. The lowest-order distribution function is the Maxwellian which is written as

$$f_{a0} = n_a \left(\frac{m_a}{2\pi T_a} \right)^{3/2} \exp - \frac{m_a (v')^2}{2T_a} = N_a \left(\frac{m_a}{2\pi T_a} \right)^{3/2} \exp - \frac{\varepsilon}{T_a} \tag{17.18}$$

where the temperature $T_a = T_a(\Psi)$ and $N_a = N_a(\Psi)$ are flux-surface functions although generally the density n_a depends on the poloidal angle θ through ξ_a and is given by Hinton and Wong (1985).

$$n_a = N_a \exp \left(- \frac{\xi_a}{T_a} \right) = N_a \exp \left(- \frac{e_a}{T_a} \widetilde{\Phi}_1 + \frac{1}{2} \frac{m_a V_0^2}{2T_a} \right). \tag{17.19}$$

This shows the dependence of the density n_a on the poloidal angle θ is the cause of several complications in the derivation of the classical and neoclassical transport coefficients for the rotating plasma.

For plasmas consisting of electrons and a single ion species with charge $e_i \equiv Z_i e$ [Hinton and Wong (1985)], we have

$$\begin{aligned} \frac{e}{T_e} \widetilde{\Phi}_1 &= \frac{m_i (V^\zeta)^2 (R^2 - \langle R^2 \rangle)}{2(Z_i T_e + T_i)}, \\ Z_i N_i(\Psi) &= N_e(\Psi) \exp - \frac{m_i (V^\zeta)^2 \langle R^2 \rangle}{2T_i} \end{aligned} \tag{17.20}$$

where the charge neutrality condition $\Sigma_a e_a n_a = 0$ is used and $m_e/m_i (\ll 1)$ is neglected.

In toroidally-rotating axisymmetric systems, the linearized drift-kinetic equation is written [Hinton and Wong (1985); Catto, *et al.* (1987); Sugama and Horton (1997)]

$$\begin{aligned} v'_{\|} \boldsymbol{b} \cdot \boldsymbol{\nabla} \overline{g_a} - C_a^L (\overline{g_a}) = \frac{\mathbf{1}}{T_a} f_{a0} \, (& W_{a1} X_{a1} + W_{a2} X_{a2} \\ & + W_{aV} X_V + W_{aE} X_E) \end{aligned} \tag{17.21}$$

where C_a^L denotes the linearized collision operator [see Eq. (8) in Sugama and Horton (1996)] and $\overline{g_a}$ is defined in terms of the first-order gyrophase-averaged distribution function f_{A1} as

$$\overline{g_a} \equiv \overline{f_{a1}} - f_{a0} \frac{e_a}{T_a} \int^l \frac{dl}{B} \left(B E_{\|}^{(2)} - \frac{B^2}{\langle B^2 \rangle} \left\langle B E_{\|}^{(2)} \right\rangle \right). \tag{17.22}$$

Here $\int^l dl$ denotes the integral along the magnetic field line and $E_{\|}^{(2)} \equiv \boldsymbol{b} \cdot (-\nabla \Phi^{(2)} - c^{-1} \partial A / \partial t)$ is the second-order parallel electric field. The thermodynamic forces

$(X_{a1}, X_{a2}, X_V, X_E)$ are flux-surface quantities defined by

$$X_{a1} \equiv \frac{1}{N_a}\frac{\partial(N_a T_a)}{\partial \Psi} - e_a \frac{\partial \langle \Psi_1 \rangle}{\partial \Psi}, \qquad X_{a2} \equiv -\frac{\partial T_a}{\partial \Psi}$$
$$X_V \equiv \frac{\partial V^\zeta}{\partial \Psi} = c\frac{\partial^2 \Phi_0}{\partial \Psi^2}, \qquad X_E \equiv \frac{\left\langle B E_\parallel^{(A)} \right\rangle}{\langle B^2 \rangle^{1/2}}. \tag{17.23}$$

The driving forces $\overline{X}_{a1} = X_{a1} = \partial \ell n\, N_a / \partial \Psi - e_a \partial \langle \Psi_1 \rangle / \partial \Psi$ and $\overline{X}_{a2} = X_{a2}$ since the diagnostics give directly the density and temperature profile.

The functions $(W_{a1}, W_{a2}, W_{aV}, W_{aE})$ are defined by

$$W_{a1} \equiv \frac{m_a}{e_a} v'_\parallel \boldsymbol{b} \cdot \boldsymbol{\nabla} \left(R^2 V^\zeta + \frac{I}{b} v'_\parallel \right) \equiv v'_\parallel \boldsymbol{b} \cdot \boldsymbol{\nabla} U_{a1}$$
$$W_{a2} \equiv W_{a1} \left(\frac{\varepsilon}{T_a} - \frac{5}{2} \right) \equiv v'_\parallel \boldsymbol{b} \cdot \boldsymbol{\nabla} U_{a2}$$
$$W_{aV} \equiv \frac{m_a}{2e_a} v'_\parallel \boldsymbol{b} \cdot \boldsymbol{\nabla} \left[m_a \left(R^2 V^\zeta + \frac{I}{B} v'_\parallel \right)^2 + \mu \frac{R^2 B_P^2}{B} \right] \tag{17.24}$$
$$\equiv v'_\parallel \boldsymbol{b} \cdot \boldsymbol{\nabla} U_{aV}$$
$$W_{aE} \equiv \frac{e_a v'_\parallel B^{1/2}}{\langle B^2 \rangle}.$$

The neoclassical entropy production [Sugama and Horton (1996); Sugama and Horton (1997)] is kinetically defined in terms of $\overline{f_{a1}}$ and C_a^L and is rewritten in the thermodynamic form by using Eq. (17.21). The surface-averaged total neoclassical entropy production is given by

$$\sum_a T_a \langle \sigma_a^{\rm ncl} \rangle \equiv -\sum_a T_a \left\langle \int d^3 v \frac{\overline{f_{a1}}}{f_{a0}} C_a^L \left(\overline{f_{a1}} \right) \right\rangle$$
$$= \sum_a \left(\Gamma_a^{\rm ncl} X_{a1} + \frac{1}{T_a} q_a^{\rm ncl} X_{a2} + \Pi_a^{\rm ncl} X_V \right) + J_e X_e \tag{17.25}$$

where the neoclassical transport fluxes $\Gamma_a^{\rm ncl}, q_a^{\rm ncl}/T_a, \Pi_a^{\rm ncl}, J_E)$ conjugate to the forces $(X_{a1}, X_{a2}, X_V, X_E)$ are defined by

$$\Gamma_a^{\rm ncl} \equiv \left\langle \int d^3 v \overline{g_a} W_{a1} \right\rangle, \frac{1}{T_a} q_a^{\rm ncl} \equiv \left\langle \int d^3 v \overline{g_a} W_{a2} \right\rangle$$
$$\Gamma_a^{\rm ncl} \equiv \left\langle \int d^3 v \overline{g_a} W_{aV} \right\rangle \tag{17.26}$$
$$J_e \equiv \frac{\langle B J_\parallel \rangle}{\langle B^2 \rangle^{1/2}} \equiv \sum_a \left\langle \int d^3 v \overline{g_a} W_{aE} \right\rangle.$$

Here Γ_a^{ncl}, q_a^{ncl}, and Π_a^{ncl} denote the surface-averaged radial fluxes of particles, heat, and toroidal (angular) momentum, respectively, and J_E represents the surface-averaged parallel current. The neoclassical transport equations connecting the conjugate pairs of the fluxes and forces are written as

$$\begin{aligned}
\Gamma_a^{\mathrm{ncl}} &= \sum_b \left(L_{11}^{ab}X_{b1} + L_{12}^{ab}X_{b2}\right) + L_{1V}^{a}X_V + L_{1E}^{a}X_E, \\
\frac{1}{T_a}q_a^{\mathrm{ncl}} &= \sum_b \left(L_{21}^{ab}X_{b1} + L_{22}^{ab}X_{b2}\right) + L_{2V}^{a}X_V + L_{2E}^{a}X_E \\
\sum_a \Pi_a^{\mathrm{ncl}} &= \sum_b \left(L_{V1}^{b}X_{b1} + L_{V2}^{b}X_{b2}\right) + L_{VV}X_V + L_{VE}X_E \\
J_e &= \sum_b \left(L_{E1}^{b}X_{b1} + L_{E2}^{b}X_{b2}\right) + L_{EV}^{a}X_V + L_{EE}X_E
\end{aligned} \tag{17.27}$$

where the transport coefficients are dependent on the radial electric field through the toroidal angular velocity $V^{\zeta} = -\Phi_0'$.

From the formal solution of the linearized drift kinetic Eq. (17.21), one proves the Onsager symmetry of the neoclassical transport matrix for the rotating plasma consisting of electrons and multi-species ions with arbitrary collision frequencies. The drift wave turbulence then adds to these collisional transport fluxes to give the total transport of the particles, heat, and toroidal momentum. In the cases of single-species ions, the explicit forms of the neoclassical transport matrices for the Pfirsch-Schlüter and banana regimes are well known. The transport observed in the laboratory, however, typically requires the inclusion of at least one low-Z impurity (e.g. B^{+5} or C^{+6} since the ion particle flux becomes much larger in the presence of an impurity due to breaking the $\Gamma_i = \Gamma_e$ constraint [Helander and Sigmar (2005)].

To first order in ϵ one shows that the angular moment flux for given rotational velocity changes sign when $v_\phi \to -v_\phi$ and is zero at $v_\phi = 0$.

When there is a single null divertor configuration there arises a significant angular momentum flux that does not vanish at zero rotation speed. This flux creates a spontaneous rotation that develops further with auxiliary heating from the strong density and temperature gradients as typically observed in the pear-shaped single null tokamaks. A kinetic theory analysis of the equilibrium of rotating toroidal plasmas with a temperature anisotropy is given by Cremashini and Tessarotto (2011). The work develops the theory of kinetic description of the collisionless rotating tokamak plasmas with anisotropic temperatures.

References

ASDEX Team. (1989). *Nucl. Fusion* **29**, p. 1959.

Balescu, R. (1988). *Transport Processes in Plasmas* **1-2**, (North-Holland, Amsterdam and New York), ISBN:0444870938.

Berk, H. L. and Pratt, J. (2011). Trapped particle stability for the kinetic stabilizer, *Nucl. Fusion* **51**, pp. 083025-08338, doi:10.1088/0029-5515/51/8/083025.

Berk, H. L., Cheng, C. Z., Rosenbluth, M. N., and Van Dam, J. W. (1983). Finite Larmor radius stability theory of ELMO Bumpy Torus plasmas, *Phys. Fluids* **26**, p. 2642, `http://link.aip.org/link/doi/10.1063/1.864456`.

Catto, P. J., Bernstein, I. B., and Tessarotto, M. (1987). *Phys. Fluids* **30**, p. 2784.

Cho, T., Kohagura, J., Hirata, M., Numakura, T., Higaki, H., Hojo, H., Ichimura, M., Ishii, K., Islam, K. M., Itakura, A., Katanuma, I., Nakashima, Y., Saito, T., Tatematsu, Y., Yoshikawa, M., Takemura, Y., Kojima, A., Kobayashi, T., Yamaguchi, Y., Miyata, Y., Yokoyama, N., Tomii, Y., Miyake, Y., Kiminami, S., Shimizu, K., Kubota, Y., Saimaru, H., Higashizono, Y., Mase, A., Yasaka, Y., Ogura, K., Sakamoto, K., Yoshida, M., Pastukhov, V. P., Imai, T., Miyoshi, S., and Gamma 10 Group. (2005). Progress in potential formation and findings in the associated radially sheared electric-field effects on suppressing intermittent turbulent vortex-like fluctuations and reducing transverse losses, *Nucl. Fusion* **45**, pp. 1650-1657, doi:10.1088/0029-5515/45/12/021.

Cooper, W. A., *et al.* (1980). Beam-induced tensor pressure tokamak equilibria, *Nucl. Fusion* **20**, p. 985.

Cremashini and Tessarotto. (2011). *Phys. Plasmas* **18**, 112502.

Diamond, P. H., Lebedev, V. B., Newman, D. E., Carreras, B. A., Hahm, T. S., Tang, W. M., Rewoldt, G., and Avinash, K. (1997). Dynamics of transition to enhanced confinement in reversed magnetic shear discharges, *Phys. Rev. Lett.* **78**, pp. 1472-1475, `http://link.aps.org/doi/10.1103/PhysRevLett.78.1472`.

Dimov, G. I. (2005). The ambipolar trap, *Physics-Uspekhi* **48**, pp. 1129-1149, doi:10.1070/PU2005v048n11ABE005804.

Field, A. R., McCone, L., and Conway, N. J., *et al.* (2009). *Plasma Phys. Control. Fusion* **51**, p. 105002, doi:10.1088/0741-3335/51/10/105002.

Fitzpatrick, R., *et al.* (1993). *Nucl. Fusion* **33**, p. 1533.

Frieman, E. and Rotenberg, M. (1960). *Rev. Mod. Phys.* **32**, pp. 898-902, `http://dx.doi.org/10.1103/RevModPhys.32.898`.

Gibson, A. and the JET Team. (1998). Deuterium-tritium plasmas in the Joint European Torus (JET): Behavior and implications, *Phys. Plasmas* **5**, p. 1839, `http://dx.doi.org/10.1063/1.872854`.

Gill, A. E. (1982). *Atmosphere-Ocean Dynamics, International Geophysics Series* (Academic Press), **30**, ISBN:0-12-283520-4.

Gladd, N. T. and Horton, W. (1973). *Phys. Fluids* **16**, p. 879.

Goedbloed, J. P., Keppens, R., and Poedts, S. (2010). *Advanced Magnetohydrodynamics with Applications to Laboratory and Astrophysical Plasmas*, pp. 269-304 (Cambridge University Press) ISBN:978-0-521-87957-5.

Helander, P. and Sigmar, D. J. (2005). *Collisional transport in magnetized plasmas* (Lavoisier).

Hinton, F. L. and Hazeltine, R. D. (1976). *Rev. Mod. Phys.* **48**, pp. 239-308, doi:10.1103/RevModPhys.48.239

Hinton, F. L. and Wong, S. K. (1985). Neoclassical ion transport in rotating axisymmetric plasmas, *Phys. Fluids* **28**, p. 3082.

Hirshman, S. P., and Sigmar, D. J. (1997). *Phys Fluids* **20**, p. 518.

Hoang, G. T., Becoulet, A., Jacquinot, J., Artaud, J. F., Bae, Y. S., Beaumont, B., Belo, J. H., Berger-By, G., João, P. S. Bizarro, Bonoli, P., Cho, M. H., Decker,

J., Delpech, L., Ekedahl, A., Garcia, J., Giruzzi, G., Goniche, M., Gormezano, C., Guilhem, D., Hillairet, J., Imbeaux, F., Kazarian, F., Kessel, C., Kim, S. H., Kwak, J. G., Jeong, J. H., Listerd, J. B., Litaudon, X., Magne, R., Milora, S., Mirizzi, F., Namkung, W., Noterdaeme, J. M., Park, S. I., Parker, R., Peysson, Y., Rasmussen, D., Sharma, P. K., Schneider, M., Synakowski, E., Tanga, A., Tuccillo, A., and Wan, Y. X. (2009). A lower hybrid current drive system for ITER, *Nucl. Fusion* **49**, 7, p. 075001, http://stacks.iop.org/0029-5515/49/i=7/a=075001.

Hooper, B., Hallock, G. A., and Foote, J. H. (1982). Low-frequency oscillations in the central cell of the TMX tandem mirror experiment, *Phys. Fluids*, http://link.aip.org/link/doi/10.1063/1.863975.

Horton, W., Hu, G., and Laval, G. (1996). Turbulent transport in mixed states of convective cells and sheared flows, http://link.aip.org/link/doi/10.1063/1.871651.

Hu, B. and Horton, W. (1997). Minimal model for transport barrier dynamics based on ion-temperature-gradient turbulence, *Phys. Plasmas* **4**, p. 3262, http://link.aip.org/link/doi/10.1063/1.872467.

Ilic, D. B., Rognlien, T. D., Self, S. A., and Crawford, F. W. (1973). Low-frequency flute instabilities of a hollow cathode arc discharge: Theory and experiment, *Phys. Fluids* **16**, p. 1042.

Janssen, Peter A. E. M. (1982). Single mode saturation of the rotational instability in a low-beta, finite Larmor radius plasma, *Phys. Fluids* **25**, pp. 316-324.

Kim, J.-H., Perez, J. C., Horton, W., Chagelishvili, G. D., Changishvili, R. G., Lominadze, J. G., and Bowman, J. C. (2006). Self-sustaining vortex perturbations in smooth shear flows, *Phys. Plasmas* **13**, p. 062304.

Kishimoto, Y., Kim, J-Y., Horton, W., Tajima, T., LeBrun, M. J., Dettrick, S. A., Li, J. Q., and Shirai, S. (2000). Discontinuity model for internal transport barrier formation in reversed magnetic shear plasmas, *Nucl. Fusion* **40**, p. 667, http://iopscience.iop.org/0029-5515/40/3Y/330/.

Knoepfel, H., and Spong, D. A. (1979). Runaway electrons in toroidal discharges, *Nucl. Fusion* **19**, p. 785.

Koide, Y., Takizuka, T., Takeji, S., Ishida, S., Kikuchi, M., Kamada, Y., Ozeki, T., Neyatani, Y., Shirai, H., Mori, M., and Tsuji-Iio, S. (1996). Internal transport barrier with improved confinement in the JT-60U tokamak, *Plasma Phys. Control. Fusion* **38**, p. 1011, doi:10.1088/0741-3335/38/7/006.

Krishnamurti, R. and Howard, L. N. (1981). Large-scale flow generation in turbulent convection, *Proc. Nat. Acad. Sci.* **78**, p. 4.

Lazzaro, E., Buttery, R. J., Hender, T. C., Zanca, P., Fitzpatrick, R., Bigi, M., Bolzonella, T., Coelho, R., DeBenedetti, M., Nowak, S., Sauter, O., and Stamp, M. (2002). Error field locked modes thresholds in rotating plasmas, anomalous braking and spin-up, *Phys. Plasmas* **9**, p. 3906, http://dx.doi.org/10.1063/1.1499495.

Levinton, F. M., Zarnstorff, M. C., Batha, S. H., Bell, M., Bell, R. E., Budny, R. V., Bush, C., Chang, Z., Fredrickson, E., Janos, A., Manickam, J., Ramsey, A., Sabbagh, S. A., Schmidt, G. L., Synakowski, E. J., and Taylor, G. (1995). Improved confinement with reversed magnetic shear in TFTR, *Phys. Rev. Lett.* **75**, pp. 4417-4420, http://link.aps.org/doi/10.1103/PhysRevLett.75.4417.

Mikhailovskii, A. B. (1992). *Electromagnetic Instabilities in an Inhomogeneous Plasma* (Institute of Physics Publishing) ISBN:0-7503-0182-1.

Newton, S. and Helander, P. (2006). *Phys. Plasma* **13**, p. 102505.

Ninomiya, H. and the JT-60 Team. (1992). *Phys. Fluids B* **4**, p. 2070.

Parra, F. I. and Catto P. J. (2010). *Phys. Plasmas* **17**, p. 056106, http://link.aip.org/link/doi/10.1063/1.3327127.

Peeters, A. G., Angioni, C., and Strintzi, D. (2007). Toroidal momentum pinch velocity due to the Coriolis drift effect on small- scale instabilities in a toroidal plasma, *Phys. Rev. Lett.* **98**, p. 265003, http://link.aps.org/doi/10.1103/PhysRevLett.98.265003.

Peeters, A. G. and Angioni, C. (2005). Linear gyrokinetic calculations of toroidal momentum transport in a tokamak due to the ion temperature gradient mode, *Phys. Plasmas* **12**, p. 072515, http://link.aip.org/link/doi/10.1063/1.1949608.

Rognlien, T. D. (1973). Low-frequency flute instabilities of a bounded plasma column, *J. Appl. Phys.* **44**, pp. 3505-3512, doi:10.1063/1.1662794.

Rosenbluth, M. N. and Simon, A. (1965). Finite Larmor radius equations with nonuniform electric fields and velocities, *Phys. Fluids* **8**, p. 1300.

Rosenbluth, M. N., Krall, N. A., and Rostoker, N. (1962). Finite Larmor radius stabilization of "weakly" unstable confined plasmas, *Nucl. Fusion Suppl.* **143**, OSTI ID:4808729.

Sugama, H. and Horton, W. (1998). Nonlinear electromagnetic gyrokinetic equation for plasmas with large mean flows, *Phys. Plasmas* **5**, p. 2560, http://link.aip.org/link/doi/10.1063/1.872941.

Sugama, H. and Horton, W. (1997). Neoclassical electron and ion transport in toroidally rotating plasmas, *Phys. Plasmas* **4**, p. 6, http://hdl.handle.net/10655/3844.

Sugama, H. and Horton, W. (1996). Entropy production and Onsager symmetry in neoclassical transport processes of toroidal plasmas, *Phys. Plasmas* **3**, pp. 304-322, http://hdl.handle.net/10655/3847.

Uzawa, K., *et al.* (2009). Global characteristics of zonal flows due to the effect of finite bandwidth in drift wave turbulence, *Phys. Plasmas* **16**, p. 042304.

Wagner, F., Becker, G., Behringer, K., Campbell, D., Eberhagen, A., Engelhardt, W., Fussmann, G., Gehre, O., Gernhardt, J., Gierke, G. V., Haas, G., Huang, M., Karger, F., Keilhacker, M., Klüber, O., Kornherr, M., Lackner, K., Lisitano, G., Lister, G. G., Mayer, H. M., Meisel, D., Müller, E. R., Murmann, H., Niedermeyer, H., Poschenrieder, W., Rapp, H., Röhr, H., Schneider, F., Siller, G., Speth, E., Staebler, A., Steuer, K. H., Venus, G., Vollmer, O., and Yü, Z. (1982). *Phys. Rev. Lett.* **49**, pp. 1408-1412, http://link.aps.org/doi/10.1103/PhysRevLett.49.1408.

Waltz, R. E., Kerbel, G. D., and Milovich, J. (1994). Toroidal gyro-Landau fluid model turbulence simulations in a nonlinear ballooning mode representation with radial modes, *Phys. Plasmas* **1**, p. 2229, http://link.aip.org/link/doi/10.1063/1.870934.

Chapter 18

Transport Simulations with Multiscale Codes for Internal Transport Barrier Formation

Local radial zones of reduced transport called internal transport barriers are now routinely produced in the large tokamaks. Internal Transport Barriers (ITBs) were first observed in the large tokamaks in the 1990s [Koide, *et al.* (1994); Mori, *et al.* (1994); Levinton, *et al.* (1995); Strait, *et al.* (1995); Gormezano and the Jet Team (1996); Söldner and the Jet Team (1997)] were first discovered in Japan Atomic Energy Research Institute Tokamak-60 Upgrade (JT-60U) [Ninomiya and the JT-60 Team (1992); Koide, *et al.* (1994); Mori, *et al.* (1994)]. An ITB is a region where a steep temperature and/or density gradient forms inside a tokamak plasma, usually within the region $\rho \lesssim 0.6$, where ρ is the normalized minor radius. In tokamaks with toroidal plasma velocity measurements, it is found that the region within an ITB is also the region where the toroidal rotation velocity gradient is steep. This high performance regime can occur with a lower power threshold than required for the high confinement mode (H-mode) regime, where the transport barriers are located at the plasma edge.

A schematic of the multiscale physics for the formation of the internal and edge transport barriers along with the particle and thermal transport is shown in Fig. 18.1. In this Chapter we give some details of these multiscale transport simulations for the transport barrier dynamics for experiments in JET and DIII-D. The name "transport barrier", while well accepted in the fusion literature, is misleading since there is no actual barrier but rather there is a marked change in the dominant transport characteristics across relatively narrow radial regions. These changes in the transport are associated with regions of steep gradients in the toroidal rotation velocity and/or magnetic shear. A local minimum in the magnetic shear enhances the effect of the rotational shear and is the method by which the small scale turbulent electron transport can be strongly reduced.

Internal transport barriers were produced in two optimized magnetic shear (OS) discharges in the Joint European Torus (JET) [Rebut, *et al.* (1985)], shot #40542 and #40847 [Gormezano and the Jet Team (1996); Söldner and the Jet Team (1997); Parail, *et al.* (1999)]. In both shots, Neutral Beam Injection (NBI) is the major auxiliary power source, although there is some RF heating as well. The q-profiles generated by current ramping in these two shots are flat and low in the core (near

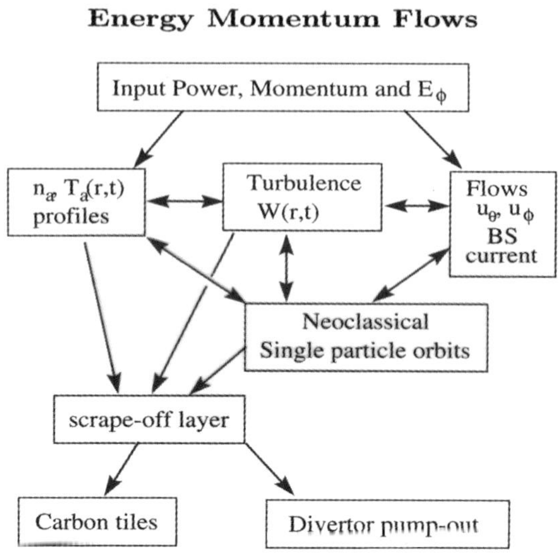

Fig. 18.1 A flow diagram of the complex weather-like dynamics of the driven toroidal plasma. The closed loops of couplings give rise to the bifurcations described in Chapter 18 in the simulations using the JETTO and MMM codes.

the magnetic axis) and steeply rising approaching the edge, with the minimum value of q greater than unity throughout the plasma. In both discharges, the ITB forms first in the core while the edge remains in the low confinement mode (L-mode). Then the ITB expands towards the edge. In one shot (#40542) an ELMy (edge localized mode) H-mode subsequently occurs and results in the co-existence of both an internal transport barrier and an edge transport barrier (ETB) until the ramping down of NBI power. In the other JET shot (#40847), the edge plasma first transits to an ELM-free H-mode and then enters an ELMy H-mode phase, during which the internal transport barrier decays. While it is an interesting feature for these two JET shots that both ITB and ETB co-exist, we focus on the ITB behavior in the simulations.

Internal transport barriers were formed in the two negative central magnetic shear (NCS) discharges in the Doublet III-D Tokamak [Luxon and Davis (1985)], shot #84682 and #87031 [Rice, *et al.* (1996); Schissel, *et al.* (1996)]. NBI is the only auxiliary power source in these two shots, which produced ion temperature profiles with large gradients in the region of the minimum of the q-profiles. In the discharge with larger negative central magnetic shear (#84682), the internal transport barrier is steeper than the one in the discharge with weaker NCS (#87031). Both of these DIII-D discharges retain L-mode behavior at the edge, which provides an opportunity for the validation of the physics of ITB in the simulations.

The weak and/or reversed magnetic shear in the plasma core is an important factor contributing to the formation of ITBs. In the Tokamak Fusion Test Reactor (TFTR) [Grove and Meade (1985)], internal transport barriers were usually produced with reversed magnetic shear (RS) in the plasma core. In JET, ITBs occur in the optimized magnetic shear scenario, with low central magnetic shear. One theoretical argument is that the toroidal coupling of the drift modes is weak in regions of low magnetic shear, makes the modes more slab-like and hence more localized [Beklemishev and Horton (1992)]. At the radial surface where the magnetic shear $dq/dr = 0$ vanishes, a large gap opens in the density of rational surfaces reducing the overlapping of the drift wave structures, as shown in Fig. 7b.

In addition to the weak or reversed magnetic shear, the $\boldsymbol{E} \times \boldsymbol{B}$ flow shear in the core region is another crucial factor for the formation of ITBs. Since the discovery of the H-mode, the ion drift mode turbulence suppression due to $\boldsymbol{E} \times \boldsymbol{B}$ flow shear has long been identified as the critical mechanism in a number of different high performance regimes [Biglari, *et al.* (1990)]. Magnetic shear and $\boldsymbol{E} \times \boldsymbol{B}$ flow shear have been incorporated into various transport models, and simulations have been carried out based on a paradigm of the ITB dynamics [Fukuyama, *et al.* (1998); Kinsey (1999)] similar to that of H-mode [Sugama and Horton (1995); Diamond, *et al.* (1997)]. In this paradigm, the build-up of the microturbulence level due to the increasing temperature or density gradient driven by auxiliary heating generates spontaneous $\boldsymbol{E} \times \boldsymbol{B}$ flow in the plasma through anomalous momentum transport. This $\boldsymbol{E} \times \boldsymbol{B}$ flow is in addition to the $\boldsymbol{E} \times \boldsymbol{B}$ flow of plasma that can be driven by input torque (*e.g.* NBI). The increase in $\boldsymbol{E} \times \boldsymbol{B}$ flow shear suppresses the microturbulence, leading to the decrease of anomalous transport of particle and energy and the increase of density and temperature gradients, namely, the formation of the transport barrier. The suppression of the turbulence level also reduces the anomalous driving force (*e.g.* "Reynolds stress") for the $\boldsymbol{E} \times \boldsymbol{B}$ flow and its shear. The balance of these two opposite trends produces a new steady state with higher plasma confinement, which is the case for H-modes when this process happens at plasma edge. In the central region of plasma, the weak magnetic shear serves as another confinement enhancing mechanism by reducing radial decorrelation length of drift modes and decreasing the growth rate of the drift mode ballooning turbulence. The decrease of the radial correlation length is due to lower density of the rational surfaces analyzed in Chapter 7.3.2. The combination of these two mechanisms, $\boldsymbol{E} \times \boldsymbol{B}$ flow and weak magnetic shear, in the inner region of plasma is expected to result in the formation of ITBs with a power threshold that is lower than if only one mechanism were present. Recently, remarkable agreement with experimental data is reported for the simulations of a DIII-D internal transport barrier discharge by Kinsey, *et al.* (1999) using the GLF23 transport model.

The Multi-Mode model (MMM) in the time-dependent transport code BALDUR simulates the profiles for the ITBs in the two JET optimized shear (OS) discharges and the two DIII-D negative central shear (NCS) discharges. The Multi-Mode model is an advanced fluid turbulent transport model that is used to compute four channels of transport. This transport model has been used in simulations of a wide range of L-mode and H-mode discharges where good agreement with experimental data is obtained [Bateman (1996, 1997) and Kinsey, *et al.* (1999)]. The $\boldsymbol{E} \times \boldsymbol{B}$ flow shear mechanism is implemented in the current version of the Multi-Mode model either by (1) subtracting the Hahm-Burrell shearing rate ω_s [Hahm and Burrell (1995)] from the drift mode growth rate, or (2) by multiplying the ion drift mode induced transport coefficients by the suppression factor $1/(1 + (\Upsilon_s/\Upsilon_{sc})^2)$, where Υ_s is the Hamaguchi-Horton shear parameter [Hamaguchi and Horton (1992)]. Both ω_s and Υ_s are computed using the radial electric field E_r, which is computed from the poloidal and toroidal flow velocities and the pressure gradient. The poloidal flow velocity is obtained from a neoclassical model [Ernst, *et al.* (1998); Zhu, *et al.* (1999)], while the toroidal flow velocity is taken from the experimental data. The weak magnetic shear mechanism works through the magnetic shear effect already present in Weiland model for drift modes in the Multi-Mode model. The magnetic shear enters the model through the Hamaguchi-Horton parameter Υ_s in combination with the $\boldsymbol{E} \times \boldsymbol{B}$ flow shear effect divided by the magnetic shear. Internal transport barriers are reproduced in the simulations for the discharges in both machines, consistent with the $\boldsymbol{E} \times \boldsymbol{B}$ flow shear and weak magnetic shear mechanisms of ITB dynamics.

18.1 Multi-Mode Models

The Multi-Mode model determines multiple channels anomalous for the transport coefficients by combining quasilinear transport contributions from several drift modes that co-exist in tokamak plasmas. These include ion temperature gradient (ITG) and trapped electron modes (TEM) in the Weiland model [Nordman, *et al.* (1990); Weiland and Hirose (1992); Nilsson and Weiland (1994)], the drift-Alfvén modes in a model developed by Scott (1997) and Bateman, *et al.* (1998), as well as a model for kinetic ballooning modes. A complete description of the model and the parameters in the MMM95 version of the Multi-Mode model is given in Bateman (1992) and Bateman, *et al.* (1997). The first work validating the model was for analysis of five L-mode TFTR discharges in PPPL-2764 with currents from 1 to 2 MA and NBI heating powers from 5 to 17 MW. The results confirmed that the degradation of the confinement with increasing heating power given by the empirical L-mode confinement law is explained by the combination of the ITG/TEM modes and the current dependence of the confinement is related to the resistive-g modes in the L-mode discharges of TFTR. The Multi-Mode Model (MMM) went through a number of upgrades over the past decade. The 1995 version, called MMM95, and the 2000 update were widely used with success.

The $\boldsymbol{E} \times \boldsymbol{B}$ flow shear has long been considered as a suppression mechanism for the microturbulence [Biglari, *et al.* (1990)]. Gyrokinetic simulations by Waltz, *et al.* (1994) show that the suppression of microturbulence occurs when the Hahm-Burrell $\boldsymbol{E} \times \boldsymbol{B}$ flow shearing rate ω_s is greater than the maximum growth rate of all the drift modes. This result is used to simulate the $\boldsymbol{E} \times \boldsymbol{B}$ flow shear mechanism in several transport models, including the IFS-PPPL and MMM95 models, by using the reduced growth rate $\gamma_{\text{eff}} = \gamma - \omega_s$ in the estimation of turbulent transport coefficients. The Hahm-Burrell shearing rate ω_s is given by Hahm and Burrell (1995)

$$\omega_s = \left| \frac{RB_\theta}{B_\phi} \frac{\partial}{\partial r} \left(\frac{E_r}{RB_\theta} \right) \right| \tag{18.1}$$

where the profiles of radial electric field E_r and safety factor q are needed. Here, R is the major radius, B_θ and B_ϕ are poloidal and toroidal components of magnetic field, and r is the minor radius (half-width). The radial electric field E_r is usually obtained by solving poloidal and toroidal momentum transport equations for poloidal and toroidal rotation velocity profiles, since

$$E_r = u_{\phi i} B_\theta - u_{\theta i} B_\phi + \frac{1}{Z_i e n_i} \frac{dp_i}{dr}. \tag{18.2}$$

where $u_{\phi i}$ and $u_{\theta i}$ are toroidal and poloidal components of ion velocity, e is the electron charge, Z_i is the ion charge number, n_i is the ion density, and p_i is the ion pressure. In the BALDUR code, the poloidal rotation velocity $u_{\theta i}$ is computed based on neoclassical model [Ernst, *et al.* (1998); Zhu, *et al.* (1999)], while the toroidal rotation velocity $u_{\phi i}$ is taken from experiment measurement, which is available in JET and DIII-D for the impurity ions. The reduced growth rate $\gamma_{\text{eff}} = \gamma - \omega_s$ is used to compute the turbulent transport coefficients for the drift waves in the current Multi-Mode model. However, a coefficient C, greater than unity in front of ω_s in the reduced growth rate $\gamma - C\omega_s$, is required in order to produce simulation profiles that match the experimental data in some cases. The need for the coefficient C has been reported by other researchers as well Kinsey, *et al.* (1999).

The alternative to the Hahm-Burrell flow shear model takes into account the effect of weak magnetic shear in addition to the $\boldsymbol{E} \times \boldsymbol{B}$ flow shear mechanism. In this model, all the turbulent transport coefficients of ion drift modes are multiplied by a factor $1/[1 + (\Upsilon_s/\Upsilon_{sc})^2]$, where Υ_s is the dimensionless Hamaguchi-Horton shear parameter [Hamaguchi and Horton (1992)]

$$\Upsilon_s \simeq \sqrt{\frac{m_i}{T_e}} \left| R\partial_\psi \frac{(E_r/RB_\theta)}{\partial_\psi \ln q} \right|. \tag{18.3}$$

Here Υ_{sc} is the critical value for the onset of strong transport suppression. Here, m_i is the ion mass, T_e is the electron temperature, and ψ is the magnetic function that is used to label magnetic surfaces. Note that when the magnetic shear in the denominator in Eq. (18.3) is small the value of Υ_s is greatly increased. In shear-slab case, Hamaguchi and Horton (1992) showed that linear ion temperature gradient mode growth rate decreases as Υ_s becomes sufficiently large and that the corresponding turbulent transport is reduced when Υ_s rises above a critical value between 1 and 2 in 3D turbulence simulations. These studies suggest the use of Υ_s as the $\boldsymbol{E} \times \boldsymbol{B}$ flow shear parameter in the weak magnetic shear region.

Turbulence suppression from toroidal flow shear in ITER has been modeled by deGrassie (2009). The work suggests that toroidal velocity profiles with core velocities greater than 120 km/s are required for the suppression of transport. They validate their code by comparisons with the experiments and scaling law derived in C-Mod, where a significant spontaneous rotation counter to the toroidal current is found [Rice, *et al.* (2007)].

The magnetic shear effect enters the Weiland model for drift modes through its influence on the form of the perturbations along the magnetic field lines [Weiland and Hirose (1992)]. In the strong magnetic shear region, the electron motions parallel to the magnetic field line are greatly impeded; hence, the perturbations are strongly ballooning-like. This ballooning feature of the perturbation is reduced in the weak magnetic shear region, along with the growth rate of the mode. This mechanism is implemented in Weiland model through the averaged ion acoustic operator $\left\langle \nabla_{\parallel}^2 \right\rangle$ and the averaged Alfvén line-bending operator $\langle \nabla_{\parallel} \nabla_{\perp}^2 \nabla_{\parallel} \rangle$, which are strong functions of the magnetic shear.

A statistical formulation of the confinement transitions based on pde models using the Langevin equation and the fluctuation dissipation theorem is given by Itoh and Itoh (2001). Phase diagrams are given for the multiple confinement and fluctuation regimes based on a five-field model with resistive-g and ITG turbulence.

18.2 Electron Thermal Transport Barriers in Tokamaks

By proper adjustment of the tangential and near perpendicular neutral beam injectors to control the injected momentum and thus the plasma flow and magnetic shear profiles, Ishida, *et al.* (1997) found regimes in JT-60U where there were strong transport barriers in both the ions and electrons. The plasma energy $W(t)$ increased linearly in time until reaching approximately 10 MJ and then collapsed. The neutron production from the D-D plasma reached 5×10^{16}/s in this discharge E27969. This D-D neutron yields is reported to be equivalent to a $Q_{\rm DT} \sim 1.0 \pm 0.05$. In these barrier experiments the ion temperature reaches 16 KeV and electron temperature at 8 KeV with profiles that are flat out to the transport barrier which occurs across the region $\rho_{\rm ITB} = 0.6 - 0.7$ just inside the minimum $q_{\rm min} \sim 2$ surface at $\rho = 0.75$. The density has a similar profile with the core electron density of $9 \times 10^{19} \mathrm{m}^{-3}$. Carbon is the dominant impurity with $Z_{\rm eff} = 3.5$ and the energy confinement time in the D-D plasma is reported as $\tau_E = 0.97$ s being an $H_{\rm ITER89-P} = 3.2$. The absorbed NBI power is 16 MW and the $B_{\rm tor} = 4.34$ T and $I_p = 2.8$ MA.

Another important record shot from JT60U is 27315 with negative ion (deuterium) beam injection driving a substantial electrical current as reported by Koide and the JT60 Team (1997). The negative ion beam energy was 400 KeV with power $P_{\rm NNB} = 2.5$ MW. The current drive efficiency parameter $\eta_{\rm CD} = \overline{n}_e R I_{\rm BD}/P_{\rm NB}$ was 8×10^8 A/W which was 70% higher than achieved with the 85 KeV positive ion beams. The core electron temperature was 4.5 KeV. More details of the record shot E27969 are given in the article [Koide and JT60 Team (1997)] along with comparison with a weakly reversed magnetic shear and monotonic magnetic shear high performance shots. The motional Stark effect (MSE) diagnostic is used to infer the safety factor profiles in these plasmas.

The article [Koide and JT60 Team (1997)] reports results from 2 MW of Lower Hybrid Current Drive from 1.74 to 2.23 GHz driving frequency transmitters. The target plasma had $\overline{n}_e = 1.2\times10^{19}\text{m}^{-3}$ and $P_{\text{NB}} = 0.5$ MW with $I_p = 1$ MA and $B_t = 3$ T. Radiative divertor experiments where neon gas was injected into an ITB high-performance plasma were carried out. The radiated power from the divertor region increased from 1.4 MW to 4 MW from the impurity radiation. The concentration of neon in the main plasma increased in 1 second to 1.8%. The increased resistivity causes the current profile to relax to a monotonic q profile in 2.4 s.

18.3 Predictive Transport Simulations of Internal Transport Barriers Using the Multimode Model

Predictive simulations involve the specification of some of the experimental conditions and the theoretical models. The experimental conditions include:

(1) Geometrical and physical parameters: tokamak size, toroidal magnetic field, and properties of the hydrogenic and impurity gases. The parameters for the JET and DIII-D shots are listed in Table 18.1.

Table 18.1 JET OS and DIII-D NCS Discharge Parameters.

Shot	JET 40542	JET 40847	DIII-D 84682	DIII-D 87031
Time (s)	47.4	46.9	1.58	1.80
R (m)/a (m)	2.9/0.94	2.9/0.95	1.7/0.62	1.7/0.65
B_T (T)	3.6	3.6	2.2	2.3
I_p (MA)	3.3	3.2	1.6	1.6
P_{NBI} (MW)	17	19	9.3	9.6
$\overline{n}_e$ (10^{19}m^{-3})	2.5	3.9	2.5	3.1
Working gas	deuterium	deuterium	deuterium	deuterium
Impurity	carbon	carbon	carbon	carbon

(2) Sources and sinks. In principle, all source profiles should be computed inside the code. In current BALDUR code, NBI power deposition is computed by solving Fokker-Planck equation based on realistic experimental settings of the beams, while the RF heating power profiles are provided by experimental data. The heating power levels are set according to the experimental data as a function of time. The plasma current ramping in the experiment is followed

in the simulations in order to obtain q-profiles close to those obtained in the experiments.

(3) Initial and boundary conditions for density and temperature profiles.

Since the toroidal velocity is not computed in the code, the time evolution of the toroidal rotation profiles are input data needed for the simulation to compute the radial electric field. All the experimental data are the TRANSP processed data obtained from the International Thermonuclear Experimental Reactor (ITER) Profile Database. The theoretical model includes the following: The eleven-equation version of Weiland model is used for the drift modes in the core [Bateman, *et al.* (1997)]. The drift-Alfvén mode by Scott (1997) and Bateman, *et al.* (1998) is taken for the edge mode model. The supershot setting of kinetic ballooning model used in previous MMM simulations is used here [Bateman, *et al.* (1997)]. To simulate the favorable effect of plasma elongation, the pressure gradient threshold for the onset of the kinetic ballooning mode is increased from that used for the circular magnetic surfaces. This shape adjustment to the critical pressure turns out to be quite important in obtaining the sharp ion temperature gradients in the simulations of the JET discharges.

The experimental and simulation ion temperature profiles at six time slices are shown in Fig. 18.2 for the two JET discharges and Fig. 18.3 for the two DIII-D discharges. In the simulations shown in Fig. 18.2 and Fig. 18.3 the Hahm-Burrell model is employed with a coefficient C as described above. Six time slices of particular interest for each shot is the focus here.

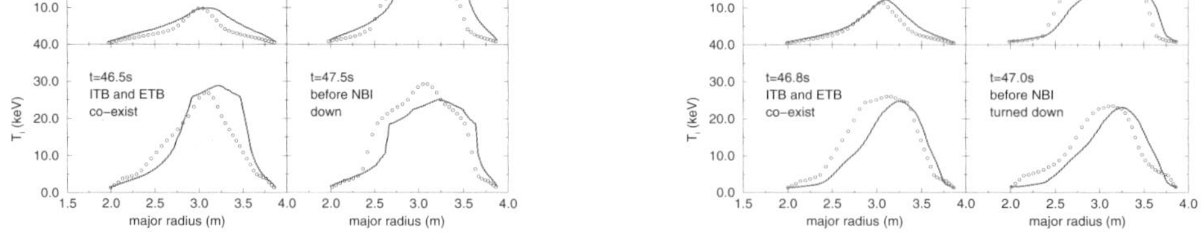

(a) (b)

Fig. 18.2 Temperature profiles for two JET optimized shear discharges that form internal transport barriers before the H-mode transition. The left panel gives JET 40542 $T_i(r,t)$ data in circles and the simulation in the solid curve. The right panel gives JET 4087 $T_i(r,t)$ data in circles and the simulation in the solid curve. The onset of the ITB is for the time slices at 45.5 s as label in the third time slice frames [Zhu, *et al.* (2000)].

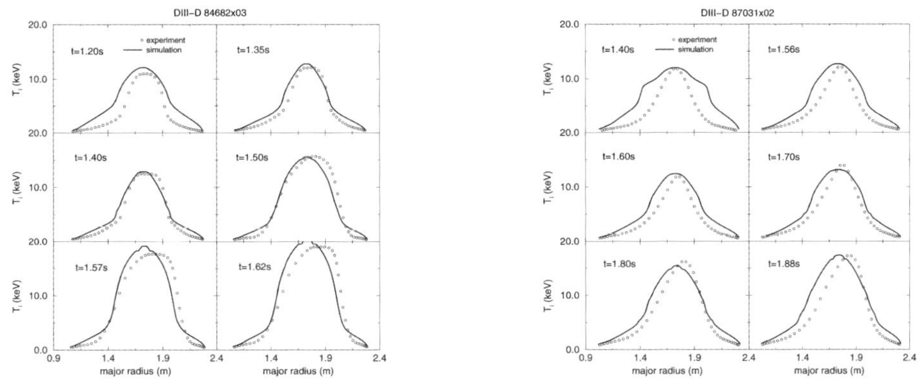

(a) (b)

Fig. 18.3 Temperature profiles for two DIII-D reversed shear discharges that form internal transport barriers before the H-mode transition. The left panel gives DIII-D 84662 $T_i(r,t)$ data in circles and the simulation in the solid curve. The right panel gives DIII-D 87031 $T_i(r,t)$ data in circles and the simulation in the solid curve is the transport simulation.

In the case of JET, the simulations start during the ohmic stage and carry on till the end of all auxiliary heating. The first time slice for the JET discharges is the in the middle of ohmic stage, and the second time slice is right before the auxiliary heating is turned on. The third and fourth time slice are taken right after the internal transport barrier forms and just before the formation of the edge transport barrier. The last two time slices are during the stage when the internal transport barrier co-exists with the ELMy H-mode at the edge and before the time when the NBI heating is turned down. In the simulation of the JET discharge #40542, the reduced growth rate $\gamma - \omega_s$ is used for the $\boldsymbol{E} \times \boldsymbol{B}$ flow shear mechanism, while in simulation of the JET discharge #40847, a factor $C = 3 \sim 5$ in front of ω_s is used in the reduced growth rate $\gamma - C\omega_s$ in order to obtain simulated ion temperature profiles in agreement with experiment. Kinsey, *et al.* (1999) found the factor $C = 2.65$ is necessary to obtain agreement with the experimental temperature profiles in their simulation of this same JET discharge in which they use the GLF23 transport model. During the early time stages, the times of formation of the ITBs in the simulations for both shots agree with the experimental data. During the evolution of ITBs at later time stages, there are some radial shifts of ITBs in the simulations as compared with the experiments.

In the case of DIII-D shown in Fig. 18.3, the simulations start at the beginning of the NBI heating stage, where the experimental data is available. In both of the DIII-D discharges, the NBI heating starts with a lower power, and steps up to about double power in the middle of the shot. Plots of the ion temperature profiles at two time slices before, one time slice during, and three time slices after the NBI power step-up are presented in Fig. 18.3. The reduced growth rate $\gamma - \omega_s$ is used for the $\boldsymbol{E} \times \boldsymbol{B}$ flow shear mechanism in the simulation of DIII-D shot #84682 and DIII-D shot #87031. The timing and location of the ITBs in the simulations agree well with the experiment data for both shots.

18.3.1 *Dynamic behavior of ITBs*

In addition to the evolution of ion temperature radial profiles shown in Fig. 18.2 and Fig. 18.3, the dynamic behavior of ITBs can also be illustrated by the ion temperature evolution curves at a sequence radii. The experimental and simulation time traces are shown in Fig. 18.4 for JET shot #40542 and in Fig. 18.5 for DIII-D shot #84682.

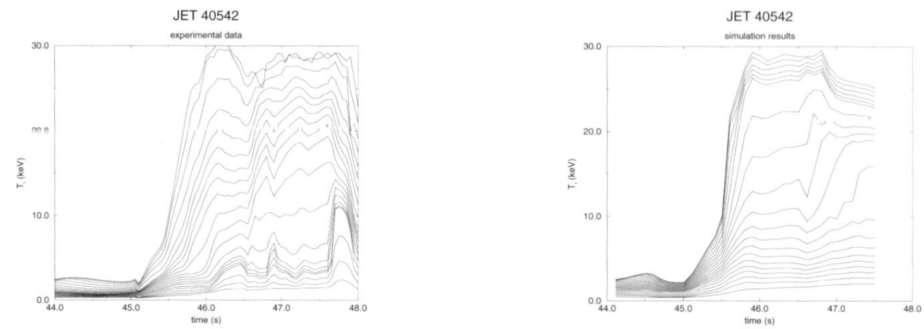

(a) (b)

Fig. 18.4 An alternative representation of the ITB formation shown in Fig. 18.2 by time traces of the ion temperature data at fixed radial positions. The formation of the steep gradient corresponds to the widening paces in the nested temperature traces. Frame (a) gives the JET 40542 data and frame (b) gives the simulation result.

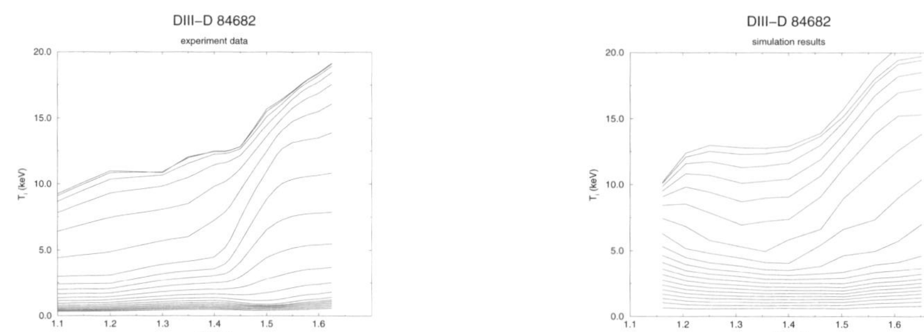

(a) (b)

Fig. 18.5 An alternative representation of the ITB formation shown in Fig. 18.3 by time traces of the ion temperature data at fixed radial positions. The formation of the steep gradient corresponds to the widening gaps in the nested temperature traces. The left panel shows the DIII-D data and the right panel the simulation.

In Fig. 18.4 and Fig. 18.5, each curve represents the ion temperature at a given normalized minor radius as a function of time. In each plot, the curve at the top is the ion temperature trace near the magnetic axis, while the curve at the bottom is the ion temperature trace at the edge of the plasma. The rest of the time traces are at equally spaced intervals of normalized minor radius. Internal transport barriers are characterized by wider spacing between adjacent curves (steeper gradients) relative to the spacing between other adjacent curves.

In the experimental data from the JET shot #40542 (Fig. 18.4a), the time when the ITB first forms is evident from the curves near the inner radii ($t = 46.0\,\text{s}$). At a later time the edge transport barrier corresponding to the H-mode is seen from the lower curves near the plasma edge. In the plot of the simulation data (Fig. 18.4b), the formation of an ITB is seen at $t = 45.4$, though the formation of an ETB is absent. From Fig. 18.4b, it can be seen that the ITB moves from the central region (10 to 27 KeV) at $t = 46.0\,\text{s}$ to a lower temperature region (9 to 18 KeV) at $t = 47.5\,\text{s}$. There is a similar motion in the data in the left frame of Fig. 18.4a from the central region (especially 20 to 28 KeV) at $t = 46.0\,\text{s}$ to a region closer to the plasma edge (5 to 17 KeV) at $t = 47.5\,\text{s}$. Hence, the simulation shown follows the trend of the outward radial movement of the ITB observed in the experimental data.

In the case of the DIII-D shot #84682, the experimental curves (Fig. 18.5a) show the formation of an ITB at early times (before $t = 1.4\,\text{s}$) and its outward expansion up to $t = 1.45\,\text{s}$. The same trend is also seen in the simulation plot (Fig. 18.5b). At $t = 1.2\,\text{s}$, the ITB is mostly concentrated between a normalized minor radius of $r/a = 0.2$ and $r/a = 0.3$ in both the experimental data and the simulation. By $t = 1.6\,\text{s}$, the radial extent of the ITB broadens by extending toward the edge of the plasma, as indicated in both experimental data and in the simulation results.

18.3.2 *Weak magnetic shear OS and reversed magnetic shear NCS confinement*

The transport simulations predict the time evolution of the magnetic q-profile by advancing the magnetic diffusion equation. The optimized magnetic shear (OS) feature of the safety factor q-profile is well reproduced in the simulation for JET. In the DIII-D case, the q-profile produced by the simulation has negative central magnetic shear in experiment (NCS). In the transport model, it is the weak magnitude of magnetic shear that is significant, while the sign of the shear is irrelevant. Therefore, the q-profile with weak shear in the simulation produces comparable transport reduction to the q-profile with negative magnetic shear. The evolution of experimentally-measured q-profiles and the simulation are compared in Fig. 18.6.

Implementation of $\boldsymbol{E} \times \boldsymbol{B}$ flow shear mechanism that includes weak magnetic shear effects through the Hamaguchi-Horton shear parameter Υ_s, Eq. (18.3). The value of Υ_s is plotted in Fig. 18.7a for JET shot #40542 and in Fig. 18.6b for DIII-D

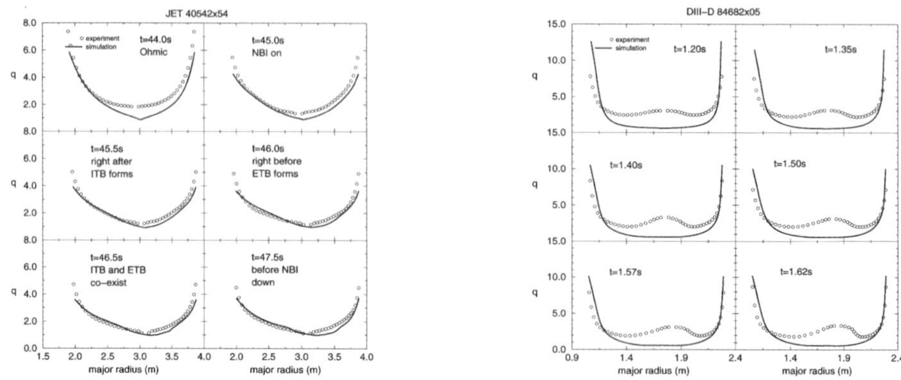

(a) (b)

Fig. 18.6 The time evolving $q(r,t)$ profiles in the left panel for the JET discharge and in the right panel for the DIII-D discharge. The flat-bottomed q-profile in the JET profile and the reversed shear profile in the DIII-D profile is called RS or negative central shear NCS. Both q-profiles produce internal transport barriers.

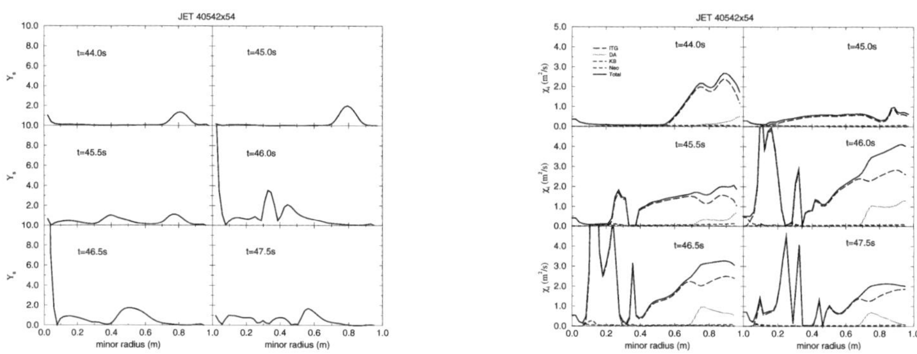

(a) (b)

Fig. 18.7 For the JET discharge the $\boldsymbol{E}\times\boldsymbol{B}$ shearing rate parameter from the Hamaguchi-Horton formula Eq. (18.3) in the left panel and the ion thermal diffusivity in the right panel.

shot #84682. The Υ_s profiles are peaked at both the maximum $\boldsymbol{E}\times\boldsymbol{B}$ flow shear rate ω_s and the low central magnetic shear regions in cases of both shots.

In Fig. 18.6a and Fig. 18.6b we see the reduction of the thermal diffusivity χ_i by Υ_s, especially in the central region of plasma. Transport barriers are also reproduced in ion temperature profiles of the simulations, where the agreement with experiment depends on the choice of the critical value Υ_{sc}. The critical shear parameter $\Upsilon_{sc}=1$ is used for the JET case in Fig. 18.6a and $\Upsilon_{sc}=0.5$ is used for the DIII-D case in Fig. 18.6b. In general, the ion temperature profiles of the simulations using this model tend to be narrowly peaked in the inner plasma. In the DIII-D case, the values of Υ_s are much greater than that in JET case, due to the nearly zero magnetic shear over broader central region of plasma and stronger $\boldsymbol{E}\times\boldsymbol{B}$ flow shear in DIII-D barrier region Fig. 18.6a and Fig. 18.6a. Hence, the flow and simulation profiles have more influence on the ion thermal diffusivity and temperature profiles in the simulations of DIII-D than in the corresponding simulations of JET.

18.3.3 *Statistical analysis of the simulation profiles*

To quantify the comparison of simulations with experiments, we compute the relative root-mean-square (rms) deviation between the two profiles for the six time slices of the four discharges we simulated. The relative rms deviation of each quantity X (T_i, T_e, and n_e) is defined as

$$\sigma_X = \sqrt{\frac{1}{N}\sum_{j=1}^{N}\left(\frac{X_j^{\exp} - X_j^{\text{sim}}}{X_{\max}^{\exp}}\right)^2} \tag{18.4}$$

in Table 18.2 for the two JET discharges.

Table 18.2 Simulation Statistics for JET Simulations.

JET 40542				JET 40847			
Time (s)	σ_{T_i}(%)	σ_{T_e}(%)	σ_{N_e}(%)	Time (s)	σ_{T_i}(%)	σ_{T_e}(%)	σ_{N_e}(%)
44.0	22	39	7	44.0	12	18	10
45.0	19	30	9	45.0	26	11	7
45.5	19	13	9	45.5	12	11	6
46.0	20	12	11	46.5	19	12	14
46.5	20	17	11	46.8	17	13	16
47.5	16	20	11	47.0	16	13	12

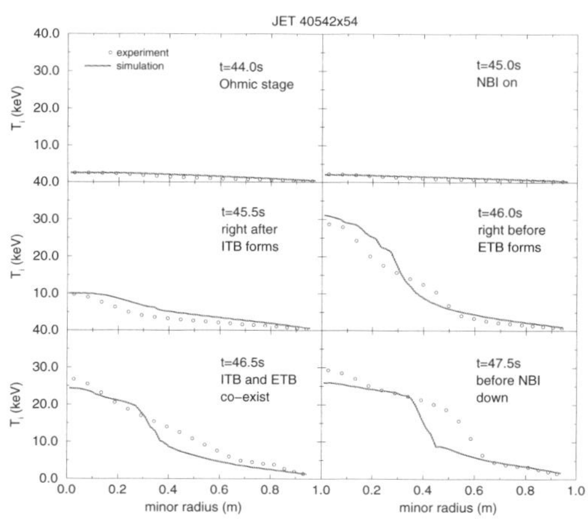

Fig. 18.8 Comparison of the evolution of the predicted (solid line) and experimental data (dashed line) for the JET 40542 discharge.

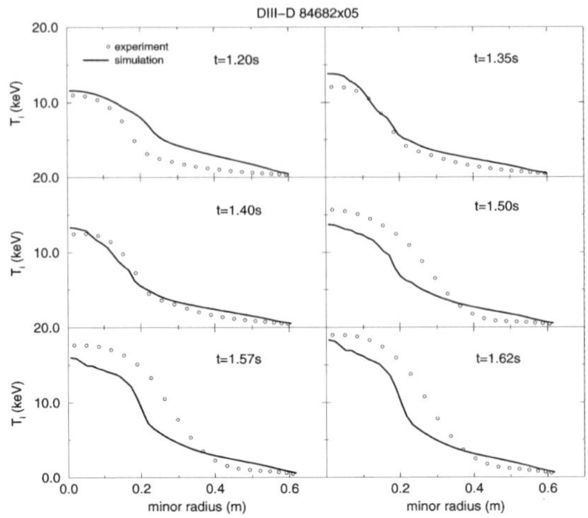

Fig. 18.9 Comparison of the simulation (solid line) with the DIII-D 84662 ion temperature profile (dashed line) for the NCS transport barrier. The $\boldsymbol{E} \times \boldsymbol{B}$ shearing rate parameter from the Hamaguchi-Horton formula, Eq. (18.3) gives the ion thermal diffusivity formula and predicts the experimental ion temperature profile.

For the JET simulations, the statistics show that the simulation profiles generally match the experimental data about equally well during all six time slices the auxiliary heating stages of the discharges. During these time stages with transport barriers, the relative rms deviations are within 20%, which is comparable with the general experimental measurement error. In the DIII-D cases, the overall relative rms deviations are smaller than for the JET cases.

In the Multi-Mode model the overall mechanisms of the ITB dynamics are implemented in three different levels: theoretical, empirical, and experimental. The theoretical models include the Weiland model for ion drift modes, the Scott and Bateman model for drift-Alfvén mode at plasma edge, and the neoclassical model for the poloidal plasma flow. The kinetic ballooning mode model and the $\boldsymbol{E}\times\boldsymbol{B}$ flow shear mechanism are implemented empirically. The toroidal velocity profiles from experimental data have been used directly instead of predicting toroidal velocity with a toroidal momentum transport model.

Simulations carried out with this transport model have reproduced the internal transport barriers observed in both JET OS and DIII-D NCS discharges. For the two JET shots, the exact location of the ITB differs from the location indicated by experimental data at later time stages after the barrier first occurs. For the two DIII-D shots, the timing and location of ITBs in the simulations agree with the experimental results. From the effects of $\boldsymbol{E} \times \boldsymbol{B}$ flow shear and q-profile on the profiles of ion thermal diffusivity and ion temperature, the formation of internal transport barrier is shown to predict to within 40% the profiles of the plasma temperatures and density.

References

Bateman, G. (1992). Theory-based transport simulations of TFTR L-mode temperature profiles, *Phys. Fluids B* **4**, p. 634, http://link.aip.org/link/doi/10.1063/1.860261.

Bateman, G., Kinsey, J. E., Kritz, A. H., *et al.* (1996). *Proceedings of the Sixteenth International Atomic Energy Agency Fusion Energy Conference* (Montréal, Canada, 7-11 October), International Atomic Energy Agency, Vienna), vol. II, pp. 559-565.

Bateman, G., Kritz, A. H., Kinsey, J. E., *et al.* (1997). *Phys. Plasmas* **5**, p. 1793.

Bateman, G., Kritz, A. H., Redd, A. J., *et al.* (1998). *Proceedings of the Seventeenth International Atomic Energy Agency Fusion Energy Conference* (Yokohama, Japan, 19-24 October, 1998), International Atomic Energy Agency, Vienna, 1999), paper IAEA-F1-CN-69/THP2-19.

Beklemishev, A. and Horton, W. (1992). Anomalous ion thermal diffusion from η_i modes, *Phys. Fluids B* **4**, p. 2176, http://link.aip.org/link/doi/10.1063/1.860022.

Biglari, H., Diamond, P. H., and Terry, P. W. (1990). Influence of sheared poloidal rotation on edge turbulence, *Phys. Fluids B* **2**, 1, http://link.aip.org/link/doi/10.1063/1.859529.

deGrassie, J. S. (2009). *Plasma Phys. Control. Fusion* **51**, p. 124047, doi:10.1088/0741-3335/51/12/124047.

Diamond, P. H., Lebedev, V. B., Newman, D. E., Carreras, B. A., Hahm, T. S., Tang, W. M., Rewoldt, G., and Avinash, K. (1997). Dynamics of transition to enhanced confinement in reversed magnetic shear discharges, *Phys. Rev. Lett.* **78**, pp. 1472-1475, http://link.aps.org/doi/10.1103/PhysRevLett.78.1472.

Ernst, D., Bell, M., Bell, R., Bush, C., *et al.* (1998). *Phys. Plasmas* **5**, p. 665.

Fukuyama, A., Itoh, K., Itoh, S.-I., and Yagi, M. (1998). *Proceedings of the Seventeenth International Atomic Energy Agency Fusion Energy Conference, Yokohama, Japan, 19-24 October* (International Atomic Energy Agency, Vienna, 1999).

Gormezano, C. and the JET Team, *Proceedings of the Sixteenth International Atomic Energy Agency Fusion Energy Conference, 1997* (Montréal, Canada, 1996). 1, p. 487.

Grove, D. J. and Meade, D. M. (1985). Initial studies of confinement, adiabatic compression, and neutral-beam heating in TFTR, *Nucl. Fusion* **25**, p. 1167, doi:10.1088/0029-5515/25/9/029.

Hahm, T. S. and Burrell, K. H. (1995) Flow shear induced fluctuation suppression in finite aspect ratio shaped tokamak plasma, *Phys. Plasmas* **2**, p. 1648, http://link.aip.org/link/doi/10.1063/1.871313.

Hamaguchi, S. and Horton, W. (1992). Effects of sheared flows on ion-temperature-gradient-driven turbulent transport, *Phys. Fluids B* **4**, p. 319, http://link.aip.org/link/doi/10.1063/1.860280.

Ishida, S., *et al.* (1997). Achievement of high fusion performance in JT60-U reversed shear discharges, *Phys. Rev. Lett.* **79**, p. 3917.

Itoh, S-I. and Itoh, K. (2001). Statistical theory and transition in multi-scale-length turbulence in plasmas, *Plasma Phys. Control Fusion* **43**, pp. 1055-1102.

Kinsey, J. E. (1999). Time dependent transport simulations of JET H mode plasmas, *Nucl. Fusion* **39**, p. 539, doi:10.1088/0029-5515/39/4/310.

Koide, Y., Kikuchi, M., Mori, M., Tsuji, S., Ishida, S., Asakura, N., Kamada, Y., Nishitani, T., Kawano, Y., Hatae, T., Fujita, T., Fukuda, T., Sakasai, A., Kondoh, T., Yoshino, R., and Neyatani, Y. (1994). *Phys. Rev. Lett.* **2**, pp. 3662-3665, http://link.aps.org/doi/10.1103/PhysRevLett.72.3662.

Koide, Y. and JT60 Team. (1997). *Progress in confinement and stability with plasma shape and profile control for steady-state operation in the Japan Atomic Energy Research Institute Tokamak-60 Upgrade*, *Phys. Plasmas* **4**, p. 1623, `http://link.aip.org/link/?PHPAEN/4/1623/1`.

Levinton, F. M., Zarnstorff, M. C., Batha, S. H., Bell, M., Bell, R. E., Budny, R. V., Bush, C., Chang, Z., Fredrickson, E., Janos, A., Manickam, J., Ramsey, A., Sabbagh, S. A., Schmidt, G. L., Synakowski, E. J., and Taylor, G. (1995). *Phys. Rev. Lett.* **75**, pp. 4417-4420, `http://link.aps.org/doi/10.1103/PhysRevLett.75.4417`.

Luxon, J. L. and Davis, L. G. (1985). Big Dee-A flexible facility operating near break-even conditions, *Fusion Technology* **8**, pp. 441-449, OSTI ID:6011619.

Mori, M., Ishida, S., Ando, T., Annoh, K., Asakura, N., Azumi, M., van Blokland, A. A. E., Frieling, G. J. Fujii, T., Fujita, T., Fukuda, T., Funahashi, A., Hatae, T., Hoek, M., Honda, M., Hosogane, N., Isei, N., Itami, K., Kamada, Y., Kawano, Y., Kikuchi, M., Kimura, H., Kimura, T., Kishimoto, H., Kitsunezaki, A., Kodama, K., Koide, Y., Kondoh, T., Kubo, H., Kuriyama, M., Matsuoka, M., Matsuzaki, Y., Miya, N., Nagami, M., Nagashima, A., Naito, O., Nakamura, H., Nemoto, M., Neyatani, Y., Ninomiya, H., Nishitani, T., Ohga, T., Ohmori, S., Saidoh, M., Sakasai, A., Sato, M., Shimada, M., Shimizu, K., Shirai, H., Sugie, T., Takeuchi, H., Tani, K., Tobita, K., Tsuji, S., Ushigusa, K., Yamada, M., Yonekawa, I., Yoshida, H., and Yoshino, R. (1994). *Nucl. Fusion* **34**, p. 1045, doi:10.1088/0029-5515/34/7/I09.

Nilsson, J. and Weiland, J. (1994). Fluid model for general collisionality and magnetic curvature, *Nucl. Fusion* **34**, p. 803, doi:10.1088/0029-5515/34/6/I04.

Ninomiya, H. and the JT-60 Team. (1992). *Phys. Fluids B* **4**, p. 2070.

Nordman, H., Weiland, J., and Jarmén, A. (1990). Simulation of toroidal drift mode turbulence driven by temperature gradients and electron trapping, *Nucl. Fusion* **30**, p. 983, doi:10.1088/0029-5515/30/6/001.

Parail, V. V., Baranov, Yu.F., Challis, C. D., Cottrell, G. A., Fischer, B., Gormezano, C., Huysmans, G. T. A., Litaudon, X., Sips, A. C. C., Söldner, F. X., Springmann, E. M., Taroni, A., and Ward, D. J. (1999). Predictive modeling of JET optimized shear discharges, *Nucl. Fusion* **39**, p. 429, doi:10.1088/0029-5515/39/3/310.

Rebut, P.-H., Bickerton, R. J., and Keen, B. E. (1985). The Joint European Torus: installation, first results and prospects, *Nucl. Fusion* **25**, p. 1011, doi:10.1088/0029-5515/25/9/003.

Rice, B. W., Lazarus, E. A., Austin, M. E., Burrell, K. H., Casper, T. A., Groebner, R. J., Gohil, P., Forest, C. B., Ikezi, H., Lao, L. L., Mauel, M. E., Navratil, G. A., Stallard, B. W., Strait, E. J., and Taylor, T. S. (1996). Observations of enhanced core confinement in negative magnetic shear discharges with an L mode edge on DIII-D, *Nucl. Fusion* **36**, p. 1271, doi:10.1088/0029-5515/36/10/I12.

Rice, J. E., Ince-Cushman, A., deGrassie, J. S., Eriksson, L.-G., Sakamoto, Y., Scarabosio, A., Bortolon, A., Burrell, K. H., Duval, B. P., Fenzi-Bonizec, C., Greenwald, M. J., Groebner, R. J., Hoang, G. T., Koide, Y., Marmar, E. S., Pochelon, A., and Podpaly, Y. (2007). Inter-machine comparison of intrinsic toroidal rotation in tokamaks, *Nucl. Fusion* **47**, p. 1618, doi:10.1088/0029-5515/47/11/025.

Schissel, D. R., Greenfield, C. M., Deboo, J. C., *et al.* (1996). *Proceedings of the Sixteenth International Atomic Energy Agency Fusion Energy Conference, International Atomic Energy Agency, Vienna, 1997)* (Montréal, Canada, 1996) **1**, p. 463, paper IAEA F1-CN-64/A5-3.

Scott, B. (1997). Three-dimensional computation of drift Alfvén turbulence,*Plasma Phys. Control. Fusion* **39**, p. 1635, doi:10.1088/0741-3335/39/10/010.

Söldner, F. and the JET Team. (1997). *Plasma Phys. Control. Fusion* **39**, p. B353.

Strait, E. J., Lao, L. L., Mauel, M. E., Rice, B. W., Taylor, T. S., Burrell, K. H.,

Chu, M. S., Lazarus, E. A., Osborne, T. H., Thompson, S. J., and Turnbull A. D. (1995). Enhanced confinement and stability in DIII-D discharges with reversed magnetic shear, *Phys. Rev. Lett.* **75**, pp. 4421-4424, http://link.aps.org/doi/10.1103/PhysRevLett.75.4421.

Sugama, H. and Horton, W. (1995). Neoclassical and anomalous transport in axisymmetric toroidal plasmas with electrostatic turbulence, *Phys. Plasmas* **2**, 8, pp. 2989-3006, http://link.aip.org/link/?PHPAEN/2/2989/1.

Waltz, R. E., Kerbel, G. D., and Milovich, J. (1994). Toroidal gyro-Landau fluid model turbulence simulations in a nonlinear ballooning mode representation with radial modes, *Phys. Plasmas* **1**, p. 2229, http://link.aip.org/link/doi/10.1063/1.870934.

Weiland, J. and Hirose, A. (1992). Electromagnetic and kinetic effects on the ion temperature gradient mode, *Nucl. Fusion* **32**, p. 151, doi:10.1088/0029-5515/32/1/I13.

Zhu, P., Bateman, G., Kritz, A. H., and Horton, W. (2000). *Phys. Plasmas* **7**, p. 2898, http://ojps.aip.org/pop/popcr.jsp.

Zhu, P., Horton, W., and Sugama, H. (1999). *Phys. Plasmas* **6**, p. 2503.

Chapter 19

Gyrokinetic Simulations of Drift Wave Turbulence

A large variety of computer simulation codes have been developed over the last decade to perform first principle numerical calculations for the nonlinear dynamics of drift waves and other low frequency plasma collective dynamics. These plasma simulation codes include the interaction of the plasma convection and vortex structures with the plasma flows treated in Chapters 13 to 18. Understanding and treating analytically, even with relatively simple models, the interactions of the turbulence, plasma flows and radially elongated convective cells is a challenge. While simple paradigm models developed by theorists such as the "predator-prey" models of Diamond, *et al.* (2001, 2008, 2009), and low-dimensional nonlinear models provide a useful framework for deriving analytical conditions for the bifurcation points between the different confinement regimes [Hu and Horton (1997); Yagi, *et al.* (1989); Kim, *et al.* (1993); Sugama and Horton (1995); Yagi and Horton (1994); Thiffeault and Horton (1996); Horton, *et al.* (1996)]. Gluhovsky and Tong (1999) develop further the conditions for the energy conserving truncations in the low-order models. Numerical simulations are required for comparisons with data as shown in Chapter 18 and to give quantitative projections. The analytic models provide guides as to what to expect in the complex, but realistic simulations, as one of the many system parameters, is varied.

In a gyrokinetic approach the dimensionality of the phase space is reduced from six to five, by averaging out the fast gyromotion. Based on the dynamics of gyrocenters, the gyrokinetic equation is formulated for each species. These gyrocenter phase-space distributions are solved either with sampled particle orbits or as the continuum Vlasov phase-space probability density together in each time step with the Poisson equation for the electric potential, and Maxwell's equations for the electromagnetic fields. A review of the method is given in Brizard and Hahm (2007); Hahm (1988); Lee, *et al.* (2001); and Lee, *et al.* (2008a,b).

Gyrokinetic simulations are generally based on two main methods: PIC and fluid methods. In the Lagrangian Particle-in-Cell (PIC) [Birdsall and Langdon (2004)] approach equations of motion are solved for each individual marker-particle taken from the plasma particle distributions. Choice of the markers is equivalent to the random sampling of the phase space. The fluid moments, such as density, flow velocity, and pressure are calculated by summing up the contributions from the individual markers. The remaining field equations are solved on the three-dimensional spatial mesh. The main disadvantage of the Lagrangian approach is statistical error caused by insufficient number of samples (number of markers is much smaller than number of physical particles), which is referred to as discrete particle noise. The problem of noise can be controlled by increasing the number of simulation particles, applying various smart sampling techniques, and by using δf method [Parker, *et al.* (1993)], where additional dynamic variable, representing particle's phase space density, is introduced. One of the advantages of PIC method is good parallelization, which is especially important with modern supercomputers. The major codes using Lagrangian approach are GEM [Parker, *et al.* (1993)], GTC [Lin, *et al.* (1998)], GT3D [Idomura (2003, 2006)], GTS [Wang, *et al.* (2007)], ORB5 [Jolliet, *et al.* (2007)], and others.

In the Eulerian (Vlasov) approach the gyrokinetic equation is solved numerically on a regular five-dimensional space-velocity grid. This approach typically requires large computational resources to achieve sufficient resolution. The major codes using Eulerian approach are GS2 [Kotschenreuther, *et al.* (1995); Dorland, *et al.* (2000)], GENE [Jenko (2000)], GYRO [Candy and Waltz (2003a)], GT5D [Idomura, *et al.* (2006)], GKV [Watanabe, *et al.* (2008)], and others.

The main drawback of the gyrokinetic simulations is that hours of wall clock time are required to run the simulations for each second of laboratory plasma discharge time. Currently, the developers are turning to using the graphical processing units, GPUs, to reduce the runs times at the cost of increased complexity of programming. In contrast, low-dimensional models (LOMs) in Chapter 12 and 13 plasma models are well suited for building real-time control systems with set of ODEs, typically or not more than ten to twelve. These low-dimensional dynamical models can be programmed in Field Programmable Gated Arrays (FPGAs) and Graphical Processing Units (GPUs) for feedback control of the fueling system, the heating and magnetic control systems, and the external ELM control coils. Predicting the plasma dynamics with low-dimensional models is well documented in the implementation of space weather forecasting tools. One example is the WINDMI model running real-time at the NASA Community Coordinated Modeling Center `http:\\ccmc.gsfc.nasa.gov`. For space weather the input signals for the computer codes are from solar wind monitors at the Lagrangian-1 point one-hour upstream on the Earth-Sun line. Data from the solar wind monitoring spacecraft is received every few minutes giving forecasts of storm and substorm with lead times of approximately one hour.

In the period the 1990s to 2007 there was a community-wide effort to benchmark a set of tokamak simulation codes for a set of reference parameters. The choice of parameters was not critical for the benchmarking but was to be typical for tokamak discharges of that era. A particular time slice and radial position in an L-mode discharge in the DIII-D tokamak was selected. Note, the chosen L-mode discharge

Table 19.1 Historical growth from 1997 to 2009 of computational power from massively parallel computers for gyrokinetic simulations.

	Computer	Core #	Speed TF	Particle #	Step #
1997	Cray C90	10^1 NERSC	10^{-2}	10^7	10^4
1998	Cray T3E	10^2 NERSC	10^{-1}	10^8	10^4
2002	IBM SP	10^3 NERSC	10^0	10^9	10^4
2007	Cray XT3	10^4 ORNL	10^2	10^{10}	10^4
2008	Cray XT4	10^4 NERSC	10^2	10^{10}	10^4
2009	Cray XT5	10^5 ORNL	10^3	10^{11}	10^5

is considerably simpler than the DIII-D discharge analyzed in Chapter 18. The benchmarking mission was called CYCLONE, and this set of discharge parameters became known as the CYCLONE parameters. The parameters are such that there is a moderately strong ion temperature gradient (ITG) driven instability. The ITG instabilities are described in Chapter 13 where their general form includes a coupling with the trapped electron mode (TEM) and electromagnetic fluctuations are developed. In the CYCLONE base test case for code comparisons, however, these complications were eliminated by defining the benchmark mission as being a comparison in which the electron dynamics is taken as adiabatic. These means that the ion dynamics is complicated and nonlinear but the electron response to the fluctuating potential is linear and analytically specified as $\delta n_e = n_e \, e\phi/T_e$ in the plasma fluctuations.

In Fig. 19.1 we show an example for one simulation code used in these comparisons of models with data. As shown in Chapter 13 all of the codes used in the CYCLONE bench marking mission were able to get the basic frequencies, growth rates and ion heat flux to agree with the linear ITG plasma dispersion relation for the reference L mode discharge in DIII-D. The differences in variations of frequency and growth rate with wavenumber over a certain range of mode numbers are relatively narrow. The nonlinear development of the instabilities show more spread in the results depending on the simulation code was used. The simulation codes use different choices for how to represent the spatial grids, the fields and the gyro-averaging of the strongly magnetized particles. There are numerous choices to be made for the algorithms for advancing the fields and the particles. These algorithms

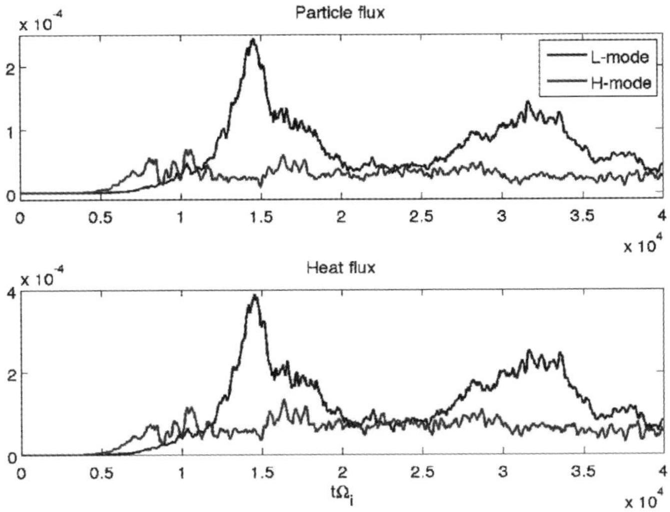

Fig. 19.1 Typical gyrokinetic simulation of a tokamak from the 2000 decade with massively parallel computers [Chang, *et al.* (2009)]. The DIII-D experimental magnetic equilibrium is used. The density, temperature and E_r profiles are computed with the XGC0 neoclassical code. The particle flux and ion energy flux normalized to the ion thermal velocity for the L- and H-mode comparison. Density, temperature and E_r profiles are obtained from the XGC0 neoclassical calculation. The DIII-D experimental magnetic equilibrium is used.

will dictate various constraints on the time steps used in the evolution. Thus, for a given physical problem there will be a spread of answers corresponding to differences produced by the different algorithms used in the numerical simulations.

19.1 Benchmark of Simulation Codes on Massively Parallel Machines of the 2000-2010 Era

In gyrokinetic simulations each time step has two parts: first, a particle push according to gyrocenter equation of motion (PIC approach), or grid solution of gyrokinetic equation (Vlasov approach) followed by (2) a calculation of the electromagnetic fields using the new distributions of charges and currents. The method of solving the self-consistent field problem using a particle dynamics where the gyro-phase angle has been averaged out varies with codes and is complex when the electromagnetic fields are included. The electrostatic dynamics is relatively simple and available in well documented codes. The problem is called gyrokinetics because the orbits have been reduced to eliminate the gyrophase angle which rotates at the cyclotron frequency. This method reduces the particle phase space to $d = 5$ dimensions allowing higher resolution for the low frequency drift waves than the original particle-in-cell simulations of Okuda, *et al.* (1979) and Horton, *et al.* (1977) that used full $d = 6$ particle dynamics. The gyrokinetic method has been developed in such detail that one could devote a book to describing the method. Texts for the particle-in-cell simulations methods include Birdsall and Langdon (2004) and Tajima (2004). The gyrokinetic simulations evolved from these PIC methods.

Gyrokinetic simulations have demonstrated parallel scalability and efficient utilization of hundreds of thousands of processors on modern parallel computers. Writing the codes requires the methods of message passing instructions (mpi) and Open MP for utilizing the shared memory on the massively parallel computers. Some machines used in a recent bench mark study are shown in Table 19.1.

The significant benefits of the parallel computing performance are two-fold. First, the parallel computing allows the simulation of large plasma volumes with few approximations using spatially global magnetic fields of the tokamak and also the helical/stellarator confinement which has no symmetry. Second, parallel computing provides the capability for investigating convergence with respect to phase space resolution for verifying convergence from discretizations and comparisons with other methods including direct discretization of the continuum partial differential equations of the phase space density. Having available both continuum (Galerkin approximation, for example) and particle-in-cell methods solving nonlinear gyrokinetic equations provides a cross-check and improved verification of the plasma simulations.

Present day advanced gyrokinetic codes include (i) kinetic electrons, (ii) electromagnetic perturbations, and (iii) collisions. Some codes are able to use complicated magnetic equilibria. The first codes used the electrostatic approximation for the electric field with no perturbations to the magnetic field. This approximation is sufficient in low-beta plasmas and results in great simplification of the problem. For the fusion grade plasmas, however, one sees from Chapters 11 and 15 that magnetic perturbations are typically critical to the dynamics. The inclusion of the electromagnetic field perturbations has proven to be a difficult problem which is still under some development.

The simpler adiabatic electron and/or electrostatic simulations have been widely reported by many codes. The differences with respect to the finite plasma beta electromagnetic simulations are reported in Chen, *et al.* (2003) and Candy and Waltz (2003b).

The PPPL code of Wang (2007, 2009), called the GTS particle code has been applied to collisionless TEM with kinetic electrons. Center for Plasma Edge Simulation (CPES) code is called the XGC1 edge turbulence code which has results with adiabatic electron physics in Chang, *et al.* (2009) and Ku, *et al.* (2009). There are also widely used and important particle codes with less physics including the PG3EQ [Dimits, *et al.* (1996)] flux tube simulation with one species treated as adiabatic and the global UCAN code [Sydora, *et al.* (1996)]. The UCAN code has results with trapped electrons and adiabatic passing electrons [Sydora, *et al.* (1996)]. The UCAN code was used without trapped electron physics in publications [Rhodes, *et al.* (2002); Kniep, *et al.* (2004)].

In addition, there are two complimentary leading U.S.-based continuum gyrokinetic codes: GYRO [Candy and Waltz (2003a)] and GS2 [Dorland, *et al.* (2000)] and several international continuum efforts as well.

There are also many outstanding international gyrokinetic particle simulation efforts including: UCAN-Canada, ORB5-Switzerland, ELMFIRE-Finland, GKV and GT3D-Japan, and GYSELA in France to name a few.

These numerical simulations have advantages for studies near the last closed flux surface regions which are regions with steep plasma gradients. These gradients regions are called the edge pedestal and are the regions of emissions of the Edge Localized Modes or ELMs shown in Chapter 14. More generally the codes have been applied in verification and validation efforts in particular, profiles taken from the NSTX, DIII-D, and Alcator C-Mod tokamaks and the basic plasma experiments LAPD and CLM.

The global axisymmetric simulations give important information on the radial electric field and toroidal plasma rotation in nearly collisionless toroidal plasmas. In the future, the gyrokinetic codes can be generalized to include the role of density fueling and impurity transport in tokamak plasmas using gyrokinetic particle simulation. Investigations by W. W. Lee showed a difference in the transport with the working gas ion species of H, D, T consistent with an explanation derived from drift wave theory for the ion isotope effect Dong (1994) on the energy confinement time.

The development and use of this gyrokinetic algorithm requires knowledge of performance optimization on present-day petascale computers and preparatory work underway for the future exascale computers. Advanced data management tools are utilized for the particle codes, as well as fast parallel I/O, with the goals of both improving performance and facilitating the sharing of data for efficient collaboration amongst the research team.

Now we give some details of particular gyrokinetic simulation codes. Benchmark runs of the gyrokinetic Vlasov code called GKV, developed by Watanabe and Sugama and the MHD code called MEGA, developed by Todo (2008), both at the National Institute for Fusion Science (NIFS) on six computing platforms is published as Watanabe, *et al.* (2008). The same large simulations are compared on the platforms of the Plasma Simulator (SX-7 vector machine with 1.4 TFlops), the Earth Simulator (ES vector machine with 40 TFlops), HPCx (Power5+ with 18.8 TFlops), Franklin (Opteron dual core processors with 101 TFlops) and Ranger (Opteron quad core processors with 503 TFlops). The compilers, the interconnect methods and the parallelized FFT libraries are given in the article. The speedup factors of five machines versus the first machine, the Plasma Simulator with the SX-7 vector processor are shown. The codes were originally developed for the Plasma Simulator but were modified and partially optimized for the other five machines. The Franklin and the Ranger are massively parallel machines with the Opteron cores. The speed-up factor is order 10 to 14 for the Franklin computer compared with the Plasma Simulator. The growth of the speed-up factor was found to saturate on the Ranger at about 16,384 cores due to the increase in the internode communication times. The results of the comparisons show that the internode-connection speed is the critical factor for speed-up on the massively parallel computers. These times begin to dominate at around 10,000 cores. Thus, to continue a near-to linear speed-up beyond this number of cores, it would be necessary to develop new programs, or new simulation algorithms, that make smaller demands on the memory throughput and internode connections.

19.2 Gyrokinetic Toroidal Code

The gyrokinetic toroidal code (GTC) [Lin, *et al.* (1998)] is a well-benchmarked full torus particle-in-cell code for simulations of microturbulence and energetic particle transport. In 2010-11 time frame integrated single version capable of nonperturbative (full-f) and perturbative (δf) simulations was made available. The GTC software includes kinetic electrons and electromagnetic fluctuations, general toroidal geometry with shaped, up-down asymmetric equilibrium (using EFIT data) and experimental plasma profiles (using TRANSP or ONETWO data), multiple ion species, neoclassical effects with Fokker-Planck collision operators conserving particle, momentum and energy, equilibrium radial electric field with toroidal and poloidal rotations, and sources/sinks and external antenna mode excitation. GTC is a platform-independent program using standard FORTRAN, MPI, and OpenMP, and achieves nearly perfect scalability on many massively parallel computers, including the world fastest petascale computer with more than 10^5 cores.

GTC has made long-time runs on the world's fastest supercomputers: the CRAY XT5 petaflop computer at ORNL, where the runs served as a benchmark code [SciDac Review (2009)].

19.2.1 *Kinetic electrons and electromagnetic turbulence*

The small electron mass presents a numerical difficulty for simultaneously treating the dynamics of ions and electrons in simulations of drift-Alfvénic turbulence in high-pressure plasmas. A fluid-kinetic hybrid electron model that overcomes this difficulty has been developed based on an expansion of the electron response using the electron-ion mass ratio as a small parameter [Lin and Chen (2001)].

Electrons are adiabatic to the lowest order and can be described by a massless fluid model. At a higher order, kinetic effects are treated by the electron drift kinetic equation [Lin, *et al.* (2007); Holod and Lin (2010)]. Excellent numerical efficiency is achieved by overcoming the electron Courant condition and suppressing high frequency modes, while retaining electron wave-particle interactions. Using this kinetic electron model, electrostatic ITG and TEM turbulence, as well as electromagnetic Alfvénic modes have been studied extensively [Xiao and Lin (2009); Deng and Lin (2009); Nishimura, *et al.* (2009); Holod and Lin (2010); Xiao, *et al.* (2010); Zhang, *et al.* (2010b)]. Shear Alfvén wave, continuum damping, and the frequency gap in toroidal geometry has been demonstrated [Nishimura, *et al.* (2007)]. Excitation of the toroidal Alfvén eigenmode by energetic particles has been verified [Nishimura (2009)]. The linear tearing mode is removed in this fluid-kinetic electron model to improve numerical properties. Because of the flexibility of the GTC algorithm and code structure, the electron split-weight scheme [Lee, *et al.* (2001)] can be easily implemented in addition to the fluid-kinetic electron model for studying tearing mode physics as needed.

19.2.2 *Perturbative (δf) and nonperturbative (full-f) simulation*

The perturbative method (δf) has been used extensively in plasma simulation to improve numerical properties [Parker, *et al.* (1993)]. However, it cannot efficiently address key aspects of turbulent transport on long-time scales, such as the plasma profile evolution. Properly addressing the long time behavior is important for multiscale global simulation with time scales ranging from the turbulence dynamical time to the plasma profile relaxation time and with spatial scales ranging from the ion gyroradius to the device size. A generalized weight-based scheme has been implemented in GTC to simulate both the perturbed distribution (δf) and the full distribution (full-f) within the same GTC code framework. Turbulence intensity and heat transport from preliminary full-f ITG simulations agree with δf simulation on the turbulence dynamical time scale.

19.2.3 *General toroidal geometry and experimental plasma profiles*

Toroidal geometry is treated rigorously in real space using magnetic coordinates. The guiding center equations of motion can be derived from a guiding center Hamiltonian or Lagrangian [White and Zakharov (2003)] which conserves phase space volume and is best for integrating particle orbits for a long period. Large device simulations only became feasible with the implementation of an efficient global field-aligned mesh [Lin, *et al.* (2002)] for the quasi-2D structure of toroidal drift and Alfvén eigenmodes. The global field-aligned mesh provides the highest computational efficiency without any simplification in physics models or toroidal geometry. A non-spectral gyrokinetic Poisson solver [Lin and Lee (1995)] calculates the polarization current locally in real space and allows realistic treatment of the equilibrium profile variations such as steep pressure gradients and poloidal variations of the magnetic field. The original iterative solver is generalized for treating kinetic electrons and electromagnetic fluctuations by using a highly scalable sparse matrix solver PETSc [Nishimura, *et al.* (2006)]. Global toroidal geometry with strong shaping and up-down asymmetry has been implemented through interface with MHD equilibrium solvers EFIT and experimental plasma profiles have been used through interface with tokamak transport codes TRANSP and ONETWO.

Non-axisymmetric equilibrium magnetic field can be added using a spline fit. Collisional transport in 3D non-axisymmetric equilibrium has been studied for stellarator design [Reimann, *et al.* (1999); Neilson, *et al.* (2000); Reimann, *et al.* (2001); Lee, *et al.* (2001)].

19.2.4 *Versions and development path of the GTC gyrokinetic code*

GTC Version 1 was developed (1996-2000) at Princeton Plasma Physics Laboratory (PPPL) with electrostatic turbulence, collision operators, general geometry, and 1D domain-decomposition [Lin, *et al.* (1998)]. Kinetic electrons, field-aligned mesh, and multi-level parallelism were implemented (2000-2004) in GTC Version 2 at PPPL and UCI [Lin, *et al.* (2002); Ethier and Lin, 2004]. GTC Version 3 was developed beginning in 2004 by UCI GTC team and collaborators and maintained at UCI with electromagnetic turbulence, nonperturbative (full-f) method, equilibrium electric field and rotation, multiple ion species, PETSc solver, and ADIOS, in addition to all capabilities in versions 1 and 2 [Holod, *et al.* (2009)]. Additional variants of Version 2 also exist and are known as GTC-P [Ethier, *et al.* (2008)] and GTS [Wang, *et al.* (2006)] both limited to electrostatic turbulence.

During the 2007-2010 SciDAC GPS funding cycle, the GTC team completed the most comprehensive upgrade of GTC, published 30 journal articles including four in PRL and delivered 14 invited and plenary talks at major conferences (APS-DPP, EPS, Sherwood, SciDAC, etc). In particular, 15 turbulence simulation papers [Holod and Lin (2007); Holod and Lin (2008); Holod and Lin (2010); Lin, *et al.* (2007); Zhang, *et al.* (2008); Zhang, *et al.* (2009); Zhang, *et al.* (2010a); Zhang, *et al.* (2010b); Nishimura, *et al.* (2007); Nishimura, *et al.* (2009); Nishimura (2009); Xiao and Lin (2009); Xiao, *et al.* (2010); Deng and Lin (2009)] first-authored by GTC members were published in peer-reviewed physics journals. In addition, during the same period, there were two turbulence simulation articles published each by XGC1 [Chang, *et al.* (2009); Ku, *et al.* (2009)] and by GTS [Wang, *et al.* (2007, 2009)].

19.2.5 *Non-diffusive transport of trapped electron mode turbulence*

The nature of the trapped electron transport is reported in Xiao and Lin (2009) and Xiao, *et al.* (2010). GTC simulations of collisionless trapped electron mode (CTEM) turbulence show that the electron heat transport exhibits a gradual transition from Bohm to gyroBohm with device size scaling. Turbulence eddies are predominantly microscopic but with a significant mesoscale component. The radial profile of the electron heat conductivity only follows the profile of fluctuation intensity on a global scale, whereas the ion transport is more sensitive to the local fluctuation intensity. This suggests the existence of a non-diffusive component in the electron heat flux, which is characterized by a probability density function that deviates from a lognormal distribution. The non-diffusive transport arises from the ballistic radial $\boldsymbol{E} \times \boldsymbol{B}$ drift of trapped electrons due to a combination of the presence of mesoscale eddies and the weak detuning of the toroidal precessional resonance that drives the CTEM instability.

In contrast, the ion radial excursion is not affected significantly by the mesoscale eddies due to the rapid parallel decorrelation in the ion dynamics. This effect is not operational for trapped electrons because of the bounce averaging process associated with the fast electron parallel motion. This difference in the ion and electron response wasconfirmed by a comprehensive analysis of kinetic and fluid time scales.

19.2.6 *Turbulence spreading and non-local transport in reversed shear plasmas*

Deng and Lin (2009) and Guo, *et al.* (2009) investigate the turbulence spreading and non-local transport in reversed shear plasmas. GTC simulations of reversed magnetic shear plasmas find that the linear eigenmode of the ITG instability exhibits a mode gap around the minimum safety factor ($q_{\min}$) region when $q_{\min}$ is an integer, and that the CTEM instability is suppressed in the negative-shear region. However, after nonlinear saturation, the ITG gap is filled up by turbulence spreading. The CTEM fluctuations propagate into the stable negative-shear region. The simulations indicate that the electrostatic drift wave turbulence itself does not support either a linear or a nonlinear mechanism for the formation of internal transport barriers in reversed magnetic shear plasmas, when the $q_{\min}$ crossing is an integer. In a related analytic study, the stationary solution of the coupled partial differential equations governing the self-consistent spatio-temporal evolution of a drift-wave (DW) radial envelope and a zonal-flow (ZF) amplitude yields the formation of DW-ZF soliton structures, which propagate radially with speed depending on the envelope peak amplitude. The propagation of solitons causes significant radial spreading of DW turbulence and therefore can affect transport scaling with the system size by broadening of the turbulent region.

19.2.7 *Vlasov (Continuum) Gyrokinetic Simulations*

The Vlasov (or continuum) gyrokinetic simulation method has been developed as an alternative to the particle (marker)-based simulation codes. Advantages of the Vlasov-type simulation using numerical grids and/or spectral expansion for solving the gyrokinetic equations are the noise-free property and the explicitness of numerical resolution.

The GKV code numerically time-integrates the gyrokinetic equation for the perturbed distribution function (δf) code was developed [Watanabe and Sugama (2006); Watanabe, *et al.* (2008b)] at the National Institute for Fusion Science and has been applied to the ITG and ETG turbulence and zonal flows in tokamak and helical systems. A GKV-X for the helical toroidal system with ion temperature gradient driven turbulence is shown in Fig. 12.2 and in color on the cover of this book.

The nonlinear $\boldsymbol{E} \times \boldsymbol{B}$ advection term is calculated by the Fourier spectral method in the flux tube coordinates [Beer, *et al.* (1995)]. The fourth-order finite difference method is applied to the field-aligned and the velocity-space coordinates. The finite collisionality is introduced by the Lenard-Bernstein model collision operator, which can dissipate fine scale structures of the perturbed distribution function [Watanabe and Sugama (2006); Watanabe, *et al.* (2008b)].

Numerical results of the GKV simulations have been carefully monitored by the entropy balance relation so as to confirm the high resolution for the configuration (x, y, z)-space and velocity space to determine the required grid size. The entropy transfer analysis also gives rise to a new insight into the turbulence and zonal flow interactions [Nakata, *et al.* (2012)]. The GKV code has been cross-benchmarked with other continuum codes and applied to non-axisymmetric systems [Xanthopoulos, *et al.* (2011)]. The code is continuously maintained with further extensions of the physical models and the numerical methods while being implemented in state-of-the-art of supercomputer systems.

19.3 Energetic Particle Transport

GTC simulations of the energetic particle diffusion by the ITG turbulence show that the diffusivity decreases drastically for high-energy particles due to the averaging effect of the large gyroradius and banana width, and also the fast wave-particle decorrelation. By performing the integration in phase space, we can calculate the diffusivity for any distribution function. The NBI ion diffusivity driven by the ITG turbulence is found to decrease rapidly for birth energies up to an order of magnitude above the background plasma thermal energy. Results from GTC simulations have been used to explain some features of the measured fast-ion transport in the DIII-D tokamak [Zhang, *et al.* (2008); Zhang, *et al.* (2010a)].

19.4 Momentum Turbulent Transport

Now we consider the works of gyrokinetic particle simulations of toroidal momentum transport [Holod and Lin (2008)] and the effects of electron dynamics in toroidal momentum transport driven by ion temperature gradient turbulence [Holod and Lin (2010)].

On the turbulence simulation time scale (many correlation times or milliseconds), a significant redistribution of toroidal momentum is observed in GTC simulations from ITG turbulence, resulting in a peaked momentum profile in the central region of the radial domain. Diffusive and off-diagonal (pinch and residual) fluxes are identified and delineated. Both simulation and quasilinear theory find that the ratio of momentum to heat conductivity is smaller than unity even after subtracting the pinch contribution when wave-particle resonance energy is larger than the thermal energy. The magnitude and structure of the momentum flux follows gyroBohm scaling for large device size. The symmetry breaking due to the shear of the radial electric field is found to be a mechanism for generating the residual momentum flux. However,the residual flux is small compared to the momentum pinch term in the case of a finite background rotation. The trapped electrons in the ITG turbulence increase the intensity and modify the spectral properties, leading to the increase of the toroidal momentum pinch.

19.5 Wave-Particle Decorrelation as the Transport Mechanism in ETG Turbulence

Comprehensive analysis of first-principle GTC simulations shows that stochastic wave particle decorrelation is the dominant mechanism responsible for electron heat transport driven by electron temperature gradient (ETG) turbulence with extended radial streamers [Fu, *et al.* (2012)]. The transport is proportional to the local fluctuation intensity, and phase-space island overlap leads to a diffusive process with a time scale comparable to the wave-particle decorrelation time, determined by the fluctuation spectral width. This kinetic time scale is much shorter than the fluid time scale of eddy mixing [Lin, *et al.* (2007)].

19.5.1 *Statistical theory of particle noise-induced transport in PIC simulation*

The problem of discrete particle noise has been studied based on direct fluctuation measurements from GTC simulations of stable plasmas. Noise-driven transport is calculated using the quasilinear theory and the noise spectrum measured from simulations. The theoretical value of the diffusivity shows good agreement with simulations. The simulations show that for the realistic parameters, the noise-driven transport depends linearly on the numerical entropy in actual turbulence simulations. This makes it possible to estimate and subtract the noise contribution to the total transport during turbulence simulations [Holod and Lin (2007)].

19.5.2 *Global simulation of ITG modes with profile effects*

The profile modification due to the ITG turbulence, using the global GTC code, has shown a weak effect on the steady-state transport. Similar simulations have found that the nonlinear saturation of ITG modes and the resulting transport are intimately related to the nonlinear velocity space effects and the long wavelength global zonal flows [Lee, *et al.* (2008a); Lee, *et al.* (2008b)].

19.6 GEM Code Development Path

The GEM code [Chen and Parker (2003, 2007a)] is a gyrokinetic δf Particle-In-Cell (PIC) simulation developed for the study of core tokamak plasma microturbulence and associated transport. GEM uses generalized field-aligned magnetic coordinates [Chen and Parker (2007a)] parametrized by the Miller equilibrium model [Miller, *et al.* (1998)]. Kinetic electrons and electromagnetic perturbations are included using the parallel canonical momentum formalism [Hahm (1988)]. An adjustable split-weight scheme [Chen and Parker (2003)] is used to enhance the time step otherwise limited by the fast electron motion along the magnetic field.

GEM has an advanced finite-β Ampére solver to allow accurate simulation of Alfvén waves or finite-β modification of ITG modes [Chen and Parker (2003)]. GEM is radially global, with multiple ion species, arbitrary density and temperature profiles and equilibrium E_r profile [Chen and Parker (2007a)]. For long-time simulation, a Coarse-Graining Procedure (CGP) has been developed to control the secular growth of the mean square of the particle weights [Chen and Parker (2007b)]. Such secular growth of the particle weights is a necessary consequence of entropy production in steady-state turbulence with non-zero transport [Krommes (1999)], but can be suppressed using a CGP algorithm without changing quantities of physical interest such as the particle and energy fluxes.

For parallelization, GEM currently uses a 1-D domain decomposition along the field line direction, with domain cloning in the perpendicular plane [Kim and Parker (2000)]. GEM has been extensively benchmarked both linearly and nonlinearly with the continuum codes GS2 and GYRO on ITG, TEM, and ETG and the kinetic ballooning mode, η_i collisional effects and flux surface shape effects [Chen and Parker (2003, 2007a), Nevins, *et al.* (2007), and Ernst, *et al.* (2009)]. GEM has been used extensively to study TEM turbulence [Lang, *et al.* (2009, 2009)], edge turbulence [Chen and Parker (2008)] and density fueling in burning plasmas [Wan, *et al.* (2010)].

GEM has also demonstrated nonlinear saturation of the collision-less and semi-collisional tearing mode and the Rutherford regime [Wan, *et al.* (2005a,b)]. A linear mass-less fluid electron model has been implemented in GEM and has been used to study energetic particle driven TAEs. GEM demonstrated the first kinetic simulations of radiative damping of TAE due to thermal ion kinetic effects [Lang, *et al.* (2009)].

19.7 Outlook for Future of Toroidal Magnetic Confinement Systems and Nuclear Fusion Power Reactors

There is now, after fifty years of research and development of the toroidal magnetic confinement machines, a large literature and database on the plasma confinement and control that continues to grow every year. Dahlburg, *et al.* (2005) give a detailed review of the accomplishments and prospects for the three major facilities at this time in the US: NSTX, Alcator C Mod and DIII-D. The Dahlburg report was prepared in response to a request from the U.S. Department of Energy Fusion Energy Sciences Advisory Committee (FESAC) for an evaluation of the major magnetic fusion experimental facilities in 2005.

The status of ITER is discussed in Chapter 1 and in Tables 1.1 and 1.6. The fusion community waits anxiously to see if that huge international effort will achieve its fusion power objectives, or perhaps exceed, the expected power amplification of a factor ten from the large injected plasma heating power planned for the facility. Meanwhile, the new Japan-Europe effort to build a new superconducting tokamak JT60SA running near break-even is being built to test the plasma facing wall components to the continuous high-power neutron fluxes that will occur in a fusion reactor. This fusion materials testing facility is proceeding rapidly and will be ready before the tritium runs in ITER are carried out.

With some serendipity, magnetic confinement fusion scientists expect to use the knowledge we gained from ITER and JT60SA to build a fusion power reactor as a demonstration power plant in the next thirty years. This length of time required to build a demo is of course discouragingly for the private sector and to political parties providing the research funds for new sources of carbon-free electric power.

The lessons from the March 2011 Fukushima Daiichi meltdown of the fission power fuel rods, and the loss of confinement of radio activity fission materials from the fuel rods has slowed the progress toward a fission reactor basis for a carbon-free electric power community, is that the promise of a fusion power electric generator system must be pursued as the long-term solution to a clean, low-radiation level, nuclear power electric economy. The Cadarache 2007 Annual Report sets forth the scale-up required from ITER to DEMO for the building of a demonstration tokamak-based fusion power electric generator [Garcia, *et al.* (2008)].

The purpose of this book is to put in better perspective the key issues of the turbulent transport of particles and their thermal energy that has limited the performance of several successive generations of toroidal confinement machines short of achieving the expected results. The progress has been considerable and the author expects the next generation of scientists and their machines to achieve the conditions of a steady nuclear reacting fusion plasma.

Mother nature's way to produce large power is clearly nuclear fusion. The carbon based fuels used in today's electric based-society arise from the nuclear fusion power of the Sun. The possibility of using the same method for producing power from a man-made machine is the grand challenge for the next generation of engineers and scientists.

References

Beer, M. A., Cowley, S. C., Hammett, G. W. (1995). *Phys Plasmas* **2**, p. 2687.

Birdsall, C. K. and Langdon, A. B. (2004). *Series in Plasma Physics* (Taylor and Francis Group), ISBN:9780750310253.

Brizard, A. J. and Hahm, T. S. (2007). Foundations of nonlinear gyrokinetic theory, *Rev. Mod. Phys.* **79**, pp. 421-468, doi:10.1103/RevModPhys.79.421.

Candy, J. and Waltz, R. (2003a). An Eulerian gyrokinetic-Maxwell solver, *J. Comput. Phys.* **186**, p. 545, doi:10.1016/S0021-9991(03)00079-2.

Candy, J. and Waltz, R. (2003b). Anomalous transport scaling in the DIII-D Tokamak matched by supercomputer simulation, *Phys. Rev. Lett.* **91**, p. 045001, `http://link.aps.org/doi/10.1103/PhysRevLett.91.045001`.

Chang, C. S., Ku, S., Diamond, P. H., Lin, Z. H., Parker, S. E., Hahm, T. S., and Samatova, N. (2009). Compressed ion temperature gradient turbulence in diverted tokamak edge, *Phys. Plasmas* **16**, p. 056108, `http://link.aip.org/link/doi/10.1063/1.3099329`.

Chen, Y. and Parker, S. E. (2003). A δf particle method for gyrokinetic simulations with kinetic electrons and electromagnetic perturbations, *J. Comput. Phys.* **189**, p. 463, doi:10.1016/S0021-9991(03)00228-6.

Chen, Y. and Parker, S. E. (2007). Electromagnetic gyrokinetic δf particle-in-cell turbulence simulation with realistic equilibrium profiles and geometry, *J. Comput. Phys.* **220**, pp. 839-855, doi:10.1016/j.jcp.2006.05.028.

Chen, Y. and Parker, S. E. (2007). Coarse-graining phase space in δf particle-in-cell simulations, *Phys. Plasmas* **14**, p. 082301, `http://link.aip.org/link/doi/10.1063/1.2751603`.

Chen, Y. and Parker, S. E. (2008). *Phys. Plasmas* **15**, p. 055905.

Dahlburg, J., Allen, S. L., Betti, R., Knowlton, S., Maingi, R, Navratil, G. A., Sabbagh, S. A., Sheffield, J.,Van Dam, J. W., and Whyte, D. (2005). Characteristics and contributions of the three major United States toroidal magnetic fusion facilities, *J. Fusion Energy* **24**, pp. 3/4, doi:10.1007/s10894-005-8543-y.

Deng, W. J. and Lin, Z. H. (2009). *Phys. Plasmas* **16**, p. 102503.

Diamond, P. H., Champeaux, S., Malkov, M., Das, A., Gruzinov, I., Rosenbluth, M. N., Holland, C., Wecht, B., Smolyakov, A. I., Hinton, F. L., Lin, Z. H., and Hahm, T. S. (2001). *Nucl. Fusion* **41**, p. 1067.

Diamond, P. H., McDevitt, C. J., Gurcan, O. D., Hahm, T. S., and Naulin, V. (2008). *Phys. Plasmas* **15**, p. 012303.

Diamond, P. H., McDevitt, C. J., Gurcan, O. D., Hahm, T. S., Wang, W. X., Yoon, E. S., Holod, I., Lin, Z., Naulin, V., and Singh, R. (2009). Large orbit effects, *Nucl. Fusion* **49**, p. 11.

Dimits, A. M., Williams, T. J., Byers, J. A., and Cohen, B. I. (1996). Scalings of Ion-Temperature-Gradient-Driven Anomalous Transport in Tokamaks, *Phys. Rev. Lett.* **77**, pp. 71-74, `http://link.aps.org/doi/10.1103/PhysRevLett.77.71`.

Dong, J.-Q., Horton, W., and Dorland, W. (1994). Isotope scaling and η_i mode with impurities in tokamak plasmas), *Phys. Plasmas* **1**, p. 3635.

Dorland, W., Jenko, F., Kotschenreuther, M., and Rogers, R. N. (2000). Electron temperature gradient turbulence, *Phys. Rev. Lett.* **85**, pp. 5579-5582, `http://link.aps.org/doi/10.1103/PhysRevLett.85.5579`.

Ernst, D., *et al.* (2009). *Phys. Plasmas* **16**, p. 055906.

Ethier, S. and Lin, Z. (2004). *Comput. Phys. Comm.* **164**, p. 456.

Ethier, S., Tang, W. M., Walkup, R., and Oliker, L. (2008). *IBM J. Res. Dev.* **52**, p. 105.

Fu, X., Horton, Xiao, Y., W., Sen, A. K., Sokolov, V., and Lin, Z. (2012). *Phys. Plasmas* **19**, p. 032303.

Garcia, J., Giruzzi, G., Artaud, J. F., Basiuk, V., Imbeaux, F., Schneider, M. (2008).

Analysis of DEMO scenarios with the CRONOS suite of codes, *Nucl. Fusion* **48**, p. 075007.
Gluhovsky, A. and Tong, C. (1999). The structure of energy conserving low-order models, *Phys. Plasmas* **11**, p. 334.
Guo, Z. H., Chen, L., and Zonca, F. (2009). *Phys. Rev. Lett.* **103**, p. 055002.
Hahm, T. S. (1988). Nonlinear gyrokinetic equations for tokamak, *Phys. Fluids* **31**, p. 2670, doi:10.1063/1.866544.
Holod, I. and Lin, Z. (2007). Statistical analysis of fluctuations and noise-driven transport in particle-in-cell simulations of plasma turbulence, *Phys. Plasmas* **14**, p. 032306, doi:10.1063/1.2673002.
Holod, I. and Lin, Z. H. (2008). Gyrokinetic particle simulations of toroidal momentum transport). *Phys. Plasmas* **15**, p. 092302, doi:10.1063/1.2977769.
Holod, I., Zhang, W. L., Xiao, Y., and Lin, Z. H. (2009). Electromagnetic formulation of global gyrokinetic particle simulation in toroidal geometry, *Phys. Plasmas* **16**, p. 122307, doi:10.1063/1.3273070.
Holod, I. and Lin, Z. H. (2010). Effects of electron dynamics in toroidal momentum transport driven by ion temperature gradient turbulence), *Plasma Phys. Control. Fusion* **52**, p. 035002, doi:10.1088/0741-3335/52/3/035002.
Horton, W., Hu, G., and Laval, G. (1996). Turbulent transport in mixed states of convective cells and sheared flows, *Phys. Plasmas* **3**, pp. 2912-2923 .
Horton, W., *et al.* (1977). Correlations between drift wave theory, particle simulations and the observed anomalous transport in tokamaks), *Plasma Phys. Control. Nucl. Fusion Res.* **II**, (International Atomic Energy agency, Vienna) pp. 467-479.
Hu, B. and Horton, W. (1997). Minimal model for transport barrier dynamics based on ion-temperature-gradient turbulence, *Phys. Plasmas* **4**, p. 3262, `http://link.aip.org/link/doi/10.1063/1.872467`.
Idomura, Y., Watanabe, T-H., Sugama, H. (2006). Kinetic simulations of turbulent fusion plasmas, *Comptes Rendus Physique* **7**, pp. 650-669.
Idomura, Y., Tokuda, S., and Kishimoto, Y. (2003). Global gyrokinetic simulation of ion temperature gradient driven turbulence in plasmas using a canonical Maxwellian distribution, *Nucl. Fusion* **43**, p. 234, doi:10.1088/0029-5515/43/4/303.
Jenko, F. (2000). *Phys. Plasmas* **7**, p. 514.
Jolliet, S., Bottino, A., Angelino, P., Hatzky, R., Tran, T. M., Mcmillan, B. F., Sauter, O., Appert, K., Idomura, Y., Villard, L. (2007). A global collisionless PIC code in magnetic coordinates, *Comp. Phys. Comm.* **177**, 5, pp. 409-425, `http://dx.doi.org/10.1016/j.cpc.2007.04.006`.
Kim, C. and Parker, S. E. (2000). *J. Comput. Phys.* **161**, p. 589.
Kim, C. B., Horton, W., and Hamaguchi, S. (1993). New fluid model for the turbulent transport due to the ion temperature gradient, *Phys. Fluids B* **5**, p. 1516.
Kniep, J., Leboeuf, J-N., and Decyk, V. (2004). *Comput. Phys. Comm.* **164**, p. 98.
Kotschenreuther, M., Rewoldt, G., and Tang, W. M. (1995). *Comp. Phys. Comm.* **88**, p. 128.
Krommes, J. (1999). *Phys. Plasmas* **6**, p. 1477.
Ku, S., Chang, C. S., and Diamond, P. (2009). *Nucl. Fusion* **49**, p. 115021.
Lang, J., Parker, S. E., and Chen, Y. (2008). *Phys. Plasmas* **15**, p. 055907.
Lang, J., Chen, Y., Parker, S. E., and Fu, G-Y. (2009). *Phys. Plasmas* **16**, p. 102101.
Lee, W. W., Lewandowski, J. L. V., Hahm, T. S., and Lin, Z. H. (2001). *Phys. Plasmas* **8**, p. 4435.
Lee, W. W., Ethier, S., Ganesh, R., Kolesnikov, R., and Wang, W. X. (2008). Multiscale Turbulence Simulation and Steady State Transport *Invited talk, Joint Varenna-*

Lausanne International Workshop on Theory of Fusion Plasmas, AIP Conf. Proc. 1069, p. 144.

Lee, W. W., Ethier, S., Kolesnikov, R., Wang, W. X., and Tang, W. M. (2008). Nonlinear turbulence transport in magnetic fusion plasmas, invited talk 13th EU-US TTF Workshop, Copenhagen, *Computational Science and Discovery* **1**, p. 015010.

Lin, Z. H. and Lee, W. W. (1995). *Phys. Rev. E* **52**, pp. 5646-5652.

Lin, Z. H., Hahm, T. S., Lee, W. W., Tang, W. M., and White, R. B. (1998). *Science* **281**, pp. 1835-1837.

Lin, Z. H. and Chen, L. (2001). *Phys. Plasmas* **8**, p. 1447.

Lin, Z. H., Ethier, S., Hahm, T. S., and Tang, W. M. (2002). *Phys. Rev. Lett.* **88**, p. 195004.

Lin, Z. H., Nishimura, Y., Xiao, Y., Holod, I., Zhang, W. L., and Chen, L. (2007). *Plasma Phys. Control. Fusion* **49**, p. B163.

Miller, R. L., Chu, M. S., Greene, J. M., Lin-Liu, Y. R., and Waltz, R. E. (1998). *Phys. Plasmas* **5**, p. 973.

Nakata, M., Watanabe, T-H., Sugama, H. (2012). *Phys. Plasmas* **19**, p. 022303.

Neilson, G. H., Reiman, A. H., Zarnstorff, M. C., Brooks, A., Fu, G.-Y., Goldston, R. J., Ku, L. P., Lin, Z., Majeski, R., Monticello, D. A., Mynick, H., Pomphrey, N., Redi, M. H., Reiersen, W. T., Schmidt, J. A., Hirshman, S. P., Lyon, J. F., Berry, L. A., Nelson, B. E., Sanchez, R., Spong, D. A., Boozer, A. H., Miner, W. H., Valanju, P. M., Cooper, W. A., Drevlak, M., Merkel, P., and Nuehrenberg, C. (2000). *Phys. Plasmas* **7**, p. 1911.

Nevins, W. M., Parker, S. E., Chen, Y., Candy, J., Dimits, A. M., Dorland, W., Hammett, G., and Jenko, F. (2007). *Phys. Plasmas* **14**, p. 084501.

Nishimura, Y., Lin, Z. H., Lewandowski, J. L. V., and Ethier, S. (2006). *J. Comp. Phys.* **214**, p. 657.

Nishimura, Y., Lin, Z. H., and Wang, W. X. (2007). *Phys. Plasmas* **14**, p. 042503.

Nishimura, Y., Lin, Z. H., and Chen, L. (2009). *Comm. Comp. Phys.* **5**, p. 183.

Nishimura, Y. (2009). *Phys. Plasmas* **16**, p. 030702.

Okuda, H., Lee, W. W., and Lin, A. T. (1979). Plasma diffusion due to magnetic field fluctuations, *Phys. Fluids* **22**, 10, pp. 1899-1906, doi:10.1063/1.862478.

Parker, S. E., Lee, W. W., and Santoro, R. A. (1993). Gyrokinetic simulation of ITG driven turbulence in 3D toroidal geometry, *Phys. Rev. Lett.* **71**, p. 2042.

Reimann, A., Fu, G-Y., Hirshman, S., Ku, L., Monticello, D., Mynick, H., Redi, M., Spong, D., Zarnstorff, M., Blackwell, B., Boozer, A., Brooks, A., Cooper, W. A., Drevlak, M., Goldston, R., Harris, J., Isaev, M., Kessel, C., Lin, Z., Lyon, J. F., Merkel, P., Mikhailov, M., Miner, W. M., Nakajima, N., Neilson, G., Nuhrenberg, C., Okamoto, M., Pomphrey, N., Reiersen, W., Sanchez, R., Schmidt, J., Subbotin, A., Valanju, P., Watanabe, K. Y., and White, R. (1999). *Plasma Phys. Control. Fusion* **41**, p. B273.

Reimann, A., Ku, L., Monticello, D., Hirshman, S., Hudson, S., Kessel, C., Lazarus, E., Mikkelsen, D., Zarnstorff, M., Berry, L. A., Boozer, A., Brooks, A., Cooper, W. A., Drevlak, M., Fredrickson, E., Fu, G-Y., Goldston, R., Hatcher, R., Isaev, M., Jun, C. , Knowlton, S., Lewandowski, J., Lin, Z., Lyon, J. F., Merkel, P., Mikhailov, M., Miner, W. H., Mynick, H., Neilson, G., Nelson, B. E., Nuhrenberg, C., Pomphrey, N., Redi, M., Reiersen, W., Rutherford, P., Sanchez, R., Schmidt, J., Spong, D. A., Strickler, D., Subbotin, A., Valanju, P., and White, R. B. (2001). *Phys. Plasmas* **8**, p. 2083.

Rhodes, T., *et al.* (2002). *Phys. Plasmas* **9**, p. 2141.

SciDac Review. (2009). *Breakthrough Fusion Simulation Sheds Light on Plasma Confinement*, GTC is selected for *Top Breakthroughs in Computational Science* and

HPCwire, *Fusion Gets Faster.*
Sugama, H. and Horton, W. (1995). L-H Confinement mode dynamics in three-dimensional state space, *Plasma Phys. Control. Fusion* **37**, pp. 345-362.
Sydora, R. D., Decyk, V., and Dawson, J. M. (1996). Fluctuation-induced heat transport results from a large global 3D toroidal particle simulation model, *Plasma Phys. Control. Fusion* **38**, p. A281, doi:10.1088/0741-3335/38/12A/021.
Tajima, T. (2004). *Computational Plasma Physics: With Applications To Fusion and Astrophysics* (Frontiers in Physics) [Paperback], ISBN-13:978-0813342115.
Thiffeault, J.-L. and Horton, W. (1996). Energy-conservign truncations for convection with shear flow, *Phys. Fluids* **8**, p. 1715.
Todo, Y., Nakajima, N., Osakabe, M., Yamamoto, S., and Spong, D. A. (2008). Simulation study of energetic ion transport due to Alfvén eigenmodes in LHD plasma, *Plasma and Fusion Research* **3**, p. S1074.
Wan, W., Chen, Y., and Parker, S. E. (2005). Gyrokinetic δf simulation of the collisionless and semicollisional tearing mode instability, *Phys. Plasmas* **12**, p. 012311, `http://link.aip.org/link/doi/10.1063/1.1827216`.
Wan, W., Chen, Y., and Parker, S. E. (2005). δf Simulation of the collisionless tearing mode instability with a gyrokinetic ion response, *IEEE Trans. Plasma Sci.* **33**, pp. 609-614, doi:10.1109/TPS.2005.844606.
Wan, W., Parker, S. E., Chen, Y., and Perkins, F. W. (2010). Natural fueling of a tokamak fusion reactor, *Phys. Plasmas* **17**, p. 040701, `http://link.aip.org/link/doi/10.1063/1.3389225`.
Wang, W. X., Lin, Z. H., Tang, W. M., Lee, W. W., Ethier, S., Lewandowski, J. L. V., Rewoldt, G., Hahm, T. S., and Manickam, J. (2006). Gyro-kinetic simulation of global turbulent transport properties in tokamak experiments, *Phys. Plasmas* **13**, p. 092505, `http://link.aip.org/link/doi/10.1063/1.2338775`.
Wang, W., *et al.* (2007). *Phys. Plasmas* **14**, p. 072306.
Wang, W., *et al.* (2009). *Phys. Rev. Lett.* **102**, p. 035005.
Watanabe, T.-H., Todo, Y., and Horton, W. (2008) Benchmark tests of fusion plasma simulation codes for studying microturbulence and energetic-particle dynamics, *Plasma Fusion Res.* **3**, p. 061, doi:10.1585/prf.3.061.
Watanabe, T.-H., Sugama, H., and Ferrando-Margalet S. (2008). *Phys. Rev. Lett.* **100**, p. 195002.
Watanabe, T.-H. and Sugama, H. (2006). *Nucl. Fusion* **46**, p. 24.
White, R. and Zakharov, L. E. (2003). Hamiltonian guiding center equations in toroidal magnetic configurations. *Phys. Plasmas* **10**, p. 573, `http://link.aip.org/link/doi/10.1063/1.1544500`.
Xanthopoulos, P., Mishchenko, A., Helander, P., Sugama, H., Watanabe, T-H. (2011). *Phys. Rev. Lett.* **107** p. 245002.
Xiao, Y. and Lin, Z. H. (2009). *Phys. Rev. Lett.* **103**, p. 085004.
Xiao, Y., Holod, I., Zhang, W., Klasky, S., and Lin, Z. H. (2010). *Phys. Plasmas* **17**, p. 022302.
Yagi, M. and Horton, W. (1994). Reduced Braginskii equations, *Phys. Plasmas* **1**, 7, p. 2135.
Yagi, M., Wakatani, M., Sugama, H., Hong, B-G., and Horton, W. (1989). Ion-temperature gradient driven drift mode coupled to the resistive interchange mode in a Heliotron/Torsatron, *Phys. Soc. Jpn.* **58**, pp. 4265-4268.
Zhang, W. L., Decyk, V., Holod, I., Xiao, Y., Lin, Z., and Chen, L. (2010). Scalings of energetic particle transport by ion temperature gradient microturbulence. *Phys. Plasmas* **17**, p. 055902, doi:10.1063/1.3379471

Zhang, H. S., Lin, Z., Holod, I., Wang, X., Xiao, Y., and Zhang, W. L. (2010). Gyrokinetic particle simulation of beta-induced Alfvén eigenmode, *Phys. Plasmas* **17**, p. 112505, doi:10.1063/1.3498761.

Zhang, H. S., Qiu, Z., Chen, L., and Lin, Z. H. (2009). *Nucl. Fusion* **49**, p. 125009.

Zhang, W. L., Lin, Z., and Chen, L. (2008). Research on the transport of energetic particles by microturbulence in magnetized Plasmas. *Phys. Rev. Lett.* **101**, p. 095001.

Index